PREFACE TO THE FIRST EDITION

The present book has the goal of setting forth the basic aspects of the algebraic theory of semigroups. A semigroup (or in other words an associative system) is a set considered with respect to a binary associative operation defined in it. The concept of a semigroup is so simple and natural that it is hard to say when it first appeared. As Klein[1] points out, there were doubts, even in the period when the theory of groups was formulated as a separate mathematical discipline, as to whether that which we call a semigroup should be taken as the fundamental concept. However, the problems facing mathematics at that stage of its development made it necessary to choose a more restrictive concept, that of a group.

One of the most important reasons that the theory of groups played so large a role in the ensuing development of mathematics was that the algebraic theory of groups by its very nature was the abstract study of invertible transformations. The need for considering such mappings occurs in such diverse fields of mathematics, that one encounters the concept and properties of groups in very many mathematical (and not only mathematical) disciplines.

In spite of the importance and fruitfulness of the general concept of an invertible transformation, the necessity of studying the general concept of a transformation became clear with the further development of mathematics. To create a corresponding algebraic theory it was necessary to form a corresponding concept, namely the concept of a semigroup. The basic role of the algebraic theory of semigroups in mathematics evidently rests on the fact that this theory is the abstract study of general transformations.

In order to establish a basic mathematical discipline, naturalness and importance of the original concept are not sufficient. The experience of the theory of groups showed that the success of its development, granted the importance of the original ideas, is explained by the fact that the axioms permitted the construction of an extremely deep and complex theory. In this connection, the poorer axioms of the theory of semigroups raised a natural doubt at the first stages of its development: can it serve as the foundation for a sufficiently wide construction of an independent theory? At the present time, these doubts have already passed into oblivion.

After the first, frequently fragmentary, studies of semigroups, carried out in the 1920's and 1930's, the last 15 years have seen the appcarance of hundreds

[1] F. Klein, *Lectures on the development of mathematics in the 19th century*, Part I, Ch. VIII

of different works on the theory of semigroups, which have completely proved the possibility of a sufficiently profuse and deep independent theory.

The exposition of the principal results of these works in the form of a connected systematic theory is the aim of the present book. Of course there is no need (or possibility, for that matter) of setting forth all of the results obtained up to the present time. But the basic aspects of the contemporary abstract algebraic theory of semigroups are set forth here with sufficient completeness. Concrete semigroups, which are met in the most diverse fields of mathematics (in particular, wherever one must study one or another transformation), are completely impossible to survey completely. The material is too profuse. Nevertheless, a mathematician versed in the general theory of semigroups could apply this theory in a given concrete problem.

The present book is designed to offer this possibility to mathematicians working in different fields of pure and applied mathematics. To this end the material is set forth fairly completely and in considerable detail.

The theory as given here depends only on well-known branches of mathematics (the few places where this is not the case may be omitted). Only the simplest, universally familiar concepts and properties from the theory of sets will be used extensively.

However, we should mention here that we denote the elements of a set by including them in curly brackets, in contrast to the common practice in the theory of groups of using curly brackets in another special way. We make no distinction between a single element A of a certain set and the subset of this set consisting of the single element $\{A\}$.

We call the union of disjoint sets a union without intersection, or a non-intersecting union. The empty set is denoted by the symbol $\varnothing$. At times we shall use the concept of the empty symbol. A formula containing empty symbols should be read as if they were not present.

Throughout we shall constantly meet pairs of statements that are symmetric with respect to "left" and "right" (cancellation on the left, cancellation on the right, left ideal, right ideal, and so on). In such cases we shall usually give only one of the two assertions, counting on the reader to supply the other assertion and to see that it is correct.

The text of the book is divided into separate sections, numbered on the decimal principle. For example, a reference to II, 4.11 refers the reader to the second chapter, fourth section, 11th subsection. References within a given chapter omit the number of the chapter.

The literature on the abstract theory of semigroups is listed (with reasonable completeness) at the end of the book. Of course it does not contain works devoted completely or partially to the study of one or another concrete semigroup. As already remarked, there are simply too many such works. The bibliography also contains certain works on topological semigroups, although the theory of topological semigroups, which has already consolidated itself

39.00 nett. Science

Translations of Mathematical Monographs

Volume 3

by

E. S. Ljapin

American Mathematical Society
Providence, Rhode Island
1963

ПОЛУГРУППЫ

Е. С. ЛЯПИН

Государственное Издательство
Физико-Математической Литературы

Москва 1960

Translated from the Russian by A. A. Brown, J. M. Danskin, D. Foley, S. H. Gould, E. Hewitt, S. A. Walker, J. A. Zilber

Publication aided by grant NSF-GN 57
from the
NATIONAL SCIENCE FOUNDATION

Library of Congress Cataloging in Publication Data

Liapin, E S

Semigroups.

(Translations of mathematical monographs, v. 3)

Translation of Polugruppy.

Bibliography: p.

1. Semigroups. I. Title. II. Series.

QA171.L7413 1974 512′.2 63-15659

ISBN 0-8218-1553-9

Third edition, 1974

Second printing, 1978

Printed in the United States of America

into a special branch of topological algebra, is not touched on in the book itself. References to various other works not part of the general bibliography of the theory of semigroups are given in footnotes.

I remark with gratitude that in the final redaction of the manuscript I frequently made use of the valuable advice and observations of L. M. Gluškin.

Leningrad, 1958. E. Ljapin

PREFACE TO THE SECOND EDITION

During the time since this book was written the theory of semigroups has continued to develop rapidly. Several hundred articles have been published on the subject and a second book has appeared, specifically devoted to semigroups: A. H. Clifford and G. B. Preston, *The algebraic theory of semigroups*. Vol. I, Math. Surveys, no. 7, Amer. Math. Soc., Providence, R. I., 1961, (Vol. II to be published in 1967). But the foundations, the general outlines and the principal directions of research have not essentially changed. Consequently, in this second edition, it has been possible to preserve the text of the first edition with only small changes, and to add a supplement containing a relatively small number of new results that are characteristic and important for an understanding of the present state of the theory.

It should be especially noted that in recent years a sharply increasing number of articles has been published on semigroups of transformations, partial transformations, and binary relations.

Also remarkable is the increase of interest in semigroups provided with an additional structure (e.g., ordered semigroups). It appears that such investigations constitute a special direction of research in algebra. This remark, which was made some time ago for topological semigroups, has been fully justified by the further development of that subject, although the theory of topological semigroups naturally remains closely connected with the general theory of semigroups.

For a clear understanding of the place of the theory of semigroups in contemporary mathematics and of its probable development in the future one must take into account the development of the following important theories, all of them so closely related in their essence to the theory of semigroups that they could from a certain point of view be included in that theory; namely, the theory of categories and the abstract theory of automata. It is also important to keep in mind the development of the general theory of binary relations. In this connection we wish to mention the extensive work of V. V. Vagner on the systematization of this theory: V. V. Vagner. *Theory of relations and algebra of partial transformations*, Theory of Semigroups and Its Applications, no. 1, Saratov Univ. Press, 1965, pp. 3-178. (Russian)

The bibliography on semigroups, which is reasonably complete for articles appearing up to 1958, has been retained for the present edition. A short supplementary bibliography, consisting of later works cited in the new material, has been added; it does not pretend to be complete.

Leningrad, 1966. E. S. Ljapin

PREFACE TO THE THIRD EDITION

During the past five years, the rate of development of the theory of semigroups has continued to increase. The theory has enjoyed a substantial enrichment through the large number of new results obtained in its various subdivisions. Nevertheless, just as remarked in the preface to the preceding edition concerning the period from 1958 to 1966, it can be said again that the general aspect and character of the theory is basically unaltered. As a consequence, the earlier presentation can be retained.

However, this proposition does require in part some amendment, in the following respects.

The theory of transformations (including partial and multivalued transformations) plays a critical role in the theory of semigroups. It has importance for the connections of the theory of semigroups with other domains of mathematics, and the progress achieved in this direction would appear to merit a more extensive treatment. But one gains the impression that there is in process of formation a separate theory, though quite closely tied in with the general theory of semigroups.

New problems have appeared as a result of the permeation into the theory of semigroups of the ideas and methods of mathematical logic. There has been an expansion in the study of algorithmic questions. But this is again without question a separate line of investigation.

Within the confines of the theory of semigroups proper, most significant has been the development of, and the position taken by, the line of investigation dealing with the study of identities. Since in the preceding edition this was left almost entirely untouched, it has now appeared necessary to enlarge the presentation. A new chapter has therefore been added to deal with semigroup identities.

Also added is one more section to Chapter XI, dealing with results relating to approximation properties.

These additions have in turn made necessary additions to the Supplementary Bibliography.

Leningrad, 1972 *E. S. Ljapin*

TABLE OF CONTENTS

CHAPTER I

THE CONCEPT OF A SEMIGROUP

1. Multiplicative Sets

1.1. The aim of the present chapter is to present the concept of a semigroup: the algebraic object to which the present book is devoted. Of course, we shall not merely give the definition, which requires but a few lines. We will also point out certain connections, defining the position of semigroups in contemporary mathematics. This will help to clarify the meaning, role, and problems of this theory, which will be set forth in later chapters.

In contemporary algebra, one studies a series of algebraic objects, defined with the aid of one or several algebraic operations in a set of elements of this or that kind. One obtains one or another algebraic theory, depending upon the collection of these operations, their properties, and the nature of the set. One of these theories is the theory of semigroups.

1.2. Suppose that there is a law given in a certain set $\mathfrak{M}$, which assigns to every pair of elements of $\mathfrak{M}$, taken in a definite order, a certain third element of $\mathfrak{M}$. Then we say that an *algebraic operation* is defined in $\mathfrak{M}$.

The third element just referred to is called the *result of the given operation* carried out on the two original elements.

1.3. Although this definition is very wide, it is not the most general one, and can be further generalized. For example, one frequently must give up the requirement that the result is defined for all pairs of elements of $\mathfrak{M}$. In the more general definition, the result of the operation is defined only for some (not necessarily all) pairs of elements of the set. Thus the operation of division in the set of all rational numbers is an example of this, since the result of the operation is undefined for pairs of numbers in which the second (the divisor) is equal to zero. There are times when one has to postulate that the result of the operation consists not of one element but of a certain subset of the original set. Other generalizations are also possible. However, at the present time, the main attention in algebra is devoted to operations in the sense defined in 1.2. Later, when we speak of algebraic operations, we shall mean them only in the sense of the definition in 1.2.

1.4. Suppose that there is given a certain algebraic operation in a set $\mathfrak{M}$. The element Z which is the result of this operation carried out on the elements

X and Y is usually denoted by symbols of the following forms. The elements X and Y are written one after the other, and between them some sort of symbol is written to indicate the given operation. For example, in the operation of addition between numbers, we use the symbol for plus, "+". The result of the operation of addition between the numbers X and Y is denoted by $X + Y$. For many ordinary operations, the symbols are generally known and their use is fixed. For any new operations, or in studying properties of an arbitrary operation belonging to a given class of operations, we use new symbols to denote the result of the operation on the elements X and Y. For example, we write $X \circ Y$ or $X * Y$, and so on. However, it is most customary in this case to use multiplicative notation, writing the result of the operation on the elements X and Y as $X \cdot Y$ or simply XY. Here the result of the operation XY is also called the product of X and Y, and the elements X and Y are called the left and right factors, respectively (this is essential, since the result of the operation will in general depend upon the order of the elements in the original pair, as happens for example in subtracting numbers). This multiplicative notation is in general use in modern algebra, and consequently we shall adhere to it in this book.

1.5. A nonvoid set, in which there is given an algebraic operation (in the sense of 1.2), and for which multiplicative notation is used, will be called a *multiplicative set*. Thus $\mathfrak{M}$ is a multiplicative set if for every pair of its elements $X, Y \in \mathfrak{M}$ there is defined a third element

$$Z = XY \in \mathfrak{M}.$$

The fact that the definition contains an inessential feature (namely, the multiplicative notation) causes no inconvenience, since the concept of a multiplicative set plays only an auxiliary role and will be used principally in the present chapter. We note that there is no fixed terminology at the present time for what we have called a multiplicative set. Different authors use different terms, for example, groupoid, etc.

1.6. The description of the operation in a multiplicative set $\mathfrak{M}$ can be carried out in various ways. The most natural is simply to list all results of the operation for arbitrary pairs of elements. This method of describing the operation can be presented as a *multiplication table*, also called a Cayley table.

For a multiplicative set $\mathfrak{M} = \{A, B, C, \ldots, R, \ldots, S, \ldots\}$, we construct a square table (see top of page 3).

At the intersection of the row corresponding to the element R and of the column corresponding to the element S, we write the element $T_{R,S}$, which is the product of the elements R and S, that is, $T_{R,S} = R \cdot S$.

Except for the order in which the elements are written (which is of no consequence for the multiplication table), the multiplication table of a multiplicative set is completely determined. Conversely, constructing such a table in a completely arbitrary way for a certain set defines in it an operation of multiplication, that is, turns it into a multiplicative set.

	A	B	C	...	S	...
A	$T_{A,A}$	$T_{A,B}$	$T_{A,C}$	...	$T_{A,S}$	...
B	$T_{B,A}$	$T_{B,B}$	$T_{B,C}$	...	$T_{B,S}$	...
C	$T_{C,A}$	$T_{C,B}$	$T_{C,C}$	...	$T_{C,S}$	...
... ...	... ...	... ...	... ...	... ...	... ...	... ...
R	$T_{R,A}$	$T_{R,B}$	$T_{R,C}$	...	$T_{R,S}$	...
...	...	...	...	...	...	...

Of course one can only actually construct a multiplication table for a finite multiplicative set that has not too many elements. However, we can imagine and use in arguments multiplication tables for arbitrary (even infinite) multiplicative sets.

1.7. As an example, we consider the set of the first six natural numbers $\mathfrak{M} = \{1, 2, 3, 4, 5, 6\}$ and the operation in this set which as a result of the "multiplication" of two numbers X and Y gives the greatest common divisor of these numbers. The multiplication table for this multiplicative set is the following.

	1	2	3	4	5	6
1	1	1	1	1	1	1
2	1	2	1	2	1	2
3	1	1	3	1	1	3
4	1	2	1	4	1	2
5	1	1	1	1	5	1
6	1	2	3	2	1	6

1.8. The abstract tendency, as it is sometimes called, in modern algebra consists in studying one or more algebraic operations in a set from the point of view of the results of these operations, without regard to any other properties of the elements of the set. From this point of view, the abstract theory of multiplicative sets is concerned only with the rule of multiplication in a given such set. Thus two multiplicative sets with the same number of elements, in which the rule of multiplication is defined in the same way, differ from each other only inessentially and can even be taken to be one and the same multiplicative set. This point of view is expressed more precisely with the aid of the concept of isomorphic multiplicative sets.

DEFINITION. *Two multiplicative sets* $\mathfrak{M}$ *and* $\mathfrak{M}'$ *are called* ISOMORPHIC *with each other if there is a one-to-one correspondence between the elements of* $\mathfrak{M}$ *and the elements of* $\mathfrak{M}'$, *let us say*

$$A \sim A', \qquad B \sim B', \ldots, \qquad S \sim S', \ldots$$

where $\mathfrak{M} = \{A, B, \ldots, S, \ldots\}$ *and* $\mathfrak{M}' = \{A', B', \ldots, S', \ldots\}$, *with the property that if*

$$XY = Z \qquad (X, Y, Z \in \mathfrak{M}),$$

then for the corresponding elements of $\mathfrak{M}'$, *we always have*

$$X'Y' = Z',$$

and conversely.

Given two isomorphic multiplicative sets, one also says that there exists an isomorphism between them.

It is clear that two isomorphic multiplicative sets have the same cardinal number and that the operations in them are perfectly "similar", apart from the nature of the elements of the sets. Thus, for all questions connected only with the character of the operation, isomorphic multiplicative sets are completely similar and hence can be considered, in the usual way, as one and the same multiplicative set with different designations of the elements (A', B', ... can be regarded simply as different designations of the elements A, B, ...). With this approach to multiplicative sets, it is customary in stating an assertion to add the words "up to an isomorphism." Thus for example, it is obvious how to interpret the assertion that up to an isomorphism there exists only one multiplicative set consisting of one element.

1.9. We note that the last condition in Definition 1.8, expressed by the concluding words "and conversely" is really unnecessary. It holds automatically. In fact, suppose that

$$X'Y' = Z' \qquad (X', Y', Z' \in \mathfrak{M}').$$

We write $XY = T$. In agreement with the preceding conditions of Definition 1.8, we have $X'Y' = T'$. Thus $Z' = T'$. But the correspondence under study is one-to-one, and hence, since $Z' = T'$, we have $Z = T$. Thus we have shown that the conditions of the definition, excluding the last one, imply that $XY = Z$ whenever $X'Y' = Z'$.

We note also that if

$$XY \neq Z \qquad (X, Y, Z \in \mathfrak{M}),$$

then we also have $X'Y' \neq Z'$ in $\mathfrak{M}'$. In fact, if $X'Y' = Z'$, then the definition implies that $XY = Z$.

1.10. We can arrive at the concept of an isomorphic set from another point of view as well.

DEFINITION. *A one-to-one mapping φ of a multiplicative set $\mathfrak{M}$ onto a multiplicative set $\mathfrak{M}'$ is called an isomorphic mapping or simply an* ISOMORPHISM *if for any two elements X, Y in $\mathfrak{M}$ the relation*

$$\varphi(X) \cdot \varphi(Y) = \varphi(XY)$$

always holds in $\mathfrak{M}'$.

1.11. Definitions 1.8 and 1.9 are connected with each other in an obvious way.

If there exists an isomorphism φ (1.10) of the multiplicative set $\mathfrak{M}$ onto the multiplicative set $\mathfrak{M}'$, then the correspondence

$$X \sim \varphi(X) \qquad (X \in \mathfrak{M}, \varphi(X) \in \mathfrak{M}')$$

clearly satisfies the property 1.8, thanks to 1.9. Hence $\mathfrak{M}$ and $\mathfrak{M}'$ are isomorphic in the sense of 1.8.

On the other hand, if there is a correspondence 1.8 between $\mathfrak{M}$ and $\mathfrak{M}'$, the mapping φ of the multiplicative system $\mathfrak{M}$ onto $\mathfrak{M}'$ defined by

$$\varphi(X) = X' \qquad (X \in \mathfrak{M}, X' \in \mathfrak{M}')$$

(where $X \sim X'$ in agreement with the correspondence 1.8) is, as one easily sees, an isomorphism of $\mathfrak{M}$ onto $\mathfrak{M}'$ in the sense of 1.10.

1.12. As is the case with every one-to-one mapping, an isomorphism φ of a multiplicative set $\mathfrak{M}$ onto a multiplicative set $\mathfrak{M}'$ always admits an inverse one-to-one mapping φ^{-1} of $\mathfrak{M}'$ onto $\mathfrak{M}$. It is easy to see that φ^{-1} is an isomorphism of $\mathfrak{M}'$ onto $\mathfrak{M}$. In fact, suppose that

$$\varphi^{-1}(X') \cdot \varphi^{-1}(Y') = \varphi^{-1}(Z') \qquad (X', Y', Z' \in \mathfrak{M}').$$

Then by the property of the isomorphism φ, we have

$$\varphi[\varphi^{-1}(X')] \cdot \varphi[\varphi^{-1}(Y')] = \varphi[\varphi^{-1}(Z')].$$

But $\varphi\varphi^{-1}$ is the identity mapping. Hence,

$$X'Y' = Z',$$

that is, for arbitrary X' and Y' in $\mathfrak{M}'$, we have

$$\varphi^{-1}(X') \cdot \varphi^{-1}(Y') = \varphi^{-1}(X'Y').$$

1.13. Suppose that φ is a one-to-one mapping of a multiplicative set $\mathfrak{M}$ into a multiplicative set $\mathfrak{M}'$ with the property that

$$\varphi(X) \cdot \varphi(Y) = \varphi(XY)$$

for all $X, Y \in \mathfrak{M}$. Then it is easy to see that the subset $\mathfrak{M}'' = \varphi(\mathfrak{M})$ of $\mathfrak{M}'$ is also a multiplicative set. One can also consider φ as an isomorphism of $\mathfrak{M}$ onto $\mathfrak{M}''$. In this case, we call the mapping φ an *isomorphism of* $\mathfrak{M}$ *into* $\mathfrak{M}'$.

In the sequel we will follow the custom in mathematical literature of using the term "isomorphism" in this sense. One should keep in mind the double meaning of the term "isomorphism," connected with Definitions 1.8 and 1.10. If we take account of 1.11, no confusion can result.

In place of *isomorphism* in the above sense, the term *monomorphism* is sometimes used, in which case *isomorphism* is restricted to a mapping onto.

1.14. Naturally one can assemble without any difficulty an unlimited number of examples of multiplicative sets, encountered in the most diverse branches of mathematics. The goal of an algebraic theory is to further research in these fields by providing the algebraic apparatus for studying the properties of the corresponding multiplicative sets. Here, however, one must distinguish two types of properties. Certain properties are connected with the character of the elements that make up the given multiplicative set. Under an isomorphism, such properties may be destroyed since only the operation is "preserved" (properties of elements not defined in terms of the operation need not be preserved). The other properties are those which, if holding for one multiplicative set, are automatically satisfied for every multiplicative set that is isomorphic with it. Properties of the first sort, connected (as one sometimes says) with the "concrete" nature of the given set, cannot, it is clear, be completely captured by an abstract theory. In their study, the abstract algebraic theory of operations can play only an auxiliary role. On the other hand, the study of properties of the second type, that is, of an "abstract" character, where by abstract we mean invariant under isomorphisms, can be completely carried over to the field of the abstract theory of algebraic operations. This observation defines in general terms the role of the abstract theory of algebraic operations in the science of mathematics.

1.15. Let Ξ be a certain class of multiplicative sets. Every isomorphism φ of a certain multiplicative set $\mathfrak{M}$ onto one of the multiplicative sets belonging to the class Ξ is called an *isomorphic representation* (one sometimes says merely a representation) of $\mathfrak{M}$ in the class Ξ.

The class Γ of all multiplicative sets which admit isomorphic representations in the class Ξ is obviously closed under isomorphisms; that is, every multiplicative set isomorphic to a certain multiplicative set belonging to Γ also belongs to Γ.

To describe the members of the class Γ for a given class is to give a necessary and sufficient condition in order that an arbitrary multiplicative set should be isomorphic to some multiplicative set belonging to Ξ. To obtain such a condition is usually called the determination of an *abstract characteristic* for the class Ξ of multiplicative sets. In connection with what was said above, the use of such a term is completely justified and natural. The determination of abstract characteristics for various important concrete classes of multiplicative sets is a matter of essential interest. It ordinarily determines the development of this or that direction of the theory of algebraic operations.

1.16. Of course every multiplicative set is isomorphic to itself, since the identity mapping is an isomorphism. Besides this, there sometimes exist other, nonidentity isomorphisms of a multiplicative set onto itself. Every such isomorphism (the identity or not) is called an *automorphism* of the multiplicative set. In agreement with what we have said above about isomorphisms, an automorphism can be regarded as a system for permuting the elements of a multiplicative set. Elements X and X', where one goes into the other under a certain automorphism, play one and the same role in the multiplicative set. Every property possessed by one of them that relates to the operation in the multiplicative set is also possessed by the other.

1.17. Along with isomorphisms, we must also in certain cases consider a related mapping. A one-to-one mapping φ of a multiplicative set $\mathfrak{M}$ onto a multiplicative set $\mathfrak{M}'$ is called an *anti-isomorphism* or an inverse isomorphism if for arbitrary elements $X, Y \in \mathfrak{M}$ the relation

$$\varphi(X) \cdot \varphi(Y) = \varphi(YX)$$

always holds in $\mathfrak{M}'$.

It is clear that anti-isomorphic multiplicative sets are very closely related. The operations in them are analogous in all respects except for "direction" (for the result of the operation is defined for pairs of elements taken in a definite order, or in other words, the direction in the pair of going from one element to the other is specified). Of course, the direction is reversed for anti-isomorphic multiplicative sets.

An anti-isomorphism of a multiplicative set onto itself is called an *anti-automorphism.* Of course not every multiplicative set admits an anti-automorphism. The existence of an anti-automorphism shows the symmetry of the operation for the two orders of pairs of elements, for which the results of the operation are defined. For certain multiplicative sets (semigroups) V. V. Vagner **[3]** has made a study of some special anti-automorphisms, the so-called involutions.

1.18. The general theory of multiplicative sets splits into a series of branches, each of which is devoted to the study of multiplicative sets for which the operations have varying basic properties. The following are the most important of these properties.

Let $\mathfrak{M}$ be a multiplicative set.

(α) The operation $\mathfrak{M}$ is called *associative* if for arbitrary A, B, C in $\mathfrak{M}$ we have

$$(AB)C = A(BC).$$

(β) The operation in $\mathfrak{M}$ is called *commutative* if for every A, B in $\mathfrak{M}$ we have

$$AB = BA.$$

(γ) The operation in $\mathfrak{M}$ has the property of *left invertibility* if for arbitrary A, B in $\mathfrak{M}$ there is always an X in $\mathfrak{M}$ such that

$$XA = B.$$

(δ) The operation in $\mathfrak{M}$ has the property of *right invertibility* if for all A, B in $\mathfrak{M}$ there is always a Y in $\mathfrak{M}$ such that

$$AY = B.$$

(ε) The operation $\mathfrak{M}$ has the property of *left cancellation* if for all X, A, B in $\mathfrak{M}$ the equality $XA = XB$ always implies that

$$A = B.$$

(ζ) The operation in $\mathfrak{M}$ has the property of *right cancellation* if for all Y, A, B in $\mathfrak{M}$ the equality $AY = BY$ always implies that

$$A = B.$$

We remark that the property of right cancellation (ζ) (and analogously for left cancellation) is sometimes called the property of uniqueness of division on the right (analogously, on the left). This is explained by the fact that the property (ζ) holds if and only if for all A, B in $\mathfrak{M}$ the equation

$$XA = B$$

has at most one solution.

The various branches of algebra are characterized by the study of multiplicative sets whose operations have various properties. Of course, it is sometimes necessary to study additional important properties. Also the problem of the cardinality of a multiplicative set may be important.

1.19. Even if a multiplicative set is not commutative, it still may be the case that $AB = BA$ for certain pairs of elements A, B in $\mathfrak{M}$.

In this case, the elements A and B are said to be *permutable*. One also says that A and B commute with each other. Singling out pairs of permutable elements is often useful in studying noncommutative operations.

1.20. For an arbitrary finite n, we set

$$X_1X_2X_3\ldots X_n = ((\ldots((X_1\cdot X_2)\cdot X_3)\cdot\ldots)\cdot X_n).$$

In case $n = 2$, we have the previously introduced notation for the product of two elements. For $n = 1$, we have the product consisting of just one factor, which we take to be equal to the element which is this single factor.

If the operation in a multiplicative set is associative, then we have

$$(X_1X_2 \dots X_m) \cdot (X_{m+1}X_{m+2} \dots X_n) = X_1X_2 \dots X_mX_{m+1} \dots X_n.$$

The proof of this is carried out with no trouble, using induction on n, since thanks to associativity we have

$$\begin{aligned}(X_1X_2 \dots X_m) &\cdot (X_{m+1}X_{m+2} \dots X_n) \\ &= (X_1X_2 \dots X_m) \cdot [(X_{m+1}X_{m+2} \dots X_{n-1}) \cdot X_n] \\ &= [(X_1X_2 \dots X_m) \cdot (X_{m+1}X_{m+2} \dots X_{n-1})] \cdot X_n.\end{aligned}$$

It follows from the equality stated that for an arbitrary sequence of elements $(X_1, X_2, \dots, X_n)$ of a multiplicative set with associative operation, the value of a product for an arbitrary placement of parentheses defining the sequence of multiplications of the elements $X_1, X_2, \dots, X_n$, will be equal to $X_1X_2 \dots X_n$. For this reason we usually omit the parentheses in writing down complicated products of several elements. They are written down only when it is desired to clarify the formation of some product formed in a complicated way.

In the case when all of the factors in a product are equal, we speak of a power of the element and use the corresponding notation:

$$\underbrace{X \cdot X \cdot \dots \cdot X}_{n} = X^n.$$

1.21. Let us look at several examples of multiplicative sets.

(α) Let n be any natural number. Let $\mathfrak{M}$ be the set of integers from 0 to $(n - 1)$:

$$\mathfrak{M} = \{0, 1, 2, \dots, n - 1\}.$$

The result of the operation in $\mathfrak{M}$ on the numbers m_1 and m_2 is defined as the remainder upon dividing $m_1 \cdot m_2$ by n.

It is easy to see that the operation just described in this multiplicative set is associative and commutative. This operation does not enjoy the property of cancellation since zero belongs to the set $\mathfrak{M}$.

If n is a prime number, then the subset $\mathfrak{M}'$ of the set $\mathfrak{M}$ consisting of the numbers $1, 2, \dots, n - 1$ will also be a multiplicative set with respect to the same operation, and the operation is associative and commutative. Also the operation in $\mathfrak{M}'$ will be invertible and enjoys the cancellation property. This follows from the fact that, as is known, there exists for arbitrary $m_1, m_2 \in \mathfrak{M}$ integers x and y such that

$$xm_1 + yn = m_2, \qquad x \in \mathfrak{M}'.$$

Thanks to this, the result of the operation we are considering for x' (x' is the remainder upon division of x by n) and m_1 will be m_2.

(β) The set $\mathfrak{M}_n$ of all complex square matrices of order n with respect to ordinary multiplication of matrices is a multiplicative set. As is known, this operation is associative, but does not have the other properties listed in 1.18. In the sequel we shall frequently return to this set, always using the notation $\mathfrak{M}_n$ for it.

(γ) The nonsingular matrices in $\mathfrak{M}_n$ form a multiplicative set in which, along with 1.18, (α), properties 1.18, (γ), (δ), (ε), and (ζ) are satisfied.

(δ) In the set of all continuous functions of two variables x and y, defined in the square $0 \leqslant x \leqslant a$, $0 \leqslant y \leqslant a$, the following operation, which plays an important role in the theory of integral equations, can be defined. The result of this operation carried out for the functions $K_1(x, y)$ and $K_2(x, y)$ is the function

$$\int_0^a K_1(x, t)K_2(t, y)\, dt.$$

As follows easily from the simplest properties of integrals, this operation is associative.

(ε) We consider the set of functions of one variable which are absolutely integrable for $0 \leqslant x < \infty$. In many branches of mathematics, one considers the operation in this set, the result of which for $f_1(x)$ and $f_2(x)$ is the function

$$\int_0^x f_1(t)f_2(x - t)\, dt.$$

One can show that this operation is associative and commutative.

(η) Let $\mathfrak{M}$ be the set of all possible formal power series with complex coefficients and constant term equal to zero such as

$$A = \sum_{k=1}^{\infty} d_k^{(A)} x^k.$$

The product of two such series A and B is defined in the following way. In the series A, we substitute the series B for the variable x, writing

$$\sum_{k=1}^{\infty} d_k^{(A)} \left(\sum_{l=1}^{\infty} d_l^{(B)} x^l \right)^k.$$

Removing the parentheses and grouping terms with the same powers of x, we obtain a new power series, called the "product" of the original power series. One can show that multiplication in the multiplicative set so obtained is associative.

2. Independence of the Conditions of Associativity

2.1. Since all of this book is devoted to the study of multiplicative sets with associative multiplication, it is natural to consider in more detail the question of the requirement of associativity of the operation. However, we should mention that the considerations of the present section, which explain a certain aspect of the condition of associativity, will not themselves be used in what follows, so that a knowledge of the present section is not needed for the sequel.

Let $\mathfrak{M} = \{A, B, C, \ldots\}$ be a multiplicative set. Associativity of the multiplication in $\mathfrak{M}$ means that the following equalities hold:

$$(AA)A = A(AA);$$

$$(AB)A = A(BA);$$

$$(AA)B = A(AB);$$

$$\cdots\cdots\cdots\cdots$$

$$(AB)C = A(BC);$$

$$\cdots\cdots\cdots\cdots$$

If the number of elements in $\mathfrak{M}$ is finite (equal to n), then the number of these relations is n^3. If $\mathfrak{M}$ is infinite with cardinal number $\mathfrak{m}$, then the cardinal number of the set of these equalities is also equal to $\mathfrak{m}$.

It is natural to call each of the equalities

$$(XY)Z = X(YZ)$$

(where X, Y, Z may be distinct or identical elements of $\mathfrak{M}$), an associativity condition corresponding to the sequence (X, Y, Z) of elements of $\mathfrak{M}$. Validity of the associativity conditions for all possible triples of elements (X, Y, Z) is what is meant by the associativity of the operation.

It is then also natural to raise the question whether all the associativity conditions are independent, that is, whether some subset of them may not imply all the rest. The investigations of Szász [**1**], which are described below, show that with trivial exceptions these associativity conditions are in fact independent.

2.2. If the number of elements in a set $\mathfrak{N}$ is greater than one, then the operation in $\mathfrak{N}$ can be defined in several different ways. In other words, several different multiplicative sets can exist with one and the same set of elements. It is, of course, just this fact which gives rise to the question under discussion.

We shall say that *in the set* $\mathfrak{N}$ *the associativity conditions are independent* if for every sequence (A, B, C) of three elements belonging to $\mathfrak{N}$ it is possible to define the operation of multiplication in $\mathfrak{N}$ in such a way that the equalities

$$(XY)Z = X(YZ)$$

will be valid for every sequence (X, Y, Z) of elements of $\mathfrak{N}$ which is different from (A, B, C), whereas for this latter sequence we have

$$(AB)C \neq A(BC).$$

2.3. It is immediately evident that the answer to the question of whether the associativity conditions for a set are independent or not depends only on the number of elements in the set.

THEOREM. *If the number of elements in a set $\mathfrak{N}$ is greater than three, then the associativity conditions in $\mathfrak{N}$ are independent* (2.2).

PROOF. Let (A, B, C) be an arbitrary sequence of three elements of $\mathfrak{N}$. Depending on which of these elements are alike and which are different, the argument now falls into five parts.

(1) $A = B = C$.

In $\mathfrak{N}$ we choose three distinct elements U, V, W different from A and define the operation of multiplication in $\mathfrak{N}$ as follows:

$$AA = U, \qquad AU = V,$$

and $XY = W$ in all the remaining cases.

We see at once that with this definition of multiplication the associativity condition for the given sequence of elements does not hold, since

$$(AA)A = UA = W,$$
$$A(AA) = AU = V.$$

Now we must show that for every sequence (X, Y, Z) different from (A, A, A) the associativity condition does hold.

Since $XY \neq A$, we always have

$$(XY)Z = W.$$

If X is different from A, we have

$$X(YZ) = W.$$

Let $X = A$ and note that YZ can never be equal to A. Then the triple product can be equal to V only if $Y = A$, $Z = A$, in which case $(X, Y, Z) = (A, A, A)$. In all other cases

$$X(YZ) = A(YZ) = W.$$

(2) $A = B \neq C$.

We choose in $\mathfrak{N}$ an element W different from A and C, and define the operation of multiplication in $\mathfrak{N}$ as follows:

$$AC = C,$$

with $XY = W$ in all other cases.

We have

$$(AA)C = WC = W,$$
$$A(AC) = AC = C.$$

For an arbitrary sequence (X, Y, Z) distinct from (A, A, C) we obtain

$$(XY)Z = W,$$

since XY will always be different from A. We also have

$$X(YZ) = W,$$

since either $X \neq A$ or $YZ \neq C$ in view of the fact that the equality $YZ = C$ is possible only for $Y = A, Z = C$, which for $X = A$ would mean that (X, Y, Z) is identical with (A, A, C).

(3) $A = C \neq B$.

We choose in $\mathfrak{N}$ an element W different from A and B and define the operation in $\mathfrak{N}$ as follows:

$$AA = A, \qquad BA = A, \qquad BB = B,$$

and $XY = W$ in all other cases.

We have

$$(AB)A = W,$$

$$A(BA) = AA = A.$$

For an arbitrary sequence (X, Y, Z) different from (A, B, A) and containing at least one element distinct from A and B we obtain

$$(XY)Z = W, \qquad X(YZ) = W.$$

Let each of the elements X, Y, Z be equal to A or to B. If X is equal to B, we have

$$(BY)Z = YZ, \qquad B(YZ) = YZ;$$

if Z is equal to A, we have

$$(XY)A = A, \qquad X(YA) = XA = A.$$

Finally, in the remaining two cases

$$(AA)B = AB = W, \qquad A(AB) = AW = W;$$

$$(AB)B = WB = W, \qquad A(BB) = AB = W.$$

(4) $A \neq B = C$.

We choose in $\mathfrak{N}$ an element W different from A and B and define the operation in $\mathfrak{N}$ as follows:

$$AB = A,$$

with $XY = W$ in all other cases.

We have

$$(AB)B = AB = A,$$

$$A(BB) = AW = W.$$

For an arbitrary sequence (X, Y, Z) different from (A, B, B) we obtain

$$(XY)Z = W,$$

since either $XY \neq A$ or $Z \neq B$ (i.e. for $X = A$, $Y = B$).

Also we have

$$X(YZ) = W,$$

since $YZ \neq B$.

(5) A, B, C are all different from one another.

We choose in $\mathfrak{N}$ an element W distinct from A, B, C and define the operation in $\mathfrak{N}$ as follows:

$$AC = C,$$

$XY = X$ in the other cases in which X and Y are elements A, B, C, and $XY = W$ in all other cases.

We have

$$(AB)C = AC = C,$$

$$A(BC) = AB = A.$$

If the sequence (X, Y, Z) contains an element distinct from A, B, C, we have

$$(XY)Z = W, \qquad X(YZ) = W.$$

Let X, Y, Z be elements from A, B, C. If $X \neq A$, we have

$$(XY)Z = XZ = X, \qquad X(YZ) = X.$$

If $X = A$ and $Y = C$, we have

$$(XY)Z = (AC)Z = CZ = C, \qquad X(YZ) = A(CZ) = AC = C.$$

If $X = A$, $Y = A$, $Z = C$, we have

$$(XY)Z = (AA)C = AC = C, \qquad X(YZ) = A(AC) = AC = C.$$

If $X = A$, $Y = A$, $Z \neq C$, we have

$$(XY)Z = (AA)Z = AZ = A, \qquad X(YZ) = A(AZ) = AA = A.$$

If $X = A$, $Y = B$, $Z \neq C$, we have

$$(XY)Z = (AB)Z = AZ = A, \qquad X(YZ) = A(BZ) = AB = A.$$

2.4. To complete our investigation of the independence of the associativity conditions it remains only to examine the cases when the set $\mathfrak{N}$ contains one, two or three elements.

If $\mathfrak{N}$ consists of a single element, then by 2.2 we must consider that the associativity conditions are not independent.

Now let $\mathfrak{N}$ consist of two or three elements. We fix one of them, the element U. Let us suppose that the operation in $\mathfrak{N}$ has been defined in such a way that the associativity conditions hold for all sequences (X, Y, Z) with the single exception of the sequence (U, U, U), for which we have $(UU)U \neq U(UU)$.

Let us set $U^2 = V$. By hypothesis it follows that $VU \neq UV$. Consequently, V is different from U, since otherwise we would have $VU = UV$.

By examining all the cases in succession we shall see that in each of them the above hypothesis leads to a contradiction. The conclusion is that in sets of two or three elements the associativity conditions are not independent.

(1) If $VU = U$, $UV = V$, we have

$$U = VU = (UV)U = U(VU) = UU = V.$$

(2) If $VU = U$, $UV = W$ (here and below W denotes an element distinct from U and V), we have

$$\begin{aligned} U = VU = (UU)U = [(VU)U]U = [V(UU)]U = (VV)U \\ = [(UU)V]U = [U(UV)]U = (UW)U = U(WU) \\ = U[(UV)U] = U[U(VU)] = U(UU) = UV = W. \end{aligned}$$

(3) If $VU = V$, $UV = W$, we have

$$\begin{aligned} V = VU = (VU)U = V(UU) = VV = (VU)V = V(UV) \\ = VW = (UU)W = U(UW) = U[U(UV)] = U[(UU)V] \\ = U(VV) = U[V(UU)] = U[(VU)U] = U(VU) = UV = W. \end{aligned}$$

The cases

(1′) $VU = V$, $UV = U$;
(2′) $VU = W$, $UV = U$;
(3′) $VU = W$, $UV = V$

are obviously analogous to the cases (1), (2), (3), respectively.

2.5. In those cases in which the associativity conditions are not independent it would be possible to continue our investigation by setting up minimal sets of associativity conditions from which all the rest would follow. But these cases, as we have just seen, include only the relatively uninteresting sets of not more than three elements. Consequently we will not spend any more time on them. Since the commutative case is especially interesting, it would be natural to study the associativity conditions for commutative operations. Such studies have also been made by Szász **[2]**.

3. General Semigroups and Semigroups of Transformations

3.1. Among the various branches of the general theory of algebraic operations there must exist, in agreement with what we have said above in § 1, a branch devoted to the study of multiplicative sets with an associative operation. Such multiplicative sets are called semigroups.

DEFINITION. *A* SEMIGROUP *is a nonempty set* $\mathfrak{A}$, *in which for every ordered pair of elements* X, $Y \in \mathfrak{A}$ *there is defined a new element called their product*

$$U = XY \in \mathfrak{A},$$

where for all X, $Y, Z \in \mathfrak{A}$ *we have*

$$(XY)Z = X(YZ).$$

Since we have included in our definition a reference to the method of writing the result of the operation in the form of a product it would be more exact to speak of a multiplicative semigroup. We will not do this here because the choice of notation is unimportant and also because we will have no occasion below to use any other notation than the multiplicative one.

It is to be kept in mind that in the literature other terms are used with exactly the same meaning as "semigroup," namely "associative system," "associative groupoid" and so forth. As for the term "semigroup," its use is limited by some writers to cases where there is left and right cancellation (1.18).

3.2. In agreement with 1.18 above, the following important special cases of semigroups are often singled out for particular attention.

(α) *A semigroup* $\mathfrak{A}$ *is said to be* COMMUTATIVE *if for all* A, $B \in \mathfrak{A}$ *we have*

$$AB = BA.$$

(β) *A semigroup* $\mathfrak{A}$ *is called a* GROUP *if for all* A, $B \in \mathfrak{A}$ *there exist* X *and* Y *in* $\mathfrak{A}$ *such that*

$$XA = B, \qquad AY = B.$$

(γ) *The semigroup* $\mathfrak{A}$ *is called a* SEMIGROUP WITH LEFT CANCELLATION *if for all* X, A, $B \in \mathfrak{A}$ *it follows from*

$$XA = XB$$

that $A = B$.

(δ) *A semigroup* $\mathfrak{A}$ *is called a* SEMIGROUP WITH RIGHT CANCELLATION *if for all* Y, A, $B \in \mathfrak{A}$ *it follows from*

$$AY = BY$$

that $A = B$.

(ε) *A semigroup* $\mathfrak{A}$ *is called a* SEMIGROUP WITH TWO-SIDED CANCELLATION *if it is both a semigroup with a left cancellation and a semigroup with a right cancellation.*

Of course these properties can be combined among themselves and also with other properties to give various important classes of semigroups, for example the class of commutative groups, the class of countable semigroups with left cancellation and so forth. An extremely important property is the cardinality of the set of all the elements of a semigroup. It is often called the order of the semigroup. The simplest of all semigroups, namely the semigroup consisting of a single element, is called *the unit semigroup* or the *unit group*, since it is obviously a group.

3.3. The theory of groups, which are a special case of semigroups, was developed considerably earlier than the general theory of semigroups. In fact the very term "semigroup" was formed from the already well-established term "group." At the present time the theory of groups is a profound and well developed algebraic theory with wide application. Since it is our purpose to give an exposition of the foundations of the general theory of semigroups we will occasionally have to deal with groups. In many cases we will reduce a question connected with some property of general semigroups to the study of a corresponding property of groups, and we will consider that in this case we have made a significant advance, in view of the comparatively well developed state of the theory of groups. But usually we will not pay any particular attention to questions in the theory of groups, since the reader may find them in books and articles specifically devoted to group theory.[1]

3.4. Semigroups are to be found in the most varied branches of mathematics. All the multiplicative sets mentioned in 1.21 are semigroups, since in every case the operation is associative. The only case among them which is in general a group is the multiplicative set of nonsingular matrices.

Without pausing to consider other examples let us merely note that it would be easy to give a very large number of them, a fact which indicates the importance of constructing a theory of semigroups in the various branches of mathematics.

3.5. The most important reason for constructing such a theory is the connection between the concept of a semigroup and the concept of a product of transformations.

Let Ω be a set. A mapping of the set Ω into itself is called a *transformation*. In what follows we shall express the effect of the transformation on the elements of Ω in the form of an operator. The element $\beta \in \Omega$ into which the element $\alpha \in \Omega$ is mapped by the transformation S will be written in the form of a product of S and α:

$$\beta = S\alpha.$$

In view of this notation one sometimes speaks of the product of the transformation (operator) S with α.

Let us also agree to use the following notation. If $\mathfrak{N}$ is a system of transformations of the set Ω and $\Gamma \subset \Omega$, then by $\mathfrak{N}\Gamma$ we will mean the set of all elements of Ω representable in the form $X\alpha$, where $X \in \mathfrak{N}$ and $\alpha \in \Gamma$.

Let us also mention the method of writing transformations in the form of permutations. In this notation all the elements of Ω are thought of as being written in one straight line, in arbitrary order. Then under each of these elements

---

[1] Cf. A. G. Kuroš, *The theory of groups*, Vol. I and II. Translated from the Russian and edited by K. A. Hirsch, Chelsea Publishing Co., New York, N.Y. (1955/56).

is written the element into which it is mapped (transformed) by the given transformation S:

$$S = \begin{pmatrix} \alpha & \beta & \gamma & \ldots \\ S\alpha & S\beta & S\gamma & \ldots \end{pmatrix}$$

(here $\Omega = \{\alpha, \beta, \gamma, \ldots\}$).

3.6. Let $\mathfrak{S}_\Omega$ be the set of all transformations of a given nonempty set Ω. In the set $\mathfrak{S}_\Omega$ it is natural to define the product of two transformations as the transformation obtained by carrying out the two original transformations one after the other. Thus for $X, Y, Z \in \mathfrak{S}_\Omega$ we have

$$XY = Z,$$

if for every $\alpha \in \Omega$ it is true that

$$X(Y\alpha) = Z\alpha.$$

It is easy to see that this rule for each of the pairs X, $Y \in \mathfrak{S}_\Omega$ determines XY uniquely. Consequently $\mathfrak{S}_\Omega$ is a multiplicative set.

The above operation is associative. In fact, for arbitrary $X, Y, Z \in \mathfrak{S}_\Omega$ and $\alpha \in \Omega$ we have

$$[(XY)Z]\alpha = (XY)(Z\alpha) = X[Y(Z\alpha)],$$

$$[X(YZ)]\alpha = X[(YZ)\alpha] = X[Y(Z\alpha)].$$

Since

$$[(XY)Z]\alpha = [X(YZ)]\alpha$$

for arbitrary $\alpha \in \Omega$, the transformations $(XY)Z$ and $X(YZ)$ are identical with each other.

Thus, *the set $\mathfrak{S}_\Omega$ of all transformations of an arbitrary set Ω is a semigroup with respect to the operation of forming the product* (that is, of successive application) *of the transformations.*

If $\mathfrak{A}$ is a multiplicative set of transformations of the set Ω (that is, $\mathfrak{A} \subset \mathfrak{S}_\Omega$) such that the product of two arbitrary transformations from $\mathfrak{A}$ always belongs to $\mathfrak{A}$, then obviously $\mathfrak{A}$ is a semigroup (since it follows at once from the associativity of the operation in $\mathfrak{S}_\Omega$ that the operation in $\mathfrak{A}$ is associative). Such a semigroup is called a *semigroup of transformations of the set* Ω.

The need to consider various classes of transformations arises very frequently in mathematics. If such a class is not a semigroup to begin with, it is always possible to extend it so as to form a semigroup by adjoining to it the transformations which are products of the original transformations.

3.7. Of course, the need to consider transformations arises not only in mathematics itself but also in other branches of science. Let us briefly consider, for example, the following very common situation in physics. Let $\mathfrak{A}$ be a physical system of states for which it is known that the system can occur in one of the states $\mathfrak{n}_\alpha, \mathfrak{n}_\beta, \ldots, \mathfrak{n}_\xi, \ldots$, the totality of which we will denote by Ω.

The system $\mathfrak{A}$ is now subjected to certain actions. Each of these actions S is defined by stating that if the system is in the state $\mathfrak{n}_\xi$, then as a result of the action in question the system will pass to a new state $\mathfrak{n}_\eta$ (which may of course coincide with the original state $\mathfrak{n}_\xi$ if the given state is stable with respect to this action). Thus every action on the system $\mathfrak{A}$ is simply a transformation in the set of states of the system $\Omega = \{\mathfrak{n}_\alpha, \mathfrak{n}_\beta, \ldots, \mathfrak{n}_\xi, \ldots\}$. The result of two successive actions can also be considered as an action. The transformation produced by the latter action is of course simply the product of the two transformations corresponding to the two successive actions. In this way the totality of the actions in question on the physical system $\mathfrak{A}$, being closed with respect to successive application of the actions (i.e., the result of successive applications of these actions to the system is itself one of these actions), is a semigroup of transformations of the set of all the states of the system under consideration. All questions connected with successive passage of the system from one state to another under the influence of successive application of these actions can be expressed by forming the product of elements of the semigroup in question.

3.8. Let us also point out that this situation can be made more complicated if we agree that as a result of the action the system may pass from the given state not necessarily to a certain well-defined other state but to one of several states, with well-defined probabilities for the passage to one or another of these states. Here we again obtain a semigroup whose elements, however, are defined not simply by means of transformations but by probability matrices, whose rows and columns are indexed by the elements of the given set of states and whose elements are determined by the probabilities $p_{\xi\eta}$ of passage of the system from the state $\mathfrak{n}_\xi$ corresponding to the row containing this element to the state $\mathfrak{n}_\eta$ corresponding to the column:

$$\begin{array}{c} \\ \mathfrak{n}_\alpha \\ \mathfrak{n}_\beta \\ \cdot \\ \cdot \\ \cdot \\ \mathfrak{n}_\xi \\ \cdot \\ \cdot \\ \cdot \end{array} \begin{array}{c} \begin{array}{cccc} \mathfrak{n}_\alpha & \mathfrak{n}_\beta & \cdots & \mathfrak{n}_\eta \quad \cdots \end{array} \\ \left[\begin{array}{ccccccc} & & & & & & \cdot \\ & & & & & & \cdot \\ & & & & & & \cdot \\ & & & & & & \cdot \\ & & & & & & \cdot \\ & \cdot & \cdot & \cdot & \cdot & \cdot & p_{\xi\eta} \\ & & & & & & \\ & & & & & & \\ & & & & & & \end{array} \right] \end{array}.$$

Such matrices are characterized by the conditions

$$\sum_\eta p_{\xi\eta} = 1, \qquad p_{\xi\eta} \geqslant 0.$$

Successive application of the actions corresponds, as is is easily seen, to successive formation of the product of the corresponding probability matrices.

In this way the study of questions about successive application of these actions and of their probabilistic consequences is reduced to a consideration of the product of elements in the corresponding multiplicative semigroup of probability matrices (which will be finite or infinite depending upon the number of possible states in the set Ω).

3.9. The importance of the concept of a semigroup for the theory of transformations consists not only in the fact that every multiplicative set of transformations is a semigroup but also in the converse fact, equally correct from the abstract point of view (1.8).

For the semigroup $\mathfrak{A} = \{A, B, \ldots, S, \ldots\}$ let us denote by $\Lambda_{\mathfrak{A}}$ the set consisting of all elements of $\mathfrak{A}$ and one new "separating" element I:

$$\Lambda_{\mathfrak{A}} = \{I, A, B, \ldots, S, \ldots\}.$$

To each $S \in \mathfrak{A}$ we associate the transformation T_S of the set $\Lambda_{\mathfrak{A}}$, which is called the left translation of $\Lambda_{\mathfrak{A}}$ corresponding to the element S and is such that

$$T_S I = S, \qquad T_S X = S \cdot X \qquad (X \in \mathfrak{A})$$

(the product of S and X is formed, of course, by the rule for the product of elements in $\mathfrak{A}$).

Written in the form of a substitution the left translation has the following appearance:

$$T_S = \begin{pmatrix} I & A & B & \ldots & R & \ldots \\ S & (SA) & (SB) & \ldots & (SR) & \ldots \end{pmatrix}.$$

Left translations corresponding to distinct elements of $\mathfrak{A}$ represent distinct transformations of $\Lambda_{\mathfrak{A}}$. For in fact,

$$T_S I = S, \qquad T_R I = R.$$

Let us denote by $\mathfrak{T}_{\mathfrak{A}}$ the set of all left translations corresponding to elements of $\mathfrak{A}$. It forms a semigroup of transformations of the set $\Lambda_{\mathfrak{A}}$ (the semigroup of left translations) since for arbitrary $S, R \in \mathfrak{A}$ we have

$$(T_R T_S)I = T_R(T_S I) = T_R S = RS = T_{(RS)} I,$$

$$(T_R T_S)X = T_R(T_S X) = T_R(SX) = RSX = (RS)X = T_{(RS)} X \qquad (X \in \mathfrak{A}),$$

i.e., $T_R T_S = T_{(RS)}$.

From this last equality it follows that the one-to-one mapping φ of the semigroup $\mathfrak{A}$ onto the semigroup of left translations $\mathfrak{T}_{\mathfrak{A}}$,

$$\varphi(S) = T_S \qquad (S \in \mathfrak{A}),$$

is an isomorphism.

Thus we have proof that *every semigroup is isomorphic to a semigroup of transformations* (3.6).

Since every group of transformations is contained in a semigroup $\mathfrak{S}_\Omega$ (3.6), this result shows the important role of the semigroups $\mathfrak{S}_\Omega$ in the general theory of semigroups. In a certain sense the study of arbitrary semigroups can be reduced to the study of the semigroups $\mathfrak{S}_\Omega$. This fact explains, in particular, the great amount of attention which is paid to the semigroups $\mathfrak{S}_\Omega$ in this book.

Of course, it must be understood that this statement does not mean that the isomorphic mapping of any semigroup under consideration onto some semigroup of $\mathfrak{S}_\Omega$ is in any way necessary in the investigation of an arbitrary property of a semigroup. Often it would be completely purposeless.

3.10. In terms of the theory of representations the theorem proved in 3.9 means that every semigroup is representable by transformations (i.e., representable in the class of semigroups of transformations). Of course, every semigroup has many other representations by means of transformations in addition to the representation of 3.9. But this is the representation used most often in various investigations. It is called the *representation of a semigroup by left translations.*

Investigations of the important problem of describing the construction and classification of all representations of semigroups have been begun by Stoll [**1**]. Also, V. V. Vagner [**6**] has described the construction of arbitrary representations. These studies have been continued by a number of other authors. Further details will be given below in Chapter V, §7 and Chapter X, §6.

Let us denote by Ξ the class of all multiplicative sets of transformations, i.e., the class of semigroups of transformations. As we have just shown, every semigroup can be represented in the class Ξ. On the other hand, every multiplicative set representable in Ξ is a semigroup (since every multiplicative set isomorphic to a semigroup is itself a semigroup). Thus we may say that in the class of all multiplicative sets the class of all semigroups is distinguished as the class of those multiplicative sets which are representable by transformations (the class Γ in the notation of 1.15). In other words the requirement of associativity of multiplication in a multiplicative set (i.e., the condition that the multiplicative set belong to the class of semigroups) forms an abstract characterization of the class of multiplicative sets of transformations (1.15).

3.11. It must be noted that in the set of all transformations of the set Ω, in addition to the operation defined in 3.6, we sometimes consider the operation symmetric to it, which we will denote by the symbol $\circ$. For two transformations S and R of the set Ω the result of this operation is the transformation

$$S \circ R = R \cdot S$$

(where the product $R \cdot S$ is to be understood in the sense of 3.6).

The operation $\circ$ naturally comes into consideration in those cases where the symbol for the transformation is written on the right, rather than on the left, of the element transformed (this is a rather widespread practice). In this notation we have $\alpha(S \circ R) = (\alpha S)R \quad (\alpha \in \Omega)$.

The theory described in 3.6 for the operation of multiplication of transformations is obviously altogether analogous to the theory of the operation $\circ$. The multiplicative set of all transformations of Ω, when considered with respect to this operation $\circ$, will be denoted by $\mathfrak{S}_\Omega^\circ$. It is easy to see that $\mathfrak{S}_\Omega^\circ$ is a semigroup. Although the semigroups $\mathfrak{S}_\Omega$ and $\mathfrak{S}_\Omega^\circ$ consist of exactly the same elements, the operations on them are different and consequently the semigroups $\mathfrak{S}$ and $\mathfrak{S}^\circ$ must be considered to be distinct.

The identity mapping of the set of all transformations of Ω defines an anti-isomorphism of $\mathfrak{S}_\Omega$ onto $\mathfrak{S}_\Omega^\circ$ (1.17). Since the semigroups $\mathfrak{S}_\Omega$ and $\mathfrak{S}_\Omega^\circ$ are anti-isomorphic, their structure and properties are very similar to each other. More precisely, their properties are symmetric. The semigroups $\mathfrak{S}_\Omega$ and $\mathfrak{S}_\Omega^\circ$ are not isomorphic (assuming that Ω contains more than one element). This is obvious, for example, from the fact that $\mathfrak{S}_\Omega$ contains an element U_α (α is an arbitrary fixed element from Ω, and U_α is the transformation such that $U_\alpha\xi = \alpha$ for arbitrary $\xi \in \Omega$) with the property that

$$U_\alpha S = U_\alpha$$

for every $S \in \mathfrak{S}_\Omega$. In $\mathfrak{S}_\Omega^\circ$ there is no element with this property, since for every $X \in \mathfrak{S}_\Omega^\circ$,

$$X \circ U_\alpha = U_\alpha X = U_\alpha,$$

$$X \circ U_\beta = U_\beta X = U_\beta,$$

$$U_\alpha \neq U_\beta \qquad (\alpha \neq \beta;\ \alpha, \beta \in \Omega),$$

and either $X \circ U_\alpha$ or $X \circ U_\beta$ is distinct from X.

3.12. A study of the operation $\circ$ (3.11) naturally gives rise to a type of operation which is symmetric to left translations. For $S \in \mathfrak{A}$ we define the transformation T_S° in the set $\Lambda_{\mathfrak{A}}$ (3.9) which is called a right translation:

$$T_S^\circ I = S, \qquad T_S^\circ X = X \cdot S,$$

$$T_S^\circ = \begin{pmatrix} I & A & B & \dots & R & \dots \\ S & (AS) & (BS) & \dots & (RS) & \dots \end{pmatrix}.$$

As in 3.9, it is easy to see that the mapping ψ of the semigroup $\mathfrak{A}$ into the semigroup $\mathfrak{S}_{\Lambda_{\mathfrak{A}}}^\circ$,

$$\psi(S) = T_S^\circ \qquad (S \in \mathfrak{A}),$$

is an isomorphism of $\mathfrak{A}$. This is the so-called representation of $\mathfrak{A}$ by right translations. The semigroup of right translations is denoted by $\mathfrak{T}_{\mathfrak{A}}^\circ$.

3.13. Among all transformations the invertible transformations form a particularly important class. The transformation S of the set Ω is called *invertible* if there exists a transformation S^{-1} of the same set which is its inverse in the sense that

$$SS^{-1} = S^{-1}S = E,$$

where E is the *identity transformation*, i.e., the transformation such that $E\alpha = \alpha$ for every $\alpha \in \Omega$.

From the definition it immediately follows that the transformation S is in its turn the inverse of the transformation S^{-1}. Consequently the inverse transformation is itself invertible.

The identity transformation E itself belongs to the class of invertible transformations, and for it we obviously have $E^{-1} = E$.

Let us here note one simple property of the identity transformation E. For an arbitrary transformation S of the same set Ω it is obvious that we will always have

$$SE = ES = S.$$

3.14. It is also possible to characterize an invertible transformation in an "intrinsic" way.

THEOREM. *For a transformation S of the set Ω to be invertible it is necessary and sufficient that*

(α) *for arbitrary $\xi, \eta \in \Omega$ ($\xi \neq \eta$) we have $S\xi \neq S\eta$,*

(β) $S\Omega = \Omega$ (3.5).

PROOF. (1) Let S be invertible. If for some $\xi, \eta \in \Omega$ we have

$$S\xi = S\eta,$$

then we also have

$$\xi = E\xi = S^{-1}S\xi = S^{-1}S\eta = E\eta = \eta.$$

Furthermore, since

$$\Omega = E\Omega = SS^{-1}\Omega = S(S^{-1}\Omega) \subset S\Omega$$

it follows that $S\Omega = \Omega$.

(2) If the two conditions (α) and (β) are satisfied for the transformation S, then we may define S^{-1} as the transformation of Ω such that for $S\xi = \eta$ (in view of (β) every $\eta \in \Omega$ can be represented in this fashion, and in view of (α) for every η the corresponding ξ is uniquely defined) we have

$$S^{-1}\eta = \xi.$$

For this S^{-1} we have

$$SS^{-1}\eta = S\xi = \eta, \qquad S^{-1}S\xi = S^{-1}\eta = \xi.$$

Since we may choose arbitrary elements from Ω for our ξ and η, this means that both SS^{-1} and $S^{-1}S$ are the identity transformation.

3.15. From this theorem it is clear that the transformation inverse to a given invertible transformation S is uniquely defined. In fact, let the transformations S' and S'' both be inverse to S. Every element $\eta \in \Omega$ is representable in the form $S\xi$. Since

$$S'(S\xi) = (S'S)\xi = E\xi = (S''S)\xi = S''(S\xi)$$

it follows that $S' = S''$.

If we write these transformations in the form of substitutions we obviously have

$$S = \begin{pmatrix} \alpha & \beta & \gamma & \cdots \\ \alpha' & \beta' & \gamma' & \cdots \end{pmatrix}, \qquad S^{-1} = \begin{pmatrix} \alpha' & \beta' & \gamma' & \cdots \\ \alpha & \beta & \gamma & \cdots \end{pmatrix}.$$

3.16. The property of invertibility of a transformation is closely connected with the concept of a group (3.2).

A semigroup of invertible transformations $\mathfrak{G}$ of a set Ω is called a *group of invertible transformations* if $\mathfrak{G}$ contains the inverse transformation S^{-1} for every transformation S contained in $\mathfrak{G}$. It is easy to see that in this case the semigroup $\mathfrak{G}$ is actually a group.

For every $A, B \in \mathfrak{G}$ with $X = BA^{-1}$ and $Y = A^{-1}B$, where A^{-1} is the transformation inverse to A, we have by 3.13 that

$$XA = BA^{-1}A = BE = B, \qquad AY = AA^{-1}B = EB = B.$$

On the other hand, we will show below (II, 2.17) that every group is isomorphic to a certain group of invertible transformations.

Thus the class of groups is characterized as the class of multiplicative sets possessing representations in the class of multiplicative sets of invertible transformations which with every transformation also contains its inverse. If we denote the class of such multiplicative sets of transformations by Ξ, then in the notation of 1.15 the class of groups is the class Γ. In other words the condition of being a group is the abstract characterization of the class of multiplicative sets of transformations Ξ.

3.17. Comparing this result with 3.9, we can make clear not only the roles of the general theory of semigroups and of the theory of groups in mathematics as a whole, but also the mutual relation between these two concepts. The algebraic theory of semigroups can be concisely characterized as the abstract study of arbitrary transformations. Analogously, the theory of groups is the abstract study of invertible transformations. In view of the important role of invertible transformations it is clear why the theory of groups occupies such an important place. Since the study of invertible transformations naturally was the first to attract the attention of mathematicians, the development of the theory of groups naturally took place first. Since, on the other hand, we obviously cannot confine ourselves to the study of invertible transformations, the problem naturally arose of constructing a general abstract theory of semigroups.

3.18. The study of various transformations and of their successive application (their product) is so customary in present day mathematics that we would have no difficulty in finding a practically limitless number of examples. Let us hastily examine just one direction in the theory of transformations which is destined to play an important role in the theory of semigroups and in its applications to various branches of mathematics.

Among the elements of a given set let certain relations be defined. For the moment we will understand the term "relation" in its general descriptive sense and will not try to make it precise (below we give a general construction for relations which obviously makes the idea in question quite precise).

The presence of certain relations in the set naturally distinguishes among the various transformations of this set those transformations which do not destroy these relations. Such transformations of the set with the given system of relations will be called *endomorphisms* (for a different, somewhat narrower, meaning of this term see below). Thus a transformation of a set with relations is an endomorphism if every system of elements from this set which are related to each other by the given relations, is transformed into a system of elements which are again related to one another by the given relations. The identity mapping of a set on itself is, of course, always an endomorphism. The result of the successive application of two endomorphisms is a transformation which obviously does not destroy the relations in question, so that it too is an endomorphism. From this it follows that for an arbitrary system of relations in an arbitrary set the aggregate of endomorphisms is always a semigroup, which is called the semigroup of endomorphisms.

Obviously the algebraic properties of a semigroup of endomorphisms are determined by the properties of the system of relations in the set. Conversely, the algebraic properties of the semigroup of endomorphisms characterize to a certain extent the original relations in the set under consideration.

Consequently the study of semigroups of endomorphisms can be utilized to classify and investigate the properties of sets with various systems of relations among their elements.

It is natural to expect that the study of semigroups of endomorphisms and the establishment of the above mentioned connections between their properties and the properties of the sets themselves should be one of the most important directions for development of the theory of semigroups and should very likely be particularly rich in applications of the algebraic theory of semigroups in various branches of mathematics. Some investigations in this direction will be discussed below in Chapter V, §8.

3.19. In illustration of these remarks let us consider some semigroups of endomorphisms.

(α) Let $\mathfrak{A}$ be a semigroup (in particular, it may be a group). As relations in $\mathfrak{A}$ let us take those relations which determine the operation in $\mathfrak{A}$,

$$XY = Z \qquad (X, Y, Z \in \mathfrak{A}).$$

In this case an endomorphism will always be a mapping of the semigroup $\mathfrak{A}$ into itself such that for the elements X, Y, Z connected by relation $XY = Z$ we always have

$$\varphi(X) \cdot \varphi(Y) = \varphi(Z).$$

Let us note that it was in this sense (and restricted to groups) that the term "endomorphism" has been used up to now. Only a few of the connections between the properties of a group and the properties of its semigroup of endomorphisms have been noted in the mathematical literature.

(β) Let Ω be the set of all monomials of the variables $x_1, x_2, \ldots, x_n$ over a certain field P:

$$u = \alpha x_1^{s_1} x_2^{s_2} \ldots x_n^{s_n} \qquad (\alpha \in P).$$

In Ω there is defined a relation among certain triplets of monomials u, v, w, which defines multiplication of these monomials in the ordinary sense: $uv = w$. Let us, moreover, distinguish the monomials of zero degree, that is, the elements of the field P. Then let us also consider the fixed choice of each of these as a relation in Ω. Then the endomorphisms of the set Ω are to be considered as those transformations X of Ω for which we always have $X(uv) = X(u) \cdot X(v)$ $(u, v \in \Omega)$ and $X(\alpha) = \alpha$ $(\alpha \in P)$. Clearly every such endomorphism X is completely determined by the images of the variables x_i:

$$\begin{aligned} Xx_1 &= \alpha_1 x_1^{k_{11}} x_2^{k_{12}} \ldots x_n^{k_{1n}}; \\ Xx_2 &= \alpha_2 x_1^{k_{21}} x_2^{k_{22}} \ldots x_2^{k_{2n}}; \\ &\ldots\ldots\ldots\ldots\ldots ; \\ Xx_n &= \alpha_n x_1^{k_{n1}} x_2^{k_{n2}} \ldots x_n^{k_{nn}}. \end{aligned}$$

Here the choice of an arbitrary sequence of elements $\alpha_1, \alpha_2, \ldots, \alpha_n$ from P and of non-negative integers k_{ij} $(1 \leqslant i, j \leqslant n)$ obviously determines an endomorphism X.

These endomorphisms are a completely natural and immediate generalization of the so-called monomial substitutions studied in the theory of groups, where they play an important role. The ordinary monomial substitutions in group theory are those endomorphisms, among the ones listed above, for which $\alpha_i \neq 0$ and every k_{ij} is equal to zero with the exception of n indices $k_{1j_1}, k_{2j_2}, \ldots, k_{nj_n}$ $(j_p \neq j_q$ for $p \neq q)$ which are equal to unity.

(γ) In the same set Ω which was considered under (β) we may define the relation of divisibility. The elements u and v are connected by this relation if there exists a $t \in \Omega$ such that $u = vt$. For this relation there exists the corresponding semigroup of endomorphisms. Obviously, every transformation of the semigroup considered in (β) belongs to the semigroup of endomorphisms which does not destroy the relation of divisibility. But this latter semigroup is wider than the semigroup in (β). For example, it includes the transformation Z,

$$Z(\alpha x_1^{s_1} x_2^{s_2} \ldots x_n^{s_n}) = \alpha x_1^{s_1+1} x_2^{s_2+1} \ldots x_n^{s_n+1},$$

which is not contained, as can easily be seen, in the semigroup of endomorphisms considered in (β).

(δ) Let $\Gamma = \{\alpha, \beta, \ldots\}$ be an arbitrary subset of a topological space Ω. We shall say that the elements $\alpha, \beta, \ldots$ which make up Γ are connected by a given relation if Γ is an open subset of Ω. The definition of this relation is equivalent to the definition of the topology in the set of all the elements of Ω.

This relation can be given equally well by the definition of the topology in the whole space Ω. The endomorphisms of Ω will in this case be the transformations of Ω which take any open subset into an open subset. Thus in the present case the semigroup of endomorphisms is the semigroup of open mappings of the space Ω into itself.

(ε) Analogously to (δ) we may consider in Ω the relation which connects the elements $\alpha, \beta, \ldots$ in the case that $\{\alpha, \beta, \ldots\}$ is a closed subset of Ω. The endomorphisms corresponding to this relation are the closed mappings of Ω into itself.

(ζ) Let Ω_1 be a field which is a normal extension of a given field Ω_2. As relations in Ω_1 we could take those which determine the operations of addition and multiplication in Ω_1, and also those which leave fixed every element belonging to Ω_2. As is well known, all transformations which preserve these relations are one-to-one invertible transformations.

The semigroup of endomorphisms in this case consists of invertible transformations, the so-called automorphisms of the extension, and form a group. The connections between the properties of the extension and the properties of its group of automorphisms have been thoroughly studied. They are of profound importance for the theory of fields. They form the subject matter of the so-called Galois theory, which itself is a magnificent example of the above mentioned direction in the development of the theory of semigroups, namely, the study of semigroups of transformations.

3.20. In general, the definition of an endomorphism given in 3.18 is not sufficiently precise, in view of the fact that we have not defined very clearly just what is to be understood by a relation in a set in the general case. This did not prevent us from considering a number of semigroups of endomorphisms, since in each concrete case we stated very definitely just which relations we were speaking about in the sets under consideration. Let us now spend a few moments on the problem of giving a precise sense to the concept of an endomorphism in the general case.

Let Ω be the set in question. Let us choose some set Σ and consider the mappings of Σ into Ω. From the set of all such mappings we select a certain system of mappings Φ. This system will define a relation in the following way. For $\varphi \in \Phi$ we will say that the elements $\varphi(\xi)$, $\varphi(\eta), \ldots$ of the set Ω (where $\Sigma = \{\xi, \eta, \ldots\}$) are related to each other by the relation defined by the system Φ.

If Ω is a set together with a system of relations, each one of which is defined in the above manner, then the endomorphism X will be defined as a transformation of Ω such that the mapping $\psi = X\varphi$ (the result of the successive application of the mapping φ followed by X) belongs to Φ for arbitrary $\varphi \in \Phi$.

It is easy to see that the examples introduced in 3.19 will come under this general definition.

For example, let there be an operation defined in the set Ω. For Σ we will take the set $\{1, 2, 3\}$. We then define the mapping φ of the set Σ into Ω as belonging to Φ if for elements $\varphi(1)$, $\varphi(2)$, $\varphi(3)$ belonging to Ω we have $\varphi(1) \cdot \varphi(2) = \varphi(3)$ in Ω. It is obvious that the relation defined in this way completely determines the original operation in Ω.

The specification of the constants constituting a subset Ω' of the set Ω can also be carried out by means of the introduction of certain relations. For every $\lambda \in \Omega'$ a relation is determined by defining the set Σ consisting of a single element, where the mapping φ of this $\Sigma = \{\sigma\}$ into Ω belongs to Φ if and only if $\varphi(\sigma) = \lambda$. If we include this relation in our system of relations, then the endomorphism X will certainly leave λ unchanged, i.e., $X(\lambda) = \lambda$.

It is also possible to define a topology in the set Ω by defining a certain relation. To do this, we take for Σ a set with cardinality not less than the cardinality of Ω. A mapping of Σ into Ω is included in the class Φ if $\varphi(\Sigma)$ is an open set (or a closed set in the example considered in 3.19, (ε)). Then the transformations which do not destroy this relation are the open mappings of the topological space into itself.

It is easy to imagine various other forms and types of relations which can be obtained as concrete examples of the above general construction. In the study of such relations the corresponding semigroup of endomorphisms is often a useful tool.

4. Partial Transformations

4.1. In certain branches of mathematics it is necessary to generalize the concept of a transformation.

Let Π_1 and Π_2 be two subsets of the set Ω. A mapping S of Π_1 onto Π_2 is called a *partial transformation of the set* Ω.

Just as for transformations, the operator notation is also used for partial transformations.

$$S\alpha = \beta$$

denotes that $\alpha \in \Pi_1$ and in the partial transformation S the element α is mapped onto $\beta \in \Pi_2$. Among partial transformations we also consider the empty one, for which Π_1 and Π_2 are empty. The sets Π_1 and Π_2 are sometimes called, respectively, the first and the second projection of S (these terms are natural when the partial mapping is written in the form of a binary relation in Ω). We will usually write these two sets with an indication of the partial transformation S in the manner $\Pi_1^{(S)}$ and $\Pi_2^{(S)}$. Let us note that the partial transformation S may also be defined by giving $\Pi_1^{(S)}$ and a rule whereby to each $\alpha \in \Pi_1^{(S)}$ the corresponding element $S\alpha \in \Omega$ is determined. The set $\Pi_2^{(S)}$ is then completely defined.

An ordinary (the term "complete" is also used) transformation S is a particular case of a partial mapping, where $\Pi_1^{(S)} = \Omega$.

4.2. In the set $\mathfrak{P}_\Omega$ of all partial transformations of the set Ω we define the operation of multiplication as being the successive application of partial transformations. Namely, the product of the partial mappings X, $Y \in \mathfrak{P}_\Omega$ is the partial mapping $Z = XY \in \mathfrak{P}_\Omega$ for which $\Pi_1^{(Z)}$ consists of all those elements $\alpha \in \Omega$ which are contained in $\Pi_1^{(Y)}$, where $Y\alpha \in \Pi_1^{(X)}$ (if there is no such α in Ω, then Z is the empty partial transformation). For each $\alpha \in \Pi_1^{(Z)}$ we set $Z\alpha = X(Y\alpha)$.

It is easy to prove that the product of partial transformations is associative. Each of the three-fold products

$$(XY)Z, \qquad X(YZ)$$

has for its Π_1 the set of all elements $\alpha \in \Omega$ contained in $\Pi_1^{(Z)}$, where $Z\alpha \in \Pi_1^{(Y)}$ and $Y(Z\alpha) \in \Pi_1^{(X)}$.

For such α we have

$$(XY)Z\alpha = X(Y(Z\alpha)),$$
$$X(YZ)\alpha = X(Y(Z\alpha)).$$

Consequently the set of all partial transformations $\mathfrak{P}_\Omega$ is a semigroup.

4.3. For a deeper study of partial transformations the following relation is of essential importance. For X, $Y \in \mathfrak{P}_\Omega$ we write $X \leqslant Y$ if $\Pi_1^{(X)} \subset \Pi_1^{(Y)}$ and $X\alpha = Y\alpha$ for every $\alpha \in \Pi_1^{(X)}$. In this case we may say that Y is an extension of the partial transformation X. In particular, an important role is played by the partially identical mappings for which the identical mapping is an extension. It is obvious that every partially identical partial mapping is completely determined by the set Π_1, which is in this case identical with Π_2.

4.4. Just as for transformations, it is often necessary in the theory of partial transformations to consider not only the semigroup of all partial transformations $\mathfrak{P}_\Omega$ but also various subsets of it which are closed with respect to multiplication and therefore are themselves obviously semigroups.

For example, let Ω be the set of all real numbers. The partial transformations of Ω are just the real functions of one variable. For such a transformation $X = f(x)$ the $\Pi_1^{(X)}$ is what is usually called the domain of definition of the real function $f(x)$, and the $\Pi_2^{(X)}$ is the range of this function. The product of two such partial transformations $X_1 = f_1(x)$ and $X_2 = f_2(x)$ is the composite function

$$\varphi(x) = f_1[f_2(x)].$$

In the theory of functions the operation of multiplication of such partial transformations is called composition of the corresponding functions.

The set $\mathfrak{P}_\Omega$ of all real functions (including the "improper function" with empty sets Π_1 and Π_2) forms a semigroup with respect to the above operation. The subset consisting of all continuous functions is easily seen to be closed with respect to this operation and therefore forms a semigroup itself. The subset consisting of all functions whose domain is included in a given interval E of the real axis is also a semigroup. The periodic real functions also form a semigroup with respect to the same operation.

4.5. Both for partial transformations in general and for the partial transformations of any given concrete set Ω, an important role is played by the one-to-one partial transformations which determine a one-to-one mapping of Π_1 onto Π_2. The entire system of such mappings we will denote by $\mathfrak{Q}_\Omega$. Evidently $\mathfrak{Q}_\Omega$ is a semigroup. The partially identical partial transformations discussed above belong to the semigroup $\mathfrak{Q}_\Omega$.

To every one-to-one partial transformation X there corresponds an inverse partial transformation $\bar{X}$ (it is often denoted by X^{-1}) for which $\Pi_1^{(\bar{X})} = \Pi_2^{(X)}$ and $\Pi_2^{(\bar{X})} = \Pi_1^{(X)}$, where for every $\alpha \in \Pi_1^{(X)}$ and $\beta \in \Pi_2^{(X)}$ such that $X\alpha = \beta$ we have

$$\bar{X}\beta = \alpha.$$

$\bar{X}X$ is the partially identical partial transformation corresponding to the set $\Pi_1^{(X)}$, and $X\bar{X}$ is the partially identical transformation corresponding to the set $\Pi_2^{(X)}$. It is easy to see that in the subgroup $\mathfrak{Q}_\Omega$ the mapping with which every $X \in \mathfrak{Q}_\Omega$ associates the element $\bar{X}$ is an anti-isomorphism (1.17).

4.6. V. V. Vagner [3] has cleared up the question of when a system of one-to-one partial transformations is a group. Of course, it is necessary to this end that the system should, in the first place, be closed with respect to multiplication, that is it must be a semigroup.

THEOREM. *A semigroup of one-to-one partial transformations $\mathfrak{A}$ of the set Ω forms a group if and only if $\mathfrak{A}$ contains the inverse transformation for every transformation contained in $\mathfrak{A}$ and in Ω there is a subset $\Omega_0 \subset \Omega$ such that*

$$\Pi_1 = \Pi_2 = \Omega_0$$

for all partial transformations in $\mathfrak{A}$.

PROOF. (1) Let $\mathfrak{A}$ be a group.

Then for arbitrary $A, B \in \mathfrak{A}$ there must exist $X, Y \in \mathfrak{A}$ such that

$$XA = B, \qquad AY = B.$$

From the definition of Π_1 and Π_2 for a product it is clear that if these equalities are to hold we must have

$$\Pi_1^{(A)} \supset \Pi_1^{(B)}, \qquad \Pi_2^{(A)} \supset \Pi_2^{(B)}.$$

But in exactly the same way we must have also the inverse inclusions. Thus, for arbitrary A and B the sets Π_1 must be identical and also the sets Π_2.

For $A \in \mathfrak{A}$ there must exist a partial transformation I in $\mathfrak{A}$, such that

$$A = AI.$$

From this equality it follows that for every $\alpha \in \Pi_1$ we have

$$A\alpha = A(I\alpha).$$

Since A is one-to-one, this can happen only for $\alpha = I\alpha$. Consequently I is a partially identical partial transformation. But for such a transformation $\Pi_1 = \Pi_2$. Consequently, for all partial transformations in $\mathfrak{A}$ the sets Π_1 and Π_2 are equal to each other for all elements.

For $A \in \mathfrak{A}$ there exists a partial transformation A' in $\mathfrak{A}$ such that

$$A'A = I.$$

Since I is partially identical, we conclude from the equality of all Π_1 and Π_2 that A', which belongs to $\mathfrak{A}$, must be the one-to-one partial transformation inverse to A.

(2) Let $\mathfrak{A}$ possess the stated properties.

Since $\Pi_1^{(A)} = \Pi_2^{(A)}$, we therefore have for the partial transformation $\bar{A}$ inverse to A the following:

$$A\bar{A} = \bar{A}A = I,$$

where I is the partially identical partial transformation corresponding to the set Ω_0. Since it is obviously true that

$$IB = BI = B$$

for arbitrary $B \in \mathfrak{A}$, therefore for $X = B\bar{A}$ and $Y = \bar{A}B$ we have

$$XA = (B\bar{A})A = B(\bar{A}A) = BI = B,$$

$$AY = A(\bar{A}B) = (A\bar{A})B = IB = B.$$

4.7. The study of semigroups of partial transformations can be reduced to the study of transformations. Let $\mathfrak{A}$ be a semigroup of partial transformations of the set Ω. Let us consider a new set Ω' obtained from Ω by the adjunction of a new element Θ. To every partial transformation A of $\mathfrak{A}$ we associate the transformation A' of the set Ω' such that

$$A'\alpha = A\alpha$$

for all $\alpha \in \Pi_1^{(A)}$ and

$$A'\beta = \Theta$$

for every element $\beta \in \Omega' \ \Pi_1^{(A)}$. If $A \neq B\,(A, B \in \mathfrak{A})$, then obviously $A' \neq B'$. It is also obvious that for arbitrary $A, B \in \mathfrak{A}$ we have

$$(AB)' = A'B'.$$

From this it follows that the aggregate $\mathfrak{A}'$ consisting of all possible transformations A' ($A \in \mathfrak{A}$) of the set Ω' is a semigroup isomorphic to the semigroup $\mathfrak{A}$. Here the correspondence $A \sim A'$ ($A \in \mathfrak{A}$, $A' \in \mathfrak{A}'$) not only defines an isomorphism but also sets up an obvious parallelism between the properties of the elements of $\mathfrak{A}$ considered as partial transformations of Ω and the corresponding properties of the elements of $\mathfrak{A}'$, considered as transformations of the set Ω'.

Incidentally we should remark that this reduction of semigroups of partial transformations to semigroups of transformations is by no means always expedient. In particular, in the study of one-to-one partial transformations it is usually inconvenient, since the corresponding transformations of the new set Ω' may very well turn out not to be one-to-one.

4.8. Let $\mathfrak{A}$ be a semigroup of partial transformations of the set Ω. From the point of view of the operation of the elements of $\mathfrak{A}$ on the elements of Ω the latter set may sometimes turn out to be unnecessarily wide. So it is sometimes desirable to construct a certain reduction of the set Ω. Let us denote by Ω' the union of all $\Pi_2^{(A)}$ $(A \in \mathfrak{A})$. Also we write $\Omega'' = \Omega \backslash \Omega'$. For $\alpha \in \Omega''$ we denote by Γ_α the set of all elements $\beta \in \Omega''$ such that $A\alpha = A\beta$ (in particular, $A\alpha$ and $A\beta$ may both be empty symbols for $\alpha, \beta \bar{\in} \Pi_1^{(A)}$). Obviously it follows from $\Gamma_\alpha \ni \beta$ that $\Gamma_\beta = \Gamma_\alpha$. Consequently, Ω'' is divided into the nonintersecting union of sets Γ_α. In each of these let us fix a certain element. The union of the set of these elements and Ω' we will denote by $\tilde{\Omega}$. To each $A \in \mathfrak{A}$ it is natural to associate a partial transformation of the set $\tilde{\Omega}$:

$$\tilde{\Pi}_1^{(\tilde{A})} = \Pi_1^{(A)} \cap \tilde{\Omega}, \qquad \tilde{\Pi}_2^{(\tilde{A})} = \Pi_2^{(A)},$$

$$\tilde{A}\alpha = A\alpha \qquad (\alpha \in \tilde{\Pi}_1^{(\tilde{A})})$$

(the tilde on Π_1 and Π_2 indicates that we are speaking of partial transformations of the set $\tilde{\Omega}$).

Assume $A_1 \neq A_2$ $(A_1, A_2 \in \mathfrak{A})$. If also $\Pi_1^{(A_1)} \not\subseteq \Pi_1^{(A_2)}$ (the case $\Pi_1^{(A_2)} \not\subseteq \Pi_1^{(A_1)}$ is completely analogous), then for some $\xi \in \Pi_1^{(A_1)}$ we have $\xi \bar{\in} \Pi_1^{(A_2)}$. If $\xi \in \Omega'$, then $\xi \in \tilde{\Omega}$ and therefore $\xi \in \tilde{\Pi}_1^{(\tilde{A}_1)}$, but $\xi \bar{\in} \tilde{\Pi}_1^{(\tilde{A}_2)}$. But if $\xi \in \Omega''$, then we choose from Γ_ξ an element ξ_0 belonging to $\tilde{\Omega}$. This ξ_0 belongs to $\tilde{\Pi}_1^{(\tilde{A}_1)}$ but not to $\tilde{\Pi}_1^{(\tilde{A}_2)}$. Thus, in this case $\tilde{A}_1 \neq \tilde{A}_2$.

If $\Pi_1^{(A_1)} = \Pi_1^{(A_2)}$, then because of $A_1 \neq A_2$ there must exist $\xi \in \Pi_1^{(A_1)}$ such that $A_1\xi \neq A_2\xi$. Since $A_1\xi, A_2\xi \in \Omega' \subset \Omega$ it follows that

$$\tilde{A}_1\xi = A_1\xi \neq A_2\xi = \tilde{A}_2\xi,$$

i.e., again $\tilde{A}_1 \neq \tilde{A}_2$.

Thus the correspondence between the partial transformations $\tilde{A}$ and the partial transformations A is one-to-one. Here we also have $\widetilde{A_1A_2} = \tilde{A}_1\tilde{A}_2$. Consequently, the semigroup $\mathfrak{A}$ is isomorphic to the semigroup $\tilde{\mathfrak{A}}$ consisting of the partial transformations $\tilde{A}$ of the set $\tilde{\Omega}$. This correspondence is even more than an isomorphism, since all the properties of the partial transformations of $\mathfrak{A}$ can be immediately deduced in an evident way from the corresponding properties of the partial transformations of $\tilde{\mathfrak{A}}$. In a certain sense it is possible to say that $\tilde{\mathfrak{A}}$ is the semigroup of the same partial transformations A in $\mathfrak{A}$, but with a shortened "reduced" domain of application. In certain cases the reduced set may turn out to be substantially narrower than Ω, and the passage to $\tilde{\Omega}$ may offer definite advantages.

A second nontrivial reduction of $\tilde{\Omega}$ is impossible. In this sense $\tilde{\Omega}$ is irreducible. In fact, it is evident that each class $\tilde{\Gamma}_\alpha$ for $\tilde{\Omega}$ (with respect to $\tilde{\mathfrak{A}}$) consists of only one element. Thus, $\widetilde{(\tilde{\Omega})} = \tilde{\Omega}$.

4.9. Just as for transformations the theory of partial transformations may prove definitely useful in the investigation of properties of various relations in a set Ω if we consider those partial transformations which preserve the given relations in Ω. These may be called partial endomorphisms of the set Ω with the given system of relations. They form a semigroup. This semigroup of partial endomorphisms obviously includes all endomorphisms (3.18).

In studying partial endomorphisms one should keep in mind that some of them can be extended to form endomorphisms. In other words they may be obtained from certain endomorphisms by narrowing the range of definition in Ω down to a part of the whole set Ω. An example of an investigation in this direction is given in a paper by B. Neumann and H. Neumann.[2]

The study of partial transformations and in particular of one-to-one partial transformations can be very useful, as was shown by V. V. Vagner in certain geometric investigations.

5. Relations

5.1. In our further study of various properties of semigroups it will often be convenient for us to make use of the concept of relations in semigroups and of some of their simplest properties. *A relation in a set* $\mathfrak{M}$ is a binary relation in $\mathfrak{M}$, that is, a rule which distinguishes certain pairs of elements of $\mathfrak{M}$. The pair (X, Y) is defined by its two elements X and Y and by their order, that is, by which of the two is the first component of the pair and which is the second. Of course we may take the relation to be simply the set itself of distinguished pairs of elements from $\mathfrak{M}$. The fact that the pair (X, Y) $(X, Y \in \mathfrak{M})$ is one of the distinguished pairs with respect to the relation $\mathfrak{n}$ (that is, that the elements X and Y *are in the relation* $\mathfrak{n}$), will usually be written in the form $X \sim Y(\mathfrak{n})$. If the elements X and Y are not in the relation $\mathfrak{n}$ (that is, if the pair (X, Y) is not one of the distinguished pairs), then we will write $X \not\sim Y(\mathfrak{n})$.

To indicate that the elements X and Y are in the relation $\mathfrak{n}$ with respect to each other the following notations are also used:

$$X \underset{\mathfrak{n}}{\sim} Y, \quad X \mathfrak{n} Y, \quad X \equiv Y(\mathfrak{n}), \quad X \equiv Y(\text{mod } \mathfrak{n}), \quad (X, Y) \in \mathfrak{n}.$$

If in the course of a specific argument a special sign is used to denote a given relation, for example "$\leqslant$," then, of course, the same sign is used between

[2] B. Neumann and H. Neumann, *Extending partial endomorphisms of groups*, Proc. London Math. Soc. (3) **2** (1952), 337–348.

elements that are in the given relation, namely, $X \leqslant Y$. We will sometimes make use of this particular notation.

It is also possible to consider relations among elements belonging to distinct sets, but we will agree to discuss a relation only within the limits of one set. In addition to binary relations we may also consider ternary and in general n-ary relations ($n = 1, 2, \ldots$). But in this book we will take the word relation to refer only to binary relations.

5.2. Let Φ be the system of all possible relations in the set $\mathfrak{M}$ (that is, the system of all subsets of the set of all pairs of elements of $\mathfrak{M}$). Among the relations belonging to Φ we also include the "empty relation," with respect to which no two given elements of $\mathfrak{M}$ are in the given relation. For relations in Φ it is natural to define the operation of multiplication. Namely, for $\mathfrak{k}, \mathfrak{l} \in \Phi$ we define as their product $\mathfrak{m} = \mathfrak{k} \cdot \mathfrak{l}$ the relation $\mathfrak{m}$ for which $X \sim Y(\mathfrak{m})$ $(X, Y \in \mathfrak{M})$ holds if and only if there is a $Z \in \mathfrak{M}$ such that $X \sim Z(\mathfrak{k})$ and $Z \sim Y(\mathfrak{l})$ (of course, $\mathfrak{m}$ may sometimes be the empty relation, but in any case it is completely determined for given $\mathfrak{k}$ and $\mathfrak{l}$).

The multiplication defined in this way is associative. For let $\mathfrak{k}_1, \mathfrak{k}_2, \mathfrak{k}_3 \in \Phi$ and

$$(\mathfrak{k}_1\mathfrak{k}_2)\mathfrak{k}_3 = \mathfrak{m}', \qquad \mathfrak{k}_1(\mathfrak{k}_2\mathfrak{k}_3) = \mathfrak{m}''.$$

Obviously, $X \sim Y(\mathfrak{m}')$ will hold if and only if there exists $Z', Z'' \in \mathfrak{M}$ such that

$$X \sim Z'(\mathfrak{k}_1), \qquad Z' \sim Z''(\mathfrak{k}_2), \qquad Z'' \sim Y(\mathfrak{k}_3).$$

But the existence of Z' and Z'' is obviously necessary and sufficient for $X \sim Y(\mathfrak{m}'')$. Consequently, $\mathfrak{m}' = \mathfrak{m}''$.

Thus the system of all relations in a given set is a semigroup.

5.3. The properties of multiplication of relations are discussed in the books of Bourbaki[3] and Dubreil [**3**]. Subsequently they were studied by a number of other authors, in particular V. V. Vagner and B. M. Šaĭn. In the present book we shall not pay any detailed attention to them. Let us confine ourselves here to examining the question of commutativity of two arbitrary relations in Φ with respect to the operation of multiplication. The formula $X \sim Y(\mathfrak{n}_1 \cdot \mathfrak{n}_2)$ implies the existence of a $U \in \mathfrak{M}$ such that $X \sim U(\mathfrak{n}_1)$, $U \sim Y(\mathfrak{n}_2)$. The formula $X \sim Y(\mathfrak{n}_2 \cdot \mathfrak{n}_1)$ implies the existence of a $V \in \mathfrak{M}$ such that $X \sim V(\mathfrak{n}_2)$, $V \sim Y(\mathfrak{n}_1)$. Consequently the condition

$$\mathfrak{n}_1 \cdot \mathfrak{n}_2 = \mathfrak{n}_2 \cdot \mathfrak{n}_1$$

means that from

$$X \sim U(\mathfrak{n}_1), \qquad U \sim Y(\mathfrak{n}_2) \qquad (X, Y, U \in \mathfrak{M})$$

follows the existence of a $V \in \mathfrak{M}$ such that

$$X \sim V(\mathfrak{n}_2), \qquad V \sim Y(\mathfrak{n}_1).$$

---

[3] N. Bourbaki, *Théorie des ensembles*, Actualités Sci. Ind. No. 846, Hermann, Paris, 1939.

Conversely, from these last relations there follows the existence of a U satisfying the above condition.

The properties connected with commutativity of relations have been considered by several different authors. For example, particular attention has been given to them in the book of Dubreil [3]. The importance of the property of commutativity of relations was first emphasized by Dubreil and Dubreil-Jacotin.[4]

5.4. The concept of a relation in a set is a generalization of the concept of a transformation. Namely, the transformation can be defined as a relation $\mathfrak{n}$ in the set $\mathfrak{M}$ satisfying the following condition. For every $Y \in \mathfrak{M}$ there exists a unique $X \in \mathfrak{M}$ such that $X \sim Y(\mathfrak{n})$. In this case X may be considered as the result of the application of the given transformation to the element Y. Obviously, the above multiplication of relations applied to the case of transformations produces the same result as the multiplication of transformations defined earlier (3.6).

From this it follows in particular that every semigroup, since it can be represented isomorphically by transformations, can also be represented isomorphically by relations.

5.5. The concept of a relation also generalizes the concept of partial transformation (4.1). Namely, a partial transformation of a set $\mathfrak{M}$ may be considered as a relation $\mathfrak{n}$ satisfying the condition, from $X \sim Y(\mathfrak{n})$ and $X' \sim Y(\mathfrak{n})$ it follows that $X = X'$. The set Π_1 here consists of all $Y \in \mathfrak{M}$ for which there exists an $X \in \mathfrak{M}$ such that $X \sim Y(\mathfrak{n})$, and the set Π_2 consists of all such $X \in \mathfrak{M}$ for all possible $Y \in \Pi_1$. The element X with $X \sim Y(\mathfrak{n})$ is the result of the application of the given partial transformation to the element Y in Π_1.

The multiplication of partial transformations is obviously a particular case of the multiplication of relations.

One-to-one partial transformations, which are of such great importance, can be characterized as those relations which satisfy the following condition. If $X \sim Y(\mathfrak{n})$ and $X' \sim Y'(\mathfrak{n})$, then from $X = X'$ follows $Y = Y'$ and from $Y = Y'$ follows $X = X'$.

5.6. In the study of relations it is extremely important to determine whether they possess certain properties.

DEFINITION. *Let $\mathfrak{n}$ be a relation in the set $\mathfrak{M}$.*

The relation $\mathfrak{n}$ is called REFLEXIVE *if for every $X \in \mathfrak{M}$ we have $X \sim X(\mathfrak{n})$.*

The relation $\mathfrak{n}$ is called TRANSITIVE *if from $X \sim Y(\mathfrak{n})$ and $Y \sim Z(\mathfrak{n})$ we always have $X \sim Z(\mathfrak{n})$.*

The relation $\mathfrak{n}$ is called SYMMETRIC *if from $X \sim Y(\mathfrak{n})$ always follows $Y \sim X(\mathfrak{n})$.*

The relation $\mathfrak{n}$ is called ANTI-SYMMETRIC *if from $X \sim Y(\mathfrak{n})$ and $Y \sim X(\mathfrak{n})$ always follows $X = Y$.*

[4] P. Dubreil and M.-L. Dubreil-Jacotin, *Théorie algébrique des relations d'équivalence*, J. Math. Pures Appl. **18** (1939), 63–95.

5.7. Among all types of relations the following two classes are most important to us. The first is the class of equivalences.

DEFINITION. *The relation* $\mathfrak{n}$ *in the set* $\mathfrak{M}$ *is called a* RELATION OF EQUIVALENCE *in* $\mathfrak{M}$ *or simply an equivalence if* $\mathfrak{n}$ *is reflexive, transitive and symmetric.*

5.8. The representation of the set $\mathfrak{M}$ in the form of a union of certain nonempty nonintersecting subsets

$$\mathfrak{M} = \bigcup_{\xi} \mathfrak{M}_{\xi}$$

is called a *decomposition* (or *partition*) *of the set* $\mathfrak{M}$.

Let $\mathfrak{n}$ be an equivalence in the set $\mathfrak{M}$. The system of all elements equivalent with respect to $\mathfrak{n}$ to a certain element $X \in \mathfrak{M}$ (that is, of elements which are in relation $\mathfrak{n}$ to X) will be called an equivalence class with respect to $\mathfrak{n}$, or more briefly, simply an $\mathfrak{n}$-*class*. All elements of one $\mathfrak{n}$-class are equivalent to one another with respect to $\mathfrak{n}$ and none of them is equivalent with respect to $\mathfrak{n}$ to any element lying in any other class. Thus equivalence with respect to $\mathfrak{M}$ in the set $\mathfrak{M}$ corresponds to the decomposition of $\mathfrak{M}$ into $\mathfrak{n}$-classes.

The converse is obvious. If we have a certain decomposition of $\mathfrak{M}$

$$\mathfrak{M} = \bigcup_{\xi} \mathfrak{M}_{\xi}$$

we can define a relation $\mathfrak{n}$ in $\mathfrak{M}$ by setting $X \sim Y(\mathfrak{n})$ if and only if the elements X and Y belong to one and the same component $\mathfrak{M}_{\xi}$ of the given decomposition. Obviously this relation will be reflexive, transitive, and symmetric, that is, it will be an equivalence.

This equivalence corresponds to the original decomposition of $\mathfrak{M}$, since the components $\mathfrak{M}_{\xi}$ of the original decomposition will be the $\mathfrak{n}$-classes of this equivalence.

Thus *to define an equivalence in* $\mathfrak{M}$ *is essentially the same as to define the corresponding decomposition of* $\mathfrak{M}$.

In the future we will make no distinction between these two points of view.

5.9. A second important class of relations is formed by partial orderings.

DEFINITION. *A relation* $\mathfrak{n}$ *in the set* $\mathfrak{M}$ *is called a* RELATION OF PARTIAL ORDERING *if* $\mathfrak{n}$ *is reflexive, transitive and anti-symmetric* (5.6).

A set considered with respect to a relation of partial ordering defined for it is called a PARTIALLY ORDERED SET.

For the relation of partial ordering it is customary to use the inequality symbol. For $X \sim Y(\mathfrak{n})$, where $\mathfrak{n}$ is a relation of partial ordering, we write $X \leqslant Y$ (the notation $Y \geqslant X$ means exactly the same).

We will make frequent use of this symbol, and will say that X *precedes* Y, or Y *follows* X with respect to the given partial ordering $\mathfrak{n}$. In particular, every element of $\mathfrak{M}$ both precedes and follows itself. The use of the symbol of inequality is convenient in view of the natural association with the inequality of real numbers (it is obvious that the relation with respect to magnitude is a partial ordering).

5.10. If with respect to the partial ordering $\mathfrak{n}$ in the set $\mathfrak{M}$, for $X, Y \in \mathfrak{M}$ we have $X \leqslant Y$, where $X \neq Y$, then we also write $X < Y$ and $Y > X$. Starting from the relation $\mathfrak{n}$, it is possible to define a relation $\mathfrak{m}$ by setting $X \sim Y(\mathfrak{m})$, if $X \sim Y(\mathfrak{n})$ and $X \neq Y$ (i.e., $X < Y$). The relation $\mathfrak{m}$ is obviously transitive and strictly anti-symmetric in the sense that $X \sim Y(\mathfrak{m})$ and $Y \sim X(\mathfrak{m})$ cannot occur at the same time for any elements $X, Y \in \mathfrak{M}$. A relation $\mathfrak{m}$ with these properties defines in its turn in a natural way a relation of partial ordering in the sense of 5.9.

5.11. If a relation of partial ordering $\mathfrak{n}$ in $\mathfrak{M}$ is such that for arbitrary $X, Y \in \mathfrak{M}$ it necessarily follows that either $X \sim Y(\mathfrak{n})$ or $Y \sim X(\mathfrak{n})$, then $\mathfrak{n}$ is called a relation of *linear ordering* (sometimes simply called an "ordering" without addition of the word "linear").

5.12. Assume that with respect to a given relation of partial ordering in the set $\mathfrak{M}$ for some $\mathfrak{M}' \subset \mathfrak{M}$ and $X \in \mathfrak{M}$ we have $Z \leqslant X$ for all $Z \in \mathfrak{M}'$. Then X is called an *upper bound* for $\mathfrak{M}'$. A *lower bound* is defined similarly. If X is an upper bound for $\mathfrak{M}'$, and at the same time X is a lower bound for the set $\mathfrak{L}$ consisting of all upper bounds for $\mathfrak{M}'$, then X is called the exact upper bound for the set $\mathfrak{M}'$ (the least upper bound for $\mathfrak{M}'$). An exact lower bound (greatest lower bound) for $\mathfrak{M}'$ is defined analogously.

Of course it may happen that the set $\mathfrak{M}'$ has no upper bounds at all or may have several of them. The set $\mathfrak{M}'$ cannot have more than one exact upper bound, and the situation is the same for lower bounds.

5.13. The following classes of partially ordered sets are particularly important.

DEFINITION. *Let $\mathfrak{M}$ be a partially ordered set.*

The set $\mathfrak{M}$ is called a SEMILATTICE (*the terms demi-treillis and Halbverband are also used*) *if for every pair of elements there exists an exact lower bound.*

The set $\mathfrak{M}$ is called a LATTICE (*treillis, Verband*) *if for every pair of elements there exists an exact upper bound and an exact lower bound.*

The set $\mathfrak{M}$ is called a COMPLETE LATTICE *if for every subset of it there exists an exact upper bound and an exact lower bound.*

The theory of lattices at the present time forms a well-developed and important branch of contemporary algebra (see, for example, Birkhoff [3]).

5.14. For the relations themselves in an arbitrary set $\mathfrak{M}$ we can define a relation of partial ordering in a natural way. Namely, let us set

$$\mathfrak{n} \leqslant \mathfrak{m}$$

if from $X \sim Y(\mathfrak{n})$ it always follows that $X \sim Y(\mathfrak{m})$.

A relation in a set is defined by the set of pairs of elements which stand in the given relation; consequently, $\mathfrak{n} \leqslant \mathfrak{m}$ denotes that the set of pairs corresponding to the relation $\mathfrak{n}$ is a subset of the set of pairs corresponding to the relation $\mathfrak{m}$.

It is obvious that the relation defined in this way among the relations in $\mathfrak{M}$ is actually a relation of partial ordering.

Let us denote by Φ the system of all equivalences in a set $\mathfrak{M}$. The meaning of the partial ordering just defined, when applied to equivalences, becomes particularly clear if we characterize the equivalences by the decompositions of $\mathfrak{M}$ which correspond to them (5.8). In fact, $\mathfrak{n} \leqslant \mathfrak{m}$ ($\mathfrak{n}, \mathfrak{m} \in \Phi$) denotes that the decomposition $\mathfrak{n}$ is finer than $\mathfrak{m}$, or as we say, it is a refinement of $\mathfrak{m}$. Every $\mathfrak{m}$-class is in this case a union of nonintersecting $\mathfrak{n}$-classes.

5.15. Let Φ' be a nonempty subset of Φ, that is, a nonempty system of equivalences in the set $\mathfrak{M}$. We define a relation $\mathfrak{k}$ by setting

$$X \sim Y(\mathfrak{k}) \qquad (X, Y \in \mathfrak{M}),$$

if for every $\mathfrak{n} \in \Phi'$ we have $X \sim Y(\mathfrak{n})$. It is easy to see that relation $\mathfrak{k}$ is an equivalence.

In the partial ordering introduced in 5.14, $\mathfrak{k}$ is a lower bound for Φ'. Let $\mathfrak{m}$ be an arbitrary lower bound for Φ'. If $X \sim Y(\mathfrak{m})$ ($X, Y \in \mathfrak{m}$), then for arbitrary $\mathfrak{n} \in \Phi'$ we must have $X \sim Y(\mathfrak{n})$ in view of the fact that $\mathfrak{m} \leqslant \mathfrak{n}$. Since $X \sim Y(\mathfrak{n})$ holds for all $\mathfrak{n} \in \Phi'$, we have $X \sim Y(\mathfrak{k})$ by the definition of $\mathfrak{k}$. Thus from $X \sim Y(\mathfrak{m})$ it always follows that $X \sim Y(\mathfrak{k})$, i.e., $\mathfrak{m} \leqslant \mathfrak{k}$. This means that $\mathfrak{k}$ *is the exact lower bound for the set of equivalences* Φ' (5.12). The relation $\mathfrak{k}$ could be called the intersection of the equivalences in Φ'. The decomposition of $\mathfrak{M}$ generated by $\mathfrak{k}$ is the combined result of all the decompositions generated by equivalences in Φ'.

5.16. Now for $\Phi' \subset \Phi$ let us define a relation $\mathfrak{l}$ by setting

$$X \sim Y(\mathfrak{l}) \qquad (X, Y \in \mathfrak{M})$$

if there exist equivalences $\mathfrak{m}_1, \mathfrak{m}_2, \ldots, \mathfrak{m}_s \in \Phi'$ and elements $Z_1, Z_2, \ldots, Z_{s+1} \in \mathfrak{M}$ such that

$$Z_i \sim Z_{i+1}(\mathfrak{m}_i),$$
$$X = Z_1, \qquad Y = Z_{s+1} \qquad (i = 1, 2, \ldots, s).$$

It is easy to see that the relation $\mathfrak{l}$ is an equivalence.

If for some $X, Y \in \mathfrak{M}$ and $\mathfrak{n} \in \Phi'$ we have $X \sim Y(\mathfrak{n})$, then $X \sim Y(\mathfrak{l})$ by the definition of $\mathfrak{l}$. From this it follows that $\mathfrak{l}$ is an upper bound for the set of equivalences of Φ'. Suppose that the equivalence $\mathfrak{m}$ is an upper bound for Φ'. Assume also for $Z_1, Z_2, \ldots, Z_{s+1} \in \mathfrak{M}$ and $\mathfrak{m}_1, \mathfrak{m}_2, \ldots, \mathfrak{m}_s \in \Phi'$ we have

$$Z_i \sim Z_{i+1}(\mathfrak{m}_i) \qquad (i = 1, 2, \ldots, s).$$

Since $\mathfrak{m} \geqslant \mathfrak{m}_i$, it follows that

$$Z_i \sim Z_{i+1}(\mathfrak{m})$$

and therefore $Z_1 \sim Z_{s+1}(\mathfrak{m})$. We have shown that from $Z_1 \sim Z_{s+1}(\mathfrak{l})$ it always follows that $Z_1 \sim Z_{s+1}(\mathfrak{m})$. This means that $\mathfrak{l} \leqslant \mathfrak{m}$. Consequently $\mathfrak{l}$ is *the exact upper bound for the set of equivalences in* Φ'.

From these arguments it follows that *the system of all equivalences in a set is a complete lattice* (5.13).

5.17. Let $\mathfrak{A}$ be an arbitrary semigroup. Among the various relations in the set of all elements of the semigroup particular interest, from the point of view of the theory of semigroups, attaches to those relations which, as we say, are consistent with the operation in $\mathfrak{A}$, i.e., are not destroyed by multiplication by an arbitrary element of the semigroup.

DEFINITION. *A relation* $\mathfrak{n}$ *in the semigroup* $\mathfrak{A}$ *is called* LEFT STABLE *if from* $X \sim Y(\mathfrak{n})$ *for arbitrary* $Z \in \mathfrak{A}$ *it always follows that* $ZX \sim ZY(\mathfrak{n})$.

The relation $\mathfrak{n}$ *is called* RIGHT STABLE *if from* $X \sim Y(\mathfrak{n})$ *always follows* $XZ \sim YZ(\mathfrak{n})$.

The relation $\mathfrak{n}$ *is called* TWO-SIDEDLY STABLE (*or simply stable*) *if it is both right and left stable.*

5.18. Many articles have been devoted to the study of stability. For example, in the book of Dubreil [3] the author introduces a number of properties of stability, which is there called regularity.

Particular attention is usually given to the two-sidedly stable equivalences, which are often called congruences or relations of congruence. We will return to this subject in Chapter VII. At the moment we will consider only the following property.

Let $\mathfrak{n}$ be a transitive and reflexive relation in a semigroup. For $\mathfrak{n}$ to be two-sidedly stable it is necessary and sufficient that from

$$X \sim Y(\mathfrak{n}), \qquad X' \sim Y'(\mathfrak{n})$$

it always follows that

$$XX' \sim YY'(\mathfrak{n}).$$

In fact if $\mathfrak{n}$ is two-sidedly stable, then from $X \sim Y(\mathfrak{n})$ and $X' \sim Y'(\mathfrak{n})$ it follows that

$$XX' \sim YX'(\mathfrak{n}), \qquad YX' \sim YY'(\mathfrak{n}).$$

Consequently, from the transitivity it follows that $XX' \sim YY'(\mathfrak{n})$.

On the other hand, if $\mathfrak{n}$ has the indicated property, then from $X \sim Y(\mathfrak{n})$ and $Z \sim Z(\mathfrak{n})$ it follows that $ZX \sim ZY(\mathfrak{n})$ and $XZ \sim YZ(\mathfrak{n})$.

5.19. Let us consider some examples of relations in the set of elements of a semigroup.

(α) Let $\mathfrak{A}$ be an arbitrary semigroup.

In this book an important role will be played by the relation of left divisibility $\mathfrak{n}_l$, which is defined as follows:

$$A \sim B(\mathfrak{n}_l) \qquad (A, B \in \mathfrak{A})$$

holds if and only if there exists an element Z in $\mathfrak{A}$ such that $A = BZ$.

It is immediately evident that relation $\mathfrak{n}_l$ is left stable.

(β) Analogously we define a relation of right divisibility $\mathfrak{n}_r$ in $\mathfrak{A}$ which is right stable.

(γ) By 5.2 there is defined in a semigroup $\mathfrak{A}$ a relation $\mathfrak{n}_l \cdot \mathfrak{n}_r$ which is the product of the relations of left and right divisibility.

The formula $A \sim B(\mathfrak{n}_l \cdot \mathfrak{n}_r)$ means that there exist X, Y, $C \in \mathfrak{A}$ such that

$$A \sim C(\mathfrak{n}_l), \qquad C \sim B(\mathfrak{n}_r),$$
$$A = CY, \qquad C = XB.$$

From this it follows that $A = XBY$. But the converse is also easy to see. In fact, from $A = XBY$ follows

$$A \sim XB(\mathfrak{n}_l), \qquad XB \sim B(\mathfrak{n}_r).$$

Thus $A \sim B(\mathfrak{n}_l \cdot \mathfrak{n}_r)$ holds if and only if there exist elements X and Y in $\mathfrak{A}$ such that $A = XBY$.

In a completely analogous way we see that the existence of X and Y such that $A = XBY$ is a necessary and sufficient condition that $A \sim B(\mathfrak{n}_r \cdot \mathfrak{n}_l)$.

Consequently, the relations of left divisibility $\mathfrak{n}_l$ and right divisibility $\mathfrak{n}_r$ are commutative (5.3):

$$\mathfrak{n}_l \cdot \mathfrak{n}_r = \mathfrak{n}_r \cdot \mathfrak{n}_l.$$

(δ) Let $\mathfrak{Q}_\Omega$ be the semigroup of all one-to-one partial transformations of a set Ω (4.5). We introduce into $\mathfrak{Q}_\Omega$ a relation $\mathfrak{p}_1$ by setting

$$X \sim Y(\mathfrak{p}_1) \qquad (X, Y \in \mathfrak{Q}_\Omega)$$

if $\Pi_1^{(X)} = \Pi_1^{(Y)}$. We also define a relation $\mathfrak{p}_2$ by setting

$$X \sim Y(\mathfrak{p}_2)$$

if $\Pi_2^{(X)} = \Pi_2^{(Y)}$.

The relation $\mathfrak{p}_1$ is right stable. In fact, for $X \sim Y(\mathfrak{p}_1)$ the set $\Pi_1^{(XZ)}$ consists of those elements $\alpha \in \Pi_1^{(Z)}$ for which $Z\alpha \in \Pi_1^{(X)}$. It coincides with $\Pi_1^{(YZ)}$ since $\Pi_1^{(X)} = \Pi_1^{(Y)}$. But $\mathfrak{p}_1$ will not be left stable if Ω consists of more than one element. In fact, let $\Pi_1^{(X)} = \Pi_1^{(Y)}$ and let α be an element of this set. Let $X\alpha = \beta$, $Y\alpha = \gamma$, $\beta \neq \gamma$. We choose $Z \in \mathfrak{Q}_\Omega$ such that $\Pi_1^{(Z)} \ni \beta$, $\Pi_1^{(Z)} \bar{\ni} \gamma$. Then $\alpha \in \Pi_1^{(ZX)}$, but $\alpha \in \Pi_1^{(ZY)}$.

By similar arguments it is easy to see that the relation $\mathfrak{p}_2$ is left stable but not right stable.

Let us now consider a relation $\mathfrak{p}_3$ such that $X \sim Y(\mathfrak{p}_3)$ holds if and only if

$$\Pi_1^{(X)} = \Pi_1^{(Y)}, \qquad \Pi_2^{(X)} = \Pi_2^{(Y)}.$$

The relation $\mathfrak{p}_3$ is seen to be the exact lower bound of the relations $\mathfrak{p}_1$ and $\mathfrak{p}_2$. It is easy to show that $\mathfrak{p}_3$ is neither left stable nor right stable.

Let $\mathfrak{m}$ be any cardinal number. We define the relation $\mathfrak{p}_4$ in $\mathfrak{Q}_\Omega$ by setting

$$X \sim Y(\mathfrak{p}_4) \qquad (X, Y \in \mathfrak{Q}_\Omega)$$

if $X = Y$ or if the cardinalities of $\Pi_1^{(X)}$ and $\Pi_1^{(Y)}$ are both less than $\mathfrak{m}$. For $X = Y$ we have $XZ \sim YZ(\mathfrak{p}_4)$, $ZX \sim ZY(\mathfrak{p}_4)$ trivially. If the cardinality of $\Pi_1^{(X)}$ is less than $\mathfrak{m}$, then the cardinalities of $\Pi_1^{(XZ)}$ and $\Pi_1^{(ZX)}$ will also obviously be less than $\mathfrak{m}$. The same will also be true for Y. So it follows immediately that the relation $\mathfrak{p}_4$ is two-sidedly stable.

Let us note that all four relations $\mathfrak{p}_1, \mathfrak{p}_2, \mathfrak{p}_3, \mathfrak{p}_4$ considered here are evidently equivalences.

5.20. Let $\mathfrak{n}$ be a relation in the set of all elements of a semigroup $\mathfrak{A}$. Beginning with $\mathfrak{n}$, we define in $\mathfrak{A}$ a new relation $\mathfrak{n}'$, which we call the first derivative relation of $\mathfrak{n}$. We set

$$X \sim Y(\mathfrak{n}') \qquad (X, Y \in \mathfrak{A})$$

if and only if there exist elements T, T', S_1, S_2 which are either elements of $\mathfrak{A}$ or empty symbols such that

$$X = TS_1T', \qquad Y = TS_2T',$$

where S_1 and S_2 are either both empty symbols or satisfy one of the conditions, $S_1 \sim S_2(\mathfrak{n})$, $S_2 \sim S_1(\mathfrak{n})$.

5.21. Beginning with $\mathfrak{n}'$ in $\mathfrak{A}$ we define *the second derivative relation* $\mathfrak{n}''$ by setting

$$X \sim Y(\mathfrak{n}'') \qquad (X, Y \in \mathfrak{A})$$

if there exists a finite sequence of elements

$$X = Z_1, Z_2, \ldots, Z_r, Z_{r+1} = Y$$

such that

$$Z_i \sim Z_{i+1}(\mathfrak{n}') \qquad (i = 1, 2, \ldots, r).$$

5.22. Let us note some of the properties of these derivative relations, the correctness of which is in most cases obvious.

(α) The relation $\mathfrak{n}'$ is reflexive.
(β) The relation $\mathfrak{n}'$ is symmetric.
(γ) The relation $\mathfrak{n}''$ is reflexive.
(δ) The relation $\mathfrak{n}''$ is symmetric.
(ε) The relation $\mathfrak{n}''$ is transitive.
(ζ) $\mathfrak{n} \leqslant \mathfrak{n}' \leqslant \mathfrak{n}''$.
(η) The relation $\mathfrak{n}'$ is two-sidedly stable.
(θ) The relation $\mathfrak{n}''$ is two-sidedly stable.

In fact, if for $Z_1, Z_2, \ldots, Z_{r+1} \in \mathfrak{A}$ we have

$$Z_i \sim Z_{i+1}(\mathfrak{n}') \qquad (i = 1, 2, \ldots, r),$$

then by (η), for arbitrary $S \in \mathfrak{A}$,

$$SZ_i \sim SZ_{i+1}(\mathfrak{n}'), \qquad Z_iS \sim Z_{i+1}S(\mathfrak{n}') \qquad (i = 1, 2, \ldots, r).$$

Thus, from $Z_1 \sim Z_{r+1}(\mathfrak{n}'')$ it follows that

$$SZ_1 \sim SZ_{r+1}(\mathfrak{n}'') \qquad \text{and} \qquad Z_1S \sim Z_{r+1}S(\mathfrak{n}'').$$

(ι) If $X \sim Y(\mathfrak{n}'')$ and $X' \sim Y'(\mathfrak{n}'')$, then $XX' \sim YY'(\mathfrak{n}'')$. This is an immediate consequence of (γ), (ε), (θ) and 5.18.

5.23. The significance of the introduction of the second derivative relation is as follows. The relation $\mathfrak{n}''$ is a two-sidedly stable equivalence such that $\mathfrak{n}'' \geqslant \mathfrak{n}$. If here $\mathfrak{m}$ is a two-sidedly stable equivalence such that $\mathfrak{m} \geqslant \mathfrak{n}$, then it is easily seen that $\mathfrak{m} \geqslant \mathfrak{n}'$ and therefore also $\mathfrak{m} \geqslant \mathfrak{n}''$. Consequently, $\mathfrak{n}''$ *is the exact lower bound of the set of all two-sidedly stable equivalences* $\mathfrak{m}$ *which follow* $\mathfrak{n}$ ($\mathfrak{m} \geqslant \mathfrak{n}$). We may say that the decomposition of $\mathfrak{A}$ generated by $\mathfrak{n}''$ is the finest of all two-sidedly stable decompositions of $\mathfrak{A}$ such that for each pair of elements which are in relation $\mathfrak{n}$ to each other the two elements are necessarily contained in the same component of the decomposition (5.8, 5.15).

If the relation $\mathfrak{n}$ is itself a two-sidedly stable equivalence, then obviously we have $\mathfrak{n}'' = \mathfrak{n}' = \mathfrak{n}$.

CHAPTER II

DIVISIBILITY OF ELEMENTS

1. Concept and Elementary Properties of Divisibility

1.1. If the semigroup $\mathfrak{A}$ is not a group, its operation is not invertible. This means that for some elements A, $B \in \mathfrak{A}$ there are no elements X, Y such that

$$XB = A, \qquad BY = A.$$

It is possible, of course, that for some pairs of elements A and B one or other of these equations will have a solution with X or Y in $\mathfrak{A}$. The question of which pairs of elements admit a solution, and which do not, is of the utmost importance in the study of the structure of semigroups and in the investigation of their properties.

1.2. DEFINITION. *An element B of the semigroup $\mathfrak{A}$ is called* A RIGHT DIVISOR OF THE ELEMENT *A of the same semigroup if there exists in $\mathfrak{A}$ an element X such that*

$$XB = A.$$

B is called a LEFT DIVISOR *of A if there exists in $\mathfrak{A}$ an element Y such that*

$$BY = A.$$

If B is a right divisor of A, we say that A is divisible on the right by B. If B is a left divisor of A, we say that A is divisible on the left by B.

1.3. It follows immediately from the definition of a group (I, 3.2) that *a semigroup $\mathfrak{A}$ is a group if and only if each of its members is divisible both on the left and on the right by every element of $\mathfrak{A}$. In other words, $\mathfrak{A}$ is a group if and only if each of its elements is both a right and a left divisor of every other element of $\mathfrak{A}$.*

The extent to which group theory has been developed, by comparison with the general theory of semigroups, provides an indirect, but convincing, demonstration of the importance of the role of divisibility in semigroups.

1.4. Certain very elementary properties of the relationship of divisibility in semigroups follow immediately from Definition 1.2.

(α) If B is a right divisor of A, and C is a right divisor of B, then C is a right divisor of A.

(β) The product A_1A_2 is divisible by A_1 on the left and by A_2 on the right.

(γ) If B is a right divisor of A, then for arbitrary $Z \in \mathfrak{A}$ the element ZA is divisible on the right by B.

(δ) If A is divided on the right by B and B is divisible on the right by A, then B is a right divisor of itself.

(ε) In order that the element B should be a right divisor of the element A, it is necessary and sufficient that in the multiplication table (I, 1.6) the column corresponding to the element B contains the element A.

(ζ) In order that the element B should be a left divisor of the element A, it is necessary and sufficient that in the multiplication table (I, 1.6) the row corresponding to the element B contains the element A.

(η) An element A of a semigroup $\mathfrak{A}$ will be a right divisor of every element of $\mathfrak{A}$ if and only if in the column of the multiplication table corresponding to the element A all elements of $\mathfrak{A}$ occur.

(θ) A will be a left divisor of every element of $\mathfrak{A}$ if and only if in the row of the multiplication table corresponding to A all elements of $\mathfrak{A}$ occur.

(ι) The element A is divisible on the right by every element of $\mathfrak{A}$ if and only if A occurs in every column of the multiplication table.

(κ) A is divisible on the left by every element of $\mathfrak{A}$ if and only if A occurs in every row of the multiplication table.

1.5. Whereas in a group every element is both a left and a right divisor of every other element, in a semigroup which is not a group, such elements may not exist or it may be that only some of the elements possess the property just mentioned. In Chapter VI we shall consider such elements in detail; here we study the collection of all such elements.

THEOREM. *The set of all elements of the semigroup $\mathfrak{A}$ that are both left and right divisors of every element of $\mathfrak{A}$ is either empty or is a group.*

PROOF. We denote the set in question by $\mathfrak{B}$. Suppose $B_1, B_2 \in \mathfrak{B}$. For every $A \in \mathfrak{A}$ there exists some X_2 in $\mathfrak{A}$ such that

$$X_2B_2 = A.$$

Corresponding to X_2 there exists in $\mathfrak{A}$ an X_1 such that

$$X_1B_1 = X_2.$$

Therefore

$$X_1B_1B_2 = X_2B_2 = A.$$

Accordingly, B_1B_2 is a right divisor of A. Similarly, one can show that B_1B_2 is a left divisor of A. Therefore, $B_1B_2 \in \mathfrak{B}$.

The multiplication in $\mathfrak{B}$ is associative, since $\mathfrak{B} \subset \mathfrak{A}$.

Suppose $B_1, B_2 \in \mathfrak{B}$. There exist in $\mathfrak{A}$ elements X and Y such that

$$XB_2 = B_1, \qquad B_2Y = B_1.$$

If we show that $X, Y \in \mathfrak{B}$ the proof that $\mathfrak{B}$ is a group will be complete.

Let A be an arbitrary element of $\mathfrak{A}$.

Since, for some $T \in \mathfrak{A}$ we have $B_1T = A$,

$$XB_2T = B_1T = A,$$

i.e., X is a left divisor of A.

There exist also in $\mathfrak{A}$ elements I, C_1, C_2, such that

$$B_2I = B_2, \qquad C_1B_1 = I, \qquad B_2C_2 = I.$$

Since for arbitrary Z in $\mathfrak{A}$ there exists an element $Z' \in \mathfrak{A}$ such that $Z = Z'B_2$, we have

$$ZI = Z'B_2I = Z'B_2 = Z,$$

i.e., I is such that $ZI = Z$ for arbitrary $Z \in \mathfrak{A}$. Making use of this property of I, we obtain

$$A = AI = AB_2C_2 = AB_2IC_2 = AB_2C_1B_1C_2$$
$$= AB_2C_1XB_2C_2 = AB_2C_1XI = AB_2C_1X,$$

i.e., X turns out to be also a right divisor of A. Therefore $X \in \mathfrak{B}$. In a wholly analogous fashion one may show that $Y \in \mathfrak{B}$.

1.6. In contrast to the elements which are both left and right divisors of every element of a semigroup, there are elements (or may be) which are divisible both on the right and on the left by all elements of the semigroup. Such elements were first considered in a paper by Clifford and Miller [**1**], where they were termed zeroid elements. We will return to the study of the properties of such elements in Chapter V.

THEOREM. *The set of all elements of a semigroup $\mathfrak{A}$ that are divisible both on the right and on the left by every element of $\mathfrak{A}$ is either empty or it forms a group.*

PROOF. We denote the set in question by $\mathfrak{C}$. Suppose C_1, $C_2 \in \mathfrak{C}$. For every $A \in \mathfrak{A}$ there is in $\mathfrak{A}$ some X such that

$$C_1 = AX.$$

Therefore $C_1C_2 = A(XC_2)$, that is, C_1C_2 is divisible on the left by an arbitrary element A. It can be shown in the same way that C_1C_2 is divisible on the right by A. Therefore $C_1C_2 \in \mathfrak{C}$.

The associativity of the multiplication in $\mathfrak{C}$ follows from the fact that $\mathfrak{C} \subset \mathfrak{A}$.

Since C_1 and C_2 belong to $\mathfrak{C}$, there exist in $\mathfrak{A}$ elements U, V, W such that

$$UC_2^2 = C_2, \qquad C_2^2V = C_2, \qquad C_1 = C_2W.$$

We have then

$$UC_2 = UC_2C_2V = C_2V.$$

This yields

$$C_2(C_2V^2C_1) = (C_2^2V)(VC_1) = C_2VC_1 = UC_2C_1 = UC_2C_2W = C_2W = C_1.$$

We have shown that the element $Y = C_2V^2C_1$ is a solution of the equation

$$C_2Y = C_1.$$

Let us show that $Y \in \mathfrak{C}$. In fact, for arbitrary $Z \in \mathfrak{A}$ there exist U' and V' in $\mathfrak{A}$ such that

$$C_1 = U'Z, \qquad C_2 = ZV'.$$

Therefore

$$Y = C_2V^2C_1 = ZV'V^2U'Z,$$

that is, Y is divisible by Z both on the right and on the left. Analogously, we may find a solution X of the equation

$$XC_2 = C_1,$$

such that $X \in \mathfrak{C}$.

1.7. Let us determine when $\mathfrak{B}$ and $\mathfrak{C}$, considered in 1.5 and 1.6, possess a common element.

THEOREM. *If a semigroup $\mathfrak{A}$ possesses an element which is both a right and left divisor of every element of $\mathfrak{A}$ and is at the same time itself divisible both on the left and on the right by every element of $\mathfrak{A}$, then $\mathfrak{A}$ is a group.*

PROOF. Let A and B be arbitrary elements of $\mathfrak{A}$ and let D be an element having the properties postulated in the hypothesis of the theorem. There exist in $\mathfrak{A}$ elements X_1, Y_1, X_2, Y_2 such that

$$A = X_1D, \qquad A = DY_1, \qquad D = X_2B, \qquad D = BY_2.$$

It follows that $X = X_1X_2$ and $Y = Y_1Y_2$ are solutions of the equations

$$A = XB, \qquad A = BY.$$

1.8. It is natural to ask what groups may constitute, for some semigroup, groups $\mathfrak{B}$ and $\mathfrak{C}$ of the type discussed in 1.5 and 1.6. First, it follows from 1.7 that the question is of interest only when the two given groups have no element in common. We shall show that in this case two arbitrary groups give rise to a semigroup in which they play the role of the pair of groups $\mathfrak{B}$ and $\mathfrak{C}$.

Let $\mathfrak{G}_1$ and $\mathfrak{G}_2$ be two arbitrary groups having no common element. The set $\mathfrak{A}$ is taken as the union of the sets $\mathfrak{G}_1$ and $\mathfrak{G}_2$

$$\mathfrak{A} = \mathfrak{G}_1 \cup \mathfrak{G}_2.$$

We define a multiplication operation in $\mathfrak{A}$. Let $A, A' \in \mathfrak{A}$. If A and A' both belong to the same group $\mathfrak{G}_i$ $(i = 1, 2)$, then in that group we have

$$AA' = A'',$$

and we agree that in $\mathfrak{A}$

$$AA' = A''.$$

If, however, $A \in \mathfrak{G}_1$, $A' \in \mathfrak{G}_2$, we set

$$AA' = A'A = A'.$$

It is easy to verify that the so defined operation is associative. We show that in $\mathfrak{A}$ the group $\mathfrak{G}_1$ plays the role of the group $\mathfrak{B}$ (1.5), and that $\mathfrak{G}_2$ plays the role of the group $\mathfrak{C}$ (1.6). In fact, if $G_1 \in \mathfrak{G}_1$ and $A \in \mathfrak{A}$, the equation

$$XG_1 = A, \qquad G_1 Y = A$$

is soluble in $\mathfrak{A}$. If $A \in \mathfrak{G}_1$, we select for X and Y the solutions of the corresponding equations in $\mathfrak{G}_1$. If, however, $A \in \mathfrak{G}_2$, we set $X = Y = A$.

For $G_2 \in \mathfrak{G}_2$ the equations

$$XA = G_2, \qquad AY = G_2$$

are also soluble. If $A \in \mathfrak{G}_2$, we select for X and Y the solutions of the corresponding equations in $\mathfrak{G}_2$. If, however, $A \in \mathfrak{G}_1$, we set $X = Y = G_2$.

2. Inverse Elements and Units

2.1. In the study of the property of divisibility of elements there arises a question which is of interest under some circumstances: what are the properties of the element that multiplies the divisor to yield the dividend?

DEFINITION. *If the element B of the semigroup $\mathfrak{A}$ is a right divisor of the element A of the same semigroup, the element X satisfying the equation*

$$XB = A$$

is called a LEFT INVERSE *of B with respect to A.*

The notion of right inverse is defined analogously.

An element which is both a right inverse and a left inverse of B with respect to A is called a TWO-SIDED INVERSE *of B with respect to A.*

2.2. The properties listed below follow from the definition.

(α) If the element X is a left inverse of B with respect to A, then for arbitrary Z belonging to the given semigroup X is a left inverse of BZ with respect to AZ.

(β) If the element X is a left inverse of B with respect to A, then for an arbitrary element Z of the given semigroup ZX is a left inverse of B with respect to ZA.

(γ) If X_1 is a left inverse of B with respect to X_2, and X_2 is a left inverse of C with respect to A, then X_1 is a left inverse of BC with respect to A.

(δ) The number of left inverses of an element B with respect to A is equal to the number of times that A appears in the column of the multiplication table of the semigroup corresponding to B (I, 1.6; II, 1.4).

(ε) The number of right inverses of an element B with respect to A is equal to the number of times that A appears in the row of the multiplication table of the semigroup corresponding to B.

2.3. The following particular case is of especial interest.

DEFINITION. *An element B of a semigroup $\mathfrak{A}$ is called a* RIGHT UNIT *of the element A of the same semigroup, and A is called a* LEFT ZERO *of B, if B is a right divisor of A and A is a left inverse of B with respect to A:*

$$AB = A.$$

Left units and right zeros are defined analogously.

An element that is both a right and a left unit of some other element is called a TWO-SIDED UNIT *of the second element. An element that is both a left zero and a right zero of some other element is called a* TWO-SIDED ZERO *of the second element.*

An element I which is its own two-sided unit is called an IDEMPOTENT:

$$I^2 = I.$$

2.4. It may happen that an element of a semigroup is a right or left unit—or zero—for several elements of the semigroup. In some cases such a property is significant.

DEFINITION. *Let $\mathfrak{N}$ be a subset of the semigroup $\mathfrak{A}$. An element B of $\mathfrak{A}$ is called a right unit of $\mathfrak{N}$ if B is a* RIGHT UNIT OF *every element N of $\mathfrak{N}$:*

$$NB = N.$$

A left unit, a right zero, and a left zero of a subset are defined analogously.

If a subset $\mathfrak{N}$ has, say, right units, it is generally of interest to ask whether one or more of these belong to $\mathfrak{N}$. Both cases, i.e., that the unit belongs or does not belong, can arise.

2.5. The definition just introduced gives rise to a number of consequences.

(α) If C is a right unit of B, and B is a right unit of A, then C is a right unit of A.

(β) If B is a right unit of A, then B is a right unit of the element ZA for any Z belonging to the given semigroup.

(γ) If B is a right divisor of A, every right unit of B is a right unit of A.

(δ) An element which is a two-sided zero of itself is an idempotent.

(ε) If $\mathfrak{N}_1 \supset \mathfrak{N}_2$ and B is a right unit of $\mathfrak{N}_1$, then B is a right unit of $\mathfrak{N}_2$.

(ζ) If $\mathfrak{N}_1 \supset \mathfrak{N}_2$ and B is a right zero of $\mathfrak{N}_1$, then B is a right zero of N_2.

2.6. An especially important position is held by those elements that are two-sided units, or two-sided zeros, of the whole semigroup.

DEFINITION. *An element that is a two-sided unit of the whole semigroup $\mathfrak{A}$ is called a* UNIT OF THE SEMIGROUP $\mathfrak{A}$.

An element that is a two-sided zero of the whole semigroup $\mathfrak{A}$ is called a ZERO OF THE SEMIGROUP.

We will usually denote a unit of a semigroup by $E_{\mathfrak{A}}$, or simply E, and a zero by $O_{\mathfrak{A}}$, or simply O.

A semigroup possessing a unit is called a semigroup with unit, and one possessing a zero is called a semigroup with zero.

2.7. THEOREM. *A semigroup possesses at most one unit and at most one zero.*

PROOF. Let E_1 and E_2 both be units of the semigroup $\mathfrak{A}$.

Since E_1 is a unit of $\mathfrak{A}$, it follows that $E_1E_2 = E_2$. But since E_2 is a unit of $\mathfrak{A}$, it also follows that $E_1E_2 = E_1$. Accordingly, $E_2 = E_1$.

Let O_1 and O_2 both be zeros of the semigroup $\mathfrak{A}$.

Since O_1 is a zero of $\mathfrak{A}$, it follows that $O_1O_2 = O_1$. But since O_2 is a zero of $\mathfrak{A}$, it also follows that $O_1O_2 = O_2$. Therefore $O_1 = O_2$.

2.8. The following properties of zeros and units are immediate consequences of the definition.

(α) A unit is a right and left divisor of every element of the semigroup.

(β) No element of the semigroup is either a right unit or a left unit of a unit element, except the unit itself.

(γ) Every element of the semigroup is both a right and left divisor of the zero of the semigroup.

(δ) The zero of a semigroup is neither a right nor a left divisor of any element except itself.

(ε) The unit of a semigroup is an idempotent.

(ζ) All elements of a semigroup are zeros of the unit.

(η) All elements of the semigroup are units of the zero.

2.9. In a number of cases the presence of a unit in a semigroup simplifies formulations and makes the reasoning easier. It is therefore often helpful to add a unit to a semigroup if it does not initially have one. This can be done in the following way.

Let $\mathfrak{A}$ be an arbitrary semigroup. We consider the set $\mathfrak{A}'$ consisting of all elements of $\mathfrak{A}$ together with one new element E. We define a multiplication in $\mathfrak{A}'$. If A_1, A_2, and A_3 are elements of $\mathfrak{A}$ and we have in $\mathfrak{A}$

$$A_1A_2 = A_3,$$

then in $\mathfrak{A}'$ we set $A_1A_2 = A_3$.

Moreover, for every $X \in \mathfrak{A}'$ we set

$$XE = EX = X.$$

The associativity of the multiplication so defined in $\mathfrak{A}'$ is easily verified. Thus $\mathfrak{A}'$ is a semigroup in which the element E is a unit of $\mathfrak{A}'$.

The semigroup $\mathfrak{A}'$ is so closely connected with the semigroup $\mathfrak{A}$, and the connection is so obvious, that we can usually study $\mathfrak{A}$ completely by studying instead the semigroup $\mathfrak{A}'$, which has a unit.

2.10. In a similar way we extend a semigroup by adding a zero.

Let $\mathfrak{A}$ be an arbitrary semigroup. We consider the set $\mathfrak{A}''$ consisting of all elements of $\mathfrak{A}$ together with a new element O. We define a multiplication in $\mathfrak{A}''$. If for $A_1, A_2, A_3 \in \mathfrak{A}$ we have

$$A_1 A_2 = A_3,$$

then in $\mathfrak{A}''$ we set

$$A_1 A_2 = A_3.$$

In addition, for every $X \in \mathfrak{A}''$ we set

$$XO = OX = O.$$

It is easy to see that $\mathfrak{A}''$ is a semigroup in which O is a zero.

2.11. In the constructions 2.9 and 2.10 it is easy to see that if we first extend the semigroup $\mathfrak{A}$ by adding the unit E, and then extend the semigroup $\mathfrak{A}'$ by adding a zero O, the elements E and O are, respectively, the unit and the zero in the semigroup consisting of all elements of $\mathfrak{A}$ together with E and O.

2.12. If, in a semigroup $\mathfrak{A}$ with unit $E_{\mathfrak{A}}$, the product of any two elements distinct from $E_{\mathfrak{A}}$ is also distinct from $E_{\mathfrak{A}}$, the unit is, so to speak, externally adjoined to the semigroup consisting of all nonunit elements of $\mathfrak{A}$. The construction of the semigroup $\mathfrak{A}'$ in 2.9 is an example of the external adjunction of a unit to an initially given semigroup.

In the same way, if in a semigroup $\mathfrak{A}$ with zero O, the product of any two elements distinct from O is always distinct from O, the zero in $\mathfrak{A}$ has been externally adjoined to the semigroup consisting of all elements of $\mathfrak{A}$ distinct from O. The construction of $\mathfrak{A}''$ in 2.10 is an example of external adjunction of a zero to an initially given semigroup $\mathfrak{A}$.

2.13. Since a unit plays an exceptionally important role in a semigroup, as we have already remarked, a natural importance attaches to those elements that are inverses of each other with respect to the unit.

DEFINITION. *If in a semigroup with unit E the element A has a two-sided inverse A' with respect to E, the element A' is called the* INVERSE *of A, and is denoted by A^{-1}.*

2.14. Let $\mathfrak{A}$ be a semigroup with unit E. We consider various properties of inverse elements.

(α) If A is both a right and left divisor of E, then A possesses an inverse. In fact, suppose that

$$AA' = E, \qquad A''A = E,$$

whence

$$A' = EA' = (A''A)A' = A''(AA') = A''E = A''.$$

Since $A' = A''$, it is evident that $A' = A'' = A^{-1}$.

(β) If the element A possesses an inverse A^{-1}, no other element of $\mathfrak{A}$, distinct from A^{-1}, is either a right or a left inverse of A with respect to E. In particular, it follows that an inverse is unique.

In fact, if $XA = E$, then

$$X = XE = XAA^{-1} = EA^{-1} = A^{-1}.$$

Similarly, $AY = E$ implies that $Y = E$.

(γ) If A possesses an inverse A^{-1}, then A^{-1} possesses an inverse, and $(A^{-1})^{-1} = A$.

(δ) If each of the elements $A_1, A_2, \ldots, A_k$ possesses an inverse, then their product possesses an inverse, and

$$(A_1A_2 \ldots A_k)^{-1} = A_k^{-1} \ldots A_2^{-1}A_1^{-1}.$$

In fact,

$$(A_1A_2 \ldots A_k) \cdot (A_k^{-1} \ldots A_2^{-1}A_1^{-1}) = E,$$
$$(A_k^{-1} \ldots A_2^{-1}A_1^{-1}) \cdot (A_1A_2 \ldots A_k) = E.$$

(ε) An idempotent distinct from E cannot possess an inverse. In fact, if $I^2 = I$ and $II^{-1} = E$, then

$$I = IE = III^{-1} = II^{-1} = E.$$

2.15. The concept and properties of units and inverse elements permit us to characterize groups.

THEOREM. *In order that the semigroup $\mathfrak{A}$ be a group it is necessary and sufficient that $\mathfrak{A}$ possess a unit and that each element possess an inverse.*

PROOF. (1) Let $\mathfrak{A}$ be a group. We choose an arbitrary element S belonging to $\mathfrak{A}$. There exist in $\mathfrak{A}$ elements E_1 and E_2 such that

$$E_1S = S, \qquad SE_2 = S.$$

Let A be an arbitrary element of $\mathfrak{A}$. There exist in $\mathfrak{A}$ elements X and Y such that

$$XS = A, \qquad SY = A.$$

Since

$$E_1A = E_1SY = SY = A,$$
$$AE_2 = XSE_2 = XS = A,$$

E_1 is a left unit for $\mathfrak{A}$, and E_2 is a right unit. In particular, it follows that

$$E_1E_2 = E_2, \qquad E_1E_2 = E_1.$$

Accordingly, $E_1 = E_2$ and $E = E_1 = E_2$ is the unit of $\mathfrak{A}$. Every element of $\mathfrak{A}$ is both a right and a left divisor of every element of $\mathfrak{A}$, and therefore of the unit element E. From 2.14, (α), it follows that every element of $\mathfrak{A}$ has an inverse.

(2) Suppose that in the semigroup $\mathfrak{A}$ with unit E every element has an inverse. Then for arbitrary $A, B \in \mathfrak{A}$ we have

$$(BA^{-1})A = BE = B,$$
$$A(A^{-1}B) = EB = B.$$

Therefore $\mathfrak{A}$ is a group.

2.16. It follows from Theorem 2.15 that a group is a semigroup with two-sided cancellation (I, 3.2, (ε)).

In fact, if we have for the elements X, A, B of a group $\mathfrak{G}$ the relation

$$XA = XB,$$

then

$$A = E_{\mathfrak{G}}A = X^{-1}XA = X^{-1}XB = E_{\mathfrak{G}}B = B.$$

A similar argument holds for cancellation on the right.

2.17. With the help of Theorem 2.15 it is also easy to prove the theorem, considered earlier (I, 3.17), on the representation of a group by invertible transformations.

THEOREM. *Every group is isomorphic to some group of invertible transformations.* (I, 3.17).

PROOF. We take the set of all elements of the group $\mathfrak{G}$ itself as the set Ω of transformed elements:

$$\Omega = \mathfrak{G} = \{A, B, \ldots, S, \ldots\}.$$

To each element $S \in \mathfrak{G}$ we associate the following transformation V_S of the set Ω:

$$V_S X = (S \cdot X). \qquad (X \in \mathfrak{G}).$$

In virtue of 2.16 the transformations corresponding to distinct elements of $\mathfrak{G}$ are themselves distinct. Since for arbitrary elements S, R, $X \in \mathfrak{G}$ we have

$$(V_S V_R)X = V_S(RX) = SRX = V_{(SR)}X,$$

we obtain

$$V_S V_R = V_{(SR)}.$$

It follows that the correspondence $S \sim V_S$ represents an isomorphism between the group $\mathfrak{G}$ and the multiplicative set $\mathfrak{G}'$ consisting of all transformations V_S. The transformation $V_{S^{-1}}$ is the inverse of the transformation V_S, since

$$V_{S^{-1}}V_S X = V_{S^{-1}}SX = S^{-1}SX = E_{\mathfrak{G}}X = X,$$

$$V_S V_{S^{-1}}X = V_S S^{-1}X = SS^{-1}X = E_{\mathfrak{G}}X = X \qquad (X \in \Omega).$$

Accordingly, all the transformations V_S are invertible, and $\mathfrak{G}'$ contains the inverse $V_{S^{-1}}$ of each of its elements V_S.

2.18. The conditions introduced in 2.15 for a semigroup to be a group may be weakened as follows.

THEOREM. *If the semigroup $\mathfrak{A}$ contains a right unit I, and every element of $\mathfrak{A}$ is a left divisor of I, then $\mathfrak{A}$ is a group.*

PROOF. For an arbitrary element $A \in \mathfrak{A}$ there exist elements A' and A'' of $\mathfrak{A}$ such that

$$AA' = I, \qquad A'A'' = I.$$

From this we obtain

$$IA = IAI = IAA'A'' = IIA'' = IA'' = AA'A'' = AI = A.$$

Accordingly, I is a left unit, and so a two-sided unit, of $\mathfrak{A}$. But in that case

$$A = AI = AA'A'' = IA'' = A''.$$

Accordingly, $AA' = A'A = I$, that is A' is a two-sided inverse of A.

By 2.15, $\mathfrak{A}$ is a group.

2.19. In connection with Theorem 2.18 it is natural to consider the properties of a semigroup having a right unit for which every element appears as a right divisor. Clearly, every group belongs to the class of such semigroups. Nevertheless, there are semigroups with this property which are not groups. For example, consider the (very simply constructed) semigroup in which the product of any two elements is always equal to the left multiplier (in which case, obviously, every element is a right unit with the property in question). In Chapter VI we turn our attention to semigroups with this property, and we shall show that every semigroup having a right unit that is divisible on the right by every element of the semigroup can be obtained from a semigroup belonging to one of two prescribed classes by a simple construction.

3. Divisibility of Transformations and Matrices

3.1. Let us now see how the concepts of the two preceding paragraphs can be realized by application to the two principal particular examples of semigroups—transformations and matrices. We first consider transformations. Let Ω be an arbitrary nonempty set, $\mathfrak{S}_\Omega$ the semigroup of all transformations of Ω. For every $\alpha \in \Omega$ and $S \in \mathfrak{S}_\Omega$ we denote by $\Gamma_\alpha^{(S)}$ the set of all $\xi \in \Omega$ such that $S\xi = S\alpha$.

In order that the transformation B of $\mathfrak{S}_\Omega$ should be a right divisor of the transformation A of $\mathfrak{S}_\Omega$ it is necessary and sufficient that for every $\alpha \in \Omega$ the relation $\Gamma_\alpha^{(B)} \subset \Gamma_\alpha^{(A)}$ should hold, that is, that $B\alpha = B\beta (\alpha, \beta \in \Omega)$ should always imply $A\alpha = A\beta$.

In fact, if for some α and β in Ω we have

$$B\alpha = B\beta, \qquad A\alpha \neq A\beta,$$

then for arbitrary $X \in \mathfrak{S}_\Omega$ $XB\alpha = XB\beta$, that is, XB is never equal to A for any X.

Now suppose that $\Gamma_\alpha^{(B)} \subset \Gamma_\alpha^{(A)}$ for every $\alpha \in \Omega$. We define a new transformation $X \in \mathfrak{S}_\Omega$, setting $X\xi = A\alpha$ for $\xi = B\alpha$ ($X\xi$ is uniquely defined by this process, independently of the choice of α in $\Gamma_\alpha^{(B)}$, since $\Gamma_\alpha^{(B)} \subset \Gamma_\alpha^{(A)}$) and for $\eta \,\bar{\in}\, B\Omega$ we define $X\eta$ arbitrarily.

For every $\alpha \in \Omega$ we have

$$(XB)\alpha = X(B\alpha) = A\alpha$$

and, therefore, $XB = A$.

3.2. *The transformation B, as an element of the semigroup $\mathfrak{S}_\Omega$, is a left divisor of the transformation A belonging to $\mathfrak{S}_\Omega$ if and only if $A\Omega \subset B\Omega$.*

In fact, since for arbitrary $Y \in \mathfrak{S}_\Omega$

$$B\Omega \supset BY\Omega,$$

$BY = A$ is possible only if $B\Omega \supset A\Omega$.

If $B\Omega \supset A\Omega$ holds, then for every $\xi \in A\Omega$ we select some $\xi' \in \Omega$ such that $B\xi' = \xi$.

We construct a new transformation Y. If $A\alpha = \beta$, we set $Y\alpha = \beta'$. For such Y we have

$$BY\alpha = B\beta' = \beta = A\alpha$$

and accordingly

$$BY = A.$$

3.3. We now indicate some of the properties of the semigroup $\mathfrak{S}_\Omega$ which follow immediately from 3.1 and 3.2. In particular, we shall dwell on the question of those elements that possess an inverse with respect to the identity transformation E, which plays a special role since it is, obviously, the unit of the semigroup $\mathfrak{S}_\Omega$ (2.6).

(α) An element $S \in \mathfrak{S}$ has in $\mathfrak{S}$ a left inverse with respect to the identity transformation E if and only if, for arbitrary $\alpha, \beta \in \Omega$, $\alpha \neq \beta$ always implies $S\alpha \neq S\beta$, that is, S is a one-to-one transformation of Ω into itself.

This follows immediately from 3.1.

(β) An element $S \in \mathfrak{S}_\Omega$ has in $\mathfrak{S}_\Omega$ a right inverse with respect to the identity transformation E if and only if $S\Omega = \Omega$, that is, S is a transformation of Ω onto the whole of Ω.

This follows immediately from 3.2.

(γ) For $\mathfrak{S}_\Omega$ the group $\mathfrak{B}$ (1.5) consisting of all transformations that are both right and left divisors of every transformation of $\mathfrak{S}_\Omega$ consists of those transformations B which satisfy the conditions

$$B\Omega = \Omega,\ B\alpha \neq B\beta \quad (\alpha, \beta \in \Omega;\ \alpha \neq \beta),$$

that is, those that are one-to-one transformations of Ω onto itself. According to I, 3.14, $\mathfrak{B}$ is the set of all invertible mappings of the set Ω.

(δ) For $\mathfrak{S}_\Omega$ the set $\mathfrak{C}$ (1.6) is empty (except for the trivial case when Ω consists of a single element). In fact, suppose that U_ξ is a transformation of Ω carrying every $\alpha \in \Omega$ into ξ. Then every transformation C that is divided by U_ξ on the left must, by 3.2, satisfy $C\Omega \subset U_\xi\Omega = \xi$ and therefore cannot be divided by U_η on the left for $\eta \neq \xi$.

3.4. In order that the transformation $B \in \mathfrak{S}_\Omega$ be a right unit of the transformation $A \in \mathfrak{S}_\Omega$, it is necessary and sufficient that for arbitrary $\alpha \in \Omega$

$$A(B\alpha) = A\alpha,$$

that is, that $B\alpha$ be contained in the set $\Gamma_\alpha^{(A)}$ (3.1).

In order that B be a left unit of A, it is necessary and sufficient that for every $\xi \in A\Omega$ we have

$$B\xi = \xi.$$

In fact, $BA = A$ signifies that for every $\alpha \in \Omega$

$$BA\alpha = B(A\alpha) = A\alpha.$$

3.5. The idempotents in $\mathfrak{S}_\Omega$ are those transformations I for which all elements in I are fixed points of the transformation I, that is, $I\xi = I$. In fact, I is an idempotent if and only if for every $\alpha \in \Omega$

$$I(I\alpha) = I^2\alpha = (I\alpha).$$

The idempotents of $\mathfrak{S}_\Omega$ may very naturally be called projections. The property that all elements of Ω are mapped by I into a subset which is left unchanged by I would seem to be the essential property required for a general definition of projection.

As we have already noted, the identity transformation of Ω is the unit element of $\mathfrak{S}_\Omega$. The zero (2.6) of the semigroup $\mathfrak{S}_\Omega$ does not exist, in virtue of 3.3, (δ), since the zero of a semigroup is divisible both on the left and on the right by every element: $OA = AO = O$.

3.6. Let us now turn to the semigroup $\mathfrak{M}_n$, consisting of all complex square matrices of order n.

In order that in $\mathfrak{M}_n$ the matrix B be a right divisor of the matrix A, where

$$A = \begin{bmatrix} a_{11} & a_{12} & \dots & a_{1n} \\ a_{21} & a_{22} & \dots & a_{2n} \\ \cdot & \cdot & \cdot & \cdot \\ a_{n1} & a_{n2} & \dots & a_{nn} \end{bmatrix}, \qquad B = \begin{bmatrix} b_{11} & b_{12} & \dots & b_{1n} \\ b_{21} & b_{22} & \dots & b_{2n} \\ \cdot & \cdot & \cdot & \cdot \\ b_{n1} & b_{n2} & \dots & b_{nn} \end{bmatrix},$$

it is necessary and sufficient that

$$\operatorname{rank} \begin{bmatrix} a_{11} & a_{12} & \dots & a_{1n} \\ a_{21} & a_{22} & \dots & a_{2n} \\ \cdot & \cdot & \cdot & \cdot \\ a_{n1} & a_{n2} & \dots & a_{nn} \\ b_{11} & b_{12} & \dots & b_{1n} \\ b_{21} & b_{22} & \dots & b_{2n} \\ \cdot & \cdot & \cdot & \cdot \\ b_{n1} & b_{n2} & \dots & b_{nn} \end{bmatrix} = \operatorname{rank} \begin{bmatrix} b_{11} & b_{12} & \dots & b_{1n} \\ b_{21} & b_{22} & \dots & b_{2n} \\ \cdot & \cdot & \cdot & \cdot \\ b_{n1} & b_{n2} & \dots & b_{nn} \end{bmatrix}.$$

In fact, the equation $XB = A$ implies that the equations

$$\begin{aligned} x_{k1}b_{11} + x_{k2}b_{21} + \ldots + x_{kn}b_{n1} &= a_{k1}, \\ x_{k1}b_{12} + x_{k2}b_{22} + \ldots + x_{kn}b_{n2} &= a_{k2}, \\ \cdots\cdots\cdots&\cdots \\ x_{k1}b_{1n} + x_{k2}b_{2n} + \ldots + x_{kn}b_{nn} &= a_{kn}, \qquad (k = 1, 2, \ldots, n) \end{aligned}$$

are satisfied.

As we know from the theory of systems of linear equations, there exists a set of numbers $x_{k1}, x_{k2}, \ldots, x_{kn}$ satisfying these equations if and only if the sequence $(a_{k1}, a_{k2}, \ldots, a_{kn})$ is linearly expressible in terms of the sequences $(b_{11}, b_{12}, \ldots, b_{1n}), (b_{21}, b_{22}, \ldots, b_{2n}), \ldots, (b_{n1}, b_{n2}, \ldots, b_{nn})$. But, this condition will be satisfied for all k from 1 to n if and only if the above rank equation is satisfied.

3.7. In a completely analogous way it can be shown that a matrix B in $\mathfrak{M}_n$ is a left divisor of the matrix A if and only if we have the following equation of ranks:

$$\operatorname{rank}\begin{bmatrix} a_{11}\ a_{12} \ldots a_{1n}\ b_{11}\ b_{12} \ldots b_{1n} \\ a_{21}\ a_{22} \ldots a_{2n}\ b_{21}\ b_{22} \ldots b_{2n} \\ \cdots\cdots\cdots\cdots \\ a_{n1}\ a_{n2} \ldots a_{nn}\ b_{n1}\ b_{n2} \ldots b_{nn} \end{bmatrix} = \operatorname{rank}\begin{bmatrix} b_{11}\ b_{12} \ldots b_{1n} \\ b_{21}\ b_{22} \ldots b_{2n} \\ \cdots\cdots \\ b_{n1}\ b_{n2} \ldots b_{nn} \end{bmatrix}.$$

3.8. Evidently, the semigroup $\mathfrak{M}_n$ has both a unit and a zero.

$$E = \begin{bmatrix} 1 & 0 & 0 \ldots 0 \\ 0 & 1 & 0 \ldots 0 \\ 0 & 0 & 1 \ldots 0 \\ \cdot & \cdot & \cdot\ \cdot\ \cdot \\ 0 & 0 & 0 \ldots 1 \end{bmatrix}, \qquad O = \begin{bmatrix} 0 & 0 & 0 \ldots 0 \\ 0 & 0 & 0 \ldots 0 \\ 0 & 0 & 0 \ldots 0 \\ \cdot & \cdot & \cdot\ \cdot\ \cdot \\ 0 & 0 & 0 \ldots 0 \end{bmatrix}.$$

3.9. The group $\mathfrak{B}$ (1.5) of $\mathfrak{M}_n$, as follows from 3.6, consists of those matrices that are of rank n, that is, of all nondegenerate matrices.

The group $\mathfrak{C}$ (1.6) of $\mathfrak{M}_n$ consists of the single null matrix O. In fact, O is divisible by every element $M \in \mathfrak{M}_n$, both on the right and on the left:

$$MO = OM = O.$$

On the other hand, no matrix distinct from O is divisible by O, and therefore cannot belong to $\mathfrak{C}$.

4. Commutative Semigroups of Idempotents

4.1. Commutative semigroups of idempotents, that is, commutative semigroups for which every element is an idempotent, have a special role in view of their connection with the concept of partial ordering. Throughout the present section we shall constantly make use of the elementary concepts of the theory of partial orderings, introduced in §5 of Chapter I.

Let us turn first to a particular class of commutative semigroups of idempotents. Let Ω be a nonempty set and $\mathfrak{B}$ a set of some of its nonempty subsets such that $\mathfrak{B}$ contains the intersection $M \cap M'$ of every pair of subsets M and M' of Ω that belong to $\mathfrak{B}$. Clearly, $\mathfrak{B}$ forms a commutative semigroup of idempotents with respect to the operation of intersection. It turns out that every commutative semigroup of idempotents is isomorphically representable as a semigroup of this type.

THEOREM. *Every commutative semigroup of idempotents is isomorphic to a semigroup whose elements are subsets of a certain set and whose operation is that of intersection.*

PROOF. In the commutative semigroup of idempotents $\mathfrak{A}$ for every $X \in \mathfrak{A}$ we denote by $\mathfrak{N}_X$ the set of all elements of $\mathfrak{A}$ that are divisible by X. For arbitrary $X, Y \in \mathfrak{A}$, it is clear that $\mathfrak{N}_{XY} \subset \mathfrak{N}_X \cap \mathfrak{N}_Y$. On the other hand, if $Z \in \mathfrak{N}_X \cap \mathfrak{N}_Y$, then $Z = UX$, $Z = VY$ for some $U, V \in \mathfrak{A}$. But $Z = Z^2 = UXVY = (XY)(UV)$ and therefore $Z \in \mathfrak{N}_{XY}$. Thus it turns out that in every case

$$\mathfrak{N}_{XY} = \mathfrak{N}_X \cap \mathfrak{N}_Y.$$

It follows that the set of subsets of the set $\mathfrak{A}$ that have the form $\mathfrak{N}_X$ $(X \in \mathfrak{A})$, which we have denoted by $\mathfrak{B}$, forms a commutative semigroup of idempotents with respect to the intersection operation.

Suppose $\mathfrak{N}_X = \mathfrak{N}_Y$; since $X = X^2 \in \mathfrak{N}_X$ and $Y = Y^2 \in \mathfrak{N}_Y$, then for some $R, S \in \mathfrak{A}$ we have $X = RY$, $Y = SX$. It follows that $X = RSX$ and $Y = RSY$. We then obtain

$$X = RSX = RSRY = RSY = Y.$$

In virtue of this the mapping φ of the semigroup $\mathfrak{A}$ on $\mathfrak{B}$ defined by

$$\varphi(X) = \mathfrak{N}_X (X \in \mathfrak{A})$$

is a one-to-one mapping. In view of what we have already proved it is in fact an isomorphism, since

$$\varphi(XY) = \mathfrak{N}_{XY} = \mathfrak{N}_X \cap \mathfrak{N}_Y = \varphi(X) \cdot \varphi(Y).$$

4.2. The representation considered above, of an arbitrary commutative semigroup of idempotents, gives rise quite naturally to the study of the following partial ordering relation on $\mathfrak{A}$. We write

$$X \leqslant Y (X, Y \in \mathfrak{A}),$$

if X is divisible by Y (in the notation of 4.1, this is equivalent to the relationship $\mathfrak{N}_X \subset N_Y$). In a commutative semigroup of idempotents this condition is equivalent to the condition that Y is a unit of X. In fact, from $X = YZ$ it follows that $XY = X$.

It is easy to show that this relation is reflexive, transitive, and antisymmetric (I, 5.6).

With respect to the partial ordering relation thus introduced in $\mathfrak{A}$ the product XY of two elements X and Y is an exact lower bound (greatest lower bound) for X and Y.

In fact,

$$(XY)X = XY, \qquad (XY)Y = XY,$$

and therefore

$$(XY) \leqslant X, \qquad (XY) \leqslant Y.$$

If $Z \leqslant X$ and $Z \leqslant Y$, that is,

$$ZX = Z, \qquad ZY = Z,$$

then

$$Z(XY) = ZX \cdot Y = ZY = Z,$$

that is, $Z \leqslant XY$.

4.3. It follows from 4.2 that with respect to the partial ordering relationship the set of all elements of a commutative semigroup of idempotents $\mathfrak{A}$ forms a semilattice (I, 5.13). We will call it the semilattice conjugate to the semigroup $\mathfrak{A}$. In turn, we call $\mathfrak{A}$ the semigroup conjugate to this semilattice.

4.4. THEOREM. *For every semilattice there exists a unique commutative semigroup of idempotents that is conjugate to it.*

PROOF. (1) In the semilattice $\mathfrak{L}$ we introduce a commutative multiplication, setting

$$X \cdot Y = Y \cdot X = Z \qquad (X, Y, Z \in \mathfrak{L}),$$

if Z is the exact lower bound of $\{X, Y\}$ in $\mathfrak{L}$. According to I, 5.12; I, 5.13 this product is uniquely defined for every pair of elements. Let us show that the multiplication operation is associative.

Suppose that

$$(XY)Z = U, \qquad X(YZ) = V.$$

We have $V \leqslant X$, $V \leqslant YZ$ and therefore $V \leqslant Y$, $V \leqslant Z$. But from $V \leqslant X$, $V \leqslant Y$ it follows that $V \leqslant XY$. Since $V \leqslant Z$, we obtain $V \leqslant (XY)Z = U$. Likewise, we show that $U \leqslant V$, so that in fact $U = V$.

The set of all elements of a semilattice $\mathfrak{L}$, together with the operation just introduced, will be denoted by $\mathfrak{A}$. $\mathfrak{A}$ is a commutative semigroup. Clearly, all of its elements are idempotents.

If in $\mathfrak{A}$ we have for some pair of elements X and Y the relationship $XY = X$, then X is an exact lower bound for X and Y, and therefore $X \leqslant Y$. In turn, if $X \leqslant Y$, the exact lower bound for X and Y is X, and therefore $XY = X$. It follows that the semilattice $\mathfrak{L}$ is conjugate to the semigroup $\mathfrak{A}$.

(2) Suppose $\mathfrak{A}'$ is an arbitrary commutative semigroup of idempotents, conjugate to the semilattice $\mathfrak{L}$. The set of elements of $\mathfrak{A}'$ coincides with the set of elements of $\mathfrak{L}$ and coincides with the set of elements of $\mathfrak{A}$ (where $\mathfrak{A}$ is the semigroup constructed in the first part of the proof).

According to 4.2, the product in $\mathfrak{A}'$ of two arbitrary elements X and Y is equal to the exact lower bound of X and Y in $\mathfrak{L}$. Thus, in the semigroups $\mathfrak{A}'$ and $\mathfrak{A}$, which consist of the same elements, the operations coincide. Therefore $\mathfrak{A}' = \mathfrak{A}$.

4.5. In virtue of Theorem 4.4, the relation of conjugacy defines a one-to-one correspondence between commutative semigroups of idempotents and semilattices. This correspondence permits us to reduce the study of semilattices to the study of commutative semigroups of idempotents.

4.6. If in a commutative semigroup of idempotents $\mathfrak{A}$ it has been determined which elements are units for which others, the partial ordering in the semilattice corresponding to $\mathfrak{A}$ has also been completely determined. But, by 4.2, this relationship completely determines the multiplication operation in $\mathfrak{A}$. Therefore, if in $\mathfrak{A}$ we determine which elements are units for which others, we completely determine the semigroup $\mathfrak{A}$.

This property may also be formulated in the following way. Let φ be a one-to-one mapping of the commutative semigroup of idempotents $\mathfrak{A}$ on the commutative semigroup of idempotents $\mathfrak{A}'$. If $XY = X\,(X, Y \in \mathfrak{A})$ always implies $\varphi(X) \cdot \varphi(Y) = \varphi(X)$, and $XY \neq X$ always implies $\varphi(X) \cdot \varphi(Y) \neq \varphi(X)$, then φ is an isomorphism of $\mathfrak{A}$ to $\mathfrak{A}'$.

4.7. In the theory of partially ordered sets, lattices (I, 5.13) play an important role. Various properties of lattices have been subjected to intensive study. The theory of lattices holds an important position in modern algebra. An essential part of the theory consists in singling out certain particular classes of lattices from the general class. We shall show below how some of these classes may be naturally characterized by the divisibility properties of the semigroups conjugate to the lattices of the classes in question.

4.8. Theorem. *A semilattice is a lattice if and only if in its conjugate semigroup $\mathfrak{A}$ the set of units of two arbitrary elements is nonempty and contains its own zero.*

Proof. (1) Suppose that $\mathfrak{L}$ is a lattice. Then for $X, Y \in \mathfrak{L}$ there is an exact upper bound Z in $\mathfrak{L}$. Since $X \leqslant Z$ and $Y \leqslant Z$, we have in $\mathfrak{A}$, $XZ = X$ and $YZ = Y$, that is, Z is a unit for the set $\{X, Y\}$.

Suppose Z' is an arbitrary unit of the set $\{X, Y\}$. This means that in $\mathfrak{L}$, Z' is an upper bound for $\{X, Y\}$. Therefore $Z' \geqslant Z$, that is, in $\mathfrak{A}$, $Z'Z = Z$. It follows that Z is the zero of the set of units of $\{X, Y\}$.

(2) Suppose that in the semigroup $\mathfrak{A}$ the set of units of $\{X, Y\}$ for arbitrary $X, Y \in \mathfrak{A}$ contains its zero Z. Since Z is a unit of X and Y, Z is an upper bound of $\{X, Y\}$ in $\mathfrak{L}$. Suppose Z' is an arbitrary upper bound of $\{X, Y\}$. Since,

in $\mathfrak{A}$, Z is a zero of Z', we have $Z \leqslant Z'$. Therefore, Z is an exact upper bound of $\{X, Y\}$.

4.9. In the commutative semigroup of idempotents $\mathfrak{A}$, conjugate to the lattice $\mathfrak{L}$, we denote by $X + Y$ the element which is the exact upper bound of $\{X, Y\}$. The set $\mathfrak{L}$ may be considered with respect to this "addition" operation. We thus obtain a new commutative semigroup of idempotents connected with the lattice $\mathfrak{L}$ in a way similar to that of the semigroup $\mathfrak{A}$ conjugate to $\mathfrak{L}$ (4.3).

4.10. A linearly ordered set (I, 5.11) is obviously a lattice. If in a linearly ordered set every nonempty subset contains its own lower bound (which is automatically its exact lower bound), the set is said to be completely ordered.

4.11. THEOREM. *A semilattice $\mathfrak{L}$ is a linearly ordered set if and only if its conjugate semigroup $\mathfrak{A}$ has the property that of two arbitrary elements one is a zero of the other.*

$\mathfrak{L}$ is a completely ordered set if and only if an arbitrary nonempty subset of $\mathfrak{A}$ contains its own zero.

PROOF. (1) Let $\mathfrak{L}$ be a linearly ordered set and $X, Y \in \mathfrak{L}$. Then either $X \leqslant Y$ or $Y \leqslant X$. In the first case X is a zero of Y in $\mathfrak{A}$, and in the second case Y is a zero of X.

If $\mathfrak{L}$ is completely ordered and $\mathfrak{M} \subset \mathfrak{L}$, $\mathfrak{M} \neq \varnothing$, then $\mathfrak{M}$ contains its lower bound Z. In $\mathfrak{A}$ Z is a zero of every element of $\mathfrak{M}$.

(2) If in $\mathfrak{A}$ for $X, Y \in \mathfrak{A}$ we always have either $XY = X$ or $XY = Y$, then in $\mathfrak{L}$ we always have either $X \leqslant Y$ or $Y \leqslant X$, i.e., $\mathfrak{L}$ is linearly ordered.

If in $\mathfrak{A}$ for $\mathfrak{B} \subset \mathfrak{A}$, $B \neq \varnothing$ we always find a $Z_{\mathfrak{B}} \in \mathfrak{B}$, which is a zero of $\mathfrak{B}$, that is $Z_{\mathfrak{B}} X = Z_{\mathfrak{B}}$ for arbitrary $X \in \mathfrak{B}$, then $Z_{\mathfrak{B}}$ is a lower bound of $\mathfrak{B}$. It follows that $\mathfrak{L}$ is completely ordered.

4.12. THEOREM. *A semilattice $\mathfrak{L}$ is a complete lattice (I,5.13) if and only if the following three conditions hold: its conjugate semigroup $\mathfrak{A}$ has a unit, for every nonempty subset there exist zeros, and in the set of these zeros there is a unit.*

PROOF. (1) Let $\mathfrak{L}$ be a complete lattice. The exact lower bound of $\mathfrak{L}$ is the unit of $\mathfrak{A}$. The exact lower bound Z of a subset $\mathfrak{B} \subset \mathfrak{A}$, $\mathfrak{B} \neq \varnothing$ is one of the zeros of $\mathfrak{B}$, and, as is easily seen, Z is a unit for the set of these zeros.

(2) Suppose the semigroup $\mathfrak{A}$ has the properties stated in the hypothesis and $\mathfrak{B} \subset \mathfrak{A}$, $\mathfrak{B} \neq \varnothing$. The zeros of the set $\mathfrak{B}$ are its lower bounds in $\mathfrak{L}$. The unit of this set of zeros is the exact lower bound of $\mathfrak{B}$.

We denote by $\mathfrak{H}$ the set of all units of $\mathfrak{B}$. It is not empty, since $E_{\mathfrak{A}} \in \mathfrak{H}$. We denote by $\mathfrak{H}'$ the set of zeros of $\mathfrak{H}$, and by I the unit of the set $\mathfrak{H}'$, belonging to $\mathfrak{H}'$. Clearly, $\mathfrak{B} \subset \mathfrak{H}'$. Accordingly, I is an upper bound for $\mathfrak{B}$ in $\mathfrak{L}$.

If U is an arbitrary upper bound for $\mathfrak{B}$, then $U \in \mathfrak{H}$. For I, just as for any element of $\mathfrak{H}'$, we must have $IU = I$. Accordingly, $I \leqslant U$. Therefore, I is an exact upper bound for $\mathfrak{B}$.

4.13. There are a number of equivalent definitions of the so-called distributive lattices. We have chosen one, not the most commonly used, but the most suitable for our purposes (Bergman's proof of its equivalence with other definitions can be found in Birkhoff [3], Chapter IX, §1).

A lattice $\mathfrak{L}$ is said to be *distributive* if the conditions

$$XY = XZ, \qquad X + Y = X + Z \qquad (X, Y, Z \in \mathfrak{L})$$

always imply $Y = Z$.

THEOREM. *The lattice $\mathfrak{L}$ is distributive if and only if its conjugate semigroup has the property that for arbitrary X, Y, Z satisfying the conditions*

$$XY = XZ, \qquad Y \neq Z,$$

there exists among the units of the element X an element that is a unit for either Y or Z but not for both.

PROOF. (1) Suppose that the lattice $\mathfrak{L}$ is distributive and suppose that in its conjugate semigroup $\mathfrak{A}$ we have for some X, Y, Z the relation

$$XY = XZ,$$

and that every unit of X that is a unit of Y or of Z is a unit of both. We denote by $\mathfrak{K}$ the set of units of the ensemble $\{X, Y\}$, which is also the set of units of the ensemble $\{X, Z\}$.

Since $\mathfrak{A}$ is conjugate to a lattice, $\mathfrak{K}$ contains its zero W. The element W is an exact upper bound for $\{X, Y\}$ and for $\{X, Z\}$. Therefore, for our elements we have

$$XY = XZ, \qquad X + Y = X + Z.$$

Since the lattice $\mathfrak{L}$ is distributive, this relation can hold only if $Y = Z$.

It follows that $\mathfrak{A}$ has the properties stated in the theorem.

(2) Suppose that the lattice $\mathfrak{L}$ is not distributive, that is, that it contains elements X, Y, Z such that

$$XY = XZ, \qquad X + Y = X + Z, \qquad Y \neq Z.$$

If I is a unit of both X and Y, that is $I \geqslant X$, $I \geqslant Y$, then $I \geqslant (X + Y)$ and therefore

$$I \cdot (X + Z) = I \cdot (X + Y) = X + Y = X + Z.$$

Accordingly, $I \geqslant X + Z \geqslant Z$, and so I is a unit of Z.

It follows that the semigroup $\mathfrak{A}$ does not have the properties stated in the theorem.

4.14. The distributive lattices considered above belong to the very wide class of the so-called Dedekind, or modular, lattices. We choose, from a number of equivalent definitions of such lattices, the following (a proof of its equivalence with others can be found in Birkhoff [3], Chapter V, §2).

A lattice $\mathfrak{L}$ is said to be a Dedekind lattice if in $\mathfrak{L}$ the conditions

$$XY = XZ, \qquad X + Y = X + Z, \qquad Y \leqslant Z$$

always imply $Y = Z$.

THEOREM. *A lattice $\mathfrak{L}$ is a Dedekind lattice if and only if in its conjugate semigroup $\mathfrak{A}$ the following condition holds: If the elements X, Y, Z satisfy the relations*

$$XY = XZ, \qquad YZ = Y, \qquad Y \neq Z,$$

there exists an element which is a unit of X and of Y, but is not a unit of Z.

PROOF. (1) Let $\mathfrak{L}$ be a Dedekind lattice, and in its conjugate semigroup $\mathfrak{A}$ let some elements X, Y, Z satisfy the relations

$$XY = XZ, \qquad YZ = Y,$$

with the further condition that every element which is a unit of X and of Y is also necessarily a unit of Z. Since $YZ = Y$ implies that every unit of Z is a unit of Y, we may carry out the proof in the same way that we did in the first part of the proof of Theorem 4.13. As a result we find that for X, Y, Z in $\mathfrak{L}$ we have

$$XY = XZ, \qquad X + Y = X + Z,$$

which, since $Y \leqslant Z$ and $\mathfrak{L}$ is a Dedekind lattice, is possible only if $Y = Z$. Therefore, $\mathfrak{A}$ satisfies the conditions stated in the theorem.

(2) Now suppose that $\mathfrak{L}$ is not a Dedekind lattice, that is, it contains some X, Y, Z such that

$$XY = XZ, \qquad X + Y = X + Z, \qquad Y < Z.$$

An argument similar to that of the second part of the proof of Theorem 4.13 leads to the conclusion that in $\mathfrak{A}$ every element which is a unit for both X and Y is necessarily a unit for Z. Accordingly, $\mathfrak{A}$ does not satisfy the conditions stated in the theorem.

4.15. A lattice $\mathfrak{L}$ is called a *complemented lattice* if the following conditions are satisfied.

(α) $\mathfrak{L}$ contains an element E such that $X \leqslant E$ for every $X \in \mathfrak{L}$.

(β) $\mathfrak{L}$ contains an element O such that $X \geqslant 0$ for every $X \in \mathfrak{L}$.

(γ) For every element $X \in \mathfrak{L}$ there exists an element X' such that $XX' = 0$, $X + X' = E$.

We derive the following conclusions immediately from the definition of the conjugate semigroup.

The lattice $\mathfrak{L}$ is a complemented lattice if and only if its conjugate semigroup has a unit E and a zero O, and has the property that for arbitrary $X \in \mathfrak{A}$ there exists an element X' such that $XX' = O$ and the ensemble $\{X, X'\}$ has no unit other than E.

4.16. Historically, the first type of lattice to be studied, one which continues to play an important role even now, was the so-called Boolean algebra.

A lattice $\mathfrak{L}$ is called a *Boolean algebra* if it is distributive (4.13) and is a complemented lattice (4.15). It follows from 4.13 and 4.15 that a lattice $\mathfrak{L}$ is a Boolean algebra if and only if its conjugate semigroup $\mathfrak{A}$ has the properties stated in Theorems 4.13 and 4.15.

The following property has been proved in the theory of lattices (see, for example, Birkhoff [3], Chapter X, §5). Let $\mathfrak{L}$ be an aggregate of subsets of a set Ω (including the empty subset) with the property that together with any two subsets $A, B \subset \Omega$ the aggregate $\mathfrak{L}$ contains their union $A \cup B$ and their intersection $A \cap B$, and also for any $A \subset \Omega$ contains the complement $A' = \Omega \backslash A$. Then, with respect to the operation of inclusion, $\mathfrak{L}$ is a Boolean algebra. In this sense, every Boolean algebra is isomorphic, as a lattice, to a Boolean algebra of subsets.

It follows that with respect to the operation of intersection every ensemble of subsets of a set Ω which possesses the properties just indicated is a commutative semigroup of idempotents with the following properties.

$\mathfrak{A}$ has a unit E and a zero O. For every X there exists an X' such that $XX' = O$, and the ensemble $\{X, X'\}$ has no unit other than E. If $XY = XZ$, where $Y \neq Z$, then among the units of X there exists an element which is a unit for one of the elements Y and Z but not for both.

Every commutative semigroup of idempotents with these properties is isomorphic to some semigroup of subsets in which the operation is that of intersection and in which the above properties hold.

4.17. Since this section is devoted to a study of the properties of partial orderings, we include here a proof of a general theorem on partially ordered sets —known as Zorn's Lemma—which in some cases is useful in the derivation of properties of semigroups. Our proof is due to Weston.[1]

Let $\mathfrak{M}$ be a nonempty partially ordered set. We denote by $\Sigma_{\mathfrak{M}}$ the class of all its subsets (sometimes called chains) characterized by the fact that in each of them the partial ordering of $\mathfrak{M}$ induces a linear ordering. The empty set is contained in $\Sigma_{\mathfrak{M}}$.

THEOREM. *If the partial ordering of the set $\mathfrak{M}$ is such that each of the subsets belonging to $\Sigma_{\mathfrak{M}}$ has an upper bound in $\mathfrak{M}$, then $\mathfrak{M}$ has a maximal element, that is, an element that precedes no other element of $\mathfrak{M}$ except itself.*

PROOF. (1) With every $\mathfrak{L} \in \Sigma_{\mathfrak{M}}$ we associate the set $\overline{\mathfrak{L}}$ consisting of all the upper bounds of $\mathfrak{L}$ that are not contained in $\mathfrak{L}$. Clearly, for some $\mathfrak{L}$ the associated set $\overline{\mathfrak{L}}$ may be empty. If $\mathfrak{L}$ itself is empty, $\overline{\mathfrak{L}} = \mathfrak{M}$. In every nonempty $\overline{\mathfrak{L}}(\mathfrak{L} \in \mathfrak{M})$ we select an arbitrary element[2] $f(\overline{\mathfrak{L}}) \in \overline{\mathfrak{L}}$ (here, if $\mathfrak{L}_1 \neq \mathfrak{L}_2$ but $\overline{\mathfrak{L}}_1 = \overline{\mathfrak{L}}_2$ we set $f(\overline{\mathfrak{L}}_1) = f(\overline{\mathfrak{L}}_2)$).

[1] J. Weston, *A short proof of Zorn's lemma*, Arch. Math. **8** (1957), 279.

[2] Here we apply the so-called axiom of choice, which is the subject of many profound and not always concurrent opinions as to its place in mathematics. It is understandable that in a text such as this there is no room to discuss the question.

(2) We denote by $\Gamma_{\mathfrak{M}}$ the class of nonempty sets $\mathfrak{L}$ belonging to $\Sigma_{\mathfrak{M}}$ which have the following property: For any $\mathfrak{N} \subset \mathfrak{L}$ such that $\overline{\mathfrak{N}} \cap \mathfrak{L} = \varnothing$ the element $f(\overline{\mathfrak{N}})$ belongs to $\overline{\mathfrak{N}} \cap \mathfrak{L}$ and is its lower bound.

We note that the class $\Gamma_{\mathfrak{M}}$ is not empty since the set consisting of the one element $f(\overline{\varnothing})$ belongs to $\Gamma_{\mathfrak{M}}$.

(3) If $\mathfrak{L} \in \Gamma_{\mathfrak{M}}$ and $\overline{\mathfrak{L}} \neq \varnothing$, then, adjoining to $\mathfrak{L}$ the element $f(\overline{\mathfrak{L}})$, we obtain a set distinct from $\mathfrak{L}$, $\mathfrak{L}' = \mathfrak{L} \cup f(\overline{\mathfrak{L}})$. We shall prove that $\mathfrak{L}'$ belongs to $\Gamma_{\mathfrak{M}}$.

Since $f(\overline{\mathfrak{L}})$ is an upper bound of $\mathfrak{L}$, clearly $\mathfrak{L}' \in \Sigma_{\mathfrak{M}}$.

Suppose that for some $\mathfrak{N} \subset \mathfrak{L}'$ we have $\overline{\mathfrak{N}} \cap \mathfrak{L}' \neq \varnothing$.

If $\mathfrak{N} \cap \mathfrak{L} \neq \varnothing$, then it follows from $\mathfrak{L} \in \Gamma_{\mathfrak{M}}$ that $f(\overline{\mathfrak{N}})$ belongs to $\overline{\mathfrak{N}} \cap \mathfrak{L}$ and is a lower bound. But then $f(\overline{\mathfrak{N}})$, obviously, is also contained in $\overline{\mathfrak{N}} \cap \mathfrak{L}'$ and is a lower bound for it.

If $\overline{\mathfrak{N}} \cap \mathfrak{L} = \varnothing$, $\overline{\mathfrak{N}} \cap \mathfrak{L}' = f(\mathfrak{L})$. Then since no $X \in \mathfrak{L} \backslash \mathfrak{N}$ is an upper bound for $\mathfrak{N}$, it follows from $\mathfrak{L} \in \Sigma_{\mathfrak{M}}$ that for $X \in \mathfrak{L} \backslash \mathfrak{N}$ we can always find $Y \in \mathfrak{N}$ such that $X \leqslant Y$. From this it follows that $\overline{\mathfrak{L}} = \overline{\mathfrak{N}}$. But then $f(\overline{\mathfrak{N}}) = f(\overline{\mathfrak{L}})$ is contained in $\overline{\mathfrak{N}} \cap \mathfrak{L}' = f(\overline{\mathfrak{L}})$ and is a lower bound.

(4) Suppose $\mathfrak{L}, \mathfrak{L}' \in \Gamma_{\mathfrak{M}}$ and $X \in \mathfrak{L}$, $X' \in \mathfrak{L}'$, $X' \overline{\in} \mathfrak{L}$. We shall show that then $X < X'$.

We denote by $\mathfrak{N}$ the set of all elements of $\mathfrak{L} \cap \mathfrak{L}'$ that are distinct from X' and precede X'.

Since $X' \in \overline{\mathfrak{N}}$, the intersection $\overline{\mathfrak{N}} \cap \mathfrak{L}'$ contains X'.

In view of the definition of $\Gamma_{\mathfrak{M}}$, $\mathfrak{L}' \in \Gamma_{\mathfrak{M}}$ implies $f(\overline{\mathfrak{N}}) \in \mathfrak{L}'$ and $f(\overline{\mathfrak{N}}) \leqslant X'$.

If $\overline{\mathfrak{N}} \cap \mathfrak{L} \neq \varnothing$ then, by definition of $\Gamma_{\mathfrak{M}}$, the element $f(\overline{\mathfrak{N}})$ belongs to $\mathfrak{L}$. But this is impossible, since otherwise $f(\overline{\mathfrak{N}})$ would belong to $\mathfrak{L} \cap \mathfrak{L}'$, which in view of the relation $f(\overline{\mathfrak{N}}) < X'$ (the equation $f(\overline{\mathfrak{N}}) = X'$ cannot hold in this case, since $X' \overline{\in} \mathfrak{L}'$), would mean that $f(\overline{\mathfrak{N}}) \in \mathfrak{N}$.

If $X \in \mathfrak{N}$, then $X < X'$ by definition of $\mathfrak{N}$.

If $X \overline{\in} \mathfrak{N}$, then it follows from $\overline{\mathfrak{N}} \cap \mathfrak{L} = \varnothing$ that X is not an upper bound of $\mathfrak{N}$. In view of the fact that $\mathfrak{L} \in \Sigma_{\mathfrak{M}}$ it then follows that to X there corresponds a $Z \in \mathfrak{N}$ such that $X \leqslant Z$. Since $Z < X'$, $X < X'$.

(5) We denote by $\mathfrak{L}_0$ the union of all $\mathfrak{L}$ belonging to $\Gamma_{\mathfrak{M}}$. We shall show that $\mathfrak{L}_0 \in \Sigma_{\mathfrak{M}}$. Suppose that $X, Y \in \mathfrak{L}_0$, which means that in $\Gamma_{\mathfrak{M}}$ we can find $\mathfrak{L}_X$ and $\mathfrak{L}_Y$ such that $X \in \mathfrak{L}_X$, $Y \in \mathfrak{L}_Y$. If $Y \in \mathfrak{L}_X$, then it follows from $\mathfrak{L}_X \in \Sigma_{\mathfrak{M}}$ that $X \leqslant Y$ or $Y \leqslant X$. If $Y \overline{\in} \mathfrak{L}_X$, then according to the fourth part of the proof, $X < Y$. All this proves that $\mathfrak{L}_0 \in \Sigma_{\mathfrak{M}}$.

(6) We now prove that $\mathfrak{L}_0 \in \Gamma_{\mathfrak{M}}$.

By the second part of the proof, $\mathfrak{L}_0$ is not empty.

Suppose $\mathfrak{N} \subset \mathfrak{L}_0$ and $\overline{\mathfrak{N}} \cap \mathfrak{L}_0 \ni X$. The element X is contained in some $\mathfrak{L} \in \Gamma_{\mathfrak{M}}$ and is an upper bound of $\mathfrak{N}$. Suppose $Z \in \mathfrak{N}$. Since $\mathfrak{N} \subset \mathfrak{L}_0$, there must exist an $\mathfrak{L}' \in \Gamma_{\mathfrak{M}}$ such that $Z \in \mathfrak{L}'$. If it were true that $Z \overline{\in} \mathfrak{L}$, then, by part 4, we would have $Z > X$, which is impossible since $Z \in \mathfrak{N}$ and $X \in \overline{\mathfrak{N}}$. Accordingly,

$\mathfrak{N} \subset \mathfrak{L}$. Since $\mathfrak{L} \in \Gamma_{\mathfrak{M}}$ and $\overline{\mathfrak{N}} \cap \mathfrak{L} \ni X$, we have $f(\overline{\mathfrak{N}}) \in \mathfrak{L}$ and $f(\overline{\mathfrak{N}}) \leqslant X$. Therefore, $f(\overline{\mathfrak{N}}) \in \overline{\mathfrak{N}} \cap \mathfrak{L}_0$ and $f(\overline{\mathfrak{N}})$ precedes an arbitrary X belonging to $\overline{\mathfrak{N}} \cap \mathfrak{L}_0$. We conclude that $\mathfrak{L}_0 \in \Gamma_{\mathfrak{M}}$.

(7) Suppose that P is an upper bound for $\mathfrak{L}_0$. We postulate that $P < Q$ for some $Q \in \mathfrak{M}$. For an arbitrary $X \in \mathfrak{L}_0$ we have $X \leqslant P < Q$. Accordingly, $Q \in \overline{\mathfrak{L}}_0$. But, if $\overline{\mathfrak{L}}_0 \neq \varnothing$, it follows from the third part of this demonstration, that $\mathfrak{L}_0$ would be contained in some $\mathfrak{L}' \in \Gamma_{\mathfrak{M}}$, distinct from $\mathfrak{L}_0$. But this is impossible, since $\mathfrak{L}_0$ contains all sets belonging to $\Gamma_{\mathfrak{M}}$. Therefore P is the desired maximal element.

4.18. We now apply the theorem just proved to the derivation of the conditions for the existence of a zero in a commutative semigroup of idempotents.

Given a commutative semigroup of idempotents $\mathfrak{A}$, consider a subset of $\mathfrak{A}$ with the property that for every two of its elements one is a zero of the other. If every such subset of $\mathfrak{A}$ contains a zero, then $\mathfrak{A}$ is a semigroup with a zero.

In fact, let us define a relation in $\mathfrak{A}$ by setting $X \leqslant Y\,(X, Y \in \mathfrak{A})$ if $XY = Y$. Evidently, this relation is a partial ordering. In view of the property postulated with respect to $\mathfrak{A}$, we may apply Theorem 4.17 to this relation. According to the theorem, there is in $\mathfrak{A}$ an element P such that $PX = X$ only if $X = P$. Let A be an arbitrary element of $\mathfrak{A}$. Since $P(PA) = PA, PA = P$. Thus P is a zero of the semigroup $\mathfrak{A}$.

4.19. We cannot obtain a result similar to that of 4.18 for units. The commutative semigroup of idempotents $\mathfrak{A}$ with a zero, $\mathfrak{A} = \{O, A, B\}$ in which $AB = O$, is an example of the absence of a unit. We can find only somewhat weaker properties.

Suppose that in a commutative semigroup of idempotents $\mathfrak{A}$ every subset has a unit if of any two of its elements one is a unit of the other. Then there exists in $\mathfrak{A}$ an element which has no unit other than itself.

The validity of this assertion follows immediately by application of Theorem 4.17 to the partial ordering in $\mathfrak{A}$ for which X precedes Y if Y is a unit of X.

5. Semigroups in Which Every Element Has a Right Zero

5.1. Among the various questions arising in the study of semigroups of transformations, one which is very important concerns the existence of fixed points, that is, elements of the set which are carried into themselves by the given transformation. In a number of branches of mathematics we consider semigroups of transformations in which each transformation has a fixed point. Such a property of semigroups of transformations, once established, generally gives rise to a further series of properties important in the study of the objects over which the transformations are defined. This property of a semigroup is, of course, not an abstract property of the kind that we defined in I, 1.8 and I, 1.14. In the first place, it relates only to semigroups of transformations. Furthermore, even when we confine ourselves to such semigroups, it is not difficult to find two isomorphic semigroups in which all the transformations of one have fixed points while the transformations of the second have none. This is true despite the fact that, as E. S. Ljapin [**11**] has shown, one can study the fixed point property of transformations from an abstract point of view.

5.2. We denote by Π the class of semigroups in which each element has a right zero.

5.3. THEOREM. *Semigroups of the class* Π, *and those only, have the property that in an arbitrary representation of the semigroup by transformations every transformation has a fixed point.*

PROOF. (1) Suppose the semigroup $\mathfrak{A}$ belongs to the class Π and φ is an arbitrary representation of $\mathfrak{A}$, that is, an isomorphism of $\mathfrak{A}$ on some semigroup of transformations of a set Ω. For $X \in \mathfrak{A}$ we can find in $\mathfrak{A}$ an element U such that $XU = U$. Let us choose an arbitrary element α belonging to Ω. We write $[\varphi(U)]\alpha = \beta$. Then

$$[\varphi(X)]\beta = [\varphi(X)] \cdot [\varphi(U)]\alpha = [\varphi(XU)]\alpha = [\varphi(U)]\alpha = \beta,$$

that is, β is a fixed point of the mapping $\varphi(X)$.

(2) Suppose the semigroup $\mathfrak{A}$ is such that for an arbitrary representation by transformations every transformation of the representation has a fixed point. We consider a representation φ of the semigroup $\mathfrak{A}$ by left displacements (I, 3.9; I, 3.10). The element $T_X = \varphi(X)$, where $X \in \mathfrak{A}$, being a transformation of this representation, has a fixed point. The exceptional element I cannot be this fixed point (since $T_X I = X$). Therefore the fixed point is some element $U \in \mathfrak{A}$ for which we have

$$U = T_X U = XU.$$

5.4. Of course, it is far from true that every semigroup of transformations for which every transformation has a fixed point is contained in the class Π. In a certain sense, however, we shall see that the class Π may be used to characterize such semigroups.

First we consider an important special case.

THEOREM. *Let* $\mathfrak{A}$ *be a semigroup of transformations of the set* Ω *containing for every* $\alpha \in \Omega$ *the transformation* U_α *which carries all elements of* Ω *into* α *(that is,* $U_\alpha \xi = \alpha$ *for every* $\xi \in \Omega$*). In order that every transformation in* $\mathfrak{A}$ *have a fixed point, it is necessary and sufficient that* $\mathfrak{A}$ *belong to the class* Π.

PROOF. (1) Suppose that every transformation of $\mathfrak{A}$ has a fixed point.

For $X \in \mathfrak{A}$ there exists an $\alpha \in \Omega$ for which

$$X\alpha = \alpha.$$

Then for arbitrary $\xi \in \Omega$

$$XU_\alpha \xi = X\alpha = \alpha = U_\alpha \xi$$

and therefore $XU_\alpha = U_\alpha$. This means that every element of $\mathfrak{A}$ has a right zero and $\mathfrak{A} \in \Pi$.

(2) If the semigroup of transformations $\mathfrak{A}$ belongs to Π then the identity transformation of $\mathfrak{A}$ on itself, being a representation of $\mathfrak{A}$ by transformations, is by 5.3 such that every transformation has a fixed point.

5.5. With the help of Theorem 5.4 it is easy to show that a number of semigroups of transformations which play an important role in many branches of mathematics belong to the class Π.

Suppose that Ω is a complete metric space[3] and that $\mathfrak{A}$ is the semigroup of all contraction operators, that is, of operators A of the set Ω such that for any one of them there exists a real number θ_A with $0 \leqslant \theta_A < 1$ and that for arbitrary $\alpha, \beta \in \Omega$

$$\rho(A\alpha, A\beta) \leqslant \theta_A \cdot \rho(\alpha, \beta)$$

(here ρ is the distance between the respective elements).

It is easy to see that $\mathfrak{A}$ is a semigroup. The transformation U_α (5.4) belongs for all $\alpha \in \Omega$ to $\mathfrak{A}$. Therefore Theorem 5.4 is applicable. As is known,[4] by the so-called Banach principle every transformation in $\mathfrak{A}$ has a fixed point. By 5.4 it follows that the semigroup $\mathfrak{A}$ belongs to the class Π.

5.6. Suppose Ω is the n-dimensional sphere (of course, we could equally well say n-dimensional simplex) in n-space. $\mathfrak{A}$ is the ensemble of all continuous transformations of Ω.

By Brouwer's theorem[5] every mapping in $\mathfrak{A}$ has a fixed point. By theorem 5.4 $\mathfrak{A}$ belongs to the class Π.

5.7. One of the strongest criteria for the existence of a fixed point is the so-called Tihonov principle.[6] Let Ω be a convex bicompact subset of a linear locally convex space. The principle says that every continuous mapping of Ω has a fixed point. From Theorem 5.4 it follows that the semigroup of all continuous mappings of Ω belongs to the class Π.

5.8. In connection with the examples cited above we come naturally to the question whether, in these instances as well as in new cases, it is not possible to prove the existence of a fixed point for each of the mappings of a semigroup of mappings $\mathfrak{A}$, by showing first that $\mathfrak{A}$ belongs to the class Π (or is contained in a semigroup of mappings of the same space that does belong to the class Π—which, by Theorem 5.9, is sufficient).

---

[3] See, for example, L. A. Ljusternik and V. P. Sobolev, *The elements of functional analysis*, GITTL, Moscow, 1951; Chapter I, § 2 and § 4. (Russian)

[4] See, for example, L. A. Ljusternik and V. P. Sobolev, *The elements of function analysis*, GITTL, Moscow, 1951; Chapter I, § 7. (Russian)

[5] See, for example, V. V. Nemitzki, *Die Fixpunktmethode in der Analysis*, Uspehi Mat. Nauk **1** (1936), 141–174; § 3. (Russian)

[6] V. V. Nemitzki, ibid. **1** (1936), 141–174; § 7. (Russian)

In principle, of course, this is entirely possible. Practically, however, such a method of proof is far from simple, since it requires an explicit representation of the structure of the semigroup of mappings $\mathfrak{A}$.

Nevertheless, without regard to the possibility of obtaining new information about fixed points in one or another branch of mathematics by the use of this method, the very fact that there exists a "purely algebraic" approach to such an important property may be of interest. We may say that the reason for the existence of fixed points in the mappings of some semigroup $\mathfrak{A}$ of mappings of a certain set Ω lies not in the nature of Ω nor in the way the mappings operate on Ω, but solely in the nature of the algebraic structure of $\mathfrak{A}$. In fact, all mappings of a semigroup of mappings $\mathfrak{A}'$ defined over a set Ω' will necessarily have fixed points whenever $\mathfrak{A}'$ is isomorphic to the semigroup $\mathfrak{A}$ of 5.6 or of 5.7, and so on, quite independently of the nature of Ω' or of the mappings that make up $\mathfrak{A}'$.

5.9. The role of the class Π in the study of the phenomena now under discussion is not exhausted by Theorem 5.4. It is possible to prove a theorem which demonstrates the role of the class Π in the general case.

THEOREM. *Let $\mathfrak{A}$ be a semigroup of mappings of the set Ω. If $\mathfrak{A} \subset \mathfrak{A}'$, where $\mathfrak{A}'$ is a semigroup of mappings of the same set Ω and $\mathfrak{A}'$ belongs to the class Π, then every mapping in $\mathfrak{A}$ has a fixed point.*

Conversely, if every mapping in $\mathfrak{A}$ has a fixed point, then there exists in the class Π a semigroup of mappings $\mathfrak{A}'$ defined on the same set Ω, and $\mathfrak{A} \subset \mathfrak{A}'$.

PROOF. (1) Suppose $\mathfrak{A} \subset \mathfrak{A}'$, where $\mathfrak{A}' \in \Pi$. Since $\mathfrak{A}$ belongs to the class Π it follows, by Theorem 5.3, that every mapping in $\mathfrak{A}'$ has a fixed point, and therefore so does every mapping in $\mathfrak{A}$ (since $\mathfrak{A} \subset \mathfrak{A}'$).

(2) For every $\alpha \in \Omega$ we denote by U_α a mapping of Ω such that for arbitrary $\xi \in \Omega$, $U_\alpha \xi = \alpha$. The set of all mappings U_α ($\alpha \in \Omega$) and of all mappings in $\mathfrak{A}$ will be denoted by $\mathfrak{A}'$. The set $\mathfrak{A}'$ is a semigroup.

In fact, and obviously

$$U_\alpha U_\beta = U_\alpha,$$
$$U_\alpha A = U_\alpha,$$
$$AU_\alpha = U_{(A\alpha)},$$
$$AB \in \mathfrak{A}, \qquad (\alpha, \beta \in \Omega;\ A, B \in \mathfrak{A}).$$

The semigroup $\mathfrak{A}'$ contains the semigroup $\mathfrak{A}$. The element $\alpha \in \Omega$ is a fixed point of the mapping U_α. Therefore, if every mapping in $\mathfrak{A}$ has a fixed point, then so does every mapping in $\mathfrak{A}'$. In this case it follows from 5.4 that the semigroup $\mathfrak{A}'$ belongs to the class Π.

5.10. The semigroup $\mathfrak{A}'$ considered in the second part of the proof of Theorem 5.9, which contains the initial semigroup $\mathfrak{A}$, can obviously be constructed for any semigroup of mappings $\mathfrak{A}$. The use of this semigroup $\mathfrak{A}'$ permits us to determine whether all mappings in $\mathfrak{A}$ have fixed points.

All mappings in the semigroup $\mathfrak{A}$ *have fixed points if and only if the semigroup of mappings* $\mathfrak{A}'$ *belongs to the class* Π.

5.11. The preceding considerations show the importance of the class Π of semigroups which we have defined. It is natural, then, to raise the question of what kind of structure semigroups of this class may have. It would be unrealistic, however, to hope to solve this question in its most general sense. In fact, every semigroup can be "converted" into a semigroup belonging to Π by the external addition of a zero (2.10; 2.12). Instead, we will return in Chapter VIII (§ 5) to the class Π and will obtain what might be called the local structure of semigroups of this class.

5.12. An important role in these investigations is played by two particular semigroups belonging to Π; we now study these as examples of semigroups in Π.

Let $\mathfrak{U}$ be the set of elements U^{β}_{α} $(\alpha, \beta = 1, 2, 3, \ldots)$ with the multiplication law

$$U^{\beta_1}_{\alpha_1} \cdot U^{\beta_2}_{\alpha_2} = U^{\beta_2}_{\alpha_2} \cdot U^{\beta_1}_{\alpha_1} = U^{\beta_2}_{\alpha_2} \qquad (\alpha_1 < \alpha_2),$$

$$U^{\beta_1}_{\alpha_1} \cdot U^{\beta_2}_{\alpha_1} = U^{\beta_1+\beta_2}_{\alpha_1}, \qquad (\alpha_1, \alpha_2, \beta_1, \beta_2 = 1, 2, 3, \ldots).$$

The associativity of this multiplication is immediately evident. We remark that U^{β}_{α} is in fact the βth power of the element $U^1_{\alpha} = U_{\alpha}$.

For every $U^{\beta_1}_{\alpha_1}$ all elements $U^{\beta_2}_{\alpha_2}$ for which $\alpha_2 > \alpha_1$ are not only right zeros but two-sided zeros. The semigroup $\mathfrak{U}$ therefore belongs to the class Π.

5.13. Let $\mathfrak{V}$ be the set of expressions of the form

$$(V^{\alpha_n}_n V^{\alpha_{n-1}}_{n-1} \ldots V^{\alpha_2}_2 V^{\alpha_1}_1),$$

where α_i is a non-negative integer and $\alpha_n > 0$. We define a multiplication in $\mathfrak{V}$:

$$(V^{\alpha_n}_n V^{\alpha_{n-1}}_{n-1} \ldots V^{\alpha_2}_2 V^{\alpha_1}_1) \cdot (V^{\beta_m}_m V^{\beta_{m-1}}_{m-1} \ldots V^{\beta_2}_2 V^{\beta_1}_1)$$

$$= \begin{cases} (V^{\beta_m}_m V^{\beta_{m-1}}_{m-1} \ldots V^{\beta_2}_2 V^{\beta_1}_1) & (m > n), \\ (V^{\alpha_n}_n \ldots V^{\alpha_{m+1}}_{m+1} V^{\alpha_m+\beta_m}_m V^{\beta_{m-1}}_{m-1} \ldots V^{\beta_2}_2 V^{\beta_1}_1) & (m \leqslant n). \end{cases}$$

The associativity of the multiplication can be verified without difficulty.

If we write

$$V_n = (V^1_n V^0_{n-1} \ldots V^0_2 V^0_1),$$

then the element $(V^{\alpha_n}_n V^{\alpha_{n-1}}_{n-1} \ldots V^{\alpha_2}_2 V^{\alpha_1}_1)$ is precisely the product $V^{\alpha_n}_n V^{\alpha_{n-1}}_{n-1} \ldots V^{\alpha_2}_2 V^{\alpha_1}_1$ (where by V^0_{β} we mean the empty symbol).

For an arbitrary element $(V^{\alpha_n}_n V^{\alpha_{n-1}}_{n-1} \ldots V^{\alpha_2}_2 V^{\alpha_1}_1)$ the right zeros are all elements $(V^{\beta_m}_m V^{\beta_{m-1}}_{m-1} \ldots V^{\beta_2}_2 V^{\beta_1}_1)$, for which $m > n$. Therefore $\mathfrak{V}$ belongs to the class Π. We note that in $\mathfrak{V}$ no element has a left zero, as may be easily verified.

5.14. The role of the semigroups $\mathfrak{U}$ and $\mathfrak{V}$ is clarified if we consider the following property.

LEMMA. *If the semigroup $\mathfrak{A}$, possessing no idempotents, belongs to* Π, *it is isomorphic to either* $\mathfrak{U}$ *or* $\mathfrak{B}$.

PROOF. (1) We remark first that for arbitrary $A \in \mathfrak{A}$, the equation $A^a = A^b$ always implies $a = b$. In fact, for $a > b$, and in virtue of the fact that $A^{b+(a-b)} = A^b$, we would have

$$(A^{(a-b)})^2 = A^{2(a-b)} = A^{b+(a-b)} \cdot A^{(a-b)-b} = A^b \cdot A^{(a-b)-b} = A^{(a-b)}.$$

We now consider separately the two possible cases.

(2) Suppose that in $\mathfrak{A}$ there exists an infinite sequence of elements A_1, A_2, $A_3, \ldots$ in which each element is a two-sided zero of its predecessor.

We define a mapping φ of the semigroup $\mathfrak{U}$ (5.12) into $\mathfrak{A}$:

$$\varphi(U_\alpha^\beta) = A_\alpha^\beta \qquad (\alpha, \beta = 1, 2, \ldots).$$

This is a one-to-one mapping. In fact, suppose

$$A_{\alpha_1}^{\beta_1} = A_{\alpha_2}^{\beta_2}.$$

For $\alpha_1 = \alpha_2$, as we showed earlier, we must have $\beta_1 = \beta_2$. If $\alpha_1 > \alpha_2$, then multiplying both sides of the equation by A_{α_1}, we obtain

$$A_{\alpha_1}^{\beta_1+1} = A_{\alpha_1},$$

which is impossible.

Since for $\alpha_1 > \alpha_2$, the element A_{α_1} is a two-sided zero of A_{α_2}, the elements A_α^β $(\alpha, \beta = 1, 2, \ldots)$ obey the same multiplication law as the elements U_α^β in the semigroup $\mathfrak{U}$ (5.12). Therefore, the mapping φ is an isomorphism from $\mathfrak{U}$ to $\mathfrak{A}$.

(3) Suppose now that in $\mathfrak{A}$ every sequence of elements $A_1, A_2, \ldots$ in which each element is a two-sided zero of its predecessor can have only a finite number of terms.

We denote by $\mathfrak{B}$ the ensemble of all elements of $\mathfrak{A}$ that have no two-sided zero, and by $\mathfrak{B}^*$ the ensemble of all those elements for which no power has a two-sided zero.

We shall show that for every $A \in \mathfrak{A}$ we can find in $\mathfrak{B}^*$ an element A^* which is a right zero of A.

We construct the sequence $X_0 = A, X_1, X_2, \ldots$ of elements of $\mathfrak{A}$. X_1 is some right zero of A. If there exist right zeros of A not belonging to $\mathfrak{B}$, then X_1 is chosen from among them. X_{n+1} $(n = 1, 2, 3, \ldots)$ is a two-sided zero of the element X_n, and, if there exist two-sided zeros of X_n not belonging to $\mathfrak{B}$, then X_{n+1} is chosen from among them. According to the hypothesis, this sequence must terminate at some X_m $(m \geqslant 1)$. This last element belongs to $\mathfrak{B}$. We shall show that, even further, $X_m \in \mathfrak{B}^*$. In fact, if this were not so, then for some k the element X_m^k would have a two-sided zero, that is, $X_m^k \,\overline{\in}\, \mathfrak{B}$. It follows, however, from $X_m \in \mathfrak{B}$ that every two-sided zero of the element X_{m-1} (or right zero of $A = X_0$, if $m = 1$) must belong to $\mathfrak{B}$. At the same time the element X_m^k

is a two-sided zero of X_{m-1} (or a right zero of $A = X_0$ if $m = 1$) and does not belong to $\mathfrak{B}$.

The element X_m, which belongs to $\mathfrak{B}^*$, is a right zero of A and therefore may be taken as the desired element A^*.

(4) Under the assumptions laid down at the beginning of the preceding portion of the proof, we construct an infinite sequence of elements of $\mathfrak{A}$,

$$A_0, A_1, A_2, \ldots .$$

A_0 is an arbitrary element of $\mathfrak{A}$; A_n $(n = 1, 2, \ldots)$ is the element A^*_{n-1} possessing with respect to A_{n-1} the properties indicated in the preceding portion of the proof.

Every element A_n $(n = 1, 2, \ldots)$ belongs to $\mathfrak{B}^*$ and is a right zero of all preceding elements.

We define a mapping φ of the semigroup $\mathfrak{B}$ (5.13) into $\mathfrak{A}$:

$$\varphi(V_n^{\alpha_n} V_{n-1}^{\alpha_{n-1}} \ldots V_2^{\alpha_2} V_1^{\alpha_1}) = A_n^{\alpha_n} A_{n-1}^{\alpha_{n-1}} \ldots A_2^{\alpha_2} A_1^{\alpha_1}$$

(A_i^0 is the empty symbol).

We shall show that φ is a one-to-one mapping. Suppose that

$$A_n^{\alpha_n} A_{n-1}^{\alpha_{n-1}} \ldots A_2^{\alpha_2} A_1^{\alpha_1} = A_n^{\beta_n} A_{n-1}^{\beta_{n-1}} \ldots A_2^{\beta_2} A_1^{\beta_1},$$

where $\alpha_n \neq 0$ and $\alpha_j - \beta_j = \delta > 0$, with $\alpha_i = \beta_i$ for all $i > j$.

We multiply this equation on the right by A_j. Since A_j is a right zero of all A_i for $i < j$, we obtain

$$A_n^{\gamma_n} A_{n-1}^{\gamma_{n-1}} \ldots A_{j+1}^{\gamma_{j+1}} A_j^{\beta_j + \delta + 1} = A_n^{\gamma_n} A_{n-1}^{\gamma_{n-1}} \ldots A_{j+1}^{\gamma_{j+1}} A_j^{\beta_j + 1}.$$

As we know, j cannot be equal to n.

Writing

$$W = A_n^{\gamma_n} A_{n-1}^{\gamma_{n-1}} \ldots A_{j+1}^{\gamma_{j+1}} A_j^{\beta_j + 1},$$

we obtain

$$A_j^{\delta} W = W, \qquad W A_j^{\delta} = W,$$

which is in contradiction with the fact that $A_j \in \mathfrak{B}^*$.

This proves that φ is a one-to-one mapping.

Since A_p is a right zero of A_q for $q < p$, the multiplication rule for elements of the form $A_n^{\alpha_n} A_{n-1}^{\alpha_{n-1}} \ldots A_2^{\alpha_2} A_1^{\alpha_1}$ is the same as that for the corresponding elements in $\mathfrak{B}$. Therefore the mapping φ is an isomorphism from $\mathfrak{B}$ to $\mathfrak{A}$.

6. Regular Elements

6.1. The concept of regularity, borrowed from the theory of rings, is useful in a number of problems in the theory of semigroups.

DEFINITION. *An element A of a semigroup $\mathfrak{A}$ is said to be* REGULAR, *if we can find in $\mathfrak{A}$ an element X such that*

$$AXA = A.$$

A semigroup consisting entirely of regular elements is said to be a REGULAR SEMIGROUP.

An element A is said to be COMPLETELY REGULAR *if we can find in $\mathfrak{A}$ an element X such that*

$$AXA = A,$$

$$AX = XA.$$

A semigroup consisting entirely of completely regular elements is said to be COMPLETELY REGULAR.

It is clear that the concepts of regularity and complete regularity coincide for commutative semigroups.

An example of a regular semigroup is the semigroup $\mathfrak{S}_\Omega$ of all mappings of a set Ω (I, 3.6). In fact, for every $S \in \mathfrak{S}_\Omega$ it is easily seen that the relationship $S\bar{S}S = S$ holds for every $\bar{S} \in \mathfrak{S}_\Omega$ such that $\bar{S}\alpha$ ($\alpha \in \Omega$) is equal to some one of the elements β for which $S\beta = \alpha$, and arbitrarily for $S\Omega \bar{\ni} \alpha$.

The concept of regularity was first introduced by J. v. Neumann[7] for elements of rings. An element of a ring is regular if it is a regular element of the multiplicative semigroup of the ring in the sense just defined. In the theory of rings, the identification of regular rings is of interest from many points of view.

In the general theory of semigroups, the regular semigroups were first studied by Thierrin **[1]** under the name of *demi-groupes inversifs*. The completely regular semigroups were introduced by Clifford **[4]** and were studied by him from the point of view which we shall consider in Chapter VIII.

6.2. The property of regularity of elements turns out to be connected with the question of whether or not they possess certain special units.

DEFINITION. *An element I of a semigroup $\mathfrak{A}$ which is a left unit of the element $A \in \mathfrak{A}$ is called a* REGULAR LEFT UNIT *if it is divisible on the left by A.*

I is called a REGULAR RIGHT UNIT *of A if it is a right unit of A and is divisible on the right by A.*

I is called a REGULAR TWO-SIDED UNIT *of A if I is a two-sided unit of A and is divisible both on the left and on the right by A.*

Thus, if $I \in \mathfrak{A}$ is a regular left unit of $A \in \mathfrak{A}$ there must exist an $X \in \mathfrak{A}$ such that

$$IA = A, \qquad AX = I.$$

The condition that I should be a right regular unit is similar to the above.

The element I is a regular unit of the element A if and only if I is simultaneously a regular left unit of A and a regular right unit of A. Although, as we shall shortly see, an element that has a regular left unit has also, of necessity, a

[7] J. v. Neumann, *On regular rings*, Proc. Nat. Acad. Sci. U.S.A. **22** (1936), 707–713.

regular right unit, there are cases in which an element possesses a two-sided unit, which is also its regular left unit, but is not a regular two-sided unit. We shall find examples of this case in III, 6.2 and III, 6.3.

6.3. Idempotents are peculiarly important in the study of the properties of regularity.

Every idempotent is completely regular. It is its own regular two-sided unit. A regular left unit of an arbitrary element is always an idempotent.

In fact, suppose that
$$IA = A, \qquad AX = I.$$
Then
$$I^2 = I \cdot AX = AX = I.$$
A similar property holds for regular right units.

6.4. THEOREM. *A nonregular element of a semigroup has neither a regular left unit nor a regular right unit.*

A regular element has both a regular left unit and a regular right unit.

An element that is not completely regular has no regular two-sided unit.

PROOF. (1) If I is a regular left unit of the element A, that is, for some X
$$IA = A, \qquad AX = I,$$
then
$$AXA = IA = A,$$
that is, A is regular.

In a similar fashion we may prove that an element having a regular right unit is regular.

(2) If $AXA = A$, then, writing $I_A = AX$ and $I^{(A)} = XA$, we obtain
$$I_A A = A, \qquad I_A = AX,$$
$$AI^{(A)} = A, \qquad I^{(A)} = XA,$$
that is, I_A is a regular left unit of A, and $I^{(A)}$ is a regular right unit.

(3) If we have for the elements $A, X \in \mathfrak{A}$
$$AXA = A, \qquad AX = XA = I,$$
then
$$AI = AXA = A, \qquad IA = AXA = A,$$
$$I = AX = XA,$$
that is, I is a regular two-sided unit of A.

(4) Let the element A have the regular two-sided unit I:
$$AI = IA = A, \qquad I = UA = AV.$$
We write $X = UAV$. We have
$$AXA = AUAVA = AIVA = AVA = IA = A,$$
$$AX = AUAV = AIV = AV = I,$$
$$XA = UAVA = UIA = UA = I,$$
$$AX = XA.$$

The relations between A and X which we have obtained show that A is completely regular.

6.5. A regular element of a semigroup may have several regular left units and several regular right units. A regular two-sided unit of a completely regular element is, however, uniquely determined.

An element of a semigroup may have no more than one regular two-sided unit.

In fact, if I_1 and I_2 are regular two-sided units of the element A, then for some X and Y

$$I_1 = XA, \qquad I_2 = AY.$$

This means that

$$I_1 = XA = XAI_2 = I_1I_2 = I_1AY = AY = I_2.$$

6.6. DEFINITION. *If the elements A and B of the semigroup $\mathfrak{A}$ are connected by the relation*

$$ABA = A, \qquad BAB = B,$$

they are said to be REGULARLY CONJUGATE.

V. V. Vagner [**1**; **2**; **3**], who studied in detail the relation of regular conjugacy, calls the regular conjugate of an element A its *generalized inverse*,[8] and denotes it by A^{-1}.

We note that an element of a semigroup may have several regular conjugates. For example, in a semigroup where the product of two elements is always equal to the left coefficient, every two elements are regular conjugates of each other.

6.7. An element possessing a regular conjugate is obviously regular. The converse also turns out to be true.

THEOREM. *Every regular element of a semigroup has a regular conjugate. A completely regular element has a regular conjugate which commutes with it.*

PROOF. Corresponding to the regular element A of the semigroup $\mathfrak{A}$ there exists in $\mathfrak{A}$ an element X such that

$$AXA = A.$$

Writing $B = XAX$, we obtain

$$ABA = A \cdot XAX \cdot A = A \cdot XA \cdot XA = AXA = A,$$

$$BAB = XAX \cdot A \cdot XAX = X \cdot AXA \cdot XAX = XAXAX = XAX = B,$$

that is, A and B are regularly conjugate.

If in addition A is completely regular, and X commutes with A, then the regular conjugate of A, that is, $B = XAX$ clearly also commutes with A.

6.8. We shall consider in detail those semigroups in which every two elements are regular conjugates.

---

[8] *Translator's note*: A. Clifford and G. Preston simply call them "inverses."

THEOREM. *Every two elements of a semigroup $\mathfrak{A}$ are regularly conjugate if and only if all elements are idempotents and for arbitrary X, Y, $Z \in \mathfrak{A}$ we have the relation*

$$XYZ = XZ.$$

PROOF. (1) Suppose that every two elements of $\mathfrak{A}$ are regularly conjugate. For arbitrary $X \in \mathfrak{A}$ we have as regular conjugates X^4 and X^2 as well as X^6 and X. Thus we obtain

$$X^8 = X^2 \cdot X^4 \cdot X^2 = X^2,$$

$$X^8 = X \cdot X^6 \cdot X = X$$

and therefore $X^2 = X$.

For arbitrary X, $Y, Z \in \mathfrak{A}$ we have

$$ZXZ = Z,$$

$$X \cdot YZ \cdot X = X.$$

Therefore

$$XYZ = XY \cdot ZXZ = (X \cdot YZ \cdot X) \cdot Z = XZ.$$

(2) If for arbitrary X, $Y, Z \in \mathfrak{A}$ we have

$$XYZ = XZ, \qquad X^2 = X,$$

then for arbitrary A, $B \in \mathfrak{A}$ we obtain

$$ABA = AA = A, \qquad BAB = BB = B.$$

6.9. McLean [**1**] has developed a quite different approach to the class of semigroups we are now considering. It turns out that this class can be characterized by the property of anti-commutativity, which is diametrically opposite to the commutativity property we have been using.

THEOREM. *All elements of a semigroup are mutually regularly conjugate if and only if for arbitrary X, Y, $\in \mathfrak{A}$ the relation $X \neq Y$ always implies*

$$XY \neq YX.$$

PROOF. (1) Suppose every two elements of $\mathfrak{A}$ are regularly conjugate. Then for arbitrary X, $Y \in \mathfrak{A}$ we have

$$(XY) \cdot (YX) = XY^2X = X,$$

$$(YX) \cdot (XY) = YX^2Y = Y.$$

It follows that for $X \neq Y$ the equality $XY = YX$ is impossible.

(2) Suppose that for all X, $Y \in \mathfrak{A}$ $X \neq Y$ implies $XY \neq YX$. Since for arbitrary $X \in \mathfrak{A}$

$$X^2 \cdot X = X \cdot X^2,$$

we must have $X^2 = X$.

In virtue of this we have for arbitrary $X, Y \in \mathfrak{A}$

$$X \cdot (XYX) = X^2YX = XYX = XYX^2 = (XYX) \cdot X,$$

and, accordingly,

$$X = XYX.$$

But, likewise,

$$Y = YXY$$

and therefore X and Y are regularly conjugate.

6.10. With the aid of the concept of regularity we can formulate a number of criteria for determining when a semigroup is a group. It can be shown without difficulty that a group always possesses the properties listed below. As Thierrin showed [**1**] these properties are not only necessary, but also sufficient, for a semigroup to be a group.

(α) *If in a semigroup with a unit the unit is a regular left unit of every element of the semigroup, then the semigroup is a group.*

This is an immediate consequence of 2.18.

(β) *If the regular semigroup $\mathfrak{A}$ has only one idempotent, it is a group.*

In fact, every element A of $\mathfrak{A}$ must have a regular left unit I_A and a regular right unit $I^{(A)}$ (6.4), which are idempotents (6.3). Therefore I_A and $I^{(A)}$ are equal to each other for every A, which means they are both equal to the unit of the semigroup $\mathfrak{A}$. It follows from (α) that $\mathfrak{A}$ is a group.

(γ) *A regular semigroup with two-sided cancellation is a group.*

In fact, let us choose in $\mathfrak{A}$ an element A. Since A is regular it has a left unit I (6.4). For arbitrary $X \in \mathfrak{A}$ we have

$$XIA = XA,$$

and since $\mathfrak{A}$ is a semigroup with right cancellation, $XI = X$, that is, I is a right unit of the entire semigroup $\mathfrak{A}$.

Let U be a regular right unit of the element X (6.4). Since

$$XU = XI,$$

and $\mathfrak{A}$ is a semigroup with left cancellation, $U = I$ and therefore I is a regular right unit of X, that is, I is divisible on the right by an arbitrary element of $\mathfrak{A}$. Similarly we demonstrate the existence of a left unit I' of the semigroup $\mathfrak{A}$. Since

$$I' = I'I = I,$$

I is a left unit of $\mathfrak{A}$. But I is divisible on the right by every element of $\mathfrak{A}$. According to 2.18 it follows that $\mathfrak{A}$ is a group.

6.11. The concept of regularity in semigroups can be generalized and sharpened with the aid of conditions first stated by Croisot [**5**].

DEFINITION. *Let m and n be non-negative integers. The* REGULARITY CLASS $\mathfrak{C}_{\mathfrak{A}}(m, n)$ *is the ensemble of all elements A of the semigroup* $\mathfrak{A}$ *for which there exists an element* $X \in \mathfrak{A}$ *such that*

$$A = A^m X A^n.$$

(A^0 is the empty symbol, referred to in the Preface).

The connection of this notion with the concept of regularity consists in the fact that regular elements belong to the class $\mathfrak{C}_{\mathfrak{A}}(1, 1)$ and the general definition of regularity classes is similar to the definition of regularity.

The significance of the concept of regularity classes emerges from the establishment of a link between this concept and certain important properties of semigroups.

6.12. Let us develop the mutual relations among regularity classes.

Suppose $\mathfrak{A}$ to be an arbitrary semigroup.

(α) $\mathfrak{C}_{\mathfrak{A}}(0, 0) = \mathfrak{A}$.

(β) *If* $m_1 \geqslant m_2$ *and* $n_1 \geqslant n_2$, *then*

$$\mathfrak{C}_{\mathfrak{A}}(m_1, n_1) \subset \mathfrak{C}_{\mathfrak{A}}(m_2, n_2).$$

In fact, if $A \in \mathfrak{C}_{\mathfrak{A}}(m_1, n_1)$, then

$$A = A^{m_1} X A^{n_1} = A^{m_2} \cdot (A^{m_1 - m_2} X A^{n_1 - n_2}) \cdot A^{n_2},$$

that is, $A \in \mathfrak{C}_{\mathfrak{A}}(m_2, n_2)$.

(γ) *If* $m_1 \geqslant m_2 \geqslant 2$, *then for arbitrary n*

$$\mathfrak{C}_{\mathfrak{A}}(m_1, n) = \mathfrak{C}_{\mathfrak{A}}(m_2, n).$$

In fact, if $A \in \mathfrak{C}_{\mathfrak{A}}(m_2, n)$, then

$$A = A^{m_2} X A^n = A^{m_2 - 1} \cdot A \cdot X A^n = A^{m_2 - 1} A^{m_2} X A^n X A^n$$
$$= A^{m_2 + 1}(A^{m_2 - 2} X A^n X) A^n,$$

that is, $A \in \mathfrak{C}_{\mathfrak{A}}(m_2 + 1, n)$.

From (β) we conclude that

$$\mathfrak{C}_{\mathfrak{A}}(m_2, n) = \mathfrak{C}_{\mathfrak{A}}(m_2 + 1, n).$$

Repeating the argument $m_1 - m_2$ times, we obtain

$$\mathfrak{C}_{\mathfrak{A}}(m_2, n) = \mathfrak{C}_{\mathfrak{A}}(m_1, n).$$

(δ) *If* $n_1 \geqslant n_2 \geqslant 2$, *then for arbitrary m*

$$\mathfrak{C}_{\mathfrak{A}}(m, n_1) = \mathfrak{C}_{\mathfrak{A}}(m, n_2).$$

The proof is similar to that of (γ).

(ε) $\mathfrak{C}_{\mathfrak{A}}(1, 2) = \mathfrak{C}_{\mathfrak{A}}(1, 1) \cap \mathfrak{C}_{\mathfrak{A}}(0, 2)$.

In fact, from (β)

$$\mathfrak{C}_{\mathfrak{A}}(1, 2) \subset \mathfrak{C}_{\mathfrak{A}}(1, 1) \cap \mathfrak{C}_{\mathfrak{A}}(0, 2).$$

But, on the other hand, if

$$A \in \mathfrak{C}_{\mathfrak{A}}(1, 1) \cap \mathfrak{C}_{\mathfrak{A}}(0, 2),$$

then

$$A = AXA, \qquad A = A^0 YA^2 = YA^2.$$

From this we obtain

$$A = AXA = AX \cdot YA^2 = A \cdot (XY)A^2$$

and, therefore, $A \in \mathfrak{C}_{\mathfrak{A}}(1, 2)$.

(ζ) $\mathfrak{C}_{\mathfrak{A}}(2, 1) = \mathfrak{C}_{\mathfrak{A}}(1, 1) \cap \mathfrak{C}_{\mathfrak{A}}(2, 0)$.

The proof follows that of (ε).

6.13. It follows from 6.12 that the classes 6.11 consist in fact of only nine classes, which satisfy the inclusion relations

$$\begin{array}{ccccccccc}
 & & & & \mathfrak{C}_{\mathfrak{A}}(0, 0) & & & & \\
 & & & \subset & & \supset & & & \\
 & & \mathfrak{C}_{\mathfrak{A}}(1, 0) & & & & \mathfrak{C}_{\mathfrak{A}}(0, 1) & & \\
 & \subset & & \supset & & \subset & & \supset & \\
\mathfrak{C}_{\mathfrak{A}}(2, 0) & & & & \mathfrak{C}_{\mathfrak{A}}(1, 1) & & & & \mathfrak{C}_{\mathfrak{A}}(0, 2) \\
 & \supset & & \subset & & \supset & & \subset & \\
 & & \mathfrak{C}_{\mathfrak{A}}(2, 1) & & & & \mathfrak{C}_{\mathfrak{A}}(1, 2) & & \\
 & & & \supset & & \subset & & & \\
 & & & & \mathfrak{C}_{\mathfrak{A}}(2, 2) & & & &
\end{array}$$

6.14. Let us consider the role of some of these classes with respect to the presence of units belonging to elements. The classes $\mathfrak{C}_{\mathfrak{A}}(0, 0)$ and $\mathfrak{C}_{\mathfrak{A}}(1, 1)$ have already been discussed in (6.11) and (6.12). $\mathfrak{C}_{\mathfrak{A}}(1, 0)$ is the class of all elements of $\mathfrak{A}$ that have right units. $\mathfrak{C}_{\mathfrak{A}}(0, 1)$ is the class of elements that have left units. The class $\mathfrak{C}_{\mathfrak{A}}(2, 0)$ consists, as is easily seen, of the elements A having right units divisible on the left by A (such a unit may be called a right quasi-regular unit). The class $\mathfrak{C}_{\mathfrak{A}}(0, 2)$ consists of elements A having left units divisible on the right by A. We will consider the other three classes when we take up other properties of semigroups.

7. Inverse Semigroups

7.1. In connection with the fact that a regular element may have several regular conjugates it is natural to consider separately the class of regular semigroups in which each element has a unique regular conjugate.

As we shall show in the next section, semigroups of this class play an important role in the theory of semigroups of partial mappings. This role was first revealed by V. V. Vagner [**1; 2**]. Somewhat later, but independently, Preston [**1**; **3**], arrived at the same results.

Semigroups of this class have been the object of several studies since then.

V. V. Vagner calls these semigroups generalized groups. This term is altogether natural since their properties arise from a weakening of the group axioms. Since, however, the term "generalized group" is often used in a wider (and nonequivalent) sense (sometimes in those cases where we used the term "multiplicative set" in Chapter I) we prefer Preston's term—inverse semigroup—which reflects the most important property of these semigroups, that is, a kind of invertibility.

7.2. DEFINITION. *An* INVERSE SEMIGROUP *is a regular semigroup in which each element has a unique regular conjugate.* (6.1 and 6.6.)

In what follows we will always, when considering inverse semigroups, denote the regular conjugate of an element A by $\bar{A}$.

7.3. Let $\mathfrak{A}$ be an inverse semigroup. We note several elementary properties of regular conjugates.

(α) *For arbitrary* $A \in \mathfrak{A}$

$$\bar{\bar{A}} = A.$$

(β) *For every idempotent* I

$$\bar{I} = I.$$

(γ) $(A\bar{A})$ *is the only regular left unit of the element* $A \in \mathfrak{A}$ (6.2).

That $(A\bar{A})$ is a regular left unit of A follows immediately from Definition 6.6. Suppose now that I is an arbitrary regular left unit of A. For some X

$$IA = A, \qquad AX = I.$$

Since I is an idempotent (6.3),

$$I = I^2 = AXI.$$

But it follows from the equation

$$A \cdot (XI) \cdot A = IA = A,$$

$$(XI) \cdot A \cdot (XI) = XI \cdot I = XI,$$

that $XI = \bar{A}$. From this we obtain $I = A \cdot (XI) = A\bar{A}$.

(δ) $(A\bar{A})$ *is the only regular right unit of the element* $A \in \mathfrak{A}$.

7.4. There are several mutually equivalent definitions of an inverse semigroup. The naturalness of these definitions, and the fact that they turn out to be mutually equivalent even though they appeal to properties of very different kinds, are convincing proof of the importance of this class.

THEOREM. *A semigroup* $\mathfrak{A}$ *is an inverse semigroup if and only if*:

(α) *every element of* $\mathfrak{A}$ *has a regular left unit* (6.2 *and* 6.4):

(β) *all idempotents of* $\mathfrak{A}$ *commute with each other.*

PROOF. (1) Let $\mathfrak{A}$ be an inverse semigroup. By 6.4, every element of $\mathfrak{A}$ has a regular left unit.

Let I_1 and I_2 be two arbitrary idempotents of $\mathfrak{A}$. Since

$$(I_1I_2)\cdot(I_2\cdot\overline{I_1I_2})\cdot(I_1I_2) = I_1I_2\cdot\overline{I_1I_2}\cdot I_1I_2 = I_1I_2,$$
$$(I_2\cdot\overline{I_1I_2})\cdot(I_1I_2)\cdot(I_2\cdot\overline{I_1I_2}) = I_2\cdot(\overline{I_1I_2})\cdot(I_1I_2)\cdot(\overline{I_1I_2}) = I_2\cdot(\overline{I_1I_2}),$$

we find, since a regular conjugate is unique, that

$$I_2\cdot\overline{I_1I_2} = \overline{I_1I_2}.$$

Similarly it can be shown that

$$\overline{I_1I_2}\cdot I_1 = \overline{I_1I_2}.$$

Since

$$\overline{I_1I_2}^2 = (\overline{I_1I_2}\cdot I_1)(I_2\cdot\overline{I_1I_2}) = \overline{I_1I_2}\cdot I_1I_2\cdot\overline{I_1I_2} = \overline{I_1I_2},$$

then, by 7.3, (α), (β),

$$I_1I_2 = \overline{\overline{I_1I_2}} = \overline{I_1I_2}.$$

We have proved that I_1I_2 is an idempotent. In like manner we may show that I_2I_1 is an idempotent. Accordingly,

$$I_1I_2\cdot I_2I_1\cdot I_1I_2 = I_1I_2I_1I_2 = I_1I_2,$$
$$I_2I_1\cdot I_1I_2\cdot I_2I_1 = I_2I_1I_2I_1 = I_2I_1,$$

which means that

$$\overline{I_1I_2} = I_2I_1.$$

But we proved earlier that $\overline{I_1I_2} = I_1I_2$. Therefore $I_2I_1 = I_1I_2$.

(2) Suppose now that a semigroup $\mathfrak{A}$ has the properties (α) and (β) stated in the theorem. By 6.4 and 6.7, every element $A \in \mathfrak{A}$ has a regular conjugate. Let us assume that for some element A both X_1 and X_2 are regular conjugates of A. Since the elements AX_1, AX_2, X_1A, X_2A are idempotents, they commute with each other, and we have

$$X_1 = X_1AX_1 = X_1\cdot AX_2A\cdot X_1 = X_1\cdot AX_2\cdot AX_1 = X_1\cdot AX_1\cdot AX_2 = X_1AX_2,$$
$$X_2 = X_2AX_2 = X_2\cdot AX_1A\cdot X_2 = X_2A\cdot X_1A\cdot X_2 = X_1A\cdot X_2A\cdot X_2 = X_1AX_2,$$

that is, $X_1 = X_2$.

7.5. COROLLARY. *A mapping of an inverse semigroup $\mathfrak{A}$ on itself, which carries every element A into its regular conjugate $\bar{A}$, is an anti-automorphism* (I, 1.17).

PROOF. Since $A\bar{A}$ and $\bar{A}A$ are idempotents, we obtain, using the commutativity of idempotents,

$$A_1A_2\cdot\bar{A}_2\bar{A}_1\cdot A_1A_2 = A_1\cdot A_2\bar{A}_2\cdot\bar{A}_1A_1\cdot A_2 = A_1\cdot\bar{A}_1A_1\cdot A_2\bar{A}_2\cdot A_2 = A_1A_2,$$
$$\bar{A}_2\bar{A}_1\cdot A_1A_2\cdot\bar{A}_2\bar{A}_1 = \bar{A}_2\cdot\bar{A}_1A_1\cdot A_2\bar{A}_2\cdot\bar{A}_1 = \bar{A}_2\cdot A_2\bar{A}_2\cdot\bar{A}_1A_1\cdot\bar{A}_1 = \bar{A}_2\bar{A}_1.$$

Therefore

$$\bar{A}_2\bar{A}_1 = \overline{A_1A_2}.$$

Since $\bar{\bar{A}} = A$, the mapping which associates A with $\bar{A}$ is a biunique mapping of $\mathfrak{A}$ on itself. It is therefore an anti-automorphism.

7.6. In studying certain properties of inverse semigroups it is useful to focus attention on a partial ordering relation which arises naturally in every inverse semigroup.

We shall say that an element A of an inverse semigroup $\mathfrak{A}$ precedes the element B.

$$A \leqslant B,$$

if the regular left unit I_A of the element A satisfies

$$I_AB = A.$$

Let us verify that this is indeed a partial ordering. Obviously

$$A \leqslant A.$$

If we have at the same time

$$A \leqslant B, \qquad B \leqslant A,$$

that is,

$$I_AB = A, \qquad I_BA = B,$$

then

$$I_AI_BA = A,$$

from which it follows that I_AI_B is a regular left unit of A (its regularity follows immediately from the regularity of I_A). By 7.3, (γ), we have $I_A \cdot I_B = I_A$. Similarly we show that $I_B I_A = I_B$. According to 7.4, this means that $I_A = I_B$. But then

$$A = I_AA = I_BA = B.$$

If

$$A \leqslant B, \qquad B \leqslant C,$$

that is,

$$I_AB = A, \qquad I_BC = B,$$

then

$$(I_AI_B)C = I_AB = A.$$

But for some X

$$I_A = AX.$$

It follows from this that

$$I_AI_B = I_BI_A = I_B \cdot AX = I_BI_ABX = I_A \cdot I_BB \cdot X = I_ABX = AX = I_A.$$

Therefore the equation $(I_AI_B)C = A$, derived earlier, means that

$$I_AC = A,$$

that is $A \leqslant C$.

7.7. By a process altogether like that of 7.6 we may introduce a partial ordering by the use of regular right units, setting

$$A \leqslant B,$$

if the regular right unit $I^{(A)}$ of the element A satisfies

$$BI^{(A)} = A.$$

By 7.3, (δ), we may write this last equation in the form

$$B\bar{A}A = A.$$

If now we apply to both sides of this equality the anti-automorphism defined in 7.5, we obtain

$$\bar{A}A\bar{B} = \bar{A}.$$

Multiplying both sides of this equality on the left by A and on the right by B, we obtain

$$A\bar{A}A\bar{B}B = A\bar{A}B,$$

that is,

$$A\bar{B}B = A\bar{A}B.$$

But $A\bar{A} = I_A$ is a regular left unit of A and $\bar{B}B = I^{(B)}$ is a regular right unit of B, and, since $BI^{(A)} = A$ and $I^{(A)}I^{(B)} = I^{(B)}I^{(A)}$, $I^{(B)}$ is also a right unit of A. From this we obtain

$$A = I_A B.$$

But this shows that A precedes B in the partial ordering introduced in 7.6. In a wholly similar fashion we may prove the converse—that the partial ordering introduced in 7.6 gives rise to the same precedence relations as does the partial ordering introduced here. Therefore, these partial orderings coincide.

7.8. The introduction of one or other partial ordering in a semigroup is especially useful when it agrees with the multiplication operation, i.e., when it is two-sidedly stable (I, 5.17).

The partial ordering we have just introduced in an inverse semigroup has this property. It is two-sidedly stable. In fact,

$$A \leqslant B$$

means that

$$I_A B = A.$$

If $I_{(AX)}$ is a regular left unit of AX, then for some Y

$$I_{(AX)} = AXY,$$

from which it follows that

$$I_{(AX)} = I_A I_{(AX)} = I_{(AX)} I_A.$$

Therefore

$$I_{(AX)}BX = I_{(AX)}I_ABX = I_{(AX)}AX = AX,$$

that is

$$AX \leqslant BX.$$

In the same way, but using the partial ordering introduced in 7.7, we obtain

$$XA \leqslant XB.$$

7.9. We now take up one possible approach to the notion of inverse semigroup, limiting ourselves to the case in which a unit exists.

Let $\mathfrak{A}$ be an inverse semigroup with unit E. We define over the set $\mathfrak{A}$ a ternary operation, that is, a rule which sets up a correspondence between every sequence of three elements of $\mathfrak{A}$, A_1, A_2, A_3 and the elements of $\mathfrak{A}$:

$$[A_1, A_2, A_3] = A_1\bar{A}_2A_3.$$

The ternary operation so defined has the following properties:

(α) $[[A_1, A_2, A_3], A_4, A_5] = [A_1, [A_4, A_3, A_2], A_5] = [A_1, A_2, [A_3, A_4, A_5]]$;

(β) $[[A_1, A_2, A_2], A_3, A_3] = [[A_1, A_3, A_3], A_2, A_2]$;

(γ) $[A_1, A_1, [A_2, A_2, A_3]] = [A_2, A_2, [A_1, A_1, A_3]]$;

(δ) $[A, A, A] = A$;

(ε) $[A, E, E] = [E, E, A] = A$.

The validity of (α) follows from the fact that, by 7.5, each of the expressions entering it is equal to

$$A_1\bar{A}_2A_3\bar{A}_4A_5.$$

The expressions in (β) are equal to

$$A_1\bar{A}_2A_2\bar{A}_3A_3, \qquad A_1\bar{A}_3A_3, \qquad \bar{A}_2A_2,$$

which are the same because of the commutativity of the idempotents $\bar{A}_2A_2$ and $\bar{A}_3A_3$.

The position is the same with respect to (γ).

The equations (δ) and (ε) are altogether obvious.

7.10. Every set admitting a ternary operation with the properties defined in (α), (β), (γ), (δ), (ε) of 7.9 is called by V. V. Vagner a *generalized aggregate with a biunitary element E*. There are reasons for studying such objects.

The study of generalized aggregates with a biunitary element is equivalent to the study of inverse semigroups with unit in virtue of what we have shown in 7.9, on the one hand, and of the following proposition on the other.

It is always possible to define a multiplication over the set $\mathfrak{A}$ of all elements of a generalized aggregate with a biunitary element E in such a way that $\mathfrak{A}$ becomes an inverse semigroup having E as its unit and the ternary operation induced by the *multiplication, as defined in 7.9, coincides with the ternary operation of the given generalized aggregate.*

In fact, we write

$$A \cdot B = [A, E, B].$$

For this multiplication we have

$$(A \cdot B) \cdot C = [[A, E, B], E, C] = [A, E, [B, E, C]] = A \cdot (B \cdot C),$$

$$A \cdot E = [A, E, E] = A,$$

$$E \cdot A = [E, E, A] = A.$$

Accordingly, $\mathfrak{A}$ is a semigroup with unit E. Since

$$(A \cdot [E, A, E]) \cdot A = [A, E, [E, A, E]] \cdot A = [[A, E, E], A, E] \cdot A$$
$$= [A, A, E] \cdot A = [[A, A, E], E, A] = [A, A, [E, E, A]] = [A, A, A] = A,$$

$A \cdot [E, A, E]$ is a regular left unit of A.

Let us denote by $\mathfrak{B}$ the ensemble of elements of the form $[E, A, A]$ ($A \in \mathfrak{A}$). Every two elements of $\mathfrak{B}$ commute. In fact,

$$[E, A, A] \cdot [E, B, B] = [[E, A, A], E, [E, B, B]] = [[[E, A, A], E, E], B, B]$$
$$= [[E, A, [A, E, E]], B, B] = [[E, A, A], B, B] = [[E, B, B], A, A]$$
$$= [[E, B, [B, E, E]], A, A] = [[[E, B, B], E, E], A, A]$$
$$= [[E, B, B], E, [E, A, A]] = [E, B, B] \cdot [E, A, A].$$

We shall show that an element of the form $[A, A, E]$ also belongs to $\mathfrak{B}$. In fact,

$$[A, A, E] = [[A, E, E], A, E] = [[[E, E, A], E, E], A, E]$$
$$= [[E, [E, A, E], E], A, E]$$
$$= [E, [E, A, E], [E, A, E]] \in \mathfrak{B}.$$

Let I be an idempotent of the semigroup in question.

$$I = [I, I, I] = [I, I \cdot I, I] = [I, [I, E, I], I] = [I, I, [E, I, I]]$$
$$= [I, [E, E, I], [E, I, I]] = [[I, I, E], E, [E, I, I]]$$
$$= [I, I, E] \cdot [E, I, I].$$

We observe that every idempotent is representable as a product of elements belonging to $\mathfrak{B}$. Since the elements of $\mathfrak{B}$ commute with each other, so also do all idempotents.

It follows, by 7.4, that our semigroup is an inverse semigroup. Let us now see what sort of ternary operation is induced by the multiplication in this semigroup. We have seen that

$$A \cdot [E, A, E] \cdot A = A.$$

Also

$$[E, A, E] \cdot A \cdot [E, A, E] = [[E, A, E], E, A] \cdot [E, A, E]$$
$$= [E, A, [E, E, A]] \cdot [E, A, E] = [E, A, A] \cdot [E, A, E]$$
$$= [[E, A, A], E, [E, A, E]] = [[[E, A, A], E, E], A, E]$$
$$= [[E, A, A], A, E] = [E, [A, A, A], E] = [E, A, E].$$

Accordingly, the element $[E, A, E]$ is regularly conjugate to A. Therefore the induced ternary operation associates to the sequence A_1, A_2, A_3 the element

$$A_1\bar{A}_2A_3 = A_1 \cdot [E, A_2, E] \cdot A_3 = [A_1, E, [E, A_2, E]] \cdot A_3$$
$$= [[A_1, E, E], A_2, E] \cdot A_3 = [A_1, A_2, E] \cdot A_3$$
$$= [[A_1, A_2, E], E, A_3] = [A_1, A_2, [E, E, A_3]] = [A_1, A_2, A_3].$$

We see that the induced ternary operation coincides with the initially given ternary operation in $\mathfrak{A}$. This completes our proof.

7.11. We shall not pursue the study of further relations between inverse semigroups and generalized aggregates. These matters are taken up in a paper by V. V. Vagner [3], where the correspondence discussed above was first derived. The investigation becomes much more complicated in the absence of a unit (V. V. Vagner [**4**; **5**]).

8. Inverse Semigroups of Partial Transformations

8.1. We introduced the concept of partial transformations in § 4 of the first chapter. In this section we will develop the role of inverse semigroups in the theory of partial transformations. The essential result is the proof of Theorem 8.6, first obtained by V. V. Vagner [**2**; **3**], and later by Preston [**3**].

8.2. Let $\mathfrak{A}$ be an inverse semigroup of partial transformations of some set Ω and let $A \in \mathfrak{A}$. $I^{(A)}$ is a regular left unit of A, which as we know (7.3, (δ)) is equal to $\bar{A}A$. We denote by A^* the partial transformation of the same set Ω which satisfies the conditions

$$\Pi_1(A^*) = \Pi_2(I^{(A)}),$$
$$\Pi_2(A^*) = \Pi_2(A),$$
$$A^*\alpha = A\alpha(\alpha \in \Pi_1(A^*)).$$

We note that $\Pi_1(A) = \Pi_1(I^{(A)})$ since A and $I^{(A)}$ divide each other on the right. Also, $\Pi_2(I^{(A)}) \subset \Pi_1(I^{(A)})$ since $I^{(A)}$ is an idempotent. It follows that $\Pi_1(A^*) \subset \Pi_1(A)$. In fact, $\Pi_1(A^*)$ consists of those elements α of $\Pi_1(A)$ for which $I^{(A)}\alpha = \alpha$. We might say that A^* is obtained from A by cancelling the set $\Pi_1(A)$.

We shall show that for every $A \in \mathfrak{A}$ there exists a partial transformation of the type A^*. In fact, Π_1 is defined for this A^*. The way in which A^* maps elements of $\Pi_1(A^*)$ is defined (since $\Pi_1(A^*) \subset \Pi_1(A)$). It remains to show that for such a partial transformation we have

$$\Pi_2(A^*) = \Pi_2(A).$$

The inclusion relation $\Pi_2(A^*) \subset \Pi_2(A)$ follows immediately from $\Pi_1(A^*) \subset \Pi_1(A)$. To verify the converse inclusion we choose an arbitrary element $\alpha = A\beta$ from $\Pi_2(A)$. Since $A = AI^A$, $\alpha = A(I^{(A)}\beta)$. Inasmuch as

$$I^{(A)}\beta \in \Pi_2(I^{(A)}) = \Pi_1(A^*),$$

we have $\alpha = A(I^{(A)}\beta) = A^*(I^{(A)}\beta)$. Accordingly $\alpha \in \Pi_2(A^*)$. Thus $\Pi_2(A^*) = \Pi_2(A)$.

We shall show that the partial transformation A^* is always one-to-one. Suppose that for some $\alpha, \beta \in \Pi_1(A^*)$

$$A^*\alpha = A^*\beta,$$

that is

$$A\alpha = A\beta, \qquad \alpha = I^{(A)}\,\bar{\alpha}, \qquad \beta = I^{(A)}\bar{\beta}.$$

Since $I^{(A)} = \bar{A}A$ we have

$$\alpha = I^{(A)}\bar{\alpha} = I^{(A)}I^{(A)}\bar{\alpha} = \bar{A}A\alpha = \bar{A}A\beta = I^{(A)}I^{(A)}\bar{\beta} = I^{(A)}\bar{\beta} = \beta.$$

8.3. Let φ be the transformation of the inverse semigroup of partial transformations $\mathfrak{A}$ on the semigroup of all one-to-one partial transformations defined by

$$\varphi(A) = A^*.$$

We shall show that this is an automorphism.

First let us convince ourselves that φ is a one-to-one transformation.

Suppose that $\varphi(A) = \varphi(B)$ $(A, B \in \mathfrak{A})$ that is, $A^* = B^*$.

If $\xi \in \Pi_1(BI^{(A)})$, then $\xi \in \Pi_1(I^{(A)}) = \Pi_1(A)$. But, on the other hand, for every $\alpha \in \Pi_1(A)$ we have

$$A\alpha = AI^{(A)}\alpha = A^*I^{(A)}\alpha = B^*I^{(A)}\alpha = BI^{(A)}\alpha.$$

Accordingly, $\alpha \in \Pi_1(BI^{(A)})$. Thus, the set Π_1 is the same for A and $BI^{(A)}$, in that every element of this set is mapped in the same way by A and $BI^{(A)}$. Therefore $A = BI^{(A)}$. According to 7.7, this means that $A \leqslant B$. Since we can show in the same way that $B \leqslant A$, $A = B$ (7.6).

We now show that the one-to-one transformation φ is an isomorphism.

If $\alpha \in \Pi_1(A^*B^*)$ then $\alpha \in \Pi_1(B^*)$ and therefore $I^{(B)}\,\alpha = \alpha$. Furthermore, $B^*_\alpha \in \Pi_1(A^*)$. Accordingly $I^{(A)}B^*_\alpha = B^*_\alpha$ and therefore $I^{(A)}B\alpha = B\alpha$.
But then

$$\alpha = I^{(B)}\alpha = \bar{B}B\alpha = \bar{B}I^{(A)}B\alpha = \bar{B}\bar{A}AB\alpha = \overline{AB}\cdot AB\alpha = I^{(AB)}\alpha.$$

Therefore $\alpha \in \Pi_1[(AB)^*]$.

On the other hand, if $\beta \in \Pi_1[(AB)^*]$ then

$$\beta = I^{(AB)}\beta = \overline{AB} \cdot AB\beta = \bar{B}\bar{A}AB\beta,$$

and since

$$\Pi_1[(AB)^*] \subset \Pi_1(AB) \subset \Pi_1(B) = \Pi_1(I^{(B)}),$$

we have

$$I^{(B)}\beta = \bar{B}B\beta = \bar{B}B \cdot \bar{B}\bar{A}AB\beta = \bar{B}B\bar{B} \cdot \bar{A}AB\beta = \bar{B}\bar{A}AB\beta = \beta.$$

Therefore $\beta \in \Pi_1(B^*)$.

In view of the commutativity of the idempotents $\bar{B}B$ and $\bar{A}A$ (7.4) we obtain

$$B\beta = B\bar{B}\bar{A}AB\beta = \bar{A}AB\bar{B}B\beta = \bar{A}AB\beta = I^{(A)}(B\beta).$$

Therefore $B^*\beta = B\beta \in \Pi_1(A^*)$. It follows that $\beta \in \Pi_1[A^*B^*]$.

We have shown that $\Pi_1(A^*B^*) = \Pi_1[(AB)^*]$. But, for every element ξ of this set the partial transformations A^*, B^*, $(AB)^*$ are respectively identical with the transformations A,B, (AB). Therefore

$$(A^*B^*)\xi = A^*(B^*\xi) = A^*(B\xi) = AB\xi,$$

$$(AB)^*\xi = AB\xi,$$

that is $A^*B^* = (AB)^*$.

Thus we have proved that

$$\varphi(AB) = (AB)^* = A^*B^* = \varphi(A) \cdot \varphi(B).$$

8.4. The construction of the isomorphism φ, undertaken in 8.2 and 8.3, helps to clarify the meaning of the inversiveness of semigroups of partial transformations. The one-to-one transformation $\varphi(A) = A^*$ is essentially the same transformation as A, but with some reduction of its domain.

It is sometimes said that A^* is the reduced partial transformation A. When we go from the partial transformation A to the partial transformation A^* we obtain essentially the same semigroup, because of the isomorphism φ. The semigroup so obtained, however, consists entirely of one-to-one transformations. This shows that an inverse semigroup of partial transformations consists in essence of one-to-one transformations (or of transformations reducible to one-to-one transformations). A little later we shall show that the converse is true, in a sense. Thus, insofar as partial transformations are concerned, the inverse semigroups are distinguished from the class of all semigroups by the fact that they contain only one-to-one transformations.

8.5. THEOREM. *Let $\mathfrak{A}$ be a semigroup of one-to-one partial transformations of a set Ω. $\mathfrak{A}$ is an inverse semigroup if and only if it contains for each element A the one-to-one transformation inverse to A* (I, 4.5).

PROOF. (1) Let $\mathfrak{A}$ be an inverse semigroup and $A \in \mathfrak{A}$. If

$$A\alpha = \beta \qquad (\alpha, \beta \in \Omega),$$

then from

$$A\bar{A}A\alpha = A\alpha = \beta$$

(where $\bar{A}$ is the regular conjugate of A (6.6)) it follows that

$$A\alpha = \beta \in \Pi_1(\bar{A}).$$

We write

$$\bar{A}\beta = \gamma.$$

We have

$$A\gamma = A\bar{A}\beta = A\bar{A}A\alpha = A\alpha.$$

Since A is a one-to-one transformation, $\alpha = \gamma$. Thus $\alpha \in \Pi_1(A)$ always implies that $A\alpha \in \Pi_1(\bar{A})$ with

$$\bar{A}(A\alpha) = \alpha.$$

It is now obvious that $\Pi_2(A) \subset \Pi_1(\bar{A})$. But from $\bar{A}A\bar{A}\alpha = \bar{A}\alpha$ ($\alpha \in \Pi_1(\bar{A})$), in view of the fact that $\bar{A}$ is a one-to-one transformation, it follows that $A\bar{A}\alpha = \alpha$, that is,

$$\Pi_1(\bar{A}) \subset \Pi_2(A)$$

and, therefore, $\Pi_1(\bar{A}) = \Pi_2(A)$.

Since $\bar{\bar{A}} = A$, the converse relation is true. It follows that the one-to-one partial transformation $\bar{A}$ belonging to $\mathfrak{A}$ is inverse to A (I, 4.5).

(2) Suppose that in $\mathfrak{A}$ we find for each A also the inverse one-to-one partial transformation A'. Then for every $\alpha \in \Pi_1(A)$ we have

$$A\alpha = \beta, \qquad A'\beta = \alpha.$$

For every $\alpha \in \Pi_1(A)$, clearly

$$(AA')A\alpha = A\alpha.$$

It follows that

$$(AA')A = A$$

and AA' is a regular left unit of the element A.

If I_1 and I_2 are two idempotents belonging to $\mathfrak{A}$, then since the partial transformations are one-to-one $I_i(I_i\alpha) = I_i\alpha$ implies that

$$I_i\alpha = \alpha, \qquad (\alpha \in \Pi_1(I_i);\ i = 1, 2).$$

It follows that I_1I_2 is such that

$$\Pi_1(I_1I_2) = [\Pi_1(I_1)] \cap [\Pi_1(I_2)],$$

with

$$I_1I_2\alpha = \alpha, \qquad (\alpha \in \Pi_1(I_1I_2)).$$

In a similar fashion we obtain the corresponding relation for I_2I_1 and thus it turns out that

$$I_1I_2 = I_2I_1.$$

By 7.4, $\mathfrak{A}$ is an inverse semigroup.

8.6. A basic result which defines the role of inverse semigroups in the theory of partial transformations is a theorem that establishes the possibility of representing any inverse semigroup by one-to-one partial transformations.

THEOREM. *Every inverse semigroup is isomorphic to a semigroup of one-to-one partial transformations.*

PROOF. Let $\mathfrak{A}$ be an inverse semigroup. To every $A \in \mathfrak{A}$ we associate a transformation $\tilde{A}$ of the set $\mathfrak{A}$, defined by multiplying elements of $\mathfrak{A}$ on the left by A:

$$\tilde{A}S = (AS), \qquad (S \in \mathfrak{A}).$$

This representation is an isomorphism of $\mathfrak{A}$ to the semigroup of all transformations of $\mathfrak{A}$. In fact, if for $A_1, A_2 \in \mathfrak{A}$ we have $\tilde{A}_1 = \tilde{A}_2$, then for every $S \subset \mathfrak{A}$

$$A_1S = A_2S.$$

In particular, setting $S = \bar{A}_1A_1$ and then $S = \bar{A}_2A_2$ we obtain

$$A_1 = A_1\bar{A}_1A_1 = A_2\bar{A}_1A_1,$$

$$A_1\bar{A}_2A_2 = A_2\bar{A}_2A_2 = A_2.$$

Because of the commutativity of the idempotents $\bar{A}_2A_2$ and $\bar{A}_1A_1$ we obtain

$$A_1 = A_2\bar{A}_1A_1 = A_1\bar{A}_2A_2 \cdot \bar{A}_1A_1 = A_1 \cdot \bar{A}_1A_1 \cdot \bar{A}_2A_2 = A_1\bar{A}_2A_2 = A_2.$$

Therefore the transformation under consideration is one-to-one. Since also

$$(\tilde{A}_1\tilde{A}_2)S = \tilde{A}_1(\tilde{A}_2S) = \tilde{A}_1(A_2S) = A_1A_2S = (\widetilde{A_1A_2})S,$$

our transformation is an isomorphism.

By the method of 8.3 we can associate to every $\tilde{A}$ a one-to-one partial transformation φ $(\tilde{A})$. Since this also, by 8.3, generates an isomorphism, the transformation which carries every $A \in \mathfrak{A}$ into the one-to-one partial transformation $\varphi(\tilde{A})$ is an isomorphism of $\mathfrak{A}$ to the semigroup of all one-to-one partial transformations of the set $\mathfrak{A}$.

8.7. It follows from 8.5 and 8.6 that inverse semigroups, and only these, are isomorphic to semigroups of one-to-one partial transformations which contain the partial transformation inverse to each of the partial transformations constituting their elements. Thus, the condition that a semigroup belong to the class of inverse semigroups is an abstract characterization of the class of semigroups of one-to-one partial transformations with the property that each contains the inverse of each of its elements.

8.8. Let us now look at the meaning of the partial ordering introduced for inverse semigroups in 7.6, 7.7, and 7.8, as it may apply to inverse semigroups of one-to-one partial transformations of $\mathfrak{A}$.

Suppose $I^{(A)}$ is a regular right unit of the element $A \in \mathfrak{A}$. Since $I^{(A)}$ is an idempotent (6.3),

$$I^{(A)}(I^{(A)}\alpha) = I^{(A)}\alpha, \qquad (\alpha \in \Pi_1(I^{(A)})),$$

whence, in view of the fact that $I^{(A)}$ is one-to-one, we obtain

$$I^{(A)}\alpha = \alpha.$$

It follows from

$$AI^{(A)} = A, \qquad I^{(A)} = XA$$

that

$$\Pi_1(I^{(A)}) = \Pi_1(A).$$

The condition $A \leqslant B$, according to 7.7, means that

$$BI^{(A)} = A.$$

From what we have said about $I^{(A)}$ it follows that this last condition is equivalent to the conditions

$$\Pi_1(A) \subset \Pi_1(B),$$

$$A\alpha = B\alpha, \qquad (\alpha \in \Pi_1(A)).$$

This relation is altogether natural and highly significant with regard to the properties of partial transformations. Since it is conserved under isomorphisms, it completely defines the multiplication rule in an inverse semigroup. In particular, it holds in the representation of inverse semigroups by semigroups of one-to-one partial transformations, since as in 8.6, or otherwise, whenever $A \leqslant B$ (in the sense of 7.6) these elements are mapped into one-to-one partial transformations $\chi(A)$ and $\chi(B)$ such that

$$\Pi_1(\chi(A)) \subset \Pi_1(\chi(B)),$$

$$\chi(A)\alpha = \chi(B)\alpha, \qquad (\alpha \in \Pi_1(\chi(A))).$$

These conditions will be satisfied only by $\chi(A)$ and $\chi(B)$ such that $A \leqslant B$.

CHAPTER III

MULTIPLICATION OF SUBSETS

1. Subsemigroups

1.1. The operation of multiplying elements of a semigroup may be extended in a natural way to the multiplication of subsets.

DEFINITION. *The* PRODUCT OF THE SUBSETS $\mathfrak{K}$ *and* $\mathfrak{L}$ *of the semigroup* $\mathfrak{A}$ *is the set consisting of all elements of the form* KL, *where* $K \in \mathfrak{K}$, $L \in \mathfrak{L}$. *It is denoted by* $\mathfrak{K} \cdot \mathfrak{L}$ *or* $\mathfrak{K}\mathfrak{L}$.

Since we have agreed to make no distinction between an element of a set and the subset of that set consisting of the element alone, the multiplication of elements in a semigroup may be considered a particular case of the multiplication of subsets.

It should be noted that we will not use the notation $\mathfrak{K}^n$ in the case when the subset $\mathfrak{K}$ consists of more than one element, since frequently the expression

$$\mathfrak{K}^n, \text{ as well as } \underbrace{\mathfrak{K}\mathfrak{K}\ldots\mathfrak{K}}_{n},$$

is used to denote the set of all elements of the form K^n $(K \in \mathfrak{K})$.

This chapter is devoted to the discussion of a number of questions connected with the multiplication of subsets of a semigroup by the operation just described. In particular, from the point of view of such a multiplication, certain subsets of a semigroup are naturally singled out—namely those that are themselves semigroups with respect to the operation defined in the initial semigroup. This brings up the question of generators, that is, the question of obtaining the elements of a semigroup as products of the elements of a certain subset of the semigroup.

1.2. The concept of multiplication of subsets allows us to characterize groups in a convenient way.

THEOREM. *The semigroup* $\mathfrak{A}$ *is a group if and only if for each* $A \in \mathfrak{A}$ *we have*

$$\mathfrak{A}A = \mathfrak{A}, \qquad A\mathfrak{A} = \mathfrak{A}.$$

PROOF. If $\mathfrak{A}$ is a group, then for arbitrary $B \in \mathfrak{A}$ we can find in $\mathfrak{A}$ elements X and Y such that

$$XA = B, \qquad AY = B.$$

This means that

$$\mathfrak{A}A \supset \mathfrak{A}, \qquad A\mathfrak{A} \supset \mathfrak{A}.$$

But $\mathfrak{A}A \subset \mathfrak{A}$ and $A\mathfrak{A} \subset \mathfrak{A}$. The necessary condition is fulfilled.

On the other hand if, for arbitrary $A \in \mathfrak{A}$,

$$\mathfrak{A}A = \mathfrak{A}, \qquad A\mathfrak{A} = \mathfrak{A},$$

then for all $B \in \mathfrak{A}$

$$\mathfrak{A}A \ni B, \qquad A\mathfrak{A} \ni B,$$

that is, for some $X, Y \in \mathfrak{A}$

$$XA = B, \qquad AY = B.$$

1.3. The associativity of the multiplication operation for subsets is easily verified. The subset $(\mathfrak{K}_1\mathfrak{K}_2)\mathfrak{K}_3$, where $\mathfrak{K}_1, \mathfrak{K}_2, \mathfrak{K}_3 \subset \mathfrak{A}$, consists of all elements of the form $(K_1K_2)K_3$, where $K_1 \in \mathfrak{K}_1$, $K_2 \in \mathfrak{K}_2$, $K_3 \in \mathfrak{K}_3$. At the same time, $\mathfrak{K}_1(\mathfrak{K}_2\mathfrak{K}_3)$ consists of the elements of the form $K_1(K_2K_3) = (K_1K_2)K_3$, that is, of the same elements as $(\mathfrak{K}_1\mathfrak{K}_2)\mathfrak{K}_3$.

It follows from this that the ensemble of all nonempty subsets of a semigroup is itself a semigroup with respect to the multiplication of subsets. If the initially given semigroup is commutative, so also is the semigroup of its subsets.

The multiplication of subsets is constantly useful in the study of properties of semigroups. For all that, there has been up to now no complete and systematic study of the semigroup of subsets of a semigroup. This is understandable, perhaps, because of the complexity of the structure of semigroups of subsets. It would seem reasonable and natural to hope that if one could overcome the difficulties connected with this complexity one would be led to important results.

Dubreil [5] has done some work in this direction. He has included the empty set in the semigroup of all subsets and in this case the empty set turns out to be a zero of this semigroup.

1.4. In the ensemble of subsets of a semigroup there is a natural partial ordering (I, 5.9). For $\mathfrak{L}, \mathfrak{K} \subset \mathfrak{A}$ the set $\mathfrak{K}$ is said to precede $\mathfrak{L}$ if $\mathfrak{K} \subset \mathfrak{L}$.

It is easy to see that if one includes the empty set in the ensemble of subsets of a set $\mathfrak{A}$, the ensemble is a complete lattice (I, 5.13).

The partial ordering we have just described for the semigroup of subsets of $\mathfrak{A}$ is two-sidedly stable (I, 5.17). In fact, for $\mathfrak{K} \subset \mathfrak{L}$ and arbitrary $\mathfrak{M}$ it is clear that

$$\mathfrak{M}\mathfrak{K} \subset \mathfrak{M}\mathfrak{L}, \qquad \mathfrak{K}\mathfrak{M} \subset \mathfrak{L}\mathfrak{M} \qquad (\mathfrak{K}, \mathfrak{L}, \mathfrak{M} \subset \mathfrak{A}).$$

It is usually helpful to pay attention to inclusion relations when studying the multiplication of subsets. We note, in particular, that for arbitrary subsets $\mathfrak{M}$, $\mathfrak{L}_\lambda$ $(\lambda \in \Gamma)$ of the semigroup $\mathfrak{A}$ we always have

$$\mathfrak{M} \cdot \Big(\bigcup_{\lambda \in \Gamma} \mathfrak{L}_\lambda\Big) = \bigcup_{\lambda \in \Gamma} (\mathfrak{M}\mathfrak{L}_\lambda).$$

In fact, both the left and right sides of this equation represent the set of all possible elements of the form XY, where $X \in \mathfrak{M}$, and Y belongs to one of the sets $\mathfrak{L}_\lambda$ $(\lambda \in \Gamma)$.

Similarly

$$\left(\bigcup_{\lambda\in\Gamma} \mathfrak{L}_\lambda\right)\cdot\mathfrak{M} = \bigcup_{\lambda\in\Gamma} (\mathfrak{L}_\lambda\mathfrak{M}).$$

The analogous equations do not necessarily hold for intersections.

As an example, we might consider the multiplicative semigroup of nonnegative integers and the subsets $\mathfrak{M} = \{1, 2\}$, $\mathfrak{L}_1 = \{0, 1\}$, $\mathfrak{L}_2 = \{0, 2\}$. We have

$$\mathfrak{M}\cdot(\mathfrak{L}_1 \cap \mathfrak{L}_2) = \{1, 2\}\cdot\{0\} = 0.$$

$$(\mathfrak{M}\cdot\mathfrak{L}_1) \cap (\mathfrak{M}\cdot\mathfrak{L}_2) = \{0, 1, 2\} \cap \{0, 2, 4\} = \{0, 2\}.$$

1.5. Just as in other parts of algebra, we find that an important role is played by those subsets which themselves satisfy the initial axioms with respect to the given operation.

DEFINITION. *A nonempty subset* $\mathfrak{B}$ *of the semigroup* $\mathfrak{A}$ *is called a* SUBSEMIGROUP *of* $\mathfrak{A}$ *if*

$$\mathfrak{B}\cdot\mathfrak{B} \subset \mathfrak{B}.$$

$\mathfrak{A}$ *is called in this case a* SUPERSEMIGROUP *of* $\mathfrak{B}$.

One should keep in mind the following terminological rule in dealing with semigroups. If the subsemigroup $\mathfrak{B}$ of the semigroup $\mathfrak{A}$ is a group, it is called a *subgroup* of the semigroup $\mathfrak{A}$. If $\mathfrak{B}$ is a commutative semigroup, it is called a *commutative subsemigroup* of $\mathfrak{A}$. If $\mathfrak{B}$ is an inverse semigroup, it is called an *inverse subsemigroup* of $\mathfrak{A}$, and so on.

A subsemigroup which is a proper subset of $\mathfrak{A}$, that is, which is distinct from $\mathfrak{A}$, is called a *proper subsemigroup.*

It should be noted that in those studies that do not require that the semigroups defined in them should be nonempty, the empty set is taken as a subsemigroup.

1.6. With respect to the multiplication defined in any semigroup $\mathfrak{A}$, the set of elements of a subsemigroup is itself a semigroup. In fact, for arbitrary $B_1, B_2 \in \mathfrak{B}$ the product B_1B_2 is defined (since it is defined for arbitrary elements of $\mathfrak{A}$) and belongs to $\mathfrak{B}$ (since $B_1B_2 \in \mathfrak{B}\mathfrak{B} \subset \mathfrak{B}$). The operation in $\mathfrak{B}$ is associative, since it is associative in $\mathfrak{A}$ and $\mathfrak{B} \subset \mathfrak{A}$.

The converse, of course, is also true. If the subset $\mathfrak{B}$ of a semigroup $\mathfrak{A}$ is itself a semigroup with respect to the multiplication defined in $\mathfrak{A}$, then the product of any two of its elements belongs to $\mathfrak{B}$, that is $\mathfrak{B}\mathfrak{B} \subset \mathfrak{B}$ and $\mathfrak{B}$ is a subsemigroup.

1.7. If $\mathfrak{A}$ is a semigroup of transformations of some set Ω and $\Omega' \subset \Omega$, the ensemble of transformations A belonging to $\mathfrak{A}$, such that

$$A\Omega' \subset \Omega',$$

is evidently a subsemigroup of $\mathfrak{A}$ (supposing, of course, that it is not empty).

On the other hand, the subsemigroups are sets composed of the elements A such that

$$A\Omega' = \Omega'.$$

In the semigroup of matrices $\mathfrak{M}_n$ the set of all singular matrices forms a subsemigroup. The set of all nonsingular matrices forms a subsemigroup, and even more, a subgroup.

In the multiplicative semigroup of all natural numbers there are several subsemigroups: the set of all even numbers, the set of all numbers larger than a given number, the set of all nonprime numbers, etc.

Let $\mathfrak{A}$ be the set of all real continuous functions defined on an interval E and having values in the same interval E. With respect to the composition of functions $\mathfrak{A}$ constitutes a semigroup. Some obvious subsemigroups of $\mathfrak{A}$ are, for example, the set of all differentiable functions, the set of all monotone increasing functions, and the set of all constants.

1.8. In the study of some particular semigroup $\mathfrak{A}$ it will sometimes be advantageous to embed it in a suitable supersemigroup $\mathfrak{A}'$. This process will be helpful either because $\mathfrak{A}'$ has desirable properties, or because a number of distinct semigroups can be embedded in a single supersemigroup and therefore the study of all of them can be reduced to the study of a single semigroup. The following identification process is often useful in such embeddings.

Let φ be an isomorphism of the semigroup $\mathfrak{A}$ to the semigroup $\mathfrak{B}$, the latter having no element in common with $\mathfrak{A}$. Consider the set $\mathfrak{A}'$ obtained from $\mathfrak{B}$ by substituting for the elements belonging to $\varphi(\mathfrak{A})$ the elements of $\mathfrak{A}$ that correspond to them under the isomorphism φ:

$$\mathfrak{A}' = \mathfrak{A} \cup (\mathfrak{B} \backslash \varphi(\mathfrak{A})).$$

In $\mathfrak{A}'$ we define an operation by setting $X'Y' = Z'$ $(X', Y', Z' \in \mathfrak{A}')$ if in $\mathfrak{B}$ we have $XY = Z$ (where $X = \varphi(X')$ if $X' \in \mathfrak{A}$ and $X = X'$ if $X' \bar{\in} \mathfrak{A}$; and similarly for Y and Z).

Clearly $\mathfrak{A}'$ is a semigroup with respect to this operation, and is isomorphic to $\mathfrak{B}$. Also, $\mathfrak{A}'$ is a supersemigroup for $\mathfrak{A}$.

The transition from $\mathfrak{B}$ to $\mathfrak{A}'$ may be thought of as a substitution that replaces the elements of $\mathfrak{B}$ belonging to $\varphi(\mathfrak{A})$ by the corresponding elements of $\mathfrak{A}$, or as an identification of each element $A \in \mathfrak{A}$ with the element $\varphi(\mathfrak{A}) \in \mathfrak{B}$.

1.9. What we have just said in 1.8 gives rise to the following important concept.

DEFINITION. *A class* Γ *of semigroups is said to be a* UNIVERSAL CLASS *for the class* Ξ *of semigroups if every semigroup belonging to* Ξ *is isomorphic to some subsemigroup of each of the semigroups belonging to* Γ.

The study of the semigroups of any class for which we know a universal class Γ can be reduced to the study of the semigroups of the universal class. An especially important case arises when the semigroups of Γ all belong to Ξ. In this case, when we reduce the study of the semigroups belonging to Ξ to the study of the semigroups in Γ, we merely restrict the task to one that does not go outside the boundaries of the first task.

1.10. Let $\mathfrak{m}$ be a finite power and Ω a set with power $\mathfrak{m}$. In virtue of I, 3.9 the semigroup $\mathfrak{S}_\Omega$ is in itself the universal class for the class of all semigroups for which the power of the corresponding element-set does not exceed $\mathfrak{m}$. Of course $\mathfrak{S}_\Omega$ itself is not contained in the latter class; this follows from the fundamental properties of cardinality.

For the class of all semigroups the universal class is the class of semigroups $\mathfrak{S}_\Omega$ for which Ω takes on all values.

Let Ω be some countable set. We denote by $\mathfrak{S}'_\Omega$ the set of all mappings of Ω such that each mapping S leaves invariant all but a finite number of the elements of Ω (that is, $S\alpha \neq \alpha$ for only a finite number of the elements $\alpha \in \Omega$). Such mappings are sometimes referred to as almost identical mappings. Since the semigroup $\mathfrak{S}_\Sigma$ of all mappings of an arbitrary finite set Σ is isomorphic to some subsemigroup of the semigroup $\mathfrak{S}'_\Omega$, every finite semigroup is isomorphic to some subsemigroup of $\mathfrak{S}'_\Omega$. Thus $\mathfrak{S}'_\Omega$ is a universal class for the class of all finite semigroups.

A subsemigroup of a group is not necessarily itself a group. For example, the multiplicative semigroup of the natural numbers is not a group, but it is a subsemigroup of the group of all positive rational numbers. Semigroups that are subsemigroups of a group are said to be *embedded in that group*. The class of all semigroups that are embedded in groups has been subjected to many investigations. The study of its properties is of interest both for the general theory of semigroups and for the special problem of the theory of groups. The definition itself implies that the class of all groups, which is a part of the class of all semigroups embedded in groups, is a universal class for the latter class.

The construction and study of universal classes can be carried out for various classes of semigroups and facilitates the study of them.

1.11. We shall now indicate some of the simpler properties of subsemigroups of an arbitrary semigroup $\mathfrak{A}$.

(α) *The semigroup $\mathfrak{A}$ is a subsemigroup of itself.*

(β) *The intersection of an arbitrary ensemble of subsemigroups of a semigroup $\mathfrak{A}$ is itself a subsemigroup of $\mathfrak{A}$.*

In fact, if X and Y belong to the intersection of some ensemble of subsemigroups, XY also belongs to the intersection, since XY belongs to each of the subsemigroups making up the ensemble.

(γ) *If $\mathfrak{A}'$ is a subsemigroup of $\mathfrak{A}$, then every subsemigroup of $\mathfrak{A}'$ is a subsemigroup of $\mathfrak{A}$ contained in $\mathfrak{A}'$.*

(δ) *The subset of $\mathfrak{A}$ consisting solely of the element X is a subsemigroup if and only if X is an idempotent.*

(ε) *If the semigroup $\mathfrak{A}$ contains idempotents and they all commute with each other, the ensemble of all idempotents is a subsemigroup.*

In fact, if

$$I_1^2 = I_1, \qquad I_2^2 = I_2, \qquad I_1 I_2 = I_2 I_1,$$

then

$$(I_1 I_2)^2 = I_1 I_2 I_1 I_2 = I_1 I_1 I_2 I_2 = I_1 I_2,$$

that is, the product of two idempotents belongs to the ensemble of all idempotents.

1.12. We now pass to the following important construction. Let Σ be a nonempty ensemble of semigroups, having the property that for any two semigroups $\mathfrak{M}$ and $\mathfrak{N}$ belonging to Σ we can find in Σ a semigroup $\mathfrak{L}$ which is a subsemigroup of both $\mathfrak{M}$ and $\mathfrak{N}$.

We write

$$\mathfrak{A} = \bigcup_{\mathfrak{N} \in \Sigma} \mathfrak{N}.$$

The operation in $\mathfrak{A}$ is defined in the natural way. If $X, Y \in \mathfrak{A}$, we have for some $\mathfrak{M}, \mathfrak{N} \in \Sigma$ the relations $X \in \mathfrak{M}$, $Y \in \mathfrak{N}$. Let $\mathfrak{L}$ be a subsemigroup of $\mathfrak{M}$ and $\mathfrak{N}$. Then we define XY in $\mathfrak{A}$ as the product of X and Y when these are considered as elements of the subsemigroup $\mathfrak{L}$. This definition of the product XY is independent of the choice of $\mathfrak{M}$, $\mathfrak{N}$, and $\mathfrak{L}$. In fact, let $\mathfrak{L}' \in \Sigma$ be a subsemigroup of both $\mathfrak{M}'$ and $\mathfrak{N}'$, where $X \in \mathfrak{M}'$, $Y \in \mathfrak{N}'$ ($\mathfrak{M}', \mathfrak{N}' \in \Sigma$). We can find in Σ a semigroup $\mathfrak{T}$ which is a subsemigroup of $\mathfrak{L}$ and $\mathfrak{L}'$. The product of X and Y in $\mathfrak{L}$ must coincide with the product of X and Y in $\mathfrak{T}$. The same is true for $\mathfrak{L}'$. Accordingly, the product of X and Y is defined in the same way by $\mathfrak{M}$, $\mathfrak{N}$, and $\mathfrak{L}$ as it is by $\mathfrak{M}'$, $\mathfrak{N}'$, and $\mathfrak{L}'$.

The associativity of the operation defined in $\mathfrak{A}$ is self-evident. Therefore, $\mathfrak{A}$ is a semigroup. $\mathfrak{A}$ is the union of the semigroups of the initial ensemble Σ.

As a particular case of the preceding construction we have the union of an arbitrary ensemble Γ of subsemigroups of a given semigroup whenever Γ has the property that of any two semigroups belonging to Γ one is necessarily a subsemigroup of the other. The union of all semigroups belonging to Γ is in this case a subsemigroup of the initially given semigroup.

1.13. We now introduce a few examples of subsemigroups of an arbitrary semigroup.

Let $\mathfrak{A}$ be an arbitrary semigroup. Each of the subsets described below is a subsemigroup of $\mathfrak{A}$, provided it is not empty.

The set of all elements that are either right or left divisors of an arbitrary element of $\mathfrak{A}$ (II, 1.5).

The set of all elements divisible both on the left and on the right by all elements of $\mathfrak{A}$ (II, 1.6).

The set of all elements that commute with every element of $\mathfrak{A}$ (this subsemigroup is sometimes referred to as the center of the semigroup $\mathfrak{A}$).

For an arbitrary subset $\mathfrak{K}$ of the semigroup $\mathfrak{A}$ we define the subset $\mathfrak{U}_{\mathfrak{K}}$ consisting of all elements $U \in \mathfrak{A}$ such that $\mathfrak{A}U \cap \mathfrak{K} = \varnothing$. In a similar fashion we define $\mathfrak{V}_{\mathfrak{K}}$ consisting of the elements $V \in \mathfrak{A}$ such that $V\mathfrak{A} \cap \mathfrak{K} = \varnothing$. It is easy to see that $\mathfrak{U}_{\mathfrak{K}}$ and $\mathfrak{V}_{\mathfrak{K}}$ are, if they are not empty, subsemigroups of $\mathfrak{A}$. The subsemigroups $\mathfrak{U}_{\mathfrak{K}}$ and $\mathfrak{V}_{\mathfrak{K}}$ were introduced by Dubreil [1] (see also [3]). These subsemigroups are important in his investigations.

1.14. The subgroups of a semigroup, that is, subsemigroups which are also groups, have an important role in some studies. Let I be an arbitrary idempotent of the semigroup $\mathfrak{A}$. We denote by $\mathfrak{G}_I$ the set of all completely regular elements (II, 6.1) of $\mathfrak{A}$ for which I is a regular two-sided unit. (II, 6.2; II, 6.4).

$\mathfrak{G}_I$ is not empty, since by II, 6.3 $\mathfrak{G}_I$ contains I.

THEOREM. *$\mathfrak{G}_I$ is a subgroup of the semigroup $\mathfrak{A}$.*

PROOF. Suppose $G_1, G_2 \in \mathfrak{G}_I$. Since I is a regular two-sided unit of G_1 and G_2 we have for some $U_i, V_i \in \mathfrak{A}$ $(i = 1, 2)$

$$I = G_iU_i, \qquad I = V_iG_i, \qquad (i = 1, 2).$$

From this follows

$$(G_1G_2)(U_2U_1) = G_1IU_1 = G_1U_1 = I,$$

$$(V_2V_1)(G_1G_2) = V_2IG_2 = V_2G_2 = I.$$

Since I is a two-sided unit of the element (G_1G_2), I is a regular two-sided unit of (G_1G_2), i.e., $(G_1G_2) \in \mathfrak{G}_I$. Therefore, $\mathfrak{G}_I$ is a semigroup. I is a unit of it.

Since I is clearly a two-sided unit for IU_1I and

$$I = II = G_1U_1I = G_1 \cdot (IU_1I),$$

$$I = V_1G_1 = V_1IIG_1 = V_1G_1U_1IG_1 = (IU_1I) \cdot G_1,$$

it follows that I is a regular two-sided unit of the element (IU_1I).

Thus, this element is contained in $\mathfrak{G}_I$ and is for G_1 a two-sided inverse of I. From the fact that an arbitrary element G_1, of the semigroup $\mathfrak{G}_I$ with unit, has a two-sided inverse in $\mathfrak{G}_I$ it follows that $\mathfrak{G}_I$ is a group (II, 2.15).

1.15. COROLLARY. *An element A of a semigroup $\mathfrak{A}$ is contained in some subgroup of the semigroup $\mathfrak{A}$ if and only if A is completely regular.*

In fact, a completely regular element A has as its regular two-sided unit an idempotent I, and is therefore contained in the subgroup $\mathfrak{G}_I$.

If an arbitrary element A is contained in some subgroup $\mathfrak{G}$ of the semigroup $\mathfrak{A}$, the group property implies that $E_{\mathfrak{G}}$ is a regular two-sided unit of A and therefore A is completely regular.

1.16. Since no element may have two regular two-sided units (II, 6.5), the groups $\mathfrak{G}_I$ are mutually disjoint. Therefore the set $\mathfrak{F}$ of all completely regular elements of a semigroup $\mathfrak{A}$ can be expressed as the union of nonintersecting groups

$$\mathfrak{F} = \bigcup_{I \in \mathfrak{H}} \mathfrak{G}_I$$

(where $\mathfrak{H}$ is the set of all idempotents of $\mathfrak{A}$).

Every subgroup $\mathfrak{G}$ of the semigroup $\mathfrak{A}$ is contained in one of the components of the above union, namely in $\mathfrak{G}_{E_{\mathfrak{G}}}$. The components $\mathfrak{G}_I$ are maximal subgroups of $\mathfrak{A}$, since none is contained in any subgroup of $\mathfrak{A}$ other than itself.

The absence of idempotents is a necessary and sufficient condition for the absence of subgroups in $\mathfrak{A}$. The presence of a single idempotent signifies that every subgroup of $\mathfrak{A}$ is a subgroup of one largest subgroup $\mathfrak{G}_I$ containing all the others and consisting of the set of all completely regular elements. When there are two or more idempotents, there exists a pair of elements each contained in a subgroup, but there is no subgroup containing both.

2. Generating Sets

2.1. Let $\mathfrak{K}$ be a nonempty subset of the semigroup $\mathfrak{A}$. The ensemble $\mathfrak{B}$ consisting of all those elements B that can be represented as products of elements of $\mathfrak{K}$ (including "products" consisting of a single element), that is, of the form

$$B = X_1 X_2 \ldots X_n \qquad (X_1, X_2, \ldots, X_n \in \mathfrak{K}),$$

is clearly a subsemigroup. We shall say that the subsemigroup $\mathfrak{B}$ *is generated by the set* $\mathfrak{K}$, and $\mathfrak{K}$ will be called the *generating set* of $\mathfrak{B}$. We adopt the notation

$$\mathfrak{B} = [\mathfrak{K}]$$

(we shall of course continue to use square brackets for their customary purpose, as separation signs, but this should arouse no particular difficulty). There is an especially important case, namely when the subset $\mathfrak{K}$ is a generating set for the semigroup $\mathfrak{A}$ itself, that is, it generates the entire semigroup $\mathfrak{A}$, or

$$[\mathfrak{K}] = \mathfrak{A}.$$

The possibility of expressing elements of a semigroup (either all or only some of the elements) in the form of products of elements of a subset is clearly interesting in itself, and moreover is often useful in studying the properties of the semigroup. We shall look into this question in the following paragraph and in several of the subsequent paragraphs.

2.2. It is worth keeping in mind that in the theory of groups a subset $\mathfrak{K}$ of a group $\mathfrak{G}$ is called a generator of the subgroup $\mathfrak{G}' \subset \mathfrak{G}$, and we write $\{\mathfrak{K}\} = \mathfrak{G}'$, if the set $\bar{\mathfrak{K}}$ consisting of all elements of $\mathfrak{K}$ and of their inverses is a generator of $\mathfrak{G}'$ in the sense of 2.1.

2.3. A subsemigroup $\mathfrak{B}$ generated by a subset $\mathfrak{K}$ of a semigroup can be symbolically expressed in terms of products in the form

$$\mathfrak{B} = [\mathfrak{K}] = \mathfrak{K} \cup \mathfrak{K}\mathfrak{K} \cup \mathfrak{K}\mathfrak{K}\mathfrak{K} \cup \dots .$$

We should note also the following self-evident properties:

(1) $[[\mathfrak{K}]] = [\mathfrak{K}]$;

(2) if $\mathfrak{K}_1 \supset \mathfrak{K}_2$, then $[\mathfrak{K}_1] \supset [\mathfrak{K}_2]$.

2.4. There is another approach to the concept of the generating set. *A set $\mathfrak{K}$ is a generator of the subsemigroup $\mathfrak{B}$ of the semigroup $\mathfrak{A}$ if and only if $\mathfrak{B}$ is the intersection of all subsemigroups of $\mathfrak{A}$ that contain $\mathfrak{K}$.*

In fact, every subsemigroup $\mathfrak{N}$ of the semigroup $\mathfrak{A}$ that contains $\mathfrak{K}$ must contain all possible products of elements of $\mathfrak{K}$, i.e., $\mathfrak{N} \supset [\mathfrak{K}]$. Therefore $[\mathfrak{K}]$ is contained in the intersection of such subsemigroups. On the other hand, $[\mathfrak{K}]$ itself is a subsemigroup of $\mathfrak{A}$ and contains $\mathfrak{K}$, and therefore contains the intersection in question.

It follows in particular that the generating sets of the semigroup itself are those sets not contained in any proper subsemigroup of the semigroup.

2.5. Every semigroup, of course, has one or more generating sets. For example, the set of all elements of a semigroup is a generating set for the semigroup itself. But there are usually several different generating sets—this is evident since every subset of $\mathfrak{A}$ containing a generating set is also a generating set of $\mathfrak{A}$.

It is generally useful to pick a generating set containing as few elements as possible. We therefore try to deal with the so-called *irreducible generating sets* of a semigroup, that is, generating sets such that no proper subset of them is itself a generating set.

2.6. Every finite semigroup has an irreducible generating set (and may have more than one). An irreducible generating set of a finite semigroup may be obtained from an arbitrary generating set by eliminating those elements of the latter that can be expressed as products of the remaining elements.

If the semigroup $\mathfrak{A}$ has a finite generating set $\mathfrak{K}$, then every generating set $\mathfrak{K}'$ has a finite subset which is an irreducible generating set for the semigroup (and it follows in particular that in this case every irreducible generating set is finite, although, of course, different irreducible generating sets may have different numbers of elements). In fact, corresponding to every element of $\mathfrak{K}$ we choose a representation in the form of a product of elements of $\mathfrak{K}'$. We denote by $\mathfrak{K}''$ the ensemble of all those elements of $\mathfrak{K}'$ contained in the selected product. Since each such product contains only a finite number of elements, and since there is only a finite number of such products—because $\mathfrak{K}$ has a finite number of elements—the set $\mathfrak{K}''$ is finite. Since $[\mathfrak{K}''] \supset \mathfrak{K}$, $\mathfrak{K}''$ is a generating set of $\mathfrak{A}$. If one of the elements of $\mathfrak{K}''$ can be expressed as a product of the remaining elements of $\mathfrak{R}''$, we exclude this element and obtain a new generating set $\mathfrak{R}'''$ with fewer elements. Repeating this elimination process we finally arrive at a finite generating set $\mathfrak{R}^* \subset \mathfrak{R}'$ in which no element can be expressed as the product of others. $\mathfrak{R}^*$ is the desired irreducible generating set of the set $\mathfrak{A}$.

Infinite semigroups may have no irreducible generating sets at all.

We may cite as an example the semigroup $\mathfrak{A}$ consisting of all natural numbers in which the product of two elements is taken as their greatest common divisor. If $\mathfrak{R}$ is a generating set for this semigroup and the number n is an arbitrary element of $\mathfrak{R}$, then $\mathfrak{R}' = \mathfrak{R}\backslash n$ is also a generating set. In fact, the numbers $2n$ and $3n$ must be representable as products of elements of $\mathfrak{R}$. They are therefore representable as products of elements of $\mathfrak{R}'$ (clearly, in the representations of $2n$ and $3n$ by elements of $\mathfrak{R}$ the integer n cannot enter). But n is the product of $2n$ and $3n$. Therefore $[\mathfrak{R}']$ contains $\mathfrak{R}' \cup n = \mathfrak{R}$.

Accordingly, $[\mathfrak{R}'] \supset [\mathfrak{R}] = \mathfrak{A}$.

2.7. If the element A of the semigroup $\mathfrak{A}$ is indecomposable, i.e., cannot be represented in the form $A = XY$ where $X, Y \in \mathfrak{A}$, then A necessarily is contained in every generating set of the semigroup $\mathfrak{A}$. If the set of all indecomposable elements generates the semigroup it is the unique irreducible generating set of the semigroup.

The set of all indecomposable elements is clearly the set

$$\mathfrak{A}\backslash\mathfrak{A}\mathfrak{A}.$$

2.8. Let $\mathfrak{R}$ be a generating set for the semigroup $\mathfrak{A}$. The representation of the elements of $\mathfrak{A}$ as a product of elements of $\mathfrak{R}$ often turns out to be very useful in studying the properties of $\mathfrak{A}$. We may use it to obtain what amounts to a visual representation of the set of elements of the semigroup by thinking of it as made up of all possible products of elements of the generating set $\mathfrak{R}$. There is one circumstance, however, that seriously complicates this approach. The fact is that in general the elements of the semigroup are not uniquely factorable in terms of elements of $\mathfrak{R}$; several distinct products may be used to express a single element. Therefore the set of all elements of the semigroup is not, strictly speaking, the set of all products, but a set of classes of products obtained by putting into the same class all products which are equal to one and the same element of $\mathfrak{A}$. Moreover, the task of determining which products of elements of $\mathfrak{R}$ are equal to each other (we mean, of course, equal as elements of $\mathfrak{A}$) is in most cases rather difficult. The degree of difficulty depends on the nature of the semigroup $\mathfrak{A}$ and on the process by which the semigroup is defined.

The customary approach to a solution of this problem is as follows. Out of all possible products of elements of the generating set $\mathfrak{R}$ one attempts to choose those satisfying the following conditions: (1) no two externally distinct products are to be equal as elements of $\mathfrak{A}$; (2) every product of elements of $\mathfrak{R}$ is to be equal as an element of $\mathfrak{A}$ to some one of the chosen products. A product belonging to the set of products satisfying both these conditions is usually known as a canonical expression or canonical form of an element of $\mathfrak{A}$. If the construction of canonical forms is realizable, then every element has a unique canonical representation. Therefore, the set of all constructed canonical forms can be looked upon as the set of all elements of the semigroup. If also we have a suitable method of multiplying these canonical forms, that is a process for finding the canonical form of an element represented as the product of two factors, both given in canonical form, then we have complete data with respect to the given semigroup: the make-up of its set of elements and the rule of composition.

Of course, there may be several different systems of canonical forms with respect to a single generating set. The wisdom of one's choice among them often predetermines one's success in applying this method of defining a semigroup to the study of its properties.

2.9. One of the simplest illustrations of these notions is to be found in the multiplicative group of the natural numbers. The set $\mathfrak{K}$ consisting of unity and of all the prime numbers is a generating set for this semigroup. No number n belonging to $\mathfrak{K}$ can be represented as a product of elements of $\mathfrak{A}$ other than n itself. Therefore $\mathfrak{K}$ is the only irreducible generating set of $\mathfrak{A}$. The representation of the number 1 as a "product" consisting of the sole factor 1, and the representation of all other numbers as a product of prime factors arranged in order of magnitude, together make up a system of canonical forms for the elements of $\mathfrak{A}$. The known simple process of multiplying numbers expressed as products of prime factors is in this case the rule for multiplying canonical forms as discussed in 2.8.

2.10. A more difficult example is afforded by the semigroup $\mathfrak{S}_\Omega$ consisting of all mappings of a finite set Ω.

Let S be an arbitrary mapping contained in $\mathfrak{S}_\Omega$. We shall say that the elements α and β of Ω are equivalent with respect to S if for some integers k and l

$$S^k\alpha = S^l\beta.$$

This equivalence relation is clearly symmetric and reflexive. It is also transitive, since

$$S^k\alpha = S^l\beta, \qquad S^p\beta = S^q\gamma$$

implies that

$$S^{k+p}\alpha = S^p(S^k\alpha) = S^p(S^l\beta) = S^{p+l}\beta = S^l(S^p\beta) = S^l(S^q\gamma) = S^{l+q}\gamma.$$

We partition all elements of Ω into pairwise disjoint classes such that all members of any one class are equivalent among themselves with respect to S. We note that the equivalence, with respect to S, of the elements α and $S\alpha$ (which follows from the equality $S^2\alpha = S(S\alpha)$) implies that for every class Γ of elements equivalent with respect to S we have $S\Gamma \subset \Gamma$.

We denote by $\mathfrak{P}$ the ensemble of transformations P belonging to $\mathfrak{S}_\Omega$ and enjoying the following property: Among the classes of elements of Ω equivalent with respect to P, only one class may have more than one member. We shall show that $\mathfrak{P}$ is a generating set for $\mathfrak{S}_\Omega$.

Let S be an arbitrary transformation belonging to $\mathfrak{S}_\Omega$. We decompose Ω into classes of elements equivalent with respect to S

$$\Omega = \Gamma_1 \cup \Gamma_2 \cup \ldots \cup \Gamma_m.$$

For every class Γ_i we construct the transformation P_i which coincides with S on Γ_i and maps the elements of all other classes into themselves:

$$P_i\alpha = S\alpha, \qquad P_i\beta = \beta \qquad (\alpha \in \Gamma_i, \beta \in \Omega \backslash \Gamma_i).$$

P_i belongs to $\mathfrak{P}$, since all elements of Γ_i are equivalent with respect to P_i and no element not belonging to Γ_i is equivalent to any element except itself.

Since for every $\alpha \in \Gamma_i$

$$S\alpha = P_i\alpha, \qquad P_j(P_i\alpha) = P_i\alpha, \qquad P_j\alpha = \alpha, \qquad (j \neq i),$$

we have

$$S\alpha = P_1P_2 \ldots P_m\alpha.$$

It follows that

$$S = P_1P_2 \ldots P_m.$$

The product we have just written is characterized by the fact that all its factors belong to $\mathfrak{P}$ and that for any two of its factors there is no $\alpha \in \Omega$ such that $P_i\alpha \neq \alpha$, $P_j\alpha \neq \alpha$ $(i \neq j)$.

The factors of such a product evidently commute among themselves. If we choose two products of this type that differ in more than just the order of the factors, they will represent distinct elements of $\mathfrak{S}_\Omega$ since at least one of the elements of Ω will be mapped in two different ways by the two products. Therefore, if we agree on some ordering of the factors in the products, the products of this type will constitute canonical forms of the elements of $\mathfrak{S}_\Omega$ with respect to the generating set $\mathfrak{P}$.

We should note that the generating set $\mathfrak{P}$ cannot be irreducible if Ω contains more than one element.

In fact, we choose in Ω two elements α and β and we consider three elements of $\mathfrak{P}$, writing them in the form of substitutions:

$$P_1 = \begin{pmatrix} \alpha & \beta \\ \alpha & \alpha \end{pmatrix}, \qquad P_2 = \begin{pmatrix} \alpha & \beta \\ \beta & \beta \end{pmatrix}, \qquad P_3 = \begin{pmatrix} \alpha & \beta \\ \beta & \alpha \end{pmatrix}.$$

(The elements of Ω not written out are represented by identity substitutions.) Since

$$P_1 = P_3P_2,$$

the set obtained from $\mathfrak{P}$ by excluding P_1 is also a generating set.

2.11. In connection with the material proved in 2.10 it is natural to clarify in somewhat more detail the construction of the transformations belonging to $\mathfrak{P}$.

We shall consider a certain special method of constructing transformations belonging to $\mathfrak{S}_\Omega$. Suppose we have chosen from Ω some elements, distinct from each other, which we have indexed according to the following scheme $\mathfrak{n}$:

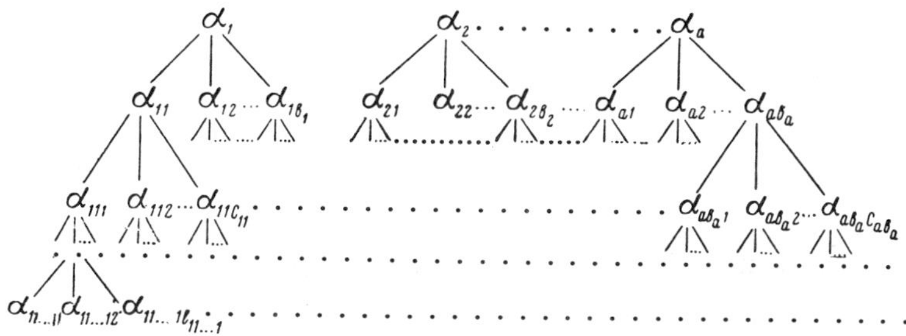

Using this scheme $\mathfrak{n}$ we construct the following transformation $P_\mathfrak{n}$:

$$P_\mathfrak{n}\alpha_1 = \alpha_2 \qquad P_\mathfrak{n}\alpha_2 = \alpha_3, \ldots, \qquad P_\mathfrak{n}\alpha_{a-1} = \alpha_a, \qquad P_\mathfrak{n}\alpha_a = \alpha_1,$$

$$P_\mathfrak{n}\alpha_{s_1 s_2 \ldots s_{r-1} s_r} = \alpha_{s_1 s_2 \ldots s_{r-1}},$$

$P_\mathfrak{n}\xi = \xi$, for all ξ not encountered in the scheme $\mathfrak{n}$.

It is easy to see that $P_\mathfrak{n}$ belongs to $\mathfrak{P}$. We shall show that every transformation $P \in \mathfrak{P}$ is constructed in this way.

Let Γ be that one equivalence class of elements of Ω with respect to P that contains more than one element (the case where all the classes have just one element is trivial). We choose from Γ an arbitrary element α and we consider the sequence

$$P\alpha, P^2\alpha, P^3\alpha, \ldots.$$

Since Ω is finite some elements of this sequence must coincide, so that

$$P^k\alpha = P^l\alpha, \qquad (k > l).$$

It follows that

$$P^{k-l}(P^l\alpha) = (P^l\alpha).$$

Thus Γ must contain elements β such that for some m, $P^m\beta = \beta$. We denote the ensemble of these by Γ'. As we have seen, for an arbitrary element $\alpha \in \Gamma$ there is an l such that $P^l\alpha \in \Gamma'$. Suppose $\beta_1 \in \Gamma'$. Let us consider the sequence

$$\beta_1, \beta_2, \ldots, \beta_a; \qquad P\beta_i = \beta_{i+1}, \qquad P\beta_a = \beta_1 \qquad (i = 1, 2, \ldots, (a-1)).$$

Since

$$P^m\beta_1 = \beta_1,$$

we have

$$P^m\beta_i = P^m(P^{i-1}\beta_1) = P^{m+i-1}\beta_1 = P^{i-1}(P^m\beta_1) = P^{i-1}\beta_1 = \beta_i,$$

that is, all elements of our sequence belong to Γ'. Γ' contains no other elements, since

$$P^r\gamma = \gamma, \qquad P^s\beta_1 = P^t\gamma$$

implies

$$\gamma = P^{rt}\gamma = P^{(r-1)t}(P^t\gamma) = P^{(r-1)t}(P^s\beta_1) = P^{(r-1)t+s}\beta_1 = \beta_l \in \Gamma'.$$

Let us now choose those elements of $\Gamma\backslash\Gamma'$ which are mapped by P into elements of Γ' and write them under the corresponding β_j:

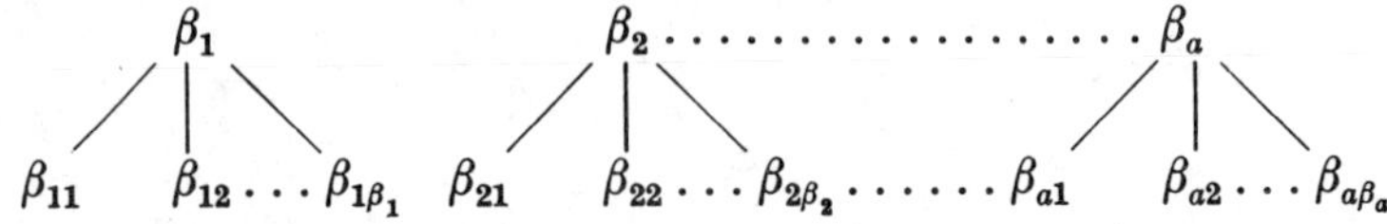

Then we choose those elements that map under P into elements of the second row, and write them under the second row to make up a third row. Continuing in this way we shall exhaust the elements of Γ, since what we have said above implies that for every $\alpha \in \Gamma$ there is some s such that $P^s\alpha = \beta_i$, that is, every $\alpha \in \Gamma$ occurs in our diagram. It is clear that the mapping P is the mapping $P_\mathfrak{n}$ for the diagram $\mathfrak{n}$ so obtained.

The representation of mappings $S \in \mathfrak{S}_\Omega$ in the form of products of elements of $\mathfrak{P}$, as indicated in 2.10, is precisely a generalization of the decomposition, well known in the theory of substitution groups, of the group substitutions into the product of independent cycles. It becomes exactly the usual cyclic substitution when the scheme $\mathfrak{n}$ consists of the top row only. An element $S \in \mathfrak{S}_\Omega$, factored into a product of the type 2.10 using elements of $\mathfrak{P}$ is a one-to-one mapping (a group substitution) if each of the factors P_i corresponds to a one-line diagram.

2.12. We shall now pass to the irreducible generating sets of the semigroup $\mathfrak{S}_\Omega$. Following the path marked out by N. N. Vorob'ev [5], we shall determine when a generating set of the semigroup $\mathfrak{S}_\Omega$ is irreducible. We denote by n the number of elements in the set Ω. For every $S \in \mathfrak{S}_\Omega$ we denote by $\chi(S)$ the number of elements in $S\Omega$. The elements S for which $\chi(S) = n$ are one-to-one mappings of the set Ω. We denote the ensemble of these by $\mathfrak{G}_\Omega$. It is a group (the so-called symmetric group of degree n).

It is easy to see that if a product of elements of S contains a factor X for which $\chi(X) = k$, the value of χ for the whole product cannot exceed k.

Now let $\mathfrak{K}$ be some irreducible generating set for the semigroup $\mathfrak{S}_\Omega$. We write

$$\mathfrak{K} \cap \mathfrak{G}_\Omega = \mathfrak{K}', \qquad \mathfrak{K}\backslash\mathfrak{K}' = \mathfrak{K}''.$$

By what we have said above, a product of elements containing even one element of $\mathfrak{K}''$ cannot belong to $\mathfrak{G}_\Omega$. It follows that every element of $\mathfrak{G}_\Omega$ is representable as a product of elements of $\mathfrak{K}'$ all belonging to $\mathfrak{G}_\Omega$. Therefore $\mathfrak{K}'$ is a generating set for $\mathfrak{G}_\Omega$. It is also an irreducible generating set for $\mathfrak{G}_\Omega$, since $\mathfrak{K}$ would otherwise not be irreducible in $\mathfrak{S}_\Omega$.

It is also obvious that for arbitrary $G \in \mathfrak{G}_\Omega$ and $S \in \mathfrak{S}_\Omega$ we have

$$\chi(GS) = \chi(SG) = \chi(S).$$

From this it follows that if $\mathfrak{R}$ were to contain no element X such that $\chi(X) = n - 1$, then no product of elements of $\mathfrak{R}$ could be equal to a mapping S for which $\chi(S) = n - 1$. Therefore the generating set $\mathfrak{R}$ must necessarily contain a mapping X for which $\chi(X) = n - 1$.

We shall show that every set $\mathfrak{R} \subset \mathfrak{S}_\Omega$ consisting of an irreducible generating set $\mathfrak{R}'$ for the group $\mathfrak{G}_\Omega$ and of one mapping X such that $\chi(X) = n - 1$, is a generating set for $\mathfrak{S}_\Omega$. It will then follow, from what we have said earlier, that such a set $\mathfrak{R}$ is an irreducible generating set and that every irreducible generating set in $\mathfrak{S}_\Omega$ is constructible in this way.

For the proof of this assertion, we show by induction on $(n - \chi(S))$ that $S \in [\mathfrak{R}' \cup X]$. If $(n - \chi(S)) = 0$, then $S \in \mathfrak{G}_\Omega$ and therefore

$$S \in [\mathfrak{R}'] \subset [\mathfrak{R}' \cup X].$$

Suppose $n - \chi(S) > 0$. We write the transformations X and S in the form of substitutions. Remembering that $\chi(X) = n - 1$ and $\chi(S) \leqslant n - 1$, these substitutions will have the form

$$X = \begin{pmatrix} \lambda_1 & \lambda_2 & \lambda_3 & \cdots & \lambda_n \\ \mu_2 & \mu_2 & \mu_3 & \cdots & \mu_n \end{pmatrix}, \qquad S = \begin{pmatrix} \nu_1 & \nu_2 & \nu_3 & \cdots & \nu_n \\ \xi_2 & \xi_2 & \xi_3 & \cdots & \xi_n \end{pmatrix}.$$

Here $\{\lambda_1, \lambda_2, \lambda_3, \ldots, \lambda_n\} = \{\nu_1, \nu_2, \nu_3, \ldots, \nu_n\} = \Omega$ and the elements $\mu_2, \mu_3, \ldots, \mu_n$ are all distinct from one another. We denote by μ_1 an element of Ω distinct from all $\mu_2, \mu_3, \ldots, \mu_n$, and by ξ_1 an element distinct from all $\xi_2, \xi_3, \ldots, \xi_n$. Since we have

$$\begin{aligned} S &= \begin{pmatrix} \nu_1 & \nu_2 & \nu_3 & \cdots & \nu_n \\ \xi_2 & \xi_2 & \xi_3 & \cdots & \xi_n \end{pmatrix} \\ &= \begin{pmatrix} \mu_1 & \mu_2 & \mu_3 & \cdots & \mu_n \\ \xi_1 & \xi_2 & \xi_3 & \cdots & \xi_n \end{pmatrix} \begin{pmatrix} \lambda_1 & \lambda_2 & \lambda_3 & \cdots & \lambda_n \\ \mu_2 & \mu_2 & \mu_3 & \cdots & \mu_n \end{pmatrix} \begin{pmatrix} \nu_1 & \nu_2 & \nu_3 & \cdots & \nu_n \\ \lambda_1 & \lambda_2 & \lambda_3 & \cdots & \lambda_n \end{pmatrix}, \end{aligned}$$

S turns out to be a product of three factors, of which one is X, another belongs to $\mathfrak{G}_\Omega$, and the third has a value of χ greater than that of $\chi(S)$. All three factors therefore belong to $[\mathfrak{R}]$, and therefore $S \in [\mathfrak{R}]$.

We have reduced the task of determining all irreducible generating sets of the semigroup $\mathfrak{S}_\Omega$ to the same task for the group $\mathfrak{G}_\Omega$. In keeping with the purpose of this book, however, we shall not go into this question, which belongs to group theory. We shall content ourselves with presenting one of the known irreducible generating sets for the group $\mathfrak{G}_\Omega$:

$$T_2 = \begin{pmatrix} \lambda_1 & \lambda_2 \\ \lambda_2 & \lambda_1 \end{pmatrix}, \qquad T_3 = \begin{pmatrix} \lambda_1 & \lambda_3 \\ \lambda_3 & \lambda_1 \end{pmatrix}, \ldots, \qquad T_n = \begin{pmatrix} \lambda_1 & \lambda_n \\ \lambda_n & \lambda_1 \end{pmatrix}.$$

With the aid of this set we obtain, as an example, the following irreducible generating set of the semigroup $\mathfrak{S}_\Omega$:

$$U = \begin{pmatrix} \lambda_1 & \lambda_2 \\ \lambda_1 & \lambda_1 \end{pmatrix}, T_2, T_3, \ldots, T_n.$$

The determination of canonical forms of the elements of $\mathfrak{S}_\Omega$ with respect to this generating set, as well as for other irreducible generating sets of $\mathfrak{S}_\Omega$, is a matter of real difficulty (cf. A. Ja. Aĭzenštat **[2]**). The elementary facts about generating sets of the group $\mathfrak{G}_\Omega$ are given in books on group theory or on Galois theory. Beyond this, there is a whole series of special works (see, for example: H. Coxeter, W. Moser, *Generators and relations for discrete groups*, Springer-Verlag, Berlin, 1957; A. Ja. Aĭzenštat **[2]** and a series of papers by S. Piccard, in particular three papers published in 1955 in the C. R. Acad. Sci. Paris).

3. Monogenic Semigroups

3.1. We shall now consider separately those semigroups that have a minimal number of elements in their generating sets.

DEFINITION. *A semigroup is called monogenic if it has a generating set consisting of a single element.*

Monogenic semigroups (the term is due to V. V. Vagner) are also called cyclic. This latter term is used in a wider sense. It is borrowed from the theory of groups, where it has a natural significance in connection with finite groups. In connection with semigroups, however, this natural connotation is lost. Since the concepts of generating set in group theory and generating set in the theory of semigroups are distinct, it seems useful to employ a different terminology in semigroup theory.

3.2. If the element X is the generating set of a monogenic semigroup

$$\mathfrak{A} = [X],$$

then

$$X, X^2, \ldots, X^n, \ldots$$

is the set of all elements of $\mathfrak{A}$. The multiplication of the elements of $\mathfrak{A}$ is commutative, since it is carried out according to the rule of addition of exponents

$$X^n \cdot X^m = X^{n+m}.$$

What we have said, however, is not an exhaustive description of the structure of $\mathfrak{A}$, since the expression of the elements of $\mathfrak{A}$ in the form X^n may not be unique. Different powers of X may represent one and the same element of $\mathfrak{A}$.

3.3. Monogenic semigroups may be either finite or infinite. We shall first consider finite monogenic semigroups.

To this end we define a function of three variables, $\bar{k} = f(k, a, b)$, with values ranging over the natural numbers:

$$\bar{k} = \begin{cases} k \text{ (if } k < a + b), \\ k - k'b, \text{ where the natural number } k' \text{ is such that } a \leqslant \bar{k} < a + b \\ \text{(if } k \geqslant a + b). \end{cases}$$

We note that, as is easily seen, for given a and b and arbitrary k_1 and k_2 we have

$$\overline{\bar{k}_1 + \bar{k}_2} = \overline{k_1 + k_2}.$$

3.4. Lemma. *If the monogenic semigroup $\mathfrak{A} = [X]$ is such that for some natural numbers a and b we have*

$$X^{a+b} = X^a,$$

then for every $X^k \in \mathfrak{A}$ we have

$$X^k = X^{\bar{k}},$$

where $\bar{k} = f(k, a, b)$.

Proof. If $k < a + b$, then $k = \bar{k}$. Suppose $k \geqslant a + b$. Then we may divide the difference $(k - a)$ by b:

$$k - a = bs + r, \qquad 0 \leqslant r < b.$$

Since X^b is a unit for the element X^a, we find that

$$X^k = X^a \cdot X^{k-a} = X^a \cdot X^{bs+r} = X^a \cdot (X^b)^s \cdot X^r = X^a \cdot X^r = X^{a+r}.$$

Since $a \leqslant a + r < a + b$ and $a + r = k - bs$, we have $a + r = f(k, a, b)$.

3.5. Lemma. *In a finite monogenic semigroup there is always at least one element having a unit.*

Proof. If $\mathfrak{A} = [X]$ is finite, then among the various powers of X some two must be equal to one another as elements:

$$X^p = X^q, \qquad (p > q).$$

In this case X^{p-q} is a unit for the element X^q:

$$X^q \cdot X^{p-q} = X^p = X^q.$$

3.6. We shall denote by $d(\mathfrak{A})$ the number of elements of the finite monogenic semigroup $\mathfrak{A}$ that have units, and by $(h(\mathfrak{A}) - 1)$ the number of elements not having a unit.

Theorem. *For every pair of natural numbers d and h there exists a finite monogenic semigroup $\mathfrak{A}$, unique except for isomorphisms, having the property that*

$$d(\mathfrak{A}) = d, \qquad h(\mathfrak{A}) = h.$$

PROOF. (1) We denote by $\mathfrak{A}$ the sequence of formal powers

$$X, X^2, X^3, \ldots, X^{h+d-1}.$$

We define over this set an operation as follows:

$$X^p \cdot X^q = X^{\overline{p+q}},$$

where

$$\overline{p+q} = f(p+q, h, d).$$

This operation is associative, since the property of the function f noted in 3.3 implies that

$$(X^p \cdot X^q) \cdot X^r = X^{\overline{p+q}} \cdot X^r = X^{\overline{\overline{p+q}+r}} = X^{\overline{p+q+r}},$$

$$X^p \cdot (X^q \cdot X^r) = X^p \cdot X^{\overline{q+r}} = X^{\overline{p+\overline{q+r}}} = X^{\overline{p+q+r}}.$$

Every formal power X^p ($p = 1, 2, \ldots, d + h - 1$) is a pth power of X with respect to this operation. Therefore $\mathfrak{A}$ is a finite monogenic semigroup with $(d + h - 1)$ elements, for which the element X is a generating set.

Since

$$X^{h+d} = X^{\overline{h+d}} = X^h,$$

the element X^d is a unit for the elements $X^h, X^{h+1}, \ldots, X^{h+d-1}$.

If $m < h$, then for arbitrary l, where $1 \leqslant l < h + d$, the definition of $f(k, h, d)$ implies that

$$X^m \cdot X^l = X^{\overline{m+l}}, \qquad h \leqslant \overline{m+l} \leqslant h + d - 1.$$

Accordingly, the elements $X, X^2, \ldots, X^{h-1}$ have no unit in $\mathfrak{A}$.

Therefore,

$$d(\mathfrak{A}) = d, \qquad h(\mathfrak{A}) = (h - 1) + 1 = h.$$

(2) Suppose $\mathfrak{A}_1 = [X_1]$ and $\mathfrak{A}_2 = [X_2]$ are two finite monogenic semigroups for which

$$d(\mathfrak{A}_1) = d(\mathfrak{A}_2) = d, \qquad h(\mathfrak{A}_1) = h(\mathfrak{A}_2) = h.$$

The elements $X_i, X_i^2, \ldots, X_i^{h+d-1}$ are all distinct among themselves, since if some two of these elements were equal to each other it would follow from the lemma of 3.4 that the number of elements of $\mathfrak{A}_i$ would be less than $h + d - 1$ ($i = 1, 2$).

Since X_i^{h+d} is an element of $\mathfrak{A}_i$, it follows that for some n, where $1 \leqslant n < h + d$, we have

$$X_i^{h+d} = X_i^n.$$

All elements $X^n, X^{n+1}, \ldots, X^{h+d-1}$ have the element X_i^{h+d-n} as a unit. The elements $X_i, X_i^2, \ldots, X_i^{n-1}$ have no unit, since for $m < n$ 3.4 implies that

$$X_i^m \cdot X_i^l = X_i^{\overline{m+l}},$$

where $\overline{m+l} = f(m+l, n, h+d-n) > m$. Accordingly,

$$d + h - n = d,$$

i.e., $n = h$.

The equation

$$X_i^{h+d} = X_i^h,$$

implies, by the lemma of 3.4, that

$$X_i^p \cdot X_i^q = X_i^{p+q} = X_i^{\overline{p+q}},$$

where $\overline{p+q} = f(p+q, h, d)$. In virtue of the correspondence

$$X_1 \sim X_2, \qquad X_1^2 \sim X_2^2, \ldots, \qquad X_1^{h+d-1} \sim X_2^{h+d-1}$$

between the semigroups $\mathfrak{A}_1$ and $\mathfrak{A}_2$, the two are isomorphic.

3.7. This theorem provides a complete classification of finite monogenic semigroups. To each such semigroup there corresponds a pair of integers

$$\chi(\mathfrak{A}) = (h, d), \qquad h = h(\mathfrak{A}), \qquad d = d(\mathfrak{A}).$$

To every pair of integers (h, d) there corresponds a finite monogenic semigroup such that

$$\chi(\mathfrak{A}) = (h, d).$$

Two finite monogenic semigroups are isomorphic if and only if

$$\chi(\mathfrak{A}_1) = \chi(\mathfrak{A}_2).$$

We call the pair of integers $\chi(\mathfrak{A}) = (h, d)$ the *type of the finite monogenic semigroup* $\mathfrak{A}$, and also the type of the element X for which $[X] = \mathfrak{A}$. In an arbitrary semigroup $\mathfrak{A}$ the element X is said to be of *finite type* (one also sometimes says "finite order") if $[X]$ is finite. The element X is said to be of *infinite type* (infinite order) if $[X]$ is infinite.

3.8. Prescribing the type $\chi(\mathfrak{A}) = (h, d)$ of the finite monogenic semigroup $\mathfrak{A} = [X]$ completely defines its structure in a convenient way. In fact, it follows from the proof of the theorem of 3.6, that $\mathfrak{A}$ consists precisely of the elements

$$X, X^2, \ldots, X^{h+d-1}.$$

Multiplication of these elements proceeds according to the law of addition of exponents, and powers with an exponent greater than $h + d - 1$ can be reduced to one of the given powers by means of the equation

$$X^{h+d} = X^h.$$

3.9. Let $\mathfrak{A} = [X]$ be a finite monogenic semigroup. We shall show how the properties of $\mathfrak{A}$ depend on the integers that make up its type $\chi(\mathfrak{A}) = (h, d)$.

If $h > 1$, then $\mathfrak{A}$ has no units, which follows from the fact that the elements $X, X^2, \ldots, X^{h-1}$ have no unit. Therefore, for $h > 1$ $\mathfrak{A}$ is not a group.

If $h = 1$, $\mathfrak{A}$ has a unit.

In fact, in this case

$$X^{d+1} = X.$$

Therefore X^d is a unit for X, and so is a unit for the entire semigroup $\mathfrak{A}$. Every element X^k $(k = 1, 2, \ldots, d)$ has an inverse:

$$X^{d-k} \cdot X^k = X^d, \qquad X^k \cdot X^{d-k} = X^d, \qquad (k = 1, 2, \ldots, d-1).$$

Accordingly, for $h = 1$, $\mathfrak{A}$ is a group.

The only possible zero in $\mathfrak{A}$ is obviously X^{h+d-1}, since every other element is changed when multiplied by X.

In order that X^{h+d-1} be a zero it is necessary and sufficient that

$$X \cdot X^{h+d-1} = X^{h+d-1},$$

i.e.,

$$X^{h+d} = X^{h+d-1}.$$

But

$$X^{h+d} = X^h.$$

Accordingly, $\mathfrak{A}$ has a zero if and only if $d = 1$.

3.10. Let us now consider the question of divisibility of the elements of a finite monogenic semigroup $\mathfrak{A} = [X]$.

If $p \geqslant h$, then X^p is divisible by every element of the semigroup $\mathfrak{A}$.

In fact, for $X^q \in \mathfrak{A}$ $(1 \leqslant q < h + d)$ we have by 3.4,

$$X^q \cdot X^{p+qd-q} = X^{p+qd} = X^{\overline{p+qd}} = X^p$$

(since $h \leqslant p < h + d$).

If $p < h$, X^p is divisible by every X^q for which $1 \leqslant q < p$.

In fact, if $1 \leqslant q < p$, then

$$X^p = X^q \cdot X^{p-q}.$$

If $q \geqslant p$, then for arbitrary s

$$X^q \cdot X^s = X^{q+s} = X^{\overline{q+s}} \neq X^p,$$

since $\overline{q+s}$ is either equal to $q + s$ or is not less than h. In either case, $\overline{q+s} > p$.

3.11. By II, 1.6, we may conclude from 3.10 that the ensemble $\{X^h, X^{h+1}, \ldots, X^{h+d-1}\}$ is a group. This is a maximal subgroup of $\mathfrak{A}$. In fact, no one of the elements $X, X^2, \ldots, X^{h-1}$ is contained in any subgroup of $\mathfrak{A}$, since it cannot have a unit.

It follows in particular that a finite monogenic semigroup always has one unique idempotent—the unit of the group just described.

3.12. If $h > 1$, that is $\mathfrak{A} = [X]$ is not a group, then X, in accordance with 3.10, is not divisible by any element. It follows that X cannot be represented as a product of any elements whatever, except for the singular representation $X = X$. But this implies that every generating set of the semigroup $\mathfrak{A}$ must contain X. Therefore, for $h > 1$ the only irreducible generating set of the finite monogenic semigroup $\mathfrak{A} = [X]$ is the set consisting of the single element X.

3.13. Suppose that $h = 1$, that is $\mathfrak{A} = [X]$ is a group. We consider an arbitrary subset of $\mathfrak{A}$:

$$\mathfrak{K} = \{X^{k_1}, X^{k_2}, \ldots, X^{k_m}\} \quad (1 \leqslant k_i < d + 1; \quad i = 1, 2, \ldots, m).$$

We denote by l the greatest common divisor of the integers $d, k_1, k_2, \ldots, k_m$. As is well known, there exist integers $\alpha_0, \alpha_1, \alpha_2, \ldots, \alpha_m$, such that

$$l = \alpha_0 d + \alpha_1 k_1 + \alpha_2 k_2 + \ldots + \alpha_m k_m.$$

Adding to both sides of this equation the integer Nd, we have for sufficiently large N the equation

$$l + Nd = \beta_0 d + \beta_1 k_1 + \beta_2 k_2 + \ldots + \beta_m k_m,$$

where $\beta_0, \beta_1, \beta_2, \ldots, \beta_m$ are also integers.

From this equation we obtain

$$\begin{aligned} X^l \cdot (X^d)^N &= X^{l+dN} = X^{\beta_0 d + \beta_1 k_1 + \beta_2 k_2 + \ldots + \beta_m k_m} \\ &= (X^d)^{\beta_0} \cdot (X^{k_1})^{\beta_1} \cdot (X^{k_2})^{\beta_2} \cdot \ldots \cdot (X^{k_m})^{\beta_m}. \end{aligned}$$

But $X^{d+1} = X$ implies that X^d is a unit of X, and therefore a unit of the entire semigroup $\mathfrak{A}$. The equation we have obtained therefore means that

$$X^l = (X^{k_1})^{\beta_1} \cdot (X^{k_2})^{\beta_2} \cdot \ldots \cdot (X^{k_m})^{\beta_m}.$$

If $l = 1$, $[\mathfrak{K}]$ contains X and therefore

$$[\mathfrak{K}] = \mathfrak{A},$$

that is, $\mathfrak{K}$ is a generating set of the semigroup $\mathfrak{A}$.

If $l > 1$, then an arbitrary element Y of $[\mathfrak{K}]$

$$Y = (X^{k_1})^{\gamma_1} \cdot (X^{k_2})^{\gamma_2} \cdot \ldots \cdot (X^{k_m})^{\gamma_m}$$

has the form

$$Y = X^{l\lambda}$$

and therefore

$$Y^{\left(\frac{d}{l}\right)} = (X^{l\lambda})^{\frac{d}{l}} = X^{d\lambda} = X^d.$$

But

$$X^{\left(\frac{d}{l}\right)} \neq X^d,$$

since otherwise it would follow from 3.4 that the number of elements of $\mathfrak{A}$ would be less than d. Therefore $[\mathfrak{K}]$ does not contain X. Thus, for $l > 1$, $\mathfrak{K}$ is not a generating set for $\mathfrak{A} = [X]$.

All this implies, in particular, that the element X^k $(1 \leqslant k < h + d)$ is a generating set of $\mathfrak{A}$ if and only if k is prime to d.

3.14. It follows from 3.12 that for $h > 1$ the semigroup $\mathfrak{A} = [X]$ admits no automorphisms other than the identity. In fact, for every automorphism of the semigroup a generating set is mapped into a generating set. Therefore X must be mapped into X, and so every power X^k is mapped into X^k.

3.15. If $h = 1$ and $d = 1$, $\mathfrak{A}$ is a unitary group and therefore admits no automorphism other than the identity.

If $h = 1$ and $d > 1$, that is, $\mathfrak{A}$ is a nonunitary group, then for arbitrary k $(1 \leqslant k < d + 1)$ relatively prime to d, the mapping φ_k of the semigroup $\mathfrak{A}$ on itself

$$\varphi_k(X^n) = X^{kn},$$

is an automorphism, as is easily seen. $\mathfrak{A}$ does not admit other automorphisms, since for every automorphism ψ the number k, where

$$\psi(X) = X^k,$$

must be prime to d, by 3.13. In this case, evidently, $\psi = \varphi_k$.

Thus, for $h = 1, d = 2$ the group $\mathfrak{A}$ admits no automorphism other than the identity. For $h = 1, d > 2$ $\mathfrak{A}$ admits the nonidentical automorphism φ_{d-1}.

3.16. We shall now pass on to the study of an infinite monogenic semigroup $\mathfrak{A} = [X]$. We shall say that its *type is infinite*.

All the elements $X, X^2, \ldots, X^n, \ldots$ of a monogenic semigroup $\mathfrak{A} = [X]$ of infinite type must be distinct, by 3.4. The operation in $\mathfrak{A}$ is defined by the rule for compounding exponents. It follows that all infinite monogenic semigroups are isomorphic among themselves. The question of divisibility is answered very simply: X^p is divisible by X^q if and only if $p > q$. This implies that $\mathfrak{A}$ has no idempotents.

The only irreducible generating set is the set consisting of the single element X. $\mathfrak{A}$ admits no automorphism other than the identity.

3.17. Finite monogenic semigroups that are also groups (that is, in the case $h = 1$) are also called finite cyclic groups. There also exist infinite cyclic groups. An infinite cyclic group $\mathfrak{A}$ is one in which the elements may be represented in the form

$$\ldots, A^{-2}, A^{-1}, A^0 = E, A, A^2, \ldots, A^n, \ldots,$$

where all these elements are distinct, and the multiplication is carried out according to the rule of addition of exponents. We note that $\mathfrak{A} = [A, A^{-1}]$.

The reason for singling out the class of cyclic groups is that, as we easily see, these are all the groups possessing a one-element set that is a generating set in the group-theoretic sense (2.2).

4. Periodic Semigroups

4.1. Every element A of the semigroup $\mathfrak{A}$ generates a monogenic subsemigroup $[A]$ of the semigroup $\mathfrak{A}$. $\mathfrak{A}$ itself is the union of all its monogenic subsemigroups. A number of properties of the semigroup $\mathfrak{A}$ are definable as properties of these monogenic subsemigroups. For example, one may distinguish the class of so-called periodic semigroups, which includes the class of finite semigroups.

DEFINITION. *A semigroup $\mathfrak{A}$ is said to be* PERIODIC *if all its monogenic subsemigroups are finite.*

Since every finite monogenic semigroup admits an idempotent, and no infinite monogenic semigroup does, the periodic semigroups may also be defined as semigroups in which every subsemigroup has an idempotent. Making use of the fact that in a finite monogenic semigroup $\mathfrak{A}$ some power of the element A is an idempotent, we may again define periodic semigroups, this time as semigroups in which every element, when raised to some power, is an idempotent.

Later we shall consider various properties of periodic semigroups, of which the basic ones were obtained by Schwarz **[5]**.

4.2. In the class of periodic semigroups we may single out the groups (which are, as we know, semigroups with two-sided cancellation) by means of the following property.

THEOREM. *A periodic semigroup with left cancellation is a group if and only if it has a single idempotent.*

PROOF. (1) If the semigroup $\mathfrak{A}$ is a group, it is known that $E_{\mathfrak{A}}$ is its only idempotent.

(2) Let I be the unique idempotent of $\mathfrak{A}$, a semigroup with left cancellation. An arbitrary element $A \in \mathfrak{A}$ generates a finite monogenic subsemigroup $[A]$ of $\mathfrak{A}$, and we denote its type (3.7) by (h, d). If it were true that $h > 1$, then

$$A^{h+d} = A^h$$

would imply

$$A \cdot (A^{h+d-1}) = A \cdot A^{h-1},$$

which contradicts the assumption that $\mathfrak{A}$ is a semigroup with left cancellation, since by definition of the type of a monogenic semigroup

$$A^{h+d-1} \neq A^{h-1}.$$

Accordingly, $h = 1$, which means that $[A]$ is a group (3.11). The unit of the group $[A]$ must also be the idempotent I. Therefore, I is a two-sided unit of an arbitrary element A of the semigroup $\mathfrak{A}$, and A has a two-sided inverse with respect to I. But this means that $\mathfrak{A}$ is a group.

4.3. We remark that all three conditions stated in the preceding theorem are necessary in order that a semigroup be a group—that is, periodicity, left cancellation, and uniqueness of the idempotent. We shall show by example that the existence of any two of these properties is not in itself sufficient.

(1) The additive semigroup of all non-negative integers admits left cancellation and has a unique idempotent (the number 0), but is not a group.

(2) A semigroup with zero in which the product of any two elements is the zero is periodic and the zero is a unique idempotent, but the semigroup is not a group if it has more than one element.

(3) A semigroup in which the product of any two elements is the right-hand factor is a periodic semigroup with left cancellation. However, if it has more than one element it is not a group.

4.4. Since every finite monogenic semigroup has a single idempotent, there corresponds to each element A of the periodic semigroup $\mathfrak{A}$ a uniquely defined idempotent, which is expressible as a power of A. This correspondence is of basic importance and is often used in the study of periodic semigroups. By use of this correspondence the ensemble of elements of a periodic semigroup may be divided into pairwise disjoint classes $\mathfrak{K}_I$ such that each class contains a single idempotent I and some power of each element of the class is equal to I:

$$\mathfrak{A} = \bigcup_{I \in \mathfrak{H}} \mathfrak{K}_I, \qquad \mathfrak{K}_I \cap \mathfrak{K}_{I'} = \varnothing, \qquad (I \neq I'),$$

(where $\mathfrak{H}$ is the set of all idempotents of the semigroup $\mathfrak{A}$).

4.5. The classes $\mathfrak{K}_I$, as we shall see in the example of 4.11, need not be subsemigroups of the semigroup $\mathfrak{A}$. The cases in which the $\mathfrak{K}_I$ are all semigroups deserve attention. We shall mention a few of them.

(α) It is easily shown that if $\mathfrak{A}$ is commutative, all $\mathfrak{K}_I$ are semigroups.

(β) Iseki **[4]** has drawn attention to the fact that all the $\mathfrak{K}_I$ will be subsemigroups when they arise from semigroups of a class introduced by Thierrin **[16]**, characterized by a kind of generalized commutativity.

Let the periodic semigroup $\mathfrak{A}$ be such that for an arbitrary pair of elements A and B there always exist integers r, s, t such that

$$(AB)^r = A^s B^t = B^t A^s.$$

Then every class $\mathfrak{K}_I$ is a subsemigroup.

In fact, suppose $A, B \in \mathfrak{K}_I$, that is

$$A^a = I, \qquad B^b = I.$$

Since for some r, s, t

$$(AB)^r = A^s B^t = B^t A^s,$$

the commutativity of A^s and B^t implies that

$$(AB)^{rab} = [(AB)^r]^{ab} = (A^sB^t)^{ab} = (A^s)^{ab} \cdot (B^t)^{ab} = (A^a)^{bs} \cdot (B^b)^{at} = I^{bs} \cdot I^{at} = I.$$

Accordingly, $A, B \in \mathfrak{K}_I$ implies that $AB \in \mathfrak{K}_I$, that is, $\mathfrak{K}_I$ is in fact a subsemigroup.

(γ) Not only commutativity and a close generalization of it, but even a kind of converse of commutativity, imply the semigroup property of the $\mathfrak{K}_I$.

If in $\mathfrak{A}$ any two distinct idempotents I and I' satisfy

$$II' \neq I'I,$$

then every class $\mathfrak{K}_I$ is a subsemigroup.

In fact, suppose that $A, B \in \mathfrak{K}_I$, that is,

$$A^a = I, \qquad B^b = I.$$

The product AB belongs to some $\mathfrak{K}_{I'}$:

$$(AB)^c = I', \qquad I'^2 = I'.$$

Since I commutes with A and with B, it must also commute with AB, and therefore with I'. But by hypothesis, this implies that $I = I'$, that is, $AB \in \mathfrak{K}_I$.

4.6. Since in the general case $\mathfrak{K}_I$ is not a semigroup, it is natural to inquire as to subsemigroups of the semigroup $\mathfrak{A}$ contained in $\mathfrak{K}_I$ and in particular as to the maximal subsemigroups of this type. In this connection we make use of a general lemma which may be useful in other circumstances as well. The lemma in question is a simple consequence of the theorem proved in II, 4.17.

LEMMA. *Let $\mathfrak{P}$ be an ensemble of certain subsets of the set Ω, with the following property. If $\mathfrak{Q}$ is a subset of $\mathfrak{P}$ such that for any two M and M' belonging to $\mathfrak{Q}$ one is necessarily a subset of the other, then the union of all sets belonging to $\mathfrak{Q}$ belongs to $\mathfrak{P}$. Then $\mathfrak{P}$ contains a maximal set M_0, that is, a set $M_0 \in \mathfrak{P}$ such that M_0 is not a proper subset of any other set M belonging to $\mathfrak{P}$.*

PROOF. The inclusion relation in $\mathfrak{P}$ is a partial ordering satisfying the conditions of Theorem II, 4.17. The validity of the lemma follows immediately.

4.7. We note an immediate consequence of the lemma, which will be of later use.

COROLLARY. *If $\mathfrak{M}$ is a subsemigroup of the semigroup $\mathfrak{A}$ and $\mathfrak{K}$ is a subset of $\mathfrak{A}$ such that $\mathfrak{M} \subset \mathfrak{K}$ then $\mathfrak{A}$ admits a subsemigroup $\mathfrak{N}$ for which $\mathfrak{M} \subset \mathfrak{N} \subset \mathfrak{K}$ and $\mathfrak{N}$ is contained in no subsemigroup $\mathfrak{N}'$ of the semigroup $\mathfrak{A}$ distinct from $\mathfrak{N}$ and satisfying the relation $\mathfrak{M} \subset \mathfrak{N}' \subset \mathfrak{K}$.*

In fact, we apply Lemma 4.6 and designate the class $\mathfrak{P}$ as the ensemble of subsemigroups of $\mathfrak{A}$ that contain $\mathfrak{M}$ and are contained in $\mathfrak{K}$. The hypotheses of 4.6 are satisfied because of 1.12.

4.8. Now let us choose, in the semigroup $\mathfrak{A}$, an arbitrary idempotent I and a subsemigroup $\mathfrak{M}$ of the semigroup $\mathfrak{A}$, containing I as a unique idempotent. Since $\mathfrak{M} \subset \mathfrak{K}_I$ (4.4), we may apply to $\mathfrak{M}$ and $\mathfrak{K}_I$ the corollary 4.7. It follows that $\mathfrak{M}$ is contained in a subsemigroup $\mathfrak{N}$ of the semigroup $\mathfrak{A}$ such that $\mathfrak{N} \subset \mathfrak{K}_I$ and $\mathfrak{N}$ is contained in no subsemigroup $\mathfrak{N}'$ of the semigroup $\mathfrak{A}$ for which $\mathfrak{N}' \subset \mathfrak{K}_I$ and $\mathfrak{N}' \neq \mathfrak{N}$.

All this means that in a periodic semigroup $\mathfrak{A}$ every subsemigroup containing a unique idempotent is contained in some maximal subsemigroup $\mathfrak{N}$ which has a single idempotent and is not itself contained in any subsemigroup of $\mathfrak{A}$ distinct from $\mathfrak{N}$ and having only one idempotent.

The class $\mathfrak{K}_I$ corresponding to an arbitrary idempotent I we shall represent as the union of subsemigroups maximal in the sense that none is contained in a subsemigroup of $\mathfrak{A}$ having a single idempotent.

4.9. Suppose that the element A of an arbitrary periodic semigroup belongs to the class $\mathfrak{K}_I$, that is, some power of A is equal to the idempotent I. If I is a left (and therefore a right) unit of $\mathfrak{A}$, I is a regular two-sided unit of $\mathfrak{A}$.

In fact,

$$A^r = I, \qquad AI = A,$$

implies

$$IA = A^rA = AA^r = AI,$$

$$A^{r-1}A = I, \qquad AA^{r-1} = I.$$

If I is not a left unit of $\mathfrak{A}$, then $\mathfrak{A}$ has no regular two-sided unit. In fact, let us assume that some idempotent $I' \in \mathfrak{A}$ satisfies

$$A = AI' = I'A, \qquad I' = AX = YA.$$

Since

$$A^{2r} = II = I = A^r, \qquad A(AX)X = AI'X = I',$$

we have

$$I = A^{r}I' = A^{2r}X^r = A^rX^r = I',$$

which contradicts our hypothesis.

4.10. If the element A of a periodic semigroup $\mathfrak{A}$ belongs to $\mathfrak{K}_I$, that is, $A^r = I$, then $AI = IA \in \mathfrak{K}_I$. The element AI has the indempotent I as a regular two-sided unit.

If A itself has a regular two-sided unit, it follows from 4.9 that that unit is I, and therefore that $AI = IA = A$. It follows also that

$$I\mathfrak{K}_I = \mathfrak{K}_I I$$

is the set of all elements of $\mathfrak{A}$ having I as a regular two-sided unit. With the notation of 1.14

$$\mathfrak{G}_I = I\mathfrak{K}_I = \mathfrak{K}_I I.$$

$\mathfrak{G}_I$ is a maximal subgroup of $\mathfrak{A}$ having I as its unit.

If we denote by $\mathfrak{H}$ the ensemble of all idempotents of the periodic semigroup $\mathfrak{A}$, we have, by 1.16, that

$$\bigcup_{I \in \mathfrak{H}} I\mathfrak{K}_I$$

is the set of all elements of $\mathfrak{A}$ contained in the subgroups of the semigroup $\mathfrak{A}$.

4.11. We now note one property of the subgroup $\mathfrak{G}_I$ of the periodic semigroup $\mathfrak{A}$. If some element $A \in \mathfrak{A}$ has the idempotent I as a right unit and I is divisible on the left by A, then $A \in \mathfrak{G}_I$.

In fact, suppose that

$$AI = A, \qquad AB = I.$$

These equations clearly imply that for arbitrary integer k we have $A^kB^k = I$. The element A, like every element of a periodic semigroup, is when raised to some power an idempotent:

$$(A^r)^2 = A^r.$$

Therefore

$$I = A^rB^r = A^rA^rB^r = A^rI = A^r.$$

By 4.9, $A \in \mathfrak{G}_I$.

4.12. As an example of the construction set forth above let us look at the multiplicative semigroup $\mathfrak{M}$ of matrices consisting of the nine matrices

$$M_{11} = \begin{bmatrix} 1 & 0 \\ 0 & 0 \end{bmatrix}, \quad M_{12} = \begin{bmatrix} 0 & 1 \\ 0 & 0 \end{bmatrix}, \quad M_{21} = \begin{bmatrix} 0 & 0 \\ 1 & 0 \end{bmatrix}, \quad M_{22} = \begin{bmatrix} 0 & 0 \\ 0 & 1 \end{bmatrix},$$

$$N_{11} = \begin{bmatrix} -1 & 0 \\ 0 & 0 \end{bmatrix}, \quad N_{12} = \begin{bmatrix} 0 & -1 \\ 0 & 0 \end{bmatrix}, \quad N_{21} = \begin{bmatrix} 0 & 0 \\ -1 & 0 \end{bmatrix}, \quad N_{22} = \begin{bmatrix} 0 & 0 \\ 0 & -1 \end{bmatrix},$$

$$O = \begin{bmatrix} 0 & 0 \\ 0 & 0 \end{bmatrix}.$$

The idempotents of the semigroup $\mathfrak{M}$ are the elements O, M_{11}, M_{22}. The classes $\mathfrak{K}_I$ are

$$\mathfrak{K}_O = \{O, M_{12}, M_{21}, N_{12}, N_{21}\},$$

$$\mathfrak{K}_{M_{11}} = \{M_{11}, N_{11}\},$$

$$\mathfrak{K}_{M_{22}} = \{M_{22}, N_{22}\}.$$

The classes $\mathfrak{K}_{M_{11}}$ and $\mathfrak{K}_{M_{22}}$ are subsemigroups. The class $\mathfrak{K}_O$ is not. It is the union of two maximal subsemigroups having a single idempotent

$$\mathfrak{K}_O = \{O, M_{12}, N_{12}\} \cup \{O, M_{21}, N_{21}\}.$$

The maximal subgroups of $\mathfrak{M}$ are

$$\mathfrak{G}_O = O, \qquad \mathfrak{G}_{M_{11}} = \mathfrak{K}_{M_{11}}, \qquad \mathfrak{G}_{M_{22}} = \mathfrak{K}_{M_{22}}.$$

5. Magnifying Elements

5.1. We may single out certain elements of a semigroup by a property, that of the so-called magnifying elements, introduced by E. S. Ljapin [**9**]. These elements will come to our attention several times in the later parts of our discussion.

DEFINITION. *An element U of a semigroup $\mathfrak{A}$ is called a right magnifier if $\mathfrak{A}$ contains a proper subset $\mathfrak{A}'$ such that*

$$\mathfrak{A}'U = \mathfrak{A} \qquad (\mathfrak{A}' \subset \mathfrak{A}, \mathfrak{A}' \neq \mathfrak{A}).$$

An element V is called a left magnifier if $\mathfrak{A}$ contains a proper subset $\mathfrak{A}''$ such that

$$V\mathfrak{A}'' = \mathfrak{A} \qquad (\mathfrak{A}'' \subset \mathfrak{A}, \mathfrak{A}'' \neq \mathfrak{A}).$$

It is not immediately obvious from the definition that there exist semigroups containing magnifiers. The proof that such semigroups exist is deferred to a later section; for the moment we merely consider the properties of the postulated magnifying elements.

5.2. THEOREM. *No element of a semigroup is simultaneously a left and a right magnifier.*

PROOF. Suppose that the element X of the semigroup $\mathfrak{A}$ is both a left and a right magnifier. Then for some $\mathfrak{A}' \in \mathfrak{A}$ and $\mathfrak{A}'' \in \mathfrak{A}$ we have

$$\mathfrak{A}'X = \mathfrak{A}, \qquad X\mathfrak{A}'' = \mathfrak{A}, \qquad \mathfrak{A}' \neq \mathfrak{A}, \qquad \mathfrak{A}'' \neq \mathfrak{A}.$$

But $\mathfrak{A}'X = \mathfrak{A}$ implies that we can find in $\mathfrak{A}$ elements Z_1 and Z_2 such that

$$Z_1X = X, \qquad Z_2X = Z_1.$$

Since $X\mathfrak{A}'' = \mathfrak{A}$, every element A of the semigroup $\mathfrak{A}$ may be represented in the form XA'' ($A'' \subset \mathfrak{A}''$). It follows that

$$Z_1A = Z_1XA'' = XA'' = A,$$

that is, Z_1 is a left unit of $\mathfrak{A}$. Therefore

$$Z_2\mathfrak{A} = Z_2X\mathfrak{A}'' = Z_1\mathfrak{A}'' = \mathfrak{A}''.$$

But, on the other hand,

$$Z_2\mathfrak{A} \supset Z_2X\mathfrak{A} = Z_1\mathfrak{A} = \mathfrak{A}.$$

This means that $\mathfrak{A}'' \supset \mathfrak{A}$, which conflicts with the hypothesis.

5.3. It goes without saying that many semigroups do not contain magnifying elements. The following are examples of those that do not.

(α) *Finite semigroups have no magnifying elements.*

In fact, if $\mathfrak{A}'$ is a proper subset of the finite semigroup $\mathfrak{A}$, the set $X\mathfrak{A}'$ (and also $\mathfrak{A}'X$) contains fewer elements than $\mathfrak{A}$, whatever the element X may be, and is therefore distinct from $\mathfrak{A}$.

(β) *Commutative semigroups have no magnifying elements.*

In a commutative semigroup, every right magnifier is also a left magnifier and, by 5.2, cannot exist.

(γ) *Semigroups with two-sided cancellation have no magnifying elements.*

Suppose that $\mathfrak{A}$ is a semigroup with two-sided cancellation and that for some $U \in \mathfrak{A}$ and $\mathfrak{A}' \subset \mathfrak{A}$ we have

$$\mathfrak{A}'U = \mathfrak{A}, \qquad \mathfrak{A}' \neq \mathfrak{A}.$$

We choose an element B of $\mathfrak{A}$ not belonging to $\mathfrak{A}'$. Since $\mathfrak{A}'U = \mathfrak{A}$, we have for some $A' \in \mathfrak{A}'$ the equation $A'U = BU$. In a semigroup with right cancellation this cannot occur unless $A' = B$, which contradicts the assumption that A' belongs to $\mathfrak{A}'$ and B does not.

(δ) *Groups do not contain magnifying elements.*

Every group is a semigroup with two-sided cancellation (II, 2.16).

5.4. THEOREM. *Suppose that some elements of a semigroup $\mathfrak{A}$ are right magnifiers and some are left magnifiers. The ensemble of all right magnifiers in $\mathfrak{A}$ is a subsemigroup of $\mathfrak{A}$, and so is its complement in $\mathfrak{A}$.*

(Similarly for left magnifiers.)

PROOF. (1) Let U_1 and U_2 be two right magnifiers belonging to $\mathfrak{A}$. For some $\mathfrak{A}_1 \subset \mathfrak{A}$ and $\mathfrak{A}_2 \subset \mathfrak{A}$ we have

$$\mathfrak{A}_1 U_1 = \mathfrak{A}, \qquad \mathfrak{A}_2 U = \mathfrak{A}, \qquad \mathfrak{A}_1 \neq \mathfrak{A}, \qquad \mathfrak{A}_2 \neq \mathfrak{A}.$$

Since

$$\mathfrak{A}_1(U_1 U_2) = \mathfrak{A}U_2 \supset \mathfrak{A}_2 U_2 = \mathfrak{A},$$

$U_1 U_2$ is also a right magnifier in $\mathfrak{A}$. The set of all right magnifiers is therefore a subsemigroup of $\mathfrak{A}$.

(2) Let X_1 and X_2 be two elements of $\mathfrak{A}$, neither being a right magnifier. If their product were a right magnifier, we would have for some $\mathfrak{A}' \subset \mathfrak{A}$

$$\mathfrak{A}'X_1X_2 = \mathfrak{A}, \qquad \mathfrak{A}' \neq \mathfrak{A}.$$

But $\mathfrak{A}'X_1 = \mathfrak{A}'' \neq \mathfrak{A}$, since X_1 is not a right magnifier. But then the relation

$$\mathfrak{A}''X_2 = \mathfrak{A}'X_1X_2 = \mathfrak{A}, \qquad \mathfrak{A}'' \neq \mathfrak{A}$$

would be incompatible with the fact that X_2 is not a right magnifier.

5.5. Let us denote by $\mathfrak{A}^{(r)}$ the ensemble of all right magnifiers belonging to the semigroup $\mathfrak{A}$, by $\mathfrak{A}^{(l)}$ the ensemble of all left magnifiers, and by $\mathfrak{A}^{(n)}$ the ensemble of all elements which are neither right nor left magnifiers belonging to $\mathfrak{A}$. According to 5.2, the intersection of $\mathfrak{A}^{(r)}$ and $\mathfrak{A}^{(l)}$ is empty. Therefore $\mathfrak{A}^{(r)} \cup \mathfrak{A}^{(n)}$ is the ensemble of elements that are not left magnifiers in $\mathfrak{A}$, and

$\mathfrak{A}^{(l)} \cup \mathfrak{A}^{(n)}$ is the ensemble of elements that are not right magnifiers. Both ensembles are subsemigroups, by 5.4, unless they are empty. It follows that their intersection $\mathfrak{A}^{(n)}$ is a subsemigroup unless it is empty. Making use of 5.2 and 5.4 we may draw the following conclusion.

The semigroup $\mathfrak{A}$ is the union of three pairwise disjoint subsets

$$\mathfrak{A} = \mathfrak{A}^{(r)} \cup \mathfrak{A}^{(l)} \cup \mathfrak{A}^{(n)},$$

each of which is a subsemigroup or is the empty set.

5.6. With respect to the existence of magnifying elements, the semigroups $\mathfrak{A}^{(r)}, \mathfrak{A}^{(l)}, \mathfrak{A}^{(n)}$ have the following properties.

(α) *The semigroup $\mathfrak{A}^{(r)}$ contains no left magnifier.*

Suppose the element $X \in \mathfrak{A}^{(r)}$ is a left magnifier belonging to $\mathfrak{A}^{(r)}$. Then in $\mathfrak{A}^{(r)}$ we must be able to find a Z such that $XZ = X$. Since $Z \in \mathfrak{A}^{(r)}$ we have that for some $\mathfrak{A}' \subset \mathfrak{A}$

$$\mathfrak{A}'Z = \mathfrak{A}, \qquad \mathfrak{A}' \neq \mathfrak{A}.$$

We choose an arbitrary element A of $\mathfrak{A}$. Since $X \in \mathfrak{A}^{(r)}$ we have for some $Y \in \mathfrak{A}$ that $YX = A$. It follows that

$$AZ = YXZ = YX = A.$$

Then Z is a right unit of the semigroup $\mathfrak{A}$. But this means that $\mathfrak{A}'Z = \mathfrak{A}'$, which conflicts with the fact that $\mathfrak{A}'$ was chosen so that $\mathfrak{A}'Z = \mathfrak{A}$.

(β) *The semigroup $\mathfrak{A}^{(l)}$ contains no right magnifier.*

The proof is similar to that of (α).

(γ) *If the semigroup $\mathfrak{A}$ has a unit, then $\mathfrak{A}^{(n)}$ is nonempty and is a semigroup without either left or right magnifiers.*

In fact, $\mathfrak{A}^{(n)}$ is not empty, since it contains the unit $E_{\mathfrak{A}}$.

Let us assume that the element $X \in \mathfrak{A}^{(n)}$ is a right magnifier belonging to the semigroup $\mathfrak{A}^{(n)}$. Then for some $\mathfrak{A}' \subset \mathfrak{A}^{(n)}$ we have

$$\mathfrak{A}'X = \mathfrak{A}^{(n)}, \qquad \mathfrak{A}' \neq \mathfrak{A}^{(n)}.$$

In particular, we can find in $\mathfrak{A}'$ a Y such that $YX = E_{\mathfrak{A}}$. This implies

$$(\mathfrak{A}Y)X = \mathfrak{A}E_{\mathfrak{A}} = \mathfrak{A}.$$

But X is not a right magnifier of $\mathfrak{A}$. Therefore we must have $\mathfrak{A}Y = \mathfrak{A}$. But this equation implies that for some $Z \in \mathfrak{A}$ we have $ZY = E_{\mathfrak{A}}$, whence

$$X = E_{\mathfrak{A}}X = ZYX = ZE_{\mathfrak{A}} = Z.$$

This yields

$$\mathfrak{A}' = \mathfrak{A}'E_{\mathfrak{A}} = \mathfrak{A}'ZY = \mathfrak{A}'XY = \mathfrak{A}^{(n)}Y \supset \mathfrak{A}^{(n)}ZY = \mathfrak{A}^{(n)}$$

in contradiction to the fact that $\mathfrak{A}'$ is a proper subset of $\mathfrak{A}^{(n)}$.

6. Magnifying Elements in Semigroups with Unit

6.1. In this section we consider a specific semigroup containing magnifiers, which is not only our first example demonstrating the existence of semigroups with magnifiers but is also useful in studying the properties of semigroups with unit and containing magnifiers.

In studying the semigroup in question, we shall incidentally obtain illustrations of certain general semigroup concepts and properties introduced earlier.

6.2. Let us denote by $h(x)$ the following real function:

$$h(x) = \begin{cases} x & (\text{if } x \geqslant 0), \\ 0 & (\text{if } x < 0). \end{cases}$$

Also let $\mathfrak{P}$ be the ensemble of all ordered pairs of non-negative integers $[a, b]$ $(a, b = 0, 1, 2, \ldots)$. We define the following multiplication rule in $\mathfrak{P}$:

$$[a_1, b_1] \cdot [a_2, b_2] = [a_1 + h(a_2 - b_1), b_2 + h(b_1 - a_2)].$$

We shall verify the associativity of this operation:

$$\begin{aligned} ([a_1, b_1] \cdot [a_2, b_2]) \cdot [a_3, b_3] &= [a_1 + h(a_2 - b_1), b_2 + h(b_1 - a_2)] \cdot [a_3, b_3] \\ &= [a_1 + h(a_2 - b_1) + h\{a_3 - b_2 - h(b_1 - a_2)\}, \\ &\qquad b_3 + h\{b_2 + h(b_1 - a_2) - a_3\}], \end{aligned}$$

$$\begin{aligned} [a_1, b_1] \cdot ([a_2, b_2] \cdot [a_3, b_3]) &= [a_1, b_1]([a_2 + h(a_3 - b_2), b_3 + h(b_2 - a_3)]) \\ &= [a_1 + h\{a_2 + h(a_3 - b_2) - b_1\}, b_3 + h(b_2 - a_3) \\ &\qquad + h\{b_1 - a_2 - h(a_3 - b_2)\}]. \end{aligned}$$

To convince oneself of the equality of these triple products it is best to consider four cases in turn:

(1) $a_2 \geqslant b_1, \quad a_3 \geqslant b_2$; (2) $a_2 \geqslant b_1, \quad a_3 < b_2$; (3) $a_2 < b_1, \quad a_3 \geqslant b_2$; (4) $a_2 < b_1, \quad a_3 < b_2$.

If we note in this connection that $h\{h(x)\} = h(x)$ and $h(x_1 + x_2) = h(x_1) + h(x_2)$ for positive x_1 and x_2, the equality of the two expanded expressions is immediately obvious.

6.3. In the semigroup $\mathfrak{P}$ we write

$$[0, 0] = U^0 = V^0 = E, \qquad [1, 0] = U, \qquad [0, 1] = V.$$

As an immediate consequence of the multiplication rule in $\mathfrak{P}$, the element E is a unit of the semigroup. Moreover, for arbitrary elements of $\mathfrak{P}$ we have

$$[a, b] = [a, 0] \cdot [0, b] = U^a V^b.$$

The elements U and V are therefore a generating set for $\mathfrak{P}$, and every element of $\mathfrak{P}$ may be expressed uniquely in the form U^aV^b (if $a = 0$, we may write simply V^b, and if $b = 0$ we write U^a). Thus, the expressions U^aV^b may be taken as the canonical forms of the elements with respect to the generating set $\{U, V\}$.

Since

$$VU = [0, 1] \cdot [1, 0] = [0 + h(1 - 1), 0 + h(1 - 1)] = [0, 0] = E,$$

the multiplication of elements expressed in canonical form is very easily carried out:

$$U^{a_1}V^{b_1} \cdot U^{a_2}V^{b_2} = \begin{cases} U^{a_1+a_2-b_1}V^{b_2} & (\text{if } a_2 \geqslant b_1), \\ U^{a_1}V^{b_2+b_1-a_2} & (\text{if } b_1 \geqslant a_2). \end{cases}$$

From the above it follows that the semigroup $\mathfrak{P}$ may also be defined as the set of all possible expressions

$$U^aV^b \qquad (a, b = 0, 1, 2, \ldots),$$

with the multiplication rule just stated.

6.4. We shall now consider the automorphisms (I, 1.16) of the semigroup $\mathfrak{P}$. A knowledge of its automorphisms will be useful when we come to applications of this semigroup. We first prove an auxiliary lemma.

LEMMA. *To each element X of the semigroup $\mathfrak{P}$ which is distinct from U and from V there corresponds an element X' distinct from E and from any power of X, which commutes with X.*

If $X = U$ or $X = V$ there exists no element X' with these properties.

PROOF. If $X = U^aV^b$, where $a > 0$, $b > 0$, $a + b > 2$, we may take as X' the element $X' = UV$, which is distinct from E.

In fact,

$$XX' = U^aV^b \cdot UV = U^aV^b,$$

$$X'X = UV \cdot U^aV^b = U^aV^b.$$

Moreover, if we assume that $X' = X^n$ we obtain

$$UV = U^aV^bU^aV^b \ldots U^aV^b.$$

But when we multiply the element U^a on the right by any element of $\mathfrak{P}$ we always obtain an element U^cV^d in which $c \geqslant a$. The analogous result holds for V^b. The above equation can therefore hold only for $a = 1$ and $b = 1$, which contradicts the assumption that $a + b > 2$.

If $X = UV$, we may take as our X' the element $X' = U^2V^2 \neq E$. We see at once that

$$U^2V^2 \cdot UV = UV \cdot U^2V^2.$$

Since $(UV)^2 = UV$, U^2V^2 is not a power of UV.

If $X = U^a$ $(a > 1)$, we take as X' the element $X' = U \neq E$. In fact,

$$UU^a = U^aU.$$

U is not a power of U^a, since $(U^a)^n = U^{an}$, which is distinct from U.

If $X = V^b$ $(b > 1)$, we take as X' the element $X' = V$.

Now suppose that some element $U' = U^aV^b \neq E$ has the postulated properties with respect to the element U. Since U' cannot be a power of U, we have $b \neq 0$. But then

$$UU' = U \cdot U^aV^b = U^{a+1}V^b,$$

$$U'U = U^aV^b \cdot U = U^aV^{b-1}$$

and U' does not commute with U.

In a similar way we could prove that there exists no element V' having the postulated properties with respect to the element V.

6.5. We now employ Lemma 6.4 to show that the semigroup $\mathfrak{P}$ admits no automorphism other than the identity transformation.

Let us suppose that φ is some nonidentical automorphism of the semigroup $\mathfrak{P}$. If the element X possesses an X' distinct from unity, commutative with X, and not a power of X, then evidently the element $\varphi(X')$ will have these properties with respect to $\varphi(X)$. It follows from Lemma 6.4 that no element distinct from U and V can be mapped by φ into either U or V. There are then only two possible cases: either

$$\varphi(U) = U, \qquad \varphi(V) = V,$$

or

$$\varphi(U) = V, \qquad \varphi(V) = U.$$

In the first case we obtain for an arbitrary element

$$\varphi(U^aV^b) = [\varphi(U)]^a[\varphi(V)]^b = U^aV^b,$$

which contradicts the assumption that φ is a nonidentical automorphism.

In the second case we obtain the contradiction:

$$\varphi(U^2V) = [\varphi(U)]^2 \cdot \varphi(V) = V^2U = V = \varphi(U), \qquad U^2V \neq U.$$

6.6. Let $\mathfrak{Q}$ be a semigroup isomorphic to the semigroup $\mathfrak{P}$. There exists only one isomorphism of $\mathfrak{P}$ to $\mathfrak{Q}$. In fact, if the mappings φ and ψ are isomorphisms of $\mathfrak{P}$ into $\mathfrak{Q}$, then $\varphi^{-1}\psi$ is an automorphism of the semigroup $\mathfrak{P}$. By 6.5, $\varphi^{-1}\psi$ can only be the identity transformation, and therefore the mappings φ and ψ coincide.

Suppose φ is an isomorphism from $\mathfrak{P}$ to $\mathfrak{Q}$. In view of the uniqueness of the isomorphism, and of the fact that U and V are fully defined in $\mathfrak{P}$ by their properties, without regard to the explicit mode of representation of $\mathfrak{P}$, we may conclude that in the semigroup $\mathfrak{Q} = \varphi(\mathfrak{P})$ the elements $\varphi(U)$ and $\varphi(V)$ are completely and uniquely defined by the semigroup $\mathfrak{Q}$ independently of its mode of representation. This property of the semigroup $\mathfrak{P}$ should be kept in mind during both the analysis and the applications of the fundamental Theorem 6.8, which we shall take up shortly.

6.7. We now consider the magnifying elements of the semigroup $\mathfrak{P}$. In the semigroup $\mathfrak{P}$ all elements of the form U^a $(a > 0)$ are right magnifiers. In fact,

$$(\mathfrak{P}V^a)U^a = \mathfrak{P}V^aU^a = \mathfrak{P}E = \mathfrak{P}.$$

Here the set $\mathfrak{P}V^a$, consisting of all elements of $\mathfrak{P}$ of the form U^cV^{d+a} $(d \geqslant 0)$, is obviously distinct from $\mathfrak{P}$, since it fails, for example, to contain the element U.

In a similar way the elements of the form V^b $(b > 0)$ turn out to be left magnifiers, since

$$V^b(U^b\mathfrak{P}) = \mathfrak{P}, \qquad \mathfrak{P} \neq U^b\mathfrak{P} \bar{\ni} V.$$

No other element of the semigroup $\mathfrak{P}$ is a magnifier. This is obvious as regards the unit E. For $U^aV^b\,(a > 0, b > 0)$ we have

$$U^aV^b\mathfrak{P} \bar{\ni} V, \qquad \mathfrak{P}U^aV^b \bar{\ni} U.$$

It follows that no proper subset of $\mathfrak{P}$ can satisfy the equation of 5.1 which characterizes an element as a magnifier.

6.8. We now cast some light on the special role of the semigroup $\mathfrak{P}$ in the theory of magnifying elements. It turns out that magnifying elements in a semigroup with unit are contained in subsemigroups isomorphic to $\mathfrak{P}$ and containing the unit of the semigroup, and that furthermore in the isomorphisms of these subsemigroups on $\mathfrak{P}$ the magnifying elements are mapped into U and V. Only magnifying elements have this property.

THEOREM. *Let $\mathfrak{A}$ be a semigroup with unit.*

An element X of the semigroup $\mathfrak{A}$ is a right magnifier if and only if there exists an isomorphism φ of the semigroup $\mathfrak{P}$ into $\mathfrak{A}$ for which $\varphi(E_{\mathfrak{P}}) = E_{\mathfrak{A}}$ and $\varphi(U) = X$.

The element X of the semigroup $\mathfrak{A}$ is a left magnifier if and only if there exists an isomorphism φ of the semigroup $\mathfrak{P}$ into $\mathfrak{A}$ such that $(E_{\mathfrak{P}}) = E_{\mathfrak{A}}$ and $\varphi(V) = X$.

PROOF. (1) Let X be a right magnifying element of the semigroup $\mathfrak{A}$. Then for some $\mathfrak{A}' \subset \mathfrak{A}$

$$\mathfrak{A}'X = \mathfrak{A}, \qquad \mathfrak{A}' \neq \mathfrak{A}.$$

It follows that $\mathfrak{A}'$ contains an element Y such that

$$YX = E_{\mathfrak{A}}.$$

We consider the semigroup

$$\mathfrak{Q} = [X, Y].$$

Since YX is the unit of $\mathfrak{A}$, every element of $\mathfrak{A}$ may be represented in the form

$$X^aY^b$$

(where X^0 and Y^0 denote $E_{\mathfrak{A}}$).

We shall now see whether two products with different exponents may represent the same element. Suppose that

$$X^{a_1}Y^{b_1} = X^{a_2}Y^{b_2},$$

where $a_1 \geqslant a_2$. Multiplying our equation on the left by Y^{a_2} we obtain

$$X^{a_1-a_2}Y^{b_1} = Y^{b_2}.$$

If we were to assume that $a_1 = a_2$ and $b_1 \neq b_2$, then, multiplying the equation on the right by X^c, where c is the larger of b_1 and b_2, we would find that some power of the right magnifier X was equal to the unit element $E_{\mathfrak{A}}$, which is not a magnifier. By 5.4 this is impossible.

Next we set $a_1 > a_2$. Then we obtain from the equation derived above:

$$X\mathfrak{A} \supset X \cdot X^{a_1-a_2-1}Y^{b_1}\mathfrak{A} = Y^{b_2}\mathfrak{A} \supset Y^{b_2}X^{b_2}\mathfrak{A} = \mathfrak{A}$$

and therefore, for some $Z \in \mathfrak{A}$

$$XZ = E.$$

But then

$$\mathfrak{A}'X = \mathfrak{A}$$

yields, after right multiplication by Z,

$$\mathfrak{A}' = \mathfrak{A}Z,$$

whence

$$\mathfrak{A}' = \mathfrak{A}Z \supset \mathfrak{A}XZ = \mathfrak{A},$$

which disagrees with the choice of $\mathfrak{A}'$.

We have shown that the semigroup $\mathfrak{Q}$ consists of all products of the form X^aY^b, which are all distinct. Since $YX = E_{\mathfrak{A}} = E_{\mathfrak{Q}}$, the multiplication of these products is carried out precisely according to the rule for multiplying elements of $\mathfrak{P}$ when these are expressed in canonical form (6.3). The mapping φ of the semigroup $\mathfrak{P}$ on $\mathfrak{Q}$

$$\varphi(U^aV^b) = X^aY^b$$

is therefore an isomorphism of $\mathfrak{P}$ into $\mathfrak{A}$. We have

$$\varphi(U) = X, \qquad \varphi E_{\mathfrak{P}} = E_{\mathfrak{A}}.$$

(2) Now suppose X is a left magnifying element of the semigroup. By a line of reasoning wholly analogous to the preceding argument, we may show that $\mathfrak{A}$ contains a subsemigroup $\mathfrak{Q}'$ which contains $E_{\mathfrak{A}}$ and is isomorphic to $\mathfrak{P}$ in such a way that the mapping φ of the semigroup $\mathfrak{P}$ on $\mathfrak{Q}'$ satisfies $\varphi(V) = X$, $\varphi(E_{\mathfrak{P}}) = E_{\mathfrak{A}}$.

(3) Let φ be an isomorphism of $\mathfrak{P}$ into $\mathfrak{A}$ for which $\varphi(E_{\mathfrak{P}}) = E_{\mathfrak{A}}$.

We have

$$\varphi(U) \cdot \varphi(V) = \varphi(UV) \neq \varphi(E_{\mathfrak{P}}) = E_{\mathfrak{A}},$$

$$\varphi(V) \cdot \varphi(U) = \varphi(VU) = \varphi(E_{\mathfrak{P}}) = E_{\mathfrak{A}}.$$

The element $\varphi(U) \in \mathfrak{A}$ has no right inverse in $\mathfrak{A}$ with respect to $E_{\mathfrak{A}}$. Otherwise, since it does have a left inverse (which is $\varphi(V)$), it would have an inverse element (II, 2.14), which would necessarily coincide with the left inverse $\varphi(V)$ (II, 2.14). But this cannot be, since the product $\varphi(U) \cdot \varphi(V)$ has been shown to be different from $E_{\mathfrak{A}}$. We may convince ourselves in a similar way that $\varphi(V)$ has no left inverse with respect to $E_{\mathfrak{A}}$. It follows that

$$[\mathfrak{A} \cdot \varphi(V)] \cdot \varphi(U) = \mathfrak{A}E_{\mathfrak{A}} = \mathfrak{A},$$

$$\varphi(V) \cdot [\varphi(U) \cdot \mathfrak{A}] = E_{\mathfrak{A}}\mathfrak{A} = \mathfrak{A},$$

while

$$\mathfrak{A} \cdot \varphi(V) \bar{\ni} E_{\mathfrak{A}}, \qquad \varphi(U) \cdot \mathfrak{A} \bar{\ni} E_{\mathfrak{A}},$$

that is,

$$\mathfrak{A} \cdot \varphi(V) \neq \mathfrak{A}, \qquad \varphi(U) \cdot \mathfrak{A} \neq \mathfrak{A}.$$

This proves that in $\mathfrak{A}$ the element $\varphi(V)$ is a left magnifier, and $\varphi(U)$ is a right magnifier.

6.9. COROLLARY. *In a semigroup with unit every magnifying element generates an infinite monogenic semigroup.*

6.10. COROLLARY. *In a semigroup with unit every magnifying element is regular.*

In fact, let X be a right magnifying element of the semigroup with unit $\mathfrak{A}$. Let us consider the isomorphism φ of the semigroup $\mathfrak{P}$ into $\mathfrak{A}$, corresponding to X as indicated in 6.8. We have

$$\varphi(U) \cdot \varphi(V) \cdot \varphi(U) = \varphi(UVU) = \varphi(U),$$

which proves the regularity of $X = \varphi(U)$.

A similar argument holds for left magnifiers.

6.11. COROLLARY. *A semigroup with unit $\mathfrak{A}$ contains magnifying elements if and only if $\mathfrak{A}$ contains a subsemigroup which contains $E_{\mathfrak{A}}$ and is isomorphic to $\mathfrak{P}$.*

6.12. COROLLARY. *If a semigroup with unit contains right magnifiers, it also contains left magnifiers, and conversely.*

In fact, by the preceding corollary, the semigroup $\mathfrak{A}$ must contain a subsemigroup $\mathfrak{P}'$ isomorphic to $\mathfrak{P}$ and containing the unit $E_{\mathfrak{A}}$. But then, by Theorem 6.8, $\mathfrak{A}$ must contain both right and left magnifiers.

6.13. We must remark that the last of these properties does not in general hold in semigroups without unit. Let us consider an example of this sort.

Let $\mathfrak{U}$ be a semigroup with the zero O, such that the product of any two elements is equal to O. Suppose $\mathfrak{U}$ has not less than two elements. Let $\mathfrak{B}$ be an arbitrary semigroup. In the set

$$\mathfrak{A} = \mathfrak{U} \cup \mathfrak{B}$$

we define an operation. If two elements of $\mathfrak{A}$ both belong to $\mathfrak{U}$, or both belong to $\mathfrak{B}$, then their product is defined by the multiplication rule of $\mathfrak{U}$ or of $\mathfrak{B}$. If $X \in \mathfrak{U}$ and $Y \in \mathfrak{B}$, we write

$$XY = O, \qquad YX = X.$$

It is easy to verify the associativity of this operation. Let

$$Z_1 = (A_1A_2)A_3, \qquad Z_2 = A_1(A_2A_3).$$

If A_1 or A_2 belongs to $\mathfrak{U}$, then $Z_1 = O$ and $Z_2 = O$. If $A_1, A_2 \in \mathfrak{B}$ and $A_3 \in \mathfrak{U}$, then $Z_1 = A_3$ and $Z_2 = A_3$. If, on the other hand, $A_1, A_2, A_3 \in \mathfrak{B}$, then $Z_1 = Z_2$ because the multiplication in $\mathfrak{B}$ is associative.

We now determine the magnifying elements in the semigroup $\mathfrak{A}$. Suppose that

$$X \in \mathfrak{U}, \qquad Y \in \mathfrak{B}, \qquad \mathfrak{A}' \subset \mathfrak{A}, \qquad \mathfrak{A}' = \mathfrak{U}' \cup \mathfrak{B}', \qquad \mathfrak{U}' \subset \mathfrak{U}, \qquad \mathfrak{B}' \subset \mathfrak{B}.$$

(Here $\mathfrak{U}'$ and $\mathfrak{B}'$ may be empty.)

We consider the products:

$$X\mathfrak{A}' = X\mathfrak{U}' \cup X\mathfrak{B}' \subset O \cup O = O,$$

$$\mathfrak{A}'X = \mathfrak{U}'X \cup \mathfrak{B}'X \subset O \cup X,$$

$$Y\mathfrak{A}' = Y\mathfrak{U}' \cup Y\mathfrak{B}' \subset \mathfrak{U}' \cup Y\mathfrak{B}',$$

$$\mathfrak{A}'Y = \mathfrak{U}'Y \cup \mathfrak{B}'Y \subset O \cup \mathfrak{B}'Y.$$

It is obvious at once that no element of $\mathfrak{U}$ can be either a right or a left magnifier of $\mathfrak{A}$. As for an element $Y \in \mathfrak{B}$, it cannot be a right magnifier: Y may be a left magnifier. It will be a left magnifier when in the third case we choose $\mathfrak{U}' = \mathfrak{U}$ and $Y\mathfrak{B}' = \mathfrak{B}$, where $\mathfrak{B}' \neq \mathfrak{B}$. Thus, the left magnifiers of the semigroup $\mathfrak{A}$ are all left magnifiers of the semigroup $\mathfrak{B}$. If for $\mathfrak{B}$ we choose some semigroup having left multipliers (for instance $\mathfrak{P}$) the semigroup $\mathfrak{A}$ will have left magnifiers but no right magnifiers.

7. The Subsemigroup Characteristic of a Semigroup

7.1. In this section we shall denote the ensemble of all subsemigroups of an arbitrary semigroup $\mathfrak{A}$ by $\Sigma(\mathfrak{A})$. We consider a multiplication in $\Sigma(\mathfrak{A})$, putting, for $\mathfrak{B}, \mathfrak{B}' \in \Sigma(\mathfrak{A})$,

$$\mathfrak{B} \circ \mathfrak{B}' = [\mathfrak{B} \cup \mathfrak{B}'].$$

It is necessary to use the sign $\circ$ for this operation, as distinct from the usual multiplication sign, since in general the above defined operation is not the same as the multiplication of subsets of a semigroup and, in particular, of subsemigroups.

This distinction may occur even in commutative semigroups. Suppose, for example, that $\mathfrak{A}'$ is an arbitrary nonempty set, that the element O does not belong to $\mathfrak{A}'$, and that $\mathfrak{A} = \mathfrak{A}' \cup O$. We define an operation in $\mathfrak{A}$, writing $XY = O$ for arbitrary $X, Y \in \mathfrak{A}$. Clearly, the subsemigroups of $\mathfrak{A}$ consist of all subsets of $\mathfrak{A}$ containing O. For two arbitrary $\mathfrak{B}, \mathfrak{B}' \in \Sigma(\mathfrak{A})$ we have:

$$\mathfrak{B} \cdot \mathfrak{B}' = O, \qquad \mathfrak{B} \circ \mathfrak{B}' = [\mathfrak{B} \cup \mathfrak{B}'] = \mathfrak{B} \cup \mathfrak{B}'.$$

7.2. It is easily seen that the operation $\circ$ in $\Sigma(\mathfrak{A})$ is associative and commutative, and that all elements of $\Sigma(\mathfrak{A})$ are idempotents with respect to it. Thus, $\Sigma(\mathfrak{A})$ is a commutative semigroup of idempotents. We may call it the semigroup of subsemigroups of the semigroup $\mathfrak{A}$. We shall refer to $\Sigma(\mathfrak{A})$ as the *subsemigroup characteristic of the semigroup* $\mathfrak{A}$.

In what follows one should keep carefully in mind that any semigroup $\mathfrak{B}$ belonging to $\Sigma(\mathfrak{A})$ is sometimes thought of as a subset of the semigroup $\mathfrak{A}$ and sometimes as an element of the semigroup $\Sigma(\mathfrak{A})$.

7.3. Since $\Sigma(\mathfrak{A})$ is a commutative semigroup of idempotents, there exists a semilattice $\overline{\Sigma}$ associated with it as in II, 4.3, II, 4.4. This semilattice consists of all elements of the semigroup $\Sigma(\mathfrak{A})$, that is, of all subsemigroups of the semigroup $\mathfrak{A}$. In this semilattice $\mathfrak{B}$ precedes $\mathfrak{B}'$ ($\mathfrak{B} \leqslant \mathfrak{B}'$) if

$$\mathfrak{B} \circ \mathfrak{B}' = \mathfrak{B} \qquad (\mathfrak{B}, \mathfrak{B}' \in \Sigma(\mathfrak{A})),$$

which means that $\mathfrak{B} \supset \mathfrak{B}'$.

It follows from II, 4.4 and II, 4.5 that a determination of either one of the associated pair—the semigroup $\Sigma(\mathfrak{A})$ and the semilattice $\overline{\Sigma}$—fully determines the other. This fact is connected with a phenomenon already noted (II, 4.6), namely that in $\Sigma(\mathfrak{A})$, as in every commutative semigroup of idempotents, the multiplication law is completely defined when it is known which elements are units of which others.

In studying the set $\Sigma(\mathfrak{A})$ one may indifferently take as fundamental either the operation defined above or the inclusion relation.

We shall use both, but begin with the multiplication. At the outset, then, $\Sigma(\mathfrak{A})$ is to be thought of as a commutative semigroup of idempotents.

We view the relation $\mathfrak{B} \subset \mathfrak{B}'(\mathfrak{B}, \mathfrak{B}' \in \Sigma(\mathfrak{A}))$ as the condition that $\mathfrak{B}$ be a unit of $\mathfrak{B}'$ in $\Sigma(\mathfrak{A})$.

7.4. The semilattice $\overline{\Sigma}$ associated with the subsemigroup character $\Sigma(\mathfrak{A})$ of the semigroup $\mathfrak{A}$ is not in general a lattice, since the intersection of two subsemigroups of $\mathfrak{A}$ may be empty. If, however, we count the empty set as a subsemigroup of $\mathfrak{A}$, the extended set $\Sigma'(\mathfrak{A})$ will be a lattice (and moreover, complete) with respect to the inclusion relation. We shall of course make use of results from lattice theory whenever it suits our purpose to do so.

7.5. In the set $\Sigma'(\mathfrak{A})$ (7.4) we may also consider the intersection operation

$$\mathfrak{B} * \mathfrak{B}' = \mathfrak{B} \cap \mathfrak{B}' \qquad (\mathfrak{B}, \mathfrak{B}' \in \Sigma'(\mathfrak{A})).$$

With respect to this operation $\Sigma'(\mathfrak{A})$ will be a commutative semigroup of idempotents. This semigroup is associated with the semilattice dual to the semilattice of all subsemigroups considered in the preceding paragraph (including the empty subset as a subsemigroup). Obviously, arguments based on the semigroup $\Sigma'(\mathfrak{A})$ are altogether symmetric to arguments based on $\Sigma(\mathfrak{A})$ and lead to the same results.

7.6. It is often possible and useful to reduce a question concerning some semigroup to a question concerning a commutative semigroup of idempotents, by going over from the semigroup itself to its subsemigroup characteristic. Corresponding to a property $\mathfrak{l}$ possessed by some semigroups and not by others, one looks for a property $\mathfrak{l}'$ of commutative semigroups of idempotents such that the semigroup $\mathfrak{A}$ possesses the property $\mathfrak{l}$ if and only if its subsemigroup character $\Sigma(\mathfrak{A})$ possesses the property $\mathfrak{l}'$.

The study of properties of semigroups is obviously equivalent to the study of classes of semigroups, since every property is characterized by the class of semigroups possessing the property, and to every class there corresponds a property, namely class membership. In connection with this idea, the reduction of questions concerning semigroup properties to questions concerning the corresponding subsemigroup characteristics may be built up around the following concept.

One may say that a certain class of semigroups Γ *is defined by a subsemigroup characteristic* (defined, of course, to within an isomorphism) if the isomorphism of the semigroups $\Sigma(\mathfrak{A}_1)$ and $\Sigma(\mathfrak{A}_2)$, where $\mathfrak{A}_1 \in \Gamma$, always implies that $\mathfrak{A}_2$ is isomorphic to some semigroup belonging to Γ. In particular, if Γ consists of a single semigroup $\mathfrak{A}$, we say that if the preceding condition is satisfied the semigroup $\mathfrak{A}$ *is defined by a semigroup characteristic.*

An isomorphism of the semigroups $\Sigma(\mathfrak{A}_1)$ and $\Sigma(\mathfrak{A}_2)$ (or, more properly, an isomorphism of the lattices $\Sigma'(\mathfrak{A}_1)$ and $\Sigma'(\mathfrak{A}_2)$ obtained by adjoining the empty set as an element—7.4) is sometimes called a lattice isomorphism of the semigroups $\mathfrak{A}_1$ and $\mathfrak{A}_2$. Thus the question of whether a semigroup $\mathfrak{A}$ is defined by a semigroup characteristic may be formulated as the question whether the existence of a lattice isomorphism with $\Sigma(\mathfrak{A})$ always implies the existence of an isomorphism with the semigroup $\mathfrak{A}$ itself.

Subsemigroup characteristics belong to the class of commutative semigroups of idempotents, which is quite simple and easily pictured. Therefore, the reduction of questions concerning properties of semigroups to questions concerning properties of their subsemigroup characteristics may always be looked at in principle as an important step along the road to a study of general semigroups.

7.7. Of course, not all classes of semigroups, and not every individual semigroup, can be defined by a subsemigroup characteristic. We consider, for example, semigroups consisting of two elements. The semigroup $\Sigma(\mathfrak{A})$ for all such semigroups $\mathfrak{A}$ consists of either two or three elements, one of which is the zero of $\Sigma(\mathfrak{A})$ (the semigroup $\mathfrak{A}$ itself).

There is only one commutative semigroup consisting of two idempotents. There are, however, two nonisomorphic semigroups each consisting of two elements and having a subsemigroup characteristic consisting of two elements. The first is the cyclic group of two elements (that is, the monogenic semigroup of type (1,2)), and the second is the monogenic semigroup of type (2,1)—cf. 3.7. These two semigroups are not isomorphic, although their subsemigroup characteristics are isomorphic to each other.

For the semigroup $\mathfrak{A} = \{A_1, A_2\}$, consisting of two elements, the semigroup $\Sigma(\mathfrak{A})$ consists of three elements if and only if the elements A_1 and A_2 are idempotents. In this case $\Sigma(\mathfrak{A}) = \{\mathfrak{A}, A_1, A_2\}$, where $\mathfrak{A}$ is the zero of $\Sigma(\mathfrak{A})$ and $A_1 \circ A_2 = \{A_1, A_2\} = \mathfrak{A}$. It is easy to see that there are three nonisomorphic semigroups each consisting of two idempotents $\mathfrak{A} = \{A_1, A_2\}$. They are defined by the following multiplication rules:

(1) $A_1A_2 = A_2A_1 = A_1^2 = A_1, \qquad A_2^2 = A_2;$

(2) $A_iA_j = A_i \qquad (i, j = 1, 2);$

(3) $A_iA_j = A_j \qquad (i, j = 1, 2).$

Finally then, we observe that the class of semigroups each consisting of two elements is not defined by a semigroup characteristic, since the cyclic group consisting of p elements, where p is an arbitrary prime number, has a subsemigroup characteristic consisting of two elements only and therefore isomorphic to the subsemigroup character of some semigroup consisting of two elements.

7.8. LEMMA. *In the subsemigroup character $\Sigma(\mathfrak{A})$ of the semigroup $\mathfrak{A}$ the element $\mathfrak{B} \in \Sigma(\mathfrak{A})$ has no unit other than itself, if and only if $\mathfrak{B}$ consists of a single idempotent of the semigroup $\mathfrak{A}$.*

PROOF. For $\mathfrak{B}, \mathfrak{B}' \in \Sigma(\mathfrak{A})$, the relation

$$\mathfrak{B} \circ \mathfrak{B}' = [\mathfrak{B} \cup \mathfrak{B}'] = \mathfrak{B}$$

holds if and only if $\mathfrak{B}' \subset \mathfrak{B}$. If $\mathfrak{B}$ consists of a single element, this set inclusion cannot hold for any $\mathfrak{B}' \neq \mathfrak{B}$. Suppose on the other hand that $\mathfrak{B}$ contains more than one element. For $X \in \mathfrak{B}$ we consider $[X] \subset \mathfrak{B}$. If $[X]$ is an infinite monogenic semigroup, $[X^2] \neq [X]$ and therefore $\mathfrak{B}' = [X^2]$ is a unit of $\mathfrak{B}$ and distinct from $\mathfrak{B}$. If $[X]$ is finite, it contains an idempotent I (3.11) and then $\mathfrak{B}' = I$ is a unit of $\mathfrak{B}$ and distinct from $\mathfrak{B}$.

7.9. We list a number of classes of semigroups definable by subsemigroup characteristics.

THEOREM. *Suppose that the semigroups $\mathfrak{A}_1$ and $\mathfrak{A}_2$ have subsemigroup characteristics that are isomorphic to each other.*

If $\mathfrak{A}_1$ *is a unitary semigroup, then* $\mathfrak{A}_2$ *is a unitary semigroup.*

If $\mathfrak{A}_1$ *is finite, then* $\mathfrak{A}_2$ *is finite.*

If $\mathfrak{A}_1$ *is infinite, then* $\mathfrak{A}_2$ *is infinite.*

If $\mathfrak{A}_1$ *is an infinite monogenic semigroup, then* $\mathfrak{A}_2$ *is also an infinite monogenic semigroup.*

If $\mathfrak{A}_1$ *is a periodic semigroup* (4.1), *then* $\mathfrak{A}_2$ *is a periodic semigroup.*

If all elements of $\mathfrak{A}_1$ *are of infinite type, then all elements of* $\mathfrak{A}_2$ *are of infinite type* (3.7).

If $\mathfrak{A}_1$ *has both elements of infinite type and elements of finite type, then so does* $\mathfrak{A}_2$.

Proof. Let φ be an isomorphism of $\Sigma(\mathfrak{A}_1)$ to $\Sigma(\mathfrak{A}_2)$.

If $\mathfrak{A}_1$ is a unitary semigroup, the unitary property of $\mathfrak{A}_2$ follows immediately from 7.8.

Let us assume that $\mathfrak{A}_1$ contains an element X of finite type. Then $\mathfrak{A}_1$ has an idempotent I (3.11). According to 7.8, it follows that in $\Sigma(\mathfrak{A}_1)$ I is an element having no unit other than itself. But then, in $\Sigma(\mathfrak{A}_2)$, $\varphi(I)$ will be an element having the same property and also in $\mathfrak{A}_2$ it follows that $\varphi(I)$ is a semigroup consisting of a single idempotent (7.8) and is of finite type.

Let us assume that $\mathfrak{A}_1$ contains an element Y of infinite type. In $\Sigma(\mathfrak{A}_1)$ the element $[Y]$ is such that every unit of it has in its turn a unit distinct from itself. It follows that $\varphi([Y])$ enjoys the same property in $\Sigma(\mathfrak{A}_2)$. But this means that in the subsemigroup $\varphi([Y])$ of the semigroup $\mathfrak{A}_2$ there are no idempotents, and therefore the semigroup $[Y']$ is infinite for every choice of $Y' \in \varphi([Y])$ $(Y' \in \mathfrak{A}_2)$.

Similarly the existence in $\mathfrak{A}_2$ of an element of finite type implies the existence of such an element in $\mathfrak{A}_1$ and the existence in $\mathfrak{A}_2$ of an element of infinite type implies the existence of such an element in $\mathfrak{A}_1$.

The last three assertions of the theorem follow immediately.

Suppose now that $\mathfrak{A}_1$ is finite. Then $\mathfrak{A}_1$ is periodic and therefore so is $\mathfrak{A}_2$. $\Sigma(\mathfrak{A}_1)$ and, therefore, $\Sigma(\mathfrak{A}_2)$ are finite. Thus $\mathfrak{A}_2$ has only a finite number of distinct monogenic subsemigroups, each of which has only a finite number of elements. Since each element of $\mathfrak{A}_2$ is contained in one of these monogenic subsemigroups, $\mathfrak{A}_2$ has only a finite number of elements.

If $\mathfrak{A}_1$ is infinite, then $\mathfrak{A}_2$ cannot be finite since the finiteness of $\mathfrak{A}_2$, via the isomorphism φ^{-1}, implies the finiteness of $\mathfrak{A}_1$, by the above argument.

Suppose $\mathfrak{A}_1$ is an infinite monogenic subsemigroup: $\mathfrak{A}_1 = [A_1]$. The subsemigroup $\mathfrak{B}_1 = \{A_1^2, A_1^3, A_1^4, \ldots\}$ contains all subsemigroups of $\mathfrak{A}_1$ except $\mathfrak{A}_1$ itself. Accordingly, $\mathfrak{B}_1$ is a zero for all elements of $\Sigma(\mathfrak{A}_1)$ except for $\mathfrak{A}_1$. It follows that $\mathfrak{B}_2 = \varphi(\mathfrak{B}_1) \in \Sigma(\mathfrak{A}_2)$ is also a zero of all elements of $\Sigma(\mathfrak{A}_2)$ except for $\mathfrak{A}_2$. Since $\mathfrak{B}_1 \neq \mathfrak{A}_1$, $\mathfrak{B}_2 \neq \mathfrak{A}_2$. We choose some element A_2 from $\mathfrak{A}_2 \backslash \mathfrak{B}_2$. Since $[A_2] \not\subseteq \mathfrak{B}_2$, $\mathfrak{B}_2$ is not a zero of $[A_2]$. Therefore $[A_2] = \mathfrak{A}_2$. The monogenic semigroup $\mathfrak{A}_2 = [A_2]$ cannot be finite, since $\mathfrak{A}_1$ is infinite, and $\Sigma(\mathfrak{A}_1)$ and $\Sigma(\mathfrak{A}_2)$ are isomorphic.

7.10. Suppose that for the semigroups $\mathfrak{A}_1$ and $\mathfrak{A}_2$ we have a given isomorphism φ of $\Sigma(\mathfrak{A}_1)$ on $\Sigma(\mathfrak{A}_2)$. We note a number of properties of this isomorphism.

(α) *For every $\mathfrak{B}_1 \in \Sigma(\mathfrak{A}_1)$ there exists an isomorphism of $\Sigma(\mathfrak{B}_1)$ on $\Sigma(\mathfrak{B}_2)$.*

In fact, $\Sigma(\mathfrak{B}_1)$ consists of all subsemigroups of the semigroup $\mathfrak{A}_1$ which are units of $\mathfrak{B}_1$ in $\Sigma(\mathfrak{A}_1)$ and $\Sigma[\varphi(\mathfrak{B}_1)]$ consists of all subsemigroups of $\mathfrak{A}_2$ which are units of $\varphi(\mathfrak{B}_1)$ in $\Sigma(\mathfrak{A}_2)$. The isomorphism φ is a one-to-one mapping of $\Sigma(\mathfrak{B}_1)$ on $\Sigma(\mathfrak{B}_2)$. This mapping is clearly an isomorphism.

(β) *If I is an idempotent of the semigroup $\mathfrak{A}_1$, then $\varphi(I)$ is an idempotent of the semigroup $\mathfrak{A}_2$.*

In fact, by (α), $\Sigma([\varphi(I)])$ is isomorphic to $\Sigma(I)$ and therefore, by 7.9, $\varphi(I)$ is a unitary semigroup.

(γ) *If $A_1 \in \mathfrak{A}_1$ is an element of infinite type, there exists in $\mathfrak{A}_2$ a unique element A_2, of infinite type, such that $\varphi([A_1]) = [A_2]$.*

In fact, by (α), $\Sigma([A_1])$ and $\Sigma(\varphi([A_1]))$ are isomorphic. It follows by 7.9 that $[A_1]$ and $\varphi([A_1])$ are isomorphic, that is, $\varphi([A_1])$ is the infinite monogenic semigroup $[A_2]$. But what we know about the structure of infinite monogenic semigroups immediately implies that $[A_2] = [A_2']$ only for $A_2 = A_2'$.

(δ) *If for $\mathfrak{B}_i, \mathfrak{B}_i^{(\alpha)}, \mathfrak{B}_i^{(\beta)}, \ldots \in \Sigma(\mathfrak{A}_i)$ we have $\mathfrak{B}_i = [\mathfrak{B}_i^{(\alpha)}, \mathfrak{B}_i^{(\beta)}, \ldots]$, $\varphi(\mathfrak{B}_1^{(\xi)}) = \mathfrak{B}_2^{(\xi)}$ $(i = 1, 2; \xi = \alpha, \beta, \ldots)$, then $\varphi(\mathfrak{B}_1) = \mathfrak{B}_2$.*

In fact, let us consider the subsemigroups $\mathfrak{B}_1' \in \Sigma(\mathfrak{A}_1)$ and $\mathfrak{B}_2' \in \Sigma(\mathfrak{A}_2)$:

$$\varphi(\mathfrak{B}_1) = \mathfrak{B}_2', \qquad \varphi(\mathfrak{B}_1') = \mathfrak{B}_2.$$

Since in $\Sigma(\mathfrak{A}_1)$ every $\mathfrak{B}_1^{(\xi)}$ is a unit of $\mathfrak{B}_1$, in $\Sigma(\mathfrak{A}_2)$ $\mathfrak{B}_2^{(\xi)}$ is a unit of $\mathfrak{B}_2'$ and therefore

$$\mathfrak{B}_2' \supset \mathfrak{B}_2.$$

Since in $\Sigma(\mathfrak{A}_2)$ every $\mathfrak{B}_2^{(\xi)}$ is a unit of $\mathfrak{B}_2$, in $\Sigma(\mathfrak{A}_1)$ $\mathfrak{B}_1^{(\xi)}$ is a unit of $\mathfrak{B}_1'$ and therefore

$$\mathfrak{B}_1 \subset \mathfrak{B}_1'.$$

Since $\mathfrak{B}_1$ is a unit of $\mathfrak{B}_1'$ it follows that $\mathfrak{B}_2$ is a unit of $\mathfrak{B}_2$:

$$\mathfrak{B}_2' \subset \mathfrak{B}_2.$$

Taken together with the inclusion relation derived earlier, this implies that $\mathfrak{B}_2 = \mathfrak{B}_2'$ and therefore $\varphi(\mathfrak{B}_1) = \mathfrak{B}_2$.

(ε) *Corresponding to the elements $X_1^{(\alpha)}, X_1^{(\beta)}, \ldots \in \mathfrak{A}_1$ of infinite type there exist elements $X_2^{(\alpha)}, X_2^{(\beta)}, \ldots \in \mathfrak{A}_2$* such that

$$\varphi[X_1^{(\alpha)}] = [X_2^{(\alpha)}], \qquad \varphi[X_1^{(\beta)}] = [X_2^{(\beta)}], \ldots;$$

$$\varphi[X_1^{(\alpha)}, X_1^{(\beta)}, \ldots] = [X_2^{(\alpha)}, X_2^{(\beta)}, \ldots].$$

This follows immediately from (γ) and (δ), since

$$[X_i^{(\alpha)}, X_i^{(\beta)}, \ldots] = [[X_i^{(\alpha)}], [X_i^{(\beta)}], \ldots].$$

(ζ) *If the element* $A_1 \in \mathfrak{A}_1$ *is of infinite type and* $\varphi([A_1]) = [A_2]$, *then for arbitrary integer n*

$$\varphi[A_1^n] = [A_2^n].$$

According to (γ), for every element $X_1 \in \mathfrak{A}_1$ of infinite type and for every integer n there exist elements X_2, $Y_2 \in \mathfrak{A}_2$ such that

$$\varphi[X_1] = [X_2], \qquad \varphi[X_1^n] = [Y_2].$$

Since $[X_1] \supset [X_1^n]$, we have $[X_2] \supset [Y_2]$, that is, $Y_2 = X_2^m$.

Let us suppose that there exist an element of infinite type X_1 and an integer n with $m \neq n$; we shall assume that n is the least of all such integers (with respect to all possible elements of the semigroup $\mathfrak{A}$ of infinite type).

Since $[X_i^{(l)}]$ $(l = 1, 2, \ldots)$ consists of all possible units of the element $[X_i]$ in $\Sigma(\mathfrak{A}_i)$, which are infinite monogenic semigroups in $\mathfrak{A}$, and $\varphi([X_1]) = [X_2]$, it follows from 7.9 that φ is a one-to-one mapping of the set $\{[X_1], [X_1^2], [X_1^3], \ldots\}$ on $\{[X_2], [X_2^2], [X_2^3], \ldots\}$, and $\varphi([X_1^s]) = [X_2^s]$ for $s = 1, 2, \ldots, n-1$ and $\varphi([X_1^n]) = [X_2^m]$, $m > n$. We write

$$\mathfrak{B}_i^{(s)} = \{X_i^{(s)}, X_i^{s+1}, X_i^{s+2}, \ldots\} \in \Sigma(\mathfrak{A}_i) \; (s = 1, 2, 3, \ldots; i = 1, 2).$$

In view of (ε) we have:

$$\varphi(\mathfrak{B}_1^{(s)}) = \mathfrak{B}_2^{(s)} \qquad (s = 1, 2, \ldots, n),$$

$$\varphi(\mathfrak{B}_1^{(n+1)}) = [X_2^n, \ldots, X_2^{m-1}, X_2^{m+1}, X_2^{m+2}, \ldots] = \mathfrak{B}_2'.$$

We suppose that $n = 2$. The equation

$$\mathfrak{B}_i^{(2)} = [X_i^p] \circ [X_i^q] = [X_i^p, X_i^q] \qquad (i = 1, 2)$$

is clearly true only when p and q are the integers 2 and 3.

Since $\varphi[X_1^2] \neq [X_2^2]$, we have

$$\varphi[X_1^2] = [X_2^3], \qquad \varphi[X_1^3] = [X_2^2].$$

But $[X_1^2] \circ [X_1^5] = [X_1^2, X_1^4, X_1^5, X_1^6, \ldots]$ contains all units of $\mathfrak{B}_1^{(2)}$, which are monogenic semigroups in $\mathfrak{A}_1$, except for one: $[X_1^3]$, since $[X_2^3] \circ [X_2^l] = [X_2^3, X_2^l]$ does not contain $[X_2^2]$ and $[X_2^4]$ for arbitrary $l > 4$ and does not contain $[X_2^2]$ and $[X_2^5]$ for $l = 4$. Therefore, we cannot have $n = 2$.

Suppose that n is even and greater than two. Since $[X_1^2] \supset [X_1^n]$ and $\varphi([X_1^2]) = [X_2^2]$, we have $[X_2^2] \supset [X_2^m]$, that is, m is even. For the elements $Y_1 = X_1^2$ and $Y_2 = X_2^2$ we have

$$\varphi[Y_1^{n/2}] = [Y_2^{m/2}], \qquad \frac{n}{2} \neq \frac{m}{2},$$

which contradicts the assumption that n is the smallest number with the stated property.

It remains to assume that n is odd.

Since $[X_1^n] \in \mathfrak{B}_1^{n+1}$, we have $[X_2^m] \in \mathfrak{B}_2'$. This implies that $m < 2n$, that is, $0 < m - n < n$. In $\Sigma(\mathfrak{A}_2)$ we have

$$[X_2^{m-n}] \circ \mathfrak{B}_2' \supset [X_2^m].$$

Therefore in $\Sigma(\mathfrak{A}_1)$

$$[X_1^{m-n}] \circ \mathfrak{B}_1^{(n+1)} \supset [X_1^n].$$

But

$$[X_1^{m-n}] \circ \mathfrak{B}_1^{(n+1)} = [X_1^{m-n}, X_1^{n+1}, X_1^{n+2}, \ldots] \ni X_1^n$$

is possible only if $X_1^n = (X_1^{m-n})^k$. This means that $m - n$ is a divisor of n. Therefore $m - n$ is odd. Since n is odd, m must be even. But then, since n is a minimum and $n > 2$, we obtain the contradiction:

$$\varphi[X_1^n] = [X_2^m] = [(X_2^{m/2})^2] = \varphi[(X_1^{m/2})^2] = \varphi[X_1^m],$$

$$[X_1^n] \neq [X_1^m].$$

7.11. As follows from 7.7, the class of all groups does not correspond to the class of semigroups defined by subsemigroup characteristics. It is natural to ask, therefore, whether there are groups, and if so which groups they may be, that have subsemigroup characteristics isomorphic to the subsemigroup characteristic of some semigroup which is not a group. The converse question also arises. Both these problems have been elaborated in detail by R. V. Petropavlovskaja [**4**; **5**; **6**]. She has determined the structure of groups with this property and moreover has discovered several classes of groups that are defined by subsemigroup characteristics. We shall examine a few cases of this sort. If all nonunit elements of a group are of infinite type, the group is said to be torsion-free. Such groups are important in group theory. We shall show that they are defined by subsemigroup characteristics.

7.12. We shall show first that an infinite cyclic group is defined by its subsemigroup characteristic.

THEOREM. *If $\mathfrak{A}_1$ is an infinite cyclic group and the semigroup $\mathfrak{A}_2$ is such that $\Sigma(\mathfrak{A}_1)$ and $\Sigma(\mathfrak{A}_2)$ are isomorphic, then $\mathfrak{A}_2$ is an infinite cyclic group.*

PROOF. $\mathfrak{A}_1 = [A_1, A_1^{-1}, E_1]$, where E_1 is the unit of $\mathfrak{A}_1$. According to 7.10, $\mathfrak{A}_2$ has a single idempotent E_2 and $\varphi(E_1) = E_2$. $\mathfrak{A}_1$ admits a decomposition into three subsemigroups:

$$\mathfrak{A}_1 = [A_1] \cup [A_1^{-1}] \cup E_1.$$

The three components of this decomposition have no unit in common in $\Sigma(\mathfrak{A}_1)$, and every infinite monogenic subsemigroup in $\mathfrak{A}_1$ is a unit in $\Sigma(\mathfrak{A}_1)$ either of $[A_1]$, or of $[A_1^{-1}]$. In virtue of 7.10, (γ), this implies that $\mathfrak{A}_2$ admits the decomposition:

$$\mathfrak{A}_2 = [A_2] \cup [A_2'] \cup E_2,$$

$$\varphi[A_1] = [A_2], \qquad \varphi[A_1^{-1}] = [A_2'], \qquad \varphi(E_1) = E_2,$$

where $[A_2]$ and $[A_2']$ are infinite.

Since $[A_1] \circ [A_1^{-1}] \supset E_1$, we have $[A_2] \circ [A_2'] \supset E_2$. It follows that $A_2^{\alpha_1} A_2'^{\beta_1} A_2^{\alpha_2} A_2'^{\beta_2} \dots A_2'^{\beta_l} = E_2$, $\alpha_1 \geqslant 1$ (if $A_2^{\alpha_1}$ is the empty symbol, the argument is the same). This implies that $A_2 X = E_2$, where $X \neq E_2$. Since $[A_1] \mathrel{\bar{\supset}} E_1$, we have $[A_2] \mathrel{\bar{\supset}} E_2$ and therefore $X \mathrel{\bar{\in}} [A_2]$. Accordingly, $X = A_2'^s$. If it were true that $A_2 A_2' \neq E_2$ we would have $A_2 A_2' = A_2^p$ or $A_2 A_2' = A_2'^q$ and from $A_2 A_2'^s = E_2$ we would have $A_2^m = E_2$ or $A_2'^m = E_2$. But neither is admissible. Therefore, $A_2 A_2' = E_2$.

Since

$$(A_2' A_2)^3 = A_2' \cdot A_2 A_2' \cdot A_2 A_2' \cdot A_2 = A_2' \cdot A_2 A_2' \cdot A_2 = (A_2' A_2)^2,$$

we have $A_2' A_2 = E_2$.

The element $A_2 E_2$ cannot belong to $[A_2']$ since otherwise $A_2 E_2 = A_2'^m$, on multiplication by A_2', would imply $E_2 = A_2'^{m+1}$. Therefore, either $A_2 E_2 = E_2$, or $A_2 E_2 = A_2^r$. Suppose $A_2 E_2 = E_2$. Then $E_2 A_2 \cdot E_2 A_2 = E_2 A_2$, that is, $E_2 A_2 = E_2$. From $A_2 E_2 = E_2$ and $E_2 A_2 = E_2$ we obtain, on multiplication by A_2', $A_2' E_2 = E_2$, $E_2 A_2' = E_2$. It follows that $[A_2^2] \circ [A_2'] = [A_2^2, A_2']$ consists of elements of the form $A_2^{2k}, A_2'^k, E_2$ $(k = 1, 2, \dots)$ and therefore does not contain A_2, which contradicts the assumption that $[A_1^2] \circ [A_1^{-1}] \supset [A_1]$. Therefore, $A_2 E_2$ cannot be equal to E_2.

From the equation $A_2 E_2 = A_2^r$ we obtain

$$A_2^r = A_2 E_2 = A_2 E_2 E_2 = A_2^r E_2 = A_2^{r-1} \cdot A_2 E_2 = A_2^{r-1} A_2^r = A_2^{2r-1}.$$

Since $[A_2]$ is infinite, this equation can hold only if $r = 1$.

In the same way that we proved that $A_2 E_2 = A_2$ we may show that $E_2 A_2 = A_2$, $A_2' E_2 = A_2'$, $E_2 A_2' = A_2'$. These equations, together with those derived earlier—$A_2 A_2' = E_2$, $A_2' A_2 = E_2$—imply that the infinite semigroup $\mathfrak{A}_2 = [A_2, A_2', E_2]$ is an infinite cyclic group.

7.13. THEOREM. *If $\mathfrak{A}_1$ is a torsion-free group* (7.11) *and $\mathfrak{A}_2$ is a semigroup such that $\Sigma(\mathfrak{A}_1)$ and $\Sigma(\mathfrak{A}_2)$ are isomorphic, then $\mathfrak{A}_2$ is a torsion-free group.*

PROOF. Since $\mathfrak{A}_1$ has a single idempotent $E_{\mathfrak{A}_1}$, we find in accordance with 7.10, that $\mathfrak{A}_2$ has a single idempotent E_2. Except for $E_{\mathfrak{A}_1}$, every subsemigroup of $\mathfrak{A}_1$ is infinite. Therefore, by 7.9 and 7.10, (α), every subsemigroup of $\mathfrak{A}_2$ distinct from E_2 is infinite.

Suppose that φ is an isomorphism of $\Sigma(\mathfrak{A}_1)$ on $\Sigma(\mathfrak{A}_2)$; A_2 is an arbitrary element of $\mathfrak{A}_2$, distinct from E_2. According to 7.10, (γ), to each $[A_2]$ corresponds an $A_1 \in \mathfrak{A}_1$ such that $\varphi([A_1]) = [A_2]$. In $\Sigma(\mathfrak{A}_1)$ there exists an infinite cyclic group $\mathfrak{B}_1$, such that $\mathfrak{B}_1 \supset [A_1]$. In virtue of 7.10, ($\alpha$) and 7.12, $\varphi(\mathfrak{B}_1)$ is an infinite cyclic group, and $\varphi(\mathfrak{B}_1) \supset [A_2]$. Therefore, an arbitrary element A_2 of the semigroup $\mathfrak{A}_2$, distinct from E_2, has E_2 as a unit and possesses an inverse with respect to E_2. It follows that $\mathfrak{A}_2$ is a group and, from what we have said earlier, is a torsion-free group.

7.14. The furthest advances in this direction have been based on the theory of commutative semigroups. R. V. Petropavlovskaja [3] has shown that every aperiodic commutative group is defined by a subsemigroup characteristic.

7.15. With the aid of the semigroup $\Sigma(\mathfrak{A})$ of all subsemigroups of a given semigroup $\mathfrak{A}$, one may pose similar problems as to the definability of $\mathfrak{A}$ itself with respect to other semigroups of subsemigroups. For example, when studying groups it is wholly natural to consider the ensemble of subgroups $\Sigma^*(\mathfrak{A})$. This is a subsemigroup of the semigroup $\Sigma(\mathfrak{A})$. Historically, the first question along the line developed in this section concerned those groups for which the isomorphism of $\Sigma^*(\mathfrak{A}_1)$ and $\Sigma^*(\mathfrak{A}_2)$ always implies the isomorphism of $\mathfrak{A}_1$ and $\mathfrak{A}_2$. The essential results were first obtained by Baer[1] and by L. E. Sadovski.[2] This line has been even further developed.[3] It is of some interest to note that Baer's fundamental result was subsequently obtained by R. V. Petropavlovskaja [3] in connection with work on the semigroup of all subsemigroups of a group.

7.16. The following theorem plays a significant part in the derivation of these results.

THEOREM. *Suppose that for the groups $\mathfrak{G}_1$ and $\mathfrak{G}_2$ the subsemigroup characters $\Sigma(\mathfrak{G}_1)$ and $\Sigma(\mathfrak{G}_2)$ are isomorphic. Then the semigroups of the subgroups $\Sigma^*(\mathfrak{G}_1)$ and $\Sigma^*(\mathfrak{G}_2)$ are isomorphic.*

PROOF. Let φ be the isomorphism of $\Sigma(\mathfrak{G}_1)$ on $\Sigma(\mathfrak{G}_2)$ and

$$\varphi(\mathfrak{B}_1) = \mathfrak{B}_2, \qquad \mathfrak{B}_1 \in \Sigma(\mathfrak{G}_1), \qquad \mathfrak{B}_2 \in \Sigma(\mathfrak{G}_2).$$

We suppose that $\mathfrak{B}_2 \in \Sigma^*(\mathfrak{G}_2)$. We choose from $\mathfrak{B}_1$ an arbitrary element X_1. If X_1 is of finite type, then $[X_1]$ contains an element inverse to X_1. Therefore $X_1^{-1} \in \mathfrak{B}_1$. If X_1 is of infinite type, then by 7.9 and 7.10, (α), $\varphi([X_1])$ is an infinite monogenic semigroup: $\varphi([X_1]) = [X_2]$. Since $[X_1] \subset \mathfrak{B}_1$, we have $[X_2] \subset \mathfrak{B}_2$. But $\mathfrak{B}_2$ is a group, and therefore $[X_2, X_2^{-1}] \subset \mathfrak{B}_2$. But then we have for arbitrary $\mathfrak{H} \in \Sigma(\mathfrak{G}_1)$,

$$[X_1] \subset \mathfrak{H} \subset \mathfrak{B}_1, \qquad \varphi(\mathfrak{H}) = [X_2, X_2^{-1}].$$

---

[1] *The significance of the system of subgroups for the structure of groups*, Amer. J. Math. **61** (1939), 1–44.

[2] *On structural isomorphisms of free groups*, Dokl. Akad. Nauk SSSR **32** (1951), 171–174. (Russian)

Structural isomorphisms of free groups and of free products, Mat. Sb. (N.S.) **14 (56)** (1944), 155–173. (Russian)

On structural isomorphisms of free products of groups, Mat. Sb. (N.S.) 21 (**63**) (1947), 63–82. (Russian)

[3] See, for example, M. Suzuki, *Structure of a group and the structure of its lattice of subgroups*, Springer, Berlin, 1956.

B. I. Plotkin, *Generalized solvable and generalized nilpotent groups*, Uspehi Mat. Nauk **13** (1958), no. 4 (82), 89–172. (Russian)

According to 7.9 and 7.10, (α), $\mathfrak{H}$ is an infinite cyclic group. Therefore $X_1^{-1} \in \mathfrak{H} \subset \mathfrak{B}_1$. Since $\mathfrak{B}_1$ contains for every one of its elements X_1 the inverse X_1^{-1}, $\mathfrak{B}_1 \in \Sigma^* (\mathfrak{G}_1)$.

In an altogether similar way we may show that $\mathfrak{B}_1 \in \Sigma^* (\mathfrak{G}_1)$ implies $\mathfrak{B}_2 \in \Sigma^* (\mathfrak{G}_2)$.

We have found, then, that the isomorphism φ of the semigroup $\Sigma (\mathfrak{G}_1)$ on $\Sigma (\mathfrak{G}_2)$ is a one-to-one mapping of $\Sigma^* (\mathfrak{G}_1)$ on $\Sigma^* (\mathfrak{G}_2)$, and is therefore an isomorphism of $\Sigma^* (\mathfrak{G}_1)$ on $\Sigma^* (\mathfrak{G}_2)$.

7.17. These results mean that for groups every isomorphism of $\Sigma (\mathfrak{G}_1)$ to $\Sigma (\mathfrak{G}_2)$ can be obtained as an extension of some isomorphism of $\Sigma^* (\mathfrak{G}_1)$ to $\Sigma^* (\mathfrak{G}_2)$. R. V. Petropavlovskaja has shown [3] that the converse is not in general true. There are groups $\mathfrak{G}_1$ and $\mathfrak{G}_2$ for which the isomorphisms of $\Sigma^* (\mathfrak{G}_1)$ to $\Sigma^* (\mathfrak{G}_2)$ exist but cannot be extended to isomorphisms of $\Sigma (\mathfrak{G}_1)$ to $\Sigma (\mathfrak{G}_2)$. Moreover, there are groups $\mathfrak{G}_1$ and $\mathfrak{G}_2$ (and even commutative groups) such that $\Sigma^* (\mathfrak{G}_1)$ and $\Sigma^* (\mathfrak{G}_2)$ are isomorphic whereas $\Sigma (\mathfrak{G}_1)$ and $\Sigma (\mathfrak{G}_2)$ are not.

Thus, the requirement that $\Sigma (\mathfrak{G}_1)$ and $\Sigma (\mathfrak{G}_2)$ be isomorphic is stronger, in the case that $\mathfrak{G}_1$ and $\mathfrak{G}_2$ are groups, than the requirement that $\Sigma^* (\mathfrak{G}_1)$ and $\Sigma^* (\mathfrak{G}_2)$ be isomorphic.

CHAPTER IV

IDEALS

1. The Concept of Ideals and Their Simplest Properties

1.1. We have had the opportunity to convince ourselves of the important role of subsemigroups, i.e., of those subsets of semigroups that are closed under multiplication. A natural strengthening of this condition is the requirement that the subset be closed under multiplication by any element of the semigroup. Subsets of semigroups with such properties play an important role in the theory of semigroups.

DEFINITION. *A nonempty subset* $\mathfrak{T}$ *of a semigroup* $\mathfrak{A}$ *is said to be a* LEFT IDEAL *of* $\mathfrak{A}$ *if*

$$\mathfrak{A}\mathfrak{T} \subset \mathfrak{T}.$$

$\mathfrak{T}$ *is said to be a* RIGHT IDEAL *if*

$$\mathfrak{T}\mathfrak{A} \subset \mathfrak{T}.$$

$\mathfrak{T}$ *is said to be a* TWO-SIDED IDEAL *if it is simultaneously a left and a right ideal, i.e., if it is nonempty and*

$$\mathfrak{A}\mathfrak{T} \subset \mathfrak{T} \quad \text{and} \quad \mathfrak{T}\mathfrak{A} \subset \mathfrak{T}.$$

$\mathfrak{T}$ *is said to be an* IDEAL *if it is either a left or a right ideal of* $\mathfrak{A}$.

In commutative semigroups the concepts of ideal, left ideal, right ideal and two-sided ideal clearly coincide.

Since

$$\mathfrak{T}\mathfrak{T} \subset \mathfrak{A}\mathfrak{T}, \qquad \mathfrak{T}\mathfrak{T} \subset \mathfrak{T}\mathfrak{A},$$

an ideal is always a subsemigroup.

One should note that the word "ideal" is often used to signify only two-sided ideals. Left and right ideals are sometimes spoken of as one-sided ideals.

The term "ideal" appeared first in the theory of algebraic numbers where its use was motivated by well-known reasons. Later it passed into other branches of algebra where it became so strongly rooted that it would be impossible to change it in spite of its being quite unsuitable.

The properties of right, left and, in particular, two-sided ideals of a semigroup are not only interesting in themselves but are closely connected with various other properties of the semigroup. For example, their role in divisibility is immediately clear. In many cases the structure of a semigroup is determined in varying degrees by the existence and interrelations of its ideals. A significant part of the work done on semigroups utilizes the concept and properties of ideals. The present chapter and the following one will be completely concerned with the investigation of various properties of ideals. Moreover, we will often employ ideals in later discussions.

1.2. In the multiplicative semigroup of the natural numbers the set of all the even numbers is an ideal.

In the semigroup of all real functions defined on the whole real axis considered with respect to the operation of composition the set of all the constants is a two-sided ideal. The set of all the periodic functions is a left ideal. The set of all the functions differing from zero for all values of the independent variable is a right ideal.

In the multiplicative semigroup $\mathfrak{M}_n$ of all complex square matrices of order n the set of matrices for which all the elements of a given fixed column are equal to zero is a left ideal. Moreover, it is not a right ideal if $n > 1$. The set of matrices for which all the elements of a given row are equal to zero is a right ideal but is not a left ideal if $n > 1$. The set of all singular matrices is a two-sided ideal.

In an arbitrary semigroup $\mathfrak{A}$ for any nonempty subset $\mathfrak{N} \subset \mathfrak{A}$ the product $\mathfrak{A}\mathfrak{N}$ is clearly a left ideal, the product $\mathfrak{N}\mathfrak{A}$ a right ideal and the product $\mathfrak{A}\mathfrak{N}\mathfrak{A}$ a two-sided ideal. The set of all factorable elements, i.e., $\mathfrak{A}\mathfrak{A}$, is obviously a two-sided ideal of $\mathfrak{A}$. If an element X is irreducible, then the set $\mathfrak{A}\backslash X$ is a two-sided ideal of $\mathfrak{A}$.

1.3. In a finite monogenic semigroup $\mathfrak{A} = [X]$ of type (h, d) (III, 3.7) the set

$$\mathfrak{T}_k = \{X^k, X^{k+1}, \ldots, X^{h+d-1}\}$$

clearly forms an ideal of $\mathfrak{A}$ for any $k = 1, 2, \ldots, h$. There are no other ideals in $\mathfrak{A}$. In fact, let X^k be the least power of X belonging to some ideal $\mathfrak{T}$ of the semigroup $\mathfrak{A}$. Then $\mathfrak{T}$ also contains $X^{k+1}, X^{k+2}, \ldots, X^{h+d-1}$.

The number k cannot be larger than h, for it follows from $X^{h+r} \in \mathfrak{T}$ $(0 < r < d)$ that

$$X^h = X^{h+d} = X^{h+r} \cdot X^{d-r} \in \mathfrak{T} \cdot X^{d-r} \subset \mathfrak{T}.$$

Incidentally, it follows from this that the first component of the pair (h, d) may be defined as the number of ideals in the semigroup $\mathfrak{A}$.

If $\mathfrak{A} = [X]$ is an infinite monogenic semigroup, then clearly all its ideals are sets of the form

$$\{X^k, X^{k+1}, X^{k+2}, \ldots\} \qquad (k = 1, 2, 3, \ldots).$$

1.4. The important role of ideals in the theory of rings and algebras is well-known.

Clearly, any left, right or two-sided ideal of a ring will be respectively a left, right or two-sided ideal of the multiplicative semigroup of this ring. The converse is in general not valid. The ring of all the integers will serve as an example of this. Its subset consisting of zero and all the integers greater in absolute value than some arbitrary fixed natural number is obviously a two-sided ideal of the multiplicative semigroup of all the integers but is not an ideal of the ring.

For a ring with an identity it is easy to give all the cases when all the ideals of the multiplicative semigroup are also ideals of the ring. (Aubert [1] formulated this result for the commutative case.)

THEOREM. *In a ring with an identity every left ideal of the multiplicative semigroup of the ring will be a left ideal of the ring itself if and only if for any two elements one is the right divisor of the other.*

PROOF. (1) Let every left ideal of the multiplicative semigroup of the ring $\mathfrak{A}$ be a left ideal of the ring itself. For any $X, Y \in \mathfrak{A}$ the set $\mathfrak{A}X \cup \mathfrak{A}Y$ is clearly a left ideal of the semigroup. Since it must also be a left ideal it must also contain the element $X + Y$. Let $(X + Y) \in \mathfrak{A}X$. Then for some $Z \in \mathfrak{A}$ we have

$$X + Y = ZX,$$

$$Y = (Z - E_{\mathfrak{A}})X.$$

(2) Assume that for any two elements of $\mathfrak{A}$ one is always the right divisor of the other and that $\mathfrak{L}$ is an arbitrary left ideal of the multiplicative semigroup of the ring $\mathfrak{A}$. Let $X, Y \in \mathfrak{L}$ and $A \in \mathfrak{A}$. We have $AX \in \mathfrak{L}$. If furthermore $Y = ZX$, then

$$X - Y = E_{\mathfrak{A}}X - ZX = (E_{\mathfrak{A}} - Z)X \in \mathfrak{L},$$

$$(Y - X) = (-E_{\mathfrak{A}})(X - Y) \in \mathfrak{L}$$

and, consequently, $\mathfrak{L}$ is a left ideal of the ring.

1.5. Although it is not true even for multiplicative semigroups of rings that the theory of ideals of rings and the theory of ideals of semigroups coincide, there is nevertheless a definite connection between them. Since the contents of a whole sequence of properties of rings are connected only with the operation of multiplication it is natural to ask whether it is possible to carry over these properties directly or, in a generalized sense, into the theory of semigroups. Among such properties the characteristics of distinct factorizations occupy an important position. They are related to the arithmetic characteristics and are related in an essential way to ideals. As examples of investigations with the indicated tendency, see the works of V. I. Arnol'd [**1**], Asano and Murata [**1**], Aubert [**1**], Weaver [**2**], Dubreil-Jacotin [**2**], Kawada and Kondô [**1**], Clifford [**2**], Lesieur [**3**], Mackenzie [**1**], L. M. Rybakoff [**1**], and Skolem [**2**; **3**; **5**].

1.6. Let us point out some of the simplest properties of ideals. Let $\mathfrak{A}$ be an arbitrary semigroup.

(α) *$\mathfrak{A}$ is a two-sided ideal of itself.*

(β) *If $\mathfrak{A}$ has a zero $O_{\mathfrak{A}}$, then $O_{\mathfrak{A}}$ is a two-sided ideal of $\mathfrak{A}$.*

(γ) *The union of any collection of left ideals is itself a left ideal.*

In fact, if $\mathfrak{B}_\lambda$ ($\lambda \in \Gamma$) are left ideals of $\mathfrak{A}$, then

$$\mathfrak{A} \cdot (\bigcup_{\lambda \in \Gamma} \mathfrak{B}_\lambda) = \bigcup_{\lambda \in \Gamma} (\mathfrak{A}\mathfrak{B}_\lambda) \subset \bigcup_{\lambda \in \Gamma} \mathfrak{B}_\lambda.$$

(δ) *The intersection of any collection of left ideals is itself a left ideal if it is not empty.*

In fact, using the notation of (γ) we get that for any $\mu \in \Gamma$

$$\mathfrak{A} \cdot (\bigcap_{\lambda \in \Gamma} \mathfrak{B}_\lambda) \subset \mathfrak{A}\mathfrak{B}_\mu \subset \mathfrak{B}_\mu.$$

Hence

$$\mathfrak{A} \cdot (\bigcap_{\lambda \in \Gamma} \mathfrak{B}_\lambda) \subset \bigcap_{\mu \in \Gamma} \mathfrak{B}_\mu.$$

(ε) *If $\mathfrak{B}$ is a subsemigroup of $\mathfrak{A}$, $\mathfrak{T}$ a left ideal of $\mathfrak{A}$ and the intersection of $\mathfrak{B}$ and $\mathfrak{T}$ is nonempty, then $\mathfrak{B} \cap \mathfrak{T}$ is a left ideal of $\mathfrak{B}$.*

In fact, since $\mathfrak{B} \cap \mathfrak{T} \subset \mathfrak{T}$, it follows that

$$\mathfrak{B} \cdot (\mathfrak{B} \cap \mathfrak{T}) \subset \mathfrak{B}\mathfrak{T} \subset \mathfrak{T},$$

and since $\mathfrak{B} \cap \mathfrak{T} \subset \mathfrak{B}$ we have

$$\mathfrak{B} \cdot (\mathfrak{B} \cap \mathfrak{T}) \subset \mathfrak{B} \cdot \mathfrak{B} \subset \mathfrak{B}.$$

Hence,

$$\mathfrak{B} \cdot (\mathfrak{B} \cap \mathfrak{T}) \subset \mathfrak{B} \cap \mathfrak{T}.$$

(ζ) *If an element X is included in some left ideal $\mathfrak{T}$ of the semigroup $\mathfrak{A}$ while an element Y is not included in $\mathfrak{T}$, then X does not divide Y on the right.*

For if X divided Y on the right, i.e.,

$$Y = ZX, \qquad Z \in \mathfrak{A},$$

then

$$Y = ZX \subset \mathfrak{A}\mathfrak{T} \subset \mathfrak{T}.$$

(η) *If $\mathfrak{T}$ is a semigroup without nonfactorable elements (i.e., $\mathfrak{T}\mathfrak{T} = \mathfrak{T}$), $\mathfrak{A}$ is a supersemigroup of $\mathfrak{T}$ such that $\mathfrak{T}$ is a left ideal of $\mathfrak{A}$ and $\mathfrak{A}'$ is a supersemigroup of $\mathfrak{A}$ such that $\mathfrak{A}$ is a left ideal of $\mathfrak{A}'$, then $\mathfrak{T}$ is a left ideal of the semigroup $\mathfrak{A}'$.*

In fact,

$$\mathfrak{A}'\mathfrak{T} = \mathfrak{A}'\mathfrak{T}\mathfrak{T} \subset \mathfrak{A}'\mathfrak{A}\mathfrak{T} \subset \mathfrak{A}\mathfrak{T} \subset \mathfrak{T}.$$

1.7. In connection with the property (η) of 1.6, it is necessary to note that in general the relationship of "being a left ideal" is nontransitive. If $\mathfrak{T}$ is a left ideal of the semigroup $\mathfrak{A}$ and $\mathfrak{A}$ is a left ideal of the semigroup $\mathfrak{A}'$, then it is not necessary that $\mathfrak{T}$ be a left ideal of $\mathfrak{A}'$. The same is true of right and two-sided ideals.

As an example we consider the semigroup consisting of the four elements

$$\mathfrak{A}' = \{A, B, C, O\}$$

in which the product of any two elements is equal to O, with the exception of the one product $AB = C$.

The associativity of the operation is verifiable without difficulty inasmuch as, for any three elements X, Y, and Z from $\mathfrak{A}'$,

$$(XY)Z = O, \qquad X(YZ) = O.$$

$\mathfrak{A} = \{B, C, O\}$ is clearly a two-sided ideal of $\mathfrak{A}'$. Its subset $\mathfrak{T} = \{B, O\}$ is a two-sided ideal of $\mathfrak{A}$. However, $\mathfrak{T}$ is not even a left ideal of $\mathfrak{A}'$ since

$$\mathfrak{A}'\mathfrak{T} \ni AB = C.$$

1.8. Semigroups are usually very rich in ideals. Moreover, the further away (in some sense) the semigroup is from a group the more ideals it has. Groups are the limiting case in this respect.

A semigroup is a group if and only if it has no proper ideals.

In fact, if $\mathfrak{A}$ is a group then each of its elements divides any other element of $\mathfrak{A}$ on both the right and the left. Thus, by 1.6 (ζ), no element of $\mathfrak{A}$ may be in any proper left or right ideal.

If $\mathfrak{A}$ contains no proper ideals, then for any $A \in \mathfrak{A}$

$$\mathfrak{A}A = \mathfrak{A}, \qquad A\mathfrak{A} = \mathfrak{A}$$

(since $\mathfrak{A}A$ and $A\mathfrak{A}$ are proper left and right ideals of $\mathfrak{A}$). By III, 1.2 it follows from this that $\mathfrak{A}$ is a group.

1.9. We will give some of the simplest properties of two-sided ideals. Here, considering the especially important role of two-sided ideals, we also include some properties whose validity follows directly from the properties of left and right ideals considered in 1.6.

Let $\mathfrak{A}$ be an arbitrary semigroup.

(α) *The union of any collection of two-sided ideals of a semigroup $\mathfrak{A}$ is itself a two-sided ideal of $\mathfrak{A}$.*

(β) *The product of two two-sided ideals of $\mathfrak{A}$ is a two-sided ideal of $\mathfrak{A}$.*

(γ) *The intersection of any collection of two-sided ideals of $\mathfrak{A}$ is a two-sided ideal of $\mathfrak{A}$ if it is nonempty.*

(δ) *The intersection of two two-sided ideals of $\mathfrak{A}$ is a two-sided ideal of $\mathfrak{A}$.*

In fact, if $\mathfrak{T}_1$ and $\mathfrak{T}_2$ are two-sided ideals of $\mathfrak{A}$, then their intersection is nonempty since it clearly contains their product $\mathfrak{T}_1 \cdot \mathfrak{T}_2$. But then, according to (γ), this intersection is a two-sided ideal of $\mathfrak{A}$.

(ε) *A subset of $\mathfrak{A}$ consisting of one element X is a two-sided ideal of $\mathfrak{A}$ if and only if X is the zero of the semigroup $\mathfrak{A}$.*

(ζ) *If $\mathfrak{A}$ has a zero $O_{\mathfrak{A}}$, then $O_{\mathfrak{A}}$ is included in every two-sided ideal of $\mathfrak{A}$.*

(η) *If $\mathfrak{B}$ is a subsemigroup of $\mathfrak{A}$ and $\mathfrak{T}$ is a two-sided ideal of $\mathfrak{A}$, then the intersection $\mathfrak{B} \cap \mathfrak{T}$, if it is not empty, is a two-sided ideal of the semigroup $\mathfrak{B}$.*

(θ) *If* $\mathfrak{T}$ *is a two-sided ideal of* $\mathfrak{A}$, *then the set* $\mathfrak{U}$ *consisting of all the elements* $U \in \mathfrak{A}$ *such that*

$$U\mathfrak{A} \subset \mathfrak{T}$$

is a two-sided ideal of $\mathfrak{A}$.

In fact, the set $\mathfrak{U}$ is nonempty since obviously $\mathfrak{U} \supset \mathfrak{T}$. Further, for any $U \in \mathfrak{U}$ and $A \in \mathfrak{A}$ we have

$$(AU)\mathfrak{A} \subset A\mathfrak{T} \subset \mathfrak{T},$$

$$(UA)\mathfrak{A} \subset U\mathfrak{A} \subset \mathfrak{T},$$

i.e., AU, $UA \in \mathfrak{U}$.

1.10. It follows from 1.9 that we may introduce naturally in the set of all two-sided ideals Γ of the semigroup $\mathfrak{A}$ several operations with respect to each of which it will be a semigroup.

It follows from 1.9 (α) that Γ is a commutative semigroup with respect to the operation of set union.

It follows from 1.9 (β) that Γ is a semigroup with respect to the operation of multiplication of subsets of semigroups.

It follows from 1.9 (δ) that Γ is a commutative semigroup with respect to the operation of set intersection.

1.11. The properties indicated in 1.10 of the semigroups of all the two-sided ideals of the semigroup $\mathfrak{A}$ are connected with the properties of $\mathfrak{A}$ itself.

Without going into these connections in detail let us consider as an illustration the question of identities and zeros in these semigroups.

(α) We consider Γ with respect to the operation of set union.

$\mathfrak{A}$ itself, being an element of Γ, will clearly be the zero of the semigroup Γ.

A two-sided ideal $\mathfrak{N}$ will be the identity of Γ if its union with any two-sided ideal $\mathfrak{T}$ is again $\mathfrak{T}$.

But this occurs only when $\mathfrak{N} \subset \mathfrak{T}$. Thus the identity of Γ is a two-sided ideal $\mathfrak{N}$ which is included in every two-sided ideal of $\mathfrak{A}$. It follows from this that the semigroup Γ possesses an identity if and only if the intersection of the set of all the two-sided ideals of $\mathfrak{A}$ is nonempty.

(β) We consider Γ with respect to the operation of multiplication of subsets of semigroups of $\mathfrak{A}$.

The zero of the semigroup Γ will be a two-sided ideal $\mathfrak{N}$ such that $\mathfrak{N}\mathfrak{T} = \mathfrak{T}\mathfrak{N} = \mathfrak{N}$ for any two-sided ideal $\mathfrak{T}$. But since $\mathfrak{N}\mathfrak{T} \subset \mathfrak{T}$, consequently, $\mathfrak{N}$ must be included in all the two-sided ideals of the semigroup $\mathfrak{A}$.

If such a two-sided ideal $\mathfrak{N}$ exists, i.e., the intersection of the set of all two-sided ideals of $\mathfrak{A}$ is nonempty, then it must be the zero of the semigroup Γ. In fact, for such an $\mathfrak{N}$ and any two-sided ideal $\mathfrak{T}$ of the semigroup $\mathfrak{A}$, the products $\mathfrak{N}\mathfrak{T}$ and $\mathfrak{T}\mathfrak{N}$ are two-sided ideals of $\mathfrak{A}$ belonging to $\mathfrak{N}$. Consequently,

$$\mathfrak{N}\mathfrak{T} = \mathfrak{T}\mathfrak{N} = \mathfrak{N}.$$

If a two-sided ideal $\mathfrak{M}$ is the identity of the semigroup Γ, then, in particular, it must satisfy $\mathfrak{A}\mathfrak{M} = \mathfrak{A}$. But $\mathfrak{A}\mathfrak{M} \subset \mathfrak{M}$ and thus $\mathfrak{M} = \mathfrak{A}$.

Thus the semigroup $\mathfrak{A}$ itself is the only possible identity in Γ. If $\mathfrak{A}$ is a semigroup with an identity, then $\mathfrak{A}$ must be the identity of the semigroup Γ. In fact, for any $\mathfrak{T} \in \Gamma$,

$$\mathfrak{T}\mathfrak{A} \subset \mathfrak{T}, \qquad \mathfrak{T}\mathfrak{A} \supset \mathfrak{T}E_{\mathfrak{A}} = \mathfrak{T},$$

i.e., $\mathfrak{T}\mathfrak{A} = \mathfrak{T}$. Analogously, $\mathfrak{A}\mathfrak{T} = \mathfrak{T}$.

If $\mathfrak{A}$ possesses no identities, then in some cases $\mathfrak{A}$ does not have to be an identity of the semigroup Γ and, consequently, in this case Γ has no identities. An example of this is the multiplicative semigroup of all the even natural numbers for which $\mathfrak{A} \cdot \mathfrak{A} \neq \mathfrak{A}$ and hence $\mathfrak{A}$ is not an identity of Γ.

(γ) We consider Γ with respect to the operation of set intersection. As in the preceding case Γ will possess a zero if and only if the intersection of the set of all the two-sided ideals of the semigroup $\mathfrak{A}$ is nonempty. This intersection is the zero of Γ.

As may be clearly seen, the semigroup $\mathfrak{A}$ itself will always be the identity of Γ.

1.12. If Γ' is the collection of all the two-sided ideals of some ring $\mathfrak{A}$, then we consider in Γ' the following operation (usually called the multiplication of ideals). If $\mathfrak{T}_1, \mathfrak{T}_2 \in \Gamma'$, then $\mathfrak{T}_3 = \mathfrak{T}_1 \circ \mathfrak{T}_2$ is the collection of all the elements of $\mathfrak{A}$ representable in the form

$$T_1^{(1)}T_2^{(1)} + T_1^{(2)}T_2^{(2)} + \ldots + T_1^{(n)}T_2^{(n)}$$
$$(T_1^{(i)} \in \mathfrak{T}_1 T_2^{(i)} \subset \mathfrak{T}_2; \qquad i = 1, 2, \ldots, n).$$

It is not difficult to see that Γ' is a semigroup with respect to this operation. The investigation of this semigroup of ideals plays an important role in the theory of rings.[1]

1.13. Many properties of semigroups may be characterized with the aid of various properties of systems of ideals of semigroups. As Iseki [**11**] showed, the important property of regularity of a semigroup (II, 6.1) may be thus characterized. Here there is a well-known similarity with the situation in the theory of rings.[2]

THEOREM. *In order for a semigroup $\mathfrak{A}$ to be regular, it is necessary and sufficient that for each of its left ideals $\mathfrak{L}$ and for each of its right ideals $\mathfrak{R}$ we have*

$$\mathfrak{R}\mathfrak{L} = \mathfrak{R} \cap \mathfrak{L}.$$

PROOF. (1) Since $\mathfrak{R}\mathfrak{L} \subset \mathfrak{R}$ and $\mathfrak{R}\mathfrak{L} \subset \mathfrak{L}$, we see that $\mathfrak{R}\mathfrak{L} \subset \mathfrak{R} \cap \mathfrak{L}$.

Let $\mathfrak{A}$ be regular. We choose in $\mathfrak{R} \cap \mathfrak{L}$ an arbitrary element A. We find for it an $X \in \mathfrak{A}$ such that $A = AXA$. But since $A \in \mathfrak{L}$ it follows that $XA \in \mathfrak{L}$.

Thus

$$A = AXA = A \cdot XA \in \mathfrak{R}\mathfrak{L}.$$

Consequently, $\mathfrak{R} \cap \mathfrak{L} \subset \mathfrak{R}\mathfrak{L}$ and thus $\mathfrak{R} \cap \mathfrak{L} = \mathfrak{R}\mathfrak{L}$.

---

[1] See, for example, N. Jacobson, *The theory of rings*, American Mathematical Society Mathematical Surveys, Vol. I, Amer. Math. Soc., New York, 1943.

[2] L. Kovács, *A note on regular rings*, Publ. Math. Debrecen **4** (1956), 465–468.

(2) Let the given property hold for the ideals in $\mathfrak{A}$. For an $A \in \mathfrak{A}$ we choose a right ideal $\mathfrak{R} = A\mathfrak{A} \cup A$. By assumption, we then get

$$\mathfrak{R} = \mathfrak{R} \cap \mathfrak{A} = \mathfrak{R}\mathfrak{A} = (A\mathfrak{A} \cup A)\mathfrak{A} \subset A\mathfrak{A}.$$

Consequently, $A \in A\mathfrak{A}$.

We can prove similarly that $A \subset \mathfrak{A}A$.

But $A\mathfrak{A}$ is a right ideal of $\mathfrak{A}$, while $\mathfrak{A}A$ is a left ideal of $\mathfrak{A}$. Thus

$$A \in A\mathfrak{A} \cap \mathfrak{A}A = A\mathfrak{A}\mathfrak{A}A \subset A\mathfrak{A}A,$$

from which it follows that for some $X \in \mathfrak{A}$ it is true that $A = AXA$.

1.14. Corollary. *In order for a commutative semigroup $\mathfrak{A}$ to be regular, it is necessary and sufficient that for each of its ideals $\mathfrak{T}$ we have*

$$\mathfrak{T}\mathfrak{T} = \mathfrak{T}.$$

Proof. (1) If $\mathfrak{A}$ is regular, then by 1.13 we have

$$\mathfrak{T}\mathfrak{T} = \mathfrak{T} \cap \mathfrak{T} = \mathfrak{T}.$$

(2) Let the indicated property of ideals hold in $\mathfrak{A}$. For an arbitrary pair of its ideals, $\mathfrak{T}_1$ and $\mathfrak{T}_2$, their intersection $\mathfrak{T}_1 \cap \mathfrak{T}_2$ will itself be an ideal. Thus

$$\mathfrak{T}_1 \cap \mathfrak{T}_2 = (\mathfrak{T}_1 \cap \mathfrak{T}_2) \cdot (\mathfrak{T}_1 \cap \mathfrak{T}_2) \subset \mathfrak{T}_1\mathfrak{T}_2,$$

and since always $\mathfrak{T}_1 \cap \mathfrak{T}_2 \supset \mathfrak{T}_1\mathfrak{T}_2$, we obtain $\mathfrak{T}_1 \cap \mathfrak{T}_2 = \mathfrak{T}_1\mathfrak{T}_2$. The regularity of $\mathfrak{A}$ follows from this by 1.13.

2. Chains of Subsets of an Arbitrary Set

2.1. We consider some general concepts and properties of subsets of an arbitrary set. Their choice is determined by our desire to introduce with their aid in the following sections a series of properties of ideals of semigroups which were obtained earlier by means of direct constructions by N. N. Vorob'ev [**3**; **6**] and Green [**1**].

Let Γ be some nonempty collection of nonempty subsets of some set $\mathfrak{M}$.

Definition. *A set $M \in \Gamma$ is said to be* MINIMAL *in Γ if none of its proper subsets belong to Γ.*

A set $M \in \Gamma$ is said to be UNIVERSALLY MINIMAL *in Γ if it is a subset of every set belonging to Γ.*

MAXIMAL *and* UNIVERSALLY MAXIMAL *sets in Γ are defined analogously.*

2.2. In the remainder of this section, unless we indicate to the contrary, we will understand by Γ an arbitrary nonempty collection of nonempty subsets of a set $\mathfrak{M}$ which satisfies the conditions:

(α) the set $\mathfrak{M}$ itself belongs to Γ;

(β) the union of any nonempty collection of sets from Γ belongs to Γ;

(γ) the intersection of any collection of sets of Γ belongs to Γ if it is nonempty.

2.3. If the intersection M_0 of all the sets of Γ is nonempty, then M_0 is clearly a universally minimal set in Γ. In this case Γ has no minimal sets other than M_0.

If the indicated intersection M_0 is empty, it is easy to see that there does not exist a universally minimal set in Γ.

2.4. Clearly, the following relation in $\mathfrak{M}$ is an equivalence.

DEFINITION. *Elements x and y of $\mathfrak{M}$ are said to be* Γ-EQUIVALENT *if any set of Γ that contains one of them must contain the other.*

A class of Γ-equivalent elements is said to be a Γ-LAYER.

2.5. *$\mathfrak{M}$ is the nonintersecting union of classes of Γ-equivalent elements, i.e., the nonintersecting union of all of its Γ-layers.*

2.6. For a nonempty subset N of the set $\mathfrak{M}$, its Γ-*envelope* is the intersection of all the sets of Γ containing N. Of course there are such sets, for $\mathfrak{M}$ itself belongs to Γ. Obviously a Γ-envelope is the universally minimal set in the collection of those sets of Γ which contain N.

2.7. THEOREM. *Two elements x and y of $\mathfrak{M}$ belong to the same Γ-layer if and only if their Γ-envelopes coincide.*

PROOF. (1) Let the Γ-envelopes of the elements x and y coincide. If some set $M \in \Gamma$ contains x, then it contains the Γ-envelope of x, and hence the Γ-envelope of y and thus y itself. The converse is analogous. Thus x and y are Γ-equivalent.

(2) Let x and y be Γ-equivalent. The Γ-envelope of x, inasmuch as it belongs to Γ and contains x, must also contain y. The converse is analogous. Hence the Γ-envelopes of x and y coincide.

2.8. There is another approach to the concept of a Γ-layer other than those which were employed in 2.4 and 2.7.

DEFINITION. *Two distinct sets from Γ are said to be* ADJACENT *in Γ if one of them M_1 contains the other M_2 and there is no set M' in Γ other than M_1 or M_2 such that*

$$M_1 \supset M' \supset M_2.$$

2.9. THEOREM. *The set $N \subset \mathfrak{M}$ will be a Γ-layer if and only if one of the following two conditions is satisfied:*

(1) *N belongs to Γ and is a minimal set in Γ;*

(2) *there are in Γ two adjacent sets $M_1 \supset M_2$ such that*

$$N = M_1 \backslash M_2.$$

PROOF. (1) Let N be some Γ-layer. We denote by M_1 the Γ-envelope of N and by M_2 the union of all the sets in Γ that are in M_1 but have no common elements with N (M_2 may be empty).

Clearly, $M_1 \backslash M_2 \supset N$. Let $x \in N$ and $y \in M_1 \backslash M_2$. If x is contained in some set $M' \in \Gamma$, then, by the definition of a Γ-layer, $M' \supset N$, i.e., $M' \supset M_1$, and thus $M' \ni y$. If y is included in some set $M'' \in \Gamma$, then $M_1 \cap M'' \ni y$. Since $y \bar{\in} M_2$, $M_1 \cap M''$ is not included in M_2. Thus $M_1 \cap M''$ possesses an element z belonging to N. But then $M_1 \cap M'' \ni x$, i.e., $x \in M''$. From what has been said it follows that x and y are Γ-equivalent. It follows from this that $M_1 \backslash M_2 \subset N$. Hence,

$$N = M_1 \backslash M_2.$$

It follows from the definition of a Γ-layer that there does not exist a set $M_3 \in \Gamma$ such that $M_1 \supset M_3 \supset M_2$, where $M_3 \neq M_1$, $M_3 \neq M_2$. If M_2 is empty, then $N = M$ is a minimal set in Γ.

(2) If M is a minimal set in Γ, then any set M' from Γ either contains M or has no elements in common with M. In fact, in the contrary case $M \cap M'$ would be a set from Γ contained in M and distinct from M. Thus any two elements of M are Γ-equivalent. Elements that do not belong to M cannot be Γ-equivalent with elements from M.

Let the sets M_1 and M_2 be adjacent in Γ, where $M_1 \supset M_2$.

Clearly, no element of $M_1 \backslash M_2$ can be Γ-equivalent with any element that is not included in $M_1 \backslash M_2$. If we can show that any two elements of $M_1 \backslash M_2$ are Γ-equivalent, this will mean that $M_1 \backslash M_2$ is a Γ-layer. Let us assume the contrary. Let two elements x and y of $M_1 \backslash M_2$ not be Γ-equivalent. This means that one of them, say x, is included in some set $M' \in \Gamma$ which does not contain the element y. But then the set from Γ,

$$M'' = (M' \cap M_1) \cup M_2,$$

is included in M_1 and contains M_2. Moreover, it is distinct from M_1 since it does not contain y, and it is distinct from M_2 since it contains x. The existence of such a set contradicts the fact that M_1 and M_2 are adjacent.

2.10. There is another approach to the concept of a Γ-layer that is associated with the concept of a Γ-chain.

DEFINITION. *A nonempty collection $\Sigma \subset \Gamma$ is said to be a* Γ-CHAIN *if, for any two sets in Σ, one must necessarily be a subset of the other.*

A Γ-chain Σ is said to be a PRINCIPAL *Γ-chain if there does not exist a Γ-chain Σ' distinct from Σ such that $\Sigma \subset \Sigma'$.*

2.11. THEOREM. *Any principal Γ-chain Σ itself possesses properties* 2.2 (α), (β) *and* (γ).

PROOF. Adjoining the set $\mathfrak{M}$ to Σ, we clearly obtain a Γ-chain containing Σ. It must coincide with Σ. Consequently, $\mathfrak{M} \in \Sigma$.

Let M be the union of some sets from Σ:

$$M = \bigcup_{\lambda} M_{\lambda}.$$

By the definition of Γ, M belongs to Γ.

We take an arbitrary set $\mathfrak{L}$ from Σ. If $\mathfrak{L}$ is contained in some component M_ξ of our union, that is, if $\mathfrak{L} \subset M_\xi$, then $\mathfrak{L} \subset M$. If $\mathfrak{L}$ is not included in any component M_λ, then, by the definition of a Γ-chain, $\mathfrak{L}$ must contain all the M_λ and hence $\mathfrak{L} \supset M$. From this it follows that the collection consisting of Σ and the set M will be a Γ-chain containing Σ. It must coincide with Σ. Consequently, $M \in \Sigma$.

Analogously, if M' is a nonempty intersection of sets from Σ, then it either contains or is contained in every set from Σ. Thus, adjoining the set M' to Σ, we obtain a Γ-chain containing Σ. It must coincide with Σ, from which it follows that $M' \in \Sigma$.

2.12. The role of principal Γ-chains is determined by the fact that any Γ-chain can always be extended to a principal Γ-chain.

THEOREM. *For any Γ-chain Σ, there is always a principal Γ-chain Σ' such that $\Sigma \subset \Sigma'$.*

PROOF. Let $\mathfrak{P}$ be the collection of all the Γ-chains containing a given Γ-chain Σ. If $\mathfrak{Q} \subset \mathfrak{P}$, where $\mathfrak{Q}$ is such that, for any two Γ-chains belonging to $\mathfrak{Q}$, one must always be contained in the other, then $\Sigma_{\mathfrak{Q}}$, the union of all the Γ-chains in $\mathfrak{Q}$, will itself be a Γ-chain. In fact, assume $M_1, M_2 \in \Sigma_{\mathfrak{Q}}$. Then there are Γ-chains Σ_1 and Σ_2 in $\mathfrak{Q}$ such that $M_1 \in \Sigma_1$ and $M_2 \in \Sigma_2$. If $\Sigma_1 \subset \Sigma_2$, then the sets M_1 and M_2 both belong to the chain Σ_2 and thus one of them is a subset of the other. This means that $\Sigma_{\mathfrak{Q}}$ is a Γ-chain.

By what has been said we may apply Lemma III, 4.6 to $\mathfrak{P}$. By this lemma there exists in $\mathfrak{P}$ a Γ-chain Σ' which is not included in any other Γ-chain of $\mathfrak{P}$. Clearly, Σ' will be the principal Γ-chain that we are looking for.

2.13. By Theorem 2.11 all the definitions and results above for the collection Γ are applicable to a principal Γ-chain Σ. Clearly, Σ is a Σ-chain for the collection Σ itself. Here it is obviously a principal Σ-chain and there can be no other principal Σ-chains different from Σ itself.

Consideration of the principal Γ-chain Σ is helpful because the concepts of the Γ-layer and the Σ-layer turn out to be *equivalent*.

THEOREM. *If Σ is a principal Γ-chain, then each Γ-layer is a Σ-layer and each Σ-layer is a Γ-layer.*

PROOF. (1) Let us assume that in some Γ-layer N there are two elements belonging to two distinct Σ-layers. This means that one of these elements, say x, is included in some set N' from Σ which does not contain y, the second of the elements. But this is impossible, for x and y are Γ-equivalent and $N' \in \Sigma \subset \Gamma$.

(2) Now let us assume that there are in some Σ-layer P two elements belonging to two distinct Γ-layers. This means that one of these elements, say u, is included in some set M from Γ which does not contain v, the second of the elements.

P cannot be a minimal set in Σ. Otherwise, adjoining to Σ the set $P \cap M$, which is nonempty since it contains u and is distinct from P since it does not contain v, we would obtain a new Γ-chain containing Σ and distinct from Σ. But this would contradict the fact that Σ is a principal Γ-chain.

Since P is not a minimal set in Σ, by 2.9 there are in Σ two sets M_1 and M_2 adjacent in Σ such that

$$P = M_1 \backslash M_2.$$

The set from Γ,

$$M'' = (M \cap M_1) \cup M_2,$$

is included in M_1 and contains M_2. It is distinct from M_2 since it contains u, and distinct from M_1 since it does not contain v. Consequently M'' does not belong to Σ. But then, adjoining the set M'' to Σ, we would clearly obtain a new Γ-chain containing Σ and distinct from Σ, which is impossible.

(3) By 2.5 the set $\mathfrak{M}$ may be represented as the union of all the Γ-layers, and by 2.11 as the union of all the Σ-layers. If we take one of the components of the first factorization N, and one of the components of the second factorization P, it follows from the discussions in the first two parts of the proof that N and P either coincide or have no common elements. It directly follows from this that the two unions considered consist of the same elements. Thus, each Γ-layer is included in the second union, i.e. is some Σ-layer, and conversely.

2.14. It follows from 2.13 that all the Γ-layers may be obtained, starting from any principal Γ-chain.

COROLLARY. *If Σ is a principal Γ-chain, then the set N will be a Γ-layer if and only if one of the following two conditions is satisfied:*

(1) *N belongs to Σ and is a minimal (and thus a universally minimal) set in Σ.*

(2) *There are in Σ two adjacent sets $M_1 \supset M_2$ such that*

$$N = M_1 \backslash M_2.$$

2.15. We note also the following corollary of 2.13.

COROLLARY. *If Σ_1 and Σ_2 are two principal Γ-chains, then the collection of all Σ_1-layers coincides with the collection of all Σ_2-layers.*

In fact, by 2.13 the collection of all Σ_1-layers and the collection of all Σ-layers both coincide with the collection of all Γ-layers.

2.16. In the study of the collection Γ the conditions of minimality and maximality often play an important role. Γ is said to satisfy the *condition of minimality* if each nonempty subset Γ' of the set Γ contains a set minimal in Γ'. It follows directly from the definition that if Γ satisfies the property of minimality, then any Γ-chain forming a descending sequence

$$M_1 \supset M_2 \supset \ldots \supset M_n \supset M_{n+1} \supset \ldots$$

possesses the property that, starting with some n, all its members coincide. It is not difficult to convince oneself of the converse. Let the indicated property about decreasing sequences be satisfied in Γ. In an arbitrary $\Gamma' \subset \Gamma$ we take an arbitrary set $M_1 \in \Gamma'$. If M_1 is not minimal in Γ', then we take some $M_2 \in \Gamma'$ such that $M_2 \subset M_1$, $M_2 \neq M_1$. Then, starting from M_2, we choose $M_3 \subset M_2$, $M_3 \neq M_2$ and so on. By assumption, the construction of such a sequence must break off at some member M_n. Obviously, $M_n \in \Gamma'$ will be minimal in Γ'.

2.17. The *condition of maximality* is defined and its equivalence with the conditions of stability of any increasing sequence is established analogously.

Another necessary and sufficient condition for the property of maximality is the requirement that in each $M \in \Gamma$ there exists a finite subset $M' \subset M$ such that the Γ'-envelope of M' is equal to M.

In fact, let the condition of maximality be satisfied. We choose in the set $M \in \Gamma$ an arbitrary element x_1 and consider its Γ-envelope X_1. If $X_1 \neq M$, we choose in $M \backslash X_1$ an element x_2 and consider the Γ-envelope X_2 of the set $\{x_1, x_2\}$. We then construct X_3, and so on. We obtain an increasing sequence of sets in Γ,

$$X_1 \subset X_2 \subset \ldots \subset X_n \subset \ldots .$$

Since $X_{i-1} \neq X_i$, the sequence must break off at some member X_m, which means that $M = X_m$, i.e., M is the Γ-envelope of the finite set $\{x_1, x_2, \ldots, x_m\}$.

Now let each set of Γ be the Γ-envelope of some finite set.

We consider an arbitrary increasing sequence of sets from Γ: $M_1 \subset M_2 \subset M_3 \subset \cdots$.

The set $N = \bigcup_n M_n$ belongs to Γ and is the Γ-envelope of some finite set $\{x_1, x_2, \ldots, x_m\}$. For some k all the x_i are included in M_k. Consequently, $M_k = N \supset M_i$ $(i = 1, 2, \ldots, m)$, i.e., $M_k = M_{k+1} = \cdots$.

2.18. The collection Γ satisfies simultaneously the condition of minimality and the condition of maximality only when it is finite. In fact, because of the condition of minimality there must exist in Γ a minimal set M_1. In the collection of all sets in Γ containing M_1 but distinct from M_1 there must exist a minimal set M_2. Repeating the argument, we obtain an increasing sequence which must break off:

$$M_1 \subset M_2 \subset \ldots \subset M_{n-1} \subset M_n.$$

It follows from the construction that we have obtained a principal Γ-chain. The number of its layers is finite. By 2.13 it follows from this that the number of all the Γ-layers is also finite. Since each set of Γ is the union of certain Γ-layers, the number of all the sets in Γ is also finite.

3. Principal Ideals and Ideal Layers

3.1. As was already mentioned, the study of properties of collections of subsets of an arbitrary set that was made in the preceding section will now be applied to collections of ideals of semigroups. Corresponding properties of ideals were obtained in part by Green [**1**], and in part by N. N. Vorob′ev [**3**; **6**].

We take as $\mathfrak{M}$ the set of all the elements of the semigroup $\mathfrak{A}$, and as Γ a collection of ideals of $\mathfrak{A}$ such that $\mathfrak{A}$ itself belongs to Γ and the union and intersection of any collection of ideals from Γ belong to Γ if they are nonempty. Then all the results obtained in § 2 are automatically satisfied for such a collection of ideals.

3.2. We will be interested in the following three collections:

(α) *the collection of all the left ideals of a semigroup*;

(β) *the collection of all the right ideals of a semigroup*;

(γ) *the collection of all the two-sided ideals of a semigroup.*

It follows from 1.6 (α), (γ) and (δ), that each of these possesses the above property.

All the results obtained in § 2 for the collection Γ are valid for each of these collections.

Each of these results, when formulated for the collection (α), (β) or (γ), is an important property of ideals of semigroups. Clearly, there is no need to reformulate in detail the results of § 2 for the indicated collections of ideals since this may be done without difficulty and in a completely automatic way.

3.3. When used in connection with ideals, the terms introduced in the preceding section are correspondingly changed.

Let Γ be the collection of all the left ideals of a semigroup $\mathfrak{A}$.

An ideal minimal in Γ is said to be a *minimal left ideal* of $\mathfrak{A}$.

Γ-equivalence becomes *left ideal equivalence.* A Γ-layer is said to be a *left ideal layer.* A Γ-envelope becomes a *left ideal envelope.*

The corresponding concepts are analogously formed when Γ is the collection of all the right ideals of $\mathfrak{A}$ or the collection of all the two-sided ideals.

3.4. THEOREM. *If $\mathfrak{N}$ is a nonempty subset of the semigroup $\mathfrak{A}$, then*

$$\mathfrak{A}\mathfrak{N} \cup \mathfrak{N}$$

is the left ideal envelope of $\mathfrak{N}$;

$$\mathfrak{N}\mathfrak{A} \cup \mathfrak{N}$$

is the right ideal envelope of $\mathfrak{N}$;

$$\mathfrak{A}\mathfrak{N}\mathfrak{A} \cup \mathfrak{A}\mathfrak{N} \cup \mathfrak{N}\mathfrak{A} \cup \mathfrak{N}$$

is the two-sided ideal envelope of $\mathfrak{N}$.

PROOF. Since

$$\mathfrak{A}(\mathfrak{A}\mathfrak{N} \cup \mathfrak{N}) = \mathfrak{A}\mathfrak{A}\mathfrak{N} \cup \mathfrak{A}\mathfrak{N} \subset \mathfrak{A}\mathfrak{N} \subset \mathfrak{A}\mathfrak{N} \cup \mathfrak{N},$$

it follows that $\mathfrak{A}\mathfrak{N} \cup \mathfrak{N}$ is a left ideal of $\mathfrak{A}$ containing $\mathfrak{N}$. But every left ideal of $\mathfrak{A}$ containing $\mathfrak{N}$ must also contain $\mathfrak{A}\mathfrak{N}$ and $\mathfrak{N}$. Thus $\mathfrak{A}\mathfrak{N} \cup \mathfrak{N}$ is the intersection of the set of all the left ideals of $\mathfrak{A}$ that contain $\mathfrak{N}$.

The argument is analogous for right ideals.

Since

$$\mathfrak{A}(\mathfrak{A}\mathfrak{N}\mathfrak{A} \cup \mathfrak{A}\mathfrak{N} \cup \mathfrak{N}\mathfrak{A} \cup \mathfrak{N}) = \mathfrak{A}\mathfrak{A}\mathfrak{N}\mathfrak{A} \cup \mathfrak{A}\mathfrak{A}\mathfrak{N} \cup \mathfrak{A}\mathfrak{N}\mathfrak{A} \cup \mathfrak{A}\mathfrak{N}$$
$$\subset \mathfrak{A}\mathfrak{N}\mathfrak{A} \cup \mathfrak{A}\mathfrak{N} \cup \mathfrak{N}\mathfrak{A} \cup \mathfrak{N},$$

we see that $\mathfrak{A}\mathfrak{N}\mathfrak{A} \cup \mathfrak{A}\mathfrak{N} \cup \mathfrak{N}\mathfrak{A} \cup \mathfrak{N}$ is a left ideal of $\mathfrak{A}$ containing $\mathfrak{N}$. Similarly, we can show that it is also a right ideal. It is easy to see that any two-sided ideal $\mathfrak{A}$ containing $\mathfrak{N}$ must also contain $\mathfrak{A}\mathfrak{N}\mathfrak{A} \cup \mathfrak{A}\mathfrak{N} \cup \mathfrak{N}\mathfrak{A} \cup \mathfrak{N}$. Thus, this two-sided ideal is the intersection of the set of all the two-sided ideals of $\mathfrak{A}$ containing $\mathfrak{N}$.

3.5. When $\mathfrak{A}$ possesses an identity, the expression for ideal envelopes is simplified, since

$$\mathfrak{A}\mathfrak{N} \supset \mathfrak{N}, \quad \mathfrak{N}\mathfrak{A} \supset \mathfrak{N}, \quad \mathfrak{A}\mathfrak{N}\mathfrak{A} \supset \mathfrak{A}\mathfrak{N}, \quad \mathfrak{A}\mathfrak{N}\mathfrak{A} \supset \mathfrak{N}\mathfrak{A}, \quad \mathfrak{A}\mathfrak{N}\mathfrak{A} \supset \mathfrak{N}.$$

COROLLARY. *If $\mathfrak{A}$ is a semigroup with an identity, then $\mathfrak{A}\mathfrak{N}$ is the left ideal envelope of $\mathfrak{N} \subset \mathfrak{A}$, $\mathfrak{N}\mathfrak{A}$ is the right ideal envelope and $\mathfrak{A}\mathfrak{N}\mathfrak{A}$ is the two-sided ideal envelope.*

3.6. Especially important is the case when the subset $\mathfrak{N}$ consists of one element.

DEFINITION. *A left ideal of the semigroup $\mathfrak{A}$ which is the left ideal envelope of one of its elements is said to be a* PRINCIPAL LEFT IDEAL.

Principal right ideals and *principal two-sided ideals* are defined analogously.

It follows directly from 3.4 and 3.5 that a principal left ideal of a semigroup $\mathfrak{A}$ has the form

$$\mathfrak{A}X \cup X,$$

where X is some element of $\mathfrak{A}$. If $\mathfrak{A}$ possesses an identity, then

$$\mathfrak{A}X \cup X = \mathfrak{A}X.$$

Analogously, a principal right ideal has the form $X\mathfrak{A} \cup X$, or $X\mathfrak{A}$ in the presence of an identity. A principal two-sided ideal has the form

$$\mathfrak{A}X\mathfrak{A} \cup \mathfrak{A}X \cup X\mathfrak{A} \cup X,$$

or $\mathfrak{A}X\mathfrak{A}$ if $\mathfrak{A}$ has an identity.

3.7. The relation of inclusion between principal left ideals of elements of a semigroup determines the important relation of divisibility on the right (and analogously of divisibility on the left).

THEOREM. *Let A and B be two distinct elements of a semigroup $\mathfrak{A}$. The element A is divisible on the right by the element B if and only if the left ideal envelope of the element A is included in the left ideal envelope of the element B.*

PROOF. If, for some $X \in \mathfrak{A}$,

$$A = XB,$$

then

$$\mathfrak{A}A \cup A = \mathfrak{A}XB \cup XB \subset \mathfrak{A}B \cup B.$$

On the other hand, if

$$\mathfrak{A}A \cup A \subset \mathfrak{A}B \cup B,$$

then, taking into account that $A \neq B$, we have

$$A \subset \mathfrak{A}B,$$

i.e., for some $X \in \mathfrak{A}$ it must be true that

$$A = XB.$$

3.8. By 2.5 a semigroup may be represented as the nonintersecting union of all of its left ideal layers, as the nonintersecting union of all of its right ideal layers and also as the nonintersecting union of all of its two-sided ideal layers. As was shown in the preceding section, an ideal layer (left, right, or two-sided) may be determined in various different ways. By 2.7 two elements A and B of a semigroup $\mathfrak{A}$ belong to the same left ideal layer if their left ideal envelopes coincide. Obviously, for this it is necessary and sufficient for A to be contained in the left ideal envelope of B while B is contained in the left ideal envelope of A. It is also possible to use Theorem 3.7.

In order for two distinct elements of a semigroup to be included in the same left ideal layer it is necessary and sufficient for each of them to be divisible on the right by the other.

It is clear that an analogous assertion is valid for right ideal layers.

3.9. If a left ideal $\mathfrak{L}$ of a semigroup $\mathfrak{A}$ contains some element X, then by definition of left ideal equivalence, $\mathfrak{L}$ will also contain all the elements left-ideally equivalent with X. It follows from this that the left ideal $\mathfrak{L}$ is the nonintersecting union of certain left ideal bands of the semigroup. Of course, in the general case not every union of left ideal bands is a left ideal.

The left ideal $\mathfrak{L}$ can also be obtained from principal left ideals. For each $X \in \mathfrak{L}$ the left ideal envelope of X is contained in $\mathfrak{L}$ and contains X. Thus $\mathfrak{L}$ is the union of the left ideal envelopes of all the elements X contained in $\mathfrak{L}$. The converse is clear: any union of principal left ideals is a left ideal. If there occur in such a union the left ideal envelope of an element X and the left ideal envelope of an element Y, where X is divisible on the right by Y, then by 3.7 it is possible to exclude from the union the left ideal envelope of the element X. Thus in certain cases (but of course not always), for example, when $\mathfrak{A}$ is finite, each left ideal may be represented as the union of principal left ideals such that no two of them are left ideal envelopes of elements where one is divisible by the other on the right. Clearly, it is impossible to discard from such an ideal any of the principal left ideals included in it. The representation of a left ideal $\mathfrak{L}$ in the form of the indicated union $\mathfrak{L} = \bigcup_\nu \mathfrak{T}_\nu$ has the following important property. *If $\mathfrak{L}$ is represented in the form of some union of principal left ideals $\mathfrak{L} = \bigcup_\xi \mathfrak{U}_\xi$, then all the $\mathfrak{T}_\nu$ must be included in its components $\mathfrak{U}_\xi$.* In fact, if $\mathfrak{T}_\nu$ is a left ideal envelope of the element T_ν, then T_ν is included in some $\mathfrak{U}_\nu$ which is the left ideal envelope of some element A_μ. But A_μ must be included in some $\mathfrak{T}_\lambda$. By the definition of left ideal envelopes we obtain $\mathfrak{T}_\nu \subset \mathfrak{U}_\mu$, $\mathfrak{U}_\mu \subset \mathfrak{T}_\lambda$, i.e., $\mathfrak{T}_\nu \subset \mathfrak{T}_\lambda$. By assumption this is possible only for $\mathfrak{T}_\nu = \mathfrak{T}_\lambda$, and then $\mathfrak{T}_\nu = \mathfrak{U}_\mu$. From what has been proven it follows in particular that every left ideal can have no more than one representation in the form of the union of principal ideals such that no two of them are left ideal envelopes of elements where one element is divisible by the other on the right.

The situation is analogous for right ideals.

Two-sided ideals are also all the possible unions of principal two-sided ideals.

3.10. As was indicated in the work of Munn and Penrose **[1]** (see also Miller and Clifford **[1]**), the existence and properties of idempotents in left ideal and right ideal layers is connected with certain important properties of semigroups.

We first consider the relationship between idempotents included in the same left ideal layer.

THEOREM. *If the idempotents I_1 and I_2 are both included in the same left ideal layer, then each of them is a left zero for the other. If they are both included in the same right ideal layer, then each of them is a right zero for the other.*

PROOF. I_2 is included in the left ideal $\mathfrak{A}I_2$. Inasmuch as I_1 belongs to the same left ideal layer as I_2, it must also belong to this left ideal, i.e., for some $X \in \mathfrak{A}$,

$$I_1 = XI_2.$$

But then

$$I_1 I_2 = XI_2 \cdot I_2 = XI_2 = I_1.$$

3.11. COROLLARY. *No two distinct idempotents included in the same left ideal layer can be commutative with each other.*

In fact, by the preceding discussion, we have for the indicated idempotents I_1 and I_2

$$I_1 I_2 = I_1, \qquad I_2 I_1 = I_2,$$

and then for $I_1 I_2 = I_2 I_1$, we obtain $I_1 = I_2$.

3.12. THEOREM. *In order for an element A of a semigroup $\mathfrak{A}$ to be regular* (II, 6.1) *it is necessary and sufficient for the left ideal layer containing A to have an idempotent.*

PROOF. (1) If A is regular, then, for some $B \in \mathfrak{A}$,

$$ABA = A, \qquad (BA)^2 = BA.$$

Since $A = ABA$ is divisible on the right by BA and BA is divisible on the right by A, it follows by 3.8 that A and the idempotent BA lie in the same left ideal layer.

(2) If the idempotent I lies in the same left ideal layer as A, then, for some B and $C \in \mathfrak{A}$,

$$I = BA, \qquad A = CI.$$

From this we obtain

$$ABA = AI = CI \cdot I = CI = A,$$

i.e., A is regular.

3.13. It follows directly from the theorem that in any left ideal (and analogously in any right ideal) layer either all the elements are regular or all of them are nonregular. If all the left ideal layers contain idempotents, then all the right ideal layers also contain idempotents, and conversely. The existence of idempotents in all the left ideal layers is a necessary and sufficient condition for a semigroup to be regular.

3.14. THEOREM. *In order for a semigroup to be inverse* (II, 7.2) *it is necessary and sufficient that there be in each of its left ideal layers and in each of its right ideal layers a unique idempotent.*

PROOF. (1) If $\mathfrak{A}$ is an inverse semigroup, then it is regular and by 3.12 each of its left ideal layers and each of its right ideal layers contains idempotents.

Since all the idempotents of an inverse semigroup are commutative (II, 7.4) it follows by 3.11 that no layer contains more than one idempotent.

(2) Let there be in each left ideal layer and in each right ideal layer of a semigroup $\mathfrak{A}$ exactly one idempotent. By 3.12, $\mathfrak{A}$ is regular.

Let us assume that the two elements B_1 and B_2 are both regularly associated with some element $A \in \mathfrak{A}$.

Since B_iA and A (equal to AB_iA) are divisible by each other on the right, by 3.8 the idempotent B_iA is included in the same left ideal layer as A. Consequently, $B_1A = B_2A$. Considering right ideal layers analogously, we obtain $AB_1 = AB_2$. Because of this

$$B_1 = B_1AB_1 = B_2AB_1 = B_2AB_2 = B_2.$$

Thus an arbitrary element A has only one element regularly associated with it.

3.15. Using the theorem just proven, it is also possible to weaken the condition that singles out the inverse semigroups in the class of all regular semigroups.

COROLLARY. *If in a regular semigroup none of the idempotents have elements regularly associated with them other than themselves, then the semigroup is inverse.*

PROOF. By 3.12 there are idempotents in each left ideal layer. Let us assume that some of these contain two idempotents I_1 and I_2. By 3.10,

$$I_1I_2 = I_1, \qquad I_2I_1 = I_2.$$

From this we get that

$$I_1I_2I_1 = I_1I_1 = I_1, \qquad I_2I_1I_2 = I_2I_2 = I_2,$$

i.e., I_1 and I_2 are regularly associated. By assumption this is possible only when $I_1 = I_2$. The argument is analogous for right ideal layers. By Theorem 3.14 the semigroup is inverse.

3.16. COROLLARY. *If, in a regular semigroup* $\mathfrak{A}$, *for any idempotent* I *the equality*

$$IXI = I \qquad (X \in \mathfrak{A})$$

is satisfied only for $X = 1$, *then* $\mathfrak{A}$ *is a group.*

PROOF. By 3.15, $\mathfrak{A}$ is an inverse semigroup. It follows from this that all its idempotents are commutative (II, 7.4).

Let I_1 and I_2 be idempotents of $\mathfrak{A}$. Since they commute, I_1I_2 is an idempotent. Obviously,

$$(I_1I_2)I_1(I_1I_2) = I_1I_2, \qquad (I_1I_2)I_2(I_1I_2) = I_1I_2.$$

Thus $I_1 = I_1I_2$ and $I_2 = I_1I_2$, i.e., $I_1 = I_2$. Consequently, there is only one idempotent in $\mathfrak{A}$ and, by II, 6.10 (β), $\mathfrak{A}$ is a group.

3.17. Theorem 3.14 allows us to clarify the structure of systems of principal left ideals and of principal right ideals of an inverse semigroup (II, 7.2).

Let $\mathfrak{A}$ be an inverse semigroup and $\mathfrak{H}$ the commutative subsemigroup of its idempotents (II, 7.4). Let $\mathfrak{L}$ be an arbitrary principal left ideal which is the left ideal envelope of an element A. By 3.14 there must exist some idempotent I in the same left ideal band that contains A. Since the left ideal envelopes of A and I must coincide,

$$\mathfrak{L} = \mathfrak{A}I \cup I = \mathfrak{A}I.$$

If the idempotents I_1 and I_2 are distinct, the equality $\mathfrak{A}I_1 = \mathfrak{A}I_2$ is impossible, for it would mean that I_1 and I_2 would lie in the same left ideal layer, which would contradict 3.14.

Thus the idempotents of $\mathfrak{A}$ and its principal left ideals may be put into one-to-one correspondence,

$$I \sim \mathfrak{A}I.$$

Similarly, the idempotents and the principal left ideals may be put into one-to-one correspondence,

$$I \sim I\mathfrak{A}.$$

These correspondences preserve the partial orderings in the sets of principal ideals and in $\mathfrak{H}$ where in the sets of principal ideals the ordering is defined as the relationship of inclusion, while in $\mathfrak{H}$ it is defined in the following manner.

We assume that $I_1 \geqslant I_2$ if $I_1I_2 = I_2I_1 = I_2$.

In fact, if

$$\mathfrak{A}I_1 \supset \mathfrak{A}I_2,$$

then, for some $X \in \mathfrak{A}$,

$$XI_1 = I_2,$$

from which it follows that

$$I_2I_1 = XI_1 \cdot I_1 = XI_1 = I_2.$$

Conversely, for $I_2I_1 = I_2$ we have

$$\mathfrak{A}I_2 = \mathfrak{A}I_2I_1 \subset \mathfrak{A}I_1.$$

3.18. Since, for $I_1, I_2 \in \mathfrak{H}$,

$$I_3 = I_1I_2 \in \mathfrak{H}, \qquad I_1I_3 = I_3, \qquad I_2I_3 = I_3,$$

it follows from what was proven in 3.17 that

$$\mathfrak{A}I_3 \subset \mathfrak{A}I_1, \qquad \mathfrak{A}I_3 \subset \mathfrak{A}I_2.$$

On the other hand, if $X \in \mathfrak{A}I_1 \cap \mathfrak{A}I_2$, then, for some $A_1, A_2 \in \mathfrak{A}$,

$$X = A_1I_1 = A_2I_2.$$

Also, since I_1 and I_2 commute (II, 7.4), we see that

$$X = A_2I_2 = A_2I_2I_2 = XI_2 = A_1I_1I_2 = A_1I_1I_1I_2 = XI_1I_2 = XI_3.$$

This means that $X \in \mathfrak{A}I_3$.
Thus we have shown that

$$\mathfrak{A}I_1 \cap \mathfrak{A}I_2 = \mathfrak{A}I_3.$$

The intersection of the principal left ideals of an inverse semigroup is itself a principal left ideal. It follows from this that the collection of all the principal left ideals forms a commutative semigroup of idempotents with respect to the operation of intersection. From the proven equality

$$\mathfrak{A}I_1 \cap \mathfrak{A}I_2 = \mathfrak{A}(I_1I_2)$$

it follows that the correspondence

$$\mathfrak{A}I \sim I$$

is an isomorphism between this semigroup and the subsemigroup of all the idempotents $\mathfrak{H}$ of the semigroup $\mathfrak{A}$.

4. Two-Sided Ideal Chains

4.1. The properties of ideal layers of a semigroup are closely connected with the properties of chains of ideals.

If Γ is the collection of all the two-sided ideals of the semigroup $\mathfrak{A}$, then a Γ-chain is said to be a *two-sided ideal chain.*

Left ideal chains and *right ideal chains* are defined analogously.

It follows from 2.12 *that every two-sided ideal chain can be extended to a principal two-sided ideal chain.* The situation is the same for left ideal and right ideal chains.

Properties concerned with the satisfaction of the conditions of minimality and maximality in the collection of all the left, right or two-sided ideals are formulated directly from the properties considered in 2.16, 2.17 and 2.18.

4.2. The study of adjacent ideals (2.8) is also related to that of ideal chains. For the fact that two two-sided ideals are adjacent is equivalent to the fact that they form a two-sided ideal chain that has the property that there is no other two-sided ideal chain for which the given ideals serve as ends.

By 2.9 a subset $\mathfrak{N}$ of a semigroup $\mathfrak{A}$ is a two-sided ideal layer if and only if $\mathfrak{N}$ is a minimal two-sided ideal of $\mathfrak{A}$ or if there exist two adjacent two-sided ideals $\mathfrak{T}_1$ and $\mathfrak{T}_2$ in $\mathfrak{A}$ such that

$$\mathfrak{N} = \mathfrak{T}_1 \backslash \mathfrak{T}_2.$$

The analogous statement is true for left ideal and right ideal layers.

4.3. If Σ is a principal two-sided ideal chain, then by 2.13 and 2.14 all the two-sided ideal layers may be obtained as Σ-layers, i.e., all the two-sided ideal layers consist of a minimal two-sided ideal belonging to Σ (if such a one exists) and of sets of the form $\mathfrak{T}_1 \backslash \mathfrak{T}_2$, where $\mathfrak{T}_1$ and $\mathfrak{T}_2$ are adjacent members in the chain Σ. The analogous proposition is true for left ideal and right ideal layers and chains.

4.4. Further discussions in this section will be concerned only with two-sided ideal chains. This stems from the great importance of two-sided ideals in comparison with one-sided ones, and also from the fact that the construction given here for ideal factors is practicable only for two-sided ideal layers.

We first note the important role of minimal two-sided ideals.[3] *A semigroup cannot possess more than one minimal two-sided ideal.* The validity of this follows from 2.2 and from the following property.

A minimal two-sided ideal of a semigroup is always a universally minimal two-sided ideal.

In fact, by 1.9 (δ) the intersection of a minimal two-sided ideal $\mathfrak{T}$ with an arbitrary two-sided ideal $\mathfrak{T}'$ is a two-sided ideal of the semigroup contained in $\mathfrak{T}$. Consequently, it must coincide with $\mathfrak{T}$, from which we have that $\mathfrak{T} \subset \mathfrak{T}'$.

The existence in a semigroup of a minimal two-sided ideal is always an important property. Its role is essential for the consideration of various properties of the semigroup. In the literature the minimal two-sided ideal is often called the kernel of the semigroup, or the ideal kernel, or the kernel of Suškevič. This latter term is explained by the fact that A. K. Suškevič [3] first drew attention to the role of such ideals and investigated their various properties. The numerous subsequent works in this direction are to a great degree connected with the development and generalization of the first results of A. K. Suškevič.

---

[3] It must be remembered that in the literature the term "minimal two-sided ideal" is sometimes given a wider sense, which includes among minimal two-sided ideals the minimal two-sided nonzero ideals discussed in § 3 and § 4 of the next chapter.

4.5. DEFINITION. *Let $\mathfrak{N}$ be a two-sided ideal layer of a semigroup $\mathfrak{A}$. By the* IDEAL FACTOR *corresponding to $\mathfrak{N}$ we will mean the following semigroup $\mathfrak{N}^*$:*

(1) *If $\mathfrak{N}$ is a semigroup, then $\mathfrak{N}^* = \mathfrak{N}$.*

(2) *If $\mathfrak{N}$ is not a semigroup, then $\mathfrak{N}^*$ consists of all the elements of $\mathfrak{N}$ and of a new element O^*. If for $X, Y, Z \in \mathfrak{N}$ it is true in $\mathfrak{A}$ that $XY = Z$, then we assume that $XY = Z$ also in $\mathfrak{N}^*$. In all the remaining cases we assume that $N_1N_2 = O^*$ ($N_1, N_2 \in \mathfrak{N}^*$).*

The associativity of the operation defined in $\mathfrak{N}^*$ may be verified without difficulty.

We note that in the literature the ideal factor is sometimes defined even in the first case by the external adjunction of a zero to the semigroup $\mathfrak{N}$. It is easy to see that the distinction here is completely immaterial. In view of 4.2 the notation sometimes used for the ideal factor, $\mathfrak{N}^* = \mathfrak{T}_1 - \mathfrak{T}_2$, is quite natural.

4.6. We will show by example that the construction of factors cannot in general be effected for left ideal layers. Let Ω be the denumerable set

$$\Omega = \{\alpha_1, \alpha_2, \ldots, \beta_1, \beta_2, \ldots, \gamma_1, \gamma_2, \ldots\}.$$

We consider in the semigroup $\mathfrak{S}_\Omega$ the following three transformations, given by the permutations:

$$A = \begin{pmatrix} \alpha_1 & \alpha_2 & \alpha_3 & \cdots & \beta_1 & \beta_2 & \beta_3 & \cdots & \gamma_1 & \gamma_2 & \gamma_3 & \cdots \\ \alpha_1 & \alpha_2 & \alpha_3 & \cdots & \gamma_2 & \gamma_3 & \gamma_4 & \cdots & \gamma_1 & \gamma_1 & \gamma_1 & \cdots \end{pmatrix};$$

$$B = \begin{pmatrix} \alpha_1 & \alpha_2 & \alpha_3 & \cdots & \beta_1 & \beta_2 & \beta_3 & \cdots & \gamma_1 & \gamma_2 & \gamma_3 & \cdots \\ \alpha_2 & \alpha_4 & \alpha_6 & \cdots & \alpha_1 & \alpha_3 & \alpha_5 & \cdots & \gamma_1 & \gamma_1 & \gamma_1 & \cdots \end{pmatrix};$$

$$C = \begin{pmatrix} \alpha_1 & \alpha_2 & \alpha_3 & \cdots & \beta_1 & \beta_2 & \beta_3 & \cdots & \gamma_1 & \gamma_2 & \gamma_3 & \cdots \\ \alpha_1 & \alpha_2 & \alpha_3 & \cdots & \gamma_1 & \gamma_1 & \gamma_1 & \cdots & \gamma_1 & \gamma_1 & \gamma_1 & \cdots \end{pmatrix}.$$

It follows from II, 3.1 that A is divisible on the right by B, and B is divisible on the right by A. Neither A nor B is divisible on the right by C. We denote by $\mathfrak{N}$ the left ideal layer of the semigroup $\mathfrak{S}_\Omega$ that contains A. By 3.8, $B \in \mathfrak{N}$, but $C \bar{\in} \mathfrak{N}$. By direct multiplication we verify that $AB = B$ and $AA = C$.

Since $AA \bar{\in} \mathfrak{N}$, $\mathfrak{N}$ is not a semigroup. We consider the set $\mathfrak{N}^*$ consisting of all the elements of $\mathfrak{N}$ and of a new element O^*. We define an operation in $\mathfrak{N}^*$ as was done in 4.5. As a result of this $\mathfrak{N}^*$ turns out to be a multiplicative set. However, $\mathfrak{N}^*$ is not a semigroup, for the property of associativity is not satisfied in $\mathfrak{N}^*$. In fact, by the operation in $\mathfrak{A}$, $AB = B$ and $AA \bar{\in} \mathfrak{N}$, so that in $\mathfrak{N}^*$,

$$A(AB) = AB = B, \qquad (AA)B = O^*B = O^*.$$

4.7. THEOREM. *Let $\mathfrak{N}^*$ be the ideal factor of a semigroup $\mathfrak{A}$ corresponding to a two-sided ideal layer $\mathfrak{N}$. If $\mathfrak{N}^* = \mathfrak{N}$, then $\mathfrak{N}^*$ is a semigroup that contains no proper two-sided ideals. If $\mathfrak{N}^* = \mathfrak{N} \cup O^*$, then either $\mathfrak{N}^*$ is a semigroup having* *no proper two-sided ideals other than the zero ideal O^*, or $\mathfrak{N}^*$ is a semigroup in which the product of any two elements is equal to the zero element O^*.*

PROOF. (1) Let $\mathfrak{N}^* = \mathfrak{N}$. By 4.2,

$$\mathfrak{N} = \mathfrak{T}_1 \backslash \mathfrak{T}_2,$$

where $\mathfrak{T}_1$ and $\mathfrak{T}_2$ are adjacent two-sided ideals (or $\mathfrak{T}_2$ is the empty set while $\mathfrak{T}_1$ is a minimal two-sided ideal).

Let $\mathfrak{H}$ be an arbitrary two-sided ideal of $\mathfrak{N}$.

We consider the set

$$\mathfrak{H}' = \mathfrak{T}_1\mathfrak{H}\mathfrak{T}_1 \cup \mathfrak{T}_2.$$

Since $\mathfrak{T}_1$ and $\mathfrak{T}_2$ are two-sided ideals of $\mathfrak{A}$, it clearly follows that $\mathfrak{H}$ is also a two-sided ideal of $\mathfrak{A}$, where

$$\mathfrak{T}_1 \supset \mathfrak{H}' \supset \mathfrak{T}_2.$$

Also, $\mathfrak{H}' \neq \mathfrak{T}_2$, since

$$\mathfrak{H}' \supset \mathfrak{H}\mathfrak{H}\mathfrak{H}, \qquad \mathfrak{H}\mathfrak{H}\mathfrak{H} \subset \mathfrak{N}, \qquad \mathfrak{N} \cap \mathfrak{T}_2 = \varnothing.$$

Consequently, $\mathfrak{H}' = \mathfrak{T}_1 = \mathfrak{N} \cup \mathfrak{T}_2$. But $\mathfrak{T}_1\mathfrak{H}\mathfrak{T}_1 \subset \mathfrak{H} \cup \mathfrak{T}_2$. Thus $\mathfrak{H}'$ will contain $\mathfrak{N}$ only when $\mathfrak{N} \subset \mathfrak{H}$. Hence $\mathfrak{H}$ cannot be a proper ideal of $\mathfrak{N}$.

(2) Let $\mathfrak{N}^* = \mathfrak{N} \cup O^*$, and, as in the first part,

$$\mathfrak{N} = \mathfrak{T}_1 \backslash \mathfrak{T}_2.$$

Let the semigroup $\mathfrak{N}^*$ have elements whose product yields an element other than the zero O^*, i.e., $\mathfrak{T}_1\mathfrak{T}_2 \,\bar{\subset}\, \mathfrak{T}_2$. Let $\mathfrak{H}^*$ be an arbitrary two-sided ideal of the semigroup $\mathfrak{N}^*$ possessing elements other than the zero O^*. We denote by $\mathfrak{H}$ the set of all the elements of $\mathfrak{A}$ that are included in $\mathfrak{H}^*$. The set $\mathfrak{H}$ is nonempty, and the set

$$\mathfrak{H}' = \mathfrak{T}_1\mathfrak{H}\mathfrak{T}_1 \cup \mathfrak{T}_2$$

is a two-sided ideal of $\mathfrak{A}$ included in $\mathfrak{T}_1$ and containing $\mathfrak{T}_2$.

We denote by $\mathfrak{F}$ the collection of those elements F in $\mathfrak{T}_1$ for which $\mathfrak{T}_1 F \subset \mathfrak{T}_2$. $\mathfrak{F}$ is a two-sided ideal of $\mathfrak{A}$, since

$$\mathfrak{T}_1\mathfrak{A}F \subset \mathfrak{T}_1 F \subset \mathfrak{T}_2, \qquad \mathfrak{T}_1 F\mathfrak{A} \subset \mathfrak{T}_2\mathfrak{A} \subset \mathfrak{T}_2.$$

Here $\mathfrak{T}_1 \supset \mathfrak{F} \supset \mathfrak{T}_2$. But $\mathfrak{T}_1\mathfrak{T}_1 \,\bar{\subset}\, \mathfrak{T}_2$. Thus $\mathfrak{F} = \mathfrak{T}_2$. It follows from this that for each $T \in \mathfrak{T}_1 \backslash \mathfrak{T}_2$ there exists a $T' \in \mathfrak{T}_1$ such that $T'T \,\bar{\in}\, \mathfrak{T}_2$. We show similarly that for each $T \in \mathfrak{T}_1 \backslash \mathfrak{T}_2$ there exists a T'' such that $TT'' \,\bar{\in}\, \mathfrak{T}_2$.

Let $H \in \mathfrak{H}$. For some $H' \in \mathfrak{T}_1$,

$$H'H \,\bar{\in}\, \mathfrak{T}_2$$

and for some $H'' \in \mathfrak{T}_1$,

$$(H'H)H'' \,\bar{\in}\, \mathfrak{T}_2.$$

It follows from this that the two-sided ideal $\mathfrak{H}'$ of the semigroup $\mathfrak{A}$ is distinct from $\mathfrak{T}_2$. Consequently, $\mathfrak{H}' = \mathfrak{T}_1$. But, from the fact that $\mathfrak{H}^*$ is a two-sided ideal of $\mathfrak{N}^*$, it follows from the rule of multiplication in $\mathfrak{N}^*$ that

$$\mathfrak{H}' \subset \mathfrak{H} \cup \mathfrak{T}_2.$$

This is possible only when $\mathfrak{H} = \mathfrak{N}$. But then $\mathfrak{H}^* = \mathfrak{N}^*$, i.e., the two-sided ideal $\mathfrak{H}^*$ of the semigroup $\mathfrak{N}^*$ cannot be a proper ideal of the semigroup.

4.8. If $\mathfrak{K}$ is the minimal two-sided ideal of a semigroup $\mathfrak{A}$, then by 4.2, $\mathfrak{K}$ is a two-sided ideal layer. Here, by 4.5, $\mathfrak{K}^* = \mathfrak{K}$. By 4.7 we obtain from this the following important property (which may also be developed independently).

COROLLARY. *If $\mathfrak{K}$ is the minimal two-sided ideal of a semigroup $\mathfrak{A}$, then $\mathfrak{K}$ is a semigroup having no proper two-sided ideals.*

4.9. A more detailed investigation of ideal factors under certain limitations on the semigroup was carried out by N. N. Vorob'ev **[6]**. He also considered some multiplicative properties of elements in left ideal layers.

4.10. We know that every semigroup is the union of all its two-sided ideal layers which pairwise have no common elements.

By 4.5, to each two-sided ideal layer there corresponds in a natural way a semigroup closely associated with it, the ideal factor. This ideal factor consists of the same elements as the layer itself, with perhaps the addition of one new zero element O^*. Thus the semigroup $\mathfrak{A}$ may in some sense be considered as consisting of sets of semigroups, namely all its ideal factors. The latter, as we saw in 4.7 is either the very simply constructed semigroup in which the product of any two elements is equal to zero, or is a semigroup having no proper two-sided ideals or having only the one proper two-sided zero ideal. This situation explains the interest in semigroups having no proper nonzero two-sided ideals. Such semigroups are often said to be simple, a term which we will not employ in this book. We will devote a considerable part of the next chapter to the investigation of these semigroups.

4.11. Let Σ be an arbitrary principal two-sided ideal chain of a semigroup $\mathfrak{A}$. All the ideal factors of $\mathfrak{A}$ are obtained from its two-sided ideal layers. But the collection of the latter coincides with the collection of all the Σ-layers. Thus, all the ideal factors of the semigroup $\mathfrak{A}$ can be obtained from the Σ-layers.

Let Σ_1 and Σ_2 be two principal two-sided ideal chains of the semigroup $\mathfrak{A}$. As we have already remarked (2.15), the collection of all the Σ_1-layers and the collection of all the Σ_2-layers are identical. Hence the collection of the ideal factors obtained from the Σ_1-layers coincides with the collection of ideal factors obtained from the Σ_2-layers. This proposition is a strengthened parallel of the results of a number of group theorems of the type of Jordan-Hölder.[4] The first

[4] See, for example, A. G. Kuroš, *Theory of groups*, 2nd ed., GITTL, Moscow, 1953; §§ 16 and 56. (Russian)

corresponding result was obtained by Rees [1] and was similar in form and content to group-theoretic discussions. Later N. N. Vorob'ev [6] strengthened this result to the degree in which it is presented here.

4.12. Some semigroups have a unique principal two-sided ideal chain. Since any two-sided ideal can be included in some principal two-sided ideal chain (4.1), this situation occurs for those and only those semigroups in which the set of all the two-sided ideals is linearly ordered with respect to inclusion.

THEOREM. *Let the collection of two-sided ideals of a semigroup* $\mathfrak{A}$ *satisfy the condition of maximality* (2.17). *In order for* $\mathfrak{A}$ *to possess a unique principal two-sided ideal chain, it is necessary and sufficient that all the two-sided ideals of* $\mathfrak{A}$ *be principal two-sided ideals.*

PROOF. (1) Let $\mathfrak{A}$ possess a unique principal two-sided ideal chain. Let $\mathfrak{T}$ be an arbitrary two-sided ideal of $\mathfrak{A}$. By 2.17, it is the two-sided ideal envelope of one of its finite subsets $\{T_1, T_2, \ldots, T_n\}$. We consider the two-sided ideal envelopes of the elements of this set: $\mathfrak{T}_1, \mathfrak{T}_2, \ldots, \mathfrak{T}_n$. Since they are all included in the same principal two-sided ideal chain, we have, for an appropriate numeration,

$$\mathfrak{T}_1 \subset \mathfrak{T}_2 \subset \ldots \subset \mathfrak{T}_n.$$

It follows from this that $\mathfrak{T}_n$ is the principal two-sided ideal containing the elements $T_1, T_2, \ldots, T_n$. Since $\mathfrak{T}_n \subset \mathfrak{T}$, it follows that $\mathfrak{T}_n = \mathfrak{T}$.

(2) Let all the two-sided ideals of $\mathfrak{A}$ be principal. We take two arbitrary two-sided ideals $\mathfrak{T}_1$ and $\mathfrak{T}_2$ which are the two-sided ideal envelopes of the elements A_1 and A_2. The two-sided ideal $\mathfrak{T}_3 = \mathfrak{T}_1 \cup \mathfrak{T}_2$ is also the two-sided ideal envelope of some element A_3. This element is included in $\mathfrak{T}_1$ or in $\mathfrak{T}_2$. Let $A_3 \in \mathfrak{T}_1$. This means that $A_3 = XA_1Y$, where X and Y are elements of $\mathfrak{A}$ or are empty symbols. In turn, $A_2 \in \mathfrak{T}_3$ and thus $A_2 = UA_3V$, where U and V are elements of $\mathfrak{A}$ or are empty symbols. Thus $A_2 = UXA_1YV$, from which it follows that $A_2 \in \mathfrak{T}_1$ and hence $\mathfrak{T}_2 \subset \mathfrak{T}_1$. We have shown that of two arbitrary two-sided ideals of $\mathfrak{A}$ one must necessarily be contained in the other. It follows from this that $\mathfrak{A}$ possesses a unique principal two-sided ideal chain.

4.13. We will show that the multiplicative semigroup $\mathfrak{M}_n$ of all the complex square matrices of order n belongs to the type of semigroup considered above. We denote by $\mathfrak{T}_k$ $(0 \leqslant k \leqslant n)$ the set of matrices from $\mathfrak{M}_n$ whose rank does not exceed k. As is well-known, the rank of a product of matrices does not exceed the rank of the factors. It follows from this that $\mathfrak{T}_k$ is a two-sided ideal of $\mathfrak{M}_n$. It turns out that $\mathfrak{M}_n$ has no other two-sided ideals.

Let M be a matrix of greatest rank r belonging to the two-sided ideal $\mathfrak{U}$ of the semigroup $\mathfrak{M}_n$, and let N be an arbitrary matrix of rank $\rho \leqslant r$. As is known from the theory of matrices, by means of so-called elementary transformations the matrices M and N may be brought to the form

$$M' = \begin{bmatrix} 1 & & & & & & & & \\ & 1 & & & & & & & \\ & & \cdot & & & & & & \\ & & & \cdot & & & & & \\ & & & & \cdot & & & & \\ & & & & & 1 & & & \\ & & & & & & 0 & & \\ & & & & & & & \cdot & \\ & & & & & & & & \cdot \\ & & & & & & & & & \cdot \\ & & & & & & & & & & 0 \end{bmatrix} \left.\vphantom{\begin{matrix}1\\1\\ \cdot \\ \cdot \\ \cdot \\ 1\end{matrix}}\right\} r ,$$

$$N' = \begin{bmatrix} 1 & & & & & & & & \\ & 1 & & & & & & & \\ & & \cdot & & & & & & \\ & & & \cdot & & & & & \\ & & & & \cdot & & & & \\ & & & & & 1 & & & \\ & & & & & & 0 & & \\ & & & & & & & \cdot & \\ & & & & & & & & \cdot \\ & & & & & & & & & \cdot \\ & & & & & & & & & & 0 \end{bmatrix} \left.\vphantom{\begin{matrix}1\\1\\ \cdot \\ \cdot \\ \cdot \\ 1\end{matrix}}\right\} \rho$$

(elements not written are equal to zero). But the elementary transformations, as is well-known, may be realized by multiplication of the matrix on the left and on the right by nonsingular matrices. Thus,

$$M' = P_1 M P_2, \qquad N' = Q_1 N Q_2,$$

where P_1, P_2, Q_1, and Q_2 are nonsingular and thus possess inverse matrices. Since clearly $N' = M'N'$,

$$N = Q_1^{-1} N' Q_2^{-1} = Q_1^{-1} M' N' Q_2^{-1} = Q_1^{-1} P_1 M P_2 N' Q_2^{-1}$$

and thus $N \in \mathfrak{U}$.

It follows from the above argument that $\mathfrak{U} = \mathfrak{T}_r$.

Thus the semigroup $\mathfrak{M}_n$ possesses only the following two-sided ideals, which form its unique principal two-sided ideal chain:

$$\mathfrak{M}_n = \mathfrak{T}_n \supset \mathfrak{T}_{n-1} \supset \ldots \supset \mathfrak{T}_1 \supset \mathfrak{T}_0 = O.$$

We note that, by 4.12, each of these ideals is a principal two-sided ideal.

4.14. For an arbitrary nonempty set Ω, the cardinality of which we denote by $\mathfrak{m}$, we consider the semigroup $\mathfrak{S}_\Omega$ of all of its transformations. For any cardinality $\mathfrak{n} > 1$ not exceeding $\mathfrak{m}$, we denote by $\mathfrak{T}_\mathfrak{n}$ the collection of all the transformations $X \in \mathfrak{S}_\Omega$ such that the cardinality of $X\Omega$ is less than $\mathfrak{n}$.

For any $S \in \mathfrak{S}_\Omega$ and $X \in \mathfrak{T}_\mathfrak{n}$

$$(XS)\Omega = X(S\Omega) \subset X\Omega,$$

and thus $XS \in \mathfrak{T}_\mathfrak{n}$. Since the cardinality of $X\Omega$ is less than $\mathfrak{n}$, the cardinality of $(SX)\Omega = S(X\Omega)$ will also be less than $\mathfrak{n}$ and hence $SX \in \mathfrak{T}_\mathfrak{n}$. Thus $\mathfrak{T}_\mathfrak{n}$ turns out to be a two-sided ideal. As A. I. Mal'cev [4] has shown, $\mathfrak{S}_\Omega$ has no other proper two-sided ideals.

Let $\mathfrak{U}$ be an arbitrary two-sided ideal of the semigroup $\mathfrak{S}_\Omega$ that is distinct from $\mathfrak{S}_\Omega$. We choose an arbitrary transformation $U \in \mathfrak{U}$ and denote the cardinality of $U\Omega$ by $\mathfrak{l}$. Let V be an arbitrary transformation for which the cardinality of $V\Omega$ does not exceed $\mathfrak{l}$.

The set Ω may be represented in the form of the nonintersecting union

$$\Omega = \bigcup_\xi \Omega_\xi^{(U)},$$

where each component $\Omega_\xi^{(U)}$ consists of all the elements of Ω that the transformation U takes into the same element. Analogously, for the transformation V,

$$\Omega = \bigcup_\eta \Omega_\eta^{(V)}.$$

Inasmuch as the cardinality of the set of components of the first union is $\mathfrak{l}$, while the second does not exceed $\mathfrak{l}$, there exists a one-to-one mapping φ of the collection of components of the second union into the first. Choosing an arbitrary element σ_ξ in each $\Omega_\xi^{(U)}$, we define a transformation $Y \in \mathfrak{S}_\Omega$ such that, for $\alpha \in \Omega_\eta^{(V)}$, it is true that $Y\alpha = \sigma_{\varphi(\eta)}$. Since $U\sigma_{\xi_1} \neq U\sigma_{\xi_2}$ (if $\xi_1 \neq \xi_2$), then from $V\mu_1 \neq V\mu_2$ (which means that μ_1 and μ_2 are included in different components $\Omega_\eta^{(V)}$) it follows that $UY\mu_1 \neq UY\mu_2$. By II, 3.1, it follows that the transformation UY is a right divisor of the transformation V, i.e., for some $X \in \mathfrak{S}_\Omega$

$$V = XUY.$$

Since $U \in \mathfrak{U}$, it follows from this that also $V \in \mathfrak{U}$.

It follows from the above argument that to a two-sided ideal $\mathfrak{U}$ there corresponds a class of cardinalities having the property that along with a particular cardinality all the smaller ones belong to the given class. These are those cardinalities for which there exists a $U \in \mathfrak{U}$ such that $U\Omega$ has the given cardinality. We may assume that for a given class there exists such a cardinality $\mathfrak{n}$ and that this class is the class of all the cardinalities less than the cardinality $\mathfrak{n}$. By the preceding argument, $\mathfrak{U}$ will be identical with $\mathfrak{T}_\mathfrak{n}$.

Since, given any two unequal cardinalities $\mathfrak{n}_1$ and $\mathfrak{n}_2$, one of them must be smaller than the other, and since it clearly follows from $\mathfrak{n}_1 > \mathfrak{n}_2$ that $\mathfrak{T}_{\mathfrak{n}_1} \supset \mathfrak{T}_{\mathfrak{n}_2}$ the set of all the two-sided ideals of the semigroup $\mathfrak{S}_\Omega$ turns out to be linearly ordered by inclusion, i.e., forms a unique principal two-sided ideal chain of the semigroup $\mathfrak{S}_\Omega$.

5. The Interrelations of Ideal Equivalences

5.1. Along with the interest presented by the independent study of each of the ideal equivalences, i.e., left ideal, right ideal, and two-sided, there is a substantial connection between these equivalences.

We first consider an important relation between left ideal equivalence and right ideal equivalence that was obtained by Green [**1**].

Let an element A_1 of a semigroup $\mathfrak{A}$ belong to a left ideal layer $\mathfrak{L}_1$ and to a right ideal layer $\mathfrak{R}_1$. Similarly, let an element A_2 belong to the layers $\mathfrak{L}_2$ and $\mathfrak{R}_2$.

If the intersection of $\mathfrak{L}_2$ and $\mathfrak{R}_2$ is nonempty, the intersection of $\mathfrak{L}_2$ and $\mathfrak{R}_2$ is also nonempty.

In fact, let

$$\mathfrak{L}_1 \cap \mathfrak{R}_2 \ni Z.$$

If $Z \neq A_1$ and $Z \neq A_2$, then by 3.8 there exist in the proper semigroup elements T, U, V and W such that

$$TA_1 = Z, \qquad UZ = A_1, \qquad ZV = A_2, \qquad A_2W = Z.$$

We take the element

$$X = A_1V = UZV = UA_2.$$

Since

$$A_1 = UZ = UA_2W = XW, \qquad X = A_1V,$$

it follows by 3.8 that X and A_1 are in the same right ideal layer, i.e., $X \in \mathfrak{R}_1$.

Similarly, it follows from

$$A_2 = ZV = TA_1V = TX, \qquad X = UA_2$$

that $X \in \mathfrak{L}_2$. Hence, in the case under consideration $\mathfrak{R}_1 \cap \mathfrak{L}_2 \ni X$.

When $Z = A_1$, we have $A_1 \in \mathfrak{R}_2$, i.e., $\mathfrak{R}_2 = \mathfrak{R}_1$, and hence

$$\mathfrak{L}_2 \cap \mathfrak{R}_1 = \mathfrak{L}_2 \cap \mathfrak{R}_2 \ni A_2.$$

Similarly, for $Z = A_2$,

$$\mathfrak{L}_2 \cap \mathfrak{R}_1 = \mathfrak{L}_1 \cap \mathfrak{R}_1 \ni A_1.$$

5.2. If we employ the concept of the product of relations (I, 5.2), then the fact that for some elements A_1 and A_2 the intersection $\mathfrak{L}_1$ and $\mathfrak{R}_2$ is nonempty means that A_1 and A_2 are in the relation $\mathfrak{l} \cdot \mathfrak{r}$, where $\mathfrak{l}$ designates left ideal equivalence and $\mathfrak{r}$ right ideal equivalence. Analogously, $\mathfrak{L}_2 \cap \mathfrak{R}_1 \neq \varnothing$ signifies that A_1 and A_2 are in the relation $\mathfrak{r} \cdot \mathfrak{l}$. Thus, the property proven in 5.1 is simply the commutativity of the relations $\mathfrak{l}$ and $\mathfrak{r}$ (I, 5.3): $\mathfrak{l} \cdot \mathfrak{r} = \mathfrak{r} \cdot \mathfrak{l}$.

5.3. The relation $\mathfrak{l} \cdot \mathfrak{r}$ (5.2) was introduced and considered similarly by Green [**1**]. Later it was investigated in the work of Clifford [9] and of Miller and Clifford [**1**]. As Green showed, in some cases this relation coincides with two-sided ideal equivalence. For example, this is true for periodic semigroups.

In fact, let $\mathfrak{A}$ be a periodic semigroup.

If two of its distinct elements A and B are associated by the relation (5.2), then there exists an element X such that A and X divide each other on the right and B and X divide each other on the left. It follows from this that X belongs to the two-sided ideal envelope of A, while A in turn belongs to the two-sided ideal envelope of X. Consequently, X and A are two-sided-ideally equivalent. X and B are also two-sided-ideally equivalent. Hence A and B are two-sided-ideally equivalent.

Now let A and B be two-sided-ideally equivalent. Then by 3.6, for some X_1, X_2, X_3 and X_4, which are elements of $\mathfrak{A}$ or empty symbols,

$$X_1AX_2 = B, \qquad X_3BX_4 = A.$$

From this we obtain

$$B = (X_1X_3)B(X_4X_2)$$

and thus also

$$B = (X_1X_3)^kB(X_4X_2)^k$$

for any natural k. We choose k so that $(X_4X_2)^k$ is an idempotent, which is possible since $\mathfrak{A}$ is periodic. Then it follows from the indicated equality that

$$B(X_4X_2)^k = B.$$

For $Z = BX_4$ we have

$$Z \cdot X_2(X_4X_2)^{k-1} = BX_4X_2(X_4X_2)^{k-1} = B(X_4X_2)^k = B$$

(or $ZX_2 = B$ if $k = 1$). It follows from this that B and Z are right-ideally equivalent. We show similarly that the elements $Z' = X_3B$ and B are left-ideally equivalent. This means that Z' and B divide each other on the right. But then $Z'X_4$ and BX_4 will also divide each other on the right, i.e., will be left-ideally equivalent. But $Z'X_4 = A$, while $BX_4 = Z$. Thus A and Z are left-ideally equivalent.

From what was proved above we thus obtain the result that A and B are connected by the relation $\mathfrak{l} \cdot \mathfrak{r}$ (5.2).

5.4. As was shown by Green **[1]** the intersection of arbitrary left ideal layers and right ideal layers of a semigroup often turns out to be a group. He obtained the following condition for this.

THEOREM. *Let $\mathfrak{L}$ be a left ideal layer and $\mathfrak{R}$ a right ideal layer of a semigroup $\mathfrak{A}$; let $\mathfrak{K} = \mathfrak{L} \cap \mathfrak{R}$. If*

$$\mathfrak{K}\mathfrak{K} \cap \mathfrak{K} \neq \varnothing,$$

then $\mathfrak{K}$ is a group.

PROOF. Let X and Y be elements of $\mathfrak{K}$ such that $XY \in \mathfrak{K}$. By 3.8 it is necessary to find a U and V which are elements of $\mathfrak{A}$ or empty symbols such that

$$U \cdot XY = Y, \qquad XY \cdot V = X.$$

For the element $I = UXYV$ we clearly have

$$I = YV, \qquad I = UX,$$
$$XI = XYV = X, \qquad IY = UXY = Y,$$
$$I^2 = I \cdot I = UX \cdot YV = I.$$

It follows from this that X and I are left-ideally equivalent, i.e., $I \in \mathfrak{L}$, while Y and I are right-ideally equivalent, i.e., $I \in \mathfrak{R}$.

Thus $I \in \mathfrak{K}$.

Let $Z \in \mathfrak{K}$. Since Z and I are left-ideally and right-ideally equivalent we see that, for some $Z_1, Z_2, Z_3, Z_4 \in \mathfrak{A}$,

$$Z_1 I = Z, \qquad IZ_2 = Z, \qquad Z_3 Z = I, \qquad ZZ_4 = I.$$

It follows from this that

$$IZ = Z, \qquad ZI = Z.$$

Thus I is a two-sided identity in the set $\mathfrak{K}$.

Since

$$IZ_3 I \cdot Z = IZ_3 Z = II = I, \qquad Z \cdot IZ_4 I = ZZ_4 I = II = I,$$
$$IZ_3 I = IZ_3 I \cdot I = IZ_3 I \cdot Z \cdot IZ_4 I = I \cdot IZ_4 I = IZ_4 I,$$

it follows that the element IZ_3I is left-ideally equivalent with I and thus belongs to $\mathfrak{L}$. Similarly, $IZ_3 I \in \mathfrak{R}$. Hence IZ_3I belongs to $\mathfrak{K}$ and is a two-sided inverse to Z with respect to I.

Let $T, Z \in \mathfrak{K}$ and $ZZ' = I$. Inasmuch as TZ is divisible by T on the left and T is divisible by TZ on the left, i.e.,

$$T = TI = T \cdot ZZ' = TZ \cdot Z',$$

both the elements T and TZ must be included in the same right ideal layer, i.e., $TZ \in \mathfrak{R}$. We show similarly that $TZ \in \mathfrak{L}$. Consequently, $TZ \in \mathfrak{K}$, which completes the proof that $\mathfrak{K}$ is a group.

5.5. In particular, if $\mathfrak{K}$ contains an idempotent, the intersection $\mathfrak{K}\mathfrak{K} \cap \mathfrak{K}$ will necessarily be nonempty.

COROLLARY. *If $\mathfrak{L}$ is a left ideal layer and $\mathfrak{R}$ a right ideal layer of a semigroup, where $\mathfrak{L} \cap \mathfrak{R}$ contains an idempotent, then $\mathfrak{L} \cap \mathfrak{R}$ is a group* (and, consequently, such an idempotent is unique).

5.6. Let I be an arbitrary idempotent of a semigroup $\mathfrak{A}$ and let $\mathfrak{G}_I$ be the set of all the completely regular elements for which I is a two-sided regular identity. As we saw in III, 1.14 and III, 1.16, $\mathfrak{G}_I$ is the maximal subgroup of $\mathfrak{A}$ having I as its identity. Since all the elements of a group divide each other on the right and on the left, all the elements of $\mathfrak{G}_I$ will be both left-ideally and right-ideally equivalent with I.

Considering 5.5, we conclude that $\mathfrak{G}_I = \mathfrak{L} \cap \mathfrak{R}$, where $\mathfrak{L}$ is a left ideal layer containing I and $\mathfrak{R}$ is a right ideal layer containing I.

5.7. Since every two-sided ideal of a semigroup is both a left and a right ideal of it, left-ideally or right-ideally equivalent elements will necessarily be two-sided-ideally equivalent. It follows from this that each two-sided ideal layer is the nonintersecting union of certain left ideal layers and also the nonintersecting union of certain right ideal layers.

5.8. The question as to what kind of left ideal layers compose a two-sided ideal layer and how they are obtained is answered in some degree by the following theorem obtained by N. N. Vorob'ev [**6**].

THEOREM. *If $\mathfrak{T}_1$ and $\mathfrak{T}_2$ are two adjacent* (2.8) *two-sided ideals of a semigroup $\mathfrak{A}(\mathfrak{T}_1 \supset \mathfrak{T}_2)$ and some left ideal layer $\mathfrak{L}$ is included in $\mathfrak{T}_1 \backslash \mathfrak{T}_2$ and is such that $\mathfrak{T}_2 \cup \mathfrak{L}$ is a left ideal of $\mathfrak{A}$, then for any left ideal layer $\mathfrak{L}'$ contained in $\mathfrak{T}_1 \backslash \mathfrak{T}_2$ the set $\mathfrak{T}_2 \cup \mathfrak{L}'$ will be a left ideal of the semigroup $\mathfrak{A}$.*

PROOF. (1) We denote by Σ the set of all the left ideal layers $\mathfrak{N}$ of the semigroup $\mathfrak{A}$ such that each of them is included in $\mathfrak{T}_1 \backslash \mathfrak{T}_2$ and such that $\mathfrak{T}_2 \cup \mathfrak{N}$ is a left ideal of $\mathfrak{A}$. Σ is nonempty since $\mathfrak{L} \in \Sigma$.

We consider the set

$$\mathfrak{P} = \Big(\bigcup_{\mathfrak{N} \in \Sigma} \mathfrak{N}\Big) \cup \mathfrak{T}_2 = \bigcup_{\mathfrak{N} \in \Sigma} (\mathfrak{N} \cup \mathfrak{T}_2).$$

Since $\mathfrak{N} \cup \mathfrak{T}_2$ for $\mathfrak{N} \in \Sigma$ is a left ideal of $\mathfrak{A}$, $\mathfrak{P}$ is also a left ideal.

$\mathfrak{P}$ is contained in $\mathfrak{T}_1$, contains $\mathfrak{T}_2$, and is distinct from $\mathfrak{T}_2$ (since $\mathfrak{L} \not\subseteq \mathfrak{T}_2$). If we succeed in showing that $\mathfrak{P}$ is a right and thus a two-sided ideal of $\mathfrak{A}$, this will mean that $\mathfrak{P} = \mathfrak{T}_1$. The validity of the theorem follows directly from this, for $\mathfrak{L}'$, as one of the left ideal layers belonging to $\mathfrak{T}_1 \backslash \mathfrak{T}_2 = \mathfrak{P} \backslash \mathfrak{T}_2$, will necessarily coincide with some $\mathfrak{N} \in \Sigma$.

(2) Let $A \in \mathfrak{A}$ and $X \in \mathfrak{P}$. For some $\mathfrak{N} \in \Sigma$ the element X belongs to the left ideal $\mathfrak{N} \cup \mathfrak{T}_2$.

We assume that there exist no fewer than two left ideal layers contained in the left ideal $(\mathfrak{N} \cup \mathfrak{T}_2)A$ but not contained in $\mathfrak{T}_2$. This means that in

$$(\mathfrak{N} \cup \mathfrak{T}_2)A \backslash [\mathfrak{T}_2 \cap (\mathfrak{N} \cup \mathfrak{T}_2)A]$$

there are no fewer than two left ideal layers. By 2.9 it follows from this that the corresponding left ideals are not adjacent, i.e., there lies between them some left ideal $\mathfrak{L}''$ different from both of them:

$$(\mathfrak{N} \cup \mathfrak{T}_2)A \supset \mathfrak{L}'' \supset \mathfrak{T}_2 \cap (\mathfrak{N} \cup \mathfrak{T}_2)A.$$

We denote by $\mathfrak{L}_1$ the collection of all the elements $Y \in \mathfrak{A}$ such that $YA \in \mathfrak{L}''$, and by $\mathfrak{L}_2$ the intersection of $(\mathfrak{N} \cup \mathfrak{T}_2)$ and $\mathfrak{L}_1$. If $T \in \mathfrak{T}_2$, then $TA \in (\mathfrak{N} \cup \mathfrak{T}_2)A$ and $TA \in \mathfrak{T}_2$, and thus $TA \in \mathfrak{T}_2 \cap (\mathfrak{N} \cup \mathfrak{T}_2)A \subset \mathfrak{L}''$. Consequently, $T \in \mathfrak{L}_1$. Thus $\mathfrak{T}_2 = \mathfrak{L}_1$, and $\mathfrak{L}_1$ and $\mathfrak{L}_2$ are left ideals.

Since $\mathfrak{L}_2 \subset \mathfrak{L}_1$ it follows by the definition of $\mathfrak{L}_1$ that $\mathfrak{L}_2 A \subset \mathfrak{L}''$. If $L'' \subset \mathfrak{L}''$, then $L'' = ZA$, where $Z \in \mathfrak{N} \cup \mathfrak{T}_2$. By the definition of $\mathfrak{L}_1$ the element Z must belong to $\mathfrak{L}_1$. Consequently, $Z \in \mathfrak{L}_2$. If it were true that $\mathfrak{L}_2 \subset \mathfrak{T}_2$, then Z would also belong to $\mathfrak{T}_2$, and thus an arbitrary element L'' from $\mathfrak{L}''$ would also belong to $\mathfrak{T}_2$, which is impossible since $\mathfrak{L}''$ is included in $(\mathfrak{N} \cup \mathfrak{T}_2)A$ but contains $\mathfrak{T}_2 \cap (\mathfrak{N} \cup \mathfrak{T}_2)A$ and is distinct from it. Thus $\mathfrak{L}_2$ cannot in fact be included in $\mathfrak{T}_2$.

Since $\mathfrak{L}_2$ is contained in $\mathfrak{N} \cup \mathfrak{T}_2$ it follows that $\mathfrak{L}_2$ has elements in common with $\mathfrak{N}$ and thus, being a left ideal, must necessarily contain the whole left ideal layer $\mathfrak{N}$.

We showed above that $\mathfrak{T}_2 \subset \mathfrak{L}_1$. Since $\mathfrak{L}_2 = (\mathfrak{N} \cup \mathfrak{T}_2) \cap \mathfrak{L}_1$, we have $\mathfrak{T}_2 \subset \mathfrak{L}_2$, and since $\mathfrak{N} = \mathfrak{L}_2$, we have $\mathfrak{N} \cup \mathfrak{T}_2 \subset \mathfrak{L}_2$. Thus by the definition of $\mathfrak{L}_2$,

$$\mathfrak{L}_2 = \mathfrak{N} \cup \mathfrak{T}_2.$$

Since $\mathfrak{L}_2 \subset \mathfrak{L}_1$ it follows that

$$(\mathfrak{N} \cup \mathfrak{T}_2)A \subset \mathfrak{L}'',$$

which contradicts the original assumption concerning $\mathfrak{L}''$. The contradiction so obtained means that in fact $(\mathfrak{N} \cup \mathfrak{T}_2)A$ cannot contain more than one left ideal layer not included in $\mathfrak{T}_2$.

(3) If $(\mathfrak{N} \cup \mathfrak{T}_2)A \subset \mathfrak{T}_2$, then

$$XA \in (\mathfrak{N} \cup \mathfrak{T}_2)A \subset \mathfrak{T}_2 \subset \mathfrak{P}.$$

If some of the elements of $(\mathfrak{N} \cup \mathfrak{T}_2)A$ do not belong to $\mathfrak{T}_2$, then all of them, by what was proven in the second part, form one left ideal layer $\mathfrak{N}'$. Since $\mathfrak{N} \in \Sigma$, therefore $\mathfrak{N} \cup \mathfrak{T}_2$ is a left ideal of $\mathfrak{A}$ and thus $(\mathfrak{N} \cup \mathfrak{T}_2)A$ is also a left ideal. Thus

$$\mathfrak{N}' \cup \mathfrak{T}_2 = (\mathfrak{N} \cup \mathfrak{T}_2)A \cup \mathfrak{T}_2$$

is a left ideal, i.e., $\mathfrak{N}' \in \Sigma$. It follows that

$$XA \in (\mathfrak{N} \cup \mathfrak{T}_2)A \subset \mathfrak{N}' \cup \mathfrak{T}_2 \subset \mathfrak{P}.$$

Inasmuch as in both possible cases $XA \in \mathfrak{P}$, the set $\mathfrak{P}$ is a right ideal.

5.9. Corollary. *In a semigroup $\mathfrak{A}$ let the collection of principal left ideals satisfy the condition of minimality. If $\mathfrak{T}_1$ and $\mathfrak{T}_2$ are two adjacent two-sided ideals of $\mathfrak{A}$, and $\mathfrak{N}$ is a left ideal layer included in $\mathfrak{T}_1 \backslash \mathfrak{T}_2$, then $\mathfrak{T}_2 \cup \mathfrak{N}$ is a left ideal of $\mathfrak{A}$.*

Proof. In the set of all principal left ideals included in $\mathfrak{T}_1$ but not included in $\mathfrak{T}_2$, there must exist an ideal $\mathfrak{L}$ that is minimal in the set. The left ideals $\mathfrak{T}_2$ and $\mathfrak{T}_2 \cup \mathfrak{L}$ are adjacent. In fact, if $\mathfrak{L}'$ is a left ideal such that

$$\mathfrak{T}_2 \subset \mathfrak{L}' \subset \mathfrak{T}_2 \cup \mathfrak{L}, \qquad \mathfrak{L}' \neq \mathfrak{T}_2,$$

then we choose in $\mathfrak{L}' \backslash \mathfrak{T}_2$ some element X and consider its left ideal envelope $\mathfrak{L}_X$. Since $X \bar{\in} \mathfrak{T}_2$, we have $X \in \mathfrak{L}$ and thus also $\mathfrak{L}_X \subset \mathfrak{L}$. Thanks to the minimality of $\mathfrak{L}$ it follows that $\mathfrak{L}_X = \mathfrak{L}$. But $\mathfrak{L}_X \subset \mathfrak{L}'$. Consequently, $\mathfrak{L}' \supset \mathfrak{L}$ and thus $\mathfrak{L}' = \mathfrak{T}_2 \cup \mathfrak{L}$.

It follows from what has been proven that $(\mathfrak{T}_2 \cup \mathfrak{L})\backslash\mathfrak{T}_2$ is a left ideal layer. Its union with $\mathfrak{T}_2$ yields a left ideal $\mathfrak{A}$. By 5.8 it follows from this that the union of the left ideal layer $\mathfrak{N}$ with $\mathfrak{T}_2$ must also yield a left ideal.

5.10. Any two-sided ideal chain of a semigroup is a left ideal chain of the semigroup. Of course, a principal two-sided ideal chain will not in general be a principal left ideal chain, but by 2.12 it may be extended to a principal left ideal chain. Usually this may be done in a number of ways.

Using the corollary of 5.9 it is possible to describe how all the possible extensions may be realized under the corresponding assumptions.

In a semigroup $\mathfrak{A}$ let the collection of principal left ideals satisfy the condition of minimality. Let Σ be an arbitrary principal two-sided ideal chain of $\mathfrak{A}$. For arbitrary adjacent two-sided ideals $\mathfrak{T}_1$ and $\mathfrak{T}_2$ of the chain Σ we consider the collection Φ of all the left ideal layers contained in $\mathfrak{T}_1\backslash\mathfrak{T}_2$.

By 5.9, for any $\mathfrak{N} \in \Phi$, the union $\mathfrak{T}_2 \cup \mathfrak{N}$ is a left ideal. It follows from this that for any set $\Phi' \subset \Phi$ the union

$$\left(\bigcup_{\mathfrak{N}\in\Phi'} \mathfrak{N}\right) \cup \mathfrak{T}_2$$

will be a left ideal of $\mathfrak{A}$ containing $\mathfrak{T}_2$ and contained in $\mathfrak{T}_1$. Clearly, all such left ideals can be obtained in this way since each of them contains a whole layer $\mathfrak{N} \in \Phi$ if at least one element of $\mathfrak{N}$ belongs to it.

We introduce in an arbitrary way a linear ordering in Φ. We consider subsets Φ' of the set Φ such that if $\mathfrak{N} \in \Phi'$, then Φ' also contains any $\mathfrak{N}'$ that precedes $\mathfrak{N}$ in the ordering introduced. By what was obtained above, $\left(\bigcup_{\mathfrak{N}\in\Phi'} \mathfrak{N}\right) \cup \mathfrak{T}_2$ is a left ideal of $\mathfrak{A}$ containing $\mathfrak{T}_2$ and contained in $\mathfrak{T}_1$. The set of all such left ideals constructed for any pair of adjacent two-sided ideals $\mathfrak{T}_1$ and $\mathfrak{T}_2$ from the chain Σ forms a left ideal chain Σ' which is an extension of the two-sided ideal chain Σ. It is a principal left ideal chain of $\mathfrak{A}$. In fact, let the addition of the left ideal $\mathfrak{L}$ to the chain Σ' give a left ideal chain of $\mathfrak{A}$. Let

$$\mathfrak{T}_2 \subset \mathfrak{L} \subset \mathfrak{T}_1,$$

where $\mathfrak{T}_1$ and $\mathfrak{T}_2$ are adjacent two-sided ideals belonging to Σ. The ideal $\mathfrak{L}$, like any left ideal, is the union of certain left ideal layers. Consequently,

$$\mathfrak{L} = \left(\bigcup_{\mathfrak{N}\in\Psi} \mathfrak{N}\right) \cup \mathfrak{T}_2.$$

Let $\mathfrak{N} \in \Psi$ and let $\mathfrak{N}'$ precede $\mathfrak{N}$ in the ordering relation indicated above. We consider the left ideal from Σ':

$$\mathfrak{L}' = \left(\bigcup_{\mathfrak{N}\in\Phi'} \mathfrak{N}\right) \cup \mathfrak{T}_2,$$

where Φ' consists of all the $\mathfrak{N}''$ preceding $\mathfrak{N}'$. If $\mathfrak{N}' \neq \mathfrak{N}$, then $\mathfrak{L}'$ does not contain $\mathfrak{L}$ and thus, by the definition of a chain, must be itself contained in $\mathfrak{L}$. It follows from this that $\mathfrak{L} \supset \mathfrak{N}'$, i.e. $\mathfrak{N}' \in \Psi$. We have shown that Ψ is such that $\mathfrak{L}$ itself turns out to be an ideal belonging to Σ'.

It is obvious that every principal left ideal chain which is the extension of a given principal two-sided ideal chain Σ can be obtained by the method described.

5.11. The results obtained, and also various other investigations of N. N. Vorob′ev **[6]** and Green **[1]**, show the importance of the condition of minimality for principal ideals. In this connection a relation between such conditions established by Green also turns out to be significant.

THEOREM. *If in a semigroup the collection of all the principal left ideals and the collection of all the principal right ideals both satisfy the condition of minimality, then the collection of all the principal two-sided ideals also satisfies the condition of minimality.*

PROOF. Let Γ' be some nonempty set of principal two-sided ideals of a semigroup $\mathfrak{A}$. We denote by $\mathfrak{K}$ the set of all the elements of $\mathfrak{A}$ the two-sided ideal envelopes of which belong to Γ'. We denote by Φ the set of left ideal envelopes of those elements X for which

$$(X \cup X\mathfrak{A}) \cap \mathfrak{K} \neq \varnothing .$$

Since Φ is obviously nonempty, by the condition of the theorem there exists in Φ a minimal ideal $\mathfrak{L}_0$ which is the left ideal envelope of some element X_0.

We denote by Ψ the set of right ideal envelopes of those elements Y for which $X_0 Y \in \mathfrak{K}$. If Ψ is nonempty it contains a minimal ideal $\mathfrak{R}_0$ which is the right ideal envelope of some element Y_0.

We denote by $\mathfrak{T}_0$ the two-sided ideal envelope of the element $X_0 Y_0$. Since $X_0 Y_0 \in \mathfrak{K}$, we have that $\mathfrak{T}_0 \in \Gamma'$.

We assume that the two-sided ideal envelope of some element Z belongs to Γ' and is included in $\mathfrak{T}_0$. Then for some A and B which are elements of $\mathfrak{A}$ or empty symbols,

$$Z = AX_0 Y_0 B.$$

Since $Z \in \mathfrak{K}$, the left ideal envelope $\mathfrak{L}'$ of the element AX_0 must, by the definition of Φ, belong to Φ. Here $\mathfrak{L}' \subset \mathfrak{L}_0$. Consequently, $\mathfrak{L}' = \mathfrak{L}_0$. This means that AX_0 and X_0 are left-ideally equivalent. But then, clearly, the elements $(AX_0) \cdot (Y_0 B)$, and $X_0 Y_0 B$ obtained from the multiplication of those elements on the right by a common factor will also be left-ideally equivalent. These elements will also be two-sided-ideally equivalent. Thus, $X_0 Y_0 B \in \mathfrak{K}$. Hence the right ideal envelope of the element $Y_0 B$ belongs to Ψ and is included in the right ideal envelope $\mathfrak{R}_0$ of the element Y_0. It follows from the minimality of the latter in Ψ that these envelopes must coincide, i.e., Y_0 and $Y_0 B$ must be right-ideally equivalent. But then the elements $X_0 Y_0$ and $X_0 Y_0 B$ will also be right-ideally equivalent. We showed above that the second of these is two-sided-ideally equivalent to the element Z. Consequently, Z is two-sided-ideally equivalent to $X_0 Y_0$, i.e., its two-sided ideal envelope is $\mathfrak{T}_0$. Having shown this, we have proven the minimality of $\mathfrak{T}_0$ in Γ'.

If Ψ is empty, then $X_0 \in \mathfrak{R}$ and analogously to the preceding (considering Y_0 as an empty symbol), we show that the two-sided ideal envelope of the element X_0 is the minimal set in Γ'.

6. Isolated Ideals

6.1. The role of a subsemigroup $\mathfrak{B}$ of some semigroup $\mathfrak{A}$ depends to a great extent on the relationship between $\mathfrak{B}$ and the elements of $\mathfrak{A}\backslash\mathfrak{B}$. Essentially, the question is: Can the products of various elements of $\mathfrak{A}\backslash\mathfrak{B}$ be equal to an element of $\mathfrak{B}$ and, in particular, can some power of an element of $\mathfrak{A}\backslash\mathfrak{B}$ be equal to some element of $\mathfrak{B}$?

DEFINITION. *A subsemigroup $\mathfrak{B}$ of a semigroup $\mathfrak{A}$ is said to be* ISOLATED *if, for any $X \in \mathfrak{A}$ and any natural n, it always follows from $X^n \in \mathfrak{B}$ that $X \in \mathfrak{B}$.*

$\mathfrak{B}$ is said to be COMPLETELY ISOLATED *if, for any X, $Y \in \mathfrak{A}$, it always follows from $XY \in \mathfrak{B}$ that X and Y belong to $\mathfrak{B}$.*

The investigation of the basic characteristics of isolation and complete isolation has been conducted up to now only for ideals. Completely isolated ideals are also called in the literature simple ideals, while isolated ideals (this term is due to P. G. Kontorovič) are sometimes said to be semisimple (for example, Croisot uses this term).

6.2. We note several properties of isolated subsemigroups of an arbitrary semigroup $\mathfrak{A}$.

(α) *A completely isolated subsemigroup of a semigroup $\mathfrak{A}$ is isolated.*

(β) *$\mathfrak{A}$ itself is a completely isolated subsemigroup of itself.*

(γ) *A subsemigroup $\mathfrak{B}$ will be completely isolated if and only if $\mathfrak{A}\backslash\mathfrak{B}$ is either a subsemigroup or is empty.*

(δ) *The intersection of any set of isolated subsemigroups is itself an isolated subsemigroup if it is nonempty.*

(ε) *The union of any set of isolated left ideals is an isolated left ideal.*

(ζ) *The union of any set of completely isolated left ideals is a completely isolated left ideal.*

(η) *In order for an ideal $\mathfrak{B}$ to be isolated it is sufficient that, for any $X \in \mathfrak{A}$, it always follows from $X^2 \in \mathfrak{B}$ that $X \in \mathfrak{B}$.*

In fact, let the ideal $\mathfrak{B}$ not be isolated. This means that for some element X in $\mathfrak{A}\backslash\mathfrak{B}$ it is true that $X^n \in \mathfrak{B}$.

Let the element X from $\mathfrak{A}\backslash\mathfrak{B}$ and the natural number n be so chosen that n is the least of all the possible numbers with this property (for any $X \in \mathfrak{A}\backslash\mathfrak{B}$). The number n cannot be even and different from two, for otherwise

$$(X^{n/2})^2 \in \mathfrak{B}$$

and both the possible cases

$$(1)\ X^{n/2} \in \mathfrak{B}; \qquad (2)\ X^{n/2} = Y \bar{\in} \mathfrak{B}, \qquad Y^2 \in \mathfrak{B}$$

contradict the minimality of n.

The number n cannot be odd (of course $n \neq 1$), for otherwise

$$X^{n+1} \in \mathfrak{B}$$

(which follows from the fact that $X^{n+1} = X \cdot X^n = X^n \cdot X$, $X^n \in \mathfrak{B}$, while $\mathfrak{B}$ is a left or a right ideal), and again both the possible cases

$$(1)\ X^{(n+1)/2} \in \mathfrak{B}; \qquad (2)\ X^{(n+1)/2} = Y \bar{\in} \mathfrak{B}, \qquad Y^2 \in \mathfrak{B}$$

contradict the minimality of n. Consequently, $n = 2$, i.e., for a nonisolated ideal there necessarily exists an element $X \in \mathfrak{A}\backslash\mathfrak{B}$ such that

$$X^2 \in \mathfrak{B}.$$

(θ) *If $\mathfrak{B}$ is an isolated two-sided ideal of a semigroup $\mathfrak{A}$ that is a two-sided ideal of one of its supersemigroups $\mathfrak{A}'$, then $\mathfrak{B}$ is a two-sided ideal of the semigroup $\mathfrak{A}'$.*

In fact, for any $X \in \mathfrak{A}'$ and $B \in \mathfrak{B}$ the product XB belongs to $\mathfrak{A}$ since $\mathfrak{B} \subset \mathfrak{A}$, and $\mathfrak{A}$ is a two-sided ideal for $\mathfrak{A}'$. Let us assume that $XB \bar{\in} \mathfrak{B}$. Since $\mathfrak{B}$ is isolated in $\mathfrak{A}$, we also have $(XB)^2 \bar{\in} \mathfrak{B}$. However, this is impossible, for

$$(XB)^2 = XBX \cdot B \in (XBX)\mathfrak{B} \in \mathfrak{B}.$$

We show analogously that $BX \bar{\in} \mathfrak{B}$ is also impossible. Consequently, $\mathfrak{B}$ is a two-sided ideal of $\mathfrak{A}'$.

6.3. We note that the property analogous to (δ) does not apply for a completely isolated subsemigroup. An example is the semigroup composed of three elements A, B, and O for which $A^2 = A$, $B^2 = B$, and $XY = O$ in all the remaining cases. Both $\{A, O\}$ and $\{B, O\}$ are completely isolated two-sided ideals. At the same time their intersection O is not a completely isolated subsemigroup since $AB = O$.

We remark also that the property analogous to (η) does not apply for subsemigroups that are not ideals. An example is the subsemigroup $\mathfrak{B}$ of the multiplicative semigroup of all the natural numbers that consists of numbers of the form 8^k ($k = 1, 2, 3, \ldots$). Clearly $N^2 = 8^k$ is possible only for $N \in \mathfrak{B}$. However, $\mathfrak{B}$ is not an isolated subsemigroup, for $2^3 \in \mathfrak{B}$, but $2 \bar{\in} \mathfrak{B}$.

6.4. As Croisot [5] noted, it is convenient to distinguish the semigroups that are characterized by various ideals by use of the concept of classes of regularity (II, 6.11).

(α) *In order for all the left ideals of a semigroup $\mathfrak{A}$ to be isolated it is necessary and sufficient that*

$$\mathfrak{A} = \mathfrak{C}_{\mathfrak{A}}(0, 2).$$

To see this, let all the left ideals of $\mathfrak{A}$ be isolated. For an arbitrary $A \in \mathfrak{A}$ the left ideal $\mathfrak{A}A^2$ is isolated. But $A^3 \in \mathfrak{A}A^2$. Thus $A \in \mathfrak{A}A^2$, i.e., for some $X \in \mathfrak{A}$

$$A = XA^2,$$

which also means that $A \in \mathfrak{C}_{\mathfrak{A}}(0, 2)$.

On the other hand, if $\mathfrak{A} = \mathfrak{C}_{\mathfrak{A}}(0, 2)$, $\mathfrak{L}$ is some left ideal of $\mathfrak{A}$ and $A^2 \in \mathfrak{L}$ ($A \in \mathfrak{A}$), then for some $X \in \mathfrak{A}$

$$A = XA^2.$$

Since $A^2 \in \mathfrak{L}$ it follows from this that also $A \in \mathfrak{L}$. By 7.2 (η) this yields the isolation of $\mathfrak{L}$.

(β) *In order for all the right ideals of a semigroup* $\mathfrak{A}$ *to be isolated it is necessary and sufficient that*

$$\mathfrak{A} = \mathfrak{C}_{\mathfrak{A}}(2, 0).$$

The proof is analogous to the proof of (α).

(γ) *In order for all the ideals (both the left and the right) of a semigroup* $\mathfrak{A}$ *to be isolated it is necessary and sufficient that*

$$\mathfrak{A} = \mathfrak{C}_{\mathfrak{A}}(0, 2) = \mathfrak{C}_{\mathfrak{A}}(2, 0).$$

This is a direct corollary of (α) and (β).

(δ) *In order for all the two-sided ideals of a semigroup* $\mathfrak{A}$ *to be isolated it is necessary and sufficient that for any* $A \in \mathfrak{A}$

$$A \in \mathfrak{A}A^2\mathfrak{A}.$$

In fact, if any two-sided ideal is isolated, then, in particular, the two-sided ideal $\mathfrak{A}A^2\mathfrak{A}$ will also be isolated. But $A^4 \in \mathfrak{A}A^2\mathfrak{A}$. Thus $A \in \mathfrak{A}A^2\mathfrak{A}$.

Now let any element A be included in $\mathfrak{A}A^2\mathfrak{A}$ and let $\mathfrak{T}$ be an arbitrary two-sided ideal of $\mathfrak{A}$. If $A^2 \in \mathfrak{T}$ ($A \in \mathfrak{A}$), then $A \in \mathfrak{A}A^2\mathfrak{A} \subset \mathfrak{A}\mathfrak{T}\mathfrak{A} \subset \mathfrak{T}$ and $\mathfrak{T}$ is an isolated ideal (6.2, (η)).

6.5. Let $\mathfrak{K}$ be some nonempty subset of a semigroup $\mathfrak{A}$. We consider the collection of all the isolated subsemigroups of $\mathfrak{A}$ that contain $\mathfrak{K}$. By 6.2 (β) it is nonempty. A minimal subsemigroup in this collection (2.1) is said to be an *isolated subsemigroup of* $\mathfrak{A}$ *minimal over* $\mathfrak{K}$. Similarly, a minimal semigroup in the collection of all the completely isolated subsemigroups of $\mathfrak{A}$ containing $\mathfrak{K}$ is said to be a *completely isolated subsemigroup of* $\mathfrak{A}$ *minimal over* $\mathfrak{K}$.

By what was said in 2.3 it follows from the properties 6.2 (α), (β), and (δ) that there exists an isolated subsemigroup of $\mathfrak{A}$ universally minimal over $\mathfrak{K}$ which is the unique isolated subsemigroup of $\mathfrak{A}$ minimal over $\mathfrak{K}$. It is often called the isolator of $\mathfrak{K}$. However, there may be a number of completely isolated subsemigroups minimal over $\mathfrak{K}$.

Clearly, it is for isolated subsemigroups and only for them that the isolated subsemigroup minimal over them coincides with the subsemigroup itself.

For completely isolated subsemigroups and only for them, a completely isolated subsemigroup minimal over them coincides with the subsemigroup itself.

6.6. The property of isolation was studied in detail by P. G. Kontorovič **[1; 2; 3]** for a class of semigroups imbedded in groups (III, 1.10). We will consider below the description obtained by him for isolated and completely isolated subsemigroups minimal over a given ideal. We relate these investigations to a class of semigroups that contains the class of semigroups imbedded in groups.

6.7. We will say that a semigroup $\mathfrak{A}$ satisfies the *commutator condition* if for any $A, B \in \mathfrak{A}$ there always exist in $\mathfrak{A}$ elements $L_{A,B}$ and $R_{A,B}$ such that

$$AB = L_{A,B}A, \qquad AB = BR_{A,B}.$$

Clearly, the commutator condition signifies the coincidence of the relations of left and right divisibility. This condition is important, in particular because it is satisfied by all commutative semigroups and also by every semigroup for which there exists a supergroup $\mathfrak{G}$ such that

$$G^{-1}\mathfrak{A}G \subset \mathfrak{A}$$

for each $G \in \mathfrak{G}$. In fact, for such a semigroup we may obviously take for $L_{A,B}$ and $R_{A,B}$

$$L_{A,B} = ABA^{-1}, \qquad R_{A,B} = B^{-1}AB,$$

where A^{-1} and B^{-1} are elements inverse to A and B in $\mathfrak{G}$, respectively (because of the definition of $\mathfrak{G}$ both $L_{A,B}$ and $R_{A,B}$ belong to $\mathfrak{A}$).

6.8. LEMMA. *Let a semigroup $\mathfrak{A}$ satisfy the commutator condition* (6.7); *then any ideal of $\mathfrak{A}$ is two-sided.*

PROOF. Let $\mathfrak{L}$ be a left ideal of $\mathfrak{A}$. If $X \in \mathfrak{L}$ and $A \in \mathfrak{A}$, then

$$XA = L_{A,X}X \in \mathfrak{L}.$$

Consequently, $\mathfrak{L}$ is at the same time a right ideal of $\mathfrak{A}$.

We show similarly that each right ideal of $\mathfrak{A}$ must necessarily be two-sided.

6.9. THEOREM. *Let $\mathfrak{A}$ be a semigroup satisfying the commutator condition and let $\mathfrak{T}$ be one of its ideals.*

The isolated subsemigroups of $\mathfrak{A}$ minimal over $\mathfrak{T}$ will be those and only those subsemigroups $\mathfrak{P}$ for which $\mathfrak{A}\backslash\mathfrak{P}$ is a subsemigroup of $\mathfrak{A}$ maximal in the set of subsemigroups of $\mathfrak{A}$ contained in $\mathfrak{A}\backslash\mathfrak{T}$. If this set is empty, then $\mathfrak{A}$ itself is a completely isolated subsemigroup of $\mathfrak{A}$ minimal over $\mathfrak{T}$.

The subsemigroup $\mathfrak{P}$ is an ideal of $\mathfrak{A}$.

PROOF. (1) Let $\mathfrak{Q}$ be a subsemigroup of $\mathfrak{A}$ belonging to $\mathfrak{A}\backslash\mathfrak{T}$ and not being contained in any subsemigroup belonging to $\mathfrak{A}\backslash\mathfrak{T}$ other than $\mathfrak{Q}$ itself (if $\mathfrak{A}\backslash\mathfrak{T}$ does not contain any subsemigroups of $\mathfrak{A}$, then $\mathfrak{Q}$ is the empty set). We denote by $\mathfrak{N}$ the ideal which is the union of all the ideals of $\mathfrak{A}$ lying in $\mathfrak{P} = \mathfrak{A}\backslash\mathfrak{Q}$. Since $\mathfrak{Q} = \mathfrak{A}\backslash\mathfrak{P} \subset \mathfrak{A}\backslash\mathfrak{T}$, we have $\mathfrak{T} \subset \mathfrak{P}$ and thus $\mathfrak{T} \subset \mathfrak{N}$. We assume that the product of some elements A_1 and A_2 of $\mathfrak{A}\backslash\mathfrak{N}$ is included in $\mathfrak{N}$. Both sets

$$\mathfrak{N}_1 = \mathfrak{A}A_1 \cup A_1, \qquad \mathfrak{N}_2 = \mathfrak{A}A_2 \cup A_2$$

are obviously left ideals. Since they are not contained in $\mathfrak{N}$, they must have elements in common with $\mathfrak{Q}$:

$$F_1 = S_1A_1 \in \mathfrak{N}_1 \cap \mathfrak{Q}, \qquad F_2 = S_2A_2 \in \mathfrak{N}_2 \cap \mathfrak{Q}$$

(the case when $F_i = A_i$ is completely analogous). But

$$F_1F_2 = S_1A_1S_2A_2 = S_1L_{A_1,S_2}A_1A_2 \in \mathfrak{N} \subset \mathfrak{P},$$

which contradicts the fact that F_1F_2 is the product of elements belonging to the semigroup $\mathfrak{Q}$ and thus must be included in $\mathfrak{Q}$. It follows from this that $\mathfrak{N}$ is a completely isolated ideal. Since $\mathfrak{N} \supset \mathfrak{T}$ it follows that $\mathfrak{A}\backslash\mathfrak{N}$ is a subsemigroup of $\mathfrak{A}$ belonging to $\mathfrak{A}\backslash\mathfrak{T}$. Because

$$\mathfrak{A}\backslash\mathfrak{N} \supset \mathfrak{A}\backslash\mathfrak{P} = \mathfrak{Q}$$

this is possible only for $\mathfrak{A}\backslash\mathfrak{N} = \mathfrak{A}\backslash\mathfrak{P}$, i.e., when $\mathfrak{N} = \mathfrak{P}$. Thus $\mathfrak{P}$ turns out to be a completely isolated subsemigroup and an ideal. Let $\mathfrak{P}'$ be a completely isolated subsemigroup of $\mathfrak{A}$ such that

$$\mathfrak{T} \subset \mathfrak{P}' \subset \mathfrak{P}.$$

Then $\mathfrak{A}\backslash\mathfrak{P}'$ is a subsemigroup of $\mathfrak{A}$ belonging to $\mathfrak{A}\backslash\mathfrak{T}$ and containing $\mathfrak{Q} = \mathfrak{A}\backslash\mathfrak{P}$. By the original assumption about $\mathfrak{Q}$ it follows that $\mathfrak{A}\backslash\mathfrak{P}' = \mathfrak{A}\backslash\mathfrak{P}$, i.e., $\mathfrak{P}' = \mathfrak{P}$. This proves that $\mathfrak{P}$ is a completely isolated subsemigroup of $\mathfrak{A}$ minimal over $\mathfrak{T}$.

(2) Let $\mathfrak{P}$ be a completely isolated subsemigroup minimal over $\mathfrak{T}$. $\mathfrak{Q} = \mathfrak{A}\backslash\mathfrak{P}$ is a subsemigroup of $\mathfrak{A}$ that belongs to $\mathfrak{A}\backslash\mathfrak{T}$ (6.2 (γ)). By III, 4.7 there exists in $\mathfrak{A}\backslash\mathfrak{T}$ a subsemigroup $\mathfrak{Q}'$ containing $\mathfrak{Q}$ that is not included in any subsemigroup belonging to $\mathfrak{A}\backslash\mathfrak{T}$ other than $\mathfrak{Q}'$. By what was shown in the first part $\mathfrak{A}\backslash\mathfrak{Q}'$ is a completely isolated subsemigroup minimal over $\mathfrak{T}$. Since $\mathfrak{A}\backslash\mathfrak{Q}' \subset \mathfrak{A}\backslash\mathfrak{Q} = \mathfrak{P}$ it follows by the definition of $\mathfrak{P}$ that $\mathfrak{A}\backslash\mathfrak{Q}' = \mathfrak{P}$, i.e., $\mathfrak{Q}' = \mathfrak{Q}$. This completes the proof since $\mathfrak{Q}'$ possesses the property required of $\mathfrak{Q}$.

6.10. THEOREM. *Let $\mathfrak{A}$ be a semigroup satisfying the commutator condition and let $\mathfrak{T}$ be an ideal of $\mathfrak{A}$. Then the isolated subsemigroup minimal over $\mathfrak{T}$ is itself an ideal of $\mathfrak{A}$, consists of all the elements some power of which is included in $\mathfrak{T}$ and is equal to the intersection of all the completely isolated subsemigroups of the semigroup $\mathfrak{A}$ minimal over $\mathfrak{T}$.*

PROOF. (1) Let $\mathfrak{T}'$ be the isolated subsemigroup minimal over $\mathfrak{T}$ and let $\mathfrak{B}$ be the set of all elements some power of which belongs to $\mathfrak{T}$. If $X^n \in \mathfrak{T}$ ($X \in \mathfrak{A}$), then $X \in \mathfrak{T}'$. Consequently, $\mathfrak{B} \subset \mathfrak{T}'$.

$\mathfrak{B}$ is an ideal of $\mathfrak{A}$. In fact, from $X^n \in \mathfrak{T}$ we obtain, for any $\mathfrak{A} \in \mathfrak{A}$,

$$(XA)^n = XAXAXA \ldots XA = XXR_{A,X}AXA \ldots XA$$
$$= XXXR_{(R_{A,X},AX)}A \ldots XA = \ldots = X^nY \in \mathfrak{T},$$

i.e., it follows from $X \in \mathfrak{B}$ that $XA \in \mathfrak{B}$.

The isolation of $\mathfrak{B}$ is obvious. It follows that $\mathfrak{B} \supset \mathfrak{T}'$. In view of the converse proved above, we obtain $\mathfrak{B} = \mathfrak{T}'$.

(2) We denote by $\mathfrak{C}$ the intersection of all the completely isolated subsemigroups of $\mathfrak{A}$ minimal over $\mathfrak{T}$. We have $\mathfrak{C} \supset \mathfrak{T}'$ since $\mathfrak{T}'$ is included in each isolated subsemigroup containing $\mathfrak{T}$ (6.2 (δ); 6.5).

Let $X \in \mathfrak{A}\backslash\mathfrak{T}'$. Since $\mathfrak{T}'$ is an isolated subsemigroup we have $[X] \cap \mathfrak{T}' = \varnothing$, i.e., $[X] \subset \mathfrak{A}\backslash\mathfrak{T}' \subset \mathfrak{A}\backslash\mathfrak{T}$. Let $\mathfrak{N}$ be a subsemigroup containing $[X]$ that belongs to $\mathfrak{A}\backslash\mathfrak{T}$ but is not contained in any subsemigroup belonging to $\mathfrak{A}\backslash\mathfrak{T}$ other than $\mathfrak{N}$ (III, 4.7). By 6.9, $\mathfrak{M} = \mathfrak{A}\backslash\mathfrak{N}$ is an ideal that is a completely isolated subsemigroup of $\mathfrak{A}$ minimal over $\mathfrak{T}$. Since $X \in \mathfrak{N}$ it follows that $X \bar{\in} \mathfrak{M}$. But $\mathfrak{C} \subset \mathfrak{M}$ and hence $X \bar{\in} \mathfrak{C}$. We have shown that $\mathfrak{A}\backslash\mathfrak{T}'$ and $\mathfrak{L}$ can have no common elements. Consequently, $\mathfrak{C} \subset \mathfrak{T}'$. In view of the above converse we have $\mathfrak{T}' = \mathfrak{C}$.

6.11. There are semigroups that have no proper isolated ideals. The condition for this for commutative semigroups was found by Thierrin **[16]**.

We introduce the relation $\mathfrak{n}$ into a commutative semigroup $\mathfrak{A}$, where $X \sim Y(\mathfrak{n})$ $(X, Y \in \mathfrak{A})$, if

$$X\mathfrak{A} \cap [Y] \neq \varnothing.$$

If $X \sim Y(\mathfrak{n})$ in $\mathfrak{A}$ for each pair of elements X and Y, then $\mathfrak{A}$ has no proper isolated ideals.

To see this, let us assume that $\mathfrak{B}$ is a proper isolated ideal of $\mathfrak{A}$. Let $B \in \mathfrak{B}$ and $A \in \mathfrak{A}\backslash\mathfrak{B}$. Since $B \sim A(\mathfrak{n})$ for some A and some natural number n it must be true that $BZ = A^n$. But $BZ \in \mathfrak{B}$, while $A^n \bar{\in} \mathfrak{B}$ in view of the isolation of $\mathfrak{B}$.

If there exist in $\mathfrak{A}$ elements X and Y such that $X \nsim Y(\mathfrak{n})$, then $\mathfrak{A}$ possesses proper completely isolated ideals.

We denote by $\mathfrak{B}$ the collection of all elements B such that $B \nsim Y(\mathfrak{n})$, and by $\mathfrak{C}$ the collection of all elements C such that $C \sim Y(\mathfrak{n})$.

$\mathfrak{B}$ is nonempty since $X \in \mathfrak{B}$, while $\mathfrak{C}$ is nonempty since $Y \in \mathfrak{C}$.

$\mathfrak{B}$ is an ideal of $\mathfrak{A}$, since for any $B \in \mathfrak{B}$ and $A \in \mathfrak{A}$ it follows from $B\mathfrak{A} \cap [Y] = \varnothing$ that $BA\mathfrak{A} \cap [Y] = \varnothing$.

We show that $\mathfrak{B}$ is completely isolated. Let us assume that $C_1C_2 \in \mathfrak{B}$, where $C_1, C_2 \in \mathfrak{C} = \mathfrak{A}\backslash\mathfrak{B}$. The latter means that, for some $Z_1, Z_2 \in \mathfrak{A}$,

$$C_1Z_1 = Y^n, \qquad C_2Z_2 = Y^m.$$

But then

$$(C_1C_2)(Z_1Z_2) = Y^{n+m},$$

which contradicts the fact that $C_1C_2 \in \mathfrak{B}$.

6.12. In particular, it follows from 6.11 that if a commutative semigroup has no proper completely isolated ideals, then it has no proper isolated ideals.

6.13. It is sometimes appropriate to consider related conditions along with the conditions of isolation and complete isolation. For example, for a subsemigroup $\mathfrak{B}$ of a semigroup $\mathfrak{A}$ it is sometimes required that $XY \in \mathfrak{B}$ be possible only when $X, Y \in \mathfrak{B}$ (MacKenzie [**1**]). Sometimes the requirement is strengthened so that from $XY \in \mathfrak{B}$ it follows that either $X^n \in \mathfrak{B}$ or $Y^n \in \mathfrak{B}$ for some natural number n (Aubert [**1**]).

Just as for the conditions of isolation and complete isolation, these conditions also may be given starting with the complement $\mathfrak{A}\backslash\mathfrak{B}$ of the semigroup $\mathfrak{A}$ to the semigroup $\mathfrak{B}$.

$\mathfrak{B}$ will be isolated if $\mathfrak{A}\backslash\mathfrak{B}$, along with each element X, also contains X^2.

$\mathfrak{B}$ will be completely isolated if $\mathfrak{A}\backslash\mathfrak{B}$ is a subsemigroup of $\mathfrak{A}$ or is empty.

$\mathfrak{B}$ satisfies the condition that $XY \in \mathfrak{B}$ implies that $X, Y \in \mathfrak{B}$ if $\mathfrak{A}\backslash\mathfrak{B}$ is a two-sided ideal of $\mathfrak{A}$.

$\mathfrak{B}$ satisfies the condition that $XY \in \mathfrak{B}$ implies $X^n \in \mathfrak{B}$ or $Y^n \in \mathfrak{B}$ for some n, if $\mathfrak{A}\backslash\mathfrak{B}$ is such that $[X], [Y] \subset \mathfrak{A}\backslash\mathfrak{B}$ always implies $XY \in \mathfrak{A}\backslash\mathfrak{B}$.

CHAPTER V

SEMIGROUPS WITH MINIMAL IDEALS

1. Two-Sided Ideals that are Groups

1.1. In the present chapter we will be concerned with the investigation of minimal left, minimal right, and minimal two-sided ideals of a semigroup (IV, 3.3). Of course, not every semigroup possesses such ideals. The fact of the existence of various minimal ideals in a semigroup is not only interesting in itself but turns out to have an important influence on certain other properties of semigroups. Since many properties of the semigroups possessing minimal ideals lend themselves to deeper investigation, while the structure of the minimal ideals themselves is for the most part relatively easy to clarify, mathematicians studying semigroups have been very much interested in minimal ideals.

The first results in this direction were obtained by A. K. Suškevič [**3**], who was mostly concerned with finite semigroups in which, of course, minimal left, right, and two-sided ideals always exist. Later the investigations were continued for infinite semigroups in the work of Rees [**1**; **2**], Clifford [**6**; **7**], Rich [**1**], Schwarz [**1**; **3**; **4**], Hashimoto [**2**; **3**], L. M. Gluskin [**3**; **6**], Teissier [**2**; **4**] and Tamura [**9**].

1.2. We first occupy ourselves with the question of the existence of a universally minimal ideal in the collection of all the ideals (left, right, and two-sided) of a semigroup. We will show later that if such an ideal exists it is two-sided. Thus, in the case of the existence of such an ideal, it is the unique minimal left, the unique minimal right and the unique minimal two-sided ideal.

As we will show, the question of the existence of a universally minimal ideal in the collection of all the ideals is equivalent to the question of the existence of two-sided ideals that are groups.

We have already seen examples of two-sided ideals that are groups (II, 1.8). Of course, not every semigroup possesses a two-sided ideal that is a group. For example, the infinite monogenic semigroup does not in general have subsemigroups that are groups.

THEOREM. *If a semigroup $\mathfrak{A}$ possesses a two-sided ideal $\mathfrak{T}$ that is a group, then $\mathfrak{T}$ is contained in every ideal of $\mathfrak{A}$ and is thus the universally minimal ideal in the collection of all the ideals of $\mathfrak{A}$.*

PROOF. Let $\mathfrak{L}$ be an arbitrary left ideal (the argument for a right ideal is completely analogous). Since $\mathfrak{T}$ is a two-sided ideal of $\mathfrak{A}$, while $\mathfrak{L}$ is a left ideal of $\mathfrak{A}$, we have

$$\mathfrak{T}\mathfrak{L} \subset \mathfrak{T}, \qquad \mathfrak{T}\mathfrak{L} \subset \mathfrak{L}.$$

Here $\mathfrak{T}\mathfrak{L}$ is obviously a left ideal of $\mathfrak{T}$. But $\mathfrak{T}$, being a group, has no ideals other than itself (IV, 1.8). Consequently,

$$\mathfrak{T} = \mathfrak{T}\mathfrak{L} \subset \mathfrak{L}.$$

Being contained in all the ideals of the semigroup $\mathfrak{A}$ and being itself an ideal, by the same token $\mathfrak{T}$ is the intersection of the collection of all the ideals of $\mathfrak{A}$, i.e., is the universally minimal set in this collection (IV, 2.3).

1.3. COROLLARY. *A semigroup cannot have more than one two-sided ideal that is a group.*

1.4. The question of the presence in a semigroup of a two-sided ideal that is a group turns out to be identical with the question of the existence in the semigroup of the zeroid elements (II, 1.6) of Clifford and Miller **[1]**.

THEOREM. *A semigroup $\mathfrak{A}$ has a two-sided ideal that is a group if and only if among its elements there are elements that are divisible on both the right and left by every element of $\mathfrak{A}$. In this case, the two-sided ideal that is a group consists of all the elements that are divisible on both the right and the left by every element of $\mathfrak{A}$.*

PROOF. (1) In a semigroup $\mathfrak{A}$ let $\mathfrak{T}$ be a two-sided ideal that is a group. Let A be an arbitrary element of $\mathfrak{A}$ and let T be an arbitrary element of $\mathfrak{T}$. Since $\mathfrak{T}$ is a two-sided ideal it follows that

$$TA = T' \in \mathfrak{T}.$$

Since $\mathfrak{T}$ is a group, for some $X \in \mathfrak{T}$ we have

$$XT' = T,$$

from which it follows that

$$(XT)A = XT' = T.$$

It may be shown similarly that T is divisible by A on the left.

(2) Let the set $\mathfrak{C}$ consisting of all those elements of the semigroup $\mathfrak{A}$ that are divisible on the left and right by every element of $\mathfrak{A}$ be nonempty. We have already shown in II, 1.6 that $\mathfrak{C}$ is a group. Let A be an arbitrary element of $\mathfrak{A}$. The identity $E_{\mathfrak{C}}$ of the group $\mathfrak{C}$ is divisible on the left by the element $AE_{\mathfrak{C}}$:

$$E_{\mathfrak{C}} = AE_{\mathfrak{C}}Y.$$

Multiplying this equality on the right by $E_{\mathfrak{C}}$, we obtain

$$E_{\mathfrak{C}} = AE_{\mathfrak{C}}YE_{\mathfrak{C}} = AC_0.$$

It follows from the fact that $E_{\mathfrak{C}}$ is divisible by every element of $\mathfrak{A}$ on both the right and the left that the element $C_0 = E_{\mathfrak{C}} Y E_{\mathfrak{C}}$ also possesses this property (II, 1.4 (α)), i.e., $C_0 \in \mathfrak{C}$. Thus

$$A\mathfrak{C} = AE_{\mathfrak{C}}\mathfrak{C} = AC_0C_0^{-1}\mathfrak{C} = E_{\mathfrak{C}}C_0^{-1}\mathfrak{C} \subset \mathfrak{C}$$

(here C_0^{-1} is the inverse of C_0 with respect to $E_{\mathfrak{C}}$). Consequently, $\mathfrak{C}$ is a left ideal of $\mathfrak{A}$. It may be shown similarly that $\mathfrak{C}$ is a right ideal.

Hence we have seen that if $\mathfrak{C}$ is nonempty, then it is a two-sided ideal of $\mathfrak{A}$ and is a group.

1.5. COROLLARY. *If a semigroup $\mathfrak{A}$ possesses a two-sided ideal that is a group, then every element of $\mathfrak{A}$ that is divisible on the left by all the elements of $\mathfrak{A}$ will also be divisible on the right by all the elements of $\mathfrak{A}$; each element that is divisible on the right by all the elements of $\mathfrak{A}$ will also be divisible on the left by all the elements of $\mathfrak{A}$.*

To see this, let T be some element of a two-sided ideal $\mathfrak{T}$ that is a group, and let A be an arbitrary element of $\mathfrak{A}$ that is divisible on the left by all the elements of $\mathfrak{A}$. For some $Y \in \mathfrak{A}$,

$$A = TY.$$

Since $\mathfrak{T}$ is a two-sided ideal of $\mathfrak{A}$, it follows from this that $A \in \mathfrak{T}$. Thus, by 1.4, A must be divisible on the right by every element of $\mathfrak{A}$.

1.6. We note that in a semigroup which does not possess a two-sided ideal that is a group the property of 1.5 may not be fulfilled. For example, in a semigroup with no less than two elements in which $XY = Y$ for each pair of elements X and Y, each element is divisible by all the elements on the left. At the same time, this semigroup contains no element which is divisible by all the elements on the right.

1.7. Using 1.4, it is not difficult to show that the sufficient condition obtained in 1.2 for the existence of a universally minimal ideal in the collection of all the left and right ideals is also a necessary condition.

THEOREM. *If the collection of all the ideals of a semigroup $\mathfrak{A}$ possesses a universally minimal ideal, i.e., the intersection of all the ideals of $\mathfrak{A}$ is nonempty* (IV, 2.3), *then this intersection is a two-sided ideal that is a group.*

PROOF. Let the intersection $\mathfrak{T}$ of all the ideals of $\mathfrak{A}$ be nonempty. We choose arbitrary elements A from $\mathfrak{A}$ and T from $\mathfrak{T}$. Since $\mathfrak{A}A$ and $A\mathfrak{A}$ are ideals of $\mathfrak{A}$ it follows that $\mathfrak{A}A \supset \mathfrak{T}$ and $A\mathfrak{A} \supset \mathfrak{T}$. This means that for some X, $Y \in \mathfrak{A}$,

$$XA = T, \qquad AY = T.$$

Consequently, T is divisible on the right and left by every element of the semigroup $\mathfrak{A}$.

On the other hand, if $Z \in \mathfrak{A}$ is divisible on the left and right by all the elements of $\mathfrak{A}$, then Z is included in each ideal of the semigroup $\mathfrak{A}$. In fact, if $\mathfrak{L}$ is a left ideal of $\mathfrak{A}$ (and analogously for a right ideal), while $L \in \mathfrak{L}$, then for some $X \in \mathfrak{A}$

$$Z = XL,$$

and since $\mathfrak{L}$ is a left ideal, $Z \in \mathfrak{L}$. It follows that $Z \in \mathfrak{T}$.

We have shown that $\mathfrak{T}$ is the set of all the elements of $\mathfrak{A}$ that are divisible on the right and the left by every element of $\mathfrak{A}$. By 1.4 the desired property of $\mathfrak{T}$ follows from this.

1.8. If a semigroup $\mathfrak{A}$ possesses a two-sided ideal $\mathfrak{T}$ that is a group, then the identity $E_{\mathfrak{T}}$ of the group is an element with the following properties:

(α) $E_{\mathfrak{T}}$ is an idempotent;

(β) $E_{\mathfrak{T}}$ is divisible on the left and right by every element of $\mathfrak{A}$ (1.4);

(γ) $E_{\mathfrak{T}}$ commutes with every element of $\mathfrak{A}$.

In fact, for any $A \in \mathfrak{A}$,

$$AE_{\mathfrak{T}} \in \mathfrak{T}, \qquad E_{\mathfrak{T}}A \in \mathfrak{T},$$

and thus

$$AE_{\mathfrak{T}} = E_{\mathfrak{T}}(AE_{\mathfrak{T}}) = (E_{\mathfrak{T}}A)E_{\mathfrak{T}} = E_{\mathfrak{T}}A.$$

(δ) For every idempotent I of the semigroup $\mathfrak{A}$, we have

$$IE_{\mathfrak{T}} = E_{\mathfrak{T}}I = E_{\mathfrak{T}}.$$

In fact, by (γ), $IE_{\mathfrak{T}} = E_{\mathfrak{T}}I$ and

$$(E_{\mathfrak{T}}I)^2 = E_{\mathfrak{T}}I \cdot E_{\mathfrak{T}}I = E_{\mathfrak{T}}^2I^2 = E_{\mathfrak{T}}I.$$

$E_{\mathfrak{T}}I$ belongs to $\mathfrak{T}$. Since $\mathfrak{T}$, being a group, has only one idempotent, $E_{\mathfrak{T}}I = E_{\mathfrak{T}}$.

By 1.4 the existence in $\mathfrak{A}$ of an element with the property (β) is also sufficient in order for $\mathfrak{A}$ to possess a two-sided ideal that is a group. A semigroup possessing an element satisfying properties (α), (β) and (γ) was called a homogroup by Thierrin [**6**]. Thus a homogroup is nothing other than a semigroup possessing a two-sided ideal that is a group. Along with the properties listed above, a whole series of other properties have been obtained for homogroups (see, for example, Thierrin [**19**]).

We note that by 1.4 every finite commutative semigroup is a homogroup since the product of all of its elements is clearly divisible on the left and the right by each of its elements. Groups are related to homogroups, a homogroup being a group if and only if it has no proper two-sided ideals.

2. Semigroups with Minimal Left Ideals

2.1. As we have already remarked, not every semigroup possesses minimal left, right, or two-sided ideals. In the present section we will consider semigroups possessing minimal left ideals. We will explain the connection between different minimal left ideals and will give their relationship to a minimal two-sided ideal. It is understood that analogous results are valid for semigroups possessing minimal right ideals. The results mentioned were obtained in part by Clifford [**6**] and in part by Schwarz [**3**].

2.2. It turns out that all the minimal left ideals may be easily obtained from one of them.

THEOREM. *Let $\mathfrak{L}$ be a minimal left ideal of a semigroup $\mathfrak{A}$. Then for any $S \in \mathfrak{A}$ the product $\mathfrak{L}S$ is also a minimal left ideal of $\mathfrak{A}$, while every minimal left ideal of $\mathfrak{A}$ may be represented in the form $\mathfrak{L}S$ for some $S \in \mathfrak{A}$.*

Distinct minimal left ideals of $\mathfrak{A}$ have no common elements.

PROOF. (1) $\mathfrak{L}S$ is clearly a left ideal of $\mathfrak{A}$. Let some ideal $\mathfrak{L}'$ of the semigroup $\mathfrak{A}$ be contained in $\mathfrak{L}S$. We denote by $\mathfrak{N}$ the collection of all the elements X of $\mathfrak{L}$ for which $XS \in \mathfrak{L}'$. Since each element of $\mathfrak{L}'$ is included in $\mathfrak{L}S$, i.e., is representable in the form XS, where $X \in \mathfrak{L}$, the set $\mathfrak{N}$ is nonempty. From the fact that $\mathfrak{L}'$ is a left ideal it follows that, for any $X \in \mathfrak{N}$ and $Z \in \mathfrak{A}$,

$$ZXS \subset Z\mathfrak{L}' \subset \mathfrak{L}',$$

i.e., $ZX \in \mathfrak{N}$. Consequently, $\mathfrak{N}$ is a left ideal of $\mathfrak{A}$. But $\mathfrak{N} \subset \mathfrak{L}$ and since $\mathfrak{L}$ is a minimal left ideal of $\mathfrak{A}$, $\mathfrak{N} = \mathfrak{L}$. Hence $\mathfrak{L}'$ is contained in $\mathfrak{N}S$. But by the definition of $\mathfrak{N}$, $\mathfrak{N}S$ is included in $\mathfrak{L}'$. Consequently,

$$\mathfrak{L}' = \mathfrak{N}S = \mathfrak{L}S.$$

We conclude from this that $\mathfrak{L}S$ is a minimal left ideal of $\mathfrak{A}$.

(2) Let $\mathfrak{M}$ be a minimal left ideal of $\mathfrak{A}$. We choose some element S in $\mathfrak{M}$. Since $\mathfrak{M}$ is a left ideal,

$$\mathfrak{L}S \subset \mathfrak{A}\mathfrak{M} \subset \mathfrak{M}.$$

$\mathfrak{L}S$ is a left ideal of $\mathfrak{A}$ contained in the minimal left ideal $\mathfrak{M}$. By definition, this is possible only when $\mathfrak{M} = \mathfrak{L}S$.

(3) Let the minimal left ideals $\mathfrak{M}_1$ and $\mathfrak{M}_2$ of a semigroup $\mathfrak{A}$ have a common element X. The intersection $\mathfrak{M}_0$ of the left ideals $\mathfrak{M}_1$ and $\mathfrak{M}_2$, being nonempty, will itself be a left ideal of $\mathfrak{A}$. It will be included in both $\mathfrak{M}_1$ and $\mathfrak{M}_2$, and since they are minimal left ideals, $\mathfrak{M}_0 = \mathfrak{M}_1$ and $\mathfrak{M}_0 = \mathfrak{M}_2$.

2.3. COROLLARY. *If a semigroup $\mathfrak{A}$ possesses minimal left ideals, then each left ideal of $\mathfrak{A}$ contains some minimal left ideal of $\mathfrak{A}$.*

PROOF. Let $\mathfrak{L}$ be a minimal left ideal of $\mathfrak{A}$ and let $\mathfrak{N}$ be an arbitrary left ideal of $\mathfrak{A}$. Then for any $N \in \mathfrak{N}$, $\mathfrak{L}N$ belongs to $\mathfrak{N}$. By 2.2, $\mathfrak{L}N$ is a minimal left ideal of $\mathfrak{A}$.

2.4. As we have already noted, the relation of being a left or right ideal is not transitive. Thus it does not follow directly from the definition that minimal left and right ideals themselves cannot have their own corresponding left and right ideals. We will show that this cannot occur. It will follow in particular that a left ideal of a semigroup will be a minimal left ideal if and only if it is a semigroup without proper left ideals.

THEOREM. *If $\mathfrak{L}$ is a minimal left ideal of a semigroup $\mathfrak{A}$, then $\mathfrak{L}$ is a semigroup without proper left (and thus also two-sided) ideals.*

PROOF. Let $\mathfrak{N}$ be an arbitrary left ideal of the semigroup $\mathfrak{L}$. Then

$$\mathfrak{L}\mathfrak{N} \subset \mathfrak{N} \subset \mathfrak{L}.$$

But since $\mathfrak{L}$ is a left ideal of a semigroup $\mathfrak{A}$,

$$\mathfrak{A}(\mathfrak{L}\mathfrak{N}) = (\mathfrak{A}\mathfrak{L})\mathfrak{N} \subset \mathfrak{L}\mathfrak{N}.$$

Consequently, $\mathfrak{L}\mathfrak{N}$ is a left ideal of $\mathfrak{A}$ included in the minimal left ideal $\mathfrak{L}$ of the semigroup $\mathfrak{A}$. This is possible only when $\mathfrak{L}\mathfrak{N} = \mathfrak{L}$. But, as we showed earlier, $\mathfrak{L}\mathfrak{N} \subset \mathfrak{N}$. Consequently,

$$\mathfrak{N} = \mathfrak{L},$$

i.e., $\mathfrak{L}$ has no proper left ideals.

2.5. Among the different ideals of a semigroup which we have already discussed and will discuss below, of special interest are the minimal two-sided ideals. As we have already remarked, if such an ideal exists, it is the universally minimal two-sided ideal (IV, 4.4). In § 1 we considered a special case of such ideals, namely, two-sided ideals that are groups. However, minimal two-sided ideals may also exist in other cases. One sufficient condition for the existence of a minimal two-sided ideal in a semigroup is the presence in the semigroup of minimal left (or right) ideals. In this case the minimal two-sided ideal is their union.

THEOREM. *If a semigroup $\mathfrak{A}$ possesses a minimal left ideal, then $\mathfrak{A}$ possesses a minimal two-sided ideal which is equal to the union of all the minimal left ideals.*

PROOF. Let $\mathfrak{L}$ be one of the minimal left ideals of a semigroup $\mathfrak{A}$. It follows from 2.2 that

$$\mathfrak{L}\mathfrak{A} = \bigcup_{S \in \mathfrak{A}} \mathfrak{L}S$$

is the union of all the minimal left ideals of $\mathfrak{A}$. Obviously $\mathfrak{L}\mathfrak{A}$ is a two-sided ideal of $\mathfrak{A}$. Let $\mathfrak{T}$ be an arbitrary two-sided ideal of $\mathfrak{A}$. The product $\mathfrak{T} \cdot (\mathfrak{L}S)$ is a left ideal of $\mathfrak{A}$ included in the minimal left ideal $\mathfrak{L}S$. Consequently,

$$\mathfrak{T}\mathfrak{L}S = \mathfrak{L}S.$$

On the other hand, it follows from the fact that $\mathfrak{T}$ is a two-sided ideal of $\mathfrak{A}$ that

$$\mathfrak{T}\mathfrak{L}S \subset \mathfrak{T}.$$

Consequently,

$$\mathfrak{L}S \subset \mathfrak{T}.$$

Thus,

$$\mathfrak{L}\mathfrak{A} \subset \mathfrak{T},$$

i.e., $\mathfrak{L}\mathfrak{A}$ is the universally minimal two-sided ideal of $\mathfrak{A}$.

2.6. For the case considered in the theorem of 2.5, it is easy to explain which are the left ideals of the minimal two-sided ideal of a semigroup.

THEOREM. *Let a semigroup $\mathfrak{A}$ possess minimal left ideals and let $\mathfrak{K}$ be the minimal two-sided ideal of $\mathfrak{A}$. Then the left ideals of $\mathfrak{A}$ included in $\mathfrak{K}$ are the only left ideals of the semigroup $\mathfrak{K}$.*

PROOF. It is obvious that every left ideal of the semigroup $\mathfrak{A}$ contained in $\mathfrak{K}$ is a left ideal of $\mathfrak{K}$.

Let $\mathfrak{T}$ be an arbitrary left ideal of the semigroup $\mathfrak{K}$ and let T be an arbitrary element of $\mathfrak{T}$. Since $\mathfrak{K}$ by 2.5 is the union of the minimal left ideals of $\mathfrak{A}$, T is included in some minimal left ideal $\mathfrak{L}$ of the semigroup $\mathfrak{A}$. Clearly, $\mathfrak{K}T$ is a left ideal of the semigroup $\mathfrak{A}$ contained in $\mathfrak{L}$. Consequently,

$$\mathfrak{L} = \mathfrak{K}T.$$

In particular, $T \subset \mathfrak{K}T$. Thus each element of $\mathfrak{T}$ is included in $\mathfrak{K}\mathfrak{T}$, i.e.,

$$\mathfrak{T} \subset \mathfrak{K}\mathfrak{T}.$$

On the other hand, it follows from the fact that $\mathfrak{T}$ is a left ideal of $\mathfrak{K}$ that

$$\mathfrak{T} \supset \mathfrak{K}\mathfrak{T}.$$

Consequently,

$$\mathfrak{T} = \mathfrak{K}\mathfrak{T}.$$

But $\mathfrak{K}\mathfrak{T}$ is obviously a left ideal of the semigroup $\mathfrak{A}$.

2.7. The question of the existence in a semigroup of minimal left, right, and two-sided ideals, and also of the way in which they are interrelated, has great importance in the study of many of the properties of semigroups. It was shown in 2.5 that if a semigroup $\mathfrak{A}$ possesses minimal left or minimal right ideals, it will also possess a minimal two-sided ideal $\mathfrak{K}$. Here $\mathfrak{K}$ by 2.5 will contain all the minimal left and minimal right ideals of the semigroup $\mathfrak{A}$, which by 2.6 are respectively minimal left and right ideals of the semigroup $\mathfrak{K}$. But $\mathfrak{K}$, as was shown in IV, 4.8, is a semigroup without proper two-sided ideals. Thus the question of the relationship between the minimal left and minimal right ideals of an arbitrary semigroup $\mathfrak{A}$ turns out to be equivalent to the question of the relationship between minimal left and minimal right ideals of the semigroup $\mathfrak{K}$, which is a semigroup without proper two-sided ideals. This situation presents another reason for the interest in semigroups without proper two-sided ideals (in passing we mention that we have already indicated the fundamental importance of such semigroups in IV, 4.10).

2.8. The condition of the absence in a semigroup of proper two-sided ideals is clearly equivalent to that of the semigroup being itself its own minimal two-sided ideal. In the presence of minimal left ideals in the semigroup, this will occur by 2.5, if and only if the semigroup is the union of all of its minimal left ideals.

2.9. THEOREM. *Let a semigroup $\mathfrak{A}$ possess minimal left ideals. In order for there to be no proper two-sided ideals in $\mathfrak{A}$ it is necessary and sufficient for each principal left ideal* (IV, 3.6) *to be a minimal left ideal.*

PROOF. (1) Each element is included in its own principal left ideal. It follows that if all the principal left ideals are minimal left ideals, then $\mathfrak{A}$ is the union of minimal left ideals and thus has no proper two-sided ideals (2.8).

(2) If $\mathfrak{A}$ has no proper two-sided ideals, then by 2.8 each element A is included in some minimal left ideal $\mathfrak{L}$. The principal left ideal of the element A is included in the left ideal $\mathfrak{L}$ containing A. It follows from the minimality of $\mathfrak{L}$ that the principal left ideal of A coincides with $\mathfrak{L}$.

2.10. THEOREM. *Let a semigroup $\mathfrak{A}$ possess minimal left ideals. In order for there to be no proper two-sided ideals in $\mathfrak{A}$ it is necessary and sufficient for each right divisor of an arbitrary element of the semigroup to be itself always divisible on the right by this element.* (This condition means the symmetry of the relation of divisibility on the right.)

PROOF. (1) Let $\mathfrak{A}$ have no proper two-sided ideals.

It is necessary for us to show that from the fact that some element A is divisible on the right by B, namely,

$$A = XB \qquad (A, B, X \in \mathfrak{A}),$$

it follows that B is also divisible on the right by A.

If $A = B$, this is clear. Let $A \neq B$. Since

$$\mathfrak{A}A \cup A = \mathfrak{A}XB \cup XB \subset \mathfrak{A}B \cup B,$$

where $\mathfrak{A}A \cup A$ and $\mathfrak{A}B \cup B$ are principal left ideals of $\mathfrak{A}$, it follows from 2.9 that

$$\mathfrak{A}A \cup A = \mathfrak{A}B \cup B.$$

Since $B \neq A$, it follows from this that $B \in \mathfrak{A}A$, i.e., for some Y,

$$B = Y\mathfrak{A}.$$

(2) Suppose that in $\mathfrak{A}$ every right divisor of an arbitrary element is itself divisible on the right by this element.

We show that an arbitrary principal left ideal $\mathfrak{L} = A \cup \mathfrak{A}A$ $(A \in \mathfrak{A})$ of $\mathfrak{A}$ is a minimal left ideal. By 2.9 it will follow from this that $\mathfrak{A}$ does not have proper two-sided ideals.

Let $\mathfrak{L}'$ be a left ideal of $\mathfrak{A}$ contained in $\mathfrak{L}$, and let $B \in \mathfrak{L}'$. If $B = A$, then obviously $\mathfrak{L}' = \mathfrak{L}$. Suppose $B \neq A$. Since $\mathfrak{L}'$ is a left ideal, $B \cup \mathfrak{A}B \subset \mathfrak{L}' \subset \mathfrak{L}$. Since $B \in \mathfrak{L}$ and $B \neq A$, it follows that $B = XA$ for some $X \in \mathfrak{A}$. Hence by the assumption on $\mathfrak{A}$ it follows that A in turn must be divisible on the right by B, i.e. $A = YB$. Thus

$$\mathfrak{L} = A \cup \mathfrak{A}A = YB \cup \mathfrak{A}YB \subset \mathfrak{A}B \subset \mathfrak{L}'$$

and, consequently, $\mathfrak{L}' = \mathfrak{L}$.

2.11. We dwell in particular on the case studied by Clifford and Miller [1], where the semigroup has a unique minimal left ideal. Such an ideal will by 2.3 be a universally minimal left ideal. There is a necessary and sufficient condition for the existence of such an ideal that may be obtained directly from the next theorem.

THEOREM. *If there are among the elements of a semigroup $\mathfrak{A}$ elements that are divisible on the right by all the elements of $\mathfrak{A}$, then the collection of all such elements is a minimal two-sided ideal of $\mathfrak{A}$ that is a universally minimal left ideal of $\mathfrak{A}$.*

If there are among the elements of $\mathfrak{A}$ no elements that are divisible on the right by all the elements of $\mathfrak{A}$, then $\mathfrak{A}$ either has no minimal left ideals or it has more than one.

PROOF. (1) Let the collection $\mathfrak{U}$ of all the elements of $\mathfrak{A}$ that are divisible on the right by all the elements of $\mathfrak{A}$ be nonempty. For any $A_1, A_2 \in \mathfrak{A}$ and $U \in \mathfrak{U}$ in $\mathfrak{A}$ there exists an element X such that

$$U = XA_1.$$

Thus

$$(A_2U) = (A_2X) \cdot A_1.$$

Consequently, A_2U is divisible on the right by any element A_1 from $\mathfrak{A}$, i.e., $A_2U \in \mathfrak{U}$. This means that $\mathfrak{U}$ is a left ideal.

If $\mathfrak{B}$ is an arbitrary left ideal of $\mathfrak{A}$, then for any $U \in \mathfrak{U}$ and $V \in \mathfrak{B}$ there exists a $Z \in \mathfrak{A}$ such that

$$U = ZV.$$

Since $\mathfrak{B}$ is a left ideal of $\mathfrak{A}$, we have $ZV \in \mathfrak{B}$. It follows that $\mathfrak{U}$ is a universally minimal left ideal. Of course, there can be no other minimal left ideals of $\mathfrak{A}$; for, by 2.5, $\mathfrak{U}$ is the minimal two-sided ideal of $\mathfrak{A}$.

(2) Let $\mathfrak{L}$ be a unique minimal left ideal of $\mathfrak{A}$. For any $A \in \mathfrak{A}$ the product $\mathfrak{L}A$ is by 2.2 also necessarily a minimal left ideal of $\mathfrak{A}$. Consequently,

$$\mathfrak{L}A = \mathfrak{L}.$$

But this means that for any $L \in \mathfrak{L}$ there exists in $\mathfrak{L}$ an X such that

$$XA = L.$$

Consequently, the elements of $\mathfrak{L}$ are such that each of them is divisible on the right by each element of $\mathfrak{A}$. It is only in the presence of such elements that $\mathfrak{A}$ can possess a unique minimal left ideal.

2.12. In 1.2 we saw that when there exists in the semigroup a two-sided ideal that is a group, the semigroup has a unique minimal left and a unique minimal right ideal. By the theorem of 2.11 it is now possible to state the converse.

COROLLARY. *If a semigroup $\mathfrak{A}$ has a unique minimal left ideal $\mathfrak{U}$ and a unique minimal right ideal $\mathfrak{B}$, then $\mathfrak{U}$ is equal to $\mathfrak{B}$, is a two-sided ideal of $\mathfrak{A}$ and is a group.*

PROOF. By 2.11, $\mathfrak{U}$ is a minimal two-sided ideal of $\mathfrak{A}$. Analogously, $\mathfrak{B}$ is a minimal two-sided ideal of $\mathfrak{A}$. Consequently, $\mathfrak{U} = \mathfrak{B}$. Here $\mathfrak{U}$, equal to $\mathfrak{B}$, is the collection of all the elements divisible on the left by all the elements of $\mathfrak{A}$. By 1.4 this collection is a two-sided ideal that is a group.

2.13. Using the above corollary we can easily discuss the question of minimal left ideals of inverse semigroups (II, 7.2) which were studied in detail by Preston [**2**]. It turns out that an inverse semigroup can have no more than one minimal left ideal, which in this case turns out to be simultaneously a minimal right ideal. This follows from the next theorem if we keep in mind the theorem of 1.2.

THEOREM. *If in the set of the idempotents of an inverse semigroup $\mathfrak{A}$ the idempotent I_0 is a two-sided zero of this set, then*

$$\mathfrak{A}I_0\mathfrak{A} = \mathfrak{A}I_0 = I_0\mathfrak{A} = I_0\mathfrak{A}I_0$$

is a two-sided ideal that is a group.

If none of the idempotents is a two-sided zero of the set, then there are neither minimal left ideals nor minimal right ideals in $\mathfrak{A}$.

PROOF. (1) Let the idempotent I_0 be a zero in the set of all the idempotents of $\mathfrak{A}$. For an arbitrary idempotent I it follows from $I_0I = I_0$ that the left ideal $\mathfrak{A}I_0$ is included in the principal left ideal $\mathfrak{A}I$. Since every principal left ideal has the form $\mathfrak{A}I$ (IV, 3.17), therefore $\mathfrak{A}I_0$ is contained in every left ideal, i.e., is a universally minimal left ideal. Similarly, $I_0\mathfrak{A}$ is a universally minimal right ideal. By 2.12,

$$\mathfrak{A}I_0 = I_0\mathfrak{A}$$

is a two-sided ideal which is a group. In particular, it follows that

$$\mathfrak{A}I_0\mathfrak{A} = \mathfrak{A}I_0 = I_0\mathfrak{A} = I_0\mathfrak{A}I_0.$$

(2) Let $\mathfrak{L}$ be a minimal left ideal of $\mathfrak{A}$. It clearly is the left ideal envelope of every element of $\mathfrak{L}$. By IV, 3.17, $\mathfrak{L}$ contains an idempotent I, where

$$\mathfrak{L} = \mathfrak{A}I.$$

Let I' be an arbitrary idempotent of $\mathfrak{A}$. The intersection of the left ideals

$\mathfrak{L} = \mathfrak{A}I$ and $\mathfrak{A}I'$ is nonempty since it contains II'. But this intersection is a left ideal included in $\mathfrak{L}$. Consequently it coincides with $\mathfrak{L}$. Thus

$$\mathfrak{A}I' \supset \mathfrak{A}I.$$

By IV, 3.17 it follows that $I'I = II' = I$. Consequently, the idempotent I is a two-sided zero in the set of all the idempotents.

2.14. In connection with the theorem of 2.13 one should note that an inverse semigroup that has only a finite number of idempotents (in particular, any finite inverse semigroup) always possesses an idempotent that is a zero in the set of all the idempotents. As easily follows from the commutativity of the idempotents in an inverse semigroup, such an idempotent is in this case the product of all the idempotents of the inverse semigroup.

3. Semigroups with both Minimal Left Ideals and Minimal Right Ideals

3.1. In the preceding section we saw the important role played by the very fact of the existence of minimal left ideals in a semigroup. Of course, analogous results hold for the existence in a semigroup of minimal right ideals. Even more important is the case when a semigroup possesses simultaneously both minimal left and minimal right ideals. Such semigroups were studied by Rees [**1**], Schwarz [3], Clifford [**6**], and Hashimoto [**2**; **3**]. In the present section we will consider some of these properties.

3.2. It is easy to see that the intersection of any left ideal with any right ideal is always nonempty, for it contains their product (in which the right ideal is taken as the left factor). When these ideals are minimal we obtain an important property for this intersection.

THEOREM. *Let $\mathfrak{L}$ be some minimal left ideal and $\mathfrak{R}$ some minimal right ideal of a semigroup $\mathfrak{A}$. Then $\mathfrak{G} = \mathfrak{R}\mathfrak{L}$ is a group, where*

$$\mathfrak{R} = E_{\mathfrak{G}}\mathfrak{A}, \qquad \mathfrak{L} = \mathfrak{A}E_{\mathfrak{G}}, \qquad \mathfrak{G} = \mathfrak{R}\mathfrak{L} = \mathfrak{R} \cap \mathfrak{L} = E_{\mathfrak{G}}\mathfrak{A}E_{\mathfrak{G}}.$$

PROOF. Since $\mathfrak{L}$ is a left ideal,

$$\mathfrak{G}\mathfrak{G} = \mathfrak{R}\mathfrak{L} \cdot \mathfrak{R}\mathfrak{L} = \mathfrak{R} \cdot \mathfrak{L}\mathfrak{R}\mathfrak{L} \subset \mathfrak{R}\mathfrak{L} = \mathfrak{G}.$$

Consequently, $\mathfrak{G}$ is a subsemigroup of $\mathfrak{A}$.

Let G be an arbitrary element of $\mathfrak{G}$. Since $\mathfrak{G} \in \mathfrak{R}$, $G\mathfrak{R}$ is a right ideal of the semigroup $\mathfrak{A}$ and belongs to $\mathfrak{R}$. Consequently,

$$G\mathfrak{R} = \mathfrak{R},$$

from which we obtain, by multiplication on the right by $\mathfrak{L}$,

$$G\mathfrak{G} = \mathfrak{G}.$$

It is shown analogously that

$$\mathfrak{G}G = \mathfrak{G},$$

from which it follows that $\mathfrak{G}$ is a group.

Since $E_{\mathfrak{G}} \in \mathfrak{G} \subset \mathfrak{L}$, $\mathfrak{A}E_{\mathfrak{G}}$ is a left ideal of $\mathfrak{A}$ contained in $\mathfrak{L}$. Consequently,

$$\mathfrak{A}E_{\mathfrak{G}} = \mathfrak{L},$$

which means that $\mathfrak{L}$ is the set of all the elements of $\mathfrak{A}$ for which $E_{\mathfrak{G}}$ is a right identity. We show analogously that

$$E_{\mathfrak{G}}\mathfrak{A} = \mathfrak{R},$$

i.e., $\mathfrak{R}$ is the set of all the elements of $\mathfrak{A}$ for which $E_{\mathfrak{G}}$ is a left identity. It follows that $\mathfrak{R} \cap \mathfrak{L}$ is the set of all the elements of $\mathfrak{A}$ for which $E_{\mathfrak{G}}$ is a two-sided identity. Consequently,

$$\mathfrak{R} \cap \mathfrak{L} = E_{\mathfrak{G}}\mathfrak{A}E_{\mathfrak{G}},$$

from which it follows in particular that

$$\mathfrak{R} \cap \mathfrak{L} \supset E_{\mathfrak{G}}\mathfrak{A} \cdot \mathfrak{A}E_{\mathfrak{G}} = \mathfrak{R}\mathfrak{L} = \mathfrak{G}.$$

But, on the other hand, if $X \in \mathfrak{R} \cap \mathfrak{L}$, then

$$X = XE_{\mathfrak{G}} \subset \mathfrak{R}\mathfrak{L} = \mathfrak{G}.$$

Consequently,

$$\mathfrak{R} \cap \mathfrak{L} = \mathfrak{G}.$$

3.3. Using the theorem of 3.2 it is possible to clear up completely the character of the collection of those idempotents of the semigroups we have been considering that are contained in a minimal two-sided ideal. The very fact that every minimal left ideal possesses idempotents is interesting in itself.

Let $\Gamma^{(l)}$ be the set of all the minimal left ideals of a semigroup $\mathfrak{A}$ and let $\Gamma^{(r)}$ be the set of all the minimal right ideals.

By 3.2 the intersection

$$\mathfrak{R} \cap \mathfrak{L} = \mathfrak{R}\mathfrak{L} = \mathfrak{G}_{\mathfrak{L},\mathfrak{R}} \qquad (\mathfrak{L} \in \Gamma^{(l)},\ \mathfrak{R} \in \Gamma^{(r)})$$

is always nonempty and contains a unique idempotent, namely, the identity of the group $\mathfrak{G}_{\mathfrak{L},\mathfrak{R}}$, which will be denoted by $E_{\mathfrak{L},\mathfrak{R}}$.

By 2.5 the minimal two-sided ideal $\mathfrak{K}$ of the semigroup $\mathfrak{A}$ may be represented in the form of the unions

$$\mathfrak{K} = \bigcup_{\mathfrak{L}\in\Gamma^{(l)}} \mathfrak{L} = \bigcup_{\mathfrak{R}\in\Gamma^{(r)}} \mathfrak{R},$$

where the components of each union are pairwise nonintersecting. It follows from this that

$$\mathfrak{L} = \bigcup_{\mathfrak{R}\in\Gamma^{(r)}} \mathfrak{G}_{\mathfrak{L},\mathfrak{R}} \qquad (\mathfrak{L} \in \Gamma^{(l)}),$$

$$\mathfrak{R} = \bigcup_{\mathfrak{L}\in\Gamma^{(l)}} \mathfrak{G}_{\mathfrak{L},\mathfrak{R}} \qquad (\mathfrak{R} \in \Gamma^{(r)}),$$

$$\mathfrak{K} = \bigcup_{\substack{\mathfrak{L}\in\Gamma^{(l)} \\ \mathfrak{R}\in\Gamma^{(r)}}} \mathfrak{G}_{\mathfrak{L},\mathfrak{R}},$$

where the components of each union are nonempty and pairwise nonintersecting.

It follows from what has been said that the idempotents of $\mathfrak{K}$ are $E_{\mathfrak{L},\mathfrak{R}}$. Thus, we obtain a one-to-one correspondence between those idempotents of the semigroup contained in its minimal two-sided ideal $\mathfrak{K}$ and the pairs of ideals

$$E_{\mathfrak{L},\mathfrak{R}} \sim (\mathfrak{L}, \mathfrak{R}) \qquad (\mathfrak{L} \in \Gamma^{(l)}, \mathfrak{R} \in \Gamma^{(r)}).$$

3.4. We note especially the following property of idempotents which we will need later.

COROLLARY. *Let a semigroup $\mathfrak{A}$ have minimal left and minimal right ideals. Then no idempotent of the minimal two-sided ideal $\mathfrak{K}$ can be a two-sided identity of any idempotent other than itself.*

PROOF. Let it be true for the idempotents I_1 and I_2 of the semigroup $\mathfrak{A}$ that

$$I_1 I_2 I_1 = I_2,$$

where $I_1 \in \mathfrak{K}$. By 3.3,

$$I_1 = E_{\mathfrak{L},\mathfrak{R}} \qquad (\mathfrak{L} \in \Gamma^{(l)}, \mathfrak{R} \in \Gamma^{(r)}).$$

Thus

$$I_2 = I_1 I_2 I_1 \subset \mathfrak{R}\mathfrak{L} I_2 \mathfrak{R}\mathfrak{L} = \mathfrak{R} \cdot \mathfrak{L} I_2 \mathfrak{R} \cdot \mathfrak{L} \subset \mathfrak{R}\mathfrak{L} = \mathfrak{G}_{\mathfrak{L},\mathfrak{R}}.$$

But there is only one idempotent, $E_{\mathfrak{L},\mathfrak{R}} = I_1$, in $\mathfrak{G}_{\mathfrak{L},\mathfrak{R}}$. Hence $I_2 = I_1$.

3.5. As we already know (2.5), the presence in a semigroup of a minimal left or a minimal right ideal implies the existence of a minimal two-sided ideal. We consider the connection between these minimal ideals.

THEOREM. *Let $\mathfrak{L}$ be a minimal left ideal, $\mathfrak{R}$ a minimal right ideal and $\mathfrak{K}$ the minimal two-sided ideal of a semigroup $\mathfrak{A}$. Then*

$$\mathfrak{K} = \mathfrak{L}\mathfrak{R} = \mathfrak{L}\mathfrak{A}\mathfrak{R} = \mathfrak{L}\mathfrak{A} = \mathfrak{A}\mathfrak{R} = \mathfrak{L}\mathfrak{K} = \mathfrak{K}\mathfrak{R}.$$

PROOF. By 2.5, $\mathfrak{L} \subset \mathfrak{K}$. Analogously, $\mathfrak{R} \subset \mathfrak{K}$. It follows from this that each of the sets $\mathfrak{L}\mathfrak{R}$, $\mathfrak{L}\mathfrak{A}\mathfrak{R}$, $\mathfrak{L}\mathfrak{A}$, $\mathfrak{A}\mathfrak{R}$, $\mathfrak{L}\mathfrak{K}$ and $\mathfrak{K}\mathfrak{R}$ is included in $\mathfrak{K}$. But they are all obviously two-sided ideals. Inasmuch as $\mathfrak{K}$ is a minimal two-sided ideal, each of these six ideals must coincide with $\mathfrak{K}$.

3.6. COROLLARY. *If $\mathfrak{L}$ and $\mathfrak{L}'$ are minimal left ideals of $\mathfrak{A}$ while $\mathfrak{R}$ and $\mathfrak{R}'$ are minimal right ideals of $\mathfrak{A}$, then*

$$\mathfrak{L}\mathfrak{R} = \mathfrak{L}'\mathfrak{R}'.$$

By 3.5 both $\mathfrak{L}\mathfrak{R}$ and $\mathfrak{L}'\mathfrak{R}$ are equal to $\mathfrak{K}$.

3.7. We note also that for a minimal left ideal $\mathfrak{L}$ and a minimal right ideal $\mathfrak{R}$ of a semigroup $\mathfrak{A}$ the group $\mathfrak{R}\mathfrak{L}$ (3.2) is included in the minimal two-sided ideal $\mathfrak{K}$ of the semigroup $\mathfrak{A}$. This is true since by 2.5 we have $\mathfrak{L} \subset \mathfrak{K}$.

3.8. In the case under consideration we can explain without difficulty how all the minimal left and right ideals of a semigroup may be obtained.

THEOREM. *If a semigroup $\mathfrak{A}$ possesses minimal left and minimal right ideals, then the left ideal envelopes of the idempotents that belong to the minimal two-sided ideal, and only they, are the minimal left ideals of the semigroup.*

The analogous statement holds for right ideals.

PROOF. (1) Let I be an idempotent included in the minimal two-sided ideal $\mathfrak{K}$. Since I is an idempotent its left ideal envelope is $\mathfrak{A}I$.

By 2.5, I is included in some minimal left ideal $\mathfrak{L}$ of the semigroup $\mathfrak{A}$. Similarly, I is included in some minimal right ideal $\mathfrak{R}$ of the semigroup $\mathfrak{A}$. By 3.2, $\mathfrak{R}\mathfrak{L}$ is a group. Since $I = I \cdot I \in \mathfrak{R}\mathfrak{L}$, I is the identity of this group and, by 3.2, $\mathfrak{A}I = \mathfrak{L}$.

(2) Let $\mathfrak{L}$ be an arbitrary minimal left ideal of $\mathfrak{A}$. We choose some minimal right ideal $\mathfrak{R}$ of the semigroup $\mathfrak{A}$. By 3.2, $\mathfrak{G} = \mathfrak{R}\mathfrak{L}$ is a group and $\mathfrak{L} = \mathfrak{A}E_{\mathfrak{G}}$. Here $\mathfrak{A}E_{\mathfrak{G}}$ is obviously the left ideal envelope of the idempotent $E_{\mathfrak{G}}$ which belongs to $\mathfrak{K}$ by 3.5.

3.9. In IV, 4.10 we were concerned with the important role of semigroups having no proper two-sided ideals in the analysis of the structure of an arbitrary semigroup. In the general case such semigroups may be quite complicated and we do not yet have a sufficiently clear conception of their structure. However, with one additional quite natural assumption, it is possible to obtain a rather clear description of the structure of such semigroups.

DEFINITION. *If a semigroup with no proper two-sided ideals possesses minimal left and minimal right ideals, it is said to be a* COMPLETELY SIMPLE SEMIGROUP WITHOUT ZERO.

The words "without zero" in this definition are justified by the fact that, with the exception of the trivial case when the semigroup consists of one element, the semigroup so defined cannot have a zero, since it would be a proper two-sided ideal. The necessity of inserting these words arises from the necessity of considering a related class of semigroups that have a zero. The next two sections will be devoted to that class.

It should be kept in mind that for the definition of a completely simple semigroup many authors make use of a set of properties which is equivalent to the above definition. The approach to completely simple semigroups which we use in the present book is due to Clifford [**6**; **7**].

This approach is more natural than the original approach of Rees [**1**], who was the first to give in the general case the description of the structure of completely simple semigroups which will be presented below. Subsequently many mathematicians (see 1.1) studied the various properties of completely simple semigroups. The starting point for all of them was the work of A. K. Suškevič [**3**], who obtained most of these properties for finite semigroups.

3.10. The first definition of a completely simple semigroup, given by Rees

[**1**; **2**], is based on the concept of a so-called primitive idempotent. A nonzero idempotent I is said to be *primitive* if it is not a two-sided identity for any other nonzero idempotent. It follows from 3.2 that a completely simple semigroup without zero that is not the identity semigroup always possesses nonzero idempotents, and by 3.4 all of them are primitive.

As we will show in 6.15, the existence of at least one primitive idempotent in a semigroup that has no proper two-sided ideals is sufficient for it to be a completely simple semigroup without zero.

3.11. We note that the various properties of minimal ideals obtained already provide us with a whole series of properties of completely simple semigroups with zero (for example, the existence for each element of a two-sided identity, which follows directly from 2.8 and 3.2, and so on). Using these properties we could already give a description of the structure of an arbitrary completely simple semigroup without zero. However, for economy of space we will leave this until § 6, where this description will be obtained from simple corollaries of theorems concerning semigroups of a related class.

3.12. If a semigroup $\mathfrak{A}$ possesses a zero, then the properties of the minimal ideals of $\mathfrak{A}$ considered in the present and preceding sections will become completely trivial. In fact, a subset of the semigroup $\mathfrak{A}$ consisting of one element $O_{\mathfrak{A}}$ will be in $\mathfrak{A}$ simultaneously a minimal left ideal, a minimal right ideal and a minimal two-sided ideal. Of course, there are no other minimal ideals of $\mathfrak{A}$. All the relationships found above of minimal ideals of different kinds will coincide in a trivial manner. A semigroup with zero that has no proper two-sided ideals turns out to be merely a semigroup consisting of one element.

Since the approach to the study of the properties of semigroups without zero from the point of view of properties of their minimal ideals gives valuable results, the desire naturally arises to try to transform this approach so that it will also be suitable for the study of semigroups with zero. A number of successful attempts have been made in this direction. Instead of the concept of an ideal, the concept of a nonzero ideal serves as a basis.

3.13. DEFINITION. *An ideal (left, right or two-sided), is said to be* NONZERO *if it includes elements that are not zeros of the semigroup.*

A minimal ideal in the set of all the left nonzero ideals is said to be a MINIMAL LEFT NONZERO IDEAL. *Minimal right nonzero ideals and minimal two-sided nonzero ideals are defined analogously.*

3.14. The theory of minimal nonzero ideals arising in connection with these concepts is to a great degree parallel with the theory of minimal ideals of a semigroup without zero. Many results may be transferred, though usually with various weakening stipulations and complications. The reason for this clearly lies in the fact that the concept of nonzero ideals is of course a less natural one.

The first significant complication arises from the fact that the intersection of two nonzero ideals can sometimes consist of a zero, i.e., not belong to the class of nonzero ideals. In this connection, a minimal two-sided nonzero ideal may not be a universally minimal two-sided nonzero ideal. Moreover, a semigroup may have several such ideals, a fact which is evident from the example of a semigroup with a zero in which the product of any two elements is equal to zero (in this case any element X together with the zero clearly forms a minimal two-sided nonzero ideal $\{X, 0\}$). This example shows that the theorem of 2.5, while important in the theory considered above, can not be transferred in full into the theory of nonzero ideals.

For both these reasons (the merely partial parallelism with the theory of minimal ideals and the relatively less complete state of the theory) we will not make any special study of the theory of minimal nonzero ideals (we refer the reader in this connection to the articles of Rees **[1]**, Clifford [7], and Schwarz **[4]**, which contain a number of properties of such ideals). We will limit ourselves to examining in the next section certain properties in this theory (essentially obtained in the articles just mentioned) which are connected with a certain important class of semigroups.

4. Completely Simple Semigroups with Zero

4.1. For later requirements it is necessary for us to pay some attention here to one of the directions indicated at the end of the preceding section for the theory of minimal nonzero ideals. Namely, we must consider semigroups with zero having no proper two-sided ideals other than zero. We considered the very important role of these semigroups in IV, 4.10. A complete study of their properties and construction is possible at the present time under certain limitations similar to those made in the preceding sections.

4.2. DEFINITION. *If a semigroup* $\mathfrak{A}$ *with zero possessing the property that*

$$\mathfrak{A}\mathfrak{A} \neq O_{\mathfrak{A}}$$

has no proper two-sided nonzero ideals and has minimal left nonzero ideals and minimal right nonzero ideals (3.13), *then it is said to be a* COMPLETELY SIMPLE SEMIGROUP WITH ZERO.

4.3. We note that the condition

$$\mathfrak{A}\mathfrak{A} \neq O_{\mathfrak{A}}$$

is made only to exclude the following two semigroups: the semigroup consisting of one element and the semigroup consisting of two elements A and O with the rule of multiplication

$$AO = OA = A^2 = O^2 = O.$$

Both these semigroups clearly satisfy the remaining conditions of the definition. However, if these two semigroups were not excluded by the condition $\mathfrak{A}\mathfrak{A} \neq O_{\mathfrak{A}}$ from the definition of completely simple semigroups with zero (for both of which, of course, $\mathfrak{A}\mathfrak{A} = O_{\mathfrak{A}}$), then in the discussion of the rest of the theory we would continually have to make tedious conventions.

For any semigroup other than these two, the property $\mathfrak{A}\mathfrak{A} \neq O_{\mathfrak{A}}$ is a consequence of the absence of proper two-sided nonzero ideals. In fact, a semigroup with more than two elements for which $\mathfrak{A}\mathfrak{A} = O_{\mathfrak{A}}$ has a proper two-sided ideal consisting of $O_{\mathfrak{A}}$ and any nonzero element.

We note in passing that in a completely simple semigroup with zero the condition under consideration implies the equality

$$\mathfrak{A}\mathfrak{A} = \mathfrak{A}.$$

In fact, since $\mathfrak{A}\mathfrak{A} \neq O_{\mathfrak{A}}$, $\mathfrak{A}\mathfrak{A}$ is a two-sided nonzero ideal of $\mathfrak{A}$ and thus must be equal to $\mathfrak{A}$.

4.4. We begin the derivation of several properties of completely simple semigroups with zero. The choice of these properties is mostly explained by the requirements of the fundamental theorem of the next section which entirely clears up the structure of completely simple semigroups with zero, and which later aids in the explanation of the structure of completely simple semigroups without zero. The properties mentioned are noticeably similar to many of the properties of ideals considered in §§ 2 and 3. However, there are quite important differences which prevent us from treating the two theories simultaneously.

4.5. THEOREM. *A completely simple semigroup $\mathfrak{A}$ without zero is equal to the union of all of its minimal left nonzero ideals that have pairwise no common element other than $O_{\mathfrak{A}}$.*

PROOF. Let $\mathfrak{L}$ be a minimal left nonzero ideal of $\mathfrak{A}$. Obviously $\mathfrak{L}\mathfrak{A}$ is a two-sided ideal of $\mathfrak{A}$. If the ideal $\mathfrak{L}\mathfrak{U}$ consisted only of zero, this $\mathfrak{L}$ would be a two-sided ideal, i.e., would coincide with the semigroup $\mathfrak{A}$ itself. But then we would have

$$O_{\mathfrak{A}} = \mathfrak{L}\mathfrak{A} = \mathfrak{A}\mathfrak{A},$$

which is impossible for a completely simple semigroup with zero.

Consequently,

$$\mathfrak{L}\mathfrak{A} = \mathfrak{A}.$$

For every $A \in \mathfrak{A}$, the product $\mathfrak{L}A$ is clearly a left ideal. Let $\mathfrak{L}'$ be a left nonzero ideal of $\mathfrak{A}$ contained in $\mathfrak{L}A$.

We denote by $\mathfrak{N}$ the collection of all the elements X of $\mathfrak{L}$ for which $XA \in \mathfrak{L}'$. Some of these elements are nonzero inasmuch as $\mathfrak{L}' \subset \mathfrak{L}A$ and $\mathfrak{L}'$ is a nonzero ideal. Since $\mathfrak{L}'$ is a left ideal, for any $Z \in \mathfrak{A}$ and $X \in \mathfrak{N}$, the element ZX will also belong to $\mathfrak{N}$. Thus $\mathfrak{N}$ is a left nonzero ideal of $\mathfrak{A}$ included in $\mathfrak{L}$. It follows from this that $\mathfrak{N} = \mathfrak{L}$, and hence

$$\mathfrak{L}A \subset \mathfrak{L}'.$$

But $\mathfrak{L}'$ in turn is included in $\mathfrak{L}A$. Consequently, $\mathfrak{L}' = \mathfrak{L}A$, i.e., $\mathfrak{L}A$ has no proper left nonzero ideals.

Thus each product $\mathfrak{L}A$ $(A \in \mathfrak{A})$, if different from $O_{\mathfrak{A}}$, is a minimal left nonzero ideal. The equality obtained earlier implies that $\mathfrak{A}$ is the union of all the sets of the form $\mathfrak{L}A$. We may exclude from this union components equal to $O_{\mathfrak{A}}$ since $O_{\mathfrak{A}}$ is included in those $\mathfrak{L}A$ that are distinct from $O_{\mathfrak{A}}$. After this we obtain a representation of $\mathfrak{A}$ in the required form.

The intersection of two distinct minimal left nonzero ideals, being a left ideal of $\mathfrak{A}$, cannot be different from $O_{\mathfrak{A}}$. In fact, it would otherwise coincide with each of these ideals, which would contradict the fact that they were assumed to be distinct.

4.6. COROLLARY. *A completely simple semigroup $\mathfrak{A}$ with zero is the union of all the possible products of the form*

$$\mathfrak{R}\mathfrak{L},$$

where $\mathfrak{R}$ is a minimal right nonzero ideal of $\mathfrak{A}$ while $\mathfrak{L}$ is a minimal left nonzero ideal of $\mathfrak{A}$. The components of this union have pairwise only the element $O_{\mathfrak{A}}$ in common.

PROOF. By the theorem of 4.5 (applying it also for right ideals) we have

$$\mathfrak{A} = \mathfrak{A}\mathfrak{A} = \Big(\bigcup_{\mathfrak{R}} \mathfrak{R}\Big)\Big(\bigcup_{\mathfrak{L}} \mathfrak{L}\Big) = \bigcup_{\mathfrak{L},\mathfrak{R}} \mathfrak{R}\mathfrak{L}.$$

If $\mathfrak{R} \neq \mathfrak{R}'$, then $\mathfrak{R}$ and $\mathfrak{R}'$ have no elements in common other than $O_{\mathfrak{A}}$ (4.5). Since, for any $\mathfrak{L}$ and $\mathfrak{L}'$,

$$\mathfrak{R}\mathfrak{L} \subset \mathfrak{R} \text{ and } \mathfrak{R}'\mathfrak{L}' \subset \mathfrak{R}',$$

it follows that $\mathfrak{R}\mathfrak{L}$ and $\mathfrak{R}'\mathfrak{L}'$ have no elements in common other than zero. We argue similarly for $\mathfrak{L} \neq \mathfrak{L}'$.

4.7. In connection with this corollary the products of the form $\mathfrak{L}\mathfrak{R}$ play a natural role. Let us consider their structure in more detail.

THEOREM. *Let $\mathfrak{A}$ be a completely simple semigroup with zero, let $\mathfrak{L}$ be a minimal left nonzero ideal of $\mathfrak{A}$ and let $\mathfrak{R}$ be a minimal right nonzero ideal of $\mathfrak{A}$. Then*

$$\mathfrak{R}\mathfrak{L} \neq O_{\mathfrak{A}}.$$

If $\mathfrak{L}\mathfrak{R} \neq O_{\mathfrak{A}}$, then $\mathfrak{L}\mathfrak{R} = \mathfrak{A}$, while $\mathfrak{H} = \mathfrak{R}\mathfrak{L}$ is a group with an externally adjoined zero (II, 2.12), *where*

$$\mathfrak{L} = \mathfrak{A}E_{\mathfrak{H}}, \qquad \mathfrak{R} = E_{\mathfrak{H}}\mathfrak{A}, \qquad \mathfrak{H} = \mathfrak{R}\mathfrak{L} = \mathfrak{R} \cap \mathfrak{L} = E_{\mathfrak{H}}\mathfrak{A}E_{\mathfrak{H}}.$$

PROOF. (1) In the process of the proof of the theorem of 4.5 we saw that $\mathfrak{L}\mathfrak{A} = \mathfrak{A}$. Analogously $\mathfrak{A}\mathfrak{R} = \mathfrak{A}$. Thus

$$\mathfrak{A} = \mathfrak{A}\mathfrak{A} = \mathfrak{A}\mathfrak{R} \cdot \mathfrak{L}\mathfrak{A},$$

from which it follows that $\mathfrak{R}\mathfrak{L} \neq O_{\mathfrak{A}}$.

(2) In the future we will assume that

$$\mathfrak{L}\mathfrak{R} \neq O_{\mathfrak{A}}.$$

Since $\mathfrak{L}\mathfrak{R}$ is obviously a two-sided ideal of $\mathfrak{A}$, we have

$$\mathfrak{L}\mathfrak{R} = \mathfrak{A}.$$

(3) $\mathfrak{H} = \mathfrak{R}\mathfrak{L}$ is a subsemigroup:

$$\mathfrak{H}\mathfrak{H} = \mathfrak{R}\mathfrak{L}\mathfrak{R}\mathfrak{L} = \mathfrak{R} \cdot [(\mathfrak{L}\mathfrak{R}) \cdot \mathfrak{L}] \subset \mathfrak{R}\mathfrak{L} = \mathfrak{H}.$$

Let G be an arbitrary nonzero element of $\mathfrak{H}$. Since $G \in \mathfrak{R}\mathfrak{L} \subset \mathfrak{L}$, we have $\mathfrak{A}G \subset \mathfrak{L}$. But $\mathfrak{A}G$ is a left ideal of $\mathfrak{A}$. Thus either $\mathfrak{A}G = \mathfrak{L}$ or $\mathfrak{A}G = O_{\mathfrak{A}}$. In the second of these cases we would have that $\{O_{\mathfrak{A}}, G\}$ is a left nonzero ideal contained in $\mathfrak{L}$, i.e., $\{O_{\mathfrak{A}}, G\} = \mathfrak{L}$. But this would contradict the fact that $\mathfrak{R}\mathfrak{L} \neq O_{\mathfrak{A}}$. Consequently, $\mathfrak{A}G = \mathfrak{L}$.

$G\mathfrak{R}$ is a right ideal of $\mathfrak{A}$ included in $\mathfrak{R}$. Thus either $G\mathfrak{R} = \mathfrak{R}$ or $G\mathfrak{R} = O_{\mathfrak{A}}$. The second of these equations would imply by the equality $\mathfrak{A}G = \mathfrak{L}$ proved above that

$$\mathfrak{A} = \mathfrak{L}\mathfrak{R} = \mathfrak{A}G\mathfrak{R} = O_{\mathfrak{A}}.$$

Consequently, $G\mathfrak{R} = \mathfrak{R}$. We obtain from this by multiplication on the right by $\mathfrak{L}$ that

$$G\mathfrak{H} = \mathfrak{H}.$$

For the second nonzero element $G' \in \mathfrak{H}$ we have analogously that $G'\mathfrak{H} = \mathfrak{H}$, from which it follows that

$$GG'\mathfrak{H} = G\mathfrak{H} = \mathfrak{H}.$$

Consequently, $GG' \neq O_{\mathfrak{A}}$, i.e., the collection of all the nonzero elements of $\mathfrak{H}$ forms a semigroup which we will denote by $\mathfrak{G}$.

Since $GO_{\mathfrak{A}} = O_{\mathfrak{A}}$, it follows from $G\mathfrak{H} = \mathfrak{H}$ that

$$G\mathfrak{G} = \mathfrak{G}.$$

We prove analogously that $\mathfrak{G}G = \mathfrak{G}$. It follows that $\mathfrak{G}$ is a group, while $\mathfrak{H}$ is thus a group with an externally adjoined zero.

Since $E_{\mathfrak{G}} = E_{\mathfrak{H}} \subset \mathfrak{L}$, it follows that $\mathfrak{A}E_{\mathfrak{H}}$ is a left ideal of $\mathfrak{A}$ included in $\mathfrak{L}$. Here $\mathfrak{A}E_{\mathfrak{H}} \ni E_{\mathfrak{H}}E_{\mathfrak{H}} = E_{\mathfrak{H}} \neq O_{\mathfrak{A}}$. Consequently,

$$\mathfrak{A}E_{\mathfrak{H}} = \mathfrak{L}.$$

Analogously,

$$E_{\mathfrak{H}}\mathfrak{A} = \mathfrak{R}.$$

We obtain from this, keeping in mind that $\mathfrak{A}\mathfrak{A} = \mathfrak{A}$,

$$\mathfrak{H} = \mathfrak{R}\mathfrak{L} = E_{\mathfrak{H}}\mathfrak{A} \cdot \mathfrak{A}E_{\mathfrak{H}} = E_{\mathfrak{H}}\mathfrak{A}E_{\mathfrak{H}}.$$

Since

$$\mathfrak{H} = \mathfrak{R}\mathfrak{L} \subset \mathfrak{R} \text{ and } \mathfrak{H} = \mathfrak{R}\mathfrak{L} \subset \mathfrak{L},$$

we have

$$\mathfrak{H} \subset \mathfrak{R} \cap \mathfrak{L}.$$

But, on the other hand,

$$\mathfrak{R} \cap \mathfrak{L} \subset \mathfrak{L} = \mathfrak{A}E_{\mathfrak{H}}, \qquad \mathfrak{R} \cap \mathfrak{L} \subset \mathfrak{R} = E_{\mathfrak{H}}\mathfrak{A}.$$

Thus

$$\mathfrak{R} \cap \mathfrak{L} = E_{\mathfrak{H}} \cdot (\mathfrak{R} \cap \mathfrak{L}) \cdot E_{\mathfrak{H}} \subset E_{\mathfrak{H}}\mathfrak{A}E_{\mathfrak{H}} = \mathfrak{H}.$$

Consequently,

$$\mathfrak{R} \cap \mathfrak{L} = \mathfrak{H}.$$

4.8. COROLLARY. *For every minimal left nonzero ideal $\mathfrak{L}$ of a completely simple semigroup $\mathfrak{A}$ with zero there always exists a minimal right nonzero ideal $\mathfrak{R}$ such that*

$$\mathfrak{L}\mathfrak{R} = \mathfrak{A}.$$

PROOF. As was shown in the proof of the theorem of 4.5,

$$\mathfrak{L}\mathfrak{A} = \mathfrak{A} \neq O_{\mathfrak{A}}.$$

By 4.5, $\mathfrak{A}$ is the union of its minimal right nonzero ideals, and thus it is impossible that, for each of the $\mathfrak{R}$,

$$\mathfrak{L}\mathfrak{R} = O_{\mathfrak{A}}.$$

If, for some of the $\mathfrak{R}$,

$$\mathfrak{L}\mathfrak{R} \neq O_{\mathfrak{A}},$$

then by 4.7 it follows that

$$\mathfrak{L}\mathfrak{R} = \mathfrak{A}.$$

4.9. It is important that all the minimal left and right nonzero ideals of a completely simple semigroup with zero can be given with the help of the idempotents of the semigroup.

THEOREM. *If I is a nonzero idempotent of a completely simple semigroup $\mathfrak{A}$ with zero, then $\mathfrak{A}I$ is a minimal left nonzero ideal of the semigroup, while $I\mathfrak{A}$ is a minimal right nonzero ideal. Here each minimal left or right nonzero ideal can be given with the aid of some nonzero idempotent of the semigroup.*

PROOF. By 4.5, I is included in some minimal left nonzero ideal $\mathfrak{L}$. Clearly, $\mathfrak{A}I$ is a left ideal of the semigroup contained in $\mathfrak{L}$. Since $\mathfrak{A}I \ni II = I$ it follows that $\mathfrak{A}I \neq O_{\mathfrak{A}}$. Thus $\mathfrak{A}I = \mathfrak{L}$.

Let $\mathfrak{L}$ be some minimal left nonzero ideal of $\mathfrak{A}$. By 4.8 there exists for $\mathfrak{L}$ a minimal right nonzero ideal $\mathfrak{R}$ such that

$$\mathfrak{L}\mathfrak{R} \neq O_{\mathfrak{A}}.$$

By 4.7 it follows that $\mathfrak{R}\mathfrak{L}$ possesses an identity E, where

$$\mathfrak{L} = \mathfrak{A}E, \qquad E^2 = E.$$

The argument for right ideals is analogous.

4.10. COROLLARY. *Each element X of a completely simple semigroup $\mathfrak{A}$ with zero has a left identity and a right identity that are idempotents.*

PROOF. By the theorem of 4.5, X must be included in some minimal left nonzero ideal $\mathfrak{L}$ and in some minimal right nonzero ideal $\mathfrak{R}$. It follows from 4.9 that there exist in $\mathfrak{A}$ idempotents I and I' such that

$$\mathfrak{L} = \mathfrak{A}I, \qquad \mathfrak{R} = I'\mathfrak{A}.$$

Obviously,

$$XI = X, \qquad I'X = X.$$

4.11. COROLLARY. *If X is a nonzero element of a completely simple semigroup $\mathfrak{A}$ with zero, then*

$$\mathfrak{A}X\mathfrak{A} = \mathfrak{A}.$$

PROOF. $\mathfrak{A}X\mathfrak{A}$ is a two-sided ideal of $\mathfrak{A}$. Let I be a left identity of X and I' a right identity of X (4.10). Then

$$\mathfrak{A}X\mathfrak{A} \ni IXI' = X \neq O_{\mathfrak{A}}.$$

Consequently, $\mathfrak{A}X\mathfrak{A}$ is a nonzero two-sided ideal of $\mathfrak{A}$ and thus may only be $\mathfrak{A}$ itself.

4.12. Let $\mathfrak{A}$ be a completely simple semigroup with zero.

For any nonzero idempotent I the set of all the elements having I as their two-sided identity is $I\mathfrak{A}I$. Such sets are very important in the study of $\mathfrak{A}$. This is connected with the fact that each subgroup $\mathfrak{G}$ of the semigroup $\mathfrak{A}$ is contained in a set of the form $E_{\mathfrak{G}}\mathfrak{A}E_{\mathfrak{G}}$. Here we limit ourselves to the proof of only two properties of such sets.

THEOREM. *If I is a nonzero idempotent of a completely simple semigroup $\mathfrak{A}$ with zero, then $I\mathfrak{A}I$ is a group with an externally adjoined zero. Here, for every nonzero idempotent I' and for $X \in I\mathfrak{A}I'$ it is always true that*

$$X \cdot I'\mathfrak{A}I = I\mathfrak{A}I, \qquad I'\mathfrak{A}I \cdot X = I'\mathfrak{A}I'.$$

PROOF. (1) Since $\mathfrak{A}\mathfrak{A} = \mathfrak{A}$, we have

$$I\mathfrak{A}I = (I\mathfrak{A}) \cdot (\mathfrak{A}I).$$

By 4.9, $\mathfrak{A}I$ is a minimal left nonzero ideal of $\mathfrak{A}$ and $I\mathfrak{A}$ is a minimal right nonzero ideal of $\mathfrak{A}$.

Here

$$(\mathfrak{A}I) \cdot (I\mathfrak{A}) \ni I,$$

and consequently

$$(\mathfrak{A}I) \cdot (I\mathfrak{A}) \neq O_{\mathfrak{A}}.$$

By the theorem of 4.7 it follows that

$$(I\mathfrak{A}) \cdot (\mathfrak{A}I) = I\mathfrak{A}I$$

is a group with an externally adjoined zero.

(2) Since $X \in I\mathfrak{A}I'$,

$$IX = X, \qquad XI' = X.$$

Thus $X\mathfrak{A}$ is a right nonzero ideal included in the minimal right nonzero ideal $I\mathfrak{A}$ (4.9). Consequently

$$X\mathfrak{A} = I\mathfrak{A}.$$

From this we obtain

$$I\mathfrak{A}I = X\mathfrak{A}I = XI'\mathfrak{A}I.$$

The second required equality is proven analogously.

4.13. COROLLARY. *Let I be a nonzero idempotent of a completely simple semigroup $\mathfrak{A}$ with zero. Then no idempotent of $\mathfrak{A}$ different from I can be a two-sided identity of I.*

In fact, if the idempotent I' is a two-sided identity of I, then $I' \neq O_{\mathfrak{A}}$ and, by 4.12, $I'\mathfrak{A}I'$ is a group with an externally adjoined zero. The nonzero idempotents I and I' both belong to it. This is possible only when $I' = I$.

4.14. The property of 4.13 implies that each nonzero idempotent of a completely simple semigroup with zero is primitive (3.10). It turns out that even in the weakened form this property, along with the requirement of the absence of proper two-sided nonzero ideals, gives a sufficient condition for a semigroup to be a completely simple semigroup with zero.

THEOREM. *If a semigroup $\mathfrak{A}$ with zero has no proper two-sided nonzero ideals and possesses a primitive idempotent, then $\mathfrak{A}$ is a completely simple semigroup with zero.*

PROOF. (1) We denote by $\mathfrak{B}$ the collection of all the elements B such that $B\mathfrak{A} = O_{\mathfrak{A}}$.

For any $X \in \mathfrak{A}$,

$$(XB)\mathfrak{A} = O_{\mathfrak{A}}, \qquad (BX)\mathfrak{A} \subset B\mathfrak{A} = O_{\mathfrak{A}}.$$

Consequently, $\mathfrak{B}$ is a two-sided ideal of $\mathfrak{A}$, and thus $\mathfrak{B} = O_{\mathfrak{A}}$ or $\mathfrak{B} = \mathfrak{A}$. The latter would mean that $\mathfrak{A}\mathfrak{A} = O_{\mathfrak{A}}$, which is impossible, for $\mathfrak{A}\mathfrak{A} \ni II = I \neq O_{\mathfrak{A}}$. Consequently, $\mathfrak{B} = O_{\mathfrak{A}}$.

(2) For every $A \in \mathfrak{A}$ the product $\mathfrak{A}A\mathfrak{A}$ is a two-sided ideal. Thus one of the following is true:

$$\mathfrak{A}A\mathfrak{A} = O_{\mathfrak{A}}, \qquad \mathfrak{A}A\mathfrak{A} = \mathfrak{A}.$$

By the first part of the proof, for a nonzero A we have

$$A\mathfrak{A} \neq O_{\mathfrak{A}}.$$

We choose in $A\mathfrak{A}$ some nonzero element C. The fact that $\mathfrak{A}A\mathfrak{A} = O_{\mathfrak{A}}$ implies that $\mathfrak{A}C = O_{\mathfrak{A}}$.

However, it is proven in a manner similar to that of the first part that the existence of nonzero elements satisfying such a condition is impossible. Consequently, for $A \neq O_{\mathfrak{A}}$ it must necessarily be true that

$$\mathfrak{A}A\mathfrak{A} = \mathfrak{A}.$$

(3) The set $\mathfrak{A}I$ containing $II = I$ is a left nonzero ideal of $\mathfrak{A}$. Let $\mathfrak{L}$ be some left nonzero ideal of $\mathfrak{A}$ included in $\mathfrak{A}I$. We choose in $\mathfrak{L}$ some nonzero element L. As was shown in the second part,

$$\mathfrak{A}L\mathfrak{A} = \mathfrak{A}.$$

Consequently, there exist in $\mathfrak{A}$ elements U and V such that

$$ULV = I.$$

We consider the element

$$I' = IVIUL.$$

Since $L \in \mathfrak{A}I$, $LI = L$. Thus

$$I'I' = IVIUL \cdot IVIUL = IVI \cdot ULV \cdot IUL = IVI \cdot I \cdot IUL = IVIUL = I',$$

i.e., I' is an idempotent. Since furthermore,

$$IULI'V = IUL \cdot IVIUL \cdot V = I \cdot ULV \cdot I \cdot ULV = I,$$

while I is a nonzero element, it follows that I' is not a zero of the semigroup either.

Since $LI = L$ and $I^2 = I$ it follows from the expression for I' that

$$I'I = I', \qquad II' = I'.$$

But by the assumption of the theorem this is possible only when $I' = I$.

But then

$$\mathfrak{A}I = \mathfrak{A}IVIUL \subset \mathfrak{L}.$$

We have shown that $\mathfrak{A}I$ contains no proper left nonzero ideal. Consequently, $\mathfrak{A}I$ is a minimal left nonzero ideal. We argue similarly for right ideals. Noting finally that $\mathfrak{A}\mathfrak{A}$ is different from zero since it contains the nonzero element $II = I$, we come to the conclusion that $\mathfrak{A}$ is a completely simple semigroup with zero.

5. The Structure of Completely Simple Semigroups with Zero

5.1. For the study of the structure of completely simple semigroups Rees **[1]** used a certain general construction recalling the construction of Wedderburn in the theory of simple algebras.

Let $\mathfrak{H}$ be a semigroup with zero and let Γ and Γ' be two arbitrary nonempty sets of indices. Let each pair (α, β), where $\alpha \in \Gamma$, $\beta \in \Gamma'$, be associated with some element of the semigroup $P_{\alpha,\beta} \in \mathfrak{H}$. The rule giving this correspondence can be represented in the form of a matrix P (finite or infinite), where Γ is the set of the rows, Γ' is the set of the columns and all the elements belong to $\mathfrak{H}$. For finite (or even denumerable) Γ and Γ', which in this case it is natural to consider as sets of natural numbers, such a representation is completely intuitive:

$$P = \begin{bmatrix} P_{11} & P_{12} & \dots & P_{1n} \\ P_{21} & P_{22} & \dots & P_{2n} \\ \cdot & \cdot & \cdot & \cdot \\ P_{m1} & P_{m2} & \dots & P_{mn} \end{bmatrix} \qquad (P_{\alpha,\beta} \in \mathfrak{H}).$$

We denote by $\mathfrak{S}(P, \mathfrak{H})$ the set consisting of all possible three-tuples

$$(H, \xi, \eta) \qquad (H \in \mathfrak{H}\backslash O_{\mathfrak{H}}, \xi \in \Gamma', \eta \in \Gamma)$$

and of an element $O_{\mathfrak{S}}$. For uniformity in subsequent calculations we will also write the latter element in the form of three-tuples $(O_{\mathfrak{H}}, \xi, \eta)$ for any $\xi \in \Gamma'$, $\eta \in \Gamma$.

In the set $\mathfrak{S}(P, \mathfrak{H})$ we define the multiplication

$$(H_1, \xi_1, \eta_1) \cdot (H_2, \xi_2, \eta_2) = (H_1 P_{\eta_1 \xi_2} H_2, \xi_1, \eta_2).$$

We prove the associativity of this operation:

$$[(H_1, \xi_1, \eta_1) \cdot (H_2, \xi_2, \eta_2)] \cdot (H_3, \xi_3, \eta_3)$$
$$= (H_1 P_{\eta_1 \xi_2} H_2, \xi_1, \eta_2) \cdot (H_3, \xi_3, \eta_3) = (H_1 P_{\eta_1 \xi_2} H_2 P_{\eta_2 \xi_3} H_3, \xi_1, \eta_3);$$
$$(H_1, \xi_1, \eta_1) \cdot [(H_2, \xi_2, \eta_2) \cdot (H_3, \xi_3, \eta_3)]$$
$$= (H_1, \xi_1, \eta_1) \cdot (H_2 P_{\eta_2 \xi_3} H_3, \xi_2, \eta_3) = (H_1 P_{\eta_1 \xi_2} H_2 P_{\eta_2 \xi_3} H_3, \xi_1, \eta_3).$$

Thus $\mathfrak{S}(P, \mathfrak{H})$ is a semigroup. The element $O_{\mathfrak{S}}$ is clearly its zero. We will call the semigroup $\mathfrak{S}(P, \mathfrak{H})$ the *matrix semigroup over* $\mathfrak{H}$ *with defining matrix P*. If $\mathfrak{H}$ is a group with an externally adjoined zero (II, 2.12) and if there are in each row and column of P elements distinct from $O_{\mathfrak{H}}$, then $\mathfrak{S}(P, \mathfrak{H})$ will be called a *completely simple matrix semigroup with zero over* $\mathfrak{H}$ *with defining matrix P*.

The justification of the last definition will be found in 5.3 and 5.4.

5.2. The significance of the law of multiplication introduced into $\mathfrak{S}(P, \mathfrak{H})$ and the name given above for such a semigroup become completely clear and intuitive if we turn to the matrix interpretation. We limit ourselves for this to finite Γ and Γ', which will be considered to be sets of natural numbers:

$$\Gamma = \{1, 2, \dots, m\}, \qquad \Gamma' = \{1, 2, \dots, n\}.$$

The three-tuple (H, ξ, η) $(H \neq O_{\mathfrak{H}})$ may be represented in the form of a matrix with n rows and m columns (in contrast to P, in this matrix Γ' is the set of rows, while Γ is the set of the columns), in which the element in the ξth row and the ηth column is equal to H while all the other elements are equal to $O_{\mathfrak{H}}$. Here $O_{\mathfrak{S}}$ is the matrix all of whose elements are equal to $O_{\mathfrak{H}}$.

If we define addition between the element $O_{\mathfrak{H}}$ and all the elements of $\mathfrak{H}$ by the rule

$$O_{\mathfrak{H}} + H = H + O_{\mathfrak{H}} = H \qquad (H \in \mathfrak{H}),$$

then multiplication of three-tuples of $\mathfrak{S}(P, \mathfrak{H})$ may be represented in the form of matrix multiplication with the insertion of the matrix P:

$(H_1, \xi_1, \eta_1) \cdot (H_2, \xi_2, \eta_2)$

$$= \begin{bmatrix} O & \dots & O & \dots & O \\ \cdot & \cdot & \cdot & \cdot & \cdot \\ O & \dots & (H_1)_{\xi_1\eta_1} & \dots & O \\ \cdot & \cdot & \cdot & \cdot & \cdot \\ O & \dots & O & \dots & O \end{bmatrix} \cdot \begin{bmatrix} P_{11} & P_{12} & \dots & P_{1n} \\ \cdot & \cdot & \cdot & \cdot \\ P_{21} & P_{22} & \dots & P_{2n} \\ \cdot & \cdot & \cdot & \cdot \\ P_{m1} & P_{m2} & \dots & P_{mn} \end{bmatrix}$$

$$\times \begin{bmatrix} O & \dots & O & \dots & O \\ \cdot & \cdot & \cdot & \cdot & \cdot \\ O & \dots & (H_2)_{\xi_2\eta_2} & \dots & O \\ \cdot & \cdot & \cdot & \cdot & \cdot \\ O & \dots & O & \dots & O \end{bmatrix} = \begin{bmatrix} O & \dots & O & \dots & O \\ \cdot & \cdot & \cdot & \cdot & \cdot \\ O & \dots & (H_1P_{\eta_1\xi_2}H_2)_{\xi_1\eta_2} & \dots & O \\ \cdot & \cdot & \cdot & \cdot & \cdot \\ O & \dots & O & \dots & O \end{bmatrix}$$

$$= (H_1P_{\eta_1\xi_2}H_2, \xi_1, \eta_2).$$

5.3. Theorem. *A completely simple matrix semigroup $\mathfrak{S}(P, \mathfrak{H})$ with zero* (5.1) *is always a completely simple semigroup with zero* (4.2).

Proof. We denote by $\mathfrak{G}$ the group of all the nonzero elements of $\mathfrak{H}$.

As follows directly from the rule for multiplication in $\mathfrak{S}(P, \mathfrak{H})$, for each $\nu \in \Gamma$ the set $\mathfrak{L}_\nu$ consisting of all the elements of the form

$$(H, \xi, \nu)$$

is a left ideal. We show that this ideal is a minimal left nonzero ideal. We choose two arbitrary nonzero elements of this ideal:

$$L_1 = (G_1, \xi_1, \nu), \qquad L_2 = (G_2, \xi_2, \nu).$$

For $X = (G_3, \sigma, \tau)$, we have

$$XL_1 = (G_3P_{\tau\xi_1}G_1, \sigma, \nu).$$

If in X we set $\sigma = \xi_2$, choose τ so that $P_{\tau\xi_1} \in \mathfrak{G}$ (such an index τ exists by definition of the matrix P) and take as G_3 an element of $\mathfrak{G}$

$$G_3 = G_2G_1^{-1}P_{\tau\xi_1}^{-1},$$

then for such an X we find that

$$XL_1 = L_2.$$

It follows from this that every left ideal of $\mathfrak{S}(P, \mathfrak{H})$ containing one of the nonzero elements of $\mathfrak{L}_\nu$ must necessarily contain every other element of $\mathfrak{L}_\nu$. This means that $\mathfrak{L}_\nu$ is a minimal left nonzero ideal of the semigroup.

We convince ourselves similarly of the existence of minimal right nonzero ideals.

The semigroup $\mathfrak{S}(P, \mathfrak{H})$ has no proper two-sided nonzero ideals. For let us choose two arbitrary nonzero elements:

$$S_1 = (G_1, \xi_1, \eta_1), \qquad S_2 = (G_2, \xi_2, \eta_2).$$

As before we may choose an element X such that

$$XS_1 = (G_2, \xi_2, \eta_1).$$

A similar argument will yield a Y such that

$$(XS_1) \cdot Y = (G_2, \xi_2, \eta_1) \cdot Y = (G_2, \xi_2, \eta_2) = S_2.$$

It follows from this that a two-sided ideal $\mathfrak{S}(P, \mathfrak{H})$ containing some nonzero element of a semigroup will also contain every other element of the semigroup. It follows that in the semigroup $\mathfrak{S}(P, \mathfrak{H})$ the semigroup itself is a unique two-sided nonzero ideal. It is easy to see that $\mathfrak{S}(P, \mathfrak{H})\mathfrak{S}(P, \mathfrak{H}) \neq O_{\mathfrak{S}}$.

5.4. As was shown by Rees **[1]** in the general case, the completely simple matrix semigroups $\mathfrak{S}(P, \mathfrak{H})$ with zero (5.1) exhaust up to isomorphism the class of completely simple semigroups with zero (4.2).

THEOREM. *Every completely simple semigroup with zero is isomorphic with some completely simple matrix semigroup* $\mathfrak{S}(P, \mathfrak{H})$ *with zero* (5.1).

PROOF. (1) Let $\mathfrak{A}$ be a completely simple semigroup with zero. We denote by $\Gamma^{(l)}$ the collection of all of its minimal left nonzero ideals and by $\Gamma^{(r)}$ the collection of all of its minimal right nonzero ideals. We fix some pair of ideals $\mathfrak{L}_0 \in \Gamma^{(l)}$ and $\mathfrak{R}_0 \in \Gamma^{(r)}$ such that $\mathfrak{L}_0\mathfrak{R}_0 = \mathfrak{A}$ (4.8). By 4.7, $\mathfrak{H}_0 = \mathfrak{R}_0\mathfrak{L}_0$ is a group with an externally adjoined zero. We denote the group of all of its nonzero elements by $\mathfrak{G}_0$ and its identity by E_0. By 4.7 we have $\mathfrak{H}_0 = E_0\mathfrak{A}E_0$. Since $\mathfrak{H}_0 \ni O_{\mathfrak{A}}$, we have $O_{\mathfrak{H}_0} = O_{\mathfrak{A}}$. By 4.9 there exist in each ideal $\mathfrak{L} \in \Gamma^{(l)}$ and in each ideal $\mathfrak{R} \in \Gamma^{(r)}$ idempotents $I_{\mathfrak{L}}$ and $I_{\mathfrak{R}}$ such that

$$\mathfrak{L} = \mathfrak{A}I_{\mathfrak{L}}, \qquad \mathfrak{R} = I_{\mathfrak{R}}\mathfrak{A}.$$

We fix them for each $\mathfrak{L} \in \Gamma^{(l)}$ and each $\mathfrak{R} \in \Gamma^{(r)}$.

Clearly $I_{\mathfrak{L}}$ is a right identity of $\mathfrak{L}$ while $I_{\mathfrak{R}}$ is a left identity of $\mathfrak{R}$.

By 4.12,

$$I_{\mathfrak{L}}\mathfrak{A}E_0 \cdot E_0\mathfrak{A}I_{\mathfrak{L}} = I_{\mathfrak{L}}\mathfrak{A}I_{\mathfrak{L}} \ni I_{\mathfrak{L}}.$$

Thus it is possible to choose and fix elements $U_{\mathfrak{L}}$, $V_{\mathfrak{L}}$ such that $U_{\mathfrak{L}}V_{\mathfrak{L}} = I_{\mathfrak{L}}$, $U_{\mathfrak{L}} \in I_{\mathfrak{L}}\mathfrak{A}E_0$, $V_{\mathfrak{L}} \in E_0\mathfrak{A}I_{\mathfrak{L}}$.

By 4.14 it follows from $U_{\mathfrak{L}} \in I_{\mathfrak{L}}\mathfrak{A}E_0$ that

$$E_0\mathfrak{A}I_{\mathfrak{L}} \cdot U_{\mathfrak{L}} = E_0\mathfrak{A}E_0 \ni E_0.$$

Thus it is possible to choose and fix an element $W_{\mathfrak{L}}$ such that

$$W_{\mathfrak{L}}U_{\mathfrak{L}} = E_0, \qquad W_{\mathfrak{L}} \in E_0\mathfrak{A}I_{\mathfrak{L}}.$$

Similarly, for $\mathfrak{R} \in \Gamma^{(r)}$ we fix elements $U_{\mathfrak{R}}$, $V_{\mathfrak{R}}$, $W_{\mathfrak{R}}$ such that

$$U_{\mathfrak{R}}V_{\mathfrak{R}} = I_{\mathfrak{R}}, \qquad U_{\mathfrak{R}} \in I_{\mathfrak{R}}\mathfrak{A}E_0, \qquad V_{\mathfrak{R}} \in E_0\mathfrak{A}I_{\mathfrak{R}},$$

$$V_{\mathfrak{R}}W_{\mathfrak{R}} = E_0, \qquad W_{\mathfrak{R}} \in I_{\mathfrak{R}}\mathfrak{A}E_0.$$

For an arbitrary pair $\mathfrak{L} \in \Gamma^{(l)}$, $\mathfrak{R} \in \Gamma^{(r)}$ we use the notation

$$P_{\mathfrak{L},\mathfrak{R}} = V_{\mathfrak{L}}U_{\mathfrak{R}} \in E_0\mathfrak{A}E_0 = \mathfrak{H}_0.$$

The matrix over the group $\mathfrak{H}_0$ with an externally adjoined zero, the set of whose rows is $\Gamma^{(l)}$, the set of whose columns is $\Gamma^{(r)}$, and whose elements are $P_{\mathfrak{L},\mathfrak{R}}$, will be denoted by P.

We note that in every row (and similarly in every column) there are elements distinct from $O_{\mathfrak{H}_0} = O_{\mathfrak{A}}$. In fact, for $\mathfrak{L} \in \Gamma^{(l)}$ we take $\mathfrak{R} = I_{\mathfrak{L}}\mathfrak{A}$. By 4.9, $\mathfrak{R} \in \Gamma^{(r)}$. Since

$$I_{\mathfrak{L}} = I_{\mathfrak{L}}I_{\mathfrak{L}} = U_{\mathfrak{L}}V_{\mathfrak{L}}I_{\mathfrak{L}} \in U_{\mathfrak{L}}V_{\mathfrak{L}}\mathfrak{R} = U_{\mathfrak{L}}V_{\mathfrak{L}}U_{\mathfrak{R}}V_{\mathfrak{R}}\mathfrak{A},$$

it follows that $V_{\mathfrak{L}}U_{\mathfrak{R}} = P_{\mathfrak{L}\mathfrak{R},} \neq O_{\mathfrak{A}}$, for otherwise it would turn out that $I_{\mathfrak{L}} = O_{\mathfrak{A}}$ and $\mathfrak{L} = \mathfrak{A}I_{\mathfrak{L}} = O_{\mathfrak{A}}$.

(2) Let $\mathfrak{A}'$ be a completely simple matrix semigroup with zero over a group $\mathfrak{H}_0$ with an externally adjoined zero, with defining matrix given by the matrix P constructed above (5.1). By 4.5, for each nonzero element $A \in \mathfrak{A}$ there are determined in a unique fashion ideals $\mathfrak{L}^{(A)} \in \Gamma^{(l)}$, $\mathfrak{R}^{(A)} \in \Gamma^{(r)}$ such that

$$A \in \mathfrak{R}^{(A)}\mathfrak{L}^{(A)}.$$

Since

$$V_{\mathfrak{R}^{(A)}} \in E_0\mathfrak{A}I_{\mathfrak{R}^{(A)}}, \qquad U_{\mathfrak{L}^{(A)}} \in I_{\mathfrak{L}^{(A)}}\mathfrak{A}E_0,$$

it follows that

$$V_{\mathfrak{R}^{(A)}}AU_{\mathfrak{L}^{(A)}} \in E_0\mathfrak{A}E_0 = \mathfrak{H}_0.$$

Thus the following mapping χ of the semigroup $\mathfrak{A}$ is a mapping of it into $\mathfrak{A}'$:

$$\chi(A) = (V_{\mathfrak{R}^{(A)}}AU_{\mathfrak{L}^{(A)}}, \mathfrak{R}^{(A)}, \mathfrak{L}^{(A)}),$$

$$\chi(O_{\mathfrak{A}}) = O_{\mathfrak{A}'}.$$

We show that the mapping χ is one-to-one. Let

$$\chi(A) = \chi(B), \qquad (A, B \in \mathfrak{A}).$$

This means that $\mathfrak{R}^{(A)} = \mathfrak{R}^{(B)} = \mathfrak{R}$, $\mathfrak{L}^{(A)} = \mathfrak{L}^{(B)} = \mathfrak{L}$ and

$$V_{\mathfrak{R}}AU_{\mathfrak{L}} = V_{\mathfrak{R}}BU_{\mathfrak{L}}.$$

Multiplying the last equation on the left by $U_{\mathfrak{R}}$ and on the right by $V_{\mathfrak{L}}$, we obtain

$$I_{\mathfrak{R}}AI_{\mathfrak{L}} = I_{\mathfrak{R}}BI_{\mathfrak{L}}.$$

But $I_{\mathfrak{R}}$ is a left identity of the ideal $\mathfrak{R}$ that contains both A and B, while $I_{\mathfrak{L}}$ is a right identity of the ideal $\mathfrak{L}$ that also contains both A and B. Thus the last equation implies that $A = B$.

The image of $\mathfrak{A}$ under the mapping χ is the whole semigroup $\mathfrak{A}'$. To see this, let

$$A' = (G, \mathfrak{R}, \mathfrak{L}) \in \mathfrak{A}' \qquad (G \in \mathfrak{G}_0,\ \mathfrak{L} \in \Gamma^{(l)},\ \mathfrak{R} \in \Gamma^{(r)}).$$

We consider the element

$$A = W_{\mathfrak{R}}GW_{\mathfrak{L}}.$$

Since

$$W_{\mathfrak{R}} \in I_{\mathfrak{R}}\mathfrak{A}E_0, \qquad W_{\mathfrak{L}} \in E_0\mathfrak{A}I_{\mathfrak{L}},$$

it follows that

$$A \in I_{\mathfrak{R}}\mathfrak{A}I_{\mathfrak{L}} = I_{\mathfrak{R}}\mathfrak{A} \cdot \mathfrak{A}I_{\mathfrak{L}} = \mathfrak{R}\mathfrak{L}.$$

Thus

$$\chi(A) = (V_{\mathfrak{R}}W_{\mathfrak{R}}GW_{\mathfrak{L}}U_{\mathfrak{L}}, \mathfrak{R}, \mathfrak{L}) = (E_0GE_0, \mathfrak{R}, \mathfrak{L}) = (G, \mathfrak{R}, \mathfrak{L}) = A',$$

$$\chi(O_{\mathfrak{A}}) = O_{\mathfrak{A}'}.$$

(3) We show that the one-to-one mapping χ of the semigroup $\mathfrak{A}$ into the completely simple matrix semigroup $\mathfrak{A}' = \mathfrak{S}(\mathfrak{H}_0, P)$ with zero is an isomorphism.

Keeping 4.6 in mind, let

$$A_1, A_2 \in \mathfrak{A}; \qquad A_1 \in \mathfrak{R}_1\mathfrak{L}_1; \qquad A_2 \in \mathfrak{R}_2\mathfrak{L}_2; \qquad \mathfrak{L}_1, \mathfrak{L}_2 \in \Gamma^{(l)}; \qquad \mathfrak{R}_1, \mathfrak{R}_2 \in \Gamma^{(r)}.$$

By the rule for multiplication in $\mathfrak{A}'$, we obtain

$$\begin{aligned}\chi(A_1) \cdot \chi(A_2) &= (V_{\mathfrak{R}_1}A_1 U_{\mathfrak{L}_1}, \mathfrak{R}_1, \mathfrak{L}_1) \cdot (V_{\mathfrak{R}_2}A_2U_{\mathfrak{L}_2}, \mathfrak{R}_2, \mathfrak{L}_2)\\ &= (V_{\mathfrak{R}_1}A_1U_{\mathfrak{L}_1}P_{\mathfrak{L}_1,\mathfrak{R}_2}V_{\mathfrak{R}_2}A_2U_{\mathfrak{L}_2}, \mathfrak{R}_1, \mathfrak{L}_2)\\ &= (V_{\mathfrak{R}_1}A_1U_{\mathfrak{L}_1}V_{\mathfrak{L}_1}U_{\mathfrak{R}_2}V_{\mathfrak{R}_2}A_2U_{\mathfrak{L}_2}, \mathfrak{R}_1, \mathfrak{L}_2)\\ &= (V_{\mathfrak{R}_1}A_1I_{\mathfrak{L}_1}I_{\mathfrak{R}_2}A_2U_{\mathfrak{L}_2}, \mathfrak{R}_1, \mathfrak{L}_2).\end{aligned}$$

Since $A_1 \in \mathfrak{L}_1$ while $I_{\mathfrak{L}_1}$ is a right identity of $\mathfrak{L}_1$, and $A_2 \in \mathfrak{R}_2$ while $I_{\mathfrak{R}_2}$ is a left identity of $\mathfrak{R}_2$, we have

$$\chi(A_1) \cdot \chi(A_2) = (V_{\mathfrak{R}_1}A_1A_2U_{\mathfrak{L}_2}, \mathfrak{R}_1, \mathfrak{L}_2) = \chi(A_1A_2).$$

5.5. The representation of completely simple semigroups with zero in the form of matrix semigroups $\mathfrak{S}(P, \mathfrak{H})$ (5.1) allows us to easily obtain some of their various properties. We will discuss, for example, how to construct the systems of left and right ideals of a completely simple matrix semigroup with zero.

We denote by Γ the set of the rows of the matrix P and by Γ' the set of its columns.

Let Σ be a nonempty subset of the set Γ. As follows directly from the rule for multiplication in completely simple matrix semigroups, the collection of all the elements of the form (G, ξ, η), where $\eta \in \Sigma$, is a left ideal of $\mathfrak{S}(P, \mathfrak{H})$. The semigroup $\mathfrak{S}(P, \mathfrak{H})$ has no left nonzero ideals other than those obtained by means of subsets Σ. This follows from the fact that if some left ideal $\mathfrak{L}$ of the semigroup $\mathfrak{S}(P, \mathfrak{H})$ contains an element (G, ξ, η) $(G \neq O_{\mathfrak{H}})$, then, as is easily seen (in essence we have already proved this in the preceding arguments), $\mathfrak{L}$ will also contain every other element of the form (G', ξ', η).

Thus, in a completely simple semigroup with zero the left ideals are in one-to-one correspondence with the nonempty subsets of the set Γ. The minimal left nonzero ideals are those which correspond to the subsets consisting of one element of the set Γ. If to the zero ideal of the semigroup $\mathfrak{S}(P, \mathfrak{H})$ we make correspond the empty subset of the set Γ, then we will obtain a one-to-one correspondence between all the left ideals of the semigroup $\mathfrak{S}(P, \mathfrak{H})$ and all the subsets (including the empty one) of the set Γ.

The situation is analogous for right ideals, which turn out to be in one-to-one correspondence with the subsets of the set Γ'

5.6. In a completely simple matrix semigroup $\mathfrak{S}(P, \mathfrak{H})$ with zero a nonzero idempotent will be an element of the form $(P_{\eta\xi}^{-1}, \xi, \eta)$, where $P_{\eta\xi} \neq O_{\mathfrak{H}}$. The maximal subgroup corresponding to this idempotent (III, 1.16) is clearly the set consisting of all the elements of the form (G, ξ, η). The mapping φ of the group $\mathfrak{G} = \mathfrak{H} \backslash O_{\mathfrak{A}}$ into this group,

$$\varphi(G) = (GP_{\eta\xi}^{-1}, \xi, \eta),$$

is one-to-one. It is an isomorphism, for

$$\begin{aligned}\varphi(G_1) \cdot \varphi(G_2) &= (G_1P_{\eta\xi}^{-1}\ \xi, \eta) \cdot (G_2P_{\eta\xi}^{-1}\ \xi, \eta) \\ &= (G_1P_{\eta\xi}^{-1} \cdot P_{\eta\xi} \cdot G_2P_{\eta\xi}^{-1}, \xi, \eta) = \varphi(G_1G_2).\end{aligned}$$

Thus all the nonzero maximal subgroups of the semigroup $\mathfrak{S}(P, \mathfrak{H})$ are isomorphic with each other, since each of them is isomorphic with the group $\mathfrak{G} = \mathfrak{H} \backslash O_{\mathfrak{H}}$.

5.7. By what has been said above, it is clear how important and convenient is the representation of completely simple semigroups with zero in the form of matrix semigroups. To complete the question of their representation in such a form it is still necessary to say whether such a representation may be realized in a unique fashion, and if not, we must give the relationship between the completely simple matrix semigroups with zero that are isomorphic with a given completely simple semigroup with zero. Clearly, this question is equivalent to the corresponding question arising when two given completely simple matrix semigroups with zero are isomorphic to each other.

THEOREM. *Let* $\mathfrak{A} = \mathfrak{S}(P, \mathfrak{H})$ *and* $\bar{\mathfrak{A}} = \mathfrak{S}(\bar{P}, \bar{\mathfrak{H}})$ *be two completely simple matrix semigroups with zero (5.1). Let* Γ *be the set of the rows of the matrix* P, *let* Γ' *be the set of its columns and let* $\bar{\Gamma}$ *and* $\bar{\Gamma}'$ *be the corresponding sets for* $\bar{P}$; $\mathfrak{G} = \mathfrak{H} \backslash O_{\mathfrak{H}}$ *and* $\bar{\mathfrak{G}} = \bar{\mathfrak{H}} \backslash O_{\mathfrak{H}}$.

In order for $\mathfrak{A}$ *and* $\bar{\mathfrak{A}}$ *to be isomorphic it is necessary and sufficient that there exist:*

(1) *a one-to-one mapping* μ *of* Γ *onto* $\bar{\Gamma}$,

$$\mu(\eta) = \bar{\eta} \ (\eta \in \Gamma, \bar{\eta} \in \bar{\Gamma});$$

(2) *a one-to-one mapping* ν *of* Γ' *onto* $\bar{\Gamma}'$,

$$\nu(\xi) = \bar{\xi}(\xi \in \Gamma', \bar{\xi}' \in \bar{\Gamma}');$$

(3) *an isomorphism* σ *of the group* $\mathfrak{G}$ *onto* $\bar{\mathfrak{G}}$;

(4) *elements* $\bar{G}_{\bar{\eta}} \in \bar{\mathfrak{G}}$ $(\bar{\eta} \in \bar{\Gamma})$;

(5) *elements* $\bar{G}'_{\bar{\xi}} \in \bar{\mathfrak{G}}$ $(\bar{\xi} \in \bar{\Gamma}')$ *such that*

$$\bar{P}_{\bar{\eta}\bar{\xi}} = \bar{G}_{\bar{\eta}} \cdot \sigma(P_{\eta\xi}) \cdot \bar{G}'_{\bar{\xi}} \quad (\eta \in \Gamma, \xi \in \Gamma').$$

PROOF. (1) Let the five items in the formulation of the theorem be satisfied by $\mathfrak{A}$ and $\bar{\mathfrak{A}}$. We define a mapping χ of the semigroup $\mathfrak{A}$ into $\bar{\mathfrak{A}}$:

$$\chi(G, \xi, \eta) = (\bar{G}'^{-1}_{\bar{\xi}} \cdot \sigma(G) \cdot \bar{G}^{-1}_{\bar{\eta}}, \bar{\xi}, \bar{\eta}) \in \bar{\mathfrak{A}}$$
$$(\bar{\xi} = \nu(\xi), \bar{\eta} = \mu(\eta)).$$

If $G = O_{\mathfrak{H}}$, then we assume that $\sigma(G) = O_{\mathfrak{H}}$ and thus $\chi(O_{\mathfrak{A}}) = O_{\bar{\mathfrak{A}}}$. For given ξ and η, as G runs through the group $\mathfrak{G}$, by the properties of an isomorphism $\sigma(G)$ will run through $\sigma(\mathfrak{G}) = \bar{\mathfrak{G}}$. Using this, we conclude that the mapping χ is a one-to-one mapping of $\mathfrak{A}$ onto $\bar{\mathfrak{A}}$.

Using the connection between P and $\bar{P}$, we show that χ is an isomorphism:

$$\begin{aligned}
\chi(G_1, \xi_1, \eta_1) \cdot \chi(G_2, \xi_2, \eta_2)
&= (\bar{G}'^{-1}_{\bar{\xi}_1} \cdot \sigma(G_1) \cdot \bar{G}^{-1}_{\bar{\eta}_1}, \bar{\xi}_1, \bar{\eta}_1) \cdot (\bar{G}'^{-1}_{\bar{\xi}_2} \cdot \sigma(G_2) \cdot \bar{G}^{-1}_{\bar{\eta}_2}, \bar{\xi}_2, \bar{\eta}_2) \\
&= (\bar{G}'^{-1}_{\bar{\xi}_1} \cdot \sigma(G_1) \cdot \bar{G}^{-1}_{\bar{\eta}_1} \cdot \bar{P}_{\bar{\eta}_1\bar{\xi}_2} \cdot \bar{G}'^{-1}_{\bar{\xi}_2} \cdot \sigma(G_2) \cdot \bar{G}^{-1}_{\bar{\eta}_2}, \bar{\xi}_1, \bar{\eta}_2) \\
&= (\bar{G}'^{-1}_{\bar{\xi}_1} \cdot \sigma(G_1) \cdot \sigma(P_{\eta_1\xi_2}) \cdot \sigma(G_2) \cdot \bar{G}'^{-1}_{\bar{\eta}_2}, \bar{\xi}_1, \bar{\eta}_2) \\
&= \chi(G_1 P_{\eta_1\xi_2} G_2, \xi_1, \eta_2) = \chi[(G_1, \xi_1, \eta_1) \cdot (G_2, \xi_2, \eta_2)].
\end{aligned}$$

(2) Let there exist an isomorphism χ of the semigroup $\mathfrak{A}$ onto $\bar{\mathfrak{A}}$.

For a fixed $\eta \in \Gamma$ the set $\mathfrak{L}_\eta$ consisting of all the elements of the form (G, ξ, η) (5.5) constitutes a minimal nonzero left ideal of the semigroup $\mathfrak{A}$. Under the isomorphism χ it must be mapped onto some minimal nonzero left ideal of the semigroup $\bar{\mathfrak{A}}$, i.e., onto a set $\bar{\mathfrak{L}}_{\bar{\eta}}$ consisting of elements of the form $(\bar{G}, \bar{\xi}, \bar{\eta})$ for some $\bar{\eta} \in \bar{\Gamma}: \chi(\mathfrak{L}_\eta) = \bar{\mathfrak{L}}_{\bar{\eta}}$. The mapping μ,

$$\mu(\eta) = \bar{\eta},$$

is clearly a one-to-one mapping of Γ onto $\bar{\Gamma}$. We determine analogously a one-to-one mapping ν of the set Γ' onto $\bar{\Gamma}'$ with the aid of minimal nonzero right ideals such that $\nu(\xi) = \bar{\xi}: \chi(\mathfrak{R}_\xi) = \bar{\mathfrak{R}}_{\bar{\xi}}$.

We fix some pair (ξ_0, η_0) $(\xi_0 \in \Gamma', \eta_0 \in \Gamma)$ such that $P_{\eta_0\xi_0} \neq O_{\mathfrak{H}}$. Such a pair exists by the definition of a completely simple matrix semigroup with zero. In this case we also have $\bar{P}_{\bar{\eta}_0\bar{\xi}_0} \neq O_{\bar{\mathfrak{H}}}$. In fact, it follows from $P_{\eta_0\xi_0} \neq O_{\mathfrak{H}}$ that the product of any two elements of the form (G, ξ_0, η_0) $(G \neq O_{\mathfrak{H}})$ is different from $O_{\mathfrak{A}}$. But then by the definitions of μ and ν $(\mu(\xi_0) = \bar{\xi}_0, \nu(\eta_0) = \bar{\eta}_0)$ the product of any two elements of the form $(\bar{G}, \bar{\xi}_0, \bar{\eta}_0)$ $(\bar{G} \neq O_{\bar{\mathfrak{H}}})$ must also be different from $O_{\bar{\mathfrak{A}}}$ (since these elements are the images of the elements indicated above from $\mathfrak{A}$ under the isomorphism χ). As directly follows from the rule for multiplication in the matrix semigroups under consideration, the latter is possible only when $\bar{P}_{\bar{\eta}_0\bar{\xi}_0} \neq O_{\bar{\mathfrak{A}}}$.

The one-to-one mapping φ of the group $\mathfrak{G}$ onto $(\mathfrak{L}_{\eta_0} \cap \mathfrak{R}_{\xi_0}) \backslash O_{\mathfrak{A}}$,

$$\varphi(G) = (GP^{-1}_{\eta_0\xi_0}, \xi_0, \eta_0) \qquad (G \in \mathfrak{G}),$$

as we have shown in 5.6, is an isomorphism.

The analogous mapping ψ of the group $\bar{\mathfrak{G}}$ onto $(\bar{\mathfrak{L}}_{\bar{\eta}_0} \cap \mathfrak{R}_{\bar{\xi}_0}) \backslash O_{\bar{\mathfrak{A}}}$ is also an isomorphism.

By our definitions of the mappings μ and ν the isomorphism χ of the semigroup $\mathfrak{A}$ onto $\bar{\mathfrak{A}}$ brings about an isomorphism of $\mathfrak{L}_{\eta_0} \cap \mathfrak{R}_{\xi_0}$ onto $\bar{\mathfrak{L}}_{\bar{\eta}_0} \cap \bar{\mathfrak{R}}_{\bar{\xi}_0}$. Thus the mapping

$$\sigma = \psi^{-1}\chi\varphi$$

is an isomorphism of $\mathfrak{G}$ onto $\bar{\mathfrak{G}}$.

We consider the elements

$$\bar{N}_{\bar{\xi}} = (\bar{P}^{-1}_{\bar{\eta}_0\bar{\xi}_0}, \bar{\xi}, \bar{\eta}_0) \in \bar{\mathfrak{A}}, \qquad \bar{M}_{\bar{\eta}} = (E_{\bar{\mathfrak{G}}}, \bar{\xi}_0, \bar{\eta}) \in \bar{\mathfrak{A}}$$

$$(\bar{\xi} \in \bar{\Gamma}', \bar{\eta} \in \bar{\Gamma}).$$

By the choice of ξ_0 and η_0 these elements are distinct from $O_{\bar{\mathfrak{A}}}$.

By the rule of multiplication in $\bar{\mathfrak{A}}$ we have

$$\bar{M}_{\bar{\eta}}\bar{N}_{\bar{\xi}} = (E_{\bar{\mathfrak{G}}}, \bar{\xi}_0, \bar{\eta}) \cdot (\bar{P}^{-1}_{\bar{\eta}_0\bar{\xi}_0}, \bar{\xi}, \bar{\eta}_0) = (E_{\bar{\mathfrak{G}}}\bar{P}_{\bar{\eta}\bar{\xi}}\bar{P}^{-1}_{\bar{\eta}_0\bar{\xi}_0}, \bar{\xi}_0, \bar{\eta}_0) = \psi(\bar{P}_{\bar{\eta}\bar{\xi}}).$$

By the choice of the mappings μ and ν we have, for some $C_\xi \in \mathfrak{G}$ and $D_\eta \in \mathfrak{G}$,

$$\chi^{-1}(\bar{N}_{\bar{\xi}}) = (C_\xi, \xi, \eta_0), \qquad \chi^{-1}(\bar{M}_{\bar{\eta}}) = (D_\eta, \xi_0, \eta).$$

Here C_ξ and D_η are not zeros, since $\bar{N}_{\bar{\xi}}$ and $\bar{M}_{\bar{\eta}}$ are not zeros.

For the C_ξ and D_η thus chosen we obtain

$$\begin{aligned}\bar{P}_{\bar{\eta}\bar{\xi}} &= \psi^{-1}\chi\varphi\varphi^{-1}\chi^{-1}\psi(\bar{P}_{\bar{\eta}\bar{\xi}}) = \psi^{-1}\chi\varphi\varphi^{-1}\chi^{-1}(\bar{M}_{\bar{\eta}}\bar{N}_{\bar{\xi}}) = \sigma\varphi^{-1}[\chi^{-1}(\bar{M}_{\bar{\eta}}) \cdot \chi^{-1}(\bar{N}_{\bar{\xi}})] \\ &= \sigma\varphi^{-1}[(D_\eta, \xi_0, \eta) \cdot (C_\xi, \xi, \eta_0)] = \sigma\varphi^{-1}(D_\eta P_{\eta\xi} C_\xi, \xi_0, \eta_0) \\ &= \sigma(D_\eta P_{\eta\xi} C_\xi P_{\eta_0\xi_0}) = \sigma(D_\eta) \cdot \sigma(P_{\eta\xi}) \cdot \sigma(C_\xi P_{\eta_0\xi_0}).\end{aligned}$$

Setting

$$\bar{G}_{\bar{\eta}} = \sigma(D_\eta), \qquad \bar{G}_{\bar{\xi}} = \sigma(C_\xi P_{\eta_0 \xi_0}) \qquad (\bar{\eta} \in \bar{\Gamma}, \bar{\xi} \in \bar{\Gamma}'),$$

we obtain the desired expression for the elements of the matrix $\bar{P}$.

5.8. The conditions for isomorphism of two completely simple matrix semigroups with zero considered in 5.7 also permit various other approaches to simple matrix semigroups with zero.

For a matrix P over a semigroup $\mathfrak{H}$ with zero the following transformations of it, i.e., transitions from P to a new matrix $\bar{P}$, will be called elementary transformations of the matrix P:

(1) one-to-one mapping of the set of the rows Γ onto some set $\bar{\Gamma}$ of equal cardinality ("permutation of the rows");

(2) one-to-one mapping of the set of the columns Γ' onto some set $\bar{\Gamma}'$ of equal cardinality ("permutation of the columns");

(3) isomorphism of the semigroup $\mathfrak{H}$ onto a new semigroup $\bar{\mathfrak{H}}$;

(4) multiplication on the left of each row of the matrix P by one of its arbitrary elements from $\mathfrak{H}\backslash O_{\mathfrak{H}}$;

(5) multiplication on the right of each column of the matrix P by one of its arbitrary elements from $\mathfrak{H}\backslash O_{\mathfrak{H}}$.

If for two completely simple matrix semigroups with zero, $\mathfrak{S}(P, \mathfrak{H})$ and $\mathfrak{S}(\bar{P}, \bar{\mathfrak{H}})$, the matrix $\bar{P}$ can be obtained from P with the aid of one of these elementary transformations (in the case of the third transformation, of course, in the elements of the semigroup $\mathfrak{S}(P, \mathfrak{H})$ itself the first component of $\mathfrak{H}$ is replaced by a corresponding element from $\bar{\mathfrak{H}}$), then by 5.7 the semigroups $\mathfrak{S}(P, \mathfrak{H})$ and $\mathfrak{S}(\bar{P}, \bar{\mathfrak{H}})$ are isomorphic. It is understood that such an isomorphism will exist only when $\bar{P}$ can be obtained from P as a result of several successive elementary transformations. In this connection, in the case of a finite number of rows or columns the last two transformations can be replaced by transformations in which only one row or one column is multiplied.

As follows from 5.7, every isomorphism of a completely simple matrix semigroup onto another completely simple matrix semigroup may be obtained as a result of successive applications of various elementary transformations (where each of the five transformations may be used only once).

6. The Structure of Completely Simple Semigroups without Zero

6.1. Our investigation of the structure of completely simple semigroups with zero allows us to easily obtain the corresponding properties for completely simple semigroups without zero (of course, these properties could be obtained by direct means with arguments similar to those of the preceding section). The connection (discussed below) between the semigroups of the two classes given below will serve as a basis for this purpose.

THEOREM. *If in a completely simple semigroup $\mathfrak{A}$ with zero the zero is externally adjoined* (II, 2.12), *then the set of all the nonzero elements of the semigroup $\mathfrak{A}$ is a completely simple semigroup without zero.*

If a zero is adjoined in an external fashion to a completely simple semigroup $\mathfrak{A}'$ without a zero, then the semigroup obtained is a completely simple semigroup with zero.

PROOF. (1) Let $\mathfrak{A}$ be a semigroup with an externally adjoined zero and let $\mathfrak{A}'$ be the subsemigroup of its nonzero elements.

If $\mathfrak{L}$ is a left nonzero ideal of $\mathfrak{A}$, then $\mathfrak{L}\backslash O_{\mathfrak{A}}$, as is easily seen, is a left ideal of the semigroup $\mathfrak{A}'$. In this manner all the left ideals of $\mathfrak{A}'$ can be obtained, since it follows from the fact that $\mathfrak{L}'$ is a left ideal of $\mathfrak{A}'$ that $\mathfrak{L}' \cup O_{\mathfrak{A}}$ is a left nonzero ideal of $\mathfrak{A}$. We argue analogously for right and two-sided ideals.

(2) If $\mathfrak{A}$ is a completely simple semigroup with zero, $\mathfrak{L}$ is one of its minimal left nonzero ideals and $\mathfrak{R}$ a minimal right nonzero ideal, then by what was said above $\mathfrak{L}\backslash O_{\mathfrak{A}}$ and $\mathfrak{R}\backslash O_{\mathfrak{A}}$will be respectively minimal left and minimal right ideals of the semigroup $\mathfrak{A}'$. Since there are no proper nonzero two-sided ideals in $\mathfrak{A}$, there are also no proper two-sided ideals in $\mathfrak{A}'$.

(3) If $\mathfrak{A}'$ is a completely simple semigroup without zero and $\mathfrak{L}'$ and $\mathfrak{R}'$ are minimal left and minimal right ideals of it, then $\mathfrak{L}' \cup O_{\mathfrak{A}}$ and $\mathfrak{R}' \cup O_{\mathfrak{A}}$ will be respectively a minimal left nonzero ideal and a minimal right nonzero ideal of $\mathfrak{A}$. From the fact that $\mathfrak{A}'$ has no proper two-sided ideals it follows that $\mathfrak{A}$ has no proper two-sided nonzero ideals.

6.2. In connection with this theorem it is necessary, as a preliminary to our study of completely simple semigroups with zero, to clear up the question concerning when the zero in a completely simple semigroup with zero is an externally adjoined zero. By the theorem of 5.4 we may limit ourselves to completely simple matrix semigroups with zero.

THEOREM. *In a completely simple matrix semigroup $\mathfrak{S}(P, \mathfrak{H})$ with zero* (5.1), *the zero will be externally adjoined if and only if all the elements of the matrix P are different from the zero of the semigroup $\mathfrak{H}$.*

PROOF. From the definition of multiplication in $\mathfrak{S}(P, \mathfrak{H})$ it follows that if all the elements of P are distinct from $O_{\mathfrak{H}}$, then the product of any nonzero elements of the semigroup $\mathfrak{S}(P, \mathfrak{H})$ is also a nonzero element, i.e., the zero $O_{\mathfrak{S}}$ is externally adjoined to $\mathfrak{S}(P, \mathfrak{H})$.

If some element of the matrix P is equal to the zero of $\mathfrak{H}$,

$$P_{\eta\xi} = O_{\mathfrak{H}},$$

then for some $H_1, H_2 \in \mathfrak{H}\backslash O_{\mathfrak{H}}$ the following product of nonzero elements turns out to be equal to $O_{\mathfrak{S}}$:

$$(H_1, \xi, \eta) \cdot (H_2, \xi, \eta) = (H_1 P_{\eta\xi} H_2, \xi, \eta) = (O_{\mathfrak{H}}, \xi, \eta) = O_{\mathfrak{S}}.$$

6.3. Let $\mathfrak{G}$ be a group and P a matrix (finite or infinite) with a set of rows Γ and a set of columns Γ', the elements $P_{\eta\xi}$ of which belong to $\mathfrak{G}$. We denote by $\mathfrak{S}'(P, \mathfrak{G})$ the set of all the three-tuples of the form

$$(G, \xi, \eta) \qquad (G \in \mathfrak{G}, \xi \in \Gamma', \eta \in \Gamma).$$

In $\mathfrak{S}'(P, \mathfrak{G})$ we define the operation

$$(G_1, \xi_1, \eta_1) \cdot (G_2, \xi_2, \eta_2) = (G_1 P_{\eta_1 \xi_2} G_2, \xi_1, \eta_2).$$

Clearly, $\mathfrak{S}'(P, \mathfrak{G})$ will be identical with the set of all the nonzero elements of the completely simple matrix semigroup $\mathfrak{S}(P, \mathfrak{H})$ with zero, where $\mathfrak{H}$ is the semigroup obtained from $\mathfrak{G}$ by adjoining a zero externally.

Since all the elements of P are distinct from $O_{\mathfrak{H}}$, by 6.2, $\mathfrak{S}(P, \mathfrak{H})$ is a semigroup with an externally adjoined zero. By 6.1 it follows from this that $\mathfrak{S}'(P, \mathfrak{G})$ is a completely simple semigroup without zero. We call $\mathfrak{S}'(P, \mathfrak{H})$ a *completely simple matrix semigroup without zero.*

6.4. The semigroups $\mathfrak{S}'(P, \mathfrak{G})$ described in 6.3 exhaust, up to isomorphism, all the completely simple semigroups without zero.

THEOREM. *Every completely simple semigroup $\mathfrak{A}$ without zero is isomorphic to some completely simple matrix semigroup without zero* (6.3).

PROOF. Adjoining a zero O to $\mathfrak{A}$ externally, we obtain a semigroup $\mathfrak{B}$, which by 6.1 is a completely simple semigroup with zero. By the theorem of 5.4, $\mathfrak{B}$ is isomorphic with some completely simple matrix semigroup $\mathfrak{S}(P, \mathfrak{H})$ with zero, where $\mathfrak{H}$ is a group $\mathfrak{G}$ with an externally adjoined zero $O_{\mathfrak{H}}$. Since $\mathfrak{B}$, and thus also $\mathfrak{S}(P, \mathfrak{H})$, is a semigroup with an externally adjoined zero, it follows by 6.2 that all the elements of the matrix P belong to the group $\mathfrak{G}$. It follows that the set of all the nonzero elements of the semigroup $\mathfrak{S}(P, \mathfrak{H})$ is identical with the semigroup $\mathfrak{S}'(P, \mathfrak{G})$ (6.3). This semigroup is isomorphic with the subsemigroup of all the nonzero elements of the semigroup $\mathfrak{B}$, and consequently is isomorphic with the original group $\mathfrak{A}$.

6.5. Thus, for completely simple semigroups without zero, we have obtained a complete description of their structure similar to that given earlier for completely simple semigroups with zero. The question of isomorphism, i.e., of how the various representations of the same completely simple semigroup without zero are connected, is also simply decided by means of the corresponding result for completely simple semigroups with zero.

Let there be given two completely simple matrix semigroups $\mathfrak{S}'(P, \mathfrak{G})$ and $\mathfrak{S}'(\bar{P}, \overline{\mathfrak{G}})$ without zero (6.3). Adjoining to them externally the zeros O and $\bar{O}$, we obtain completely simple semigroups with zero $\mathfrak{S}(P, \mathfrak{H})$ and $\mathfrak{S}(\bar{P}, \overline{\mathfrak{H}})$, where $\mathfrak{H}$ is the group $\mathfrak{G}$ with an externally adjoined zero $O_{\mathfrak{H}}$, and $\overline{\mathfrak{H}}$ is the group $\overline{\mathfrak{G}}$ with an externally adjoined zero $O_{\overline{\mathfrak{H}}}$. Clearly, the semigroups $\mathfrak{S}'(P, \mathfrak{G})$ and $\mathfrak{S}'(\bar{P}, \overline{\mathfrak{G}})$ will be isomorphic to each other if and only if the semigroups $\mathfrak{S}(P, \mathfrak{H})$ and $\mathfrak{S}(\bar{P}, \overline{\mathfrak{H}})$ are isomorphic to each other. A necessary and sufficient condition for the isomorphism of such semigroups was given in 5.7. Using it, we obtain directly a necessary and sufficient condition for the isomorphism of the semigroups $\mathfrak{S}'(P, \mathfrak{G})$ and $\mathfrak{S}(\bar{P}, \overline{\mathfrak{G}})$. It is not even necessary to reformulate it, since it coincides word for word with the condition of 5.7. It consists of the five points indicated there, which all refer to the groups $\mathfrak{G}$ and $\overline{\mathfrak{G}}$ and the matrices P and $\bar{P}$.

6.6. Let $\mathfrak{S}'(P, \mathfrak{G})$ be a completely simple matrix semigroup without zero (6.3) and let $\mathfrak{S}(P, \mathfrak{H})$ be the completely simple matrix semigroup with zero obtained from $\mathfrak{S}'(P, \mathfrak{G})$ by the external adjoining of a zero O. In the proof of the theorem of 6.1 we noted how the ideals of the semigroup were associated. By this relationship, using the fact that in 5.5 we gave the construction of the systems of left and right ideals of a semigroup of the form $\mathfrak{S}(P, \mathfrak{H})$, we directly obtain the systems of left and right ideals of the semigroup $\mathfrak{S}'(P, \mathfrak{G})$.

Let Γ be the set of the rows of the matrix P and let Γ' be the set of its columns. For any nonempty subset $\Sigma \subset \Gamma$, the set of all the elements of the form (G, ξ, η) $(G \in \mathfrak{G})$ for which $\eta \in \Sigma$ is a left ideal of the semigroup $\mathfrak{S}'(P, \mathfrak{G})$. This semigroup has no left ideals other than ideals of the same form for different $\Sigma \subset \Gamma$. The minimal left ideals will be those left ideals for which Σ consists of one element.

If $\Sigma' \subset \Gamma'$, then the set of all the elements of the form (G, ξ, η) $(G \in \mathfrak{G})$ for which $\xi \in \Sigma'$ is a right ideal of $\mathfrak{S}'(P, \mathfrak{G})$, and such ideals exhaust all the right ideals of $\mathfrak{S}'(P, \mathfrak{G})$.

6.7. We examine one more important property of completely simple semigroups without zero. Let Γ be the set of the rows and Γ' the set of the columns of the matrix P of a completely simple matrix semigroup $\mathfrak{S}'(P, \mathfrak{G})$ without zero (6.3). For $\xi \in \Gamma'$, $\eta \in \Gamma$ we denote by $\mathfrak{G}_{\xi\eta}$ the collection of all the elements of the form (G, ξ, η) $(G \in \mathfrak{G})$. By 5.6, $\mathfrak{G}_{\xi\eta}$ is a group isomorphic with the group $\mathfrak{G}$. Thus all the $\mathfrak{G}_{\xi\eta}$ turn out to be isomorphic with one another. Since $\mathfrak{S}'(P, \mathfrak{G})$ is their nonintersecting union it follows from 6.4 that we have the following properties for an arbitrary completely simple semigroup without zero.

(α) *Every completely simple semigroup without zero is the nonintersecting union of groups that are isomorphic to one another.*

(β) *Every completely simple semigroup without zero is completely regular.*

This follows from (α) and from III, 1.15.

(γ) *Any two distinct idempotents of a completely simple semigroup without zero commute.*

In fact, the idempotents of $\mathfrak{S}'(P, \mathfrak{G})$ are identities of the groups $\mathfrak{G}_{\xi\eta}$, but by the rule for multiplication in $\mathfrak{S}'(P, \mathfrak{G})$ elements from $\mathfrak{G}_{\xi_1\eta_1}$ and $\mathfrak{G}_{\xi_2\eta_2}$ for $\xi_1 \neq \xi_2$ or for $\eta_1 \neq \eta_2$ clearly do not commute.

(δ) *If a completely simple semigroup without zero has only one idempotent, it is a group.*

This follows directly from (α).

6.8. It follows from these properties that each element of a completely simple semigroup without zero has a two-sided identity that is an idempotent.

In the semigroup $\mathfrak{S}'(P, \mathfrak{G})$ such an identity for the element (G, ξ, η) will obviously be $(P_{\eta\xi}^{-1}, \xi, \eta)$.

We note in passing that with the exception of a group, completely simple semigroups without zero never have a two-sided identity of the whole semigroup. In fact, if the set of the rows of the matrix P of a completely simple semigroup $\mathfrak{S}'(P, \mathfrak{G})$ contains even two elements, then

$$(G_1, \xi_1, \eta_1) \cdot (G_2, \xi_2, \eta_2) = (G_1 P_{\eta_1 \xi_2} G_2, \xi_1, \eta_2) \neq (G_1, \xi_1, \eta_1) \qquad (\eta_1 \neq \eta_2)$$

and (G_1, ξ_1, η_1) cannot be an identity of the semigroup. The analogous statement is true when the set of the columns of the matrix P contains no less than two elements. If both the set of the rows and the set of the columns of the semigroup $\mathfrak{S}'(P, \mathfrak{G})$ consist of only one element, then $\mathfrak{S}'(P, \mathfrak{G})$ is isomorphic to the group $\mathfrak{G}$ (6.7).

6.9. In spite of the significant similarity between completely simple semigroups without zero and completely simple semigroups with zero, there are also differences between them. As an example, we consider the following completely simple matrix semigroup with zero.

Let $\mathfrak{G} = \{E\}$ be the identity of the group, and let the sets of the indices Γ and Γ' each consist of two elements $\{1, 2\}$. As the matrix P we take the identity matrix

$$P = \begin{bmatrix} E & O \\ O & E \end{bmatrix}$$

(as a result of which the multiplication in our semigroup is obviously the usual matrix multiplication). The corresponding completely simple matrix semigroup with zero consists of the five elements

$$E_{11} = (E, 1, 1) = \begin{bmatrix} E & O \\ O & O \end{bmatrix}, \qquad E_{12} = (E, 1, 2) = \begin{bmatrix} O & E \\ O & O \end{bmatrix},$$

$$E_{21} = (E, 2, 1) = \begin{bmatrix} O & O \\ E & O \end{bmatrix}, \qquad E_{22} = (E, 2, 2) = \begin{bmatrix} O & O \\ O & E \end{bmatrix},$$

$$O = \begin{bmatrix} O & O \\ O & O \end{bmatrix}$$

and, as may be directly verified, possesses the following multiplication table:

	E_{11}	E_{12}	E_{21}	E_{22}	O
E_{11}	E_{11}	E_{12}	O	O	O
E_{12}	O	O	E_{11}	E_{12}	O
E_{21}	E_{21}	E_{22}	O	O	O
E_{22}	O	O	E_{21}	E_{22}	O
O	O	O	O	O	O

We see that the nonzero elements E_{12} and E_{21} not only are not included in the subgroups, but also have in general no two-sided identities.

6.10. It is not difficult to explain how the property of 6.7 may be transformed to apply to a completely simple semigroup with zero. We consider a completely simple matrix semigroup $\mathfrak{S}(P, \mathfrak{H})$ with zero (5.1). Its arbitrary nonzero element $S = (G, \xi, \eta)$ is included in the minimal right ideal $\mathfrak{R}_\xi$ consisting of all the elements of the form (H, ξ, λ) $(H \in \mathfrak{H}, \lambda \in \Gamma)$ (5.5), and in the minimal left ideal $\mathfrak{L}_\eta$ consisting of all the elements of the form (H, μ, η) $(H \in \mathfrak{H}, \mu \in \Gamma')$ (5.5). If $P_{\eta\xi} \neq O_{\mathfrak{S}}$, then

$$\mathfrak{L}_\eta \mathfrak{R}_\xi \ni (E_{\mathfrak{H}}, \xi, \eta) \cdot (E_{\mathfrak{H}}, \xi, \eta) = (P_{\eta\xi}, \xi, \eta) \neq O_{\mathfrak{S}},$$

from which it follows by 4.7 that $\mathfrak{R}_\xi \mathfrak{L}_\eta$ is a group with an externally adjoined zero. Our element S is included in the group $\mathfrak{R}_\xi \mathfrak{L}_\eta \backslash O_{\mathfrak{S}}$ since it belongs to $\mathfrak{R}_\xi \cap \mathfrak{L}_\eta = \mathfrak{R}_\xi \mathfrak{L}_\eta$ (4.7).

If $P_{\eta\xi} = O_{\mathfrak{H}}$, then

$$S^2 = (G, \xi, \eta) \cdot (G, \xi, \eta) = (GP_{\eta\xi}G, \xi, \eta) = O_{\mathfrak{S}}$$

and S cannot be included in any group of $\mathfrak{S}(P, \mathfrak{H})$.

From what has been said it follows that in a completely simple semigroup with zero all the elements whose squares are different from zero are included in subgroups and thus are completely regular (III, 1.18). Elements whose squares are equal to zero are, of course, not completely regular.

6.11. The consideration of completely simple semigroups without zero allows us to give the structure of completely regular semigroups in which every two elements are regularly associated with each other. We considered such semigroups in II, 6.8 and II, 6.9.

Let Γ' and Γ be two arbitrary nonempty sets. We denote by $\mathfrak{B}$ the collection of all the possible pairs of the form (ξ, η) where $\xi \in \Gamma'$, $\eta \in \Gamma$. By defining in $\mathfrak{B}$ the multiplication

$$(\xi_1, \eta_1) \cdot (\xi_2, \eta_2) = (\xi_1, \eta_2)$$

we make $\mathfrak{B}$ into a semigroup. Obviously $\mathfrak{B}$ is the completely simple matrix semigroup $\mathfrak{S}'(P, \mathfrak{G})$ without zero, where $\mathfrak{G}$ is the identity group, P is the matrix the set of whose rows is Γ, the set of whose columns is Γ', and all of whose elements are equal to the identity E, the only element of the group $\mathfrak{G}$.

If either $\xi_1 \neq \xi_2$ or $\eta_1 \neq \eta_2$, then

$$(\xi_1, \eta_1) \cdot (\xi_2, \eta_2) \neq (\xi_2, \eta_2) \cdot (\xi_1, \eta_1).$$

Thus, by II, 6.9, $\mathfrak{B}$ is a semigroup in which every two elements are regularly associated with each other.

6.12. We will show that every semigroup $\mathfrak{A}$ in which each two elements are regularly associated with each other is isomorphic with some semigroup $\mathfrak{B}$ of the type considered in 6.11.

Let X and Y be two arbitrary elements of $\mathfrak{A}$. Since $XYX = X$, every two-sided ideal of $\mathfrak{A}$ containing Y will contain X. It follows that $\mathfrak{A}$ has no proper two-sided ideals.

For any $A \in \mathfrak{A}$ the set of elements of the form $XA (X \in \mathfrak{A})$ is a left ideal. Moreover, it is a minimal left ideal since every left ideal containing an element XA will also contain (by II, 6.8) the element $YA = YXA$ for any $Y \in \mathfrak{A}$. Analogously, $A\mathfrak{A}$ is a minimal right ideal. Thus $\mathfrak{A}$ is a completely simple semigroup without zero. By 6.4, $\mathfrak{A}$ is isomorphic with some semigroup $\mathfrak{S}'(P, \mathfrak{G})$. The group $\mathfrak{G}$ must be the identity group, since on the one hand by 6.6, $\mathfrak{S}'(P, \mathfrak{G})$ is the union of groups isomorphic to $\mathfrak{G}$, and on the other hand all the elements of $\mathfrak{A}$ are idempotent (II, 6.8). The semigroup $\mathfrak{S}'(P, \mathfrak{G})$ for the identity group $\mathfrak{G}$ is in an obvious way isomorphic to a semigroup of the type considered in 6.11.

6.13. The discussion in 6.11 and 6.12 gives an exhaustive description of the structure of semigroups in which every two elements are regularly associated with each other. It also gives a classification of such semigroups, since it is easy to show directly (this also follows from the conditions for isomorphism of completely simple semigroups of matrix type) that two semigroups $\mathfrak{B}_1$ and $\mathfrak{B}_2$ of the type of 6.11 are isomorphic if and only if the corresponding sets Γ_1' and Γ_2' are of equal cardinality and Γ_1 and Γ_2 are of equal cardinality.

6.14. The semigroups of the class considered here can also be characterized as completely simple semigroups without zero, all the elements of which are idempotent. It has already been shown that they all possess these properties. Conversely, let $\mathfrak{A}$ be a completely simple semigroup without zero, all the elements of which are idempotent. It is isomorphic with some matrix semigroup $\mathfrak{S}'(P, \mathfrak{S})$ (6.4). Since all the elements of $\mathfrak{A}$ are idempotent, all the subgroups of $\mathfrak{A}$ will be the identity group. By 6.4 it follows from this that $\mathfrak{G}$ is the identity group. But we have already noted that when $\mathfrak{G}$ is the identity group, the semigroup $\mathfrak{S}'(P, \mathfrak{G})$

is isomorphic with some semigroup $\mathfrak{B}$ of the type of 6.10.

6.15. In conclusion we will note how, with the aid of the concept of a primitive idempotent (3.10), the completely simple semigroups without zero and the completely simple semigroups with zero can be naturally combined in one common class.

THEOREM. *In order for a nonidentity semigroup without proper two-sided nonzero ideals to be completely simple it is necessary and sufficient that it contain a primitive idempotent.*

PROOF. It follows from 4.10 that every completely simple semigroup with zero possesses a nonzero idempotent. All of them, as was mentioned in 4.14, are primitive.

By 3.10 a completely simple semigroup without zero possesses primitive idempotents.

If a semigroup with zero that does not have proper two-sided nonzero ideals possesses primitive idempotents, then by 4.14 it is a completely simple semigroup with zero.

If a semigroup without zero $\mathfrak{A}$ has no proper two-sided ideals and contains a primitive idempotent I, then we adjoin externally a zero. We obtain a semigroup $\mathfrak{B}$ which, as is easy to see, will have no proper two-sided nonzero ideals and in which I will also be a primitive idempotent. Thus $\mathfrak{B}$ must be a completely simple semigroup with zero. By 6.1 the set of its nonzero elements $\mathfrak{A}$ will be a completely simple semigroup without zero.

6.16. The above description of the structure of completely simple semigroups without zero and the introduction of the above properties, as well as various other properties on which we did not dwell, may be realized without difficulty, starting from the representation of completely simple semigroups without zero with the aid of the representation of them as matrix semigroups. This approach was used by Rees **[1]** and by a number of other authors in later works. However, many of these properties may also be introduced directly without the use of this representation. Schwarz **[3]** uses such an approach.

7. Minimal Ideals of Semigroups of Transformations

7.1. Let Ω be a nonvoid set, and suppose $\alpha \in \Omega$. We have already, in Chapter II, §5, considered the transformation U_α taking all elements of Ω into α, i.e., $U_\alpha \xi = \alpha$ for every $\xi \in \Omega$. Such transformations are sometimes called contractions, sometimes transformations of rank one. Using functional terminology, it is natural to call them constant transformations or simply constants.

Constant transformations have a very simple structure and so, in themselves, can scarcely be the object of a substantial investigation. Nevertheless, in various semigroups of transformations they often play an especially important role and as elements of such semigroups deserve attention. This can be fully accounted for from the point of view developed in the preceding sections of this chapter.

The set of all transformations U_α $(\alpha \in \Omega)$ will be denoted by $\mathfrak{U}_\Omega$.

7.2. **Theorem.** *Let* $\mathfrak{A}$ *be a semigroup of transformations of the set* Ω *such that* $\mathfrak{U}' = \mathfrak{A} \cap \mathfrak{U}_\Omega \neq \emptyset$. *Then*

(α) *Every element* $U_\alpha \in \mathfrak{U}'$ *constitutes a minimal right ideal of the semigroup* $\mathfrak{A}$, *and* $\mathfrak{A}$ *has no other minimal right ideals*;

(β) $\mathfrak{U}'$ *is the only minimal left ideal of* $\mathfrak{A}$;

(γ) $\mathfrak{U}'$ *is the minimal two-sided ideal of* $\mathfrak{A}$.

Proof. By the multiplication rule for transformations, we have $U_\alpha A = U_\alpha$ for any $U_\alpha \in \mathfrak{U}'$ and $A \in \mathfrak{A}$. Hence $U_\alpha \mathfrak{A} = U_\alpha$, and U_α constitutes a right ideal of $\mathfrak{A}$ (obviously minimal).

Let $\mathfrak{R}$ be a minimal right ideal of $\mathfrak{A}$. Take any elements $X \in \mathfrak{R}$ and $U_\alpha \in \mathfrak{U}'$. Let $X\alpha = \beta$.

For any $\xi \in \Omega$, we have

$$XU_\alpha \xi = X\alpha = \beta.$$

Consequently $XU_\alpha = U_\beta$. But $XU_\alpha \in \mathfrak{R}$. Hence $U_\beta \in \mathfrak{R}$. Since U_β by itself constitutes a minimal right ideal, we have $\mathfrak{R} = U_\beta$.

It now follows, by V, 2.5, that $\mathfrak{U}'$ is the minimal two-sided ideal of the semigroup $\mathfrak{A}$.

Let $\mathfrak{L}$ be any left ideal of $\mathfrak{A}$. For any $X \in \mathfrak{L}$ and $U_\alpha \in \mathfrak{U}'$, we have $U_\alpha = U_\alpha X \in \mathfrak{L}$. It follows that $\mathfrak{L} \supset \mathfrak{U}'$. Thus $\mathfrak{U}'$ is the only minimal left ideal of $\mathfrak{A}$.

7.3. Among the various semigroups of transformations, a special role is played by those semigroups which contain all constant transformations of the set being transformed. Such semigroups will be called *complete with respect to constants.*

7.4. Many important semigroups of transformations arising in the development of one or another mathematical theory are complete with respect to constants. We shall make the acquaintance of one such class in the next section.

7.5. The question of constant transformations is generally very important in studying representations of semigroups by transformations. We consider below some results in this direction obtained by E. S. Ljapin [S3].

A representation φ of a semigroup $\mathfrak{A}$ by transformations of the set Ω (i.e., an isomorphism of $\mathfrak{A}$ into the semigroup $\mathfrak{S}_\Omega$) will be called complete with respect to constants if $\mathfrak{U}_\Omega \subset \varphi(\mathfrak{A})$.

7.6. Let $\mathfrak{A}$ be a semigroup with minimal right ideals, the totality of which will be denoted, for a while, by Γ.

If $\mathfrak{R} \in \Gamma$, we know (V, 2.2) that $A\mathfrak{R} \in \Gamma$ for every $A \in \mathfrak{A}$. Hence for a given $A \in \mathfrak{A}$, if we assign to each $\mathfrak{R} \in \Gamma$ the ideal $A\mathfrak{R} \in \Gamma$ we obtain a transformation of the set Γ. The transformation so determined by the element A will be denoted by χ_A.

We shall denote by $\chi_{\mathfrak{A}}$ the mapping of the semigroup $\mathfrak{A}$ into $\mathfrak{S}_\Gamma$ such that $\chi_{\mathfrak{A}}(A) = \chi_A \in \mathfrak{S}_\Gamma$ $(A \in \mathfrak{A})$.

7.7. A family Σ of subsets of a semigroup will be called right-separating if for every pair $A \neq B (A, B \in \mathfrak{A})$ there exists an $\mathfrak{M} \in \Sigma$ such that $A\,\mathfrak{M} \neq B\mathfrak{M}$.

7.8. **Lemma.** *If a semigroup $\mathfrak{A}$ has minimal right ideals of which the totality constitutes a right-separating family, then every minimal right ideal of $\mathfrak{A}$ consists of a single element.*

Proof. Suppose a minimal right ideal $\mathfrak{R}$ of $\mathfrak{A}$ contains two distinct elements X and Y. Then there must exist a minimal right ideal $\mathfrak{R}'$ such that $X\mathfrak{R}' \neq Y\mathfrak{R}'$. But $X \in \mathfrak{R}$ implies $X\mathfrak{R}' \subset \mathfrak{R}$, and since $X\mathfrak{R}'$ is a right ideal, we have $X\mathfrak{R}' = \mathfrak{R}$. Similarly, $Y\mathfrak{R}' = \mathfrak{R}$, and this contradicts $X\mathfrak{R}' \neq Y\mathfrak{R}'$.

7.9. In connection with 7.8, note that, by V, 2.2, if one minimal right ideal of a semigroup consists of a single element, then every minimal right ideal consists of a single element.

7.10. **Theorem.** *A semigroup $\mathfrak{A}$ has a representation by transformations which is complete with respect to constants* (7.5) *if and only if $\mathfrak{A}$ has minimal right ideals whose totality Γ constitutes a right-separating family.*

Proof. 1) Let φ be a representation of $\mathfrak{A}$, by transformations of a set Ω, which is complete with respect to constants.

By 7.2 the individual transformations U_α $(\alpha \in \Omega)$, and only these, constitute all the minimal right ideals of the semigroup $\varphi(\mathfrak{A})$. Since $\varphi(\mathfrak{A})$ has minimal right ideals, the same must be true of the isomorphic semigroup $\mathfrak{A}$.

Let $A' = \varphi(A)$ and $B' = \varphi(B)$ $(A, B \in \mathfrak{A})$ be any two distinct elements of the semigroup $\varphi(\mathfrak{A})$. To say that A' and B' are distinct is to say that for some $\xi \in \Omega$ we have

$$A'\xi = \alpha, \quad B'\xi = \beta, \quad \alpha \neq \beta.$$

Since, for every $\lambda \in \Omega$,

$$A'U_\xi\lambda = A'\xi = \alpha, \quad B'U_\xi\lambda = B'\xi = \beta,$$

it follows that

$$A'U_\xi = U_\alpha, \quad B'U_\xi = U_\beta, \quad U_\alpha \neq U_\beta.$$

Thus, the minimal right ideals of the semigroup $\varphi(\mathfrak{A})$ constitute a right-separating family. The same is therefore true of the minimal right ideals of the semigroup $\mathfrak{A}$.

2) Suppose the totality Γ of minimal right ideals of a semigroup $\mathfrak{A}$ is nonvoid and constitutes a right-separating family. Consider the mapping

$\chi_{\mathfrak{A}}$ of $\mathfrak{A}$ into the semigroup $\mathfrak{S}_\Gamma$ defined in 7.6.

For any pair $A \neq B$ $(A, B \in \mathfrak{A})$, there exists an $\mathfrak{R} \in \Gamma$ such that $A\mathfrak{R} \neq B\mathfrak{R}$. This implies that $\chi_{\mathfrak{A}}(A) \neq \chi_{\mathfrak{A}}(B)$. Thus, the mapping $\chi_{\mathfrak{A}}$ is one-to-one.

For any $A, B \in \mathfrak{A}$, and every $\mathfrak{R} \in \Gamma$, we have

$$[\chi_{\mathfrak{A}}(AB)]\mathfrak{R} = (AB)\mathfrak{R} = A(B\mathfrak{R}) = [\chi_{\mathfrak{A}}(A)](B\mathfrak{R})$$
$$= [\chi_{\mathfrak{A}}(A)]\{[\chi_{\mathfrak{A}}(B)]\mathfrak{R}\} = \{[\chi_{\mathfrak{A}}(A)][\chi_{\mathfrak{A}}(B)]\}\mathfrak{R}.$$

This proves that $\chi_{\mathfrak{A}}$ is an isomorphism.

Consider any $\mathfrak{R} \in \Gamma$. By 7.8, $\mathfrak{R}$ consists of a single element X. The transformation $\chi_{\mathfrak{A}}(X)$ takes an arbitrary element $\mathfrak{R}' \in \Gamma$ into $[\chi_{\mathfrak{A}}(X)]\mathfrak{R}' = X\mathfrak{R} = X$. Consequently, $\chi_{\mathfrak{A}}(X) = U_X = U_{\mathfrak{R}}$. We have therefore shown that an arbitrary constant transformation $U_{\mathfrak{R}}$ of the set Γ belongs to $\chi_{\mathfrak{A}}(\mathfrak{A})$. Thus $\chi_{\mathfrak{A}}$ is a representation of $\mathfrak{A}$ of the required sort.

7.11. Having determined what semigroups have representations which are complete with respect to constants (7.5), it is natural to try to describe all such representations for any such semigroup.

We introduce first the following notion.

For a semigroup $\mathfrak{A}$, let φ_1 be a representation by transformations of a set Ω_1, and φ_2 a representation by transformations of a set Ω_2. We shall say that the representations φ_1 and φ_2 are not essentially distinct if there exists a one-to-one mapping ψ of Ω_1 onto Ω_2 such that, for any $\alpha_1 \in \Omega_1$ and $A \in \mathfrak{A}$,

$$\psi\{[\varphi_1(A)](\alpha_1)\} = [\varphi_2(A)][\psi(\alpha_1)].$$

This means, obviously, that the representations differ at most only in the nature of, or the notation for, the elements of Ω_1 and Ω_2, i.e., that they are in a sense essentially the same representation.

7.12. Now, it turns out that every semigroup has, up to inessential distinction, at most one representation which is complete with respect to constants. Indeed, just such a representation is the representation $\chi_{\mathfrak{A}}$ of the semigroup $\mathfrak{A}$, as was seen in the proof of Theorem 7.10.

7.13. **Theorem.** *If a representation φ of a semigroup $\mathfrak{A}$ is complete with respect to constants, then φ is not essentially distinct from the representation $\chi_{\mathfrak{A}}$.*

Proof. Let φ be a representation by transformations of a set Ω. By 7.2 the minimal right ideals in $\varphi(\mathfrak{A})$ are the elements of $\mathfrak{U}_\Omega$. The mapping φ effects a one-to-one correspondence between the elements U_ξ $(\xi \in \Omega)$ and those elements R_ξ of $\mathfrak{A}$ which constitute the minimal right ideals in $\mathfrak{A}$:

$$\varphi(R_\xi) = U_\xi \quad (\xi \in \Omega).$$

Let ψ be the one-to-one mapping of Ω onto the set of all R_ξ $(\xi \in \Omega)$: $\psi(\xi) = R_\xi$.

Now consider any $A \in \mathfrak{A}$. Let $[\varphi(A)]\xi = \xi'$. Since

$$\varphi(AR_\xi) = \varphi(A) \cdot \varphi(R_\xi) = \varphi(A) \cdot U_\xi = U_{\xi'} = \phi(R_{\xi'}),$$

we have $AR_\xi = R_{\xi'}$.

It follows that

$$[\chi_{\mathfrak{A}}(A)][\psi(\xi)] = [\chi_{\mathfrak{A}}(A)]R_\xi = AR_\xi = R_{\xi'} = \psi(\xi') = \psi\{[\varphi(A)]\xi\}.$$

This means that φ is not essentially distinct from $\chi_{\mathfrak{A}}$.

7.14. The notion introduced in 7.11 of inessential distinction between representations by transformations is related to the following important notion.

Let $\mathfrak{A}_1$ be a semigroup of transformations of a set Ω_1, and $\mathfrak{A}_2$ a semigroup of transformations of a set Ω_2. One says that the semigroups $\mathfrak{A}_1$ and $\mathfrak{A}_2$ are not essentially distinct (or are similar), if there exists a one-to-one mapping χ of Ω_1 onto Ω_2 and a one-to-one mapping φ of $\mathfrak{A}_1$ onto $\mathfrak{A}_2$ such that, for any $\xi \in \Omega_1$ and $A \in \mathfrak{A}_1$,

$$\varphi(A)\chi(\xi) = \chi(A\xi).$$

The meaning of this notion is obvious. The semigroups $\mathfrak{A}_1$ and $\mathfrak{A}_2$ consist of essentially the same transformations, differing only in the notation for, or the intrinsic nature of, the elements of Ω_1 and Ω_2 undergoing the transformations. From our point of view this latter distinction is entirely immaterial.

It is immediately evident that semigroups of transformations which are not essentially distinct are isomorphic. But it is of course far from the case that all isomorphic semigroups of transformations are not essentially distinct.

Thus the important relation, between semigroups of transformations, of being only inessentially distinct, does not reduce to the question of their isomorphism. In this connection it is natural to raise the question of singling out classes of semigroups of transformations within which the two relations coincide, i.e., within which isomorphism implies inessential distinction.

7.15. If two representations φ_1 and φ_2 of a semigroup $\mathfrak{A}$ are not essentially distinct, it is obvious that the semigroups of transformations $\varphi_1(\mathfrak{A})$ and $\varphi_2(\mathfrak{A})$ are likewise not essentially distinct.

But it is important to keep in mind that the inference cannot be made in the opposite direction. It may well happen that the two semigroups of transformations $\varphi_1(\mathfrak{A})$ and $\varphi_2(\mathfrak{A})$ are not essentially distinct, but that the representations φ_1 and φ_2 are.

7.16. By means of the theorems proved above we may now obtain some results regarding the problem mentioned in 7.14.

Theorem. *Suppose two semigroups of transformations $\mathfrak{A}_1$ and $\mathfrak{A}_2$ are each complete with respect to constants. Then if $\mathfrak{A}_1$ and $\mathfrak{A}_2$ are isomorphic, they are not essentially distinct semigroups of transformations.*

Proof. Let φ be an isomorphism of $\mathfrak{A}_1$ onto $\mathfrak{A}_2$, and ϵ the identity isomorphism of $\mathfrak{A}_1$ onto itself. Then ϕ and ϵ are representations of $\mathfrak{A}_1$ which are complete with respect to constants. By 7.13, φ and ϵ are not essentially distinct. It follows, as remarked in 7.15, that the semigroups of transformations $\varphi(\mathfrak{A}_1) = \mathfrak{A}_2$ and $\epsilon(\mathfrak{A}_1) = \mathfrak{A}_1$ can be only inessentially distinct.

8. Semigroups of Endomorphisms

8.1. In Chapter I, §3 we spoke of a contemplated direction in the development of the theory of semigroups, namely that dealing with the semigroups of endomorphisms of various mathematical systems. This direction has recently drawn considerable attention. The greatest study has been devoted to endomorphisms of graphs. By a graph $(\Omega, \mathfrak{n})$ is meant a set Ω with a given binary relation $\mathfrak{n}$. An important case of a graph is a partially ordered set (when $\mathfrak{n}$ is a partial order relation).

8.2. For two graphs $(\Omega_1, \mathfrak{n}_1)$ and $(\Omega_2, \mathfrak{n}_2)$, a one-to-one mapping ψ of the set Ω_1 onto the set Ω_2 is called an isomorphism of the first graph onto the second if $\alpha \sim \beta$ $(\mathfrak{n}_1)$ $(\alpha, \beta \in \Omega_1)$ always implies $\psi(\alpha) \sim \psi(\beta)$ $(\mathfrak{n}_2)$ and conversely.

A one-to-one mapping ψ is called an anti-isomorphism if $\alpha \sim \beta$ $(\mathfrak{n}_1)$ $(\alpha, \beta \in \Omega_1)$ always implies $\psi(\beta) \sim \psi(\alpha)$ $(\mathfrak{n}_2)$ and conversely.

A graph $(\Omega_2, \mathfrak{n}_2^*)$ is called the dual (or conjugate) of a graph $(\Omega_2, \mathfrak{n}_2)$ if $\alpha \sim \beta$ $(\mathfrak{n}_2)$ always implies $\beta \sim \alpha$ $(\mathfrak{n}_2^*)$ and conversely.

An anti-isomorphism of a graph $(\Omega_1, \mathfrak{n}_1)$ onto a graph $(\Omega_2, \mathfrak{n}_2)$ is obviously the same thing as an isomorphism of $(\Omega_1, \mathfrak{n}_1)$ onto $(\Omega_2, \mathfrak{n}_2^*)$.

8.3. By I, 3.18 and 3.20, an endomorphism of a graph $(\Omega, \mathfrak{n})$ means a transformation X of the set Ω such that $\alpha \sim \beta$ $(\mathfrak{n})$ $(\alpha, \beta \in \Omega)$ always implies $X\alpha \sim X\beta$ $(\mathfrak{n})$.

The totality of all endomorphisms of a graph $(\Omega, \mathfrak{n})$ is obviously a semigroup and will be denoted by $\mathfrak{B}(\Omega, \mathfrak{n})$.

8.4. The general idea in the study of semigroups of endomorphisms is to determine the relations between the properties of the graph $(\Omega, \mathfrak{n})$ itself and the semigroup properties of its semigroup of endomorphisms $\mathfrak{B}(\Omega, \mathfrak{n})$. Here, the semigroup $\mathfrak{B}(\Omega, \mathfrak{n})$ can be regarded from the "concrete point of view," i.e. as a semigroup of transformations, or from the "abstract point of view," i.e. as determined up to isomorphism. Both approaches are important from the point of view of the general theory of semigroups. The second is of particular interest. To understand the interrelations between the two points of view one must keep in mind the remarks at the end of the preceding section.

8.5. It is easy to see that if two graphs $(\Omega_1, \mathfrak{n}_1)$ and $(\Omega_2, \mathfrak{n}_2)$ are either isomorphic or anti-isomorphic, then the semigroups $\mathfrak{B}(\Omega_1, \mathfrak{n}_1)$ and $\mathfrak{B}(\Omega_2, \mathfrak{n}_2)$ are isomorphic. What is of interest and of prime importance is to discover those cases where, conversely, the isomorphism of the semigroups $\mathfrak{B}(\Omega_1, \mathfrak{n}_1)$ and $\mathfrak{B}(\Omega_2, \mathfrak{n}_2)$ implies the isomorphism or anti-isomorphism of the graphs $(\Omega_1, \mathfrak{n}_1)$ and $(\Omega_2, \mathfrak{n}_2)$.

If we grant that a graph and its dual are basically the same thing, then for the cases just mentioned the semigroup of endomorphisms entirely describes the graph, and the study of such graphs reduces entirely to the study of their semigroups.

The first basic step in this direction was made by L. M. Gluskin [**S2**]. Below we consider the most important case of his result. A number of investigations have subsequently been made in the same direction.

8.6. A graph $(\Omega, \mathfrak{n})$ is said to be reflexive if the relation $\mathfrak{n}$ is reflexive.

Theorem. *A graph* $(\Omega, \mathfrak{n})$, *where* $\mathfrak{n}$ *is nonvoid, is reflexive if and only if its semigroup of endomorphisms* $\mathfrak{B}(\Omega, \mathfrak{n})$ *is complete with respect to constants* (7.3).

Proof. 1) Suppose $\mathfrak{n}$ is reflexive. Consider any constant transformation U_α $(\alpha \in \Omega)$.

Let $\xi \sim \eta$ $(\mathfrak{n})$. Since $\mathfrak{n}$ is reflexive, $\alpha \sim \alpha$ $(\mathfrak{n})$, i.e., $U_\alpha \xi \sim U_\alpha \eta$ $(\mathfrak{n})$. Consequently, $U_\alpha \in \mathfrak{B}(\Omega, \mathfrak{n})$ $(\alpha \in \Omega)$.

2) Suppose $\mathfrak{U}_\Omega \subset \mathfrak{B}(\Omega, \mathfrak{n})$.

Since $\mathfrak{n}$ is nonvoid, we have $\xi \sim \eta$ $(\mathfrak{n})$ for some $\xi, \eta \in \Omega$. For any $\alpha \in \Omega$, consider the corresponding constant transformation U_α. Since $U_\alpha \in \mathfrak{B}(\Omega, \mathfrak{n})$, the fact that $\xi \sim \eta$ $(\mathfrak{n})$ implies that $U_\alpha \xi \sim U_\alpha \eta$ $(\mathfrak{n})$, i.e., $\alpha \sim \alpha$ $(\mathfrak{n})$.

8.7. For reflexive graphs the question of the interrelation between the two approaches to semigroups of endomorphisms, indicated in 8.4, becomes, by 8.6 and 7.16, completely transparent, since for that case the semigroups $\mathfrak{B}(\Omega_1, \mathfrak{n}_1)$ and $\mathfrak{B}(\Omega_2, \mathfrak{n}_2)$ are inessentially distinct (i.e. are essentially the same, as semigroups of transformations) if and only if they are isomorphic.

8.8. We consider now the most important case of reflexive graphs, viz. ordered sets, and denote the relation between elements, as usual, by the symbol $\leqq$.

In a partially ordered set Ω, select a fixed element α_0. For each pair $\alpha, \beta \in \Omega$, define a transformation $F_{\alpha\beta}$ of the set Ω by the requirement that $F_{\alpha\beta}\xi = \alpha$ if $\xi \leqq \alpha_0$ and $F_{\alpha\beta}\xi = \beta$ otherwise.

We assert that for $\alpha < \beta$, $F_{\alpha\beta}$ is an endomorphism.

To prove this, suppose $\xi \leqq \eta$. If $\eta \leqq \alpha_0$, then also $\xi \leqq \alpha_0$, and therefore $F_{\alpha\beta}\xi = \alpha_0$ and $F_{\alpha\beta}\eta = \alpha_0$. If $\eta \nleqq \alpha_0$, then $F_{\alpha\beta}\eta = \beta$, and since either $F_{\alpha\beta}\xi = \alpha$ or $F_{\alpha\beta}\xi = \beta$, we have $F_{\alpha\beta}\xi \leqq F_{\alpha\beta}\eta$.

8.9. For a wide class of cases the problem of 8.5, keeping in mind the remarks made there, has its solution in the following theorem.

Theorem. *Suppose that the partially ordered sets* Ω *and* Ω' *have isomorphic semigroups of endomorphisms* $\mathfrak{B}$ *and* $\mathfrak{B}'$. *Then as partially ordered sets* Ω *and* Ω' *are either isomorphic or anti-isomorphic.*

Proof. 1) By 7.16 and 8.6 the isomorphism of the semigroups $\mathfrak{B}$ and $\mathfrak{B}'$ implies that they are not essentially distinct (7.14). Let ψ be the corresponding one-to-one mapping of Ω onto Ω', and φ the induced isomorphism of $\mathfrak{B}$ onto $\mathfrak{B}'$. For every $\xi \in \Omega$ we shall write $\psi(\xi) = \xi'$.

If the partial orderings in both Ω and Ω' are just the identity relations, then Ω and Ω' are trivially isomorphic. We shall therefore suppose that in, say, Ω there exists a pair of elements $\alpha_0 < \beta_0$.

For every pair $\alpha < \beta$ $(\alpha, \beta \in \Omega)$, construct the endomorphism $F_{\alpha\beta}$, as in 8.8, with respect to α_0. Then $\varphi(F_{\alpha\beta}) = F'_{\alpha\beta} \in \mathfrak{B}'$, where, for any $\xi' \in \Omega'$, if $\xi \leqq \alpha_0$ we have $F'_{\alpha\beta}\xi' = \alpha'$ and if $\xi \nless \alpha_0$ we have $F'_{\alpha\beta}\xi' = \beta'$.

2) Next, we assert that either $\alpha'_0 \leqq \beta'_0$ or $\beta'_0 \leqq \alpha'_0$.

Suppose that neither inequality holds. For any $\alpha' \leqq \beta'$ $(\alpha', \beta' \in \Omega')$, we have

$$F'_{\alpha_0\beta_0}\alpha' \leqq F'_{\alpha_0\beta_0}\beta'.$$

But the elements $F'_{\alpha_0\beta_0}\alpha'$ and $F'_{\alpha_0\beta_0}\beta'$ can each be equal only to either α'_0 or β'_0, neither of which, by assumption, precedes the other. Hence it must be the case that

$$F'_{\alpha_0\beta_0}\alpha' = F'_{\alpha_0\beta_0}\beta'.$$

This implies that for $\alpha' < \beta'$ one of two possibilities holds: either α and β both precede α_0, or else neither one precedes α_0.

Consider the transformation H' of Ω' such that $H'\xi' = \beta'_0$ for $\xi \leqq \alpha_0$ and $H'\xi' = \alpha'_0$ for $\xi \nleqq \alpha_0$.

If $\alpha' < \beta'$, then, by what has just been proved, $H'\alpha' = H'\beta'$. Consequently, $H' \in \mathfrak{B}'$.

Since $H'\alpha'_0 = \beta'_0$ and $H'\beta'_0 = \alpha'_0$, we find, for $H = \varphi^{-1}(H') \in \mathfrak{B}$, that $H\alpha_0 = \beta_0$ and $H\beta_0 = \alpha_0$, which is impossible, since $\alpha_0 < \beta_0$ and H is an endomorphism of Ω.

3) Suppose, then, that $\alpha'_0 < \beta'_0$. For any pair $\alpha < \beta$, the equalities $\alpha = F_{\alpha\beta}\alpha_0$ and $\beta = F_{\alpha\beta}\beta_0$ imply that $\alpha' = F'_{\alpha\beta}\alpha'_0$ and $\beta' = F'_{\alpha\beta}\beta'_0$. Since $\alpha'_0 < \beta'_0$ and $F'_{\alpha\beta} \in \mathfrak{B}'$, we have $\alpha' \leqq \beta'$.

On the other hand, if $\alpha' < \beta'$, consider the endomorphism $F_{\alpha'\beta'}$, in the semigroup $\mathfrak{B}'$, constructed with respect to α'_0 as in 8.8.

We have

$$\alpha' = F_{\alpha'\beta'}\alpha'_0, \quad \beta' = F_{\alpha'\beta'}\beta'_0.$$

This implies, for $F = \varphi^{-1}(F_{\alpha'\beta'})$ in Ω, that

$$\alpha = F\alpha_0, \quad \beta = F\beta_0.$$

Since $F \in \mathfrak{B}$ and $\alpha_0 < \beta_0$, we conclude that $\alpha \leqq \beta$.

We have thus shown that the one-to-one mapping ψ of Ω onto Ω' is such that whenever $\alpha < \beta$ $(\alpha, \beta \in \Omega)$, $\psi(\alpha) \leqq \psi(\beta)$, and whenever $\alpha' < \beta'$ $(\alpha', \beta' \in \Omega')$, $\psi^{-1}(\alpha') \leqq \psi^{-1}(\beta')$. This says that ψ is an isomorphism of the partially ordered set Ω onto the partially ordered set Ω'.

4) Suppose, finally, that $\alpha'_0 > \beta'_0$.

Then between the partially ordered set Ω on the one hand, and, on the other hand, the set Ω' provided with the partial ordering dual to the original ordering on Ω', there exists the same relation as has already been considered in part 3). These two sets are therefore isomorphic. This implies that Ω and Ω', as originally given, are anti-isomorphic.

8.10. We remark that the constructions in 8.9, and the result itself, can be extended in a natural fashion to a wider class of graphs than that of partially ordered sets, as is done in the paper of Gluskin referred to above.

8.11. Aside from setting up against a graph $(\Omega, \mathfrak{n})$ (and in particular a partially ordered set) the semigroup $\mathfrak{B}(\Omega, \mathfrak{n})$ of *all* its endomorphisms, investigations have been made of the connections between properties of the graph $(\Omega, \mathfrak{n})$ and those of certain subsemigroups of $\mathfrak{B}(\Omega, \mathfrak{n})$. In a number of cases it has been possible to retain the fundamental relationship: namely that such semigroups be isomorphic only when the graphs are either isomorphic or anti-isomorphic.

In this connection may be mentioned papers of E. S. Ljapin [**S 2**] and L. D. Zybina [**S 1**], examining semigroups of certain special simple endomorphisms of partially ordered sets, which in turn determine partially ordered sets themselves and are comparatively simple to set up.

On the other hand, L. M. Popova [**S 1, 2**] and A. M. Kalmanovič [**S 1**] have studied semigroups of partial endomorphisms (I, 4.9), both those of all partial endomorphisms as well as those of certain special types. They describe the structure of a very wide class of graphs.

8.12. Of significant interest is the determination of those semigroups which are the semigroups of endomorphisms of some graph. Z. Hedrlín and A. Pultr [**S 1, 2**] have shown that for any semigroup $\mathfrak{A}$ of transformations of a set Ω that contains the identity transformation, there exists a relation $\mathfrak{n}$ in Ω such that $\mathfrak{A} = \mathfrak{B}(\Omega, \mathfrak{n})$ (8.3).

Since every semigroup with a unit is isomorphic to a semigroup of transformations containing the identity transformation, every such semigroup is therefore isomorphic to the semigroup of all endomorphisms of some graph.

The corresponding question for partial endomorphisms and arbitrary relations (I, 3.20, 4.9) has been examined by E. S. Ljapin [**S 4**].

CHAPTER VI

INVERTIBILITY

1. Invertibility of the Product of Elements

1.1. In this chapter we return to the question which was already touched on lightly in the second chapter.

DEFINITION. *An element S of a semigroup $\mathfrak{A}$ is said to be* RIGHT INVERTIBLE *if it is a left divisor of every element of $\mathfrak{A}$.*

An element S is said to be LEFT INVERTIBLE *if it is a right divisor of every element of $\mathfrak{A}$.*

An element which is both left and right invertible is said to be a TWO-SIDEDLY INVERTIBLE ELEMENT.

In a commutative semigroup the properties of right invertibility, left invertibility, and two-sided invertibility coincide, and we may simply speak of invertibility.

The task of distinguishing the invertible elements is a quite natural one. In particular, the abundance of elements with one or another property of invertibility characterizes the degree of closeness of the semigroup to a group. In semigroups of transformations, as we shall see, the invertibility property is essential for certain important properties of the transformations.

1.2. By definition, the element S is right invertible if for any $A \in \mathfrak{A}$ the equation

$$SY = A$$

for the unknown Y is solvable in $\mathfrak{A}$. It therefore follows that S will be right invertible if and only if

$$S\mathfrak{A} = \mathfrak{A}.$$

Analogously S is left invertible if for any $A \in \mathfrak{A}$ the equation

$$XS = A$$

for the unknown X is solvable in $\mathfrak{A}$, which is equivalent to the validity of the equation

$$\mathfrak{A}S = \mathfrak{A}.$$

The element S is two-sidedly invertible if and only if

$$S\mathfrak{A}S = \mathfrak{A}.$$

1.3. If the semigroup $\mathfrak{A}$ possesses a unit element E, then for the right invertibility of S it is necessary and sufficient that S should have a right inverse relative to E.

Indeed, from

$$S\mathfrak{A} = \mathfrak{A},$$

it follows that for some S' we have $SS' = E$. In turn, from $SS' = E$ follows

$$\mathfrak{A} = E\mathfrak{A} = SS'\mathfrak{A} \subset S\mathfrak{A},$$

i.e., $S\mathfrak{A} = \mathfrak{A}$.

In a quite analogous way one proves that for left invertibility of S it is necessary and sufficient that S should have a left inverse relative to E.

By II, 2.14, (α) it follows from what has been said that an element S, belonging to a semigroup $\mathfrak{A}$ with unit element, will be two-sidedly invertible in $\mathfrak{A}$ if and only if it has an inverse element S^{-1}:

$$SS^{-1} = S^{-1}S = E.$$

Here S^{-1} evidently will also be two-sidedly invertible.

1.4. THEOREM. *The set of all two-sidedly invertible elements of a semigroup, if it is not empty, forms a group whose unit is the unit of the whole semigroup.*

PROOF. The fact that the set $\mathfrak{G}$ of all two-sidedly invertible elements, i.e., the set of all elements which are both right and left divisors of every element of the semigroup $\mathfrak{A}$, forms a group, was proved in II, 1.5. In II, 2.15 we proved that a group always has a unit E.

Suppose that A is any element of $\mathfrak{A}$. Since $E \in \mathfrak{G}$ there must be in $\mathfrak{A}$ elements A' and A'' such that

$$A = EA', \qquad A = A''E.$$

Since $E^2 = E$, evidently $EA = A$ and $AE = A$.

1.5. COROLLARY. *In order that the semigroup $\mathfrak{A}$ should possess two-sidedly invertible elements it is necessary and sufficient that $\mathfrak{A}$ should have a unit.*

PROOF. If $\mathfrak{A}$ has two-sidedly invertible elements, then by 1.4 the unit of the group of all two-sidedly invertible elements will be the unit of $\mathfrak{A}$.

If $\mathfrak{A}$ has a unit E, then for any $A \in \mathfrak{A}$

$$EA = A, \qquad AE = A,$$

i.e., E is both a right and left divisor of A, i.e., a two-sidedly invertible element.

1.6. For any semigroup $\mathfrak{A}$ we denote by $\mathfrak{G}$ the set of all of its two-sidedly invertible elements. The set of all elements which are right invertible but not left invertible will be denoted by $\mathfrak{R}$; the set of all the elements which are left invertible but not right invertible will be denoted by $\mathfrak{L}$.

The set of all elements which are neither right nor left invertible will be denoted by $\mathfrak{K}$.

From the definitions it follows that the sets $\mathfrak{G}$, $\mathfrak{R}$, $\mathfrak{L}$, $\mathfrak{K}$ are disjoint, and that

$$\mathfrak{A} = \mathfrak{G} \cup \mathfrak{R} \cup \mathfrak{L} \cup \mathfrak{K}.$$

It is clear that for some semigroups some of the subsets $\mathfrak{G}$, $\mathfrak{R}$, $\mathfrak{L}$, $\mathfrak{K}$ will be empty.

Under the notations thus adopted $\mathfrak{G} \cup \mathfrak{R}$ is the set of all right invertible elements of $\mathfrak{A}$ and $\mathfrak{G} \cup \mathfrak{L}$ is the set of all left elements.

1.7. Let us see what can be said about the invertibility of the product of two elements. As E. S. Ljapin **[15]** proved, the invertibility of a product depends on the invertibility of the factors, although it is not determined by this alone. The results of the investigation may be put in the form of a table, to be interpreted as follows: For two given subsets $\mathfrak{X}, \mathfrak{B} \subset \mathfrak{A}$, their product is contained in the subset $\mathfrak{Z}$ indicated at the intersection of the row corresponding to $\mathfrak{X}$ and the column corresponding to $\mathfrak{B}$ (however, $\mathfrak{X}\mathfrak{B}$ may fail to coincide with $\mathfrak{Z}$).

THEOREM. *For every semigroup* $\mathfrak{A}$ *we have*:

	$\mathfrak{G}$	$\mathfrak{R}$	$\mathfrak{L}$	$\mathfrak{K}$
$\mathfrak{G}$	$\mathfrak{G}$	$\mathfrak{R}$	$\mathfrak{L}$	$\mathfrak{K}$
$\mathfrak{R}$	$\mathfrak{R}$	$\mathfrak{R}$	$\mathfrak{G} \cup \mathfrak{R} \cup \mathfrak{L} \cup \mathfrak{K}$	$\mathfrak{R} \cup \mathfrak{K}$
$\mathfrak{L}$	$\mathfrak{L}$	$\mathfrak{K}$	$\mathfrak{L}$	$\mathfrak{K}$
$\mathfrak{K}$	$\mathfrak{K}$	$\mathfrak{K}$	$\mathfrak{L} \cup \mathfrak{K}$	$\mathfrak{K}$

PROOF. First of all we observe that if $XY \in \mathfrak{G}$, then $X \in \mathfrak{G} \cup \mathfrak{R}$ and $Y \in \mathfrak{G} \cup \mathfrak{L}$.

Indeed, by 1.2,

$$XY\mathfrak{A}XY = \mathfrak{A}.$$

Hence we obtain

$$X\mathfrak{A} \supset X(Y\mathfrak{A}XY) = \mathfrak{A}, \qquad \mathfrak{A}Y \supset (XY\mathfrak{A}X)Y = \mathfrak{A},$$

which means that X is right invertible and Y is left invertible.

Now we shall consider successively all sixteen cells of our table. In what follows, G, R, L, K (also with primes) will mean arbitrary elements, respectively, of $\mathfrak{G}$, $\mathfrak{R}$, $\mathfrak{L}$, $\mathfrak{K}$.

(1) That $GG' \in \mathfrak{G}$ follows from 1.4.

(2) Since

$$GR\mathfrak{A} = G\mathfrak{A} = \mathfrak{A},$$

then $GR \in \mathfrak{G} \cup \mathfrak{R}$. Here $GR \in \mathfrak{G}$, by what was said at the beginning, is impossible, since $R \bar{\in} \mathfrak{G} \cup \mathfrak{L}$.

(3) Since

$$\mathfrak{A}GL = \mathfrak{A}L = \mathfrak{A},$$

therefore $GL \in \mathfrak{G} \cup \mathfrak{L}$. If we had $GL \in \mathfrak{G}$, then, because $G^{-1} \in \mathfrak{G}$,

$$\mathfrak{A} = G^{-1}\mathfrak{A} = G^{-1}GL\mathfrak{A} = L\mathfrak{A},$$

which would contradict the fact that L is not right invertible.

(4) If we had

$$GK \in \mathfrak{G} \cup \mathfrak{R},$$

then because $G^{-1} \in \mathfrak{G}$ we would obtain

$$\mathfrak{A} = G^{-1}\mathfrak{A} = G^{-1}GK\mathfrak{A} = K\mathfrak{A},$$

which would contradict the fact that K is not right invertible.

If we had

$$GK \in \mathfrak{G} \cup \mathfrak{L},$$

then we would obtain

$$\mathfrak{A} = \mathfrak{A}GK = \mathfrak{A}K,$$

which would contradict the fact that K is not left invertible. Accordingly, $GK \in \mathfrak{K}$.

(5) Since

$$RG\mathfrak{A} = R\mathfrak{A} = \mathfrak{A},$$

therefore $RG \in \mathfrak{G} \cup \mathfrak{R}$.

If we had $RG \in \mathfrak{G}$, then we would obtain

$$\mathfrak{A} = \mathfrak{A}G^{-1} = \mathfrak{A}RG \cdot G^{-1} = \mathfrak{A}R,$$

which would contradict the fact that R is not left invertible.

(6) Since

$$RR'\mathfrak{A} = R\mathfrak{A} = \mathfrak{A},$$

therefore $RR' \in \mathfrak{G} \cup \mathfrak{R}$. But $RR' \in \mathfrak{G}$ would contradict the fact, which follows from what was proved at the beginning that $R' \bar{\in} \mathfrak{G} \cup \mathfrak{L}$.

(7) Since $\mathfrak{G} \cup \mathfrak{R} \cup \mathfrak{L} \cup \mathfrak{K} = \mathfrak{A}$, therefore RL obviously belongs to that set.

(8) If we had

$$RK \in \mathfrak{G} \cup \mathfrak{L},$$

then we would obtain

$$\mathfrak{A} = \mathfrak{A}RK \subset \mathfrak{A}K,$$

which contradicts the fact that K is not left invertible.

(9) Since

$$\mathfrak{A}LG = \mathfrak{A}G = \mathfrak{A},$$

then $LG \in \mathfrak{G} \cup \mathfrak{L}$. Here $LG \in \mathfrak{G}$, from what was proved at the beginning, would contradict the fact that

$$L \bar{\in} \mathfrak{G} \cup \mathfrak{R}.$$

(10) If we had

$$LR \in \mathfrak{G} \cup \mathfrak{R},$$

then we would obtain

$$\mathfrak{A} = LR\mathfrak{A} = L\mathfrak{A},$$

which would contradict the fact that L is not right invertible.

Analogously one proves the impossibility of $LR \in \mathfrak{G} \cup \mathfrak{L}$.

Accordingly, $LR \in \mathfrak{K}$.

(11) Since

$$\mathfrak{A}LL' = \mathfrak{A}L' = \mathfrak{A},$$

then $LL' \in \mathfrak{G} \cup \mathfrak{L}$. But $LL' \in \mathfrak{G}$ would contradict the fact stated originally that $L \bar{\in} \mathfrak{G} \cup \mathfrak{R}$.

(12) If we had

$$LK \in \mathfrak{G} \cup \mathfrak{R},$$

then we would obtain

$$\mathfrak{A} = LK\mathfrak{A} \subset L\mathfrak{A}.$$

But $\mathfrak{A} = L\mathfrak{A}$ contradicts the fact that L is not right invertible.

If we had

$$LK \in \mathfrak{G} \cup \mathfrak{L},$$

then we would obtain

$$\mathfrak{A} = \mathfrak{A}LK = \mathfrak{A}K,$$

which would contradict the fact that K is not left invertible. Accordingly, $LK \in \mathfrak{K}$.

(13) If we had

$$KG \in \mathfrak{G} \cup \mathfrak{R},$$

then we would obtain

$$\mathfrak{A} = KG\mathfrak{A} = K\mathfrak{A},$$

which would contradict the fact that K is not right invertible.

If we had

$$KG \in \mathfrak{G} \cup \mathfrak{L},$$

then, since we have $\mathfrak{A} = \mathfrak{A}G^{-1}$, we would obtain

$$\mathfrak{A} = \mathfrak{A}G^{-1} = \mathfrak{A}KGG^{-1} = \mathfrak{A}K,$$

which would contradict the fact that K is not left invertible.

(14) If we had

$$KR \in \mathfrak{G} \cup \mathfrak{R},$$

then we would obtain

$$\mathfrak{A} = KR\mathfrak{A} = K\mathfrak{A},$$

which would contradict the fact that K is not right invertible.

If we had

$$KR \in \mathfrak{G} \cup \mathfrak{L},$$

then we would obtain

$$\mathfrak{A} = \mathfrak{A}KR \subset \mathfrak{A}R,$$

which would contradict the fact that R is not left invertible.

(15) If we had

$$KL \in \mathfrak{G} \cup \mathfrak{R},$$

then we would obtain

$$\mathfrak{A} = KL\mathfrak{A} \subset K\mathfrak{A},$$

which would contradict the fact that K is not right invertible.

(16) If we had

$$KK' \in \mathfrak{G} \cup \mathfrak{R},$$

then we would obtain

$$\mathfrak{A} = KK'\mathfrak{A} \subset K\mathfrak{A},$$

which would contradict the fact that K is not right invertible. Analogously, the hypothesis that $KK' \in \mathfrak{G} \cup \mathfrak{L}$ would lead us to a contradiction with the fact that K' is not left invertible.

1.8. From 1.7 it is immediately clear which of the unions of the sets $\mathfrak{G}$, $\mathfrak{R}$, $\mathfrak{L}$, $\mathfrak{K}$ form subsemigroups.

COROLLARY. *The following subsets are subsemigroups of* $\mathfrak{A}$:

$\mathfrak{A} = \mathfrak{G} \cup \mathfrak{R} \cup \mathfrak{L} \cup \mathfrak{K}$;
$\mathfrak{G}$—*the set of two-sided invertible elements*;
$\mathfrak{R}$—*the set of right invertible but not left invertible elements*;
$\mathfrak{L}$—*the set of left invertible but not right invertible elements*;
$\mathfrak{K}$—*the set of elements neither right nor left invertible*;
$\mathfrak{G} \cup \mathfrak{R}$—*the set of elements which are right invertible*;
$\mathfrak{G} \cup \mathfrak{L}$—*the set of elements which are left invertible*;
$\mathfrak{G} \cup \mathfrak{K}$—*the set of two-sided invertible elements and those elements which are neither right nor left invertible*;
$\mathfrak{R} \cup \mathfrak{K}$—*the set of elements not left invertible*;
$\mathfrak{L} \cup \mathfrak{K}$—*the set of elements not right invertible*;
$\mathfrak{G} \cup \mathfrak{R} \cup \mathfrak{K}$—*the set of elements which are right invertible and elements which are neither right nor left invertible*;
$\mathfrak{G} \cup \mathfrak{L} \cup \mathfrak{K}$—*the set of elements which are left invertible and elements invertible neither to the right nor to the left.*

1.9. Several of the cells of Table 1.7 contain not just one of the sets $\mathfrak{G}$, $\mathfrak{R}$, $\mathfrak{L}$, $\mathfrak{K}$ but the union of several of them. We shall show by an example that it is not possible to decrease the number of sets in those cells, i.e., that there exist semigroups for which the various products of elements of the corresponding sets belong to all the sets indicated in each such cell.

Let Ω be the set of all natural numbers. We consider the semigroup $\mathfrak{S}_\Omega$ of all transformations of Ω. Select the following elements of this semigroup, written in the form of substitutions:

$$E = \begin{pmatrix} 1 & 2 & 3 & 4 & \ldots & n & \ldots \\ 1 & 2 & 3 & 4 & \ldots & n & \ldots \end{pmatrix};$$

$$R_1 = \begin{pmatrix} 1 & 2 & 3 & 4 & \ldots & n & \ldots \\ 1 & 1 & 2 & 3 & \ldots & (n-1) & \ldots \end{pmatrix};$$

$$R_2 = \begin{pmatrix} 1 & 2 & 3 & 4 & \ldots & n & \ldots \\ 1 & 1 & 1 & 2 & \ldots & (n-2) & \ldots \end{pmatrix};$$

$$R_3 = \begin{pmatrix} 1 & 2 & 3 & 4 & \ldots & n & \ldots \\ 1 & 2 & 2 & 3 & \ldots & (n-1) & \ldots \end{pmatrix};$$

$$L_1 = \begin{pmatrix} 1 & 2 & 3 & 4 & \ldots & n & \ldots \\ 2 & 3 & 4 & 5 & \ldots & (n+1) & \ldots \end{pmatrix};$$

$$L_2 = \begin{pmatrix} 1 & 2 & 3 & 4 & \ldots & n & \ldots \\ 3 & 4 & 5 & 6 & \ldots & (n+2) & \ldots \end{pmatrix};$$

$$K_1 = \begin{pmatrix} 1 & 2 & 3 & 4 & \ldots & n & \ldots \\ 2 & 2 & 3 & 4 & \ldots & n & \ldots \end{pmatrix};$$

$$K_2 = \begin{pmatrix} 1 & 2 & 3 & 4 & \ldots & n & \ldots \\ 1 & 1 & 1 & 1 & \ldots & 1 & \ldots \end{pmatrix}.$$

By II, 3.3, R_1, R_2, R_3 considered as elements of $\mathfrak{S}_\Omega$, are left divisors of the permutation E, the identity of $\mathfrak{S}_\Omega$, and are not right divisors of E. Therefore, by 1.3, $R_1, R_2, R_3 \in \mathfrak{R}$. The elements L_1, L_2 are right divisors of E and are not left divisors of E. Therefore, by 1.3 we have $L_1, L_2 \in \mathfrak{L}$. The elements K_1 and K_2 belong to $\mathfrak{K}$ and E belongs to $\mathfrak{G}$.

We consider the seventh cell of Table 1.7. The set corresponding to this cell contains the elements:

$$R_1 L_1 = E \in \mathfrak{G};$$

$$R_2 L_1 = \begin{pmatrix} 1 & 2 & 3 & 4 & \ldots & n & \ldots \\ 1 & 1 & 2 & 3 & \ldots & (n-1) & \ldots \end{pmatrix} = R_1 \in \mathfrak{R};$$

$$R_3 L_2 = \begin{pmatrix} 1 & 2 & 3 & 4 & \ldots & n & \ldots \\ 2 & 3 & 4 & 5 & \ldots & (n+1) & \ldots \end{pmatrix} = L_1 \in \mathfrak{L};$$

$$R_3 L_1 = \begin{pmatrix} 1 & 2 & 3 & 4 & \ldots & n & \ldots \\ 2 & 2 & 3 & 4 & \ldots & n & \ldots \end{pmatrix} = K_1 \in \mathfrak{K}.$$

Thus, for the semigroup $\mathfrak{S}_\Omega$ the product $\mathfrak{R}\mathfrak{L}$ indeed contains elements from each of the four sets $\mathfrak{G}$, $\mathfrak{R}$, $\mathfrak{L}$, $\mathfrak{K}$, whose union is written in the seventh cell of the table.

Now we consider the eighth cell. In the set $\mathfrak{R}\mathfrak{K}$ corresponding to this cell there are contained the elements:

$$R_1K_1 = \begin{pmatrix} 1 & 2 & 3 & 4 & \dots & n & \dots \\ 1 & 1 & 2 & 3 & \dots & (n-1) & \dots \end{pmatrix} = R_1 \in \mathfrak{R};$$

$$R_1K_2 = \begin{pmatrix} 1 & 2 & 3 & 4 & \dots & n & \dots \\ 1 & 1 & 1 & 1 & \dots & 1 & \dots \end{pmatrix} = K_2 \in \mathfrak{K}.$$

Again we have discovered in $\mathfrak{R}\mathfrak{K}$ elements of both the sets $\mathfrak{R}$ and $\mathfrak{K}$, whose union is written in the eighth cell.

Finally, consider the fifteenth cell. In the set $\mathfrak{K}\mathfrak{L}$ corresponding to this cell there are contained the elements:

$$K_1L_1 = \begin{pmatrix} 1 & 2 & 3 & 4 & \dots & n & \dots \\ 2 & 3 & 4 & 5 & \dots & (n+1) & \dots \end{pmatrix} = L_1 \in \mathfrak{L};$$

$$K_2L_1 = \begin{pmatrix} 1 & 2 & 3 & 4 & \dots & n & \dots \\ 1 & 1 & 1 & 1 & \dots & 1 & \dots \end{pmatrix} = K_2 \in \mathfrak{K}.$$

$\mathfrak{K}\mathfrak{L}$ contains elements both of $\mathfrak{L}$ and of $\mathfrak{K}$, whose union is written in the fifteenth cell.

Finally, we observe that as the desired example we can of course take not the entire semigroup $\mathfrak{S}_\Omega$ but a countable subsemigroup of it generated by the elements E, R_1, R_2, R_3, L_1, L_2, K_1, K_2. It is easy to see that each of these elements will lie in the corresponding set $\mathfrak{G}$, $\mathfrak{R}$, $\mathfrak{L}$, $\mathfrak{K}$ of this semigroup.

1.10. We observe that the consideration of the properties of invertibility of elements is closely connected with the question as to whether elements lie in proper ideals of the semigroup.

THEOREM. *In order that the element S of the semigroup $\mathfrak{A}$ should be right invertible it is necessary and sufficient that S should have a right identity and that it should not be contained in any proper right ideal of the semigroup $\mathfrak{A}$.*

PROOF. (1) Suppose that S is right invertible:

$$S\mathfrak{A} = \mathfrak{A}.$$

Then $S\mathfrak{A} \ni S$, i.e., for some $Z \in \mathfrak{A}$ we must have $SZ = S$. Here, if $\mathfrak{R}$ is a right ideal of $\mathfrak{A}$, containing S, then

$$\mathfrak{A} = S\mathfrak{A} \subset \mathfrak{R}\mathfrak{A} \subset \mathfrak{R},$$

i.e., $\mathfrak{R} = \mathfrak{A}$.

(2) Suppose that S has both properties indicated in the formulation of the theorem. Since S has a right identity, therefore $S \in S\mathfrak{A}$. Thus the set $S\mathfrak{A}$ which is a right ideal of $\mathfrak{A}$ must obviously coincide with $\mathfrak{A}$, which means that S is right invertible.

1.11. Taking 1.5 into account, we immediately obtain from 1.10 the corollary:

COROLLARY. *In order that the element S of the semigroup $\mathfrak{A}$ should be two-sided invertible in $\mathfrak{A}$ it is necessary and sufficient that $\mathfrak{A}$ should have a unit and that S should not be contained in any proper left or right ideal of $\mathfrak{A}$.*

2. Invertibility of Magnifying Elements

2.1. As E. S. Ljapin [**14**] showed, the property of an element of being magnifying, considered in §§ 5 and 6 of the third chapter, is closely connected with the property of invertibility of that element.

THEOREM. *Every right magnifying element of a semigroup is left invertible but not right invertible.*

Every left magnifying element is right invertible but not left invertible.

PROOF. If U is a right magnifying element of the semigroup $\mathfrak{A}$, then for some $\mathfrak{A}' \subset \mathfrak{A}$ we have

$$\mathfrak{A}'U = \mathfrak{A}, \qquad \mathfrak{A}' \neq \mathfrak{A}.$$

Hence it follows that $\mathfrak{A}U = \mathfrak{A}$, i.e., U is left invertible. Suppose that U is also right invertible. Then U is a two-sidedly invertible element. By 1.4, from the existence in $\mathfrak{A}$ of two-sidedly invertible elements it follows that $\mathfrak{A}$ has a unit $E\mathfrak{A}$, and the two-sided invertible element U has an inverse U^{-1}. But from this follows

$$\mathfrak{A}' = \mathfrak{A}'E_{\mathfrak{A}} = \mathfrak{A}'UU^{-1} = \mathfrak{A}U^{-1} = \mathfrak{A}UU^{-1} = \mathfrak{A}E_{\mathfrak{A}} = \mathfrak{A},$$

which contradicts the choice of $\mathfrak{A}'$.

The reasoning for left magnifying elements is analogous.

2.2. The converse is not valid without some additional stipulations. Indeed, in a semigroup $\mathfrak{A}$ containing more than one element, in which for arbitrary elements X and Y we have $XY = Y$, it is obvious that every element is right invertible:

$$X\mathfrak{A} = \mathfrak{A}.$$

However, no element whatever is left magnifying, for

$$X\mathfrak{A}' = \mathfrak{A}'$$

for any $X \in \mathfrak{A}$ and $\mathfrak{A}' \subset \mathfrak{A}$.

2.3. The converse to Theorem 2.1 is nevertheless valid in semigroups with unity. It turns out that (in the notation of § 1) $\mathfrak{R}$ is the set of all left magnifying elements and $\mathfrak{L}$ is the set of all right magnifying elements.

THEOREM. *In a semigroup with unity the right magnifying elements, and only those, are left invertible and not right invertible.*

The left magnifying elements, and only those, are right invertible and not left invertible.

PROOF. The fact that a right magnifying element is left invertible and not right invertible was proved in 2.1.

Suppose that the element L of the semigroup $\mathfrak{A}$ with a unit is left invertible and not right invertible:

$$\mathfrak{A}L = \mathfrak{A}, \qquad L\mathfrak{A} \neq \mathfrak{A}.$$

For some $R \in \mathfrak{A}$ we have

$$RL = E_{\mathfrak{A}}.$$

From this follows

$$R(L\mathfrak{A}) = \mathfrak{A},$$

which means that R is a left magnifying element of $\mathfrak{A}$.

Consider the subsemigroup

$$\mathfrak{Q} = [L, R].$$

Each element of this semigroup

$$Q = L^{\alpha_1}R^{\beta_1}L^{\alpha_2}R^{\beta_2} \dots L^{\alpha_k}R^{\beta_k},$$

if we replace RL by the identity, may evidently be brought into the form

$$Q = L^{\alpha}R^{\beta}$$

(here and in the sequel several exponents may be equal to zero, which means $L^0 = R^0 = E_{\mathfrak{A}}$).

Let us indicate the natural method by which each element of $\mathfrak{Q}$ may be brought into this form.

Let

$$L^{\alpha_1}R^{\beta_1} = L^{\alpha_2}R^{\beta_2}.$$

If $\alpha_1 = \alpha_2$, then by multiplying the equations on the left by R^{α_1} and recalling that $RL = E_{\mathfrak{A}}$ we obtain

$$R^{\beta_1} = R^{\beta_2}.$$

If $\beta_1 > \beta_2$, then by multiplying this equation on the left by R^{β_2} we obtain

$$R^{\beta_1 - \beta_2} = E_{\mathfrak{A}}.$$

Because of III, 5.4 this is impossible, since R is a left magnifying element of $\mathfrak{A}$ and $E_{\mathfrak{A}}$ is not a left magnifying element of $\mathfrak{A}$.

If $\alpha_1 > \alpha_2$, then by multiplying the equation

$$L^{\alpha_1}R^{\beta_1} = L^{\alpha_2}R^{\beta_2}$$

on the left by R^{α_2} we obtain

$$L^{\alpha_1-\alpha_2}R^{\beta_1} = R^{\beta_2}.$$

But R is a left magnifying element. Therefore $R\mathfrak{A} = \mathfrak{A}$, and from the equality so obtained there results

$$L\mathfrak{A} \supset L^{\alpha_1-\alpha_2}R^{\beta_1}\mathfrak{A} = R^{\beta_2}\mathfrak{A} = \mathfrak{A},$$

which contradicts the fact that $L\mathfrak{A} \neq \mathfrak{A}$.

Thus we have shown that the subsemigroup $\mathfrak{Q}$ may be represented as the set of products of the form

$$L^{\alpha}R^{\beta},$$

while two products differing in either one of the exponents are distinct. The multiplication of these products, thanks to the fact that $RL = E_{\mathfrak{A}}$, is carried out according to the rule

$$L^{\alpha_1}R^{\beta_1} \cdot L^{\alpha_2}R^{\beta_2} = \begin{cases} L^{\alpha_1+\alpha_2-\beta_1}R^{\beta_2} & (\text{if } \alpha_2 \geqslant \beta_1); \\ L^{\alpha_1}_1 R^{\beta_1-\alpha_2+\beta_2} & (\text{if } \alpha_2 \leqslant \beta_1). \end{cases}$$

From all of this it follows that the semigroup $\mathfrak{Q}$ is isomorphic to the semigroup $\mathfrak{P}$ (III, 6.3). Since under the isomorphism of $\mathfrak{P}$ onto $\mathfrak{Q}$ the element $L \in \mathfrak{Q}$ corresponds to the element $U \in \mathfrak{P}$, it follows from III, 6.8 that L is a right magnifying element of the semigroup $\mathfrak{A}$.

The discussion for left magnifying elements is analogous.

2.4. Because of III, 6.8 the following corollary follows from the theorem just proved.

COROLLARY. *Suppose that the semigroup $\mathfrak{A}$ has a unity.*

The element X of the semigroup $\mathfrak{A}$ is left invertible but not right invertible if and only if there exists an isomorphism φ of the semigroup $\mathfrak{P}$ (III, 6.2, III, 6.3) *into $\mathfrak{A}$ under which $\varphi(E_{\mathfrak{P}}) = E_{\mathfrak{A}}$ and $\varphi(U) = X$.*

The element X is right invertible but not left invertible if and only if there exists an isomorphism φ of the semigroup $\mathfrak{P}$ into $\mathfrak{A}$ under which $\varphi(E_{\mathfrak{P}}) = E_{\mathfrak{A}}$ and $\varphi(V) = X$.

2.5. COROLLARY. *If, in the semigroup $\mathfrak{A}$ with unity, for some element $X \in \mathfrak{A}$ we have*

$$X\mathfrak{A} = \mathfrak{A},$$

but

$$X\mathfrak{A}' \neq \mathfrak{A}$$

for every $\mathfrak{A}' \subset \mathfrak{A}$ distinct from $\mathfrak{A}$, then the element X is two-sidedly invertible. (Analogously for right multiplication.)

Indeed, X is right invertible but not left magnifying, and therefore, from 2.3 it must be left invertible.

We observe that in the formulation of Corollary 2.5 one may not drop the requirement of the existence of an identity. This is seen from the following example. Suppose that $\mathfrak{A}$ is a semigroup with a number of elements not less than two and such that $XY = Y$ for every $X, Y \in \mathfrak{A}$. Then in $\mathfrak{A}$, for each X, we have

$$X\mathfrak{A} = \mathfrak{A}, \qquad X\mathfrak{A}' = \mathfrak{A}'.$$

However, it is clear that no element of the semigroup $\mathfrak{A}$ is two-sidedly invertible.

3. Semigroups with One-sided Invertibility

3.1. In connection with the decomposition in § 1 above of any semigroup $\mathfrak{A}$ into four subsemigroups $\mathfrak{G}$, $\mathfrak{L}$, $\mathfrak{R}$, $\mathfrak{K}$, it is natural to distinguish semigroups coinciding with their subsemigroup $\mathfrak{G} \cup \mathfrak{L}$ (or analogously $\mathfrak{G} \cup \mathfrak{R}$). Such semigroups were considered in his time by A. K. Suškevič [3; **12**], and therefore N. N. Vorob′ev [**6**] has called them Suškevič systems. They are also called semigroups with left division and left simple semigroups.

DEFINITION. *If all the elements of a semigroup are left invertible, then it is called a* SEMIGROUP WITH LEFT INVERTIBILITY.

Analogously one defines a semigroup with right invertibility.

The only semigroups with left and right invertibility at the same time are groups.

3.2. *The semigroup $\mathfrak{A}$ is a semigroup with left invertibility if and only if it does not have any proper left ideals.*

Indeed, if $\mathfrak{A}$ is a semigroup with left invertibility, then, for any $A \in \mathfrak{A}$, $\mathfrak{A}A = \mathfrak{A}$, and therefore A cannot be contained in a proper left ideal of $\mathfrak{A}$.

If $\mathfrak{A}$ has no proper left ideals, then for any $A \in \mathfrak{A}$ we have $\mathfrak{A}A = \mathfrak{A}$.

We observe that, from V, 2.4 and what has just been said, minimal left ideals of any semigroup are semigroups with left invertibility.

3.3. Semigroups with left invertibility essentially divide into two classes depending on whether there are idempotents among their elements or not.

THEOREM. *If in a semigroup with left invertibility there are idempotents, then each of them is a right unit of the semigroup which is divisible on the right by every element of the semigroup.*

If in a semigroup with left invertibility there are no idempotents, then none of its elements has a right unit.

PROOF. (1) Let I be any idempotent of a semigroup $\mathfrak{A}$ with left invertibility. Since $\mathfrak{A}$ is a semigroup with left invertibility it follows that for any $A \in \mathfrak{A}$ there will exist U and V such that

$$UA = I, \qquad VI = A.$$

Because of the second equation we obtain

$$AI = VII = VI = A.$$

(2) Suppose that in a semigroup with left invertibility $\mathfrak{A}$ the element X has a right unit:

$$XY = X \qquad (X, Y \in \mathfrak{A}).$$

For some $Z \in \mathfrak{A}$ we must have $ZX = Y$. Hence we immediately obtain the existence in $\mathfrak{A}$ of the idempotent

$$Y^2 = ZX \cdot Y = Z \cdot XY = ZX = Y.$$

3.4. As follows from Theorem 3.3, the class of semigroups with left invertibility and having idempotents coincides with the class of semigroups having right units which are divisible on the right by every element of the semigroup. Indeed, suppose that some semigroup $\mathfrak{A}$ has such a right unit I. For any $A \in \mathfrak{A}$ there is an $X \in \mathfrak{A}$ such that $XA = I$. Hence for every $B \in \mathfrak{A}$ we have

$$(BX)A = BI = B.$$

Here $\mathfrak{A} \ni I$ and $I^2 = I$.

From what has been said it follows that when we investigate below the structure of semigroups with left invertibility and with idempotents, we at the same time obtain an answer to the question in the second section of the second chapter; namely, which semigroups have right units which are divisible on the right by all the elements of the semigroup.

The investigation of a series of properties of semigroups with left invertibility and with idempotents was carried out by A. K. Suškevič [**12**]. Later, developing this direction further, Munn [**1**] gave an exhaustive description of the construction of such semigroups and carried out their complete classification.

3.5. Let $\mathfrak{H}$ be a semigroup in which for any $U, V \in \mathfrak{H}$ we have $UV = U$, and let $\mathfrak{G}$ be any group. Denote by $\mathfrak{T}$ the set of all possible pairs

$$(U, G) \qquad (U \in \mathfrak{H}, G \in \mathfrak{G}).$$

Define in $\mathfrak{T}$ the multiplication

$$(U, G) \cdot (U', G') = (UU', GG') = (U, GG')$$
$$(U, U' \in \mathfrak{H}; G, G' \in \mathfrak{G}).$$

The associativity of the operation is evident. $\mathfrak{T}$ is a semigroup with left invertibility, since for any $(U, G), (U', G') \in \mathfrak{T}$ we have

$$(U', G'G^{-1}) \cdot (U, G) = (U', G').$$

Elements of the form $(U, E_{\mathfrak{G}})$, as is easily seen, are idempotents, while $\mathfrak{T}$ has no other idempotents.

3.6. The structure of the semigroup just described is completely determined by the cardinality of the set $\mathfrak{H}$ and the group $\mathfrak{G}$.

The semigroup $\mathfrak{T}$ may be considered as a completely simple semigroup without zero of the matrix type $\mathfrak{S}'(P, \mathfrak{G})$ (V, 6.3), in which P consists of one column, with all its elements equal to $E_{\mathfrak{G}}$.

3.7. The significance of semigroups of the indicated type is explained by the fact that, as we shall show below, *every semigroup with left invertibility and with idempotents is isomorphic to some semigroup of type* 3.5.

3.8. Suppose that $\mathfrak{A}$ is any semigroup with left invertibility and with idempotents, whose union we shall denote by $\mathfrak{H}$. We shall deduce a series of properties of the semigroup $\mathfrak{A}$.

(α) If $U, V \in \mathfrak{H}$, then $UV = U$.

This follows immediately from 3.3.

(β) If $U \in \mathfrak{H}$, $X \in \mathfrak{A}$ and $UX = U$, then $X \in \mathfrak{H}$.

Indeed, because of 3.3,

$$X^2 = XUX = XU = X.$$

(γ) If $U \in \mathfrak{H}$, then the set $U\mathfrak{A}$ is a group with U as unit.

Indeed, if $X, Y \in U\mathfrak{A}$, then evidently $XY \in U\mathfrak{A}$. The element $U \in U\mathfrak{A}$ is evidently a left identity of the subsemigroup $U\mathfrak{A}$. For each $X \in U\mathfrak{A}$ there must exist in $\mathfrak{A}$ an element X' such that $U = X'X$. Since

$$U = UU = (UX') \cdot X, \qquad UX' \in U\mathfrak{A},$$

therefore U is divisible on the right in $U\mathfrak{A}$ by any element of $U\mathfrak{A}$. Hence, from II, 2.18 it follows that $U\mathfrak{A}$ is a group.

(δ) For every $X \in \mathfrak{A}$ in $\mathfrak{H}$ there is a unique element I_X which is a left unit for X.

For any $U \in \mathfrak{H}$ in $\mathfrak{A}$ one can find an X' such that $U = XX'$. For $I_X = XX'$ we obtain, using 3.4,

$$I_X X = XX'X = XU = X,$$

$$I_X^2 = XX'XX' = XUX' = XX' = I_X.$$

If V is any left unit of X which is an idempotent, then from (α) we obtain

$$V = VI_X = VXX' = XX' = I_X.$$

3.9. The properties of semigroups $\mathfrak{A}$ with left invertibility and with idempotents make it possible to prove the validity of assertion 3.7.

Fix one of the idempotents $U \in \mathfrak{H}$. Using 3.8, (δ) we associate to each element $X \in \mathfrak{A}$ the pair

$$X \sim (I_X, UX).$$

We shall show that to distinct elements of the semigroup $\mathfrak{A}$ there correspond distinct pairs. Suppose that, for $X, Y \in \mathfrak{A}$,

$$(I_X, UX) = (I_Y, UY),$$

i.e.,

$$I_X = I_Y, \qquad UX = UY.$$

Then

$$X = I_X X = I_X UX = I_Y UY = I_Y Y = Y.$$

Now we shall show that for any pair (V, G), where $V \in \mathfrak{H}$ and $G \in U\mathfrak{A}$, there exists in $\mathfrak{A}$ an element to which this pair corresponds. As such an element we choose $Z = VG$. Since

$$VZ = VVG = VG = Z, \qquad V \in \mathfrak{H},$$

then $I_Z = V$. Since $G \in U\mathfrak{A}$, then

$$UZ = UVG = UG = G.$$

Accordingly,

$$Z \sim (I_Z, UZ) = (V, G).$$

3.10. We denote by $\mathfrak{T}$ the semigroup of type 3.5 constructed for the semigroup $\mathfrak{H}$ (in which, because of 3.8, (α), the product of two arbitrary elements is equal to the left factor) and the group $U\mathfrak{A}$ (3.8, (γ)). In 3.9 there was established between the elements $\mathfrak{A}$ and $\mathfrak{T}$ a one-to-one correspondence. We shall show that this correspondence has the isomorphism property, and at the same time we shall prove the validity of assertion 3.7.

Let $XY = Z$ and

$$X \sim (I_X, UX), \qquad Y \sim (I_Y, UY), \qquad Z \sim (I_Z, UZ).$$

Since $I_X I_Y = I_X \in \mathfrak{H}$ (3.8, (α)) and

$$(I_X I_Y) \cdot Z = (I_X I_Y) \cdot (XY) = I_X XY = XY = Z,$$

then, because of 3.8, (δ), $I_X I_Y = I_Z$.

Further,

$$(UX) \cdot (UY) = U \cdot XU \cdot Y = UXY = UZ.$$

Thus, because of the operation between pairs of elements of $\mathfrak{T}$ set up in 3.5, we find that to the element $Z = XY$ there corresponds the following pair:

$$(I_Z, UZ) = (I_X I_Y, (UX) \cdot (UY)) = (I_X, UX) \cdot (I_Y, UY).$$

3.11. The isomorphic representation of the semigroups considered here in the form of semigroups of pairs not only elucidates their structure and makes it possible to deduce various properties of them but also amounts to a classification of semigroups with left invertibility and with idempotents. This follows from the fact that two semigroups of pairs, the semigroup $\mathfrak{T}_1$ constructed from $\mathfrak{H}_1$ and $\mathfrak{G}_1$ and the semigroup $\mathfrak{T}_2$ constructed from $\mathfrak{H}_2$ and $\mathfrak{G}_2$, are isomorphic if and only if the semigroups $\mathfrak{H}_1$ and $\mathfrak{H}_2$ are isomorphic (for this, evidently, it suffices that $\mathfrak{H}_1$ and $\mathfrak{H}_2$ have the same cardinality) and the groups $\mathfrak{G}_1$ and $\mathfrak{G}_2$ are isomorphic. For the proof of this fact it is evidently sufficient to prove that in the semigroup $\mathfrak{T}$ of pairs of type 3.5 the semigroup $\mathfrak{H}$ and the group $\mathfrak{G}$ are defined up to an isomorphism in a unique way. To prove this we note that $\mathfrak{H}$ has the same cardinality as the set of all idempotents in $\mathfrak{T}$, since only pairs of the form $(H, E_{\mathfrak{G}})$ $(H \in \mathfrak{H})$ are idempotents. Accordingly, $\mathfrak{H}$ is defined for $\mathfrak{T}$ up to an isomorphism in a unique way. $\mathfrak{G}$ is isomorphic to the group consisting of all the elements of $\mathfrak{T}$ having a given (arbitrary) idempotent $(H, E_{\mathfrak{G}})$ as its two-sided unit (such elements are, evidently, pairs of the form (H, G)). Accordingly, also $\mathfrak{G}$ is defined for $\mathfrak{T}$ up to an isomorphism in a unique way.

3.12. It is quite evident that the consideration of properties, description of structure, and classification, for semigroups with right invertibility which have idempotents, may be carried out in a way completely analogous to the preceding.

3.13. Now we turn to the consideration of semigroups with left invertibility which do not have idempotents. Let $\mathfrak{A}$ be one of these semigroups. It is necessarily infinite since every finite semigroup has idempotents. We recall further that in $\mathfrak{A}$, because of 3.3, no element can have a right unit.

3.14. We consider the following important example of semigroups with the indicated properties. This example was constructed by Teissier [5]. It represents a certain development of the construction of Baer and Levi [1] and plays an important role for the whole class of semigroups under consideration.

Suppose that in an infinite set Ω there is given arbitrarily an equivalence $\mathfrak{n}$ such that the number of $\mathfrak{n}$-classes is infinite.

In the semigroup $\mathfrak{S}_\Omega$ consisting of all transformations of Ω we distinguish the transformations X having the following three properties:

(α) if $\alpha \sim \beta(\mathfrak{n})$ $(\alpha, \beta \in \Omega)$, then $X\alpha = X\beta$;

(β) if $\alpha \nsim \beta(\mathfrak{n})$ $(\alpha, \beta \in \Omega)$, then $X\alpha \nsim X\beta(\mathfrak{n})$;

(γ) the set of all those $\mathfrak{n}$-classes Ω' for which $X\Omega \cap \Omega' = \varnothing$ has the same cardinality as the set of all $\mathfrak{n}$-classes.

It is easy to see that the collection of transformations X of $\mathfrak{S}_\Omega$ having all three properties is a subsemigroup, which we shall denote by $\mathfrak{T}_\Omega^{(\mathfrak{n})}$.

3.15. We shall deduce a series of properties of the semigroups $\mathfrak{T}_\Omega^{(\mathfrak{n})}$ thus defined.

(α) $\mathfrak{T}_\Omega^{(\mathfrak{n})}$ has no idempotents.

Indeed, suppose that for some $X \in \mathfrak{T}_\Omega^{(\mathfrak{n})}$ we had $X^2 = X$. By 3.14, (γ) there is an $\mathfrak{n}$-class Ω' such that Ω' has no elements in common with $X\Omega$. Suppose that $\alpha \in \Omega'$. Since

$$X(X\alpha) = X^2\alpha = X\alpha = X(\alpha),$$

therefore, because of 3.14, (β), $X\alpha \sim \alpha(\mathfrak{n})$, and consequently $X\alpha \in \Omega'$, which contradicts the fact that $X\Omega \cap \Omega' = \varnothing$.

(β) Every element of $\mathfrak{T}_\Omega^{(\mathfrak{n})}$ is a right magnifying element of this semigroup.

Suppose that $X, Y \in \mathfrak{T}_\Omega^{(\mathfrak{n})}$. We write $\mathfrak{T}_\Omega^{(\mathfrak{n})} \backslash X = \mathfrak{P}$. The set of all $\mathfrak{n}$-classes Ω' satisfying the condition $X\Omega \cap \Omega' \neq \varnothing$ will be denoted by Γ_1, and those satisfying the condition $Y\Omega \cap \Omega' = \varnothing$ by Γ_2. We break Γ_2 into two disjoint subsets $\Gamma_2 = \Gamma_3 \cup \Gamma_4$, each having a cardinality equal to the cardinality of the set of all $\mathfrak{n}$-classes. In each $\mathfrak{n}$-class $\Omega^{(\xi)} \in \Gamma_1$ we fix some element $\lambda_\xi \in \Omega^{(\xi)}$. We choose some one-to-one mapping φ of the set Γ_1 onto Γ_3, and to each $\lambda_\xi \in \Omega^{(\xi)}$ we put into correspondence some element $\lambda_\xi^\varphi \in \varphi(\Omega^{(\xi)})$ such that at least for one λ_ξ we have $\lambda_\xi^\varphi \neq X\lambda_\xi$. We introduce a new transformation $S \in \mathfrak{S}_\Omega$, putting:

(1) $S\alpha = \beta$ if there exists an element $\gamma \in \Omega$ such that $X\gamma \sim \alpha(\mathfrak{n})$ and $Y\gamma = \beta$;

(2) $S\alpha = \lambda_\xi^\varphi$ if $\alpha \in \Omega^{(\xi)}$, where $\Omega^{(\xi)} \in \Gamma_1$.

It is easy to see that these two conditions determine $S \in \mathfrak{S}_\Omega$ uniquely. By the definition of φ we have $S \in \mathfrak{P}$. It is immediate that S satisfies condition 3.14, (α), (β). Since for each $\Omega^{(\tau)} \in \Gamma_4$ we have $S\Omega \cap \Omega^{(\tau)} = \varnothing$, therefore S also has the property 3.14, (γ).

From the definition of S, for each $\nu \in \Omega$ we have $SX\nu = Y\nu$. Thus, $SX = Y$, and since Y is an arbitrary element of $\mathfrak{T}_\Omega^{(\mathfrak{n})}$ and $S \in \mathfrak{P}$, therefore $\mathfrak{P}X = \mathfrak{T}_\Omega^{(\mathfrak{n})}$.

(γ) $\mathfrak{T}_\Omega^{(\mathfrak{n})}$ is a semigroup with left invertibility. This follows immediately from (β) and 2.1.

(δ) In $\mathfrak{T}_\Omega^{(\mathfrak{n})}$ there are no left magnifying elements. This immediately follows from (β) and III, 5.2.

(ε) No element in $\mathfrak{T}_\Omega^{(\mathfrak{n})}$ has a right unit in $\mathfrak{T}_\Omega^{(\mathfrak{n})}$.

This immediately follows from (α), (γ) and 3.3.

(ζ) If every $\mathfrak{n}$-class consists of one element, then $\mathfrak{T}_\Omega^{(\mathfrak{n})}$ is a semigroup with left cancellation.

Indeed, suppose $X \neq Y (X, Y \in \mathfrak{T}_\Omega^{(\mathfrak{n})})$. For some $\alpha \in \Omega$ we must have $X\alpha \neq Y\alpha$. But this means that $X\alpha \nsim Y\alpha(\mathfrak{n})$, and therefore, by 3.14, (β), $SX\alpha \neq SY\alpha$ for any $S \in \mathfrak{T}_\Omega^{(\mathfrak{n})}$.

(η) If even one $\mathfrak{n}$-class contains more than one element, then $\mathfrak{T}_\Omega^{(\mathfrak{n})}$ is not a semigroup with left cancellation.

Indeed, if $\alpha \sim \beta(\mathfrak{n})$ and $\alpha \neq \beta$, then we choose some transformation $X \in \mathfrak{T}_\Omega^{(\mathfrak{n})}$ such that $X\Omega \ni \alpha$. We construct a transformation $X' \in \mathfrak{S}_\Omega$ such that $X'\nu = X\nu$ if $X\nu \neq \alpha$, and $X'\nu = \beta$ if $X\nu = \alpha$. It is easy to see that $X' \in \mathfrak{T}_\Omega^{(\mathfrak{n})}$. Since $S\alpha = S\beta$ for any $S \in \mathfrak{T}_\Omega^{(\mathfrak{n})}$ (since $\alpha \sim \beta(\mathfrak{n})$), then $SX = SX'$, although $X \neq X'$.

(θ) $\mathfrak{T}_\Omega^{(\mathfrak{n})}$ is not a semigroup with right cancellation.

Indeed, choose in $\mathfrak{T}_\Omega^{(\mathfrak{n})}$ two elements $X \neq X'$ for which $X\alpha = X'\alpha$ for all $\alpha \in \Omega$, with the exclusion of α, belonging to one fixed $\mathfrak{n}$-class $\Omega^{(\xi)}$. We choose Z in $\mathfrak{T}_\Omega^{(\mathfrak{n})}$ such that $Z\Omega \cap \Omega^{(\xi)} = \varnothing$. For such elements, evidently, we obtain $XZ = X'Z$.

3.16. The properties derived for $\mathfrak{T}_\Omega^{(\mathfrak{n})}$ characterize it as a semigroup which is very interesting from several points of view. In particular, we have for the first time obtained an example of a semigroup, all the elements of which are right magnifying elements for it. Moreover, as Teissier proved [5], the significance of the semigroup $\mathfrak{T}_\Omega^{(\mathfrak{n})}$ goes far beyond the bounds of an interesting example. This follows from the considerations presented below.

Suppose that $\mathfrak{A}$ is a semigroup with left invertibility which does not have idempotents. Denote by Ω the set consisting of all elements of $\mathfrak{A}$ and one further "separating" element I (I, 3.9). In the set Ω we define an equivalence relation $\mathfrak{n}$, putting $X \sim Y(\mathfrak{n})$ if $X = Y = I$ or if $X, Y \in \mathfrak{A}$ and $AX = AY$ for every $A \in \mathfrak{A}$. Symmetry, reflexivity and transitivity of $\mathfrak{n}$ are evident. We note further that if X and Y of $\mathfrak{A}$ are such that for some $Z \in \mathfrak{A}$ we have $ZX = ZY$, then $X \sim Y(\mathfrak{n})$. Indeed, let A be an arbitrary element of $\mathfrak{A}$. Since $\mathfrak{A}$ is a semigroup with left invertibility, then for some $S \in \mathfrak{A}$, $SZ = A$. Therefore

$$AX = SZX = SZY = AY.$$

For the equivalence $\mathfrak{n}$ in Ω we construct the semigroup $\mathfrak{T}_\Omega^{(\mathfrak{n})}$. It is a subsemigroup of the semigroup $\mathfrak{S}_\Omega$. We consider the isomorphism φ (I, 3.9) of the semigroup $\mathfrak{A}$ into $\mathfrak{S}_\Omega$ which is a representation of it by left translations (I, 3.10). We shall prove that $\varphi(\mathfrak{A}) \subset \mathfrak{T}_\Omega^{(\mathfrak{n})}$.

For any $A \in \mathfrak{A}$ we consider the transformation $\bar{A} = \varphi(A) \in \mathfrak{S}_\Omega$.

Let X and Y be two distinct elements of Ω belonging to one and the same $\mathfrak{n}$-class.

Since $X, Y \in \mathfrak{A}$ and $X \sim Y(\mathfrak{n})$, therefore

$$\bar{A}(X) = AX = AY = \bar{A}(Y).$$

Thus $\bar{A}$ has the property 3.14, (α).

If X and Y belong to different $\mathfrak{n}$-classes while both are elements of $\mathfrak{A}$, then, according to the remarks above, the element $\bar{A}(X) = AX$ and the element $\bar{A}(Y) = AY$ cannot stand in the relation $\mathfrak{n}$ to one another.

If $X \in \mathfrak{A}$ and $Y = I$, then

$$\bar{A}(X) = AX, \qquad \bar{A}(I) = A$$

and $AX \sim A(\mathfrak{n})$ is impossible, since otherwise, by the definition of $\mathfrak{n}$, we would obtain $AAX = AA$, which contradicts the lack of a right unit for A^2 (3.3).

Thus $\bar{A}$ has property 3.14, (β).

For each $S \in \mathfrak{A}$ there is in $\mathfrak{A}$ an element $R^{(S)}$ such that $A = R^{(S)}S$. We denote by $\Omega^{(S)}$ the set of all elements of $\mathfrak{A}$, equivalent mod $\mathfrak{n}$ to the element $R^{(S)}$.

The sets $\bar{A}\Omega$ and $\Omega^{(S)}$ have no common elements. Indeed, in the contrary case, for some $X \in \Omega$ we would have $AX \sim R^{(S)}(\mathfrak{n})$. Hence it would follow that $(SR^{(S)})SX = SR^{(S)}$, which would contradict the lack of a right unit for the element $SR^{(S)}$ (3.3). We have associated to each $S \in \mathfrak{A}$ an $\mathfrak{n}$-class $\Omega^{(S)}$ satisfying the condition $\bar{A}\Omega^{(S)} \cap \Omega^{(S)} = \varnothing$. Now we shall show that if $S_1 \not\sim S_2(\mathfrak{n})$, i.e., if S_1 and S_2 are chosen from different $\mathfrak{n}$-classes, then $\Omega^{(S_1)} \neq \Omega^{(S_2)}$. Hence it will result immediately that property 3.14, (γ) is satisfied for the transformation $\bar{A}$.

Since

$$A = R^{(S_1)}S_1, \qquad A = R^{(S_2)}S_2,$$

then for each $B \in \mathfrak{A}$ we have

$$BR^{(S_1)}S_1 = BR^{(S_2)}S_2.$$

But $S_1 \not\sim S_2(\mathfrak{n})$; therefore $BR^{(S_1)} \neq BR^{(S_2)}$. This means that $R^{(S_1)} \not\sim R^{(S_2)}(\mathfrak{n})$ and therefore the $\mathfrak{n}$-classes $\Omega^{(S_1)}$ and $\Omega^{(S_2)}$ are distinct.

3.17. The considerations presented show that *the subgroups of type* $\mathfrak{T}_\Omega^{(\mathfrak{n})}$ *form a universal class* (III, 1.9) *for the class of all subgroups with left invertibility and without idempotents.* Here they themselves belong to that class (3.15, (α), (γ)). If one restricts oneself to subgroups, the cardinalities of the sets of whose elements do not exceed some cardinality $\mathfrak{m}$, then it is not difficult to construct a semigroup $\mathfrak{T}_\Omega^{(\mathfrak{n})}$ which is a universal semigroup for all of that class (true, it will not belong itself to that class, as it has a higher cardinality).

The obtaining of the indicated universal class for the class of semigroups with left invertibility without idempotents in a certain sense clears up the character of semigroups of that class. In studying one of the semigroups of that class we may, without leaving the limits of the class, imbed in the corresponding semigroup $\mathfrak{T}_\Omega^{(\mathfrak{n})}$, the study now being continued for the latter. On the other hand, of course, our present result is far from being a complete description of the construction of semigroups with left invertibility and with idempotents, a circumstance which is due to the greater complexity, in principle, of the structure of our semigroups in the case of absence of idempotents.

3.18. In conclusion we shall consider the semigroups $\mathfrak{T}_\Omega^{(\mathfrak{n})}$ in which the equivalence $\mathfrak{n}$ is that of identity, i.e., each $\mathfrak{n}$-class consists of one element only. We obtain in this case the construction of Baer and Levi **[1]**, which served as the initial point of the considerations presented above. In this case the semigroup $\mathfrak{T}_\Omega^{(\mathfrak{n})}$ will be a semigroup with left invertibility and left cancellation and without idempotents (3.15, (ζ)). As is immediately evident from the considerations of 3.16, any semigroup having the indicated three properties maps isomorphically into some such semigroup $\mathfrak{T}_\Omega^{(\mathfrak{n})}$. Thus, the semigroups $\mathfrak{T}_\Omega^{(\mathfrak{n})}$ with the identity equivalence $\mathfrak{n}$ not only belong to the class of semigroups with the indicated three properties but also form the universal class of that class.

3.19. In the general case the properties of cancellation and invertibility are independent. An infinite monogenic semigroup has the property of two-sided cancellation. However, it is evidently neither a semigroup with left invertibility nor a semigroup with right invertibility. On the other hand, the semigroup $\mathfrak{T}_\Omega^{(\mathfrak{n})}$ is a semigroup with left invertibility (3.15, (γ)) but is never a semigroup with right cancellation (3.15, (θ)), and, with the exclusion of the case mentioned under 3.15, (η), is never a semigroup with left cancellation. The semigroup anti-isomorphic to the semigroup $\mathfrak{T}_\Omega^{(\mathfrak{n})}$ is a semigroup with right invertibility but is neither a semigroup with left cancellation nor a semigroup with right cancellation (with the exclusion of the case corresponding to 3.15, (η)).

There is an interesting connection, indicated in the work of Hewitt and

Zuckerman [1], between these important properties in the class of periodic semigroups (III, 4.1).

THEOREM. *A periodic semigroup $\mathfrak{A}$ has the property of right cancellation if and only if it is a semigroup with left invertibility.*

PROOF. (1) Let $\mathfrak{A}$ be a semigroup with right cancellation. For any idempotents U and V of the semigroup $\mathfrak{A}$ we have

$$(UV)V = UV^2 = UV,$$

from which it follows that $UV = U$.

Any element A of the semigroup $\mathfrak{A}$ generates a finite monogenic semigroup $[A]$, whose type (III, 3.7) will be denoted by (h, d). If now $h > 1$, then we would obtain

$$A^{h+d} = A^h, \qquad A^{h+d-1} \cdot A = A^{h-1} \cdot A, \qquad A^{h+d-1} \neq A^{h-1},$$

which contradicts the condition of right cancellation. From this it necessarily follows that $h = 1$, which means that $[A]$ is a group (III, 3.11).

From what has been proved it follows that for any two elements A and B of the semigroup $\mathfrak{A}$ there exist A', B', U', V' such that

$$AU' = A, \qquad A'A = U', \qquad B'B = V',$$

where U' and V' are idempotents. Because of this we obtain

$$A = AU' = AU'V' = AU'B'B = (AU'B') \cdot B.$$

Hence it follows that $\mathfrak{A}$ is a semigroup with left invertibility.

(2) Suppose that $\mathfrak{A}$ is a semigroup with left invertibility. Since $\mathfrak{A}$ is a periodic semigroup it has idempotents. Hence, from 3.7 and 3.9, it follows that $\mathfrak{A}$ is isomorphic to the semigroup described in 3.5. But the semigroup $\mathfrak{T}$ is a semigroup with right cancellation since, for any $T_i = (U_i, G_i) \in \mathfrak{T}$ $(i = 1, 2, 3)$, from

$$T_1T_3 = T_2T_3$$

it follows that $T_1 = T_2$. Indeed,

$$T_1T_3 = (U_1, G_1) \cdot (U_3, G_3) = (U_1, G_1G_3),$$
$$T_2T_3 = (U_2, G_2) \cdot (U_3, G_3) = (U_2, G_2G_3)$$

and $T_1T_3 = T_2T_3$ implies $U_1 = U_2$ and $G_1G_3 = G_2G_3$, i.e., $G_1 = G_3$ (since G_1, G_2, G_3 are elements of a group).

4. Subsemigroups Regular with Respect to Invertibility

4.1. Suppose that $\mathfrak{B}$ is a subsemigroup of the semigroup $\mathfrak{A}$ and B is some element of $\mathfrak{B}$. The invertibility of B in $\mathfrak{B}$ may differ from the invertibility of B in $\mathfrak{A}$. The case is possible that B is a right invertible element of $\mathfrak{B}$ but is not right invertible in $\mathfrak{A}$. On the other hand, the case is possible that B is not a right invertible element of $\mathfrak{B}$ but is right invertible in $\mathfrak{A}$. Analogous cases may arise in the relations of left and two-sided invertibility.

4.2. We present an example of the possibilities mentioned in 4.1.

We choose a commutative semigroup

$$\mathfrak{A} = \{\ldots, A^{-1}, A^0, A, A^2, \ldots, I\},$$

in which A^n and A^m multiply by the rule of adding exponents, $IA^n = A^nI = A^n$ ($n = \ldots, -1, 0, 1, 2, \ldots$), and $I^2 = I$. One may say that $\mathfrak{A}$ is obtained from an infinite cyclic group by external adjunction of the unit I (II, 2.12).

We consider the semigroups

$$\mathfrak{A}_1 = \{A, A^2, A^3, \ldots\}, \qquad \mathfrak{A}_2 = \{\ldots, A^{-1}, A^0, A, A^2, \ldots\}.$$

Since $A\mathfrak{A}_1 \bar{\ni} A$, then A is not an invertible element of $\mathfrak{A}_1$. At the same time A is evidently an invertible element for $\mathfrak{A}_2$, which is a supersemigroup of the semigroup $\mathfrak{A}_1$.

At the same time we observe that A is not an invertible element of $\mathfrak{A}$, since evidently

$$A\mathfrak{A} \bar{\ni} I.$$

Thus A is a supersemigroup of the semigroup $\mathfrak{A}_2$ in which A is not invertible though it is an invertible element of the semigroup $\mathfrak{A}_2$.

4.3. In connection with the phenomenon mentioned above there naturally arises the concept following below, introduced and discussed by E. S. Ljapin [**14**].

DEFINITION. *The subsemigroup $\mathfrak{B}$ of the semigroup $\mathfrak{A}$ is said to be* A REGULAR SUBSEMIGROUP WITH RESPECT TO RIGHT INVERTIBILITY *if every element B of $\mathfrak{B}$ is right invertible in $\mathfrak{A}$ if and only if it is right invertible in $\mathfrak{B}$.*

Analogously one defines subsemigroups regular with respect to left invertibility and regular with respect to two-sided invertibility.

4.4. It follows from the definition that the semigroup $\mathfrak{B}$ will be regular with respect to right invertibility if for every $B \in \mathfrak{B}$ it follows from $B\mathfrak{B} = \mathfrak{B}$ that $B\mathfrak{A} = \mathfrak{A}$ and, conversely, from $B\mathfrak{A} = \mathfrak{A}$ that $B\mathfrak{B} = \mathfrak{B}$. Analogously, for left invertibility, one of the equalities

$$\mathfrak{B}B = \mathfrak{B}, \qquad \mathfrak{A}B = \mathfrak{A}$$

must follow from the other.

4.5. We note several of the simplest properties of subsemigroups regular with respect to right invertibility (in an obvious way the corresponding properties hold for semigroups regular with respect to left invertibility and regular with respect to two-sided invertibility).

(α) The semigroup itself, regarded as a subsemigroup of itself, is regular with respect to right invertibility.

(β) If $\mathfrak{A}_1$ is a subsemigroup regular with respect to right invertibility of $\mathfrak{A}_2$, and $\mathfrak{A}_2$ is a subsemigroup regular with respect to right invertibility of $\mathfrak{A}_3$, then $\mathfrak{A}_1$ is a regular, with respect to right invertibility, subsemigroup of $\mathfrak{A}_3$.

(γ) Suppose that the semigroup $\mathfrak{A}$ has a unit E and that the subsemigroup $\mathfrak{B}$ of the semigroup $\mathfrak{A}$ contains E.

In order that $\mathfrak{B}$ be a regular, with respect to right invertibility, subsemigroup of $\mathfrak{A}$, it is necessary and sufficient that, for every $B \in \mathfrak{B}$ for which the equation $BY = E$ is solvable in $\mathfrak{A}$, that equation should be solvable in $\mathfrak{B}$.

In order that $\mathfrak{B}$ should be a regular, with respect to left invertibility, subsemigroup of $\mathfrak{A}$, it is necessary and sufficient that, for every $B \in \mathfrak{B}$ for which the equation $XB = E$ is solvable in $\mathfrak{A}$, that equation be solvable in $\mathfrak{B}$.

(δ) Suppose that the semigroup $\mathfrak{A}$ has an identity E and that $\mathfrak{B}_1$ and $\mathfrak{B}_2$ are subsemigroups of $\mathfrak{A}$, while

$$E \in \mathfrak{B}_1 \subset \mathfrak{B}_2 \subset \mathfrak{A}.$$

If $\mathfrak{B}_1$ is a subsemigroup of $\mathfrak{A}$, regular with respect to right invertibility, then $\mathfrak{B}_1$ is a regular subsemigroup of $\mathfrak{B}_2$ with respect to right invertibility.

(ε) Suppose that the semigroup $\mathfrak{A}$ has an identity E and that the subsemigroup $\mathfrak{B}$ of the semigroup $\mathfrak{A}$ contains E. In order that $\mathfrak{B}$ should be a subsemigroup of $\mathfrak{A}$, regular with respect to right invertibility, it is necessary and sufficient that every element $B \in \mathfrak{B}$ which is right invertible in $\mathfrak{A}$ should be right invertible also in $\mathfrak{B}$.

4.6. If the subsemigroup $\mathfrak{B}$ of interest to us in the semigroup $\mathfrak{A}$ is not regular with respect to right invertibility, and the presence of this property for the semigroup in question is essential to us (soon we shall encounter a question in which that property becomes essential), then one may imbed $\mathfrak{B}$ in a supersemigroup $\mathfrak{B}'$, which as a subsemigroup of the semigroup $\mathfrak{A}$ is regular with respect to right invertibility. That this is always possible follows already from the fact that one may take $\mathfrak{A}$ itself to be $\mathfrak{B}'$. But usually it turns out to be convenient to pass from $\mathfrak{B}$ to $\mathfrak{B}'$ by the addition of fewer new elements not belonging to $\mathfrak{B}$.

In certain cases one succeeds in constructing a minimal supersemigroup for $\mathfrak{B}$ which is a subsemigroup of $\mathfrak{A}$ that is regular with respect to right invertibility; however this is not always possible.

4.7. We shall present an example for the case indicated above.

Consider a commutative semigroup $\mathfrak{A}$ consisting of the elements $I_1, I_2, \ldots, I_n$, with the operation between them defined by the rule

$$I_i I_k = I_k I_i = I_i \qquad (i, k = 1, 2, \ldots ; i \leqslant k).$$

We shall show which of the subsemigroups of $\mathfrak{A}$ are regular with respect to invertibility. If the subsemigroup $\mathfrak{B} = \{I_{\alpha_1}, I_{\alpha_2}, \ldots, I_{\alpha_m}\}$ $(\alpha_1 < \alpha_2 < \ldots < \alpha_m)$ is finite, then from 4.4 it is not regular, since

$$\mathfrak{B} I_{\alpha_m} = \mathfrak{B}, \qquad \mathfrak{A} I_{\alpha_m} \neq \mathfrak{A}.$$

The last inequality follows from the fact that for no I_β is it possible that the product $I_\beta I_{\alpha_m}$ equal J_{α_m+1}.

If $\mathfrak{B} = \{I_{\alpha_1}, I_{\alpha_2}, \ldots\}$ $(\alpha_1 < \alpha_2 < \ldots)$ is infinite, then it is regular with respect to invertibility. This follows from the fact that for no I_{α_k} does one have either $\mathfrak{B}I_{\alpha_k} = \mathfrak{B}$ or $\mathfrak{A}I_{\alpha_k} = \mathfrak{A}$. For any finite subsemigroup $\mathfrak{B} = \{I_{\alpha_1}, I_{\alpha_2}, \ldots, I_{\alpha_m}\}$ $(\alpha_1 < \alpha_2 < \ldots < \alpha_m)$ of our semigroup $\mathfrak{A}$, regular with respect to invertibility, its supersemigroup can be any infinite subsemigroup of $\mathfrak{A}$ containing it:

$$\mathfrak{B}' = \{\mathfrak{B}, I_{\beta_1}, I_{\beta_2}, \ldots, I_{\beta_j}, \ldots\} \quad (\beta_j \neq \alpha_i;\, i = 1, 2, \ldots, m;\, j = 1, 2, \ldots).$$

Among such subsemigroups $\mathfrak{B}'$ there is none which is minimal.

Indeed, for any $\mathfrak{B}'$ one may always choose a subsemigroup $\mathfrak{B}'' = \{\mathfrak{B}, I_{\beta_2}, I_{\beta_3}, \ldots\}$ which is regular with respect to invertibility, contains $\mathfrak{B}$ and at the same time is a subsemigroup of $\mathfrak{B}'$ distinct from $\mathfrak{B}'$.

5. Semigroups of Transformations Regular with Respect to Invertibility

5.1. As we have already mentioned, the necessity for considering the various properties of this or that transformation arises in quite diverse branches of mathematics. Depending on the subject matter and on the purpose of the investigation, interest may lie in different properties of the transformations. Very frequently it is necessary to investigate the existence and uniqueness of a solution of the following equation.

Suppose that S is some transformation of the set Ω, α some given element of Ω and ξ an unknown desired element of Ω. The equation in question has the form

$$S\xi = \alpha.$$

In view of the arbitrariness of the set Ω and the fact that S is any of its transformations (operators), many forms of concrete equations are special cases of this very general equation. Therefore E. S. Ljapin **[14]**, in studying certain properties of this equation, discussed below, called it an equation of general form. In any discussion of this equation there usually arises at the outset the necessity of considering the two following principal questions. First, is this equation solvable with respect to the unknown ξ (problem of existence)? And secondly, if it is solvable, then is its solution unique or not (problem of uniqueness)? E. S. Ljapin **[14]** considered a series of properties of semigroups which turned out to be connected with the properties of the indicated equation. The following section of the present chapter will be devoted to the presentation of the corresponding results.

5.2. Most frequently in practice one needs to consider not just one or another separate equation of the type indicated in 5.1, but rather a collection of equations with the same transformation, i.e., the same operator S, but with distinct right sides α.

In this connection the problem may be formulated as follows. For a given transformation-operator S we need to clear up, on the one hand, whether for all

α of Ω the equation $S\xi = \alpha$ has a solution in Ω and, on the other hand, whether for all those α for which this equation is solvable it has a unique solution.

The questions thus formulated may be put differently. We need to clear up whether the equation $S\Omega = \Omega$ holds (which is equivalent to $S\xi = \alpha$ being solvable for all $\alpha \in \Omega$) and whether $S\lambda$ and $S\mu$ are different for different but arbitrary λ and μ (which is equivalent to the uniqueness of the solution of the equation $S\xi = \alpha$ for all those α for which the equation is solvable).

5.3. An affirmative or negative answer to the questions put in 5.2 is an important property of the operator-transformation S. If one considers S as an element of the semigroup $\mathfrak{S}_\Omega$ of all mappings of the set Ω, then the answers to the indicated questions are obtained immediately.

From II, 3.2 it follows that in order to have $S\Omega = \Omega$ it is necessary and sufficient that S should be a left divisor of all elements of $\mathfrak{S}_\Omega$, i.e., from 1.1, that it should be right invertible in $\mathfrak{S}_\Omega$. Inasmuch as $\mathfrak{S}_\Omega$ possesses a unit E, it follows from 1.3 that it suffices for this that S should have a right inverse with respect to E in $\mathfrak{S}_\Omega$.

It follows from II, 3.1 that in order that $S\lambda \neq S\mu$ for any $\lambda \neq \mu$ it is necessary and sufficient that S should be a right divisor of all elements of $\mathfrak{S}_\Omega$, i.e., by 1.1, that it should be left invertible in $\mathfrak{S}_\Omega$. It follows from 1.3 that it is sufficient for this that S should have a left inverse in $\mathfrak{S}_\Omega$ relative to E.

5.4. The approach indicated in 5.3 to the solution of the two fundamental questions for the equations 5.1, as originally put, is too superficial and is therefore insufficiently fruitful. In the investigation of a given collection of operator-transformations it is usually natural and fundamentally important to consider them as elements of a semigroup of transformations with certain properties which the given operator-transformation also possesses. However, if one passes from the semigroup $\mathfrak{S}_\Omega$ to one of its subsemigroups $\mathfrak{A} \subset \mathfrak{S}_\Omega$, the equivalence of the properties of the transformation S which are of interest to us with properties of invertibility of S may be disturbed. This follows from the fact that S might be right invertible in $\mathfrak{S}_\Omega$ but not right invertible in $\mathfrak{A}$, and conversely. We have spoken in the preceding section of the possibility of such phenomena. The considerations of this section will suggest a way of escape from the indicated difficulty.

5.5. DEFINITION. *A semigroup consisting of some of the transformations of the set Ω is said to be* REGULAR WITH RESPECT TO RIGHT INVERTIBILITY *if it is regular with respect to right invertibility as a subsemigroup of the semigroup of all transformations* $\mathfrak{S}_\Omega$ (4.3).

Analogously, one may define regularity with respect to left invertibility and regularity with respect to two-sided invertibility.

5.6. From this definition and from what was said in 5.3 we have the following conclusion.

Suppose that $\mathfrak{A}$ is a semigroup of transformations Ω, regular with respect to right invertibility, and that the transformation S belongs to $\mathfrak{A}$. In order that the equation

$$S\xi = \alpha$$

in the unknown $\xi \in \Omega$ should be solvable for all values $\alpha \in \Omega$ it is necessary and sufficient that S should be a right invertible element of the semigroup $\mathfrak{A}$.

Suppose that $\mathfrak{A}$ is a semigroup of transformations of the set Ω, regular with respect to left invertibility, and that the transformation S belongs to $\mathfrak{A}$. In order that the equation

$$S\xi = \alpha$$

with respect to the unknown $\xi \in \Omega$ should have for each $\alpha \in \Omega$ no more than one solution it is necessary and sufficient that S should be a left invertible element of the semigroup $\mathfrak{A}$.

5.7. It follows from 5.6 that the study of these important properties of operator-transformations belonging to some semigroup of transformations $\mathfrak{A}$, comes down to the study of a purely algebraic question on the divisibility of elements in the semigroup $\mathfrak{A}$. It is important to observe that such an approach reduces the investigation of a question related to a concrete semigroup to the study of an abstract property (in the sense of I, 1.8 and I, 1.14). Indeed, the properties of right and left invertibility are evidently preserved for isomorphic semigroups.

The indicated reduction may be of fundamental interest. However, it holds only in the case when $\mathfrak{A}$ is a regular semigroup of transformations with respect to invertibility. The question as to which semigroups of transformations are regular with respect to the invertibility properties may be resolved, of course, only on the basis of a study of the concrete properties of the given set Ω and of the properties of those transformations which make up the given semigroup of transformations. The nature of such a study will depend on the specific branch of mathematics which is under investigation. By way of illustration we present such investigations for several important cases.

5.8. Let Ω be a linear space over the field Γ.[1] In what follows, in order to avoid obvious stipulations, we shall suppose that Ω consists of more than just the null element θ.

The transformation S of the set Ω is said to be *linear* if for any $\alpha_1, \alpha_2, \ldots, \alpha_n \in \Omega$ and $\lambda_1, \lambda_2, \ldots, \lambda_n \in \Gamma$ we have

$$S(\lambda_1\alpha_1 + \lambda_2\alpha_2 + \ldots + \lambda_n\alpha_n) = \lambda_1(S\alpha_1) + \lambda_2(S\alpha_2) + \ldots + \lambda_n(S\alpha_n).$$

[1] Definitions of basis, linear space and linear dependence, and their elementary properties are taken in the same sense as in the book by A. I. Mal'cev, *Foundations of linear algebra,* OGIZ, Moscow, 1948 (see Chapter II, §§ 1, 2). (Russian)

We shall denote the set of all linear transformations by $\mathfrak{M}$. It is not hard to see that the product of two linear transformations is again a linear transformation, so that $\mathfrak{M}$ is a semigroup.

5.9. LEMMA. *Suppose that Ω' is a linearly independent subset of the linear space Ω and that φ is any mapping of Ω' into Ω. Then there exists a linear transformation S such that for all ξ on Ω' we have*

$$S\xi = \varphi(\xi).$$

PROOF. The usual proof of the existence of a basis in a linear space[2] may without difficulty be sharpened in the sense that any linearly independent subset may be included in some basis. Suppose that Ω'' is a basis of Ω, containing Ω'. The mapping φ of Ω' into Ω'' may be extended to a mapping ψ of the basis Ω'' by putting $\psi(\xi) = \varphi(\xi)$ for $\xi \in \Omega'$ and $\psi(\xi) = \theta$ for $\xi \in \Omega''\backslash\Omega'$.

For every $\alpha \in \Omega$ there exists a unique linear expression in terms of the basis elements

$$\alpha = \lambda_1\xi_1 + \lambda_2\xi_2 + \dots + \lambda_n\xi_n$$

$$(\xi_1, \xi_2, \dots, \xi_n \in \Omega''; \lambda_1, \lambda_2, \dots, \lambda_n \in \Gamma).$$

We define the following transformation S of the space Ω:

$$S\alpha = S(\lambda_1\xi_1 + \lambda_2\xi_2 + \dots + \lambda_n\xi_n) = \lambda_1\psi(\xi_1) + \lambda_2\psi(\xi_2) + \dots + \lambda_n\psi(\xi_n).$$

It is easy to see that $S \in \mathfrak{M}$. Also, for any ξ of Ω',

$$S(\xi) = \psi(\xi) = \varphi(\xi).$$

5.10. THEOREM. *The semigroup $\mathfrak{M}$ of all linear transformations of the linear space Ω is regular with respect to both right invertibility and left invertibility.*

PROOF. Since the identity transformation E, which is the unit of the semigroup of all transformations $\mathfrak{S}_\Omega$, is linear, the proof of our theorem may be carried out with the use of property 4.5, (γ).

Suppose that $AY = E$, where $A \in \mathfrak{M}$ and $Y \in \mathfrak{S}_\Omega$. We choose in Ω some basis Ω'. Using Lemma 5.9, we construct a linear transformation $B \in \mathfrak{M}$ such that for every $\xi \in \Omega'$ we have

$$B\xi = Y\xi.$$

For any element $\alpha \in \Omega$ one finds its linear expansion in terms of the elements of the basis Ω':

$$\alpha = \lambda_1\xi_1 + \lambda_2\xi_2 + \dots + \lambda_n\xi_n$$

$$(\xi_1, \xi_2, \dots, \xi_n \in \Omega'; \lambda_1, \lambda_2, \dots, \lambda_n \in \Gamma).$$

---

[2] See, for example, A. I. Mal'cev, *Foundations of linear algebra*, OGIZ, Moscow, 1948 (see Chapter II, § 2, Theorem 2). (Russian)

Since the transformations A and B are linear and $AY = E$, then

$$AB\alpha = A[\lambda_1(B\xi_1) + \lambda_2(B\xi_2) + \ldots + \lambda_n(B\xi_n)] = A[\lambda_1(Y\xi_1) + \lambda_2(Y\xi_2) + \ldots + \lambda_n(Y\xi_n)] = \lambda_1(AY\xi_1) + \lambda_2(AY\xi_2) + \ldots + \lambda_n(AY\xi_n) = \lambda_1\xi_1 + \lambda_2\xi_2 + \ldots + \lambda_n\xi_n = \alpha,$$

i.e., $AB = E$. Thus, we have proved that if the equation $AY = E$ is solvable in $\mathfrak{S}_\Omega$ it is solvable in $\mathfrak{M}$. From 4.5, (γ) it therefore follows that $\mathfrak{M}$ is a subsemigroup of $\mathfrak{S}_\Omega$ that is regular with respect to right invertibility, i.e., is a regular, with respect to right invertibility, semigroup of transformations.

Now suppose that $XA = E$, where $A \in \mathfrak{M}$ and $X \in \mathfrak{S}_\Omega$. Choose some basis Ω' of the linear space Ω. We shall prove the linear independence of the set $A\Omega'$. Let

$$\lambda_1(A\xi_1) + \lambda_2(A\xi_2) + \ldots + \lambda_n(A\xi_n) = \theta \qquad (\lambda_1, \lambda_2, \ldots, \lambda_n \in \Gamma),$$

where $\xi_1, \xi_2, \ldots, \xi_n$ are mutually distinct elements of Ω'. Then we have

$$\lambda_1\xi_1 + \lambda_2\xi_2 + \ldots + \lambda_n\xi_n = E(\lambda_1\xi_1 + \lambda_2\xi_2 + \ldots + \lambda_n\xi_n) = XA(\lambda_1\xi_1 + \lambda_2\xi_2 + \ldots + \lambda_n\xi_n) = X[\lambda_1(A\xi_1) + \lambda_2(A\xi_2) + \ldots + \lambda_n(A\xi_n)] = X\theta = \theta.$$

For the elements of the basis $\xi_1, \xi_2, \ldots, \xi_n$ this is possible only if

$$\lambda_1 = 0, \qquad \lambda_2 = 0, \ldots, \lambda_n = 0.$$

For elements μ of $A\Omega'$ we define a mapping of φ into Ω:

$$\varphi(\mu) = X\mu \qquad (\mu \in A\Omega').$$

Since the set $A\Omega'$ is linearly independent, it follows from Lemma 5.9 that there exists a transformation $C \in \mathfrak{M}$ such that

$$C\mu = X\mu$$

for all $\mu \in A\Omega'$.

Suppose that α is any element of Ω. It is linearly expressible in terms of the elements of the basis Ω':

$$\alpha = \lambda_1\xi_1 + \lambda_2\xi_2 + \ldots + \lambda_n\xi_n$$

$$(\xi_1, \xi_2, \ldots, \xi_n \in \Omega'; \lambda_1, \lambda_2, \ldots, \lambda_n \in \Gamma).$$

Since $CA \in \mathfrak{M}$ and $XA = E$, therefore

$$CA\alpha = CA[\lambda_1\xi_1 + \lambda_2\xi_2 + \ldots + \lambda_n\xi_n] = \lambda_1C(A\xi_1) + \lambda_2C(A\xi_2) + \ldots + \lambda_nC(A\xi_n) = \lambda_1X(A\xi_1) + \lambda_2X(A\xi_2) + \ldots + \lambda_nX(A\xi_n) = \lambda_1\xi_1 + \lambda_2\xi_2 + \ldots + \lambda_n\xi_n = \alpha,$$

i.e., $CA = E$.

By 4.5, (γ) it follows from what has been proved that $\mathfrak{M}$ is a regular, with respect to left invertibility, semigroup of transformations.

5.11. From 5.6 and Theorem 5.10 we have the following corollary.

Suppose that S is a linear transformation of the linear space Ω. In order that the equation

$$S\xi = \alpha$$

in the unknown $\xi \in \Omega$ should have a solution for all $\alpha \in \Omega$ it is necessary and sufficient that S should be a right invertible element of the semigroup $\mathfrak{M}$ of all linear transformations Ω.

In order that the equation

$$S\xi = \alpha$$

in the unknown $\xi \in \Omega$ should have no more than one solution for every $\alpha \in \Omega$ it is necessary and sufficient that S should be a left invertible element of the semigroup $\mathfrak{M}$ of all linear transformations Ω.

5.12. Now suppose that $\mathfrak{L}$ is any partially ordered set (I, 5.9). Denote by $\mathfrak{B}$ the set of all those transformations B of that set which do not violate the partial order. Moreover, B belongs to $\mathfrak{B}$ if and only if from $\alpha \leqslant \beta$ $(\alpha, \beta \in \mathfrak{L})$ and $B\alpha \geqslant B\beta$ it always follows that $B\alpha = B\beta$. Thus, in $\mathfrak{B}$ there fail to enter only those transformations S of the set $\mathfrak{L}$ for which in $\mathfrak{L}$ there are $\alpha < \beta$ such that $S\alpha > S\beta$. Of course, $\mathfrak{L}$ will not always be a semigroup. The further discussion refers to the case that $\mathfrak{L}$ is a semigroup.

5.13. THEOREM. *The semigroup $\mathfrak{B}$ of all transformations of the partially ordered set $\mathfrak{L}$ which do not violate the partial order* (5.12) *is regular with respect to right invertibility.*

PROOF. Since the identity transformation E, which is the unit of the semigroup $\mathfrak{S}_{\mathfrak{L}}$ of all transformations of $\mathfrak{L}$, is evidently in $\mathfrak{B}$, we may make use of property 4.5, (γ).

Let $BY = E$, where $B \in \mathfrak{B}$, $Y \in \mathfrak{S}_{\mathfrak{L}}$. If we had $Y \bar{\in} \mathfrak{B}$, then for some α, $\beta \in \mathfrak{L}$ we would have

$$\alpha < \beta, \qquad Y\alpha > Y\beta.$$

However the relations

$$Y\alpha > Y\beta,$$
$$B(Y\alpha) = \alpha < \beta = B(Y\beta)$$

contradict the fact that $B \in \mathfrak{B}$.

5.14. As to regularity with respect to left invertibility, in the general case the semigroup $\mathfrak{B}$ of all transformations not violating the partial order does not possess that property.

We cite the following example. Let $\mathfrak{L}$ be the set of all positive rational numbers less than unity. We shall say that the number α precedes β if the value of α does not exceed that of β. Consider a transformation B such that $B\alpha = (1/2)\alpha(\alpha \in \mathfrak{L})$. In $\mathfrak{S}_{\mathfrak{L}}$ the equation

$$XB = E$$

is solvable (for X one may take any transformation of $\mathfrak{L}$ such that $X\alpha = 2\alpha$ for all α with $0 < \alpha < 1/2$).

We choose any solution X of that equation. We write $X(1/2) = \rho$. We choose a rational number ρ' such that $\rho < \rho' < 1$.

We have

$$\tfrac{1}{2} > \tfrac{1}{2}\rho', \qquad X(\tfrac{1}{2}) = \rho < \rho' = E\rho' = XB\rho' = X(\tfrac{1}{2}\rho'),$$

from which it follows that X does not belong to $\mathfrak{B}$. From this, according to 4.5, (γ), it follows that $\mathfrak{B}$ is not a regular, with respect to left invertibility, subsemigroup of $\mathfrak{S}_{\mathfrak{L}}$.

5.15. Now we shall show that under some very wide restrictions one may nevertheless assert the regularity with respect to left invertibility of the semigroup of all transformations of a partially ordered set not violating the partial order.

We shall say that the partially ordered set $\mathfrak{L}$ has *separating elements* if $\mathfrak{L}$ has the following property.

Let $\mathfrak{N}'$ and $\mathfrak{N}''$ be any two arbitrary (in particular, possibly empty) subsets of $\mathfrak{L}$ such that the relation $\alpha' \geqslant \alpha''$, where $\alpha' \in \mathfrak{N}'$, $\alpha'' \in \mathfrak{N}''$, is possible only if $\alpha' = \alpha''$. Then there exists a separating element γ such that (1) $\alpha' \geqslant \gamma$ for $\alpha' \in \mathfrak{N}'$ is possible only if $\alpha' = \gamma$ and (2) $\alpha'' \leqslant \gamma$ for $\alpha'' \in \mathfrak{N}''$ is possible only if $\alpha'' = \gamma$.

5.16. THEOREM. *If the partially ordered set $\mathfrak{L}$ has separating elements* (5.15), *then the semigroup $\mathfrak{B}$ of all of its transformations which do not violate the partial order* (5.12) *is regular with respect to left invertibility.*

PROOF. (1) We suppose that for some $B \in \mathfrak{B}$ and $C \in \mathfrak{S}_{\mathfrak{L}}$ we have

$$CB = E.$$

By 4.5, (γ) it suffices for us to prove that there then follows the existence in $\mathfrak{B}$ of an element B' such that $B'B = E$.

(2) We shall consider partial transformations of the set $\mathfrak{L}$ (I, 4.1). In particular, we denote by $\mathfrak{R}$ the set of all those partial transformations X for which the relations $\alpha < \beta$ and $X\alpha > X\beta$ cannot hold simultaneously for any $\alpha, \beta \in \Pi_1 X$. We observe immediately that $\mathfrak{B} = \mathfrak{S}_{\mathfrak{L}} \cap \mathfrak{R}$.

(3) In the set of all partial transformations of the set $\mathfrak{L}$ we define a partial ordering relation by setting $X \leqslant Y$ if $\Pi_1 X \subset \Pi_1 Y$ and $X\xi = Y\xi$ for every $\xi \in \Pi_1 X$.

(4) We shall denote by P_0 a partial transformation of $\mathfrak{L}$ for which $\Pi_1 P_0 = B\mathfrak{L}$ and $P_0\xi = C\xi$ for every $\xi \in B\mathfrak{L}$. We shall denote by $\mathfrak{M}$ the set of all partial transformations of $\mathfrak{R}$ which follow P_0.

We shall prove that $P_0 \in \mathfrak{M}$, for which it suffices to prove that $P_0 \in \mathfrak{R}$. Suppose that $\alpha < \beta$ and $P_0\alpha > P_0\beta$ for some $\alpha, \beta \in \Pi_1 P_0 = B\mathfrak{L}$. Then $\alpha = B\alpha'$, $\beta = B\beta'$. We have

$$\alpha' = E\alpha' = CB\alpha' = P_0B\alpha' = P_0\alpha > P_0\beta = P_0B\beta' = CB\beta' = E\beta' = \beta',$$
$$B\alpha' = \alpha < \beta = B\beta'.$$

But this contradicts the fact that $B \in \mathfrak{B}$.

(5) Let $\mathfrak{N}$ be a subset of $\mathfrak{M}$ such that for any X and Y of $\mathfrak{N}$ one always precedes the other. We denote by N_0 a partial transformation for which $\Pi_1 N_0$ is the union of all the $\Pi_1 X$ for $X \in \mathfrak{N}$, and $N_0\alpha = \beta$ if for some $X \in \mathfrak{N}$ we have $X\alpha = \beta$. Evidently β does not depend on the choice of X in $\mathfrak{N}$ since the hypothesis stated relative to $\mathfrak{N}$ is satisfied. It is immediately evident that N_0 is an upper bound for $\mathfrak{N}$. Of course, $N_0 \geqslant P_0$. Suppose that $\alpha < \beta$ and $N_0\alpha > N_0\beta$ for some $\alpha, \beta \in \mathfrak{L}$. Then for some $X, Y \in \mathfrak{N}$ we have $X\alpha = N_0\alpha$ and $Y\beta = N_0\beta$. Suppose Z is that one of X and Y which follows both X and Y. Then $\alpha, \beta \in \Pi_1 Z$ and $Z\alpha = N_0\alpha$, $Z\beta = N_0\beta$. Since the relations $\alpha < \beta$ and $Z\alpha > Z\beta$ contradict the fact that $Z \in \mathfrak{R}$ the stated hypothesis is invalid. Accordingly, $N_0 \in \mathfrak{R}$, and therefore $N_0 \in \mathfrak{M}$.

(6) From the fact that the properties of $\mathfrak{M}$ just derived are satisfied, one may apply Theorem II, 4.17 to $\mathfrak{M}$. By that theorem there is in $\mathfrak{M}$ an element P_1 which is followed in $\mathfrak{M}$ by no element distinct from P_1.

Suppose that in $\mathfrak{L}$ there is an element λ not contained in $\Pi_1 P_1$. We denote by $\mathfrak{L}'$ the set of all elements of $\mathfrak{L}$ preceding λ, and by $\mathfrak{L}''$ the set of all elements following λ. Then $\lambda' \leqslant \lambda''$ for every $\lambda' \in \mathfrak{L}'$ and $\lambda'' \in \mathfrak{L}''$. Since $P_1 \in \mathfrak{R}$ it follows that for no $\mu' \in P_1\mathfrak{L}'$ and $\mu'' \in P_1\mathfrak{L}''$ can we have $\mu' > \mu''$. Because of the condition on the existence of separating elements (5.15), for $P_1\mathfrak{L}'$ and $P_1\mathfrak{L}''$ there exists a separating element γ. We construct a new partial transformation P_2 such that $P_2 \geqslant P_1$ and $\Pi_1 P_2 = \Pi_1 P_1 \cup \lambda$, while $P_2\lambda = \gamma$.

We shall show that $P_2 \in \mathfrak{M}$, for which it suffices to show that $P_2 \in \mathfrak{R}$. Suppose that $\alpha < \beta$ and $P_2\alpha > P_2\beta$ for some $\alpha, \beta \in \Pi_1 P_2$. If $\alpha, \beta \in \Pi_1 P_1$, we would have a contradiction with the fact that $P_2 \geqslant P_1$ and $P_1 \in \mathfrak{R}$. If $\alpha = \lambda$, we would have $\gamma > P_2\beta$, where $\beta \in \mathfrak{L}''$ (since $\beta > \alpha = \lambda$), which is impossible for γ. If we had $\beta = \lambda$, then $P_2\alpha > \gamma$, where $\alpha \in \mathfrak{L}'$ ($\alpha < \beta = \lambda$), which is impossible for γ. Accordingly, $P_2 \in \mathfrak{R}$.

Thus the assumption $\lambda \mathrel{\overline{\in}} \Pi_1 P_1$ has led us to the existence in $\mathfrak{M}$ of an element P_2 distinct from P_1 and following it. The impossibility of this means that in fact $\Pi_1 P_1 = \mathfrak{L}$, i.e., $P_1 \in \mathfrak{S}_{\mathfrak{L}}$. Since we also have $P_1 \in \mathfrak{R}$, therefore $P_1 \in \mathfrak{B}$.

(7) For any $\xi \in \mathfrak{L}$ we have $B\xi \in \Pi_1 P_0$.

Since also $P_1 \geqslant P_0$, therefore

$$P_1B\xi = P_0B\xi = CB\xi = E\xi.$$

Hence it follows that $P_1B = E$, while $P_1 \in \mathfrak{B}$.

5.17. From Theorems 5.13 and 5.16 we have the following consequence. *Suppose that* $\mathfrak{L}$ *is a partially ordered set having separating elements* (5.15), $\mathfrak{B}$ *is the semigroup of all of the transformations of* $\mathfrak{L}$ *not violating the partial ordering* (5.12), *and* $B \in \mathfrak{B}$.

In order that the equation

$$B\xi = \alpha$$

in the unknown $\xi \in \mathfrak{L}$ *should have a solution for all* $\alpha \in \mathfrak{L}$ *it is necessary and sufficient that* B *should be a right invertible element of the semigroup* $\mathfrak{B}$.

In order that the equation

$$B\xi = \alpha$$

in the unknown $\xi \in \mathfrak{L}$ *should have not more than one solution for every* $\alpha \in \mathfrak{L}$ *it is necessary and sufficient that* B *should be a left invertible element of the semigroup* $\mathfrak{B}$.

6. Groups with Separating Group Part

6.1. We have already turned our attention to the important role of two-sidedly invertible elements of a semigroup, i.e., of elements which are both right and left divisors of every element of the semigroup. In the rest of the present section the set of two-sidedly invertible elements of the semigroup $\mathfrak{A}$ will be denoted by $\mathfrak{G}(\mathfrak{A})$. In 1.4 we have proved, and now we must always keep this in view, that $\mathfrak{G}(\mathfrak{A})$, if not empty, is a group, while the unit of the group $\mathfrak{G}(\mathfrak{A})$ is a unit for the whole semigroup $\mathfrak{A}$.

We shall also introduce a notation for the set of all elements of the semigroup $\mathfrak{A}$ not lying in $\mathfrak{G}(\mathfrak{A})$:

$$\mathfrak{H}(\mathfrak{A}) = \mathfrak{A} \backslash \mathfrak{G}(\mathfrak{A}).$$

Thus, in the notation used in § 1, we have

$$\mathfrak{G}(\mathfrak{A}) = \mathfrak{G}, \qquad \mathfrak{H}(\mathfrak{A}) = \mathfrak{R} \cup \mathfrak{L} \cup \mathfrak{K}.$$

6.2. While the set $\mathfrak{G}(\mathfrak{A})$ is always closed relative to multiplication, the set $\mathfrak{H}(\mathfrak{A})$ does not have the analogous property in every semigroup. For example, for the semigroup of all transformations of a countable set, as we have seen in 1.9, the collection of elements which are not two-sidedly invertible is not a subsemigroup. At the same time, for a number of other important semigroups $\mathfrak{H}(\mathfrak{A})$ is a subsemigroup. As an example of this consider the semigroup $\mathfrak{S}_\Omega$ of all transformations of a finite set Ω. As follows immediately from II, 3.1 and II, 3.2, $\mathfrak{G}(\mathfrak{S}_\Omega)$ consists of all transformations which effect a one-to-one mapping of Ω onto itself (sometimes they are called proper permutations). $\mathfrak{H}(\mathfrak{S}_\Omega)$, consisting of all the remaining transformations (improper permutations), is evidently a subsemigroup of $\mathfrak{S}_\Omega$.

In the semigroup $\mathfrak{M}_n$ of all complex square matrices of order n, $\mathfrak{G}(\mathfrak{M}_n)$, as follows from II, 3.9, is the set of all nonsingular matrices, and $\mathfrak{H}(\mathfrak{M}_n)$ is the set of all singular matrices, which is also a subsemigroup.

Aside from the fact that many important semigroups have the indicated property, the class of semigroups for which $\mathfrak{H}(\mathfrak{A})$ is a subsemigroup is characterized by certain other important properties. Among these properties are certain important ones relating to the abstract theory of transformations.

6.3. That the collection $\mathfrak{H}(\mathfrak{A})$ is closed under multiplication means, in other words, that no two-sidedly invertible element can be represented in the form of the product of elements of $\mathfrak{H}(\mathfrak{A})$. Moreover, as we shall soon show, in this case the product of elements of $\mathfrak{A}$ belongs to $\mathfrak{G}(\mathfrak{A})$ only if all the multipliers belong to $\mathfrak{G}(\mathfrak{A})$. Thus the case in question is characterized by the setting apart, the isolation, of the group of two-sidedly invertible elements $\mathfrak{G}(\mathfrak{A})$.

DEFINITION. *The semigroup $\mathfrak{A}$ is said to be* A SEMIGROUP WITH SEPARATING GROUP PART *if it has an identity (i.e., from 1.5, if $\mathfrak{G}(\mathfrak{A})$ is nonempty) and if the product of any two of its elements not two-sidedly invertible is itself not a two-sidedly invertible element (i.e., if $\mathfrak{H}(\mathfrak{A})$ is a subsemigroup of $\mathfrak{A}$ or an empty set).*

We observe that the concept of semigroup with separating group part is closely connected with the concept of hypergroups introduced by Rauter [**1**] (see also § 51 of the book of A. K. Suškevič [**12**]).

6.4. THEOREM. *A semigroup $\mathfrak{A}$ with unit which is not a group is a semigroup with separating group part if and only if it can be represented in the form of two nonintersecting subsemigroups*

$$\mathfrak{A} = \mathfrak{A}_1 \cup \mathfrak{A}_2, \qquad \mathfrak{A}_1 \cap \mathfrak{A}_2 = \varnothing,$$

such that $\mathfrak{A}_1$ is a subgroup and $\mathfrak{A}_2$ is an ideal. In this case

$$\mathfrak{G}(\mathfrak{A}) = \mathfrak{A}_1, \qquad \mathfrak{H}(\mathfrak{A}) = \mathfrak{A}_2,$$

while the ideal $\mathfrak{A}_2$ is two-sided.

PROOF. (1) Suppose that $\mathfrak{A}$ is a semigroup with separating group part. $\mathfrak{A}$ is the union of $\mathfrak{G}(\mathfrak{A})$ and $\mathfrak{H}(\mathfrak{A})$, while $\mathfrak{G}(\mathfrak{A})$, from 1.4, is a group.

Suppose that $H \in \mathfrak{H}(\mathfrak{A})$, $A \in \mathfrak{A}$. If $A \in \mathfrak{H}(\mathfrak{A})$, then by the definition of a semigroup with separating group part, $HA \in \mathfrak{H}(\mathfrak{A})$ and $AH \in \mathfrak{H}(\mathfrak{A})$. If $A \in \mathfrak{G}(\mathfrak{A})$, then neither AH nor HA can belong to $\mathfrak{G}(\mathfrak{A})$. Indeed, in the case $HA = G \in \mathfrak{G}(\mathfrak{A})$ we would obtain $H = GA^{-1} \in \mathfrak{G}(\mathfrak{A})$. We reason analogously for AH.

Thus for any $A \in \mathfrak{A}$ we have $HA \in \mathfrak{H}(\mathfrak{A})$ and $AH \in \mathfrak{H}(\mathfrak{A})$, i.e., $\mathfrak{H}(\mathfrak{A})$ is a two-sided ideal.

(2) Suppose that $\mathfrak{A}$ decomposes in the way indicated in the theorem:

$$\mathfrak{A} = \mathfrak{A}_1 \cup \mathfrak{A}_2, \qquad \mathfrak{A}_1 \cap \mathfrak{A}_2 = \varnothing.$$

Since $\mathfrak{A}_2$ is an ideal of $\mathfrak{A}$, the unit E of the semigroup $\mathfrak{A}$ must lie in $\mathfrak{A}_1$. Since $\mathfrak{A}_1$ is a group, for any $A_1 \in \mathfrak{A}_1$ there exist $X, Y \in \mathfrak{A}_1$ such that

$$XA_1 = E, \qquad A_1 Y = E.$$

Hence from 1.3 it follows that A_1 is a two-sidedly invertible element of the semigroup $\mathfrak{A}$. On the other hand, elements of $\mathfrak{A}_2$ cannot be two-sidedly invertible elements of $\mathfrak{A}$ since $\mathfrak{A}_2$ is an ideal.

Therefore $\mathfrak{G}(\mathfrak{A}) = \mathfrak{A}_1$, from which it follows that $\mathfrak{H}(\mathfrak{A}) = \mathfrak{A}_2$, i.e., $\mathfrak{A}$ turns out to be a semigroup with separating group part. From what was proved in the first part, the ideal $\mathfrak{A}_2$ is two-sided.

6.5. COROLLARY. *In a semigroup $\mathfrak{A}$ with separating group part, $\mathfrak{H}(\mathfrak{A})$ is a two-sided ideal of $\mathfrak{A}$ or else an empty set.*

6.6. COROLLARY. *In a semigroup $\mathfrak{A}$ with separating group part the subsemigroups $\mathfrak{G}(\mathfrak{A})$ and $\mathfrak{H}(\mathfrak{A})$ (if the latter is nonempty) are completely isolated* (IV, 6.1).

6.7. In the class of semigroups with unit the semigroups with separating group part may be distinguished by the requirement that the subsets $\mathfrak{R}$ and $\mathfrak{L}$ considered by us in § 1 should be empty.

THEOREM. *Suppose that $\mathfrak{A}$ is a semigroup with unit.*

If $\mathfrak{A}$ is a semigroup with separating group part, then in $\mathfrak{A}$ every right invertible element is left invertible, and conversely.

If $\mathfrak{A}$ is not a semigroup with separating group part, then in $\mathfrak{A}$ there are elements right but not left invertible and also elements left but not right invertible.

PROOF. (1) If $\mathfrak{A}$ is a semigroup with separating group part, then no element H of $\mathfrak{H}(\mathfrak{A})$ can be either right or left invertible. This follows from the fact that $\mathfrak{H}(\mathfrak{A})$ is a two-sided ideal of $\mathfrak{A}$ (6.5). So for any $A \in \mathfrak{A}$ the elements AH and HA lie in $\mathfrak{H}(\mathfrak{A})$ and are therefore necessarily distinct from $E_{\mathfrak{A}}$.

(2) If $\mathfrak{A}$ is not a semigroup with separating group part, then in $\mathfrak{A}$ there are elements $X, Y \in \mathfrak{H}(\mathfrak{A})$ such that $XY = G \in \mathfrak{G}(\mathfrak{A})$. Because of 1.2,

$$X\mathfrak{A} \supset XY\mathfrak{A} = G\mathfrak{A} = \mathfrak{A},$$

i.e., X is right invertible. At the same time X is not left invertible, since otherwise X would belong to $\mathfrak{G}(\mathfrak{A})$. Analogously one proves that Y is left invertible but not right invertible.

6.8. Theorem 6.7 explains the role of semigroups of transformations which are semigroups with separating group part.

For transformations from such a semigroup, if they are regular with respect to invertibility, the condition that the equation $S\xi = \alpha$ should be solvable, which we discussed in § 5, turns out to be equivalent to the condition of uniqueness of the solution.

THEOREM. *Let $\mathfrak{A}$ be a semigroup of transformations of the set Ω* (5.5) *which is regular with respect to both right invertibility and left invertibility and which contains the identity transformation E.*

If $\mathfrak{A}$ *is a semigroup with separating group part, then for each* $S \in \mathfrak{A}$ *it follows from the solvability of the equation in the unknown* ξ,

$$S\xi = \alpha,$$

for any $\alpha \in \Omega$, *that the solution is always unique. In turn, in the case that the indicated equation is not solvable for some* $\alpha \in \Omega$ *we have the result that for some* $\alpha \in \Omega$ *it has more than one solution.*

If $\mathfrak{A}$ *is not a semigroup with separating group part, then there are* S_1, S_2 *in* $\mathfrak{A}$ *such that the equation*

$$S_1\xi = \alpha$$

is solvable for all $\alpha \in \Omega$, *but some* α *has more than one solution. The equation*

$$S_2\xi = \alpha$$

for some $\alpha \in \Omega$ *is not solvable, but for every* $\alpha \in \Omega$ *has not more than one solution.*

Proof. (1) Suppose that $\mathfrak{A}$ is a semigroup with separating group part. From 5.6, from the solvability of our equation it follows that S is right invertible in $\mathfrak{A}$. Hence it follows from 6.6 that S is also left invertible in $\mathfrak{A}$. Therefore, from 5.6, the equation for each $\alpha \in \Omega$ has no more than one solution.

(2) Suppose that $\mathfrak{A}$ is not a semigroup with separating group part. From 6.7 there is an element S_1 in $\mathfrak{A}$ which is right but not left invertible, and an element S_2, left but not right invertible. From 5.6 the equation

$$S_1\xi = \alpha$$

is solvable for all $\alpha \in \Omega$, but for some of these has more than one solution. The equation

$$S_2\xi = \alpha$$

has at most one solution for any $\alpha \in \Omega$. For certain α it is not solvable.

6.9. Suppose that $\mathfrak{A}$ is a semigroup with separating group part, while $\mathfrak{A}$ is not a group. Since two-sidedly invertible elements cannot be contained in any proper ideal of the semigroup $\mathfrak{A}$ it follows that $\mathfrak{H}(\mathfrak{A})$, from 6.5, is a universal maximal proper ideal of $\mathfrak{A}$.

6.10. It ought to be observed that the presence in a semigroup of a universal maximal proper ideal in the general case is not sufficient for it to be a semigroup with separating group part. We present an appropriate example.

Let $\mathfrak{G}$ be a group and G_0 a fixed element of it. We consider a set $\mathfrak{A}$ obtained by adjoining to $\mathfrak{G}$ a new element X. Elements of $\mathfrak{G}$ multiply in $\mathfrak{A}$ according to the multiplication law of the group $\mathfrak{G}$. For X we put

$$XG = G_0G, \qquad GX = GG_0, \qquad X^2 = G_0^2 \qquad (G \in \mathfrak{G}).$$

The associativity of the operation is evident. $\mathfrak{G}$ is a two-sided ideal of $\mathfrak{A}$. If $\mathfrak{T}$ is a left ideal of $\mathfrak{A}$ containing X, then for some $G \in \mathfrak{G}$ we have

$$(GG_0^{-1}) \cdot X = GG_0^{-1}G_0 = G.$$

Accordingly $\mathfrak{T}$, containing X, contains also any element of $\mathfrak{G}$, and therefore coincides with $\mathfrak{A}$. In the same way we can show the absence of proper right ideals containing X.

Thus $\mathfrak{G}$ turns out to be the unique proper ideal in $\mathfrak{A}$ and thus a universal maximal proper ideal. Nevertheless, $\mathfrak{A}$ is not a semigroup with separating group part, since $\mathfrak{A}$ does not have a unit.

6.11. Questions on the existence and properties of maximal proper left ideals, and analogously of right and two-sided ideals, were considered by Št. Schwarz [**6**; 7]. He gave particular consideration to the question of the structure of the set of elements not contained in one or another maximal ideal. From the results of Schwarz follow in particular certain sufficiency tests for the semigroup to be a semigroup with separating group part.

7. Subsemigroups of a Semigroup with Separating Group Part

7.1. Now we shall take up subsemigroups of semigroups with separating group part. First of all we observe that every semigroup may be imbedded as a subsemigroup in some semigroup with separating group part. Moreover, it can even be done in such a way that the imbedded semigroup coincides with the subsemigroup of all elements which are not two-sidedly invertible elements. For let $\mathfrak{G}$ be any group and $\mathfrak{A}$ any semigroup not having common elements with $\mathfrak{G}$. Consider the set

$$\mathfrak{A}' = \mathfrak{G} \cup \mathfrak{A}.$$

We shall define an operation in it. If both elements of $\mathfrak{A}'$ are contained simultaneously in the same semigroup $\mathfrak{G}$ or $\mathfrak{A}$, then their product is defined as the product, respectively, in $\mathfrak{G}$ or $\mathfrak{A}$. For $G \in \mathfrak{G}$ and $A \in \mathfrak{A}$ we put

$$GA = AG = A.$$

It follows from 6.4 that $\mathfrak{A}'$ is a semigroup with separating group part, while

$$\mathfrak{A} = \mathfrak{H}(\mathfrak{A}').$$

7.2. Semigroups with separating group parts do not have subsemigroups of right and left magnifying elements (III, 5.4).

THEOREM. *Suppose that $\mathfrak{A}$ is a semigroup with unit.*

If $\mathfrak{A}$ is a semigroup with separating group part, then $\mathfrak{A}$ has neither right nor left magnifying elements (III, 5.1).

If $\mathfrak{A}$ is not a semigroup with separating group part, then $\mathfrak{A}$ has right and left magnifying elements.

PROOF. By 2.1 each magnifying element is invertible on one side but not invertible on the other. From 6.7 such elements cannot lie in a semigroup with separating group part.

If $\mathfrak{A}$ is not a semigroup with separating group part, then, from 6.7, $\mathfrak{A}$ must contain elements right invertible but not left invertible, and vice-versa. By 2.3 these elements are left and right magnifying elements of the semigroup.

7.3. Because of III, 6.8 one may immediately draw a consequence of Theorem 7.2.

COROLLARY. *In order that a semigroup $\mathfrak{A}$ with unit should be a semigroup with separating group part it is necessary and sufficient that among its subsemigroups containing $E_{\mathfrak{A}}$ there are none isomorphic to the semigroup $\mathfrak{P}$* (III, 6.2; III, 6.3).

7.4. From Theorem 7.2 and III, 5.3 it follows that the semigroups belonging to various classes of semigroups are semigroups with separating group part.

(α) *Every finite semigroup with identity is a semigroup with separating group part.*

(β) *Every commutative semigroup with identity is a semigroup with separating group part.*

(γ) *Every semigroup with two-sided cancellation with an identity is a semigroup with separating group part.*

7.5. Using 7.3, it is easy to indicate still another class of semigroups with separating group part.

Suppose we are given a sequence of semigroups with separating group parts in which each term is a subsemigroup of the next one:

$$\mathfrak{A}_1 \subset \mathfrak{A}_2 \subset \ldots \subset \mathfrak{A}_n \subset \mathfrak{A}_{n+1} \subset \ldots .$$

Their union is

$$\mathfrak{B} = \bigcup_{n=1}^{\infty} \mathfrak{A}_n.$$

As we already have observed in III, 1.2, $\mathfrak{B}$ is a semigroup. If $\mathfrak{B}$ has an identity, then $\mathfrak{B}$ is a semigroup with separating group part.

Indeed, if $\mathfrak{B}$ were not a semigroup with separating group part, then from 7.3 $\mathfrak{B}$ would have a subsemigroup isomorphic to $\mathfrak{P}$ and containing $E_{\mathfrak{B}}$. Both generators of this semigroup would be contained in some $\mathfrak{A}_n$, in which they would generate a subsemigroup isomorphic to $\mathfrak{P}$ and containing $E_{\mathfrak{B}} = E_{\mathfrak{A}_n}$. But this contradicts the fact, from 7.3, that $\mathfrak{A}_n$ is a semigroup with separating group part.

7.6. THEOREM. *Let $\mathfrak{A}$ be a semigroup with separating group part. If the subsemigroup $\mathfrak{A}'$ of the semigroup $\mathfrak{A}$ contains its unit $E_{\mathfrak{A}}$, then it is itself a semigroup with separating group part.*

PROOF. $E_{\mathfrak{A}}$ is evidently the unit of $\mathfrak{A}'$. Since in $\mathfrak{A}$, from 7.3, there are no subsemigroups isomorphic to $\mathfrak{P}$ and containing $E_{\mathfrak{A}}$, there can be no such subsemigroups in the subsemigroup $\mathfrak{A}'$ either. But then, from 7.3, $\mathfrak{A}'$ is a semigroup with separating group part.

7.7. Making use of 7.4, we may indicate a still further extensive class of semigroups with separating group parts. Recall that the semigroup $\mathfrak{A}$ is said to be a semigroup representable by matrices if for some n there is an isomorphism of $\mathfrak{A}$ into the semigroup $\mathfrak{M}_n$ of all complex square matrices of order n.

THEOREM. *Every semigroup with identity, representable by matrices, is a semigroup with separating group part.*

PROOF. Suppose that n is the smallest of the integers for which there exists an isomorphism φ of the given semigroup $\mathfrak{A}$ into $\mathfrak{M}_n$. We reduce the matrix $\varphi(E_{\mathfrak{A}})$ to the normal Jordan form.[3] This means that for some nonsingular matrix $P \in \mathfrak{M}_n$ the matrix

$$P^{-1}\varphi(E_{\mathfrak{A}})P$$

has the normal Jordan form. Consider a new mapping ψ of the semigroup $\mathfrak{A}$ into $\mathfrak{M}_n$:

$$\psi(A) = P^{-1}[\varphi(A)]P.$$

Evidently ψ is a representation of $\mathfrak{A}$ by matrices. Since the element $E_{\mathfrak{A}}$ of the semigroup $\mathfrak{A}$ is idempotent, the matrix $\psi(E_{\mathfrak{A}})$, being in normal Jordan form, satisfies the condition

$$[\psi(E_{\mathfrak{A}})]^2 = \psi(\mathfrak{A}_{\mathfrak{A}}).$$

As follows immediately from the definition of the normal Jordan form, this condition may be satisfied only in the case when the matrix $\psi(E_{\mathfrak{A}})$ is diagonal and all its diagonal elements are equal to zero or unity. Suppose that the ith diagonal element of this matrix is equal to zero. Since for any element $A \in \mathfrak{A}$ we have

$$AE_{\mathfrak{A}} = A, \qquad E_{\mathfrak{A}}A = A,$$

it follows immediately that

$$[\psi(A)] \cdot [\psi(E_{\mathfrak{A}})] = \psi(A), \qquad [\psi(E_{\mathfrak{A}})] \cdot [\psi(A)] = \psi(A).$$

Taking into account that the matrix $\psi(E_{\mathfrak{A}})$ is diagonal and that its ith diagonal element is equal to zero, we conclude that all the elements of the ith row and the ith column of the matrix $\psi(A)$ are equal to zero. It is easy to verify that the matrix $\chi(A)$, obtained from the matrix $\psi(A)$ by striking out the ith row and the ith column, will again give a matrix representation of $\mathfrak{A}$ (the case when $n = 1$ and $\psi(E_{\mathfrak{A}})$ is the null matrix does not have to be considered because of its triviality). But the matrix $\chi(A)$ is contained in $\mathfrak{M}_{n-1}$, which contradicts the initial hypothesis concerning n. Hence it follows that $\varphi(E_{\mathfrak{A}})$ is the identity matrix.

[3] See, for example, A. I. Mal'cev, *Foundations of linear algebra*, OGIZ, Moscow, 1948 (see Chapter IV, § 3). (Russian)

Thus we have come to the conclusion that ψ is an isomorphism of $\mathfrak{A}$ onto some semigroup $\psi(\mathfrak{A}) \subset \mathfrak{M}_n$ containing the unit of the matrix $\psi(E_{\mathfrak{A}}) = E$. In 6.2, we indicated that $\mathfrak{M}_n$ is a semigroup with separating group part. From 7.6 it therefore follows that $\psi(\mathfrak{A})$, and therefore $\mathfrak{A}$ itself, are semigroups with separating group parts.

7.8. As to semigroups of infinite matrices, the considerations presented above do not extend to them. For example, the set of all countable matrices, in each row and column of which there is only a finite number of elements distinct from zero, evidently forms a semigroup whose identity is the countable identity matrix E. This semigroup is not a semigroup with separating group part. Indeed,

$$\begin{bmatrix} 0 & 1 & 0 & 0 & \cdots \\ 0 & 0 & 1 & 0 & \cdots \\ 0 & 0 & 0 & 1 & \cdots \\ 0 & 0 & 0 & 0 & \cdots \\ \cdot & \cdot & \cdot & \cdot & \cdot \end{bmatrix} \cdot \begin{bmatrix} 0 & 0 & 0 & 0 & \cdots \\ 1 & 0 & 0 & 0 & \cdots \\ 0 & 1 & 0 & 0 & \cdots \\ 0 & 0 & 1 & 0 & \cdots \\ \cdot & \cdot & \cdot & \cdot & \cdot \end{bmatrix} = \begin{bmatrix} 1 & 0 & 0 & 0 & \cdots \\ 0 & 1 & 0 & 0 & \cdots \\ 0 & 0 & 1 & 0 & \cdots \\ 0 & 0 & 0 & 1 & \cdots \\ \cdot & \cdot & \cdot & \cdot & \cdot \end{bmatrix} = E;$$

$$\begin{bmatrix} 0 & 0 & 0 & 0 & \cdots \\ 1 & 0 & 0 & 0 & \cdots \\ 0 & 1 & 0 & 0 & \cdots \\ 0 & 0 & 1 & 0 & \cdots \\ \cdot & \cdot & \cdot & \cdot & \cdot \end{bmatrix} \cdot \begin{bmatrix} 0 & 1 & 0 & 0 & \cdots \\ 0 & 0 & 1 & 0 & \cdots \\ 0 & 0 & 0 & 1 & \cdots \\ 0 & 0 & 0 & 0 & \cdots \\ \cdot & \cdot & \cdot & \cdot & \cdot \end{bmatrix} = \begin{bmatrix} 0 & 0 & 0 & 0 & \cdots \\ 0 & 1 & 0 & 0 & \cdots \\ 0 & 0 & 1 & 0 & \cdots \\ 0 & 0 & 0 & 1 & \cdots \\ \cdot & \cdot & \cdot & \cdot & \cdot \end{bmatrix} \neq E,$$

from which, taking account of 1.3 and II, 2.14, (β), it follows that the matrices multiplied above are elements of our semigroup invertible on one side but not invertible on the other. In semigroups with separating group part such elements, from 6.7, cannot exist.

CHAPTER VII

HOMOMORPHISMS

1. Homomorphisms and Their Divisibility

1.1. In the first chapter we spoke of mappings preserving this or that relation between elements of the set being mapped. In the construction of the theory of semigroups it is to a high degree natural to distinguish mappings of one semigroup into another under which the relations of the operation are "preserved." In other words, one is concerned with mappings such that if in the first semigroup one has $AB = C$ for certain elements A, B, C, then in the second semigroup the relation $A'B' = C'$ will hold for the elements A', B', C' onto which the respective elements A, B, C are mapped.

Isomorphisms are an example of such mappings.

DEFINITION. *The mapping φ of the semigroup $\mathfrak{A}$ into the semigroup $\mathfrak{B}$ is said to be a* HOMOMORPHISM, *if for any elements X and Y of $\mathfrak{A}$ and $\mathfrak{B}$ one always has*

$$\varphi(X) \cdot \varphi(Y) = \varphi(XY).$$

If φ is a mapping of $\mathfrak{A}$ onto $\mathfrak{B}$, then one speaks of a homomorphism of $\mathfrak{A}$ onto $\mathfrak{B}$ (sometimes in this case one uses a special term, epimorphism). If $\mathfrak{A}$ and $\mathfrak{B}$ coincide, then the homomorphism is called an endomorphism of the semigroup $\mathfrak{A}$. This use of the term does not contradict the wider meaning of it indicated in I, 3.18, since a homomorphism of a semigroup into itself is a transformation of it which preserves relations between its elements which have the form

$$AB = C.$$

1.2. It is easy to see that if the semigroup $\mathfrak{A}$ is a group and if φ is a homomorphism of $\mathfrak{A}$ onto the semigroup $\mathfrak{B} = \varphi(\mathfrak{A})$, then also $\mathfrak{B}$ will be a group. Indeed, for arbitrary elements of the semigroup $\mathfrak{B}$,

$$\varphi(A_1), \qquad \varphi(A_2) \qquad (A_1, A_2 \in \mathfrak{A}),$$

the elements $\varphi(X)$ and $\varphi(Y)$ (X and Y being elements of $\mathfrak{A}$ such that $XA_1 = A_2$ and $A_1Y = A_2$) satisfy the conditions

$$\varphi(X) \cdot \varphi(A_1) = \varphi(XA_1) = \varphi(A_2),$$

$$\varphi(A_1) \cdot \varphi(Y) = \varphi(A_1Y) = \varphi(A_2).$$

The concept of homomorphism for groups is one of the most important in the theory of groups. In most of the sections of this well-developed theory one uses directly or indirectly (through normal divisors) the properties of homomorphisms of groups. This use is greatly facilitated by the relative simplicity of defining homomorphisms of groups by means of the so-called normal divisors. In the theory of semigroups the structure of homomorphisms is incomparably more complicated. A complete study of homomorphisms has been carried out only for various special classes of semigroups. Of the numerous different general properties only a few have been considered. No doubt further study of homomorphisms will have an essential influence on the general theory.

1.3. The concept of homomorphism may be considered as a natural generalization of the concept of isomorphism. Indeed, it is evident that *an isomorphism is simply a one-to-one homomorphism.*

However in its intrinsic character this generalization goes beyond the scope of the idea of isomorphism. As distinct from an isomorphism of one semigroup onto another, with homomorphisms there is no way of thinking of both semigroups as identical in some sense. Evidently the semigroups may be essentially different with respect to very diverse properties.

1.4. There are many reasons calling for the consideration of homomorphisms, some of which, connected with the theory of transformations, we shall now mention.

In regard to various questions in mathematics and physics, particularly in the theory of differential equations, it frequently becomes necessary to consider the so-called one-parameter semigroups of transformations. Suppose that Σ is some additive semigroup of numbers and that to each number t of Σ we associate some transformation A_t of a given set Ω. If, in addition, for any $t_1, t_2 \in \Sigma$ we have

$$A_{t_1} \cdot A_{t_2} = A_{t_1+t_2},$$

then the collection $\mathfrak{A}$ of all transformations A_t $(t \in \Sigma)$ evidently is a semigroup, a one-parameter semigroup of transformations. The association to each number t of Σ of the transformation A_t represents a homomorphism of the semigroup of numbers Σ into the semigroup of all transformations of the set Ω.

The number of different important one-parameter semigroups of transformations is exceedingly large. Their study frequently turns out to be very useful in various mathematical theories. The extensive book of Hille **[2]** (see also Hille and Phillips **[1]**) is basically devoted to the study of extremely varied classes of one-parameter semigroups of transformations.

Of course, along with one-parameter transformations one sometimes has to consider transformations given by systems of parameters.

1.5. In regard to certain questions involving the transformations of sets it is convenient to take a point of view rather different from the one to which we basically adhere in the present book.

Suppose that we are given two sets $\mathfrak{A}$ and Ω such that for each pair of elements (A, α) chosen from these sets ($A \in \mathfrak{A}$ and $\alpha \in \Omega$) their product is defined to be an element of Ω, that is,

$$A\alpha = \beta \in \Omega.$$

In this case elements of the set $\mathfrak{A}$ are called *operators* on the set Ω. The operator $A \in \mathfrak{A}$ effects in Ω a certain transformation. In its turn, each transformation may be considered as an operator on the set Ω. Nevertheless, the consideration of operators is not fully identical with the consideration of transformations. The point is that two operators given as different elements of the set of operators $\mathfrak{A}$ may realize the same transformation in Ω. That in certain cases this is convenient may be seen, for example, from the following. Suppose we are given a certain family of transformations $\mathfrak{B}$ of the set Ω. Suppose that Γ is a subset of Ω such that $X\Gamma \subset \Gamma$ for every transformation $X \in \mathfrak{B}$. Then transformations of $\mathfrak{B}$ realize certain transformations of the set Γ. However, it can well happen that two different transformations of $\mathfrak{B}$ realize in Γ one and the same transformation. Thus, considering the action of the elements of $\mathfrak{B}$ on Γ, we have to take into account the possibility that certain of them which we cannot regard as identical (since they carry out different operations in Ω) effect in Γ one and the same transformation.

Suppose that in a set of operators $\mathfrak{A}$ on the set Ω one is given in some way a multiplication operation such that $\mathfrak{A}$ is a semigroup with respect to it. One says that the multiplication in the semigroup $\mathfrak{A}$ is consistent with the multiplication of operators of $\mathfrak{A}$ on elements of Ω if for any X, $Y \in \mathfrak{A}$ and $\alpha \in \Omega$ the following condition, having an associative character, is satisfied:

$$(XY)\alpha = X(Y\alpha).$$

We shall denote by $\bar{A}$ that transformation in the set Ω which is effected in Ω as a result of multiplication by the operator $A \in \mathfrak{A}$. Thus, we obtain a mapping of the semigroup of operators $\mathfrak{A}$ into the semigroup $\mathfrak{S}_\Omega$ of all transformations of the set Ω.

Since $A\alpha = \bar{A}\alpha$, it follows that for any X, $Y \in \mathfrak{A}$ and $\alpha \in \Omega$ we have

$$(\bar{X}\bar{Y})\alpha = \bar{X}(\bar{Y}\alpha) = \bar{X}(Y\alpha) = X(Y\alpha) = (XY)\alpha = (\overline{XY})\alpha.$$

Since α is arbitrary, this means

$$\bar{X} \cdot \bar{Y} = (\overline{XY}).$$

Thus, the association to each operator $A \in \mathfrak{A}$ of the transformation $\bar{A} \in \mathfrak{S}_\Omega$ yields a homomorphism of the semigroup $\mathfrak{A}$ into the semigroup $\mathfrak{S}_\Omega$.

1.6. We have already repeatedly spoken of isomorphic representations of semigroups. In connection with the concept of homomorphism one may speak of *homomorphic representations*. Suppose that $\mathfrak{A}$ is a semigroup and Ξ is a certain class of semigroups. *A homomorphism of* $\mathfrak{A}$ *into some semigroup of* Ξ *is said to be a homomorphic representation of* $\mathfrak{A}$ *by semigroups of* Ξ.

In connection with what was said in 1.2 on the fundamental difference in approach to the concepts of isomorphism and homomorphism, it is natural to outline the difference in the meaning of isomorphic representations and homomorphic representations.

The use of the homomorphic image for the study of the properties of the original semigroup is made difficult by the fact that different elements of the original semigroup may map into one and the same element under the homomorphism. This circumstance must always be borne in mind in the study of homomorphic representations. In this connection, the following concept arises. A collection of homomorphisms Φ of the semigroup $\mathfrak{A}$ into semigroups of the class Ξ is said to be a complete system of representations of $\mathfrak{A}$ by semigroups of Ξ if for any two distinct elements X and Y of $\mathfrak{A}$ there is always a homomorphism φ of Φ for which $\varphi(X) \neq \varphi(Y)$.

1.7. In § 5 of Chapter II we introduced a class of semigroups Π, having the property that each element of a semigroup of Π has a right zero. If $\mathfrak{A} \in \Pi$, then for any homomorphism φ on it the semigroup $\varphi(\mathfrak{A})$ will also belong to Π. Indeed, if U is a right zero of the element A of $\mathfrak{A}$, then evidently $\varphi(U)$ will be a right zero of the element $\varphi(A)$. Accordingly, every element of $\varphi(A)$ has in $\varphi(\mathfrak{A})$ a right zero.

Suppose that φ is any homomorphic representation by transformations of some semigroup $\mathfrak{A} \in \Pi$. Since $\varphi(\mathfrak{A}) \in \Pi$ and the identity mapping of the semigroup $\varphi(\mathfrak{A})$ is an isomorphic representation of it by transformations, it follows from II, 5.3 that every transformation of $\varphi(\mathfrak{A})$ must have an invariant point. Thus, for any homomorphic representation of semigroups of Π, all the transformations of the representation have invariant points.

1.8. In connection with what has been said, one encounters also the following property. Suppose that Γ is a subset of the set Ω such that for a semigroup $\mathfrak{A}$ of certain transformations one has $\mathfrak{A}\Gamma \subset \Gamma$. If each transformation of $\mathfrak{A}$ has an invariant point, then that point might not belong to Γ. In Γ, for some transformations in $\mathfrak{A}$, there may not exist an invariant point. However, if $\mathfrak{A}$ belongs to the class Π, then from 1.5, transformations of $\mathfrak{A}$ induce transformations of the set Γ forming a semigroup $\mathfrak{B}$ which is a homomorphic image of $\mathfrak{A}$. But, as we showed in 1.7, a homomorphic image of a semigroup of Π must also belong to Π. Hence it follows that every transformation of $\mathfrak{B}$ must have an invariant point, which, of course, is an element of Γ. But in Γ transformations of $\mathfrak{A}$ and the transformations of $\mathfrak{B}$ induced by them operate in the same way. Accordingly, each transformation of $\mathfrak{A}$, in the case when $\mathfrak{A}$ belongs to Π, has an invariant point in every set $\Gamma \subset \Omega$ such that $\mathfrak{A}\Gamma \subset \Gamma$.

1.9. Suppose that ψ is a mapping of the set Ω_1 into the set Ω_2 and that φ is a mapping of Ω_2 into Ω_3. Then one defines in the natural way a mapping χ of the set Ω_1 into Ω_3:

$$\chi(\alpha) = \varphi[\psi(\alpha)] \qquad (\alpha \in \Omega_1).$$

The mapping χ is called the product of the mappings φ and ψ and is written $\chi = \varphi \cdot \psi$ or $\chi = \varphi\psi$. It should be noted that in considering such a multiplication of mappings we in fact pass beyond the limits indicated in § 1 of the first chapter for the concept of an algebraic operation.

It is immediately clear that if for the mappings $\varphi_1, \varphi_2, \varphi_3$ the products $\varphi_1\varphi_2$ and $\varphi_2\varphi_3$ are defined, then also $(\varphi_1\varphi_2)\varphi_3$ and $\varphi_1(\varphi_2\varphi_3)$ are defined, while

$$(\varphi_1\varphi_2)\varphi_3 = \varphi_1(\varphi_2\varphi_3).$$

1.10. Suppose that Ω_1 and Ω_2 are any two nonempty sets. Denote by $\mathfrak{R}$ the set of all mappings of Ω_1 into Ω_2. The operation of multiplication of mappings considered in 1.9 is as a general rule simply inapplicable. Possible natural immediate generalizations of the operation of 1.9 lead to trivial results. However, in $\mathfrak{R}$ one may introduce a nontrivial operation in the following way. We fix some mapping π of the set Ω_2 into Ω_1. For $X, Y \in \mathfrak{R}$ we define the product $X \circ Y$ by means of the multiplication of mappings of 1.9:

$$(X \circ Y)\alpha = (X \cdot \pi \cdot Y)\alpha = X\{\pi(Y\alpha)\} \qquad (\alpha \in \Omega_1).$$

(The product $X \cdot \pi \cdot Y$ is evidently always defined in the sense of 1.9, and belongs to $\mathfrak{R}$.)

The set $\mathfrak{R}$, considered relative to this operation, will be denoted by $\mathfrak{R}_\pi$. The operation in $\mathfrak{R}_\pi$ is associative, since evidently

$$(X \circ Y) \circ Z = X\pi Y\pi Z,$$

$$X \circ (Y \circ Z) = X\pi Y\pi Z.$$

Thus $\mathfrak{R}_\pi$ is a semigroup. It sometimes is as useful for the study of mappings of $\mathfrak{R}$ as the consideration of the semigroup $\mathfrak{S}_\Omega$ is in the study of transformations of the set Ω. Apropos, $\mathfrak{S}_\Omega$ is a special case of the semigroup $\mathfrak{R}_\pi$. In fact, $\mathfrak{S}_\Omega$ is $\mathfrak{R}_\pi$ for $\Omega_1 = \Omega_2 = \Omega$ in the case when one takes for π the identity transformation.

We assign to each $X \in \mathfrak{R}_\pi$ a transformation of the set Ω_1:

$$\varphi_\pi(X) = \pi \cdot X.$$

It is clear that φ_π is a mapping of $\mathfrak{R}_\pi$ into $\mathfrak{S}_\Omega$. This mapping is a homomorphism:

$$\varphi_\pi(X) \cdot \varphi_\pi(Y) = (\pi \cdot X) \cdot (\pi \cdot Y) = \pi(X\pi Y) = \varphi_\pi(X \circ Y).$$

If the mapping π is not one-to-one, then it is easily seen that the homomorphism φ_π is not an isomorphism. If π is one-to-one, then φ_π is an isomorphism. If in addition π is a one-to-one mapping of Ω_2 onto Ω_1, then φ_π is an isomorphism of $\mathfrak{R}_\pi$ onto $\mathfrak{S}_\Omega$, so that the semigroups $\mathfrak{R}_\pi$ and $\mathfrak{S}_\Omega$ turn out in this case to be isomorphic.

In considering transformations of some set Ω, interest usually centers not only in the semigroup $\mathfrak{S}_\Omega$ of all its transformations, but in subsemigroups distinct from it, consisting of transformations having certain given properties.

Just the same, in considering the multiplication of mappings of $\mathfrak{R}_\pi$ introduced above, the interest lies not only in the semigroup $\mathfrak{R}_\pi$ itself but also in its various subsemigroups. Here, moreover, there arises further the question of clarifying the mutual relations of the semigroups $\mathfrak{R}_\pi$ for various π.

An analogous construction may of course be effected for partial mappings of one set into another.

1.11. Let $\mathfrak{M}_{m,n}$ be the set of all complex matrices having m rows and n columns, and P some fixed matrix having n rows and m columns. In $\mathfrak{M}_{m,n}$ one may consider an associative operation

$$M_1 \circ M_2 = M_1 \cdot P \cdot M_2 \qquad (M_1, M_2 \in \mathfrak{M}_{m,n})$$

(where the point denotes the usual multiplication of rectangular matrices). This is, of course, just the operation considered in 1.10, since matrices of $\mathfrak{M}_{m,n}$ may be regarded as linear mappings of vectors of n-dimensional complex space into m-dimensional complex linear space.

From the point of view of such an operation over matrices it becomes natural to consider the operation in the semigroup of matrices of type $\mathfrak{S}(P, \mathfrak{H})$, which played such an important role in §§ 5 and 6 of the fifth chapter (with the difference that the matrices there could be infinite as well, and their elements were not complex numbers).

1.12. Suppose that ψ is a homomorphism of the semigroup $\mathfrak{A}$ onto the semigroup $\mathfrak{B}$ and that φ is a homomorphism of $\mathfrak{B}$ onto the semigroup $\mathfrak{C}$. The product of transformations $\chi = \varphi\psi$, in the sense of the product of transformations described in 1.9, is defined and as one easily sees is a homomorphism of $\mathfrak{A}$ onto $\mathfrak{C}$. The homomorphism χ is called *the product of the homomorphisms* φ *and* ψ. Only in cases of the type described will we say that the multiplication of the homomorphisms φ and ψ is possible (or defined). The homomorphism ψ in this case is called a *right divisor* of the homomorphism χ. Concerning χ one says that *it is divided on the right by* ψ. We shall in this case write $\psi \sim \chi(\mathfrak{p})$. If for the homomorphisms ψ_1 and ψ_2 we have simultaneously $\psi_1 \sim \psi_2(\mathfrak{p})$ and $\psi_2 \sim \psi_1(\mathfrak{p})$, then we will write $\psi_1 \sim \psi_2(\mathfrak{q})$. We note that only homomorphisms of one and the same semigroup can lie in the relation of right divisibility $\mathfrak{p}$.

1.13. We shall consider several properties of the multiplication of homomorphisms.

(α) *If for the homomorphisms* $\varphi_1, \varphi_2, \varphi_3$ *the products* $\varphi_1\varphi_2$ *and* $\varphi_2\varphi_3$ *are defined, then also the products* $(\varphi_1\varphi_2)\varphi_3$ *and* $\varphi_1(\varphi_2\varphi_3)$ *are defined and equal to each other.*

(β) *If for the homomorphisms* $\varphi_1, \varphi_2, \varphi_3$ *one has* $\varphi_1 \sim \varphi_2(\mathfrak{p})$ *and* $\varphi_2 \sim \varphi_3(\mathfrak{p})$, *then one has also* $\varphi_1 \sim \varphi_3(\mathfrak{p})$ (*transitivity of* $\mathfrak{p}$).

Indeed, suppose

$$\varphi_3 = \psi\varphi_2, \qquad \varphi_2 = \psi'\varphi_1.$$

Then, evidently,

$$\varphi_3 = (\psi\psi')\varphi_1.$$

(γ) *If for the homomorphisms* $\varphi_1, \varphi_2, \varphi_3$ *one has* $\varphi_1 \sim \varphi_2(\mathfrak{q})$ *and* $\varphi_2 \sim \varphi_3(\mathfrak{q})$, *then there holds also* $\varphi_1 \sim \varphi_3(\mathfrak{q})$ *(transitivity of* $\mathfrak{q}$*).*

(δ) *If for the homomorphisms* φ_1, φ_2 *and the isomorphism* ε *one has* $\varphi_1 = \varepsilon\varphi_2$, *then there exists an isomorphism* ε' *such that* $\varphi_2 = \varepsilon'\varphi_1$.

Indeed, if φ_2 is a homomorphism of the semigroup $\mathfrak{A}$, then ε is an isomorphism of the semigroup $\varphi_2(\mathfrak{A})$ onto the semigroup $\varphi_1(\mathfrak{A})$. Denote by ε' the isomorphism inverse to this of the semigroup $\varphi_1(\mathfrak{A})$ onto $\varphi_2(\mathfrak{A})$. Since

$$\varepsilon'\varphi_1 = \varepsilon'\varepsilon\varphi_2$$

and since $\varepsilon'\varepsilon$ is the identity isomorphism it follows that $\varphi_2 = \varepsilon'\varphi_1$.

(ε) *For the homomorphisms* φ_1 *and* φ_2 *of the same semigroup, the relation* $\varphi_1 \sim \varphi_2(\mathfrak{q})$ *holds if and only if there exists an isomorphism* ε *for which* $\varphi_1 = \varepsilon\varphi_2$.

Indeed, if $\varphi_1 \sim \varphi_2(\mathfrak{q})$, then, for certain homomorphisms ψ and ψ',

$$\varphi_1 = \psi\varphi_2, \qquad \varphi_2 = \psi'\varphi_1,$$

i.e., $\varphi_1 = (\psi\psi')\varphi_1$.

The product $\psi\psi'$ is the identity mapping. This is possible only in the case when the homomorphisms ψ and ψ' are one-to-one mappings, i.e., isomorphisms.

Now suppose that for some isomorphism ε we have $\varphi_1 = \varepsilon\varphi_2$. This means that $\varphi_2 \sim \varphi_1(\mathfrak{p})$. But then, from ($\delta$), there must exist an isomorphism ε' such that $\varphi_2 = \varepsilon'\varphi_1$, and therefore $\varphi_1 \sim \varphi_2(\mathfrak{p})$.

(ζ) *If for the homomorphisms* $\varphi_1, \varphi_1', \varphi_2, \varphi_2'$ *of the semigroup* $\mathfrak{A}$ *we have*

$$\varphi_1 \sim \varphi_1'(\mathfrak{q}), \qquad \varphi_2 \sim \varphi_2'(\mathfrak{q}), \qquad \varphi_2 \sim \varphi_1(\mathfrak{p}),$$

then

$$\varphi_2' \sim \varphi_1'(\mathfrak{p}).$$

Indeed, for certain isomorphisms ε_1 and ε_2 and a homomorphism ψ we must have

$$\varphi_1' = \varepsilon_1\varphi_1, \qquad \varphi_2' = \varepsilon_2\varphi_2, \qquad \varphi_1 = \psi\varphi_2.$$

Denoting by ε_2' the inverse isomorphism, relative to ε_2, of the semigroup $\varphi_2'(\mathfrak{A})$ onto $\varphi_2(\mathfrak{A})$, we have

$$\varphi_1' = \varepsilon_1\varphi_1 = \varepsilon_1\psi\varphi_2 = \varepsilon_1\psi\varepsilon_2'\varepsilon_2\varphi_2 = (\varepsilon_1\psi\varepsilon_2')\varphi_2',$$

i.e., $\varphi_2' \sim \varphi_1'(\mathfrak{p})$.

(η) *If* φ *is some homomorphism of the semigroup* $\mathfrak{A}$ *and* ε *is some isomorphism of it, then* $\varepsilon \sim \varphi(\mathfrak{p})$.

Indeed, denoting by ε' the isomorphism of $\varepsilon(\mathfrak{A})$ onto $\mathfrak{A}$ inverse to ε, we obtain

$$\varphi = \varphi \cdot \varepsilon'\varepsilon = (\varphi\varepsilon') \cdot \varepsilon.$$

(θ) *If* ε *is an isomorphism of the semigroup* $\mathfrak{A}$ *and* φ *a homomorphism of* $\mathfrak{A}$, *with* $\varphi \sim \varepsilon(\mathfrak{p})$, *then* φ *is also an isomorphism.*

Indeed, for some homomorphism ψ we have

$$\varepsilon = \psi\varphi.$$

Since ε is a one-to-one mapping, the mapping φ must also be one-to-one, i.e., an isomorphism.

(ι) *Let φ and ψ be two homomorphisms of the semigroup $\mathfrak{A}$. For $\varphi \sim \psi(\mathfrak{p})$ it is necessary and sufficient that if $\varphi(A) = \varphi(B)$ for any $A, B \in \mathfrak{A}$, then*

$$\psi(A) = \psi(B).$$

Indeed, if for some homomorphism χ, $\psi = \chi\varphi$, then from $\varphi(A) = \varphi(B)$ it follows that

$$\psi(A) = (\chi \cdot \varphi)(A) = \chi[\varphi(A)] = \chi[\varphi(B)] = (\chi \cdot \varphi)(B) = \psi(B).$$

On the other hand, if φ and ψ are such that from $\varphi(A) = \varphi(B)$ it always follows that $\psi(A) = \psi(B)$, then for the semigroup $\varphi(\mathfrak{A})$ one may define the following mapping χ of it onto the semigroup $\psi(\mathfrak{A})$. Put

$$\chi[\varphi(A)] = \psi(A).$$

This mapping is uniquely defined, independently of the choice of the representative A in the class of those elements $X \in \mathfrak{A}$ for which $\varphi(X) = \varphi(A)$, since, if $\varphi(A) = \varphi(A')$, we have $\psi(A) = \psi(A')$ in $\psi(\mathfrak{A})$. From the fact that ψ and φ are homomorphisms, it follows immediately that the mapping χ is also a homomorphism. Thus $\chi\varphi = \psi$, i.e., $\varphi \sim \psi(\mathfrak{p})$.

1.14. From 1.13, (γ), (ι), it follows that in the class of all homomorphisms of a semigroup the relation $\mathfrak{q}$ is an equivalence, and thus that this class decomposes into disjoint classes of homomorphisms which are equivalent to each other relative to the relation $\mathfrak{q}$. One of these classes is formed by all the isomorphisms of the semigroup. If for the homomorphisms φ_1 and φ_2 we have $\varphi_1 \sim \varphi_2(\mathfrak{p})$, then also for any homomorphisms φ_1' and φ_2', taken from the corresponding classes, the relation $\varphi_1' \sim \varphi_2'(\mathfrak{p})$ is satisfied (1.13, (ζ)). Therefore, for those same classes Γ_1 and Γ_2 one may introduce the relation $\mathfrak{p}$, holding if for $\varphi_1 \in \Gamma_1$ and $\varphi_2 \in \Gamma_2$ one has $\varphi_1 \sim \varphi_2(\mathfrak{p})$.

Because of 1.13, (β) this relation is a partial ordering relation, and we may write $\Gamma_1 \leqslant \Gamma_2$ in place of $\Gamma_1 \sim \Gamma_2(\mathfrak{p})$. The class of all isomorphisms, because of 1.13, (η) and 1.13, (θ), turns out to be a predecessor of all the other classes. It is easy to see that the class consisting of homomorphisms mapping the semigroup onto the identity group (consisting of the unit element alone) will follow all the remaining classes.

Since homomorphisms of one and the same class differ, by 1.13, (ε), only by a multiplier which is an isomorphism, it is usual to say that they are *identical up to multiplication by an isomorphism* or that they are *inessentially distinct from one another.*

1.15. Suppose we are given some collection Γ of homomorphisms of the semigroup $\mathfrak{A}$. It is natural to call a homomorphism φ of the semigroup $\mathfrak{A}$ *the greatest common right divisor for* Γ if φ is a right divisor of every homomorphism of Γ and if it is divided on the right by every homomorphism which is a right divisor of every homomorphism of Γ.

The homomorphism ψ is said to be the *least common right multiple for* Γ if ψ is divided on the right by every homomorphism of Γ and if it is a right divisor of every homomorphism which is divided on the right by every homomorphism of Γ.

Evidently the class of homomorphisms which are the greatest common right divisors of the homomorphisms of Γ, relative to the partial ordering 1.14 of the classes, is the greatest lower bound for the classes of homomorphisms containing the homomorphisms of Γ. The class of least common right multiples is a least upper bound.

From 1.13, (ε) and 1.14 it follows that every two least common right divisors of the class Γ of homomorphisms are inessentially distinct from each other. Also every two least common right multiples are inessentially distinct.

1.16. The question as to the existence of a greatest common right divisor for a collection of homomorphisms, and of a least common right multiple, will receive a positive solution in the following section (2.11). One must, however, also keep in mind that, for a given class Γ of homomorphisms, there is usually interest not only as to whether there exist for it greatest common right divisors, but also as to whether some of them belong themselves to the class Γ. This is particularly important when the class Γ is such that if a certain homomorphism φ lies in it, then any homomorphism divided on the right by φ also lies in Γ. If the class Γ is such, and ψ is the greatest common right divisor of the homomorphisms of Γ which belongs to Γ, then Γ may be characterized as the class of all homomorphisms divided on the right by ψ. Prescribing ψ in this case accurately describes the structure of the class Γ.

As an example we mention the class of all homomorphisms of an arbitrary semigroup $\mathfrak{A}$ onto a commutative semigroup. Evidently, if the homomorphisms φ, ψ, χ are such that $\varphi\psi = \chi$ and the semigroup $\psi(\mathfrak{A})$ is commutative, then also $\chi(\mathfrak{A})$ will be commutative. In 6.7 we construct the greatest common right divisor of such homomorphisms, which will in addition lie in that class.

As an example of a negative solution of the question posed above we consider an infinite monogenic semigroup $\mathfrak{A} = [X]$ and the class of all of its homomorphisms onto groups.

For any finite cyclic group $\mathfrak{G}_d = [G]$ with a number of elements equal to d (i.e., of type $(1, d)$) there exists a homomorphism φ_d of the semigroup $\mathfrak{A}$ onto $\mathfrak{G}_d$:

$$\varphi_d = (A^k) = G^k \qquad (k = 1, 2, 3, \ldots).$$

If ψ is some homomorphism of $\mathfrak{A}$ onto the group, then, for some m, $\psi(A^m)$ will be the identity of that group. Hence we easily find that

$$\psi(\mathfrak{A}) = \{\psi(A), \psi(A^2), \ldots, \psi(A^m)\}.$$

Since the number of elements in $\psi(\mathfrak{A})$ does not exceed m it follows that ψ cannot be a right divisor for those φ_d for which $d > m$.

1.17. We have already mentioned that the connection between the properties of a semigroup and its homomorphic image in the case of an arbitrary homomorphism is far from being as elementary as with an isomorphism. Suppose that φ is a homomorphism of the semigroup $\mathfrak{A}$ onto the semigroup $\varphi(\mathfrak{A}) = \mathfrak{A}'$. If in $\mathfrak{A}$ the element A_1 is divided on the left by A_2, then obviously in $\mathfrak{A}'$ the element $\varphi(A_1)$ will be divided on the left by $\varphi(A_2)$. The same is true for divisibility on the right. In particular, if C is an element of $\mathfrak{A}$ divisible both left and right by any element of $\mathfrak{A}$ (II, 1.6), then $\varphi(C)$ will be divisible in $\mathfrak{A}'$ left and right by all elements of $\mathfrak{A}'$. However, if some element C' of $\mathfrak{A}$ is divided both left and right by every element of $\mathfrak{A}'$, then it is not necessarily an image of some element of $\mathfrak{A}$ having the same property in $\mathfrak{A}$. Indeed, the semigroup $\mathfrak{A}$ may have no such elements at all. Nevertheless, under a homomorphism of it onto the identity group the single element of the image has the property in question.

In connection with what has been said, we note that for a homogroup $\mathfrak{A}$, i.e., in the case when $\mathfrak{A}$ has a two-sided ideal $\mathfrak{T}$ which is a group (V, 1.8), the following property is satisfied. For any homomorphism φ of the semigroup $\mathfrak{A}$, if in $\mathfrak{A}' = \varphi(\mathfrak{A})$ the element C' is divided by all the elements of $\mathfrak{A}'$ both left and right, then there is in $\mathfrak{A}$ an element C divided both left and right by every element of $\mathfrak{A}$ and such that $\varphi(C) = C'$.

Indeed, it is obvious that $\varphi(\mathfrak{T})$ is a two-sided ideal of $\mathfrak{A}'$, and from 1.2 it is a group. From V, 1.4, $C' \in \varphi(\mathfrak{T})$, i.e., for some $C \in \mathfrak{T}$ we have $\varphi(C) = C'$. But from V, 1.4, $\mathfrak{T}$ consists of elements divided both left and right by every element of $\mathfrak{A}$.

2. Factor-semigroups

2.1. Suppose that φ is any homomorphism of the semigroup $\mathfrak{A}$. We define in $\mathfrak{A}$ a relation $\mathfrak{n}$ by putting $A \sim B(\mathfrak{n})$ if $\varphi(A) = \varphi(B)$. Reflexivity, symmetry and transitivity of this relation are evident. The equivalence $\mathfrak{n}$ will be called *the equivalence corresponding to the given homomorphism* φ.

Since from $\varphi(A) = \varphi(B)$ it follows for any $X \in \mathfrak{A}$ that

$$\varphi(XA) = \varphi(X) \cdot \varphi(A) = \varphi(X) \cdot \varphi(B) = \varphi(XB),$$

$$\varphi(AX) = \varphi(A) \cdot \varphi(X) = \varphi(B) \cdot \varphi(X) = \varphi(BX),$$

the equivalence $\mathfrak{n}$, *corresponding to the homomorphism* φ, *is two-sidedly stable.*

2.2. Now suppose that $\mathfrak{n}$ is any two-sidedly stable equivalence in the semigroup $\mathfrak{A}$.

Denote by $\mathfrak{A}/\mathfrak{n}$ the set of all $\mathfrak{n}$-classes (I, 5.8).

Observe that for all $\mathfrak{n}$-classes $\mathfrak{K}_1, \mathfrak{K}_2, \mathfrak{K}_3$ it always follows from

$$(\mathfrak{K}_1 \cdot \mathfrak{K}_2) \cap \mathfrak{K}_3 \neq \varnothing$$

that

$$\mathfrak{K}_1 \cdot \mathfrak{K}_2 \subset \mathfrak{K}_3.$$

Indeed, suppose $X_1, Y_1 \in \mathfrak{K}_1$; $X_2, Y_2 \in \mathfrak{K}_2$ and $Y_1Y_2 \in \mathfrak{K}_3$. In accordance with I, 5.18 it follows from $X_1 \sim Y_1(\mathfrak{n})$, $X_2 \sim Y_2(\mathfrak{n})$ that

$$X_1X_2 \sim Y_1Y_2(\mathfrak{n}).$$

Since $Y_1Y_2 \in \mathfrak{K}_3$, any element X_1X_2 of the set $\mathfrak{K}_1\mathfrak{K}_2$ must also belong to $\mathfrak{K}_3$.

Since the $\mathfrak{n}$-classes form a decomposition of the set of all elements of the semigroup (I, 5.8), it follows from the property just proved that for two arbitrary $\mathfrak{n}$-classes $\mathfrak{K}_1$ and $\mathfrak{K}_2$ there will always exist a unique $\mathfrak{n}$-class $\mathfrak{K}_3$ such that

$$(\mathfrak{K}_1 \cdot \mathfrak{K}_2) \subset \mathfrak{K}_3.$$

2.3. Because of 2.2 one may define in a natural way an operation in $\mathfrak{A}/\mathfrak{n}$. Since this operation does not coincide with the operation of multiplication of subsets, we shall temporarily use the sign $\circ$ for denoting the result of this operation. Of course, one may in this case use the usual multiplication notation if one regards the $\mathfrak{n}$-classes not as subsets of $\mathfrak{A}$ but simply as elements of a new multiplicative set $\mathfrak{A}/\mathfrak{n}$. In the mathematical literature one usually does this. In the sequel we shall also go over to the usual multiplicative notation, dropping the usage $\circ$.

Suppose that $\mathfrak{K}_1, \mathfrak{K}_2, \mathfrak{K}_3$ are three $\mathfrak{n}$-classes such that

$$\mathfrak{K}_1 \cdot \mathfrak{K}_2 \subset \mathfrak{K}_3.$$

We put in $\mathfrak{A}/\mathfrak{n}$

$$\mathfrak{K}_1 \circ \mathfrak{K}_2 = \mathfrak{K}_3.$$

Since, for any $\mathfrak{K}_1, \mathfrak{K}_2, \mathfrak{K}_3 \in \mathfrak{A}/\mathfrak{n}$,

$$(\mathfrak{K}_1 \circ \mathfrak{K}_2) \circ \mathfrak{K}_3 \supset (\mathfrak{K}_1 \cdot \mathfrak{K}_2) \cdot \mathfrak{K}_3 = \mathfrak{K}_1\mathfrak{K}_2\mathfrak{K}_3,$$

$$\mathfrak{K}_1 \circ (\mathfrak{K}_2 \circ \mathfrak{K}_3) \supset \mathfrak{K}_1 \cdot (\mathfrak{K}_2 \cdot \mathfrak{K}_3) = \mathfrak{K}_1\mathfrak{K}_2\mathfrak{K}_3,$$

the operation $\circ$ is associative in $\mathfrak{A}/\mathfrak{n}$.

2.4. DEFINITION. *For a two-sidedly stable equivalence* $\mathfrak{n}$ *in the semigroup* $\mathfrak{A}$, *the set of all* $\mathfrak{n}$-*classes* $\mathfrak{A}/\mathfrak{n}$, *considered relative to the operation* 2.3, *is a semigroup, called the* FACTOR-SEMIGROUP *of the semigroup* $\mathfrak{A}$ modulo $\mathfrak{n}$.

This construction is a generalization of the corresponding construction of the theory of groups. If $\mathfrak{N}$ is a normal divisor of the group $\mathfrak{G}$ and $\mathfrak{n}$ is a relation in $\mathfrak{G}$ according to which $X \sim Y(\mathfrak{n})$ if and only if $X^{-1}Y \in \mathfrak{N}$, then, as one easily verifies, $\mathfrak{n}$ is a two-sidedly stable equivalence and $\mathfrak{G}/\mathfrak{n}$ is exactly the factor-group of $\mathfrak{G}$ modulo $\mathfrak{N}$ of the theory of groups: $\mathfrak{G}/\mathfrak{N}$.

2.5. Suppose that $\mathfrak{n}$ is a two-sidedly stable decomposition of the semigroup $\mathfrak{A}$. Assigning to each element $A \in \mathfrak{A}$ the component $\mathfrak{K}$ of this decomposition which contains it, we obtain a mapping of $\mathfrak{A}$ onto the factor-semigroup $\mathfrak{A}/\mathfrak{n}$. This mapping is a homomorphism, since from

$$A_1A_2 = A_3 \qquad (A_1, A_2, A_3 \in \mathfrak{A}),$$

for $\mathfrak{K}_i \ni A_i$, where $\mathfrak{K}_i \in \mathfrak{A}/\mathfrak{n}$ $(i = 1, 2, 3)$, it follows that the intersection $(\mathfrak{K}_1\mathfrak{K}_2) \cap \mathfrak{K}_3$ is nonempty and that therefore

$$\mathfrak{K}_1 \circ \mathfrak{K}_2 = \mathfrak{K}_3.$$

Such a homomorphism will be called *the natural homomorphism of* $\mathfrak{A}$ *onto* $\mathfrak{A}/\mathfrak{n}$.

The obtaining of the factor-semigroup $\mathfrak{A}/\mathfrak{n}$ from $\mathfrak{A}$ by means of the natural homomorphism may be considered as the result of identifying to one another the elements of the semigroup $\mathfrak{A}$ which enter into one and the same $\mathfrak{n}$-class.

2.6. In 2.1 we showed that to each homomorphism of the semigroup $\mathfrak{A}$ there corresponds some two-sidedly stable equivalence. The construction of the factor-semigroup $\mathfrak{A}/\mathfrak{n}$ shows that for any two-sidedly stable equivalence $\mathfrak{n}$ there exists a homomorphism (in fact the natural homomorphism of $\mathfrak{A}$ onto $\mathfrak{A}/\mathfrak{n}$) to which that equivalence $\mathfrak{n}$ corresponds.

2.7. Suppose that φ is any homomorphism of the semigroup $\mathfrak{A}$ and $\mathfrak{n}$ the equivalence corresponding to it (2.1).

We denote by ψ the natural homomorphism of $\mathfrak{A}$ onto $\mathfrak{A}/\mathfrak{n}$. Assign to each $\mathfrak{n}$-class (i.e., to each element of the factor-semigroup $\mathfrak{A}/\mathfrak{n}$) that element of the semigroup $\varphi(\mathfrak{A})$ into which all the elements of the given $\mathfrak{n}$-class map under the homomorphism φ. This one-to-one mapping is easily seen to be an isomorphism. Thus

$$\varphi = \varepsilon\psi,$$

i.e., $\varphi \sim \psi(\mathfrak{q})$ (1.13, (ε)).

On the other hand, if φ is any homomorphism and ψ is the natural homomorphism onto some factor-semigroup, and if they are connected by the relation $\varphi \sim \psi(\mathfrak{q})$, i.e., for some isomorphism ε,

$$\varphi = \varepsilon\psi,$$

then evidently the decompositions corresponding to the homomorphisms φ and ψ are identical.

From what has been said it follows that to each homomorphism of the semigroup $\mathfrak{A}$ there corresponds a natural homomorphism onto the factor-semigroup $\mathfrak{A}/\mathfrak{n}$, where $\mathfrak{n}$ is the equivalence corresponding to the given homomorphism. The same natural homomorphism corresponds to different homomorphisms if and only if these homomorphisms divide one another to the right (1.12), i.e., from 1.13, (δ) and 1.13, (ε), if they differ only by a multiplier which is an isomorphism. Thus, natural homomorphisms onto factor-semigroups may be considered as representatives, taken one from each class of homomorphisms, where we unite into one class all homomorphisms which are inessentially distinct from one another (1.14), i.e., which are equivalent to one another relative to $\mathfrak{q}$ (1.12; 1.14), or in other words differ by isomorphism multipliers (1.13, (ε)).

Consequently we may regard the set of all homomorphisms of a semigroup as being exhausted, up to inessential distinctions, by the natural homomorphisms of the semigroup onto its factor-semigroups.

2.8. Suppose we are given two homomorphisms φ_1 and φ_2 of the semigroup $\mathfrak{A}$, and that $\mathfrak{n}_1$ and $\mathfrak{n}_2$ are the corresponding equivalences. If $\varphi_2 \sim \varphi_1(\mathfrak{p})$ holds for the homomorphisms, i.e., if φ_2 is a right divisor of φ_1 (1.12), i.e., $\varphi_1 = \psi\varphi_2$, then, from 1.13, ($\iota$) the equation $\varphi_2(A) = \varphi_2(B)$ $(A, B \in \mathfrak{A})$ always implies $\varphi_1(A) = \varphi_1(B)$, and, accordingly, for the equivalence relations we have $\mathfrak{n}_2 \leqslant \mathfrak{n}_1$ (I, 1.14).

Conversely, suppose that $\mathfrak{n}_2 \leqslant \mathfrak{n}_1$ is given. We assign to each $\mathfrak{n}_2$-class that $\mathfrak{n}_1$-class which contains it. It is easy to see that we obtain a homomorphism ψ of the factor-semigroup $\mathfrak{A}/\mathfrak{n}_2$ onto the factor-semigroup $\mathfrak{A}/\mathfrak{n}_1$. If ξ_1 and ξ_2 are the natural homomorphisms of $\mathfrak{A}$ onto the factor-semigroups $\mathfrak{A}/\mathfrak{n}_1$ and $\mathfrak{A}/\mathfrak{n}_2$, then we evidently have $\xi_1 = \psi\xi_2$. But ξ_1 and ξ_2 differ from φ_1 and φ_2 by multipliers which are isomorphisms. Therefore, for the homomorphisms φ_1 and φ_2 themselves, we find that for some homomorphism ψ' we have $\varphi_1 = \psi'\varphi_2$, i.e., $\varphi_2 \sim \varphi_1(\mathfrak{p})$.

It has turned out that the relation of right divisibility between homomorphisms, $\mathfrak{p}$ (1.12), is induced by the partial ordering of the corresponding equivalences.

2.9. Suppose that Ψ is some nonempty set of two-sidedly stable equivalences of the semigroup $\mathfrak{A}$. In the complete lattice of all equivalences in $\mathfrak{A}$ (I, 5.16) we take the greatest lower bound $\mathfrak{f}$ of the set Ψ (I, 5.15) and the least upper bound $\mathfrak{l}$ (I, 5.16). If for $A, B \in \mathfrak{A}$ we have $A \sim B(\mathfrak{f})$, then for any $\mathfrak{n} \in \Psi$ we have $A \sim B(\mathfrak{n})$. But $\mathfrak{n}$ is two-sidedly stable and therefore, for any $X \in \mathfrak{A}$, $AX \sim BX(\mathfrak{n})$ and $XA \sim XB(\mathfrak{n})$. Since the last relation is valid for any $\mathfrak{n} \in \Psi$, we have $AX \sim BX(\mathfrak{f})$ and $XA \sim XB(\mathfrak{f})$. Thus the equivalence $\mathfrak{f}$ itself turns out to be two-sidedly stable. Since $\mathfrak{f}$ is the greatest lower bound of Ψ in the set of all equivalences, $\mathfrak{f}$ in the same way is a least upper bound for Ψ in the set of all two-sidedly stable equivalences.

Now suppose that $A \sim B(\mathfrak{l})$. Then there are $Z_1 = A, Z_2, \ldots, Z_{2+1} = B$ in $\mathfrak{A}$ such that $Z_i \sim Z_{i+1}(\mathfrak{n}_i)$, $\mathfrak{n}_i \in \Psi$ $(i = 1, 2, \ldots, s)$. Since all the $\mathfrak{n}_i$ are two-sidedly stable, we have, for any $X \in \mathfrak{A}$,

$$AX = Z_1X, \qquad BX = Z_{s+1}X, \qquad Z_iX \sim Z_{i+1}X(\mathfrak{n}_i),$$

$$XA = XZ_1, \qquad XB = XZ_{s+1}, \qquad XZ_i \sim XZ_{i+1}(\mathfrak{n}_i) \qquad (i = 1, 2, \ldots, s).$$

Thus $AX \sim BX(\mathfrak{l})$, $XA \sim XB(\mathfrak{l})$ and the equivalence $\mathfrak{l}$ turn out to be two-sidedly stable.

Here $\mathfrak{l}$ is the least upper bound for Ψ in the set of all two-sidedly stable equivalences in $\mathfrak{A}$.

It follows from what has been said that relative to the partial ordering I, 5.14, *the set of all two-sidedly stable equivalences is a complete lattice.*

2.10. We note that sometimes the characterization of the set of all two-sidedly stable decompositions of a semigroup is realized by distinguishing in it a collection of relations (the so-called basis) such that every two-sidedly stable relation may be obtained as a least upper bound (sum) of certain relations belonging to that collection.

In this way A. E. Liber **[1]** dealt with the problem of characterizing two-sidedly stable decompositions of the semigroup $\mathfrak{Q}_\Omega$ of all one-to-one partial transformations of the set Ω (I, 4.5).

2.11. Suppose that Φ is some nonempty set of homomorphisms of the semigroup $\mathfrak{A}$. Denote by Ψ the set of two-sidedly stable equivalences corresponding to these homomorphisms (2.1). Let $\mathfrak{f}$ be their greatest lower bound and $\mathfrak{l}$ their least upper bound (2.9). From 2.8, the natural homomorphism of $\mathfrak{A}$ onto $\mathfrak{A}/\mathfrak{f}$ is the greatest common right divisor of the homomorphisms of Φ, and the natural homomorphism of $\mathfrak{A}$ onto $\mathfrak{A}/\mathfrak{l}$ is the least common right multiple (1.15).

2.12. Let $\mathfrak{n}_1$ and $\mathfrak{n}_2$ be two-sidedly stable equivalence relations on the semigroup $\mathfrak{A}$, with $\mathfrak{n}_1 \leqslant \mathfrak{n}_2$. In the factor-semigroup $\mathfrak{A}/\mathfrak{n}_1$ it is natural to define an equivalence relation $\mathfrak{n}_3 = \mathfrak{n}_2/\mathfrak{n}_1$ according to which two $\mathfrak{n}_3$-classes $\mathfrak{K}_1$ and $\mathfrak{K}_2$ are equivalent when the elements of the semigroup $\mathfrak{A}$, $X_1 \in \mathfrak{K}_1$ and $X_2 \in \mathfrak{K}$, are equivalent with respect to $\mathfrak{n}_2$ (evidently this does not depend on the choice of the representatives X_1 and X_2 in $\mathfrak{K}_1$ and $\mathfrak{K}_2$). It is easily seen that $\mathfrak{n}_3$ is a two-sidedly stable equivalence relation in $\mathfrak{A}/\mathfrak{n}_1$ and that the factor-semigroups $\mathfrak{A}/\mathfrak{n}_2$ and $(\mathfrak{A}/\mathfrak{n}_1)/(\mathfrak{n}_2/\mathfrak{n}_1)$ are isomorphic (the isomorphism is defined by assigning to the $\mathfrak{n}_2$-class $\mathfrak{K}$ of $\mathfrak{A}/\mathfrak{n}_2$ the $(\mathfrak{n}_2/\mathfrak{n}_1)$-class of $(\mathfrak{A}/\mathfrak{n}_1)/(\mathfrak{n}_2/\mathfrak{n}_1)$ which contains the $\mathfrak{n}_1$-classes $\mathfrak{K}'$ contained in $\mathfrak{K}$ as elements).

2.13. Suppose that $\mathfrak{n}$ is any decomposition of the semigroup $\mathfrak{A}$,

$$\mathfrak{A} = \bigcup_\xi \mathfrak{B}_\xi,$$

having the property that for any two of its components $\mathfrak{B}_\lambda$ and $\mathfrak{B}_\mu$ there is always a component $\mathfrak{B}_\nu$ such that $\mathfrak{B}_\lambda \cdot \mathfrak{B}_\mu \subset \mathfrak{B}_\nu$.

This decomposition may be considered as an equivalence in $\mathfrak{A}$ (I, 5.8). If $A \sim B(\mathfrak{n})$, i.e., if A and B are both contained in the same component $\mathfrak{B}_\lambda$ of our decomposition, then for $X \in \mathfrak{B}_\mu$ we have

$$AX \in \mathfrak{B}_\lambda \mathfrak{B}_\mu \subset \mathfrak{B}_\nu, \qquad BX \in \mathfrak{B}_\lambda \mathfrak{B}_\mu \subset \mathfrak{B}_\nu,$$

i.e., $AX \sim BX(\mathfrak{n})$. Analogously one proves stability to the left. Thus, $\mathfrak{n}$ turns out to be a two-sidedly stable equivalence, and the set of components $\mathfrak{B}_\xi$ of the decomposition $\mathfrak{n}$ is a set of elements of a factor-semigroup $\mathfrak{A}/\mathfrak{n}$. This shows that the preceding considerations may be carried over in their entirety to the language of decompositions of semigroups having the property indicated above.

2.14. For groups, the characterization of the structure of an arbitrary two-sidedly stable decomposition presents no difficulty. In any book on the theory of groups there is a full exposition on the question of the structure of arbitrary homomorphisms of groups. But for arbitrary semigroups the corresponding question presents significantly greater difficulty and is still far from a complete solution. Only for separate special classes of semigroups has the question of the complete characterization of all the two-sidedly stable decompositions been settled. In addition, the obtaining of the corresponding result presents significant difficulties. Below, in §§ 3 and 6, we present several of these results.

As an example, we present a construction considered by Pierce [**1**]. Suppose that $\mathfrak{K}$ is any subset of the semigroup $\mathfrak{A}$. The elements $A, B \in \mathfrak{A}$ are equivalent if for any $X, Y \in \mathfrak{A}$ it follows from $XAY \in \mathfrak{K}$ that $XBY \in \mathfrak{K}$ and conversely.

The two-sided stability of this relation is verified without difficulty. A particularly interesting case occurs when $\mathfrak{K}$ is an ideal. The consideration of such equivalences for a commutative semigroup of idempotents, adjoint to the distributive lattice $\mathfrak{L}$ (II, 4.3; II, 4.13), turns out to be useful in the study of lattice homomorphisms of the lattice $\mathfrak{L}$ (Pierce [**1**]).

3. Homomorphisms of Inverse Semigroups

3.1. In this section we consider homomorphisms of inverse semigroups, i.e., the semigroups which we defined and whose properties we considered in § 7 of the second chapter. The significance of these semigroups in connection with their role in the theory of partial transformations was considered by us in § 8 of the same chapter. We shall succeed in obtaining a number of properties of homomorphisms of inverse semigroups and in giving a complete description of the structure of arbitrary homomorphisms of these semigroups. The fundamental property of homomorphisms was obtained first by V. V. Vagner [**3**]. Somewhat later a detailed description of the structure of homomorphisms was published by Preston [**1**].

3.2. As usual, in the consideration of inverse semigroups we shall use the bar to indicate an element regularly adjoint to a given element.

LEMMA. *If φ is a homomorphism of an inverse semigroup $\mathfrak{A}$, and if $\varphi(A)$ ($A \in \mathfrak{A}$) is an idempotent of the semigroup $\varphi(\mathfrak{A})$, then $\mathfrak{A}$ contains an idempotent I for which $\varphi(I) = \varphi(A)$.*

PROOF. Since the element $\varphi(A)$ is an idempotent, we have

$$\varphi(A) \cdot \varphi(A\bar{A}) = \varphi(A) \cdot \varphi(A) \cdot \varphi(\bar{A}) = \varphi(A) \cdot \varphi(\bar{A}) = \varphi(A\bar{A}).$$

Because $A\bar{A}$ and $\bar{A}A$ are idempotents of the inverse semigroup $\mathfrak{A}$ (III, 7.3), they commute (II, 7.4). Therefore, using the equation just obtained, we find

$$\begin{aligned}\varphi(A\bar{A}) \cdot \varphi(\bar{A}A) &= \varphi(A) \cdot \varphi(A\bar{A}) \cdot \varphi(\bar{A}A) = \varphi(A) \cdot \varphi(\bar{A}A) \cdot \varphi(A\bar{A}) \\ &= \varphi(A\bar{A}A) \cdot \varphi(A\bar{A}) = \varphi(A) \cdot \varphi(A\bar{A}) = \varphi(A\bar{A}).\end{aligned}$$

On the other hand,

$$\begin{aligned}\varphi(A\bar{A}) \cdot \varphi(\bar{A}A) &= \varphi(\bar{A}A) \cdot \varphi(A\bar{A}) = \varphi(\bar{A}) \cdot \varphi(A) \cdot \varphi(A) \cdot \varphi(A) \\ &= \varphi(\bar{A}) \cdot \varphi(A) \cdot \varphi(\bar{A}) = \varphi(\bar{A}A\bar{A}) = \varphi(\bar{A}).\end{aligned}$$

Thus,
$$\varphi(A\bar{A}) = \varphi(\bar{A}).$$
Hence for the idempotent $I = \bar{A}A$ we obtain
$$\varphi(I) = \varphi(\bar{A}A) = \varphi(\bar{A}) \cdot \varphi(A) = \varphi(A\bar{A}) \cdot \varphi(A) = \varphi(A\bar{A}A) = \varphi(A).$$

3.3. With the aid of this lemma it is easy to see that the class of inverse semigroups is closed with respect to the operation of taking homomorphisms.

THEOREM. *If φ is a homomorphism of the inverse semigroup $\mathfrak{A}$, then $\varphi(\mathfrak{A})$ is also an inverse semigroup.*

PROOF. If I is a regular left identity of the element A (II, 6.2), then evidently $\varphi(I)$ is a regular left identity of the element $\varphi(A)$. In order to prove that $\varphi(\mathfrak{A})$ is an inverse semigroup it is sufficient, in view of Theorem II, 7.4, to show that any two of its idempotents commute with each other.

If λ_1 and λ_2 are idempotents of $\varphi(\mathfrak{A})$, then from 3.2 there are idempotents I_1 and I_2 in $\mathfrak{A}$ such that
$$\varphi(I_1) = \lambda_1, \qquad \varphi(I_2) = \lambda_2.$$
Because of the commutativity of I_1 and I_2 we obtain
$$\lambda_1 \cdot \lambda_2 = \varphi(I_1) \cdot \varphi(I_2) = \varphi(I_1 I_2) = \varphi(I_2 I_1) = \varphi(I_2) \cdot \varphi(I_1) = \lambda_2 \cdot \lambda_1.$$

3.4. COROLLARY. *If φ is a homomorphism of an inverse semigroup $\mathfrak{A}$, then for each $X \in \mathfrak{A}$ the equation*
$$\overline{\varphi(X)} = \varphi(\bar{X})$$
holds in the inverse semigroup $\varphi(\mathfrak{A})$.

Indeed,
$$\varphi(X) \cdot \varphi(\bar{X}) \cdot \varphi(X) = \varphi(X\bar{X}X) = \varphi(X),$$
$$\varphi(\bar{X}) \cdot \varphi(X) \cdot \varphi(\bar{X}) = \varphi(\bar{X}X\bar{X}) = \varphi(\bar{X}),$$
which means that $\varphi(\bar{X}) = \overline{\varphi(X)}$.

3.5. COROLLARY. *Suppose that φ is a homomorphism of the inverse semigroup $\mathfrak{A}$ and that $\lambda = \varphi(A)$ $(A \in \mathfrak{A})$ is an idempotent of the semigroup $\varphi(\mathfrak{A})$. Then the family $\mathfrak{B}_\lambda$ of all those elements B of $\mathfrak{A}$ for which $\varphi(B) = \lambda$ is an inverse semigroup.*

PROOF. If $B, B' \in \mathfrak{B}_\lambda$, then
$$\varphi(BB') = \varphi(B) \cdot \varphi(B') = \lambda \cdot \lambda = \lambda,$$
i.e., $BB' \in \mathfrak{B}_\lambda$. Thus, $\mathfrak{B}_\lambda$ is a subsemigroup of $\mathfrak{A}$.

Because of 3.4,
$$\varphi(\bar{B}) = \overline{\varphi(B)} = \bar{\lambda} = \lambda,$$
i.e., $\bar{B} \in \mathfrak{B}_\lambda$. The element $B\bar{B} \in \mathfrak{B}_\lambda$ is a regular left identity for B. Also, since idempotents of $\mathfrak{B}_\lambda$, as well as more generally the idempotents of $\mathfrak{A}$, commute with each other, it follows that $\mathfrak{B}_\lambda$, from II, 7.4, is an inverse semigroup.

3.6. COROLLARY. *Let φ be a homomorphism of the inverse semigroup $\mathfrak{A}$. If, for certain $A, B \in \mathfrak{A}$,*

$$\varphi(A\bar{A}) = \varphi(B\bar{B}) = \varphi(A\bar{B}),$$

then

$$\varphi(A) = \varphi(B), \qquad \varphi(\bar{A}) = \varphi(\bar{B}).$$

PROOF. Since

$$\varphi(A)\cdot\varphi(\bar{B})\cdot\varphi(A) = \varphi(A\bar{B})\cdot\varphi(A) = \varphi(A\bar{A})\cdot\varphi(A) = \varphi(A\bar{A}\bar{A}) = \varphi(A),$$

$$\varphi(\bar{B})\cdot\varphi(A)\cdot\varphi(\bar{B}) = \varphi(\bar{B})\cdot\varphi(A\bar{B}) = \varphi(\bar{B})\cdot\varphi(B\bar{B}) = \varphi(\bar{B}B\bar{B}) = \varphi(\bar{B}),$$

in the inverse semigroup $\varphi(\mathfrak{A})$ we have

$$\overline{\varphi(A)} = \varphi(\bar{B}).$$

Hence, from 3.4, we obtain

$$\varphi(\bar{A}) = \varphi(\bar{B}),$$

$$\varphi(A) = \overline{\overline{\varphi(A)}} = \overline{\varphi(\bar{A})} = \overline{\varphi(\bar{B})} = \overline{\overline{\varphi(B)}} = \varphi(B).$$

3.7. Suppose that in the inverse semigroup $\mathfrak{A}$ we are given a family Σ of some of its inverse subsemigroups. In what follows we shall be interested in the case when Σ has the following properties.

(α) *Distinct semigroups of Σ are disjoint.*

(β) *Each idempotent of $\mathfrak{A}$ is in one of the semigroups of Σ.*

(γ) *Suppose that $A\bar{A}, B\bar{B}, A\bar{B} \in \mathfrak{B}$ ($A, B \in \mathfrak{A}$), where $\mathfrak{B}$ is in Σ. If $A \in \mathfrak{B}$, then also $B \in \mathfrak{B}$.*

(δ) *Suppose that $A\bar{A}, B\bar{B}, A\bar{B} \in \mathfrak{B}$ ($A, B \in \mathfrak{A}$), where $\mathfrak{B}$ is in Σ. Then for each $\mathfrak{B}'$ of Σ there is in Σ a $\mathfrak{B}''$ such that*

$$A\mathfrak{B}'\bar{A} \subset \mathfrak{B}'', \qquad A\mathfrak{B}'\bar{B} \subset \mathfrak{B}''.$$

3.8. The appearance of conditions 3.7 is explained in the first place by the following.

Suppose that φ is any homomorphism of the inverse semigroup $\mathfrak{A}$ and that Γ is the set of idempotents of the semigroup $\varphi(\mathfrak{A})$. Because of 3.5 one may associate with the homomorphism φ a collection Σ of inverse subsemigroups $\mathfrak{B}_\lambda$ ($\lambda \in \Gamma$) of the semigroup $\mathfrak{A}$, which are complete preimages of the idempotents of the semigroup $\varphi(\mathfrak{A})$. It turns out that this family Σ has all four properties indicated in 3.7.

Property (α) is valid, since different $\mathfrak{B}_\lambda$ of Σ are complete preimages of different idempotents of Γ. (β) is valid, since under a homomorphism each idempotent maps onto an idempotent. (γ) immediately follows from 3.6. Property (δ) requires rather more detailed analysis.

Let $A\bar{A}, B\bar{B}, A\bar{B} \in \mathfrak{B}_\lambda$ ($\lambda \in \Gamma$). For $\mathfrak{B}_\mu$ ($\mu \in \Gamma$), using 3.6, we obtain

$$\varphi(A\mathfrak{B}_\mu\bar{A}) = \varphi(A)\cdot\varphi(\mathfrak{B}_\mu)\cdot\varphi(\bar{A}) = \varphi(A)\cdot\mu\cdot\varphi(\bar{A}) = \nu,$$

$$\varphi(A\mathfrak{B}_\mu\bar{B}) = \varphi(A)\cdot\varphi(\mathfrak{B}_\mu)\cdot\varphi(\bar{B}) = \varphi(A)\cdot\mu\cdot\varphi(\bar{A}) = \nu.$$

Because of the commutativity of idempotents in $\varphi(\mathfrak{A})$ (3.3),

$$\nu^2 = \varphi(A)\cdot\mu\cdot\varphi(\bar{A})\cdot\varphi(A)\cdot\mu\cdot\varphi(\bar{A}) = \varphi(A)\cdot\mu\cdot\varphi(\bar{A}A)\cdot\mu\cdot\varphi(\bar{A})$$
$$= \varphi(A)\cdot\varphi(\bar{A}A)\cdot\mu\cdot\mu\cdot\varphi(\bar{A}) = \varphi(A\bar{A}A)\cdot\mu\cdot\varphi(\bar{A}) = \varphi(A)\cdot\mu\cdot\varphi(\bar{A}) = \nu.$$

Accordingly, $\nu \in \Gamma$ and therefore

$$A\mathfrak{B}_\mu\bar{A} \subset \mathfrak{B}_\nu, \qquad A\mathfrak{B}_\mu\bar{B} \subset \mathfrak{B}_\nu.$$

3.9. Let us associate with the homomorphism φ of the inverse semigroup $\mathfrak{A}$ a relation $\mathfrak{n}$ such that

$$A \sim B(\mathfrak{n}) \qquad (A, B \in \mathfrak{A})$$

holds if and only if all three products are contained in one and the same subsemigroup $\mathfrak{B}_\lambda$ $(\lambda \in \Gamma)$ of Σ (3.8).

We shall show that $\mathfrak{n}$ is the equivalence corresponding to the homomorphism φ (2.1).

If $\varphi(A) = \varphi(B)$ $(A, B \in \mathfrak{A})$, then, from 3.4, $\varphi(\bar{A}) = \overline{\varphi(A)} = \overline{\varphi(B)} = \varphi(\bar{B})$, from which it easily follows that

$$\varphi(A\bar{A}) = \varphi(B\bar{B}) = \varphi(A\bar{B}),$$

i.e., $A\bar{A}$, $B\bar{B}$, $A\bar{B}$ lie in one and the same subsemigroup of Σ. Accordingly, $A \sim B(\mathfrak{n})$.

If $A \sim B(\mathfrak{n})$, then $\varphi(A\bar{A}) = \varphi(B\bar{B}) = \varphi(A\bar{B})$, and, from 3.6, $\varphi(A) = \varphi(B)$.

3.10. Now suppose that Σ is any collection of inverse subsemigroups of the inverse semigroup $\mathfrak{A}$, having the four properties 3.7, (α), (β), (γ), (δ).

Using Σ we define in $\mathfrak{A}$ a relation $\mathfrak{n}_\Sigma$ such that

$$A \sim B(\mathfrak{n}_\Sigma) \qquad (A, B \in \mathfrak{A})$$

if and only if all three products $A\bar{A}$, $B\bar{B}$, $A\bar{B}$ are contained in one and the same subsemigroup of Σ.

We shall prove that *such a relation* $\mathfrak{n}_\Sigma$ *is a two-sidedly stable equivalence.*

Since $A\bar{A}$ $(A \in \mathfrak{A})$ is an idempotent it follows from 3.7, (β) that $A \sim A$, i.e., $\mathfrak{n}_\Sigma$ is reflexive.

We shall prove the symmetry of $\mathfrak{n}_\Sigma$. If $A \sim B(\mathfrak{n}_\Sigma)$, that is,

$$A\bar{A},\ B\bar{B},\ A\bar{B} \in \mathfrak{B} \qquad (\mathfrak{B} \in \Sigma),$$

then from the fact that $\mathfrak{B}$ is an inverse subsemigroup of $\mathfrak{A}$ it follows that the element $\overline{A\bar{B}} = B\bar{A}$ also must be contained in $\mathfrak{B}$. Since $A\bar{A}$ and $B\bar{B}$ are contained in $\mathfrak{B}$, we have $B \sim A(\mathfrak{n}_\Sigma)$.

We shall prove the transitivity of $\mathfrak{n}_\Sigma$. Suppose that

$$A \sim B(\mathfrak{n}_\Sigma), \qquad B \sim C(\mathfrak{n}_\Sigma).$$

Then, by the definition of $\mathfrak{n}_\Sigma$, using also the symmetry property already proved, we find that for some $\mathfrak{B}' \in \Sigma$,

$$A\bar{A},\ B\bar{B},\ C\bar{C},\ A\bar{B},\ B\bar{A},\ B\bar{C},\ C\bar{B} \in \mathfrak{B}'.$$

From $B\bar{C} \cdot C\bar{B} \in \mathfrak{B}'$ and 3.7, (δ) we obtain

$$B\bar{C}C\bar{A} \in \mathfrak{B}',$$

$$A\bar{C}C\bar{B} = \overline{B\bar{C}C\bar{A}} \in \mathfrak{B}',$$

$$A\bar{C}C\bar{A} \in \mathfrak{B}'.$$

From the inclusions just obtained and 3.5, (γ) applied to $B\bar{C}$ and $A\bar{C}$, we obtain $AC \in \mathfrak{B}'$. Taken along with $A\bar{A}$, $C\bar{C} \in \mathfrak{B}'$ this means that $A \sim C(\mathfrak{n}_\Sigma)$.

We shall prove the two-sided stability of $\mathfrak{n}_\Sigma$. Suppose $A \sim B(\mathfrak{n}_\Sigma)$. In view of the symmetry of $\mathfrak{n}_\Sigma$ this means that for some $\mathfrak{B}_1 \in \Sigma$,

$$A\bar{A},\ B\bar{B},\ A\bar{B},\ B\bar{A} \in \mathfrak{B}_1.$$

Let $X \in \mathfrak{A}$. From condition 3.7, (δ) it follows that for some $\mathfrak{B}_2 \in \Sigma$ we have $X\mathfrak{B}_1\bar{X} \in \mathfrak{B}_2$. Hence

$$XA \cdot \overline{XA} = XA\bar{A}\bar{X} \in X\mathfrak{B}_1\bar{X} \subset \mathfrak{B}_2;$$

$$XB \cdot \overline{XB} = XB\bar{B}\bar{X} \in X\mathfrak{B}_1\bar{X} \subset \mathfrak{B}_2;$$

$$XA \cdot \overline{XB} = XA\bar{B}\bar{X} \in X\mathfrak{B}_1\bar{X} \subset \mathfrak{B}_2.$$

But this means that $XA \sim XB(\mathfrak{n}_\Sigma)$.

The idempotent $X\bar{X}$ is contained in some $\mathfrak{B}_3 \in \Sigma$ (3.7,(β)). Because of 3.7, (δ), for some $\mathfrak{B}_4$ and $\mathfrak{B}_5$ of Σ we have

$$A\mathfrak{B}_3\bar{A} \subset \mathfrak{B}_4, \qquad A\mathfrak{B}_3\bar{B} \subset \mathfrak{B}_4, \qquad B\mathfrak{B}_3\bar{B} \subset \mathfrak{B}_5, \qquad B\mathfrak{B}_3\bar{A} \subset \mathfrak{B}_5.$$

$\mathfrak{B}_4$ contains $AX\bar{X}\bar{B}$, and therefore also $\overline{AX\bar{X}\bar{B}} = BX\bar{X}\bar{A}$. But $BX\bar{X}\bar{A} \in \mathfrak{B}_5$. Therefore, from 3.7, ($\alpha$), $\mathfrak{B}_4 = \mathfrak{B}_5$. Hence we obtain

$$AX \cdot \overline{AX} = AX\bar{X}\bar{A} \in A\mathfrak{B}_3\bar{A} \subset \mathfrak{B}_5,$$

$$BX \cdot \overline{BX} = BX\bar{X}\bar{B} \in B\mathfrak{B}_3\bar{B} \subset \mathfrak{B}_5,$$

$$AX \cdot \overline{BX} = AX\bar{X}\bar{B} \in A\mathfrak{B}_3\bar{B} \subset \mathfrak{B}_5,$$

which means that $AX \sim BX(\mathfrak{n}_\Sigma)$.

3.11. The arguments in 3.8 and 3.9 show that an equivalence corresponding to a homomorphism of the inverse semigroup $\mathfrak{A}$ lies in the family of relations of type $\mathfrak{n}_\Sigma$ defined in 3.10. Since each two-sidedly stable equivalence is an equivalence corresponding to some homomorphism (2.6), it follows that *the two-sidedly stable equivalences of type* $\mathfrak{n}_\Sigma$ *described in* 3.10 *exhaust all the two-sidedly stable equivalences of an inverse semigroup.* Each of them is given by a collection of inverse subsemigroups satisfying the conditions 3.7, (α), (β), (γ), (δ).

Having characterized the structure of any two-sidedly stable equivalence of an inverse semigroup, we may say that we have described up to an inessential distinction (1.14) all homomorphisms of inverse semigroups.

3.12. In connection with the above-mentioned role of the relations of type $\mathfrak{n}_\Sigma$ described in 3.10, we shall go further into some of their properties.

(ε) *If* $A \sim B(\mathfrak{n}_\Sigma)$ $(A, B \in \mathfrak{A})$ *and* $A \in \mathfrak{B}$ $(\mathfrak{B} \in \Sigma)$, *then* $B \in \mathfrak{B}$.

Indeed, from $A \in \mathfrak{B}$ it follows that $\bar{A} \in \mathfrak{B}$ and $A\bar{A} \in \mathfrak{B}$, and therefore also $B\bar{B}, A\bar{B} \in \mathfrak{B}$. Hence from 3.7, ($\gamma$) we obtain $B \in \mathfrak{B}$.

(ζ) *For* $\mathfrak{B}_1, \mathfrak{B}_2 \in \Sigma$ *one can always find a* $\mathfrak{B}_3 \in \Sigma$ *such that*

$$\mathfrak{B}_1\mathfrak{B}_2 \subset \mathfrak{B}_3, \qquad \mathfrak{B}_2\mathfrak{B}_1 \subset \mathfrak{B}_3.$$

Indeed, since $\mathfrak{B}_1$ is an inverse semigroup, it has an idempotent I_1. Also $\mathfrak{B}_2$ has an idempotent I_2. In $\mathfrak{A}$ the idempotents commute. Therefore $I_3 = I_1I_2 = I_2I_1$ is also an idempotent and, from 3.7, (β), is contained in some subsemigroup $\mathfrak{B}_3 \in \Sigma$.

Any B_1 of $\mathfrak{B}_1$ is evidently equivalent to I_1, i.e., $B_1 \sim I_1(\mathfrak{n}_\Sigma)$. Also $B_2 \sim I_2(\mathfrak{n}_\Sigma)$. In view of the two-sided stability of $\mathfrak{n}_\Sigma$, we have $B_1B_2 \sim I_1I_2(\mathfrak{n}_\Sigma)$. But $I_1I_2 = I_3 \in \mathfrak{B}_3$, and therefore, on the basis of (ε), $B_1B_2 \in \mathfrak{B}_3$. Accordingly, $\mathfrak{B}_1\mathfrak{B}_2 \subset \mathfrak{B}_3$. Analogously we prove that $\mathfrak{B}_2\mathfrak{B}_1$ is contained in $\mathfrak{B}_4$, which contains I_2I_1. But

$$I_2I_1 = I_1I_2 = I_3.$$

Accordingly, $\mathfrak{B}_4 = \mathfrak{B}_3$ and $\mathfrak{B}_2\mathfrak{B}_1 \subset \mathfrak{B}_3$.

(η) *The union* $\mathfrak{N}$ *of all* $\mathfrak{B}$ *of* Σ *is an inverse subsemigroup of* $\mathfrak{A}$.

Indeed, from (ζ) it follows immediately that the product of two arbitrary elements of $\mathfrak{N}$ is contained in $\mathfrak{N}$. Each X of $\mathfrak{N}$ lies in some $\mathfrak{B}$ of Σ and therefore has in $\mathfrak{B}$, and accordingly in $\mathfrak{N}$ as well, a left regular identity. All the idempotents of $\mathfrak{N}$ commute, since they are idempotents of the inverse semigroup $\mathfrak{A}$. Accordingly the semigroup $\mathfrak{A}$ is inverse (II, 7.4).

3.13. Since the equivalence $\mathfrak{n}_\Sigma$ is two-sidedly stable, it is an equivalence corresponding to some homomorphism φ. From 3.2, for any idempotent λ of $\varphi(\mathfrak{A})$ there is an idempotent $I \in \mathfrak{A}$ such that $\varphi(I) = \lambda$. But I, from 3.7, (β), is contained in some $\mathfrak{B} \in \Sigma$. From (3.12), (ε), $\mathfrak{B}$ is a complete preimage of the idempotent λ under the homomorphism φ. Thus $\mathfrak{B}$ is $\mathfrak{B}_\lambda$ in the notation of 3.5. In addition, each $\mathfrak{B}$ of Σ is a complete preimage of some idempotent, namely, the idempotent $\varphi(I)$, where I is some idempotent contained in $\mathfrak{B}$ ($\mathfrak{B}$, as an inverse semigroup, necessarily has idempotents). Thus the family Σ is the collection of all complete preimages of idempotents of the semigroup $\varphi(\mathfrak{A})$, where φ is the homomorphism to which corresponds the equivalence $\mathfrak{n}_\Sigma$. In other words, *every family* Σ *of type* 3.7 *is a family of type* 3.8, *defined by some homomorphism of the semigroup* $\mathfrak{A}$.

From what has been said it follows in particular that for two different families Σ_1 and Σ_2 the homomorphisms φ_1 and φ_2 to which the equivalences $\mathfrak{n}_{\Sigma_1}$ and $\mathfrak{n}_{\Sigma_2}$ correspond are essentially distinct (1.14).

In view of 3.11 one may state that *every homomorphism of an inverse semigroup, up to inessential distinctions* (1.14), *is completely defined by giving the collection of complete preimages of the idempotents.*

3.14. We note that giving the family Σ of type 3.7 may be done by giving the inverse subsemigroup $\mathfrak{N}$ (3.12, (η)) of the semigroup $\mathfrak{A}$ containing all the idempotents of $\mathfrak{A}$ and by indicating a decomposition into inverse subsemigroups which have properties 3.7, (γ), (δ), (there is always such a decomposition, if perhaps trivial, i.e., consisting of only one component). Taking into account 3.12, (ζ), the indicated decomposition $\mathfrak{N}$ forms what in the following chapter we shall call a commutative band.

3.15. One can indicate a different approach to the decomposition $\mathfrak{N}$ just considered. Denote by $\mathfrak{H}$ the collection of all idempotents of the semigroup $\mathfrak{A}$. Since every $\mathfrak{B}$ of Σ (i.e., component in the indicated decomposition of the semigroup $\mathfrak{N}$) is an inverse semigroup, it must have idempotents. In this connection the decomposition of $\mathfrak{N}$ into the union of all the $\mathfrak{B}$ ($\mathfrak{B} \in \Sigma$) induces a corresponding decomposition of $\mathfrak{H}$ into the union of all $\mathfrak{B}'$, where $\mathfrak{B}' = \mathfrak{H} \cap \mathfrak{B} \neq \varnothing$. This decomposition of a commutative semigroup of idempotents of $\mathfrak{H}$, along with the giving of the semigroup $\mathfrak{N}$, completely defines the collection Σ. Indeed, any $\mathfrak{B}$ of Σ is evidently just the set of all of those elements X of $\mathfrak{N}$ for which $X\bar{X} \in \mathfrak{B}'$.

4. Normal Complexes

4.1. In connection with the concepts introduced in § 2, there naturally arises the question as to which subsets of a semigroup may be equivalence classes of elements for some two-sidedly stable equivalence relation. In other words, when is a subset a complete preimage of one element under some homomorphism of the semigroup? For the solution of this question E. S. Ljapin [5] introduced the following concept.

DEFINITION. *A nonempty subset $\mathfrak{K}$ of the semigroup $\mathfrak{A}$ is said to be a* NORMAL COMPLEX *if, for any X and Y which are elements of $\mathfrak{A}$ or empty symbols and for any $K, K' \in \mathfrak{K}$, $XKY \in \mathfrak{K}$ always implies $XK'Y \in \mathfrak{K}$.*

We note in passing that A. I. Mal'cev [7] discovered that it was possible to generalize this concept using the following link, discovered by him, with the theory of homomorphisms for general algebraic systems.

4.2. Among the normal complexes there evidently lie the semigroup itself, each of its separate elements, and every two-sided ideal of the semigroup. If the subset $\mathfrak{K} \subset \mathfrak{A}$ is such that the intersections $\mathfrak{K}\mathfrak{A} \cap \mathfrak{K}$, $\mathfrak{A}\mathfrak{K} \cap \mathfrak{K}$, $\mathfrak{A}\mathfrak{K}\mathfrak{A} \cap \mathfrak{K}$ are empty (the so-called two-sided anti-ideal), then $\mathfrak{K}$ is evidently a normal complex.

A nonempty intersection of any set of normal complexes is a normal complex.

If $\mathfrak{A}$ is a group and $\mathfrak{N}$ a normal divisor for $\mathfrak{A}$ (i.e., a subgroup such that $X^{-1}NX \in \mathfrak{N}$ for any $X \in \mathfrak{A}$ and $N \in \mathfrak{N}$), then a set of form $\mathfrak{K} = A\mathfrak{N}$ (the so-called coset modulo the normal divisor) is a normal complex. Indeed, suppose that $K = AN_1$, $K' = AN_2$ ($N_1, N_2 \in \mathfrak{N}$) and $K \in \mathfrak{K}$. Then

$$XK'Y = XAN_2Y = XAN_1Y \cdot Y^{-1}N_1^{-1}N_2Y \in XKY\mathfrak{N},$$

and therefore $XKY \in \mathfrak{K} = A\mathfrak{N}$ implies $XK'Y \in \mathfrak{K}$.

It is easy to see that every normal complex $\mathfrak{K}$ of a group has the following structure. Choosing in $\mathfrak{K}$ some element $K_0 \in \mathfrak{K}$, we write $\mathfrak{N} = K_0^{-1}\mathfrak{K}$. For any

$$N_1 = K_0^{-1}K_1 \in \mathfrak{N}, \quad N_2 = K_0^{-1}K_2 \in \mathfrak{N}, \qquad X \in \mathfrak{A},$$

since

$$(K_1K_0^{-1}K_0), \quad (K_0K_1^{-1}K_1), \qquad (K_0X^{-1}K_0^{-1}K_0X) \in \mathfrak{K},$$

using the properties of a normal complex, we obtain

$$N_1N_2 = K_0^{-1}K_1K_0^{-1}K_2 = K_0^{-1} \cdot (K_1K_0^{-1}K_2) \in K_0^{-1}\mathfrak{K} = \mathfrak{N},$$

$$N_1^{-1} = K_1^{-1}K_0 = K_0^{-1} \cdot (K_0K_1^{-1}K_0) \in K_0^{-1}\mathfrak{K} = \mathfrak{N},$$

$$X^{-1}N_1X = K_0^{-1} \cdot (K_0X^{-1}K_0^{-1}K_1X) \in K_0^{-1}\mathfrak{K} = \mathfrak{N}.$$

4.3. Suppose that $\mathfrak{n}$ is some relation in the semigroup $\mathfrak{A}$ and that φ is a homomorphism of $\mathfrak{A}$ such that $A \sim B(\mathfrak{n})$ always implies $\varphi(A) = \varphi(B)$. The relation $\mathfrak{n}_\varphi$ is the equivalence corresponding to the homomorphism φ (2.1). Evidently $\mathfrak{n}_\varphi \geqslant \mathfrak{n}$. Hence from I, 5.23 it follows that for a second arbitrary relation $\mathfrak{n}''$ we have $\mathfrak{n}_\varphi \geqslant \mathfrak{n}''$. The relation $\mathfrak{n}''$ is the greatest lower bound of all the two-sidedly stable equivalences following $\mathfrak{n}$. If ξ is the homomorphism to which the equivalence $\mathfrak{n}''$ corresponds, then, from 2.8, ξ is a right divisor of φ.

From what has been said it follows that ξ, belonging to the set of homomorphisms such that $A \sim B(\mathfrak{n})$ always implies $\varphi(A) = \varphi(B)$, is the greatest common right divisor of that set.

4.4. Suppose that $\mathfrak{K}$ is some nonempty subset of the semigroup $\mathfrak{A}$. Define in $\mathfrak{A}$ a relation $\mathfrak{n}_\mathfrak{K}$, putting $A \sim B(\mathfrak{n}_\mathfrak{K})$ if $A = B$ or if A and B both belong to $\mathfrak{K}$. The relation $\mathfrak{n}''_\mathfrak{K}$ is the greatest lower bound of the two-sidedly stable equivalences following $\mathfrak{n}_\mathfrak{K}$ (I, 5.32). From 4.3, the homomorphism to which $\mathfrak{n}''_\mathfrak{K}$ corresponds is the greatest common right divisor for those homomorphisms φ for which $\varphi(A) = \varphi(B)$ for any $A, B \in \mathfrak{K}$.

4.5. The requirement that the subset $\mathfrak{K}$ of a semigroup $\mathfrak{A}$ should be a complete preimage of some element under some homomorphism of $\mathfrak{A}$ is equivalent, from 2.1 and 2.6, to the existence in $\mathfrak{A}$ of a two-sidedly stable equivalence $\mathfrak{n}$ under which $\mathfrak{K}$ is an $\mathfrak{n}$-class.

4.6. Theorem. *In order that the subset $\mathfrak{K}$ of the semigroup $\mathfrak{A}$ should be a complete preimage of one element under some homomorphism of $\mathfrak{A}$ it is necessary and sufficient that $\mathfrak{K}$ should be a normal complex* (4.1).

Proof. (1) Suppose that, for some homomorphism of $\mathfrak{A}$, $\mathfrak{K}$ is a complete preimage of one of the elements, i.e., that $\mathfrak{K}$ is an $\mathfrak{n}$-class for some two-sidedly stable equivalence $\mathfrak{n}$ (4.5). If $XKY \in \mathfrak{K}$, where $K \in \mathfrak{K}$ and X and Y belong to $\mathfrak{A}$ or are empty symbols, then $XKY \sim XK'Y(\mathfrak{n}')$ (I, 5.20), and since $\mathfrak{n}' = \mathfrak{n}$ (I, 5.23), therefore $XK'Y \in \mathfrak{K}$.

(2) Suppose that $\mathfrak{K}$ is a normal complex of $\mathfrak{A}$. By the definition of a normal complex, it follows from $A \sim B(\mathfrak{n}'_{\mathfrak{K}})$ (where $\mathfrak{n}_{\mathfrak{K}}$ is the relation defined in 4.4) and from $A \in \mathfrak{K}$ that $B \in \mathfrak{K}$. Therefore it also follows from $A \sim B(\mathfrak{n}''_{\mathfrak{K}})$ (I, 5.21) and $A \in \mathfrak{K}$ that $B \in \mathfrak{K}$. Since, in addition, all the elements of $\mathfrak{K}$ stand in the relation $\mathfrak{n}_{\mathfrak{K}}$ to each other, $\mathfrak{K}$ turns out to be a $\mathfrak{n}''_{\mathfrak{K}}$-class for the two-sidedly stable equivalence $\mathfrak{n}''_{\mathfrak{K}}$ (I, 5.22, θ). From 4.5, $\mathfrak{K}$ is the complete preimage of one element under some homomorphism.

4.7. Suppose that $\mathfrak{K}$ is a normal complex of the semigroup $\mathfrak{A}$. Denote by Ψ the set of all those homomorphisms ψ of the semigroup $\mathfrak{A}$ for which $\mathfrak{K}$ is a complete preimage of one of the elements of the semigroup $\psi(\mathfrak{A})$. For the relation $\mathfrak{n}_{\mathfrak{K}}$ (4.4) the natural homomorphism of $\mathfrak{A}$ onto $\mathfrak{A}/\mathfrak{n}''_{\mathfrak{K}}$ (and more generally any homomorphism to which the equivalence $\mathfrak{n}''_{\mathfrak{K}}$ corresponds) is, from 4.4, the greatest common right divisor of Ψ. In addition, it belongs to Ψ itself.

One may say that this homomorphism induces the minimum of identifications in $\mathfrak{A}$ necessary for the identification of all of the elements of $\mathfrak{K}$ to each other (2.5).

4.8. While clearing up the question of complete preimages of arbitrary elements under a homomorphism, we shall dwell particularly on idempotents.

If the normal complex $\mathfrak{K}$ of the semigroup $\mathfrak{A}$ is a subsemigroup of $\mathfrak{A}$, then for each homomorphism φ under which all the elements of $\mathfrak{K}$ map into one element, that element $\varphi(\mathfrak{K})$ is an idempotent of the semigroup $\varphi(\mathfrak{A})$. Indeed,

$$\varphi(\mathfrak{K}) \cdot \varphi(\mathfrak{K}) = \varphi(\mathfrak{K}\mathfrak{K}),$$

and since $\mathfrak{K}\mathfrak{K} \subset \mathfrak{K}$, therefore $\varphi(\mathfrak{K}) \cdot \varphi(\mathfrak{K}) = \varphi(\mathfrak{K})$.

If now $\mathfrak{K}$ is a normal complex of $\mathfrak{A}$ for which, under some homomorphism ψ, $\mathfrak{K}$ is a complete preimage of an idempotent of the semigroup $\psi(\mathfrak{A})$, then $\mathfrak{K}$ is a subsemigroup of the semigroup $\mathfrak{A}$. Indeed,

$$\psi(\mathfrak{K}) = \psi(\mathfrak{K}) \cdot \psi(\mathfrak{K}) = \psi(\mathfrak{K}\mathfrak{K})$$

and therefore $\mathfrak{K}\mathfrak{K} \subset \mathfrak{K}$.

4.9. Among the idempotents a particular role is played by the identity and by the zero of a semigroup. In order to determine what are their complete preimages under homomorphisms, we must consider two-sided ideals and the so-called normal subsemigroups (or normal subsystems) introduced by E. S. Ljapin [**2**; **5**].

DEFINITION. *A nonempty subset $\mathfrak{K}$ of the semigroup $\mathfrak{A}$ is said to be a* NORMAL SUBSEMIGROUP *if for any X and Y which are elements of $\mathfrak{A}$ or empty symbols, and for any K and K' lying in $\mathfrak{K}$ or being empty symbols, $XKY \in \mathfrak{K}$ always implies $XK'Y \in \mathfrak{K}$ (given only that X, K', Y are not all empty symbols).*

If $\mathfrak{A}$ is a group, then its normal subsemigroups are called normal divisors. Their numerous important properties are investigated in the theory of groups.

4.10. As is immediately clear, normal subsemigroups are subsemigroups and normal complexes. However, one must keep in mind that a subsemigroup which is a normal complex is not always a normal subsemigroup. If the semigroup has an identity, then its normal subsemigroups are those normal complexes which contain the identity.

A nonempty intersection of any set of normal subsemigroups is a normal subsemigroup.

4.11. THEOREM. *In order that the subset $\mathfrak{K}$ of the semigroup $\mathfrak{A}$ should under some homomorphism φ of the semigroup $\mathfrak{A}$ be a complete preimage of the identity of the semigroup $\varphi(\mathfrak{A})$ it is necessary and sufficient that $\mathfrak{K}$ be a normal subsemigroup.*

PROOF. (1) If, under the homomorphism φ, the set $\mathfrak{K} \subset \mathfrak{A}$ is a complete preimage of the identity of the semigroup $\varphi(\mathfrak{A})$, then $\mathfrak{K}$ is a normal complex. Therefore, if K and K' both belong to $\mathfrak{K}$, then $XKY \in \mathfrak{N}$ implies $XK'Y \in \mathfrak{K}$. If K' is an empty symbol, and $K \in \mathfrak{K}$, i.e., $\varphi(K) = E_{\varphi(\mathfrak{A})}$, then

$$\varphi(XKY) = \varphi(X) \cdot \varphi(K) \cdot \varphi(Y) = \varphi(X) \cdot \varphi(Y) = \varphi(XY).$$

Therefore if one of the elements XKY or XY belongs to $\mathfrak{K}$, i.e., maps onto $E_{\varphi(\mathfrak{A})}$, then also the second will belong to $\mathfrak{K}$.

(2) Suppose that $\mathfrak{K}$ is a normal subsemigroup of $\mathfrak{A}$. We define in $\mathfrak{A}$ a relation $\mathfrak{m}_{\mathfrak{K}}$ by putting $A \sim B(\mathfrak{m}_{\mathfrak{K}})$ if for some X and Y belonging to $\mathfrak{A}$ or being empty symbols one has $A, B \in X\mathfrak{K}Y \cup XY$. It follows immediately from the definition of a normal subsemigroup that all the elements of $\mathfrak{K}$ stand to each other in the relation $\mathfrak{m}_{\mathfrak{K}}$, and from $A \sim B(\mathfrak{m}_{\mathfrak{K}})$ and $A \in \mathfrak{K}$ follows $B \in \mathfrak{K}$.

Evidently, the relation $\mathfrak{m}_{\mathfrak{K}}$ is symmetric, reflexive, and two-sidedly stable. Also it is evident that $\mathfrak{m}'_{\mathfrak{K}} = \mathfrak{m}_{\mathfrak{K}}$ (I, 5.20). Therefore $A \sim B(\mathfrak{m}''_{\mathfrak{K}})$ and $A \in \mathfrak{K}$ imply $B \in \mathfrak{K}$. Thus, $\mathfrak{K}$ is an $\mathfrak{m}''_{\mathfrak{K}}$-class. Since, evidently, for any $A \in \mathfrak{A}$ and $K \in \mathfrak{K}$,

$$AK \sim A(\mathfrak{m}_{\mathfrak{K}}), \qquad KA \sim A(\mathfrak{m}_{\mathfrak{K}}),$$

then under the homomorphism φ to which the equivalence $\mathfrak{m}''_{\mathfrak{K}}$ corresponds we have

$$\varphi(A) \cdot \varphi(\mathfrak{K}) = \varphi(A), \qquad \varphi(\mathfrak{K}) \cdot \varphi(A) = \varphi(A).$$

Accordingly, $\varphi(\mathfrak{K})$ is the identity of $\varphi(\mathfrak{A})$, and $\mathfrak{K}$ is the complete preimage of that identity.

4.12. Let $\mathfrak{K}$ be a normal subsemigroup of $\mathfrak{A}$. Suppose that φ is any homomorphism under which all the elements of $\mathfrak{K}$ map onto the identity of the semigroup $\varphi(\mathfrak{A})$. If $A \sim B(\mathfrak{m}_{\mathfrak{K}})$, where $\mathfrak{m}_{\mathfrak{K}}$ is the relation defined in the second part of the proof of Theorem 4.11, then evidently $\varphi(A) = \varphi(B)$. Hence it follows that the homomorphism ξ to which $\mathfrak{m}''_{\mathfrak{K}}$ corresponds is the greatest common right divisor of all those homomorphisms under which $\mathfrak{K}$ maps onto the identity.

The relations $\mathfrak{m}''_{\mathfrak{K}}$ and $\mathfrak{n}''_{\mathfrak{K}}$ (4.4) are evidently connected by the condition $\mathfrak{n}''_{\mathfrak{K}} \leqslant \mathfrak{m}''_{\mathfrak{K}}$. In addition, in special cases they may fail to coincide. This means that for certain normal subsemigroups there exist homomorphisms under which all the elements map into one and the same element which is not the identity.

As an example we mention a commutative semigroup $\mathfrak{A}$ consisting of the three elements A, B, O for which $A^2 = A$, $B^2 = B$ and $XY = O$ in all remaining cases.

The subset $\mathfrak{K}$, consisting of the single element A, is a normal subsemigroup. The isomorphisms are the greatest common right divisors of those homomorphisms under which $\mathfrak{K}$ is a complete preimage of one element. At the same time, under the isomorphism ψ, $\psi(A)$, of course, is not the identity of $\psi(\mathfrak{A})$. Since $AB = AO = O$, it follows that for every homomorphism φ under which $\varphi(A)$ is the identity of $\varphi(\mathfrak{A})$ we must have $\varphi(B) = \varphi(O)$. Hence we conclude that the homomorphism χ of the semigroup onto its subsemigroup $\{A, O\}$,

$$\chi(A) = A, \qquad \chi(B) = O, \qquad \chi(O) = O$$

is the greatest common right divisor of those homomorphisms for which A is a complete preimage of the identity.

4.13. Since every group has an identity, the question as to homomorphisms of a semigroup onto a group is connected with the properties of normal complexes which are preimages of the identity under homomorphisms. This problem was taken up by Dubreil [**1**], Levi [3], E. S. Ljapin [**2; 5**], L. M. Gluskin [**2**] and Stoll [**2**].

THEOREM. *If the semigroup $\mathfrak{A}$ has a homomorphism onto the group $\mathfrak{G}$, then the normal subsemigroup $\mathfrak{K}$ which is the complete preimage of $E_{\mathfrak{G}}$ has for any $A \in \mathfrak{A}$ the property*

$$A\mathfrak{A} \cap \mathfrak{K} \neq \varnothing, \qquad \mathfrak{A}A \cap \mathfrak{K} \neq \varnothing.$$

In turn, for every normal subsemigroup $\mathfrak{K}$ having the indicated property, there exist homomorphisms φ of the semigroup $\mathfrak{A}$ onto the group under which $\mathfrak{K}$ is the complete preimage of the identity of the group $\varphi(\mathfrak{A})$. In addition, all the groups $\varphi(\mathfrak{A})$, under all such possible homomorphisms φ, are isomorphic to one another.

PROOF. (1) Suppose that φ is a homomorphism of $\mathfrak{A}$ onto the group $\mathfrak{G}$ and that $\mathfrak{K}$ is the complete preimage of $E_{\mathfrak{G}}$. The element $\varphi(A)$ has in $\mathfrak{G} = \varphi(\mathfrak{A})$ an inverse element $\varphi(A')$ $(A' \in \mathfrak{A})$. Since

$$\varphi(AA') = \varphi(A) \cdot \varphi(A') = E_{\mathfrak{G}},$$

$$\varphi(A'A) = \varphi(A') \cdot \varphi(A) = E_{\mathfrak{G}},$$

it follows that $AA' \in \mathfrak{K}$ and $A'A \in \mathfrak{K}$, i.e.,

$$AA' \in A\mathfrak{A} \cap \mathfrak{K}, \qquad A'A \in \mathfrak{A}A \cap \mathfrak{K}.$$

(2) Suppose that the normal subsemigroup $\mathfrak{K}$ of the semigroup $\mathfrak{A}$ has the property stated in the formulation of the theorem and that φ is any homomorphism under which $\mathfrak{K}$ is a complete preimage of the identity of the semigroup

$\varphi(\mathfrak{A})$. Suppose that $\varphi(A)$ $(A \in \mathfrak{A})$ is any element of the semigroup $\varphi(\mathfrak{A})$. Since the intersection of $A\mathfrak{A}$ and $\mathfrak{K}$ is nonempty it follows that for some $A' \in \mathfrak{A}$,

$$AA' = K \in \mathfrak{K}.$$

This means that

$$\varphi(A) \cdot \varphi(A') = \varphi(K) = E_{\varphi(\mathfrak{A})},$$

i.e., $\varphi(A)$ has a right inverse with respect to the identity of the semigroup $\varphi(\mathfrak{A})$. From II, 2.18 it therefore follows that $\varphi(\mathfrak{A})$ is a group.

Suppose that φ and ψ are two homomorphisms of $\mathfrak{A}$ onto the group under each of which $\mathfrak{K}$ is a complete preimage of the identity. Suppose that for some $X, Y \in \mathfrak{A}$, $\varphi(X) = \varphi(Y)$. Denote by $\varphi(Z)$ the element inverse to $\varphi(X)$ in the group $\varphi(\mathfrak{A})$. Since

$$\varphi(XZ) = \varphi(X) \cdot \varphi(Z) = E_{\varphi(\mathfrak{A})},$$

$$\varphi(ZY) = \varphi(Z) \cdot \varphi(Y) = \varphi(Z) \cdot \varphi(X) = E_{\varphi(\mathfrak{A})}$$

it follows that $XZ \in \mathfrak{K}$ and $ZY \in \mathfrak{K}$. But $\psi(\mathfrak{K}) = E_{\psi(\mathfrak{A})}$, so that

$$\psi(X) = \psi(X) \cdot \psi(ZY) = \psi(XZY) = \psi(XZ) \cdot \psi(Y) = \psi(Y).$$

Analogously one proves that $\psi(X') = \psi(Y')$ always implies $\varphi(X') = \varphi(Y')$. Hence from 1.9, (ε), (ι) it follows that φ differs from ψ by a multiplier which is an isomorphism, so that the groups $\varphi(\mathfrak{A})$ and $\psi(\mathfrak{A})$ are isomorphic.

4.14. Let the normal subsemigroup $\mathfrak{K}$ of the semigroup $\mathfrak{A}$ be such that there exists a homomorphism of $\mathfrak{A}$ onto some group, under which $\mathfrak{K}$ is the complete preimage of unity. From the second part of the proof of Theorem 4.13 it is easy to see that for an arbitrary homomorphism φ of the semigroup $\mathfrak{A}$, where $\varphi(\mathfrak{K})$ is the identity of $\varphi(\mathfrak{A})$, the image $\varphi(\mathfrak{A})$ will be a group. Nevertheless, there may exist homomorphisms ψ such that $\mathfrak{K}$ is the complete preimage of one of the elements of $\psi(\mathfrak{A})$ but $\psi(\mathfrak{A})$ is not a group and $\psi(\mathfrak{K})$ is not the identity of the semigroup $\psi(\mathfrak{A})$.

As an example we give the commutative semigroup $\mathfrak{A}$ consisting of three elements with the following multiplication table:

	I	A	B
I	I	A	A
A	A	I	I
B	A	I	I

The subset $\mathfrak{K}$ consisting of the one element I is obviously a normal subsemigroup of $\mathfrak{A}$ satisfying the condition 4.13, but $\mathfrak{A}$ is not a group and I is not the identity of $\mathfrak{A}$ and the identity isomorphism is an example of a homomorphism ψ of the above-mentioned type. Together with this homomorphism let us consider the following homomorphism φ of the semigroup $\mathfrak{A}$ onto a subsemigroup $\{I, A\}$:

$$\varphi(I) = I, \quad \varphi(A) = \varphi(B) = A.$$

The set $\varphi(\mathfrak{A}) = (I, A)$ is a group and I is the complete preimage of the identity of this group under the homomorphism φ.

4.15. The same role as played by normal subsemigroups relative to unity is played by two-sided ideals relative to zero. The corresponding so-called ideal homomorphisms are constantly studied in various investigations of semigroups.

Suppose that $\mathfrak{T}$ is a two-sided ideal of the semigroup $\mathfrak{A}$. The relation $\mathfrak{n}_{\mathfrak{T}}$ (4.4), as one easily verifies, coincides with $\mathfrak{n}'_{\mathfrak{T}}$ and $\mathfrak{n}''_{\mathfrak{T}}$ (I, 5.20; I, 5.21). This is frequently called an *ideal equivalence*, corresponding to the given two-sided ideal $\mathfrak{T}$.

The factor-semigroup of $\mathfrak{A}$ modulo an ideal equivalence $\mathfrak{n}_{\mathfrak{T}}$ is, just like $\mathfrak{A}/\mathfrak{n}_{\mathfrak{T}}$, often denoted by $\mathfrak{A}/\mathfrak{T}$ or $\mathfrak{A}\backslash\mathfrak{T}$. It is called the *ideal factor-semigroup*, corresponding to the two-sided ideal $\mathfrak{T}$. Its elements are all the individual elements of $\mathfrak{A}$ not belonging to $\mathfrak{T}$ and the ideal $\mathfrak{T}$ itself, which is of course the zero of the semigroup $\mathfrak{A}/\mathfrak{n}_{\mathfrak{T}}$. One may represent $\mathfrak{A}/\mathfrak{n}_{\mathfrak{T}}$ as a semigroup obtained from $\mathfrak{A}$ by identifying with one another all the elements of the ideal $\mathfrak{T}$.

4.16. Suppose that $\mathfrak{T}$ is a two-sided ideal of the semigroup $\mathfrak{A}$ and $\mathfrak{K}$ is a subset of $\mathfrak{A}$ such that it contains either no elements or all elements of the ideal $\mathfrak{T}$. We denote by $\mathfrak{A}'$ the ideal factor-semigroup $\mathfrak{A}/\mathfrak{n}_{\mathfrak{T}}$ and by $\mathfrak{K}'$ a subset of it consisting of all the elements of $\mathfrak{K}$ not contained in $\mathfrak{T}$ (such elements are also elements of the semigroup $\mathfrak{A}'$), and of the element $\mathfrak{T}$ ($\mathfrak{T}$ is one of the elements of the semigroup $\mathfrak{A}/\mathfrak{n}_{\mathfrak{T}}$) if $\mathfrak{K} \supset \mathfrak{T}$. It is easy to convince oneself of the following:

(α) If $\mathfrak{K}$ is a subsemigroup of $\mathfrak{A}$, then $\mathfrak{K}'$ is a subsemigroup of $\mathfrak{A}'$.

(β) If $\mathfrak{K}$ is a left, right, or two-sided ideal of $\mathfrak{A}$, then $\mathfrak{K}'$ is a left, right, or two-sided ideal of $\mathfrak{A}'$.

(γ) If $\mathfrak{K}$ is an isolated or completely isolated subsemigroup in $\mathfrak{A}$ (IV, 6.1), then $\mathfrak{K}'$ is respectively isolated or completely isolated in $\mathfrak{A}'$.

(δ) If $\mathfrak{K}$ is a normal complex of $\mathfrak{A}$, then $\mathfrak{K}'$ is a normal complex of $\mathfrak{A}'$.

It is not difficult to see that no other subsemigroups, ideals, or normal complexes, except those obtained in a similar way from the various sets $\mathfrak{K}$, lie in the factor-semigroup $\mathfrak{A}'$.

4.17. Using the concept of ideal factor-semigroup, one may define ideal factors, as considered in § 4 of the fourth chapter.

Suppose that $\mathfrak{N}$ is a two-sided ideal layer of the semigroup $\mathfrak{A}$ (IV, 3.3), which is not a minimal two-sided ideal. From IV, 4.2 there is in $\mathfrak{A}$ a pair of two-sided ideals $\mathfrak{T}_1$ and $\mathfrak{T}_2$ such that

$$\mathfrak{N} = \mathfrak{T}_1\backslash\mathfrak{T}_2.$$

$\mathfrak{T}_2$ is a two-sided ideal also for $\mathfrak{T}_1$. Consider the ideal factor-semigroup $\mathfrak{T}_1/\mathfrak{n}_{\mathfrak{T}_2}$. If $\mathfrak{N}$ is not itself a semigroup, then it is immediately clear that the ideal factor $\mathfrak{N}^*$ will be the ideal factor-semigroup $\mathfrak{T}_1/\mathfrak{n}_{\mathfrak{T}_2}$ (if one identifies the element O^* of the factor with the element $\mathfrak{T}_2$ of the factor-semigroup $\mathfrak{T}_1/\mathfrak{n}_{\mathfrak{T}_2}$).

If $\mathfrak{N}$ is itself a semigroup (which means, of course, simply that $\mathfrak{T}_2$ is completely isolated in $\mathfrak{T}_1$), then $\mathfrak{T}_1/\mathfrak{n}_{\mathfrak{T}_2}$ is evidently the semigroup $\mathfrak{N}$ with the exterior adjunction of a zero (which is $\mathfrak{T}_2$). Discarding that exteriorly adjoined zero, we obtain the ideal factor $\mathfrak{N}^*$.

In using such an approach to ideal factors, one must bear in mind that the pair of adjacent ideals $\mathfrak{T}_1$ and $\mathfrak{T}_2$ are not in general defined uniquely for a given layer $\mathfrak{N}$. Nevertheless, the ideal factor $\mathfrak{N}^*$ is completely defined by the layer $\mathfrak{N}$ itself. Therefore, the approach used in § 4 of the fourth chapter gives a well-defined meaning to the assertion that the sets of factors of two principal two-sided ideal chains are identical (IV, 4.11).

4.18. If $\mathfrak{T}$ is a two-sided ideal of the semigroup $\mathfrak{A}$, then $\mathfrak{A}$ may be in a natural way regarded as made up of the two semigroups $\mathfrak{T}$ and $\mathfrak{A}/\mathfrak{n}_{\mathfrak{T}}$. The question as to what are the connections between the properties of these two semigroups and the semigroup $\mathfrak{A}$ itself, in analogy with the similar question in the theory of groups (relative to normal divisors) may be called the problem of ideal extensions. A paper of Clifford [**8**] and a paper of A. M. Kaufman [3] have been devoted to this question. As an illustration we consider the property of being inverse (II, 7.2).

THEOREM. *Suppose that $\mathfrak{T}$ is a two-sided ideal of the semigroup $\mathfrak{A}$. The semigroup $\mathfrak{A}$ will be inverse if and only if both of the semigroups $\mathfrak{T}$ and $\mathfrak{A}/\mathfrak{n}_{\mathfrak{T}}$ are inverse.*

PROOF. (1) Suppose that $\mathfrak{T}$ and $\mathfrak{A}/\mathfrak{n}_{\mathfrak{T}}$ are inverse. If $A \in \mathfrak{T}$, then since $\mathfrak{T}$ is inverse it contains a unique element regularly adjoint to $\mathfrak{A}$. At the same time, no element Z of $\mathfrak{A}\backslash\mathfrak{T}$ can be regularly adjoint to $\mathfrak{A}$, since

$$ZAZ \in \mathfrak{T}, \qquad ZAZ \neq Z.$$

If $A \in \mathfrak{A}\backslash\mathfrak{T}$, then since $\mathfrak{A}/\mathfrak{n}_{\mathfrak{T}}$ is inverse there is in $\mathfrak{A}\backslash\mathfrak{T}$ a unique element regularly adjoint to A (the element $\mathfrak{T}$ of the semigroup $\mathfrak{A}/\mathfrak{n}_{\mathfrak{T}}$ is the zero of that semigroup and therefore is regularly adjoint only to itself). As to the elements of $\mathfrak{T}$, none of them, as we have already seen, can be regularly adjoint to $A \in \mathfrak{A}\backslash\mathfrak{T}$.

(2) Suppose that the semigroup $\mathfrak{A}$ is inverse.

For $T \in \mathfrak{T}$ in $\mathfrak{A}$ there exists a unique regularly adjoint element. This element belongs to $\mathfrak{T}$ since no element of $\mathfrak{A}\backslash\mathfrak{T}$ can be regularly adjoint with an element of the two-sided ideal $\mathfrak{T}$. Accordingly, $\mathfrak{T}$ is an inverse semigroup.

For $Z \in \mathfrak{A}\backslash\mathfrak{T}$ there exists in $\mathfrak{A}$ a unique element regularly adjoint to it, which must belong to $\mathfrak{A}\backslash\mathfrak{T}$. This element will be the only one regularly adjoint to Z in the factor-semigroup $\mathfrak{A}/\mathfrak{n}_{\mathfrak{T}}$ since the zero $\mathfrak{T}$ of that semigroup is regularly adjoint only with itself. Accordingly, $\mathfrak{A}/\mathfrak{n}_{\mathfrak{T}}$ is inverse.

4.19. The natural homomorphism of $\mathfrak{A}$ onto the ideal factor-semigroup $\mathfrak{A}/\mathfrak{n}_{\mathfrak{T}}$ is frequently called the *ideal homomorphism corresponding to the two-sided ideal* $\mathfrak{T}$. Evidently it is an isomorphism (and in fact the identity isomorphism) if and only if $\mathfrak{T}$ consists of one element. Here one must recall that one element constitutes a two-sided ideal if and only if it is the zero of the semigroup.

4.20. The fact that two-sided ideals are normal complexes is evident. Their special role is determined by the connection (recalled above) with the zero of the homomorphic image.

THEOREM. *Suppose that $\mathfrak{K}$ is a subset of the semigroup $\mathfrak{A}$. In order that there should exist a homomorphism φ of the semigroup $\mathfrak{A}$ under which $\mathfrak{K}$ is the complete preimage of the zero of the semigroup $\varphi(\mathfrak{A})$ it is necessary and sufficient that $\mathfrak{K}$ should be a two-sided ideal of $\mathfrak{K}$.*

PROOF. (1) Suppose that, for some homomorphism φ, $\mathfrak{K}$ is the complete preimage of the zero $O_{\varphi(\mathfrak{A})}$ of the semigroup $\varphi(\mathfrak{A})$. Then for any $A \in \mathfrak{A}$, $X \in \mathfrak{K}$,

$$\varphi(AX) = \varphi(A) \cdot \varphi(X) = \varphi(A) \cdot O_{\varphi(\mathfrak{A})} = O_{\varphi(\mathfrak{A})};$$

$$\varphi(XA) = \varphi(X) \cdot \varphi(A) = O_{\varphi(\mathfrak{A})} \cdot \varphi(A) = O_{\varphi(\mathfrak{A})},$$

i.e., AX, $XA \in \mathfrak{K}$.

(2) If $\mathfrak{K}$ is a two-sided ideal of $\mathfrak{A}$, then under the natural homomorphism of $\mathfrak{A}$ onto $\mathfrak{A}/\mathfrak{n}_{\mathfrak{K}}$ (4.15), $\mathfrak{K}$, an element of $\mathfrak{A}/\mathfrak{n}_{\mathfrak{K}}$, is the zero of that semigroup, while the complete preimage of that zero is the set of all elements of the semigroup $\mathfrak{A}$ belonging to $\mathfrak{K}$.

4.21. It is necessary to observe that the natural homomorphism of the semigroup $\mathfrak{A}$ onto the ideal factor-semigroup $\mathfrak{A}/\mathfrak{n}_{\mathfrak{T}}$ corresponding to the two-sided ideal $\mathfrak{T}$ is, from 4.15, the greatest common right divisor for all the homomorphisms mapping all the elements of the ideal $\mathfrak{T}$ into one and the same element.

4.22. Suppose that $\mathfrak{G}_0$ is the unit group, i.e., a group consisting of one element. Obviously, a mapping of any semigroup $\mathfrak{A}$ onto $\mathfrak{G}_0$ is a homomorphism. Such a homomorphism is frequently called annihilating. Every semigroup possesses isomorphisms. These homomorphisms, annihilating and isomorphic, existing necessarily in every semigroup, are usually called trivial (or improper) homomorphisms. Other homomorphisms are called nontrivial (or proper). There exist semigroups having no homomorphisms other than trivial ones. Such semigroups are frequently called simple, or simple relative to homomorphisms.

Conditions for the absence of nontrivial homomorphisms in a semigroup may evidently be approached from the point of view of normal complexes. It follows immediately from 4.6 that a semigroup has no nontrivial homomorphisms if and only if it has no nontrivial normal complexes, i.e., does not have any normal complexes other than itself and the individual elements. In particular, in this case the semigroup cannot have proper nonzero two-sided ideals.

Among the semigroups having no nontrivial normal complexes evidently belong all semigroups consisting of two elements.

4.23. The condition that a semigroup with zero should not have nontrivial homomorphisms was obtained by L. M. Gluskin [**3**].

THEOREM. *Suppose that the number of elements of a semigroup $\mathfrak{A}$ with zero is larger than two and that $\mathfrak{A}$ has no proper nonzero two-sided ideals. In order that $\mathfrak{A}$ should not have any normal complexes other than $\mathfrak{A}$ itself and the individual elements it is necessary and sufficient that for any $A, B \in \mathfrak{A}$ $(A \neq B)$ there should always exist $X, Y \in \mathfrak{A}$ such that one of the products XAY and XBY should be different from $O_{\mathfrak{A}}$ and the other equal to $O_{\mathfrak{A}}$.*

PROOF. (1) For any $S \in \mathfrak{A}$ the set $\mathfrak{A}S\mathfrak{A}$ is a two-sided ideal of $\mathfrak{A}$.

The set of all $S \in \mathfrak{A}$ such that $\mathfrak{A}S\mathfrak{A} = O_{\mathfrak{A}}$ is evidently a two-sided ideal of $\mathfrak{A}$. If this collection contains elements distinct from $O_{\mathfrak{A}}$, then it must coincide with $\mathfrak{A}$, which means that $S_1S_2S_3 = O_{\mathfrak{A}}$ for any $S_1, S_2, S_3 \in \mathfrak{A}$. In this case, for any $U, V \in \mathfrak{A}$ the set $\{UV, O_{\mathfrak{A}}\}$ is evidently a proper two-sided ideal. Accordingly, one always has $UV = O_{\mathfrak{A}}$. But then $\{W, O_{\mathfrak{A}}\}$ for any $W \neq O_{\mathfrak{A}}$ turns out to be a proper nonzero two-sided ideal.

Thus $\mathfrak{A}S\mathfrak{A}$ must be equal to $\mathfrak{A}$ for any $S \in \mathfrak{A}$, $S \neq O_{\mathfrak{A}}$.

(2) Suppose that $\mathfrak{A}$ has no nontrivial normal complexes. For $A \in \mathfrak{A}$ $(A \neq O_{\mathfrak{A}})$ we denote by $\mathfrak{K}$ the collection of all the elements K such that $XKY = O_{\mathfrak{A}}$ if and only if $XAY = O_{\mathfrak{A}}$. $\mathfrak{K}$ is nonempty since $\mathfrak{K} \ni A$, and distinct from $\mathfrak{A}$ since $O_{\mathfrak{A}} \bar{\in} \mathfrak{K}$. This last follows from the fact that $XO_{\mathfrak{A}}Y = O_{\mathfrak{A}}$ holds for any $X, Y \in \mathfrak{A}$, while $\mathfrak{A}A\mathfrak{A} = \mathfrak{A}$ and therefore $XAY = O_{\mathfrak{A}}$ does not hold for all $X, Y \in \mathfrak{A}$.

$\mathfrak{K}$ is a normal complex. Indeed, suppose $K_1, K_2, K_3 \in \mathfrak{K}$ and let U and V be elements of $\mathfrak{A}$ or empty symbols, while $UK_1V = K_2$. Because of the definition of $\mathfrak{K}$ the following conditions for $X, Y \in \mathfrak{A}$ are equivalent:

$$XAY = O_{\mathfrak{A}}, \qquad XK_2Y = O_{\mathfrak{A}}, \qquad XUK_1VY = O_{\mathfrak{A}},$$
$$XUAVY = O_{\mathfrak{A}}, \qquad X(UK_3V)Y = O_{\mathfrak{A}},$$

which means that $UK_1V \in \mathfrak{K}$ implies $UK_3V \in \mathfrak{K}$.

From what has been said above, $\mathfrak{K}$ must consist of one element A. Accordingly, there is no element B of $\mathfrak{K}$ distinct from $\mathfrak{A}$. This means that for any $X, Y \in \mathfrak{A}$ one has either

$$XAY = O_{\mathfrak{A}}, \qquad XBY \neq O_{\mathfrak{A}},$$

or

$$XAY \neq O_{\mathfrak{A}}, \qquad XBY = O_{\mathfrak{A}}.$$

(3) Suppose that $\mathfrak{A}$ has the properties indicated in the formulation of the theorem and that $\mathfrak{K}$ is some normal complex of $\mathfrak{A}$ containing two distinct elements A and B. Suppose that for some $X, Y, Z \in \mathfrak{A}$,

$$XAY = O_{\mathfrak{A}}, \qquad XBY = Z \neq O_{\mathfrak{A}}.$$

Since $\mathfrak{A}Z\mathfrak{A} = \mathfrak{A}$ we have for some $P, Q \in \mathfrak{A}$,

$$PZQ = B, \qquad PXBYQ = B.$$

Since $B \in \mathfrak{K}$ it follows from the definition of a normal complex that

$$PXAYQ \in \mathfrak{K}.$$

But $XAY = O_{\mathfrak{A}}$ and therefore $O_{\mathfrak{A}} \in \mathfrak{K}$. Since for any $S \in \mathfrak{A}$ we have $SO_{\mathfrak{A}} = O_{\mathfrak{A}} \in \mathfrak{K}$, for any $K \in \mathfrak{K}$ we obtain $SK \in \mathfrak{K}$. Thus $\mathfrak{K}$ is a left ideal. One also proves that $\mathfrak{K}$ is a right ideal. Accordingly, $\mathfrak{K} = \mathfrak{A}$.

4.24. If a semigroup with zero, having no nontrivial homomorphisms, has a minimal left ideal and a minimal right ideal, then it must be a completely simple semigroup with zero (V, 4.1). However, this necessary condition for the absence of nontrivial homomorphisms is not sufficient. A completely simple semigroup may have nontrivial homomorphisms. A complete description of the structure of homomorphisms of completely simple semigroups was carried through by L. M. Gluskin **[6]**.

4.25. It is easy, using Theorem 4.23, to clear up the question as to when a completely simple semigroup with zero has no nontrivial normal complexes. From V, 5.4, we may confine ourselves to the consideration of a completely simple semigroup with zero of matrix type $\mathfrak{S}(\mathfrak{H}, P)$. If the group $\mathfrak{G}$ of nonzero elements of the semigroup $\mathfrak{H}$ is not a unit group, then, for $G, G' \in \mathfrak{G}$ $(G \neq G')$, in accordance with the right multiplication in $\mathfrak{S}(\mathfrak{H}, P)$, for any $S_1 = (X, \xi_1, \eta_1)$, $S_2 = (Y, \xi_2, \eta_2)$ the products

$$S_1 \cdot (G, \xi_0, \eta_0) \cdot S_2, \qquad S_1 \cdot (G', \xi_0, \eta_0) \cdot S_2$$

are either simultaneously equal to $O_{\mathfrak{S}}$ or simultaneously different from $O_{\mathfrak{S}}$. Accordingly, from 4.23, $\mathfrak{S}(\mathfrak{H}, P)$ has nontrivial normal complexes.

Suppose that $\mathfrak{G}$ is the unit group. If the matrix P has two identical rows, the μth and the νth, then for (G, ξ_0, μ) and (G, ξ_0, ν) $(G \in \mathfrak{G})$ we again evidently find that the products

$$S_1 \cdot (G, \xi_0, \mu) \cdot S_2, \qquad S_1 \cdot (G, \xi_0, \nu) \cdot S_2$$

are either simultaneously equal to $O_{\mathfrak{S}}$ or simultaneously different from $O_{\mathfrak{S}}$. An analogous discussion works in the case of two identical columns.

The condition that in $\mathfrak{S}(\mathfrak{H}, P)$ the group $\mathfrak{G}$ should be a unit group and that P should have no two identical rows or columns is not only sufficient but also necessary for $\mathfrak{S}(\mathfrak{H}, P)$ to have no nontrivial normal complexes.

Indeed, let us make use of 4.23. If the indicated condition is satisfied for any $(E_{\mathfrak{H}}, \lambda, \mu)$ and (H, λ', μ'), where $\lambda \neq \lambda'$ (analogously for $\mu \neq \mu'$), there exists a $\zeta \in \Gamma'$ such that $P_{\zeta\lambda} = E_{\mathfrak{H}}$, $P_{\zeta\lambda'} = O_{\mathfrak{H}}$ and a $\nu \in \Gamma$ such that $P_{\mu\nu} = E_{\mathfrak{H}}$. Therefore

$$(E_{\mathfrak{H}}, \lambda, \zeta) \cdot (E_{\mathfrak{H}}, \lambda, \mu) \cdot (E_{\mathfrak{H}}, \nu, \mu) = (E_{\mathfrak{H}}, \lambda, \mu) \neq O_{\mathfrak{S}};$$

$$(E_{\mathfrak{H}}, \lambda, \zeta) \cdot (H, \lambda', \mu') \cdot (E_{\mathfrak{H}}, \nu, \mu) = (O_{\mathfrak{H}}, \lambda, \mu) = O_{\mathfrak{S}}.$$

4.26. As for completely simple semigroups without zero, it is easy to see, from the results of § 6 of Chapter V, that such semigroups always have nontrivial normal complexes, excluding the case when it is a group without nontrivial normal divisors.

4.27. A commutative semigroup $\mathfrak{A}$ without nontrivial normal complexes cannot have nontrivial proper ideals (since these ideals are all two-sided). If it has more than two elements, then, as we proved in the first part of 4.23, for each $S \in \mathfrak{A}$ we have $\mathfrak{A}S\mathfrak{A} = \mathfrak{A}$. Accordingly, $S\mathfrak{A}$ is not the zero of $\mathfrak{A}$, so that $S\mathfrak{A} = \mathfrak{A}$. But then $\mathfrak{A}$ has to be a group. As is proved in the theory of groups, a commutative group has no nontrivial homomorphisms if and only if it is cyclic and the number of its elements is equal to one or a prime number.

5. Extension of Homomorphisms

5.1. If $\mathfrak{B}$ is a subsemigroup of the semigroup $\mathfrak{A}$, then every homomorphism φ of the semigroup $\mathfrak{A}$ induces in an obvious way a homomorphism of the semigroup $\mathfrak{B}$ onto the semigroup $\varphi(\mathfrak{B})$, whose action on the elements of $\mathfrak{B}$ coincides with the action of the homomorphism φ. Of course, it is quite possible that essentially different (1.14) homomorphisms of the semigroup $\mathfrak{A}$ may induce in $\mathfrak{B}$ one and the same homomorphism.

The same phenomenon may be approached from the opposite point of view. A homomorphism ψ of the semigroup $\mathfrak{B}$ may sometimes be extended to a homomorphism φ of one of its supersemigroups $\mathfrak{A}$, which means that ψ is induced by the homomorphism φ. However, as we shall show below by a simple example, this is not always possible.

In accordance with what has been said above, one may also have a case when the homomorphism ψ has essentially distinct extensions. Our interest then lies in the relations between the properties of φ and ψ. For example, it is obvious that if φ is an isomorphism, then ψ is also. However the inverse conclusion is not valid in the general case; isomorphisms of a semigroup frequently admit extensions to some supersemigroup which turn out to be homomorphisms, not isomorphisms.

5.2. Suppose that the homomorphisms φ_1 and φ_2 of the subsemigroup $\mathfrak{B}$ of the semigroup $\mathfrak{A}$ are essentially distinct (1.14) and that the homomorphism φ_1 may be extended to a homomorphism ψ_1 of the whole semigroup $\mathfrak{A}$. Then the homomorphism φ_2 also admits an extension to a homomorphism of $\mathfrak{A}$. Indeed, in the semigroup $\psi_1(\mathfrak{A})$ we replace, by the method of III, 1.8, all the elements appearing in $\psi_1(\mathfrak{B}) = \varphi_1(\mathfrak{B})$ by the corresponding elements of $\varphi_2(\mathfrak{B})$ ($\varphi_1(\mathfrak{B})$ and $\varphi_2(\mathfrak{B})$ are isomorphic). We obtain a new semigroup $\mathfrak{A}'$, the set of whose elements consists of $\varphi_2(\mathfrak{B})$ and of $\psi_1(\mathfrak{A})\backslash\psi_1(\mathfrak{B})$. Define a mapping ψ_2 of the semigroup $\mathfrak{A}$ onto $\mathfrak{A}'$ by putting $\psi_2(B) = \varphi_2(B)$ for $B \in \mathfrak{B}$ and $\psi_2(X) = \psi_1(X)$ for $X \in \mathfrak{A}\backslash\mathfrak{B}$. It is quite obvious that ψ_2 is a homomorphism. This homomorphism of the semigroup $\mathfrak{A}$ extends the homomorphism φ_2 of its subsemigroup $\mathfrak{B}$.

Recall that the question of extendability of homomorphisms admits a natural treatment from the point of view of relations between two-sidedly stable equivalences in $\mathfrak{B}$ and in $\mathfrak{A}$.

5.3. A simple example of extendability is given by a supersemigroup $\mathfrak{T}$ of the semigroup $\mathfrak{A}$, consisting of all subsets of the semigroup $\mathfrak{A}$. Evidently every homomorphism ψ of the semigroup $\mathfrak{A}$ may be extended to a homomorphism φ of its supersemigroup $\mathfrak{T}$. To this end, as $\varphi(\mathfrak{T})$ we choose the semigroup of all subsets of the semigroup $\psi(\mathfrak{A})$ and for $\mathfrak{N} \in \mathfrak{T}$, where $\mathfrak{N} = \{A, B, \ldots\} \subset \mathfrak{A}$, we put $\varphi(\mathfrak{N}) = \{\psi(A), \psi(B), \ldots\} \subset \psi(\mathfrak{A})$.

As an example of the impossibility of extending a homomorphism one may take the infinite monogenic semigroup $\mathfrak{A}_1 = [A_1]$ (III, 3.16), for which the mapping φ onto an arbitrary finite monogenic semigroup $\mathfrak{A}_2 = [A_2]$,

$$\varphi(A_1^k) = A_2^k \qquad (k = 1, 2, \ldots),$$

is easily seen to be a homomorphism. If $\mathfrak{A}_2$ is not a group, φ cannot be extended to the infinite cyclic group $\mathfrak{A}_3 = [A_1, A_1^{-1}]$ (III, 3.17), a supersemigroup of the semigroup $\mathfrak{A}_1$. This follows from the fact that every homomorphic image of a group must itself be a group (1.2), while a finite monogenic semigroup, not being a group, is evidently not a subsemigroup of any group.

5.4. There is still no general investigation of the question of extendability of homomorphisms. We shall consider two special cases examined by L. M. Gluskin [5].

Suppose that $\mathfrak{A}$ is a periodic semigroup (III, 4.10) and that $\mathfrak{G}_I$ is one of its maximal subgroups (III, 4.10) with unity I. We shall show that any homomorphism φ of the group $\mathfrak{G}_I$ extends to a homomorphism of $\mathfrak{A}$.

Denote by $\mathfrak{K}$ the collection of those elements K of $\mathfrak{G}_I$ for which $\varphi(K) = \varphi(I)$. $\mathfrak{K}$ is a normal complex of the group $\mathfrak{G}_I$, corresponding to the homomorphism φ. We shall show that $\mathfrak{K}$ is a normal complex of the semigroup $\mathfrak{A}$. Suppose that for some $K_1, K_2 \in \mathfrak{K}$ and X and Y which are elements of $\mathfrak{A}$ or empty symbols,

$$XK_1Y = K_2$$

and K_3 is an arbitrary element of $\mathfrak{K}$. Denote by K_2^{-1} the element inverse to K_2 relative to I in $\mathfrak{G}_I$. We have

$$XI \cdot K_1YK_2^{-1} = XK_1YK_2^{-1} = K_2K_2^{-1} = I, \quad XI \cdot I = XI.$$

Hence, from III, 4.11, $XI \in \mathfrak{G}_I$. Analogously we see that $IY \in \mathfrak{G}_I$. Therefore

$$XK_3Y = XI \cdot K_3 \cdot IY \in \mathfrak{G}_I$$

and

$$\begin{aligned}\varphi(XK_3Y) = \varphi(XI) \cdot \varphi(K_3) \cdot \varphi(IY) &= \varphi(XI) \cdot \varphi(K_1) \cdot \varphi(IY) \\ &= \varphi(XIK_1IY) = \varphi(XK_1Y) = \varphi(K_2) = \varphi(I),\end{aligned}$$

i.e., $XK_3Y \in \mathfrak{K}$.

Denote by ψ the homomorphism of the semigroup $\mathfrak{A}$ which is the greatest common right divisor of those homomorphisms for which $\mathfrak{K}$ is the complete preimage of one of the elements. The homomorphism ψ induces in $\mathfrak{G}_I$ a homomorphism ψ'. The homomorphisms ψ and ψ' of the group $\mathfrak{G}_I$ are such that the complete preimages of the identity under these homomorphisms are equal to $\mathfrak{K}$. From 4.13, it therefore follows that $\varphi(\mathfrak{G}_I)$ and $\psi'(\mathfrak{G}_I)$ are isomorphic. Since ψ' extends to the homomorphism ψ of the semigroup $\mathfrak{A}$, it follows from 5.2 that φ will extend to a homomorphism of the semigroup $\mathfrak{A}$.

5.5. Suppose that $\mathfrak{B}$ is a two-sided ideal of the semigroup $\mathfrak{A}$ and that φ is some homomorphism of $\mathfrak{B}$ such that in the semigroup $\varphi(\mathfrak{B})$ each two elements have a common left identity and a common right identity. We shall show that under this condition the homomorphism φ may be extended to a homomorphism of the semigroup $\mathfrak{A}$.

We shall define in $\mathfrak{A}$ a relation $\mathfrak{n}$, putting $A \sim B(\mathfrak{n})$ $(A, B \in \mathfrak{A})$ if $A = B$ or $A, B \in \mathfrak{B}$ and $\varphi(A) = \varphi(B)$. Obviously $\mathfrak{n}$ is an equivalence. We shall prove the right stability of $\mathfrak{n}$. Suppose that $B_1 \neq B_2$ $(B_1, B_2 \in \mathfrak{B})$, while $B_1 \sim B_2(\mathfrak{n})$, i.e., $\varphi(B_1) = \varphi(B_2)$. Since $\mathfrak{B}$ is a two-sided ideal of $\mathfrak{A}$, for any $X \in \mathfrak{A}$ we have $(B_1X), (B_2X) \in \mathfrak{B}$. By hypothesis, there is in $\mathfrak{B}$ an element $N \in \mathfrak{B}$ such that

$$\varphi(B_1X)\cdot\varphi(N) = \varphi(B_1X), \qquad \varphi(B_2X)\cdot\varphi(N) = \varphi(B_2X).$$

Because of this we get

$$\begin{aligned}\varphi(B_1X) &= \varphi(B_1X)\cdot\varphi(N) = \varphi(B_1XN) = \varphi(B_1)\cdot\varphi(XN)\\ &= \varphi(B_2)\cdot\varphi(XN) = \varphi(B_2XN) = \varphi(B_2X)\cdot\varphi(N) = \varphi(B_2X),\end{aligned}$$

i.e., $B_1X \sim B_2X(\mathfrak{n})$. Analogously we prove that $\mathfrak{n}$ is stable to the left.

Evidently the natural homomorphism of $\mathfrak{A}$ onto $\mathfrak{A}/\mathfrak{n}$ induces in $\mathfrak{B}$ a homomorphism differing from φ only by a multiplier which is an isomorphism. Since that homomorphism extends to a homomorphism of the whole semigroup $\mathfrak{A}$, it follows that φ also, from 5.2, extends to a homomorphism of $\mathfrak{A}$.

We note in passing that it is easy to see that the natural homomorphism of $\mathfrak{A}$ onto $\mathfrak{A}/\mathfrak{n}$ is the greatest common right divisor of those homomorphisms of $\mathfrak{A}$ which extend the homomorphism φ.

In particular, it follows from what has been proved that *if the two-sided ideal* $\mathfrak{B}$ *of the semigroup* $\mathfrak{A}$ *has an identity, then each of its homomorphisms may be extended to a homomorphism of the semigroup* $\mathfrak{A}$.

5.6. The properties of extendability of the identity isomorphism may be used for an abstract characterization of certain important concrete semigroups. We consider an abstract characterization of the semigroup $\mathfrak{S}_\Omega$ of all transformations of an arbitrary set Ω, obtained by E. S. Ljapin **[13]** (another abstract characterization of this same semigroup was obtained by A. I. Mal'cev **[4]**). To this end we shall need the following concept.

DEFINITION. *Suppose that Ξ is some class of semigroups. We say that the subsemigroup $\mathfrak{T}$ of the semigroup $\mathfrak{A}$ is densely embedded in $\mathfrak{A}$ relative to Ξ if the following conditions are satisfied:*

(1) *Every homomorphism of $\mathfrak{A}$ onto some semigroup of Ξ which is an extension of the identity isomorphism of $\mathfrak{T}$ must be an isomorphism;*

(2) *For any semigroup $\mathfrak{A}'$ of Ξ which is a supersemigroup of $\mathfrak{A}$ distinct from $\mathfrak{A}$, there exists a homomorphism of $\mathfrak{A}'$ onto some semigroup of Ξ which is not an isomorphism but which is an extension of the identity isomorphism of $\mathfrak{T}$.*

5.7. In the semigroup $\mathfrak{S}_\Omega$ of all transformations of the set Ω we distinguish those elements U_α ($\alpha \in \Omega$) such that $U_\alpha \xi = \alpha$ for any $\xi \in \Omega$. Their union $\mathfrak{U}_\Omega$ is a two-sided ideal of $\mathfrak{S}_\Omega$. Indeed, for any $S \in \mathfrak{S}_\Omega$, for $S\alpha = \beta$ we have

$$U_\alpha S\xi = U_\alpha(S\xi) = \alpha = U_\alpha \xi,$$

$$SU_\alpha \xi = S\alpha = U_\beta \xi$$

and accordingly

$$U_\alpha S = U_\alpha \in \mathfrak{U}_\Omega, \qquad SU_\alpha = U_\beta \in \mathfrak{U}_\Omega.$$

We shall show that the ideal $\mathfrak{U}_\Omega$ is densely embedded in $\mathfrak{S}_\Omega$ relative to the class of semigroups having the semigroup $\mathfrak{U}_\Omega$ as a two-sided ideal.

Suppose that φ is a homomorphism of $\mathfrak{S}_\Omega$ which is not an isomorphism. For some $X, Y \in \mathfrak{S}_\Omega$ we have

$$\varphi(X) = \varphi(Y), \qquad X \neq Y.$$

Since $Y \neq X$, for some $\alpha \in \Omega$ we have

$$X\alpha = \beta, \qquad Y\alpha = \gamma, \qquad \beta \neq \gamma.$$

Since

$$U_\beta = XU_\alpha, \qquad U_\gamma = YU_\alpha,$$

we have

$$\varphi(U_\beta) = \varphi(XU_\alpha) = \varphi(X) \cdot \varphi(U_\alpha) = \varphi(Y) \cdot \varphi(U_\alpha) = \varphi(YU_\alpha) = \varphi(U_\gamma).$$

Since $U_\beta \neq U_\gamma$, the equality just obtained means that any homomorphism φ of the semigroup $\mathfrak{S}_\Omega$ which is not an isomorphism cannot be an extension of the identity isomorphism of the semigroup $\mathfrak{U}_\Omega$.

Now suppose that $\mathfrak{S}'$ is some supersemigroup of $\mathfrak{S}_\Omega$, different from $\mathfrak{S}_\Omega$ and such that $\mathfrak{U}_\Omega$ is a two-sided ideal of $\mathfrak{S}'$. Define a mapping ψ of the semigroup $\mathfrak{S}'$ into $\mathfrak{S}_\Omega$ by putting, for any $X \in \mathfrak{S}'$,

$$\psi(X) = S_X, \qquad S_X\alpha = (XU_\alpha)\alpha \qquad (\alpha \in \Omega).$$

The last expression has a meaning since the product XU_α is some element of $\mathfrak{U}_\Omega$ (since $\mathfrak{U}_\Omega$ is a two-sided ideal of $\mathfrak{S}'$).

We shall show that the mapping ψ is a homomorphism.

Suppose that $X, Y \in \mathfrak{S}'$. Since $\mathfrak{U}_\Omega$ is a two-sided ideal of $\mathfrak{S}'$, for any $\alpha \in \Omega$ there exist $\beta, \gamma \in \Omega$ such that

$$XU_\alpha = U_\beta, \qquad YU_\beta = U_\gamma.$$

We have

$$S_X\alpha = (XU_\alpha)\alpha = U_\beta\alpha = \beta,$$

$$S_Y\beta = (YU_\beta)\beta = U_\gamma\beta = \gamma,$$

$$S_YS_X\alpha = S_Y(XU_\alpha)\alpha = S_YU_\beta\alpha = S_Y\beta = (YU_\beta)\beta = U_\gamma\beta = \gamma.$$

On the other hand, for $Z = YX$,

$$S_Z\alpha = ZU_\alpha\alpha = YXU_\alpha\alpha = YU_\beta\alpha = U_\gamma\alpha = \gamma.$$

Since for any $\alpha \in \Omega$ and for the mappings S_Y, S_X, S_Z we have

$$S_YS_X\alpha = S_Z\alpha,$$

it follows that $S_YS_X = S_Z$, i.e.,

$$\psi(Y)\cdot\psi(X) = \psi(Z) = \psi(YX).$$

The homomorphism ψ induces in $\mathfrak{S}_\Omega$ the identity isomorphism. Indeed, for any $\alpha \in \Omega$ and for $A \in \mathfrak{S}_\Omega$ we obtain

$$\psi(A) = S_A, \qquad S_A\alpha = (AU_\alpha)\alpha = A\alpha.$$

Accordingly,

$$\psi(A) = S_A = A.$$

As to the semigroup $\mathfrak{S}'$ itself, ψ is not an isomorphism for it. Indeed, for any $B \in \mathfrak{S}'\backslash\mathfrak{S}_\Omega$ we have

$$\psi(B) = C \in \mathfrak{S}_\Omega.$$

But, as we have proved, $\psi(C) = C$. Accordingly,

$$\psi(B) = \psi(C), \qquad B \neq C.$$

It results that ψ is a homomorphism of the semigroup $\mathfrak{S}'$ onto the semigroup $\mathfrak{S}_\Omega$, which belongs to the class in question. In addition ψ is not an isomorphism, but serves as an extension of the identity isomorphism of the semigroup $\mathfrak{U}_\Omega$.

5.8. The semigroup $\mathfrak{U}_\Omega$ of $\mathfrak{S}_\Omega$ studied in 5.7, is a two-sided ideal of $\mathfrak{S}_\Omega$ and is such that the product of any two elements is equal to the left factor. We observe that no semigroup can have two distinct two-sided ideals with that property. Indeed, the intersection of those two ideals would be a two-sided ideal of each of them (IV, 1.9, (γ), (η)), but evidently a semigroup for which the product of two elements is always equal to the left factor has no proper two-sided ideals.

5.9. The property of the semigroup $\mathfrak{S}_\Omega$ derived in 5.7 may be used to obtain an abstract characterization of it.

THEOREM. *In order that the semigroup $\mathfrak{A}$ should be isomorphic to the semigroup $\mathfrak{S}_\Omega$ of all transformations of the set Ω it is necessary and sufficient that $\mathfrak{A}$ should contain a two-sided ideal $\mathfrak{T}$ having the following properties:*

(α) $\mathfrak{T}$ has the same cardinality as Ω;

(β) In $\mathfrak{T}$ the product of any two elements is equal to the left factor;

(γ) $\mathfrak{T}$ is densely imbedded in $\mathfrak{A}$ relative to the class of semigroups containing $\mathfrak{T}$ as a two-sided ideal.

PROOF. (1) The semigroup $\mathfrak{S}_\Omega$ has a two-sided ideal $\mathfrak{U}_\Omega$, having the properties (α), (β), (γ) stated in the formulation of the theorem (5.7). Accordingly, any semigroup isomorphic to $\mathfrak{S}_\Omega$ will have such a two-sided ideal.

(2) Suppose that the semigroup $\mathfrak{A}$ contains a two-sided ideal $\mathfrak{T}$ having the properties (α), (β), (γ) stated in the formulation of the theorem. Since it has the same cardinality as the semigroup $\mathfrak{U}_\Omega$, the semigroup $\mathfrak{T}$ is obviously isomorphic to it, since in both semigroups the product of any two elements is always equal to a left multiplier. Therefore, in the semigroup $\mathfrak{A}$, one may in a natural way replace the elements of $\mathfrak{T}$ by the elements of $\mathfrak{U}_\Omega$ corresponding to them under some isomorphism (III, 1.8). As a result we obtain a new semigroup $\mathfrak{A}'$ isomorphic to the semigroup $\mathfrak{A}$ and such that $\mathfrak{U}_\Omega$ is a two-sided ideal of $\mathfrak{A}'$, while evidently $\mathfrak{U}_\Omega$ is densely imbedded in $\mathfrak{A}'$ relative to the class of semigroups containing $\mathfrak{U}_\Omega$ as a two-sided ideal.

We shall assign to each $A \in \mathfrak{A}'$ a transformation $\bar{A}$ of the set Ω in the following way. If $AU_\alpha = U_\beta$, then we put

$$\bar{A}\alpha = \beta.$$

The mapping φ of the semigroup $\mathfrak{A}'$ into $\mathfrak{S}_\Omega$, putting the transformation $\bar{A}$ into correspondence with each A, is a homomorphism. Indeed, suppose that for $A, B \in \mathfrak{A}'$ and $\alpha \in \Omega$,

$$BU_\alpha = U_\beta, \qquad AU_\beta = U_\gamma,$$

$$ABU_\alpha = AU_\beta = U_\gamma.$$

Then

$$(\overline{AB})\alpha = \gamma, \qquad \bar{B}\alpha = \beta, \qquad \bar{A}\beta = \gamma,$$

$$(\overline{AB})\alpha = \bar{A}\bar{B}\alpha,$$

i.e., $(\overline{AB}) = \bar{A} \cdot \bar{B}$ and therefore

$$\varphi(AB) = (\overline{AB}) = \bar{A} \cdot \bar{B} = \varphi(A) \cdot \varphi(B).$$

The homomorphism φ induces the identity isomorphism of $\mathfrak{U}_\Omega$. Indeed, for any $U_\xi \in \mathfrak{U}_\Omega$ and $\alpha \in \Omega$ we have

$$U_\xi U_\alpha = U_\xi, \qquad \varphi(U_\xi) = \bar{U}_\xi, \qquad \bar{U}_\xi\alpha = \xi, \qquad \bar{U}_\xi = U_\xi.$$

Since $\mathfrak{U}_\Omega$ is a two-sided ideal of $\mathfrak{S}_\Omega$, then $\mathfrak{U}_\Omega$ is a two-sided ideal also for $\varphi(\mathfrak{A}') \subset \mathfrak{S}_\Omega$. Thus φ is a homomorphism which is an extension of the identity isomorphism of $\mathfrak{U}_\Omega$ onto a semigroup belonging to the class of semigroups containing $\mathfrak{U}_\Omega$ as a two-sided ideal. It therefore follows that φ must be an isomorphism. Since $\mathfrak{U}_\Omega$ is densely embedded in $\mathfrak{A}'$ relative to the class of semigroups containing $\mathfrak{U}_\Omega$ as a two-sided ideal, $\mathfrak{U}_\Omega$ will be densely embedded relative to the same class in $\varphi(\mathfrak{A}')$ (since φ is an isomorphism). But simultaneously $\mathfrak{U}_\Omega$ is densely embedded relative to the same class in $\mathfrak{S}_\Omega$. Since $\varphi(\mathfrak{A}') \subset \mathfrak{S}_\Omega$ it follows, as one sees directly from the definition, that this is possible only if $\varphi(\mathfrak{A}') = \mathfrak{S}_\Omega$.

Thus $\mathfrak{S}_\Omega$ has turned out to be isomorphic to $\mathfrak{A}'$ and thus to $\mathfrak{A}$.

5.10. Observe that in a similar way one may find an abstract characterization also for some other important semigroups. For example, this was done for the semigroup of all one-to-one partial transformations of a set (E. S. Ljapin **[13]**) and for the semigroup of all binary relations on a set (K. A. Zareckiĭ **[1]**). In both cases the semigroups isomorphic to the given semigroup are characterized by the presence of a two-sided ideal with a definite structure, densely embedded in the semigroup relative to the class of all semigroups containing it as a two-sided ideal.

As L. M. Gluskin **[9]** proved, in the semigroup of all square matrices of the same order over any field, the collection of all matrices of rank one or zero forms a two-sided ideal, densely embedded in the semigroup relative to the class of all semigroups containing it as a two-sided ideal. The presence in a certain semigroup of an ideal isomorphic to this ideal and having the indicated properties is sufficient for isomorphism with the matrix semigroup mentioned above. In this connection, we recall that some time ago L. M. Gluskin **[1]** also found a different abstract characterization of the semigroup of square matrices.

6. Certain Special Kinds of Homomorphisms

6.1. Although in the theory of groups the description of homomorphisms of groups by means of their normal divisors is to the highest degree clear and completely describes each homomorphism up to a multiplication by an isomorphism, in the general theory of semigroups the situation is much more complicated and is still very far from a complete clarification. In this direction significant attention has been devoted to the study of various special kinds of homomorphisms. Related to the same order of ideas is the study of two-sidedly stable equivalences, which is just another approach to the same question. Below we shall present some of the results of such investigations.

6.2. For our discussion of homomorphisms of commutative semigroups, we define the following relation $\mathfrak{k}$ in an arbitrary semigroup $\mathfrak{A}$. Put $X \sim Y(\mathfrak{k})$ $(X, Y \in \mathfrak{A})$ if and only if there are elements U and V in $\mathfrak{A}$ such that $X = UV$, $Y = VU$.

If $\mathfrak{n}$ is any two-sidedly stable equivalence such that the semigroup $\mathfrak{A}/\mathfrak{n}$ is commutative, then evidently $\mathfrak{k} \leqslant \mathfrak{n}$ (I, 5.14). Conversely, for any two-sidedly stable equivalence such that $\mathfrak{k} \leqslant \mathfrak{n}$ the semigroup $\mathfrak{A}/\mathfrak{n}$ is evidently commutative.

Because of I, 5.23, it therefore follows that the second derivative relation $\mathfrak{k}''$ (I, 5.21) will be the greatest lower bound for all two-sidedly stable equivalences corresponding to homomorphisms of $\mathfrak{A}$ onto commutative semigroups. Thus, $\mathfrak{k}''$ is the finest of the decompositions $\mathfrak{A}$ corresponding to homomorphisms of $\mathfrak{A}$ onto commutative semigroups. The natural homomorphism of $\mathfrak{A}$ onto $\mathfrak{A}/\mathfrak{k}''$ is the greatest common right divisor of the set of all homomorphisms $\mathfrak{A}$ onto commutative semigroups (1. 15). In addition, it itself belongs to that set.

In the case when $\mathfrak{A}$ is a group, the decomposition $\mathfrak{k}''$ is just the decomposition of $\mathfrak{A}$ by the commutant. The natural homomorphism of $\mathfrak{A}$ onto $\mathfrak{A}/\mathfrak{k}''$ is the natural homomorphism of the group $\mathfrak{A}$ onto its factor-group modulo the commutant.

6.3. Define in $\mathfrak{A}$ a relation $\mathfrak{l}$, putting $X \sim Y(\mathfrak{l})$ $(X, Y \in \mathfrak{A})$ if and only if there is in $\mathfrak{A}$ an element Z such that $ZX = ZY$. It is easy to see that for the two-sidedly stable equivalence $\mathfrak{n}$, the semigroup $\mathfrak{A}/\mathfrak{n}$ will have the property of left cancellation if and only if $\mathfrak{n} \geqslant \mathfrak{l}$. In the set of such equivalences $\mathfrak{l}''$ will be the greatest lower bound, and will itself belong to that set. Therefore by a method analogous to that of 6.2 one may characterize the role of the decomposition $\mathfrak{l}''$ and of the natural homomorphism $\mathfrak{A}/\mathfrak{l}''$ for homomorphisms of $\mathfrak{A}$ onto semigroups with left cancellation.

Analogously one defines a relation $\mathfrak{r}$, connected with homomorphisms of $\mathfrak{A}$ onto semigroups with right cancellation. The relation $\mathfrak{m}$, the least upper bound for the relations $\mathfrak{l}$ and $\mathfrak{r}$, is easily seen to be such that $\mathfrak{m}''$ is the greatest lower bound in the set of all two-sidedly stable equivalences $\mathfrak{n}$ such that $\mathfrak{A}/\mathfrak{n}$ is a semigroup with two-sided cancellation. The natural homomorphism of $\mathfrak{A}$ onto $\mathfrak{A}/\mathfrak{m}''$ is the greatest common right divisor of the set of all homomorphisms onto semigroups with two-sided cancellation.

6.4. The relation $\mathfrak{l}$, defined in 6.3, has the following property. If $SX \sim SY(\mathfrak{l})$, then $X \sim Y(\mathfrak{l})$ $(X, Y, S \in \mathfrak{A})$. Indeed, in this case we have $ZSX = ZSY$ for some $Z \in \mathfrak{A}$, so that $X \sim Y(\mathfrak{l})$. The study of various relations with this property as well as the symmetric property for right cancellation was taken up in the papers of Dubreil [**1**; **3**; **4**; **5**]. The interest in these properties is explained by the fact that for two-sidedly stable equivalence $\mathfrak{n}$ the semigroup $\mathfrak{A}/\mathfrak{n}$ is a semigroup with left cancellation if and only if $SX \sim SY(\mathfrak{n})$ always implies $X \sim Y(\mathfrak{n})$ $(X, Y, S \in \mathfrak{A})$. For suppose that $\mathfrak{A}/\mathfrak{n}$ is a semigroup with left cancellation and that ψ is the natural homomorphism of $\mathfrak{A}$ onto $\mathfrak{A}/\mathfrak{n}$. $SX \sim SY(\mathfrak{n})$ implies $\psi(S) \cdot \psi(X) = \psi(S) \cdot \psi(Y)$, and therefore $\psi(X) = \psi(Y)$, i.e., $X \sim Y(\mathfrak{n})$. On the other hand, if the indicated property is satisfied, it follows from $\psi(S) \cdot \psi(X) = \psi(S) \cdot \psi(Y)$ that $\psi(SX) = \psi(SY)$, i.e., $SX \sim SY(\mathfrak{n})$, and therefore $X \sim Y(\mathfrak{n})$, i.e., $\psi(X) = \psi(Y)$. The situation is analogous for right cancellation.

6.5. An important relation in semigroups is the relation of regular adjointness (II, 6.6). In the general case it is not two-sidedly stable. However, in semigroups of idempotents, as follows from the investigations of McLean [**1**], two-sided stability of this relation holds. It is convenient to approach this result by considering the following two relations, which are interesting in themselves.

For the elements X and Y of the semigroup $\mathfrak{A}$ we put

$$X \sim Y(\mathfrak{n}_1)$$

if

$$XY = Y, \qquad YX = X.$$

Symmetrically we define a relation $\mathfrak{n}_2$ by putting

$$X \sim Y(\mathfrak{n}_2)$$

if

$$XY = X, \qquad YX = Y.$$

These relations are obviously symmetric. If

$$X_1X_2 = X_2; \qquad X_2X_1 = X_1, \qquad X_2X_3 = X_3; \qquad X_3X_2 = X_2\,(X_1, X_2, X_3 \in \mathfrak{A}),$$

then

$$X_1X_3 = X_1X_2X_3 = X_2X_3 = X_3, \qquad X_3X_1 = X_3X_2X_1 = X_2X_1 = X_1,$$

from which it follows that the relation $\mathfrak{n}_1$ is transitive. Analogously one proves that $\mathfrak{n}_2$ is transitive.

6.6. In the general case the equivalence relations $\mathfrak{n}_1$ and $\mathfrak{n}_2$ indicated above are not necessarily stable either to the left or to the right. As an example we consider a semigroup $\mathfrak{A}$ whose elements are pairs of integers, with the operation carried out according to the rule

$$(n_1, m_1) \cdot (n_2, m_2) = (n_1 + n_2, m_2).$$

That this operation is associative is obvious.

In this semigroup we have

$$(0, m_1) \sim (0, m_2)(n_1).$$

At the same time

$$(1, 0) \cdot (0, m_1) = (1, m_1), \qquad (1, 0) \cdot (0, m_2) = (1, m_2);$$

$$(1, m_1) \cdot (1, m_2) = (2, m_2) \neq (1, m_2);$$

$$(0, m_1) \cdot (1, 0) = (1, 0), \qquad (0, m_2) \cdot (1, 0) = (1, 0);$$

$$(1, 0) \cdot (1, 0) = (2, 0) \neq (1, 0),$$

i.e.,

$$(1, 0) \cdot (0, m_1) \nsim (1, 0) \cdot (0, m_2)(n_1);$$

$$(0, m_1) \cdot (1, 0) \nsim (0, m_2) \cdot (1, 0)(n_1).$$

6.7. If all the elements of the semigroup $\mathfrak{A}$ are idempotents, then the relations $\mathfrak{n}_1$ and $\mathfrak{n}_2$ are evidently reflexive, and therefore are equivalences. In addition,

$\mathfrak{n}_1$ is stable to the left and $\mathfrak{n}_2$ to the right. Indeed, if $X \sim Y(\mathfrak{n}_1)$, then for any $Z \in \mathfrak{A}$,

$$(ZX)(ZY) = (ZX)(Z \cdot XY) = (ZX)(ZX)Y = ZXY = ZY.$$

Analogously one proves the second of the equations, meaning that

$$ZX \sim ZY(\mathfrak{n}_1).$$

The considerations for $\mathfrak{n}_2$ are analogous.

6.8. As we have already recalled, in semigroups of idempotents the relations $\mathfrak{n}_1$ and $\mathfrak{n}_2$ may be used for the study of the relation of regular adjointness.

Suppose that $\mathfrak{A}$ is a semigroup of idempotents. Define in $\mathfrak{A}$ a relation $\mathfrak{n}_3$, putting $X \sim Y(\mathfrak{n}_3)$ if and only if X and Y are regularly adjoint elements of the semigroup $\mathfrak{A}$ (II, 6.6).

(α) *If*

$$X \sim Y(\mathfrak{n}_1), \qquad Y \sim Z(\mathfrak{n}_2),$$

then

$$X \sim Z(\mathfrak{n}_3).$$

Indeed, because $\mathfrak{n}_1$ is stable to the left and $\mathfrak{n}_2$ stable to the right,

$$ZX \sim ZY(\mathfrak{n}_1), \qquad YX \sim ZX(\mathfrak{n}_2).$$

But $ZY = Z$ and $YX = X$, so that

$$ZXZ = ZX \cdot Z = ZX \cdot ZY = ZY = Z.$$

Analogously we prove that $XZX = X$.

(β) *If $X \sim Y(\mathfrak{n}_1)$ or $X \sim Y(\mathfrak{n}_2)$, then for any $Z \in \mathfrak{A}$,*

$$X \sim Y(\mathfrak{n}_3), \qquad ZX \sim ZY(\mathfrak{n}_3), \qquad XZ \sim YZ(\mathfrak{n}_3).$$

Indeed, suppose $X \sim Y(\mathfrak{n}_1)$. Then

$$XYX = X \cdot YX = X \cdot X = X, \qquad YXY = YY = Y;$$

$$\begin{aligned}(ZX) \cdot (ZY) \cdot (ZX) &= ZX \cdot ZY \cdot ZYX \\ &= ZX \cdot (ZY)^2 X = ZXZYX = ZXZX = ZX.\end{aligned}$$

Analogously one proves: the second equation, necessary and sufficient for ZX and ZY to be regularly adjoint; the regular adjointness of XZ and YZ; the corresponding equations for the case when $X \sim Y(\mathfrak{n}_2)$.

(γ) *If $X \sim Y(\mathfrak{n}_3)$, then*

$$X \sim XY(\mathfrak{n}_1), \qquad X \sim YX(\mathfrak{n}_2).$$

Indeed,

$$X \cdot XY = X^2 Y = XY, \qquad XY \cdot X = XYX = X;$$

$$X \cdot YX = X, \qquad YX \cdot X = YX.$$

(δ) $XY \sim YX(\mathfrak{n}_3)$.

Indeed,

$$(XY)(YX)(XY) = XY^2X^2Y = XY \cdot XY = XY,$$

$$(YX)(XY)(YX) = YX^2Y^2X = YX \cdot YX = YX.$$

6.9. THEOREM. *If all the elements of $\mathfrak{A}$ are idempotents, then $\mathfrak{n}_3$ is a two-sidedly stable equivalence relation in $\mathfrak{A}$.*

PROOF. (1) The relation $\mathfrak{n}_3$ is symmetric. Reflexivity of $\mathfrak{n}_3$ follows from the fact that all the elements are idempotents. We shall prove the transitivity of $\mathfrak{n}_3$. Suppose that

$$X \sim Y(\mathfrak{n}_3), \qquad Y \sim Z(\mathfrak{n}_3).$$

Then, from 6.8, (γ),

$$Y \sim YX(\mathfrak{n}_1), \qquad Y \sim YZ(\mathfrak{n}_1),$$

$$Y \sim XY(\mathfrak{n}_2), \qquad Y \sim ZY(\mathfrak{n}_2).$$

Since $\mathfrak{n}_1$ and $\mathfrak{n}_2$ are symmetric and transitive,

$$YX \sim YZ(\mathfrak{n}_1), \qquad XY \sim ZY(\mathfrak{n}_2).$$

Because of 6.7,

$$ZYX \sim ZYZ(\mathfrak{n}_1), \qquad XYX \sim ZYX(\mathfrak{n}_2).$$

Since $ZYZ = Z$ and $XYX = X$ it follows from 6.8, (α), that

$$Z \sim X(\mathfrak{n}_3).$$

(2) Suppose that $X \sim Y(\mathfrak{n}_3)$. From 6.8, (γ),

$$X \sim XY(\mathfrak{n}_1), \qquad Y \sim XY(\mathfrak{n}_2).$$

Hence from 6.8, (β), for any $Z \in \mathfrak{A}$ we obtain

$$ZX \sim ZXY(\mathfrak{n}_3), \qquad ZY \sim ZXY(\mathfrak{n}_3),$$

$$XZ \sim XYZ(\mathfrak{n}_3), \qquad YZ \sim XYZ(\mathfrak{n}_3).$$

Because of the transitivity of $\mathfrak{n}_3$ it therefore follows that

$$ZX \sim ZY(\mathfrak{n}_3), \qquad XZ \sim YZ(\mathfrak{n}_3),$$

i.e., the equivalence $\mathfrak{n}_3$ is two-sidedly stable.

6.10. If $\mathfrak{K}$ and $\mathfrak{L}$ are two subsets of the semigroup $\mathfrak{A}$, then the set of elements $X \in \mathfrak{A}$ such that

$$\mathfrak{K} \cap X\mathfrak{L} \neq \varnothing$$

will be denoted by $(\mathfrak{K}:\mathfrak{L})_l$ and called the set of left quotients on the division of $\mathfrak{K}$ by $\mathfrak{L}$. This nomenclature is explained by the fact that X belongs to $(\mathfrak{K}:\mathfrak{L})_l$ if and only if there exist $K \in \mathfrak{K}$ and $L \in \mathfrak{L}$ such that $XL = K$.

Analogously one defines the set of right quotients $(\mathfrak{K}:\mathfrak{L})_r$. The introduction of these concepts and the study of their properties is due to Dubreil [**1**; **3**; **4**; **5**].

Using these concepts, he constructed certain equivalences, which have also been studied in detail. Further investigations in this same direction were made by Croisot [3]. In approximately the same direction are the investigations of Thierrin [**12**; **19**].

Suppose that $\mathfrak{K}$ is a fixed subset of the semigroup $\mathfrak{A}$. One defines in $\mathfrak{K}$ the relations $\mathfrak{n}^l_{\mathfrak{K}}$ and $\mathfrak{n}^r_{\mathfrak{K}}$. For $A, B \in \mathfrak{A}$ one has $A \sim B(\mathfrak{n}^l_{\mathfrak{K}})$ if and only if

$$(\mathfrak{K}:A)_l = (\mathfrak{K}:B)_l.$$

The relation $\mathfrak{n}^r_{\mathfrak{K}}$ is defined analogously, using the collection of right quotients.

The relations $\mathfrak{n}^l_{\mathfrak{K}}$ and $\mathfrak{n}^r_{\mathfrak{K}}$, as is easily seen, are reflexive, symmetric and transitive. The first of them is stable to the left. Indeed, suppose that $A \sim B(\mathfrak{n}^l_{\mathfrak{K}})$ and $X \in (\mathfrak{K}:SA)_l$. For some $K \in \mathfrak{K}$,

$$X \cdot SA = K.$$

This means that $XS \in (\mathfrak{K}:A)_l$ and therefore $XS \in (\mathfrak{K}:B)_l$, i.e., for some $K' \in \mathfrak{K}$,

$$XS \cdot B = K'.$$

But then $X \in (\mathfrak{K}:SB)_l$. Carrying out a symmetric discussion we see that $(\mathfrak{K}:SA)_l = (\mathfrak{K}:SB)_l$.

Analogously one proves that the equivalence $\mathfrak{n}^r_{\mathfrak{K}}$ is stable to the right.

The set of all $A \in \mathfrak{A}$ for which $(\mathfrak{K}:A)_l$ is empty forms one of the $\mathfrak{n}^l_{\mathfrak{K}}$-classes of the equivalence $\mathfrak{n}^l_{\mathfrak{K}}$. This $\mathfrak{n}^l_{\mathfrak{K}}$-class is just the subsemigroup $\mathfrak{U}_{\mathfrak{K}}$ indicated in III, 1.13. The analogously defined $\mathfrak{n}^r_{\mathfrak{K}}$-class is the subsemigroup $\mathfrak{V}_{\mathfrak{K}}$ defined in the same section. The properties of these subsemigroups in many respects determine the properties of the equivalences $\mathfrak{n}^l_{\mathfrak{K}}$ and $\mathfrak{n}^r_{\mathfrak{K}}$ themselves.

For certain classes of subsets $\mathfrak{K}$ the equivalences $\mathfrak{n}^l_{\mathfrak{K}}$ and $\mathfrak{n}^r_{\mathfrak{K}}$ coincide. In this case the equivalence $\mathfrak{n}_{\mathfrak{K}} = \mathfrak{n}^l_{\mathfrak{K}} = \mathfrak{n}^r_{\mathfrak{K}}$ is two-sidedly stable. Homomorphisms to which these equivalences correspond have a particular interest. As was shown in the above-mentioned papers of Dubreil and in the later work of Stoll [2], their investigation is essential for a more detailed study of homomorphisms of semigroups onto groups and onto groups with an exteriorly-adjoined zero.

6.11. Suppose again that $\mathfrak{K}$ is some fixed subset of the semigroup $\mathfrak{A}$. In some questions the relations $\mathfrak{m}^l_{\mathfrak{K}}$ and $\mathfrak{m}^r_{\mathfrak{K}}$ are of interest. For $A, B \in \mathfrak{A}$ one has $A \sim B(\mathfrak{m}^l_{\mathfrak{K}})$ if there exist $X, Y \in \mathfrak{K}$ such that $XA = YB$. The relation $\mathfrak{m}^r_{\mathfrak{K}}$ is defined symmetrically. In the sequel we shall encounter in particular a case in which, relative to $\mathfrak{K} = \mathfrak{A}$, all the elements are $\mathfrak{m}^l_{\mathfrak{K}}$-equivalent to one another.

The relation $\mathfrak{m}^l_{\mathfrak{K}}$ is obviously reflexive, symmetric and stable to the right. In the general case it is not necessarily transitive. However, if $\mathfrak{K}$ is a subsemigroup such that all its elements are in the relation $\mathfrak{m}^l_{\mathfrak{K}}$ to each other, then $\mathfrak{m}^l_{\mathfrak{K}}$ is transitive. Indeed, suppose that

$$A \sim B(\mathfrak{m}^l_{\mathfrak{K}}), \qquad B \sim C(\mathfrak{m}^l_{\mathfrak{K}}) \qquad (A, B, C \in \mathfrak{A}),$$

$$XA = YB, \qquad X'B = Y'C, \qquad UY = VX' \qquad (X, Y, X', Y', U, V \in \mathfrak{K}).$$

Then

$$(UX)A = (UY)B = (VX')B = (VY')C$$

and therefore $A \sim C(\mathfrak{m}^l_{\mathfrak{K}})$.

The situation for the relation $\mathfrak{m}^r_{\mathfrak{K}}$ is analogous. In certain cases $\mathfrak{m}^l_{\mathfrak{K}}$ and $\mathfrak{m}^r_{\mathfrak{K}}$ coincide. Then, if the indicated conditions are satisfied, we obtain a two-sidedly stable equivalence in $\mathfrak{A}$. As Dubreil **[1]** showed in introducing the relations in question, there is a certain connection and interdependence between these relations and the relations indicated in 6.10. See also Dubreil **[3; 4; 5]**, and Thierrin **[12; 19]**.

6.12. Along with the consideration of various forms of homomorphisms in arbitrary semigroups, a certain number of investigations have been devoted to the consideration of homomorphisms of important individual semigroups.

Suppose that $\mathfrak{S}_\Omega$ is the semigroup of all transformations of the set Ω. We have already mentioned the important role of such semigroups.

For $A \in \mathfrak{S}_\Omega$ the cardinality of the set $A\Omega$ will be denoted by $\rho(A)$ (A. I. Mal'cev calls $\rho(A)$ the rank of the transformation A).

Suppose that n is some integer and that $\mathfrak{N}$ is some normal divisor (4.2) of the group $\mathfrak{G}$ of all the one-to-one mappings onto itself of the set $\{1, 2, \ldots, n\}$ (these are well known: for $n = 3$ and $n > 4$, aside from the two trivial divisors, there is only one normal divisor; for $n = 4$ there are two nontrivial normal divisors; and for $n = 1, 2$ there are no nontrivial normal divisors).

For $\mathfrak{N}$ we define in $\mathfrak{S}_\Omega$ a relation $\mathfrak{n}_{\mathfrak{N}}$ by putting $A \sim B(\mathfrak{n}_{\mathfrak{N}})$ $(A, B \in \mathfrak{S}_\Omega)$, if one of the three following conditions is satisfied:

(α) $A = B$.

(β) $\rho(A) < n$, $\rho(B) < n$.

(γ) $\rho(A) = \rho(B) = n$; $A\Omega = B\Omega$; $A\xi = A\eta$ implies $B\xi = B\eta$ $(\xi, \eta \in \Omega)$, and conversely; and if $\xi_1, \xi_2, \ldots, \xi_n \in \Omega$ is such that $A\xi_i \neq A\xi_j$ $(i, j = 1, 2, \ldots, n;\ i \neq j)$ and φ is some mapping $A\Omega = \{A\xi_1, A\xi_2, \ldots, A\xi_n\}$ onto $\{1, 2, \ldots, n\}$, then

$$\begin{pmatrix} \varphi(A\xi_1)\varphi(A\xi_2) \ldots \varphi(A\xi_n) \\ \varphi(B\xi_1)\varphi(B\xi_2) \ldots \varphi(B\xi_n) \end{pmatrix} \in \mathfrak{N}.$$

We may at once observe that in case (γ) the fact that the latter permutation belongs to $\mathfrak{N}$ does not depend, as one easily sees, either on the choice of the elements $\xi_1, \xi_2, \ldots, \xi_n$ or on the choice of the mapping φ (this last follows from the fact that for any automorphism of the group $\mathfrak{G}$ its normal divisor $\mathfrak{N}$ always maps onto itself).

The relation $\mathfrak{n}_{\mathfrak{N}}$ just defined is obviously symmetric and reflexive. It is easy to see that it is transitive as well. Let

$$A \sim B(\mathfrak{n}_{\mathfrak{N}}), \qquad B \sim C(\mathfrak{n}_{\mathfrak{N}}).$$

In cases (α) and (β) the situation is evident. In case (γ) we have

$$A\Omega = B\Omega = C\Omega,$$

$$S = \begin{pmatrix} \varphi(A\xi_1)\varphi(A\xi_2)\ldots\varphi(A\xi_n) \\ \varphi(B\xi_1)\varphi(B\xi_2)\ldots\varphi(B\xi_n) \end{pmatrix} \in \mathfrak{N};$$

$$S' = \begin{pmatrix} \varphi(B\xi_1)\varphi(B\xi_2)\ldots\varphi(B\xi_n) \\ \varphi(C\xi_1)\varphi(C\xi_2)\ldots\varphi(C\xi_n) \end{pmatrix} \in \mathfrak{N},$$

from which follows

$$\begin{pmatrix} \varphi(A\xi_1)\varphi(A\xi_2)\ldots\varphi(A\xi_n) \\ \varphi(C\xi_1)\varphi(C\xi_2)\ldots\varphi(C\xi_n) \end{pmatrix} = S'S \in \mathfrak{N}.$$

It is clearly impossible to have the "mixed cases," namely, those where $A \sim B(\mathfrak{n}_{\mathfrak{N}})$ holds as a result of one of the conditions (α), (β), (γ), while $B \sim C(\mathfrak{n}_{\mathfrak{N}})$ holds as the result of another of these conditions.

We shall prove that the equivalence $\mathfrak{n}_{\mathfrak{N}}$ is two-sidedly stable. Suppose that

$$A \sim B(\mathfrak{n}_{\mathfrak{N}}) \qquad (A, B \in \mathfrak{S}_{\Omega}).$$

If case (α) or (β) holds, then evidently for any $X \in \mathfrak{A}$,

$$XA \sim XB(\mathfrak{n}_{\mathfrak{N}}), \qquad AX \sim BX(\mathfrak{n}_{\mathfrak{N}}).$$

Now suppose that case (γ) holds. If for some ξ_i and ξ_j ($i \neq j$) we have $XA\xi_i = XA\xi_j$, then $\rho(XA) < n$, $\rho(XB) < n$ and therefore $XA \sim XB(\mathfrak{n}_{\mathfrak{N}})$. If the $XA\xi_i$ are all distinct ($i = 1, 2, \ldots, n$), we define a mapping ψ of the set $XA\Omega = XB\Omega$ onto $\{1, 2, \ldots, n\}$:

$$\psi(XA\alpha) = \varphi(A\alpha).$$

We observe that it follows from $A\Omega = B\Omega$ that for $\alpha \in \Omega$ there exists $\beta \in \Omega$ such that $B\alpha = A\beta$. Hence

$$\psi(XB\alpha) = \psi(XA\beta) = \varphi(A\beta) = \varphi(B\alpha).$$

All the requirements of condition (γ) for XA and XB are satisfied. For the last of these we have relative to ψ,

$$\begin{pmatrix} \psi(XA\xi_1)\psi(XA\xi_2)\ldots\psi(XA\xi_n) \\ \psi(XB\xi_1)\psi(XB\xi_2)\ldots\psi(XB\xi_n) \end{pmatrix} = \begin{pmatrix} \varphi(A\xi_1)\varphi(A\xi_2)\ldots\varphi(A\xi_n) \\ \varphi(B\xi_1)\varphi(B\xi_2)\ldots\varphi(B\xi_n) \end{pmatrix} \in \mathfrak{N}.$$

If for some ξ_i we have $AX\Omega \not\ni A\xi_i$, then $\rho(AX) < n$, $\rho(BX) < n$ and therefore $AX \sim BX(\mathfrak{n}_{\mathfrak{N}})$. In the opposite case we fix ξ'_i ($i = 1, 2, \ldots, n$) so that $AX\xi'_i = A\xi_i$. We have

$$AX\Omega = \{A\xi_1, A\xi_2, \ldots, A\xi_n\} = A\Omega.$$

Relative to the sequence $\xi'_1, \xi'_2, \ldots, \xi'_n$ and the mapping φ we have satisfied for AX and BX the requirements of condition (γ). For the last of these conditions we have

$$\begin{pmatrix} \varphi(AX\xi_1')\varphi(AX\xi_2')\dots\varphi(AX\xi_n') \\ \varphi(BX\xi_1')\varphi(BX\xi_2')\dots\varphi(BX\xi_n') \end{pmatrix} = \begin{pmatrix} \varphi(A\xi_1)\varphi(A\xi_2)\dots\varphi(A\xi_n) \\ (\varphi B\xi_1)\varphi(B\xi_2)\dots\varphi(B\xi_n) \end{pmatrix} \in \mathfrak{N}.$$

The two-sidedly stable equivalences $\mathfrak{n}_{\mathfrak{N}}$ which we have been considering were discovered by A. I. Mal'cev [**4**]. He showed that in the case of finite Ω the semigroup $\mathfrak{S}_\Omega$ has no other two-sidedly stable equivalences.

However, in the case of infinite Ω, aside from $\mathfrak{n}_{\mathfrak{N}}$ and $\mathfrak{S}_\Omega$ there are other two-sidedly stable equivalences. Denote by Σ the sequence of infinite cardinals

$$\Sigma = (\mathfrak{m}_1, \mathfrak{m}_2, \dots, \mathfrak{m}_k, \mathfrak{m}_k', \mathfrak{m}_{k-1}', \dots, \mathfrak{m}_1', \mathfrak{m}_0'),$$

where each of the cardinals is larger than its predecessor, and only $\mathfrak{m}_k'$ can be equal to $\mathfrak{m}_k$. In addition, if $\mathfrak{m}_0'$ is larger than the cardinality of the set Ω, then $\mathfrak{m}_0'$ is the smallest of the cardinals larger than the cardinality of Ω.

Define in $\mathfrak{S}_\Omega$ a relation $\mathfrak{n}^{(\Sigma)}$ by putting

$$A \sim B(\mathfrak{n}^{(\Sigma)}) \qquad (A, B \in \mathfrak{S}_\Omega)$$

if one of the three following conditions is satisfied:

(α) $A = B$.

(β) $\rho(A) < \mathfrak{m}_k'$, $\rho(B) < \mathfrak{m}_k'$.

(γ) $\rho(A) = \rho(B) = \rho$, where ρ is larger than or equal to some $\mathfrak{m}_{i+1}'$ and less than $\mathfrak{m}_i'$, while, if one denotes by Ω_0 the set of $\alpha \in \Omega$ for which $A\alpha \neq B\alpha$, then the cardinalities of $A\Omega_0$ and $B\Omega_0$ are less than the cardinal $\mathfrak{m}_{i+1}$.

Mal'cev showed that relations of the type $\mathfrak{n}^{(\Sigma)}$ are two-sidedly stable equivalences. In addition $\mathfrak{S}_\Omega$, aside from these equivalences and the equivalences described above of type $\mathfrak{n}_{\mathfrak{N}}$, has no other two-sidedly stable equivalences. Because of these results, all the homomorphisms of the semigroup $\mathfrak{S}_\Omega$ are characterized up to isomorphisms.

6.13. The discussion of two-sidedly stable equivalences of the semigroup of all one-to-one partial transformations of any set was carried out by A. E. Liber [**1**].

A series of investigations (A. I. Mal'cev [**5**], E. A. Halezov [**1; 2**], L. M. Gluskin [**4; 7; 9**]) have been devoted to homomorphisms of multiplicative semigroups of matrices.

6.14. Aside from the investigations in which homomorphisms of semigroups are studied up to multiplication by isomorphisms (which are equivalent to the study of two-sidedly stable equivalences), in some cases there is interest in properties of the homomorphic image. For example, the case of the automorphisms of a semigroup (I, 1.16) is of interest (and more generally the case of endomorphisms, though in fact, except for automorphisms, endomorphisms have not been studied at all).

If the semigroup $\mathfrak{A}$ has an identity and the elements X and X^{-1} are mutually inverse relative to $E_{\mathfrak{A}}$,

$$XX^{-1} = X^{-1}X = E_{\mathfrak{A}},$$

then the mapping φ_X of the semigroup $\mathfrak{A}$ into itself,

$$\varphi_X(A) = XAX^{-1} \qquad (A \in \mathfrak{A}),$$

as one easily sees, is a one-to-one mapping of $\mathfrak{A}$ onto itself. In addition

$$\varphi_X(AB) = XABX^{-1} = XAX^{-1}XBX^{-1} = \varphi_X(A) \cdot \varphi_X(B).$$

Accordingly, φ_X is an automorphism. Such automorphisms are called inner automorphisms.

The set of all automorphisms of any semigroup $\mathfrak{A}$ is a group. The set of inner automorphisms, if $\mathfrak{A}$ has an identity, is easily seen to be a semigroup and in fact a normal divisor of that group.

6.15. Some semigroups have only inner automorphisms; for example, the semigroup $\mathfrak{S}_\Omega$ of all transformations of a given set Ω.

Several proofs of this theorem are known (I. Schreier **[1]**, A. I. Mal'cev **[4]**, E. S. Ljapin **[13]**). The argument presented here will be based on the results of 5.7.

In the semigroup $\mathfrak{S}_\Omega$ the two-sided ideal $\mathfrak{U}_\Omega$ (5.7) is a minimal two-sided ideal. Therefore for any automorphism of $\mathfrak{S}_\Omega$, the ideal $\mathfrak{U}_\Omega$ maps onto itself. This means that any automorphism φ of the semigroup $\mathfrak{S}_\Omega$ induces an automorphism $\bar{\varphi}$ in $\mathfrak{U}_\Omega$.

We shall show that if the automorphisms φ_1 and φ_2 of the semigroup $\mathfrak{S}_\Omega$ are distinct, then the automorphisms $\bar{\varphi}_1$ and $\bar{\varphi}_2$ induced by them are also distinct.

Suppose that for some $X \in \mathfrak{S}_\Omega$ we have

$$\varphi_1(X) = Y_1, \qquad \varphi_2(X) = Y_2, \qquad Y_1 \neq Y_2.$$

For some $\alpha \in \Omega$,

$$Y_1\alpha = \beta_1, \qquad Y_2\alpha = \beta_2, \qquad \beta_1 \neq \beta_2.$$

Suppose that $\bar{\varphi}_1$ and $\bar{\varphi}_2$ are identical. Then the automorphism $\psi = (\varphi_2\varphi_1^{-1})$ maps each element of $\mathfrak{U}_\Omega$ onto itself, and, in particular,

$$\psi(U_\alpha) = U_\alpha, \qquad \psi(U_\beta) = U_\beta.$$

Since

$$\psi(Y_1) = \varphi_2\{\varphi_1^{-1}(Y_1)\} = \varphi_2(X) = Y_2,$$

by applying the automorphism ψ to the obvious equality $Y_1U_\alpha = U_\beta$ we obtain $Y_2U_\alpha = U_{\beta_1}$. The hypothesis $\bar{\varphi}_1 = \bar{\varphi}_2$ would lead us to a contradiction since

$$Y_2U_2\alpha = Y_2\alpha = \beta_2, \qquad U_{\beta_1}\alpha = \beta_1, \qquad \beta_1 \neq \beta_2.$$

For any automorphism φ of the semigroup $\mathfrak{S}_\Omega$, the automorphism $\bar{\varphi}$ induced by it is realized by the one-to-one mapping of U_Ω onto itself:

$$\bar{\varphi}(U_\alpha) = U_{\alpha_\varphi} \qquad (\alpha \in \Omega).$$

Consider the transformations $S, S' \in \mathfrak{S}_\Omega$, which realize one-to-one mappings of Ω onto itself, i.e.,

$$S\alpha = \alpha_\varphi, \qquad S'\alpha_\varphi = \alpha \qquad (\alpha \in \Omega).$$

Evidently S and S' are mutually inverse elements in $\mathfrak{S}_\Omega$, i.e., $S' = S^{-1}$. Therefore in $\mathfrak{S}_\Omega$ there is defined an interior automorphism χ,

$$\chi(A) = SAS^{-1} \qquad (A \in \mathfrak{S}_\Omega).$$

Let us compare the automorphisms $\bar{\chi}$ and $\bar{\varphi}$ of the semigroup $\mathfrak{U}_\Omega$. Since evidently $U_\alpha S' = U_\alpha$, we have

$$[\bar{\chi}(U_\alpha)]\alpha = (SU_\alpha S')\alpha = SU_\alpha\alpha = S\alpha = \alpha_\varphi,$$
$$[\bar{\varphi}(U_\alpha)]\alpha = U_{\alpha_\varphi}\alpha = \alpha_\varphi.$$

Thus $\bar{\chi}(U_\alpha) = \bar{\varphi}(U_\alpha)$. Since $\bar{\chi}$ and $\bar{\varphi}$ coincide, it follows from what has been proved above that $\chi = \varphi$, i.e., φ is an interior automorphism.

6.16. With the aid of 6.15 it is easy to determine the structure of the group of automorphisms of the semigroup $\mathfrak{S}_\Omega$. As follows from II, 3.1 and II, 3.2, mutually inverse elements of $\mathfrak{S}_\Omega$ are transformations which realize one-to-one mappings of Ω onto itself. Any two distinct transformations S_1 and S_2 generate distinct automorphisms of $\mathfrak{S}_\Omega$. For let

$$S_1\alpha = \beta_1, \qquad S_2\alpha = \beta_2, \qquad \beta_1 \neq \beta_2 \qquad (\alpha, \beta_1, \beta_2 \in \Omega).$$

Then

$$(S_1U_\alpha S_1^{-1})\alpha = (S_1U_\alpha)\alpha = S_1(U_\alpha\alpha) = S_1\alpha = \beta_1,$$
$$(S_2U_\alpha S_2^{-1})\alpha = (S_2U_\alpha)\alpha = S_2(U_\alpha\alpha) = S_2\alpha = \beta_2,$$

i.e., $S_1U_\alpha S_1^{-1} \neq S_2U_\alpha S_2^{-1}$.

Thus, between the automorphisms of $\mathfrak{S}_\Omega$ and the one-to-one mappings of Ω onto itself, we have established a one-to-one correspondence. It is easy to see that this is an isomorphism between the group of all automorphisms and the group of all transformations of Ω which realize one-to-one mappings of Ω onto itself.

6.17. Along with the above construction of inner automorphisms (6.14) by a method borrowed from the theory of groups, various generalizations of this construction have been considered by some authors (Dubreil [**1**; 3], Croisot [**6**]) in the general theory of semigroups. One of the simplest generalizations, for example, is the following. Suppose that $\mathfrak{A}$ is a semigroup with two-sided cancellation. Denote by $\mathfrak{B}$ the set of elements $X \in \mathfrak{A}$ satisfying the condition $X\mathfrak{A} = \mathfrak{A}X$. Suppose that $X \in \mathfrak{B}$. Then for any $A \in \mathfrak{A}$ there is a $\psi_X(A)$ such that $XA = \psi_X(A) \cdot X$.

The element $\psi_X(A)$, in view of the cancellation property in $\mathfrak{A}$, is uniquely defined. Since for any $B \in \mathfrak{A}$ there exists a $B' \in \mathfrak{A}$ such that $XB' = BX$, we have $B = \psi_X(B')$. Thus X defines a mapping ψ_X of the semigroup onto itself. It is one-to-one, as immediately follows from the cancellation property in $\mathfrak{A}$. Since the mapping has the homomorphism property

$$X \cdot (AB) = \psi_X(AB) \cdot X,$$

$$X \cdot (AB) = \psi_X(A)XB = \psi_X(A) \cdot \psi_X(B) \cdot X,$$

$$\psi_X(AB) = \psi_X(A) \cdot \psi_X(B)$$

it follows that ψ_X is an automorphism of $\mathfrak{A}$.

A subset $\mathfrak{B}$ of the semigroup $\mathfrak{A}$, if nonempty, is evidently a subsemigroup of $\mathfrak{A}$. If we assign to each X of $\mathfrak{B}$ an automorphism ψ_X, we easily obtain a homomorphism of $\mathfrak{B}$ into the group of all automorphisms of the semigroup $\mathfrak{A}$. Indeed, if $X, Y \in \mathfrak{B}$, then

$$XYA = X \cdot \psi_Y(A) \cdot Y = \psi_X\{\psi_Y(A)\} \cdot XY,$$

$$XYA = \psi_{XY}(A) \cdot XY,$$

$$\psi_X\{\psi_Y(A)\} = \psi_{XY}(A).$$

Since A is arbitrary in $\mathfrak{A}$, we have $\psi_X\psi_Y = \psi_{XY}$.

It follows in particular that the collection of automorphisms ψ_X form a subsemigroup of the group of all automorphisms.

6.18. If the semigroup $\mathfrak{A}$ has an identity and $X \in \mathfrak{A}$ has an inverse element, then

$$X\mathfrak{A} = X\mathfrak{A}X^{-1}X \subset \mathfrak{A}X,$$

$$\mathfrak{A}X = XX^{-1}\mathfrak{A}X \subset X\mathfrak{A},$$

$$X\mathfrak{A} = \mathfrak{A}X,$$

i.e., $X \in \mathfrak{B}$ (6.17). The automorphism ψ_X, defined in the above way with the element X, turns out to be identical with the inner automorphism generated by X (6.14):

$$XA = XAX^{-1} \cdot X, \qquad XA = \psi_X(A) \cdot X,$$

$$XAX^{-1} = \psi_X(A).$$

6.19. The concept "generalized inner automorphism" (6.17) is wider than the concept of inner automorphism. Even a semigroup with identity can have automorphisms of type 6.17 which are not inner. We shall give an example of this. Suppose that $\mathfrak{A}$ is the set of all pairs of integers (a, b) such that either $a = b = 0$ or $a \geqslant 1$. Multiplication in $\mathfrak{A}$ is defined as follows:

$$(a_1, b_1) \cdot (a_2, b_2) = (a_1 + a_2, (-1)^{a_2} b_1 + b_2).$$

One immediately verifies the associative law. If

$$(a_1, b_1) \cdot (a_2, b_2) = (a_3, b_3),$$

then for given a_1, b_1, a_3, b_3 the numbers a_2 and b_2 are evidently uniquely defined. Thus $\mathfrak{A}$ is a semigroup with left cancellation. Analogously one has right cancellation.

(0, 0) is evidently the identity of the semigroup $\mathfrak{A}$. No two nonidentity elements have their product equal to (0, 0). Hence it follows that $\mathfrak{A}$ has no inner automorphisms (6.14) distinct from the identity automorphism. Moreover, for $X = (1, 0)$ it is easy to see that $X\mathfrak{A}$ consists of (1, 0) and all possible pairs of the form (a, b), where $a \geqslant 2$. The same is true for the set $\mathfrak{A}X$. Thus $X \in \mathfrak{B}$ (6.17). Let us construct an automorphism ψ_X using X. It is not the identity automorphism since, as is easily seen,

$$(1, 0) \cdot (a, b) = (a, -b) \cdot (1, 0)$$

and therefore

$$\psi_X(a, b) = (a, -b).$$

6.20. With the use of certain concrete homomorphisms for commutative semigroups one may construct a theory which could be called a generalized theory of characters. Let $\mathfrak{A}$ and $\mathfrak{B}$ be any two commutative semigroups. Denote by $\Phi(\mathfrak{A}, \mathfrak{B})$ the set of all homomorphisms of $\mathfrak{A}$ into $\mathfrak{A}$. In $\Phi(\mathfrak{A}, \mathfrak{B})$ we define an operation by putting $\chi_1 \cdot \chi_2 = \chi_3 (\chi_1, \chi_2, \chi_3 \in \Phi(\mathfrak{A}, \mathfrak{B}))$ if for any $A \in \mathfrak{A}$ the equation

$$\chi_1(A) \cdot \chi_2(A) = \chi_3(A)$$

is satisfied in $\mathfrak{B}$.

The result of this operation is defined for any two χ_1 and χ_2 of $\Phi(\mathfrak{A}, \mathfrak{B})$ since

$$\begin{aligned} \chi_1(AB) \cdot \chi_2(AB) &= \chi_1(A) \cdot \chi_1(B) \cdot \chi_2(A) \cdot \chi_2(B) \\ &= [\chi_1(A) \cdot \chi_2(A)] \cdot [\chi_1(B) \cdot \chi_2(B)] \qquad (A, B \in \mathfrak{A}) \end{aligned}$$

and therefore the mapping χ_3 of the semigroup $\mathfrak{A}$ into $\mathfrak{B}$ defined by the formula given above is a homomorphism, i.e., belongs to $\Phi(\mathfrak{A}, \mathfrak{B})$.

In view of the method of defining the operation in $\Phi(\mathfrak{A}, \mathfrak{B})$, it is natural to call elements of $\Phi(\mathfrak{A}, \mathfrak{B})$ generalized characters of $\mathfrak{A}$ in $\mathfrak{B}$. Since their multiplication is evidently associative it follows that $\Phi(\mathfrak{A}, \mathfrak{B})$, if nonempty, is a semigroup. In view of the commutativity of $\mathfrak{B}$ this semigroup is commutative.

This method of assigning to the pair of commutative semigroups $\mathfrak{A}$ and $\mathfrak{B}$ a third commutative semigroup $\Phi(\mathfrak{A}, \mathfrak{B})$ gives rise to a number of problems concerning the mutual relations of various properties of these three semigroups. It is of interest to make a study of $\Phi(\mathfrak{A}, \mathfrak{B})$ based on the properties of the initial semigroups $\mathfrak{A}$ and $\mathfrak{B}$. Another approach is possible: namely to characterize the properties of $\mathfrak{A}$ by means of $\mathfrak{B}$ and $\Phi(\mathfrak{A}, \mathfrak{B})$. In fact, such a way of putting the problem is most natural, since it is a method of studying $\mathfrak{A}$ by means of its semigroup of generalized characters in the given semigroup $\mathfrak{B}$. Finally, one can conceive of an investigation of $\mathfrak{B}$ beginning with the known properties of $\mathfrak{A}$ and $\Phi(\mathfrak{A}, \mathfrak{B})$.

6.21. The general study of semigroups of generalized characters was begun recently by M. M. Lesohin [**1**]. Up to that time, all that had been done in this direction was for the case of complex characters, where $\mathfrak{B}$ is the multiplicative semigroup of all complex numbers. For finite commutative semigroups Schwarz [**9**; **10**; **11**] had developed a systematic theory. It turned out in the case considered that between the properties of $\mathfrak{A}$ and the properties of $\Phi(\mathfrak{A}, \mathfrak{B})$ there exist deep and meaningful connections. In particular, the relation between the lattice of ideals of $\mathfrak{A}$ and the lattice of ideals of $\Phi(\mathfrak{A}, \mathfrak{B})$ is very interesting. Also the paper of Hewitt and Zuckerman [**1**] and a paper of Iseki [**9**] were devoted to complex characters.

6.22. In the definition of generalized characters (6.20) the commutativity of the semigroup $\mathfrak{B}$ had to be required in order that a mapping of $\mathfrak{A}$ into $\mathfrak{B}$, defined as the product of two generalized characters, would itself have the homomorphism property. Of course, this goal may be achieved by laying down a weaker requirement. Indeed, it is sufficient to require that in $\mathfrak{B}$ for any $X, U, V, Y \in \mathfrak{B}$ we have

$$XUVY = XVUY.$$

However, it is not clear how in passing from the class of commutative semigroups to semigroups with such a property one may produce an essentially generalized theory.

As far as $\mathfrak{A}$ is concerned, there is in fact no direct necessity of imposing any restrictions whatever on $\mathfrak{A}$. However, for a commutative semigroup $\mathfrak{B}$, all the generalized characters of $\mathfrak{A}$ in $\mathfrak{B}$ will be divided on the right by a homomorphism φ which is the greatest common right divisor of the homomorphisms of $\mathfrak{A}$ onto commutative semigroups (6.2). Thus $\Phi(\mathfrak{A}, \mathfrak{B})$ essentially coincides with $\Phi(\varphi(\mathfrak{A}), \mathfrak{B})$ and all results obtained from the corresponding considerations for $\mathfrak{A}$ are in essence related to $\varphi(\mathfrak{A})$, which is commutative.

Analogously, in considering a $\mathfrak{B}$ with the generalized commutativity condition noted above, it again makes sense to restrict $\mathfrak{A}$ to satisfying the same condition.

CHAPTER VIII

DECOMPOSITION OF SEMIGROUPS INTO UNIONS OF SUBSEMIGROUPS

1. Bands of Semigroups

1.1. One of the methods of studying the structure and the properties of various semigroups consists in decomposing them into unions of subgroups belonging to some better-known class. Success in the application of such a method depends first on how well we know the properties of the semigroups which are components of such a union, and, second, on the character of the interrelations between the components in that union. The first statement above is clear enough. As an illustration of the second we note that, as we already know, every semigroup is the union of its monogenic subsemigroups. These latter have extremely simple structures and their properties have been thoroughly studied. However, the mutual relations among the monogenic subsemigroups in a semigroup are complicated and for the most part are still not clear, so that the indicated decomposition taken by itself cannot be used for the study of arbitrary semigroups. In the present chapter we consider some general properties for certains kinds of decompositions of the sort suggested above.

1.2. Among the various decompositions into unions of subsemigroups, the decompositions into disjoint unions are particularly interesting, i.e., *partitions of a semigroup by subsemigroups*. Recall that we have agreed to make no distinction in meaning between the terms "equivalence" and "partition" (or "decomposition" (I, 5.8)). In the present chapter we shall usually prefer to use the word "partition" as a more descriptive term for our present investigations.

Evidently any partition $\mathfrak{n}$ of the semigroup $\mathfrak{A}$ will be a partition into subsemigroups if from $X \sim Y(\mathfrak{n})$ $(X, Y \in \mathfrak{A})$ it always follows that $XY \sim X(\mathfrak{n})$.

1.3. As an example we examine the question of partitioning a commutative semigroup into semigroups with cancellation (I, 3.2). This condition was found by Schwarz **[17]**.

THEOREM. *In order that the commutative semigroup $\mathfrak{A}$ be decomposable into a nonintersecting union of subsemigroups which are semigroups with cancellation it is necessary and sufficient that for any X, $Y \in \mathfrak{A}$ the condition*

$$X^2 = XY = Y^2$$

should be satisfied only if $X = Y$.

PROOF. (1) If $\mathfrak{A}$ is decomposed in the required way and if for $X, Y \in \mathfrak{A}$ we have $X^2 = XY = Y^2$, then X and Y cannot lie in different components of the decomposition since otherwise X^2 and Y^2 would be contained in different components and could not be equal. Since X and Y lie in the same component, and this last is a semigroup with cancellation, it follows from $X^2 = XY$ that $X = Y$.

(2) Suppose that the stated condition is satisfied in $\mathfrak{A}$. We define a relation $\mathfrak{n}$ in $\mathfrak{A}$ by writing $X \sim Y(\mathfrak{n})$ if for some natural numbers n and m we have $X^n = Y^m$. The relation $\mathfrak{n}$ is symmetric and reflexive. We shall verify its transitivity. Suppose that

$$X \sim Y(\mathfrak{n}), \qquad Y \sim Z(\mathfrak{n}),$$
$$X^n = Y^m, \qquad Y^k = Z^l.$$

Then

$$X^{nk} = Y^{mk} = (Y^k)^m = (Z^l)^m = Z^{lm},$$

i.e., $X \sim Z(\mathfrak{n})$.

Thus $\mathfrak{n}$ may be regarded as a partition of $\mathfrak{A}$ into $\mathfrak{n}$-classes (I, 5.8).

If

$$X \sim Y(\mathfrak{n}), \qquad X^m = Y^l,$$

then

$$(XY)^l = X^l Y^l = X^{l+m},$$

i.e., $XY \sim X(\mathfrak{n})$, and, from 1.2, $\mathfrak{n}$ is a partition of $\mathfrak{A}$ into subsemigroups. We shall show that each such subsemigroup $\mathfrak{B}$, consisting of elements equivalent to each other modulo $\mathfrak{n}$, is a semigroup with cancellation.

Suppose that

$$XZ = YZ \qquad (X, Y, Z \in \mathfrak{B}).$$

For certain n, m, s, t,

$$X^n = Z^m, \qquad Y^s = Z^t.$$

Suppose that p is the smallest natural number such that $X^{p+1} = X^p Y$. That such numbers exist follows from

$$X^{n+1} = X \cdot X^n = XZ^m = XZ \cdot Z^{m-1} = YZ \cdot Z^{m-1} = YZ^m = X^n Y.$$

If now p were even we would see from $X^{p+1} = X^p Y$ that

$$(X^{p/2} Y)^2 = (X^{p/2} Y) \cdot (X^{p/2+1}) = (X^{p/2+1})^2,$$

from which, because of the original hypothesis, it must follow that $X^{p/2+1} = X^{p/2} Y$. But this contradicts the definition of p. Accordingly, p is odd. But from $X^{p+1} = X^p Y$ we obtain $X^{(p+1)+1} = X^{p+1} Y$, from which by an argument analogous to that presented above we obtain $X^{[(p+1)/2]+1} = X^{(p+1)/2} Y$. Hence from the definition of p we conclude that $(p+1)/2 \geqslant p$, i.e., $p = 1$.

We have shown that $X^2 = XY$. Quite analogously one proves that $Y^2 = YX$. Thus it turns out that in $\mathfrak{B}$ it necessarily follows from $XZ = YZ$ that

$$X^2 = XY = Y^2.$$

But then, by hypothesis, it must follow that $X = Y$.

1.4. We note that the relation $\mathfrak{n}$ which we constructed in the proof of Theorem 1.3 for a commutative semigroup $\mathfrak{A}$, satisfying the condition indicated in Theorem 1.3, is the greatest lower bound for all partitions of $\mathfrak{A}$ into subsemigroups. In other words, $\mathfrak{n}$ is the finest of all possible partitions of $\mathfrak{A}$ into nonintersecting subsemigroups ($\mathfrak{n}$ is a refinement of every such partition). Indeed, if $\mathfrak{m}$ is one of these partitions and $X \sim Y(\mathfrak{n})$, i.e., $X^n = Y^m$, then evidently $X \sim Y(\mathfrak{m})$ in view of the fact that X and Y, because of $X^n = Y^m$, cannot lie in different $\mathfrak{m}$-classes, since different $\mathfrak{m}$-classes cannot have common elements.

Among the properties of the partition $\mathfrak{n}$ we note that each of its components $\mathfrak{K}$ has no more than one idempotent. Indeed, if $I_1, I_2 \in \mathfrak{K}$ and $I_1^2 = I_1$, $I_2^2 = I_2$, then from $I_1^m = I_2^n$ follows $I_1 = I_2$.

We observe further that, from IV, 6.11, $\mathfrak{K}$ is a semigroup having no proper isolated ideals. Indeed, if $X \sim Y(\mathfrak{n})$, then it immediately follows that X and Y stand to each other in the relation considered in IV, 6.11. A similar decomposition was accomplished by Thierrin **[16]**, not only for commutative semigroups but also for semigroups satisfying a certain generalized commutativity condition considered in III, 4.5, (β).

1.5. It is easy to see that the condition indicated in Theorem 1.3 is a necessary condition for a partition into subsemigroups with cancellation, not only for commutative but also for arbitrary semigroups. The question as to its sufficiency for certain classes of noncommutative semigroups was also considered by Schwarz **[16]**.

1.6. It is quite natural that among the various partitions of a semigroup into subsemigroups a particular interest lies in the two-sidedly stable partitions (I, 5.17).

DEFINITION. *A two-sidedly stable partition of a semigroup $\mathfrak{A}$, all of whose components are subsemigroups of it, is called a* BAND.

From 1.2 a two-sidedly stable partition $\mathfrak{n}$ of the semigroup $\mathfrak{A}$ will be a band if and only if for any $X \in \mathfrak{A}$ we have $X \sim X^2(\mathfrak{n})$.

If the semigroup $\mathfrak{A}$ has a band whose components are its subsemigroups $\mathfrak{B}_\alpha, \mathfrak{B}_\beta, \ldots$, then one also says that $\mathfrak{A}$ represents the band of its subsemigroups $\mathfrak{B}_\alpha, \mathfrak{B}_\beta, \ldots$, or, more simply, that $\mathfrak{A}$ is itself the band of these subsemigroups. Such a usage of the same term in different but evidently related senses will usually not lead to a misunderstanding.

1.7. As follows immediately from Definition 1.2 and I, 5.18, a band of the semigroup $\mathfrak{A}$ is defined by a collection of pairwise disjoint subsemigroups $\mathfrak{B}_\alpha, \mathfrak{B}_\beta, \ldots$ of that semigroup, whose union is equal to $\mathfrak{A}$, while for each pair of these subsemigroups $(\mathfrak{B}_\xi, \mathfrak{B}_\eta)$ one may always find a subsemigroup $\mathfrak{B}_\sigma$ of this collection such that

$$\mathfrak{B}_\xi \cdot \mathfrak{B}_\eta \subset \mathfrak{B}_\sigma.$$

Incidentally it is clear that this $\mathfrak{B}_\sigma$ is uniquely defined for a given pair $(\mathfrak{B}_\xi, \mathfrak{B}_\eta)$.

1.8. In VII, 2.1 and VII, 2.6 it was shown that every homomorphism of a semigroup defines in it a corresponding two-sidedly stable partition while each two-sidedly stable partition is the result of some homomorphism. If one restricts himself to the natural homomorphisms (VII, 2.5), then between these and all possible two-sidedly stable partitions one may establish a one-to-one correspondence.

Suppose that φ is a homomorphism of the semigroup $\mathfrak{A}$ onto the semigroup $\varphi(\mathfrak{A})$. If all the elements of the latter are idempotents, then the two-sidedly stable partition $\mathfrak{n}$ of the semigroup $\mathfrak{A}$ corresponding to this homomorphism (VII, 2.1) will satisfy the condition $X^2 \sim X(\mathfrak{n})$ for every $X \in \mathfrak{A}$ (since $\varphi(X^2) = \varphi(X) \cdot \varphi(X) = \varphi(X)$), and consequently $\mathfrak{n}$ is a band (1.2). The converse is also easily verified. If $\mathfrak{n}$ is a band and φ the homomorphism of $\mathfrak{A}$ to which the two-sidedly stable partition $\mathfrak{n}$ corresponds, then all the elements of the semigroup $\varphi(\mathfrak{A})$ are idempotents. Indeed, from $X^2 \sim X(\mathfrak{n})$ (1.6),

$$\varphi(X) \cdot \varphi(X) = \varphi(X^2) = \varphi(X).$$

Suppose that Γ is some class of semigroups. From what has been said it follows that for the semigroup $\mathfrak{A}$ to be represented in the form of a band of semigroups belonging to Γ it is necessary and sufficient that there should exist a homomorphism φ of the semigroup $\mathfrak{A}$ under which the complete preimage of each element of $\varphi(\mathfrak{A})$ is a semigroup belonging to Γ.

1.9. Suppose that Ψ is a nonempty collection of bands of the semigroup $\mathfrak{A}$. From VII, 2.9 there exists a two-sidedly stable partition $\mathfrak{k}$, the greatest lower bound of Ψ, and a two-sidedly stable partition $\mathfrak{l}$, the least upper bound of Ψ. From the constructions of $\mathfrak{k}$ and $\mathfrak{l}$ themselves (I, 5.15; I, 5.16) it is clear that since for every $\mathfrak{n} \in \Psi$ we have $X \sim X^2(\mathfrak{n})$ $(X \in \mathfrak{A})$, we must have $X \sim X^2(\mathfrak{k})$, $X \sim X^2(\mathfrak{l})$ both for $\mathfrak{k}$ and for $\mathfrak{l}$. Accordingly, $\mathfrak{k}$ and $\mathfrak{l}$ themselves are bands. It follows from this that relative to the partial ordering I, 5.14 *the set of all bands of a semigroup is a complete lattice.*

1.10. The least upper bound of all the bands of a semigroup is evidently the trivial band consisting of the one-component partition consisting of the semigroup itself.

The greatest lower bound of all bands is a band $\mathfrak{k}_0$, which is the finest two-sidedly stable decomposition of the semigroup into subsemigroups. All other bands are gotten from it by "amalgamation" of components, i.e., by taking the unions of some components of the partition $\mathfrak{k}_0$.

1.11. The structure of the band $\mathfrak{k}_0$ indicated in 1.10 was clarified for commutative semigroups in the paper of Tamura and Kimura [1]. Let us define in a commutative semigroup $\mathfrak{A}$ a relation $\mathfrak{k}$ by writing $X \sim Y(\mathfrak{k})$ $(X, Y \in \mathfrak{A})$ if there exist natural numbers p, q and elements $U, V \in \mathfrak{A}$ such that

$$X^p = YU, \qquad Y^q = XV.$$

Evidently $\mathfrak{k}$ is symmetric and reflexive. If

$$X \sim Y(\mathfrak{k}), \qquad Y \sim Z(\mathfrak{k}),$$

$$X^{p_1} = YU_1, \qquad Y^{q_1} = XV_1, \qquad Y^{p_2} = ZU_2, \qquad Z^{q_2} = YV_2,$$

then

$$X^{p_1p_2} = Y^{p_2}U_1^{p_2} = Z(U_2U_1^{p_2}), \qquad Z^{q_1q_2} = Y^{q_1}V_2^{q_1} = X(V_1V_2^{q_1}),$$

i.e., $X \sim Z(\mathfrak{k})$, and accordingly $\mathfrak{k}$ is transitive. The equivalence $\mathfrak{k}$ is two-sidedly stable, since from

$$X \sim Y(\mathfrak{k}),$$

$$X^p = YU, \qquad Y^q = XV$$

for any $S \in \mathfrak{A}$ we obtain

$$(XS)^p = X^pS^p = (YS)\cdot(US^{p-1}), \qquad (YS)^q = Y^qS^q = (XS)(VS^{q-1}),$$

which means that $XS \sim YS(\mathfrak{k})$ (S^0 here denotes the empty symbol).

Using 1.2, we verify that the partition $\mathfrak{k}$ is a band. Indeed, for any $X \in \mathfrak{A}$ we have

$$(X^2)^1 = X \cdot X, \qquad X^4 = X^2 \cdot X^2,$$

i.e., $X^2 \sim X(\mathfrak{k})$.

We shall show that the band $\mathfrak{k}$ coincides with the band $\mathfrak{k}_0$ (1.10).

Suppose that φ is some homomorphism of $\mathfrak{A}$ onto the semigroup $\varphi(\mathfrak{A})$, all elements of which are idempotents, and suppose that to φ corresponds the band $\mathfrak{k}_0$ (1.8). We note that for each $A \in \mathfrak{A}$, for any natural number s,

$$\varphi(A) = \varphi(A)^s = \varphi(A^s).$$

If for some $X, Y \in \mathfrak{A}$ we have

$$X \sim Y(\mathfrak{k}),$$

$$X^p = YU, \qquad Y^q = XV,$$

then

$$\varphi(X) = \varphi(X^p) = \varphi(YU) = \varphi(Y)\cdot\varphi(U) = \varphi(Y^q)\cdot\varphi(U)$$
$$= \varphi(XV)\cdot\varphi(U) = \varphi(X)\cdot\varphi(V)\cdot\varphi(U).$$

Analogously, one proves also

$$\varphi(Y) = \varphi(Y)\cdot\varphi(V)\cdot\varphi(U).$$

Using the equalities just obtained, we find that

$$\varphi(X) = \varphi(X)\cdot\varphi(V)\cdot\varphi(U) = \varphi(X)\cdot\varphi(V)\cdot\varphi(V)\cdot\varphi(U)$$
$$= \varphi(XV)\cdot\varphi(V)\cdot\varphi(U) = \varphi(Y^q)\cdot\varphi(V)\cdot\varphi(U) = \varphi(Y)\cdot\varphi(V)\cdot\varphi(U) = \varphi(Y).$$

The fact that $\varphi(X) = \varphi(Y)$ means that $X \sim Y(\mathfrak{k}_0)$.

We have shown that from $X \sim Y(\mathfrak{k})$ it always follows that $X \sim Y(\mathfrak{k}_0)$. Accordingly, $\mathfrak{k} \leqslant \mathfrak{k}_0$. But on the other hand $\mathfrak{k}_0$, being the greatest lower bound of all bands, precedes $\mathfrak{k}$: $\mathfrak{k}_0 \leqslant \mathfrak{k}$. Thus $\mathfrak{k} = \mathfrak{k}_0$.

1.12. As we have seen, giving a band for a semigroup is essentially equivalent to giving a homomorphism of it onto a semigroup all of whose elements are idempotents. The case when this last semigroup is commutative, i.e., when the homomorphism φ to which the band corresponds satisfies $\varphi(X)\cdot\varphi(Y) = \varphi(Y)\cdot\varphi(X)$, i.e., $\varphi(XY) = \varphi(YX)$, for any X and Y of the semigroup, merits particular attention. Bands with such properties may be characterized in the following way.

DEFINITION. *A band* $\mathfrak{n}$ *of the semigroup* $\mathfrak{A}$ *is said to be a* COMMUTATIVE BAND *if for any* $X, Y \in \mathfrak{A}$,

$$XY \sim YX(\mathfrak{n}).$$

The term "band" was introduced by Clifford **[11]**. Instead of commutative band, Clifford and certain other authors use the term "semilattice" of semigroups.

For a commutative semigroup every band is evidently a commutative band.

We recall that in considering homomorphisms of inverse semigroups (VII, § 3) we in fact have already used the construction of commutative bands.

1.13. In a way quite analogous to what was done in 1.9, one may verify that, for any nonempty collection of commutative bands, their greatest lower bound and least upper bound are commutative bands. Thus, *relative to the partial ordering* I, 5.14, *the set of commutative bands of a semigroup is a complete lattice.* Hence in particular it follows that for a semigroup there exists a commutative band $\mathfrak{n}_0$ (the greatest lower bound of all the commutative bands of the semigroup) which is the most refined partition of those partitions that are commutative bands. All the remaining bands can be gotten from $\mathfrak{n}_0$ by "amalgamation" of components, i.e., by taking unions of certain components of $\mathfrak{n}_0$.

1.14. DEFINITION. *If to each component of some band of the semigroup* $\mathfrak{A}$ *one can assign a pair of indices* $\mathfrak{B}_{\xi\eta}$, *where* ξ *runs through some set of elements* Γ', *and* η *runs through a set* Γ, *such that for any* $\xi_1, \xi_2 \in \Gamma'$ *and* $\eta_1, \eta_2 \in \Gamma$,

$$\mathfrak{B}_{\xi_1\eta_1}\cdot\mathfrak{B}_{\xi_2\eta_2} \subset \mathfrak{B}_{\xi_1\eta_2},$$

then the given band is called a MATRIX BAND.

1.15. Suppose that $\mathfrak{n}$ is some band of the semigroup $\mathfrak{A}$. If this band is a matrix band, then, assigning to all the elements of the component $\mathfrak{B}_{\xi\eta}$ the pair (ξ, η) of indexes for that component, we obviously obtain a homomorphism of the semigroup $\mathfrak{A}$ to which the given matrix band corresponds. This homomorphism maps $\mathfrak{A}$ onto a semigroup of the type considered in V, 6.11. This is a semigroup all of whose elements are regularly adjoint to one another. The inverse is also evident: if the homomorphism φ to which the given band corresponds is such that $\varphi(\mathfrak{A})$ is a semigroup all of whose elements are regularly adjoint to one another (V, 6.12), then $\mathfrak{n}$ is a matrix band.

1.16. The important role of the concept of matrix band becomes particularly clear when we prove in the following section that every band of semigroups of a certain class Σ may be represented in the form of a commutative band of semigroups which are matrix bands of semigroups of the class Σ.

1.17. We observe that we have already encountered a construction of the type described. A completely simple semigroup without zero of matrix type $\mathfrak{S}'(P, \mathfrak{G})$ (V, 6.3) is a matrix band of groups $\mathfrak{G}_{\xi\eta}$ (V, 6.7), where $\mathfrak{G}_{\xi\eta}$ consists of all elements of the form (G, ξ, η) $(G \in \mathfrak{G})$. In the same way, from V, 6.4, every completely simple semigroup without zero turns out to be a matrix band of groups.

1.18. As Clifford remarked **[11]**, the indicated property 1.17 of completely simple semigroups without zero is characteristic. Every semigroup $\mathfrak{A}$ which is a matrix band of groups is a completely simple semigroup without zero. The validity of this assertion results from the more general property which follows.

THEOREM. *A matrix band of completely simple semigroups without zero is itself a completely simple semigroup without zero.*

PROOF. Suppose that $\mathfrak{A}$ is a matrix band of completely simple semigroups without zero, namely, $\mathfrak{B}_{\alpha\beta}, \ldots, \mathfrak{B}_{\xi\eta}, \ldots$.

If the two-sided ideal $\mathfrak{T}$ of the semigroup $\mathfrak{A}$ contains the element $X \in \mathfrak{B}_{\xi\eta}$, then it also contains the set $\mathfrak{B}_{\xi\eta}X\mathfrak{B}_{\xi\eta}$. But this last is a two-sided ideal of the completely simple semigroup $\mathfrak{B}_{\xi\eta}$ without zero. Therefore

$$\mathfrak{B}_{\xi\eta} = \mathfrak{B}_{\xi\eta}X\mathfrak{B}_{\xi\eta} \subset \mathfrak{T}.$$

If $\mathfrak{T} \supset \mathfrak{B}_{\xi\eta}$, then, because of the property of a matrix band, for any $\mathfrak{B}_{\sigma\tau}$ we have

$$\mathfrak{B}_{\sigma\xi}\mathfrak{B}_{\xi\eta}\mathfrak{B}_{\eta\tau} \subset \mathfrak{B}_{\sigma\tau}.$$

Since $\mathfrak{B}_{\xi\eta} \subset \mathfrak{T}$, then also $\mathfrak{B}_{\sigma\xi}\mathfrak{B}_{\xi\eta}\mathfrak{B}_{\eta\tau} \subset \mathfrak{T}$, so that the intersection of $\mathfrak{B}_{\sigma\tau}$ and $\mathfrak{T}$ is nonempty. Hence, in accordance with the above it follows that $\mathfrak{B}_{\sigma\tau} \subset \mathfrak{T}$. Thus we have proved that any two-sided ideal $\mathfrak{T}$ of the semigroup $\mathfrak{A}$ necessarily coincides with $\mathfrak{A}$.

In one of the semigroups $\mathfrak{B}_{\xi\eta}$ we choose an idempotent I which is not a two-sided unit for any other idempotent of $\mathfrak{B}_{\xi\eta}$. Such an I exists because of V, 3.10, V, 6.15, since $\mathfrak{B}_{\xi\eta}$ is a completely simple semigroup without zero. We shall show that I cannot also be a two-sided unit for any other idempotent $I' \in \mathfrak{B}_{\sigma\tau}$, where $\sigma \neq \xi$ or $\tau \neq \eta$. Indeed, from the property of a matrix band we obtain for $\sigma \neq \xi$,

$$II' \in \mathfrak{B}_{\xi\eta}\mathfrak{B}_{\sigma\tau} \subset \mathfrak{B}_{\xi\tau}$$

and therefore II' differs from I', which is contained in $\mathfrak{B}_{\sigma\tau}$.

If $\tau \neq \eta$ we analogously prove that $I'I$ is contained in $\mathfrak{B}_{\sigma\eta}$ and therefore differs from I'.

The existence in $\mathfrak{A}$ of an idempotent A which is not a two-sided unit for any other idempotent I' means, from the above and from Theorem V, 6.15, that $\mathfrak{A}$ is a completely simple semigroup without zero.

2. Completely Regular Semigroups

2.1. We have defined a completely regular semigroup as a semigroup all of whose elements are completely regular (II, 6.1). But, as we know (III, 1.15), the only elements of the semigroup which are completely regular are those contained in certain of its subgroups. Therefore it follows that a semigroup is completely regular if and only if it is a union of groups. If one takes into account what was said in III, 1.6, one may strengthen this condition. *Every completely regular semigroup is a disjoint union of groups.* One may also say that *a semigroup is completely regular if and only if it has a partition all of whose elements are groups.* There exists a whole series of criteria for complete regularity of semigroups. Semigroups of this class were first considered by Clifford [**4**] under the name of semigroups with relative inverses. This nomenclature is explained by the fact that, as follows from II, 6.5, there exist in completely regular semigroups, and only in them, for any A, elements A' and I such that

$$IA = AI = A, \qquad AA' = A'A = I.$$

2.2. As Croisot proved [**5**], the class of completely regular semigroups may be described with the aid of regularity classes (II, 6.1).

THEOREM. *A semigroup $\mathfrak{A}$ is completely regular if and only if*

$$\mathfrak{A} = \mathfrak{C}_{\mathfrak{A}}(2, 0) = \mathfrak{C}_{\mathfrak{A}}(0, 2).$$

PROOF. (1) If $\mathfrak{A}$ is a union of groups, then any element A of $\mathfrak{A}$ is contained in some group $\mathfrak{G} \subset \mathfrak{A}$. In this group A has a two-sided inverse element A^{-1} relative to $E_{\mathfrak{G}}$. Therefore

$$A = A^2A^{-3}A^2$$

and accordingly

$$A \in \mathfrak{C}_{\mathfrak{A}}(2, 0), \qquad A \in \mathfrak{C}_{\mathfrak{A}}(0, 2),$$

i.e.,

$$\mathfrak{A} = \mathfrak{C}_{\mathfrak{A}}(2, 0) = \mathfrak{C}_{\mathfrak{A}}(0, 2).$$

(2) Suppose that

$$\mathfrak{A} = \mathfrak{C}_{\mathfrak{A}}(2, 0) = \mathfrak{C}_{\mathfrak{A}}(0, 2).$$

For any element $A \in \mathfrak{A}$ there exist $X, Y \in \mathfrak{A}$ such that

$$A = XA^2, \qquad A = A^2Y.$$

Since

$$XA = XA^2Y = AY,$$

then

$$(XA) \cdot A = A, \qquad A \cdot (XA) = AAY = A,$$

$$XA = X \cdot A = A \cdot Y.$$

XA turns out to be a regular two-sided unit for the element A, from which, using II, 6.4, it follows that A is completely regular.

2.3. COROLLARY. *The semigroup* $\mathfrak{A}$ *is completely regular if and only if*

$$\mathfrak{A} = \mathfrak{C}_{\mathfrak{A}}(2, 2).$$

Indeed, if $\mathfrak{A}$ is completely regular, then, as we proved in the first part of the preceding theorem, every element $A \in \mathfrak{A}$ is contained in $\mathfrak{C}_{\mathfrak{A}}(2, 2)$.

On the other hand, from II, 6.12, (β),

$$\mathfrak{C}_{\mathfrak{A}}(2, 2) \subset \mathfrak{C}_{\mathfrak{A}}(2, 0) \subset \mathfrak{A},$$

$$\mathfrak{C}_{\mathfrak{A}}(2, 2) \subset \mathfrak{C}_{\mathfrak{A}}(0, 2) \subset \mathfrak{A}.$$

Therefore from the equality indicated in the formulation of the corollary it follows that the conditions of the preceding theorem are fulfilled.

2.4. The sufficiency of criterion 2.3 may be strengthened.

THEOREM. *The semigroup* $\mathfrak{A}$ *is completely regular if and only if*

$$\mathfrak{A} = \mathfrak{C}_{\mathfrak{A}}(1, 1) = \mathfrak{C}_{\mathfrak{A}}(0, 2).$$

PROOF. If $\mathfrak{A}$ is completely regular, then the indicated equality follows directly from 2.3 by using II, 6.12, (β).

Suppose that

$$\mathfrak{A} = \mathfrak{C}_{\mathfrak{A}}(1, 1) = \mathfrak{C}_{\mathfrak{A}}(0, 2).$$

From II, 6.12, (ε) it therefore follows that

$$\mathfrak{A} = \mathfrak{C}_{\mathfrak{A}}(1, 2).$$

For any element $A \in \mathfrak{A}$ there exists an $X \in \mathfrak{A}$ such that

$$A = AXA^2,$$

and for XA there exists a $Y \in \mathfrak{A}$ such that

$$XA = (XA) \cdot Y \cdot (XA)^2.$$

Therefore we obtain

$$\begin{aligned} A = AXA^2 = A \cdot XA \cdot A = A \cdot (XA) \cdot Y \cdot (XA)^2 \cdot A \\ = AXAYX \cdot (AXA^2) = AXAYXA; \end{aligned}$$

$$A = AXAY \cdot XA = AXAY \cdot (XA) \cdot Y \cdot (XA)^2 = AY(XA)^2;$$

$$\begin{aligned} A = AY(XA)^2 = AYX \cdot A \cdot XA = AYX \cdot AXA^2 \cdot XA \\ = AY(XA)^2 \cdot AXA = A \cdot AXA = A^2 \cdot (XA). \end{aligned}$$

Accordingly, $A \in \mathfrak{C}_{\mathfrak{A}}(2, 0)$, i.e., $\mathfrak{A} = \mathfrak{C}_{\mathfrak{A}}(2, 0)$. Since furthermore by hypothesis, $\mathfrak{A} = \mathfrak{C}_{\mathfrak{A}}(0, 2)$, it follows from 2.2 that $\mathfrak{A}$ is completely regular.

2.5. Quite analogously one proves a symmetric theorem, according to which $\mathfrak{A}$ is completely regular if and only if $\mathfrak{A} = \mathfrak{C}_{\mathfrak{A}}(1, 1) = \mathfrak{C}_{\mathfrak{A}}(0, 2)$.

2.6. Comparing 2.3, 2.4, and 2.5, we immediately obtain a corollary on the respective regularity classes.

COROLLARY. *If in the semigroup* $\mathfrak{A}$ *one of the three classes* $\mathfrak{C}_{\mathfrak{A}}(1, 2)$, $\mathfrak{C}_{\mathfrak{A}}(2, 1)$, $\mathfrak{C}_{\mathfrak{A}}(2, 2)$ *coincides with* $\mathfrak{A}$, *then the two remaining classes also coincide with* $\mathfrak{A}$.

2.7. The criteria just obtained for regularity make it possible to characterize completely regular semigroups from the point of view of the isolation property for their ideals (IV, 6.1; IV, 6.2, (η)). Indeed, comparing 2.2 and IV, 6.4, (γ), we immediately obtain such a condition.

COROLLARY. *For completely regular semigroups, and only for such semigroups, all the ideals are isolated.*

2.8. Using 2.4 and IV, 6.4, (α), we may modify the last condition.

COROLLARY. *A semigroup is completely regular if and only if it is a regular semigroup all of whose left ideals are isolated.*

2.9. We consider a series of other properties of ideals of completely regular semigroups.

(α) *If* $\mathfrak{T}_1$ *and* $\mathfrak{T}_2$ *are two-sided ideals of* $\mathfrak{A}$, *then*

$$\mathfrak{T}_1\mathfrak{T}_2 = \mathfrak{T}_1 \cap \mathfrak{T}_2.$$

Indeed, the inclusion

$$\mathfrak{T}_1\mathfrak{T}_2 \subset \mathfrak{T}_1 \cap \mathfrak{T}_2$$

is evident. On the other hand, suppose that $X \in \mathfrak{T}_1 \cap \mathfrak{T}_2$. Since X is completely regular, we have for some $Y \in \mathfrak{A}$,

$$XYX = X.$$

Since $X \in \mathfrak{T}_1$ and $YX \in \mathfrak{T}_2$ it follows that

$$X = X \cdot YX \in \mathfrak{T}_1\mathfrak{T}_2.$$

Accordingly,

$$\mathfrak{T}_1 \cap \mathfrak{T}_2 \subset \mathfrak{T}_1\mathfrak{T}_2.$$

(β) *If* $\mathfrak{T}_1$ *and* $\mathfrak{T}_2$ *are two-sided ideals of* $\mathfrak{A}$, *then*

$$\mathfrak{T}_1\mathfrak{T}_2 = \mathfrak{T}_2\mathfrak{T}_1.$$

This immediately follows from (α).

(γ) *If* $\mathfrak{T}$ *is a two-sided ideal of* $\mathfrak{A}$, *then*

$$\mathfrak{T}\mathfrak{T} = \mathfrak{T}.$$

This immediately follows from (α).

(δ) *The two-sided ideal envelope of the element* $A \in \mathfrak{A}$ (IV, 3.3; IV, 3.6) *is* $\mathfrak{A}A\mathfrak{A}$.

Indeed, since A has a two-sided identity I, then

$$\mathfrak{A}A = \mathfrak{A}AI \subset \mathfrak{A}A\mathfrak{A};$$

$$A\mathfrak{A} = IA\mathfrak{A} \subset \mathfrak{A}A\mathfrak{A};$$

$$A = IAI \subset \mathfrak{A}A\mathfrak{A}$$

and therefore

$$\mathfrak{A}A\mathfrak{A} \cup \mathfrak{A}A \cup A\mathfrak{A} \cup A = \mathfrak{A}A\mathfrak{A}.$$

(ε) *The principal two-sided ideals of* $\mathfrak{A}$ *form a semigroup with respect to the operation of multiplication of subsets, and the assignment to each element of* $\mathfrak{A}$ *of its two-sided ideal envelope is a homomorphism of* $\mathfrak{A}$ *onto that semigroup.*

Indeed, for any $A, B, C \subset \mathfrak{A}$,

$$(BCA)^2 = BC \cdot AB \cdot CA \subset \mathfrak{A}AB\mathfrak{A}.$$

But $\mathfrak{A}AB\mathfrak{A}$ is an ideal. From 2.7 it is isolated and therefore from IV, 6.2, (η),

$$BCA \in \mathfrak{A}AB\mathfrak{A}.$$

Because of this inclusion and from (β),

$$\mathfrak{A}A\mathfrak{A} \cdot \mathfrak{A}B\mathfrak{A} = \mathfrak{A}B\mathfrak{A} \cdot \mathfrak{A}A\mathfrak{A} \subset \mathfrak{A} \cdot B\mathfrak{A}A \cdot \mathfrak{A} \subset \mathfrak{A} \cdot \mathfrak{A}AB\mathfrak{A} \cdot \mathfrak{A} \subset \mathfrak{A}AB\mathfrak{A}.$$

Since furthermore it is evident that the inverse inclusion holds, we obtain

$$\mathfrak{A}A\mathfrak{A} \cdot \mathfrak{A}B\mathfrak{A} = \mathfrak{A}AB\mathfrak{A}.$$

Hence it immediately follows that the assertion to be proved is valid.

2.10. Using the property of complete regularity we may give a new approach to the completely simple semigroups without zero that were considered in §§ 3 and 6 of Chapter V.

THEOREM. *The semigroup* $\mathfrak{A}$ *is a completely simple semigroup without zero if and only if it is completely regular and has no proper two-sided ideals.*

PROOF. (1) If $\mathfrak{A}$ is a completely simple semigroup without zero, then we know from (V, 6.7) that each of its elements is contained in some subgroup. From 2.1 this means that $\mathfrak{A}$ is completely regular. A completely simple semigroup without zero has by definition no proper two-sided ideals.

(2) Suppose that $\mathfrak{A}$ is completely regular and has no proper two-sided ideals. Let I_1 and I_2 be idempotents of $\mathfrak{A}$ such that

$$I_1I_2 = I_2I_1 = I_1.$$

From V, 3.10 and V, 6.15, in order to prove that $\mathfrak{A}$ is a completely simple semigroup without zero it is sufficient to show that this equation is possible only if $I_1 = I_2$.

Since $\mathfrak{A}I_1\mathfrak{A}$ is a two-sided ideal of $\mathfrak{A}$ it must coincide with $\mathfrak{A}$, i.e., for some $X, Y \in \mathfrak{A}$,

$$I_2 = XI_1Y.$$

Writing

$$A = I_2XI_2, \qquad B = I_2YI_2,$$

we obtain

$$AI_1B = I_2XI_2I_1I_2YI_2 = I_2XI_1YI_2 = I_2I_2I_2 = I_2.$$

Suppose that A' is an element regularly adjoint to A and commuting with it (II, 6.7). Since

$$I_2 = AI_1B = AA'AI_1B = AA'I_2 = A'AI_2 = A'A = AA',$$

it follows that

$$I_2 = I_2I_2 = A'AA'A = A'I_2A = A'AI_1BA = I_2I_1BA = I_1BA.$$

Thus I_2 turns out to be a regular left unit for I_1. Analogously we show that I_2 is also a regular right unit for I_1. But, as we know (II, 6.5), an element can have only one regular two-sided unit, which for I_1 is itself. Accordingly, $I_2 = I_1$.

2.11. We shall clear up here what two-sided ideal layers (IV, 3.3) of completely regular semigroups amount to.

LEMMA. *Suppose that $\mathfrak{N}$ is a two-sided ideal layer of the completely regular semigroup $\mathfrak{N}$. For any $N \in \mathfrak{N}$ the set*

$$\mathfrak{N}' = \mathfrak{A}N\mathfrak{A}\backslash\mathfrak{N}$$

is empty or a two-sided ideal of $\mathfrak{A}$.

PROOF. Suppose that $\mathfrak{A}\mathfrak{N}'\mathfrak{A}$ contains an element Z not contained in $\mathfrak{N}'$. For some $U, V \in \mathfrak{A}$ and $T \in \mathfrak{N}'$ we have

$$Z = UTV.$$

Since

$$Z \in U\mathfrak{N}'V \subset \mathfrak{A}\mathfrak{A}N\mathfrak{A}\mathfrak{A} \subset \mathfrak{A}N\mathfrak{A},$$

$$Z \,\bar{\in}\, \mathfrak{A}N\mathfrak{A}\backslash\mathfrak{N}$$

we have $Z \in \mathfrak{N}$. By the definition of a two-sided ideal layer, taking into account 2.9, (δ), we obtain

$$\mathfrak{A}Z\mathfrak{A} = \mathfrak{A}N\mathfrak{A}.$$

But then, on the one hand

$$\mathfrak{A}N\mathfrak{A} = \mathfrak{A}Z\mathfrak{A} = \mathfrak{A}UTV\mathfrak{A} \subset \mathfrak{A}T\mathfrak{A}$$

and on the other hand

$$\mathfrak{A}T\mathfrak{A} \subset \mathfrak{A}\mathfrak{N}'\mathfrak{A} \subset \mathfrak{A}\mathfrak{A}N\mathfrak{A}\mathfrak{A} \subset \mathfrak{A}N\mathfrak{A}.$$

However the equation $\mathfrak{A}N\mathfrak{A} = \mathfrak{A}T\mathfrak{A}$ is impossible since $N \in \mathfrak{N}$ and $T \,\bar{\in}\, \mathfrak{N}$. We have shown that $\mathfrak{A}\mathfrak{N}'\mathfrak{A} \subset \mathfrak{N}'$. Since each element of $\mathfrak{N}'$ has both a left and right unit this means that $\mathfrak{N}'$ is a two-sided ideal of $\mathfrak{A}$.

2.12. THEOREM. *The two-sided ideal layer $\mathfrak{N}$ of the completely regular semigroup $\mathfrak{A}$ is a completely regular semigroup without zero.*

PROOF. Because of 2.10 it suffices to prove that $\mathfrak{N}$ is a completely regular semigroup with no proper two-sided ideals.

If $N_1, N_2 \in \mathfrak{N}$, then from 2.9, ($\delta$),

$$\mathfrak{A}N_1\mathfrak{A} = \mathfrak{A}N_2\mathfrak{A}.$$

From 2.9, (γ), (δ), (ε),

$$\mathfrak{A}N_1N_2\mathfrak{A} = \mathfrak{A}N_1\mathfrak{A} \cdot \mathfrak{A}N_2\mathfrak{A} = \mathfrak{A}N_1\mathfrak{A} \cdot \mathfrak{A}N_1\mathfrak{A} = \mathfrak{A}N_1\mathfrak{A}.$$

From this and $N_1 \in \mathfrak{N}$ it follows that $N_1N_2 \in \mathfrak{N}$. Thus $\mathfrak{N}$ turns out to be a subsemigroup of $\mathfrak{A}$.

The set

$$\mathfrak{A}N_1\mathfrak{A} \cdot N_1 \cdot \mathfrak{A}N_1\mathfrak{A}$$

is a two-sided ideal of $\mathfrak{A}$. It contains N_1^3. Therefore, from 2.7, it must also contain N_1. But a two-sided ideal containing one of the elements of a two-sided ideal layer must also contain every other element of it. Accordingly,

$$N_2 = XN_1Y,$$

where X and Y belong to $\mathfrak{A}N_1\mathfrak{A}$. Neither X nor Y can be contained in $\mathfrak{A}N_1\mathfrak{A}\backslash\mathfrak{N}$, which does not contain N_2 and from 2.11 is a two-sided ideal. Accordingly, $X, Y \in \mathfrak{N}$. Thus, it follows from $N_2 = XN_1Y$ that every two-sided ideal of the subsemigroup $\mathfrak{N}$ that contains one of its elements necessarily contains also any other element, i.e., $\mathfrak{N}$ has no proper two-sided ideals.

Since $N_1 \in \mathfrak{N}$ is completely regular in $\mathfrak{A}$ it follows that for some $X \in \mathfrak{A}$,

$$NXN = N, \qquad XNX = N, \qquad XN = NX.$$

Evidently every two-sided ideal of $\mathfrak{A}$ that contains one of the elements N or X must also contain the other one. Accordingly, N and X belong to the same two-sided ideal layer, i.e., $X \in \mathfrak{N}$. Hence it follows that N is a completely regular element of the semigroup $\mathfrak{N}$.

2.13. The properties just obtained of completely regular semigroups make it possible to find an important decomposition for them.

THEOREM. *A completely regular semigroup is a commutative band* (1.12) *of completely simple semigroups without zero.*

PROOF. From IV, 3.8 a completely regular semigroup $\mathfrak{A}$ is the union without intersections of two-sided ideal layers, each of which, from 2.12, is a completely simple semigroup without zero, i.e.,

$$\mathfrak{A} = \bigcup_{\xi} \mathfrak{N}_{\xi}.$$

We choose two arbitrary two-sided ideal layers $\mathfrak{N}_\alpha$ and $\mathfrak{N}_\beta$ and select from them pairs of arbitrary elements

$$A_\alpha, A'_\alpha \in \mathfrak{N}_\alpha, \qquad A_\beta, A'_\beta \in \mathfrak{N}_\beta.$$

Elements of one and the same layer have the same two-sided ideal envelopes, i.e.,

$$\mathfrak{A}A_\alpha\mathfrak{A} = \mathfrak{A}A'_\alpha\mathfrak{A}, \qquad \mathfrak{A}A_\beta\mathfrak{A} = \mathfrak{A}A'_\beta\mathfrak{A}.$$

Using 2.9, (ε), we obtain

$$\mathfrak{A}A_\alpha A_\beta\mathfrak{A} = \mathfrak{A}A_\alpha\mathfrak{A}\cdot\mathfrak{A}A_\beta\mathfrak{A} = \mathfrak{A}A'_\alpha\mathfrak{A}\cdot\mathfrak{A}A'_\beta\mathfrak{A} = \mathfrak{A}A'_\alpha A'_\beta\mathfrak{A}.$$

Accordingly, $A_\alpha A_\beta$ and $A'_\alpha A'_\beta$ are contained in one and the same two-sided ideal layer $\mathfrak{N}_\gamma$. This means that

$$\mathfrak{N}_\alpha\mathfrak{N}_\beta \subset \mathfrak{N}_\gamma.$$

But from 2.9, (β),

$$\mathfrak{A}A_\beta A_\alpha\mathfrak{A} = \mathfrak{A}A_\beta\mathfrak{A}\cdot\mathfrak{A}A_\alpha\mathfrak{A} = \mathfrak{A}A_\alpha\mathfrak{A}\cdot\mathfrak{A}A_\beta\mathfrak{A} = \mathfrak{A}A_\alpha A_\beta\mathfrak{A}.$$

Therefore also $A_\beta A_\alpha \in \mathfrak{N}_\gamma$, and we obtain

$$\mathfrak{N}_\beta\mathfrak{N}_\alpha \subset \mathfrak{N}_\gamma.$$

2.14. COROLLARY. *A semigroup all of whose elements are idempotents is a commutative band of matrix bands of unit groups.*

PROOF. Since each idempotent of a semigroup forms a unit subgroup of it (consisting of a unit alone) it follows from 2.1 and 2.13 that a semigroup $\mathfrak{A}$, all of whose elements are idempotents, is a commutative band of completely simple semigroups without zero, all of whose elements are idempotents. But those semigroups, as we remarked in V, 6.14, are isomorphic to semigroups of the type considered in V, 6.11, i.e., they may in an evident way be regarded as matrix bands of unit groups.

2.15. The corollary just obtained makes it possible to derive an important result, already mentioned, on general bands due to Clifford [**11**].

THEOREM. *A semigroup which is a band of semigroups belonging to some class Γ may be represented in the form of a commutative band of semigroups which are in turn matrix bands of semigroups belonging to the class Γ.*

PROOF. Let $\mathfrak{n}$ be a band of the semigroup $\mathfrak{A}$, all of whose components $\mathfrak{B}_\alpha, \mathfrak{B}_\beta, \ldots$ belong to the class Γ. Consider the factor-semigroup $\overline{\mathfrak{A}} = \mathfrak{A}/\mathfrak{n}$ whose elements are $\mathfrak{B}_\alpha, \mathfrak{B}_\beta, \ldots$ (VII, 2.4). Since $\mathfrak{n}$ corresponds to the natural homomorphism φ of the semigroup $\mathfrak{A}$ onto $\overline{\mathfrak{A}}$ (VII, 2.5) it follows from 1.8 that all the elements of the semigroup $\varphi(\mathfrak{A}) = \overline{\mathfrak{A}}$ are idempotents. From 2.14, $\overline{\mathfrak{A}}$ is a commutative band of matrix bands of unit groups, and therefore there exists a homomorphism ψ of the semigroup $\overline{\mathfrak{A}}$ onto a commutative semigroup $\overline{\overline{\mathfrak{A}}}$, all of whose elements are idempotents, such that to the homomorphism ψ there corresponds the given commutative band of the semigroup $\overline{\mathfrak{A}}$ (1.12). The product $\psi\varphi$ is a homomorphism of $\mathfrak{A}$ onto $\overline{\overline{\mathfrak{A}}}$. The two-sidedly stable partition

$\mathfrak{m}$ corresponding to it must be a commutative band of $\mathfrak{A}$ (1.12). The complete preimage of any element of $\overline{\overline{\mathfrak{A}}}$ under the homomorphism $\psi\varphi$ is the union of certain of the $\mathfrak{B}_\alpha, \mathfrak{B}_\beta, \ldots$, which form a matrix band in $\overline{\mathfrak{A}}$. This means that they may be assigned pairs of indices $\mathfrak{B}_{\xi\eta}, \mathfrak{B}_{\sigma\tau}, \ldots$, such that, under the operation in $\overline{\mathfrak{A}} = \mathfrak{A}/\mathfrak{n}$ which we shall denote by $\circ$, we have

$$\mathfrak{B}_{\xi\eta} \circ \mathfrak{B}_{\sigma\tau} = \mathfrak{B}_{\xi\tau}$$

(recall that $\mathfrak{B}_{\xi\eta}, \mathfrak{B}_{\sigma\tau}, \ldots$ are each separate elements of the factor-semigroup $\overline{\mathfrak{A}} = \mathfrak{A}/\mathfrak{n}$). From the connection between operations in the semigroup itself and in the factor-semigroup (VII, 2.3), for multiplication of $\mathfrak{B}_{\xi\eta}, \mathfrak{B}_{\sigma\tau}, \ldots$ as subsets of $\mathfrak{A}$ we obtain

$$\mathfrak{B}_{\xi\eta} \cdot \mathfrak{B}_{\sigma\tau} \subset \mathfrak{B}_{\xi\tau}.$$

But this in turn means that the complete preimage of an element of the semigroup $\overline{\overline{\mathfrak{U}}}$ under the homomorphism $\psi\varphi$, i.e., a component of the band $\mathfrak{m}$, is a matrix band of semigroups $\mathfrak{B}_{\xi\eta}$, which all belong to the class Γ.

3. Completely Regular Inverse Semigroups

3.1. We shall pay particular attention to completely regular semigroups which are inverse semigroups. This class of semigroups was studied by Clifford [4] and then by A. E. Liber [2], who called such semigroups the simplest generalized groups.

As always, in considering inverse semigroups we shall use a bar to denote a regularly adjoint element, which in an inverse semigroup is uniquely defined. We shall use properties II, 7.3 and II, 7.5 without specifically referring to them.

If an inverse semigroup is completely regular, then from the commutativity of A and $\bar{A}$ for any element of A we have

$$\bar{A}AA = A\bar{A}A = A.$$

It results that the condition

$$\bar{A}A^2 = A,$$

which is satisfied for all elements of an inverse semigroup $\mathfrak{A}$, is also sufficient for $\mathfrak{A}$ to be completely regular.

Indeed, applying this equation to $\bar{A}$,

$$\overline{(\bar{A})}\bar{A}\bar{A} = \bar{A}, \qquad A\bar{A}\bar{A} = \bar{A}.$$

Hence, turning to the regular adjoint, we obtain

$$AA\bar{A} = A.$$

Using this equation and the fact that $\bar{A}A^2 = A$, we obtain

$$A\bar{A} = \bar{A}AA\bar{A} = \bar{A}A,$$

which proves the complete regularity of the semigroup.

Analogously, the condition

$$A^2\bar{A} = A,$$

which is also satisfied for all elements, is necessary and sufficient for complete regularity of an inverse semigroup.

3.2. If in a completely regular semigroup all the idempotents commute with one another, then it is an inverse semigroup (II, 7.4). It turns out that in this case one can make a stronger assertion concerning the commutability of idempotents.

THEOREM. *In a completely regular inverse semigroup every idempotent commutes with every element of the semigroup.*

PROOF. Using the known properties of regularly adjoint elements, we find that for any element A and any idempotent I of our semigroup,

$$(AI)\overline{(AI)} = AI\bar{I}\bar{A} = AI\bar{A},$$

$$\overline{(AI)}(AI) = \bar{I}\bar{A}AI = I\bar{A}AI.$$

Since the regularly adjoint elements (AI) and $\overline{(AI)}$ commute and since the idempotents I and $\bar{A}A$ commute, from these equations we get

$$AI\bar{A} = I\bar{A}A.$$

From this it follows that

$$AI = A\bar{A}AI = AI\bar{A}A = I\bar{A}AA = IA.$$

3.3. We denote by $\mathfrak{H}$ the set of all idempotents of the completely regular inverse semigroup $\mathfrak{A}$. $\mathfrak{H}$ is a commutative semigroup of idempotents. Since a completely simple semigroup without zero with the idempotent which commutes with all of its elements is a group (V, 6.7, (γ), (δ)), then the result of 2.13 applied to $\mathfrak{A}$ means that *a completely regular inverse semigroup $\mathfrak{A}$ is a commutative band of groups*

$$\mathfrak{A} = \bigcup_{I \in \mathfrak{H}} \mathfrak{G}_I.$$

Here $\mathfrak{G}_I$ is a group with identity I.

The converse is also true. *Every commutative band of groups is a completely regular inverse semigroup.* Indeed, its complete regularity follows from the fact that each of its elements is contained in some subgroup (III, 1.15). Suppose that I_1 and I_2 are two of its idempotents. Since the semigroup is a commutative band of groups, for some subgroup $\mathfrak{G}$ of our semigroup we have

$$I_1I_2 = G_1 \in \mathfrak{G}, \qquad I_2I_1 = G_2 \in \mathfrak{G}.$$

Hence it follows that

$$G_1G_2G_1 = I_1I_2 \cdot I_2I_1 \cdot I_1I_2 = I_1I_2I_1I_2 = G_1G_1.$$

For elements of a group this is possible only in the case when $G_2 = E_{\mathfrak{G}}$. Analogously one verifies that also $G_1 = E_{\mathfrak{G}}$. From the commutativity of any two idempotents we conclude that our completely regular semigroup is inverse.

We have shown that in a commutative band of groups the idempotents commute. It therefore follows that the set of these idempotents $\mathfrak{H}$ forms a commutative semigroup. As we proved in § 4 of Chapter II, the structure of $\mathfrak{H}$ may be given by the corresponding partial ordering in $\mathfrak{H}$.

We observe that the multiplication of idempotents in the semigroup $\mathfrak{H}$ defines the multiplication of components in our commutative band. Indeed, if

$$UV = W, \qquad (U, V, W \in \mathfrak{H}),$$

then

$$\mathfrak{G}_U \mathfrak{G}_V \cap \mathfrak{G}_W \neq \varnothing,$$

and accordingly

$$\mathfrak{G}_U \mathfrak{G}_V \subset \mathfrak{G}_W, \qquad \mathfrak{G}_V \mathfrak{G}_U \subset \mathfrak{G}_W.$$

3.4. If $UV = U$ for $U, V \in \mathfrak{H}$, we consider the mapping $\varphi_{U,V}$ of the group $\mathfrak{G}_V$ into the group $\mathfrak{G}_U$, namely,

$$\varphi_{U,V}(X) = UXU \qquad (X \in \mathfrak{G}_V).$$

Since $UX \in \mathfrak{G}_U$ and $XU \in \mathfrak{G}_U$, we have

$$UX \cdot U = UX, \qquad U \cdot XU = XU$$

and therefore

$$\varphi_{U,V}(X) = UXU = UX = XU.$$

Because of this it is easily verified that the mapping $\varphi_{U,V}$ is a homomorphism, i.e.,

$$\varphi_{U,V}(X) \cdot \varphi_{U,V}(Y) = UX \cdot YU = U(XY)U = \varphi_{U,V}(XY),$$
$$(X, Y \in \mathfrak{G}_V).$$

We note two properties of these homomorphisms.

(1) $\varphi_{U,U}$ is the identity isomorphism.

(2) If

$$UV = U, \qquad VW = V \qquad (U, V, W \in \mathfrak{H}),$$

then

$$\varphi_{U,V} \cdot \varphi_{V,W} = \varphi_{U,W}.$$

The first of these is evident since U is the unit of the group $\mathfrak{G}_U$.

The second is also easily proved. Let $X \in \mathfrak{G}_W$.

Then $VXV \in \mathfrak{G}_V$ and

$$(\varphi_{U,V} \cdot \varphi_{V,W})(X) = \varphi_{U,V}[\varphi_{V,W}(X)] = \varphi_{U,V}(VXV)$$
$$= U(VXV)U = UXU = \varphi_{U,W}(X).$$

3.5. Thus we have discovered that to a completely regular inverse semigroup $\mathfrak{A}$ there is associated a system consisting of the following elements.

(α) A commutative semigroup of idempotents $\mathfrak{H}$ (the set of all idempotents of $\mathfrak{A}$).

(β) A collection of pairwise disjoint groups $\mathfrak{G}_I$, whose units form the set $\mathfrak{H}$ ($E_{\mathfrak{G}_I} = I$).

(γ) A collection of homomorphisms $\varphi_{U,V}$ defined for each pair $U, V \in \mathfrak{H}$ such that $UV = U$. Here $\varphi_{U,V}$ is a homomorphism of the group $\mathfrak{G}_V$ into the group $\mathfrak{G}_U$. The homomorphisms $\varphi_{U,V}$ have the properties 3.4, (1), (2).

3.6. The system 3.5 completely determines the structure of a completely regular inverse semigroup $\mathfrak{A}$. Indeed, the set of elements of $\mathfrak{A}$ is the union of all the groups $\mathfrak{G}_I$ indicated in (β). The multiplication of the elements of $\mathfrak{A}$ is carried out according to the following rule.

Suppose
$$X \in \mathfrak{G}_U, \qquad Y \in \mathfrak{G}_V, \qquad UV = W, \qquad (U, V, W \in \mathfrak{H}).$$
Then $XY \in \mathfrak{G}_W$, $UW = W$, $VW = W$, and therefore
$$XY = W(XY)W = WX \cdot YW = \varphi_{W,U}(X) \cdot \varphi_{W,V}(Y).$$
This last product is completely determined by the group $\mathfrak{G}_W$, since both of its multiplicands are elements of $\mathfrak{G}_W$.

3.7. The association with a completely regular inverse semigroup of its system 3.5 gives a complete classification of such semigroups, first obtained by Clifford [4]. This follows from the fact that, as we shall now prove, each system of type 3.5 in its turn determines some completely regular inverse semigroup.

Now suppose that we are given any system consisting of the elements indicated in 3.5. In the set $\mathfrak{A}$, the union of all the groups $\mathfrak{G}_I$, we define an operation.

Let
$$X \in \mathfrak{G}_U, \qquad Y \in \mathfrak{G}_V, \qquad UV = W \qquad (U, V, W \in \mathfrak{H}).$$
Put
$$XY = T = \varphi_{W,U}(X) \cdot \varphi_{W,V}(Y).$$
T is a definite element of $\mathfrak{G}_W$ since, by the definition of the homomorphisms $\varphi_{W,U}$ and $\varphi_{W,V}$, it is the product of elements of this group.

From the fact that $\varphi_{I,I}$ ($I \in \mathfrak{H}$) is the identity isomorphism it immediately follows that for elements of $\mathfrak{G}_I$ the operation defined in $\mathfrak{A}$ coincides with the operation defined in $\mathfrak{G}_I$ itself.

The same holds for elements of $\mathfrak{H}$. The operation defined in $\mathfrak{A}$ coincides with the operation defined in $\mathfrak{H}$. This follows from the fact that if in $\mathfrak{H}$,
$$UV = W, \qquad (U, V, W \in \mathfrak{H}),$$
then in $\mathfrak{A}$,
$$U \cdot V = \varphi_{W,U}(U) \cdot \varphi_{W,V}(V).$$

But U is the unit of the group $\mathfrak{G}_U$, and W the unit of the group $\mathfrak{G}_W$. Therefore $\varphi_{W,U}(U) = W$. Analogously, $\varphi_{W,V}(V) = W$. Thus, also in $\mathfrak{A}$ we obtain

$$U \cdot V = W \cdot W = W$$

We shall show that the operation in $\mathfrak{A}$ which we have just defined is associative. Indeed, if

$$X_i \in \mathfrak{G}_{U_i}, \qquad U_1U_2 = V, \qquad VU_3 = W; \qquad U_i, V, W \in \mathfrak{H} \qquad (i = 1, 2, 3),$$

then, making use of the property of our homomorphisms, we obtain

$$\begin{aligned}(X_1X_2)X_3 &= [\varphi_{V,U_1}(X_1) \cdot \varphi_{V,U_2}(X_2)] \cdot X_3 \\ &= \varphi_{W,V}[\varphi_{V,U_1}(X_1) \cdot \varphi_{V,U_2}(X_2)] \cdot \varphi_{W,U_3}(X_3) \\ &= [(\varphi_{W,V} \cdot \varphi_{V,U_1})(X_1)] \cdot [(\varphi_{W,V} \cdot \varphi_{V,U_2})(X_2)] \cdot \varphi_{W,U_3}(X_3) \\ &= \varphi_{W,U_1}(X_1) \cdot \varphi_{W,U_2}(X_2) \cdot \varphi_{W,U_3}(X_3).\end{aligned}$$

Quite analogously one verifies that also the product $X_1(X_2X_3)$ is equal to the same element of $\mathfrak{G}_W$.

Since $\mathfrak{A}$ is the union of the groups $\mathfrak{G}_I$ $(I \in \mathfrak{H})$, it is completely regular (III, 1.15). Its idempotents are elements of $\mathfrak{H}$ which form a commutative semigroup. Therefore $\mathfrak{A}$ is inverse (II, 7.4).

It is easily verified that the system 3.5, corresponding to the completely regular inverse semigroup just obtained, coincides with the original system. $\mathfrak{H}$ is the commutative semigroup of idempotents of $\mathfrak{A}$, while $\mathfrak{A}$ is the union of the groups $\mathfrak{G}_I$ $(I \in \mathfrak{H})$.

For

$$X \in \mathfrak{G}_V, \qquad UV = V \qquad (U, V \in \mathfrak{H})$$

we have

$$UXU = \varphi_{U,U}(U) \cdot \varphi_{U,V}(X) \cdot \varphi_{U,U}(U) = U \cdot \varphi_{U,V}(X) \cdot U = \varphi_{U,V}(X),$$

i.e. the homomorphisms of the system corresponding to the semigroup just constructed coincide with the corresponding homomorphisms of the original system.

3.8. The set of idempotents of a completely regular inverse semigroup is a commutative semigroup of idempotents. Such a semigroup, as we have proved in § 4 of Chapter II, is adjoint to some sublattice. Defining the semigroup is equivalent to defining the sublattice adjoint to it. Among the sublattices those lattices which form a completely ordered set are particularly simple. In this connection, in the class of all inverse completely regular semigroups it is natural to distinguish the class of those subgroups whose idempotents are completely ordered relative to the relation of one being the unit for the other. Attention was first turned to that class of semigroups by N. N. Vorob'ev **[4; 8]**, who discovered the special role of semigroups of this class in connection with the property of existence of identities for the ideals of the semigroup.

3.9. Suppose that $\mathfrak{A}$ is a completely regular semigroup for which every non-empty set of idempotents contains its two-sided identity.

We shall consider a series of properties of $\mathfrak{A}$.

(α) *$\mathfrak{A}$ is inverse.*

Indeed, suppose that I_1 and I_2 are arbitrary idempotents of $\mathfrak{A}$. Since the set $\{I_1, I_2\}$ contains its two-sided identity, we have

$$I_1 I_2 = I_2 I_1 = I_1.$$

Accordingly the idempotents of $\mathfrak{A}$ commute, from which it follows from II, 7.4 that $\mathfrak{A}$ is inverse.

(β) *$\mathfrak{A}$ is a disjoint union of a completely ordered set of groups*, i.e.,

$$\mathfrak{A} = \mathfrak{G}_1 \cup \mathfrak{G}_2 \cup \ldots \cup \mathfrak{G}_\xi \cup \mathfrak{G}_{\xi+1} \cup \ldots \qquad (\xi < \mu),$$

while for any $\eta < \xi < \mu$,

$$E_{\mathfrak{G}_\eta} E_{\mathfrak{G}_\xi} = E_{\mathfrak{G}_\xi} E_{\mathfrak{G}_\eta} = E_{\mathfrak{G}_\xi}.$$

Indeed, if for the idempotents I and I' of the semigroup $\mathfrak{A}$ one writes $I \leqslant I'$ in the case when $II' = I'I = I'$, then from our hypothesis on $\mathfrak{A}$ the set of all idempotents relative to this ordering will be completely ordered, i.e.,

$$I_1 < I_2 < \ldots < I_\xi < I_{\xi+1} < \ldots \qquad (\xi < \mu).$$

Since $\mathfrak{A}$ is completely regular, it is a disjoint union of groups $\mathfrak{G}_\xi$ $(1 \leqslant \xi < \mu)$, where $\mathfrak{G}_\xi$ is the set of all the elements of $\mathfrak{A}$ having I_ξ as their regular two-sided identity (III, 1.14; III, 1.16).

(γ) *The mappings $\varphi_{\xi,\eta}$ for $\eta \leqslant \xi < \mu$ of the groups $\mathfrak{G}_\eta$, namely,*

$$\varphi_{\xi,\eta}(X) = E_\xi X E_\xi \qquad (X \in \mathfrak{G}_\eta)$$

are homomorphisms of $\mathfrak{G}_\eta$ into $\mathfrak{G}_\xi$.

This follows from (α), (β), and 3.4.

(δ) *The homomorphisms $\varphi_{\xi,\eta}$ satisfy the conditions*:

(1) *$\varphi_{\xi,\xi}$ is the identity isomorphism*;

(2) *for $\eta \leqslant \xi \leqslant \zeta < \mu$ we have*

$$\varphi_{\zeta,\xi} \cdot \varphi_{\xi,\eta} = \varphi_{\zeta,\eta}.$$

This follows from (γ) and 3.4.

(ε) *The structure of $\mathfrak{A}$ is completely determined by giving the completely ordered sequence of groups*

$$\mathfrak{G}_1, \mathfrak{G}_2, \ldots, \qquad \mathfrak{G}_\xi, \mathfrak{G}_{\xi+1}, \ldots \qquad (\xi < \mu),$$

and homomorphisms $\varphi_{\xi,\eta}$ $(\eta \leqslant \xi < \mu)$.

This follows from 3.6.

(ζ) *For any $\nu < \mu$ the set*

$$\mathfrak{T}_\nu = \bigcup_{\nu \leqslant \xi < \mu} \mathfrak{G}_\xi$$

is a two-sided ideal of $\mathfrak{A}$, and $\mathfrak{A}$ has no other ideals.

The fact that $\mathfrak{T}_\nu$ is a two-sided ideal of $\mathfrak{A}$ follows from the fact that the operation 3.6 is the one acting in $\mathfrak{A}$.

Suppose that $\mathfrak{L}$ is any left ideal of $\mathfrak{A}$. If $\mathfrak{L} \ni I_\xi$, then $\mathfrak{L}$ also contains all the elements of the group $\mathfrak{G}_\xi$, since $XI_\xi = X$ for any $X \in \mathfrak{G}_\xi$.

In the set of those I_ξ which are contained in $\mathfrak{L}$, there is a unit I_λ. Since all the I_ξ for $\xi \geqslant \lambda$ are contained in $\mathfrak{L}$, we obtain

$$\mathfrak{L} \supset \mathfrak{T}_\lambda.$$

Since I_σ for $\sigma < \lambda$ is not contained in $\mathfrak{L}$, no element of $\mathfrak{G}_\sigma$ belongs to $\mathfrak{L}$. Thus it results that $\mathfrak{L} = \mathfrak{T}_\lambda$.

(η) *All the ideals of* $\mathfrak{A}$ *are two-sided.*

This follows from (ζ).

(θ) *Every ideal of* $\mathfrak{A}$ *contains its unit.*

Indeed, in $\mathfrak{T}_\nu$ the element I_ν is a two-sided unit for each I_ξ ($\xi \geqq \nu$), and therefore also for any $X \in \mathfrak{G}_\xi$, since I_ξ is a two-sided unit for X.

3.10. Recall that it follows from 3.7 that a completely ordered set of arbitrary groups $\mathfrak{G}_\xi$ $(1 \leqslant \xi < \mu)$ and homomorphisms $\varphi_{\xi,\eta}$ $(\eta \leqslant \xi)$, satisfying properties 3.9, (δ), determines a certain completely regular semigroup which evidently will have the property that any nonempty set of its idempotents contains its unit.

3.11. The significance of the semigroups under discussion is determined by the following theorem.

Theorem. *If in a semigroup* $\mathfrak{A}$ *each left ideal contains its unit, then* $\mathfrak{A}$ *is a completely regular semigroup in which each nonempty set of its idempotents contains its unit.*

Proof. (1) To each idempotent I of the semigroup $\mathfrak{A}$ we assign a set $\mathfrak{A}I$, consisting of all the elements of $\mathfrak{A}$ for which I is a right unit. $\mathfrak{A}I$ is a left ideal, and therefore it must contain its unit I'. Since $I'I = I'$ and $I'I = I$ it follows that $I' = I$, i.e., I is a two-sided unit in $\mathfrak{A}I$.

There are no left ideals other than those of the type mentioned above in $\mathfrak{A}$. Indeed, suppose that $\mathfrak{L}$ is any left ideal of $\mathfrak{A}$. It must contain its unit I and therefore

$$\mathfrak{L} = \mathfrak{L}I \subset \mathfrak{A}I, \qquad \mathfrak{A}I \subset \mathfrak{A}\mathfrak{L} \subset \mathfrak{L},$$

i.e., $\mathfrak{L} = \mathfrak{A}I$. Since each left ideal contains only one of its units it follows that between the idempotents of $\mathfrak{A}$ and its left ideals one may establish a one-to-one relation.

Suppose that I_1 and I_2 are two arbitrary idempotents of $\mathfrak{A}$. Since $\mathfrak{A}I_1 \cup \mathfrak{A}I_2$ is a left ideal of $\mathfrak{A}$, for some idempotent I_3 we have

$$\mathfrak{A}I_1 \cup \mathfrak{A}I_2 = \mathfrak{A}I_3.$$

I_3 must be contained in $\mathfrak{A} I_1$ or in $\mathfrak{A} I_2$. In the first case I_3, as the unit of $\mathfrak{A} I_3$, will be the unit also for $\mathfrak{A} I_1$ and therefore $I_3 = I_1$, from which it follows that $\mathfrak{A}I_2 \subset \mathfrak{A}I_1$. In the second case $I_3 = I_2$ and $\mathfrak{A}I_1 \subset \mathfrak{A}I_2$. Thus, of any two left ideals $\mathfrak{A}I_1$ and $\mathfrak{A}I_2$, one always is contained in the other; in addition

$$\mathfrak{A}I_2 \subset \mathfrak{A}I_1$$

holds in the case when

$$I_2 I_1 = I_1 I_2 = I_2.$$

Suppose that $\mathfrak{H}$ is any nonempty set of idempotents of $\mathfrak{A}$. We consider the left ideal $\mathfrak{A}\mathfrak{H}$. For some idempotent I,

$$\mathfrak{A}\mathfrak{H} = \mathfrak{A}I.$$

Accordingly, for some $I' \in \mathfrak{H}$,

$$I \subset \mathfrak{A}I'.$$

Since

$$\mathfrak{A}I' \subset \mathfrak{A}\mathfrak{H} = \mathfrak{A}I, \qquad \mathfrak{A}I \subset \mathfrak{A}\mathfrak{A}I' \subset \mathfrak{A}I',$$

we have $\mathfrak{A}I' = \mathfrak{A}I$ and, from what has been said above, $I' = I$. Thus the idempotent $I' = I$, being the unit of the ideal of $\mathfrak{A} I = \mathfrak{A} \mathfrak{H}$, is also a two-sided unit in the set $\mathfrak{H}$ to which it belongs.

(2) Select any element A of the semigroup $\mathfrak{A}$. $A \cup \mathfrak{A}A$ is a left ideal of $\mathfrak{A}$, and therefore for some idempotent

$$\mathfrak{A}A \cup A = \mathfrak{A}I,$$

while I is the unit of $\mathfrak{A} I$.

Denote by $\mathfrak{L}$ the union of all those left ideals of the semigroup $\mathfrak{A}I$ which do not contain I. Evidently $\mathfrak{L}$ is a proper left ideal of the semigroup $\mathfrak{A}I$ or $\mathfrak{L} = \varnothing$. Suppose that $\mathfrak{L} \neq \varnothing$. Since

$$\mathfrak{A}\mathfrak{L} = \mathfrak{A}I\mathfrak{L} = (\mathfrak{A}I)\mathfrak{L} \subset \mathfrak{L},$$

it follows that $\mathfrak{L}$ is a left ideal of $\mathfrak{A}$. Consequently, for some idempotent I',

$$\mathfrak{L} = \mathfrak{A}I'.$$

$\mathfrak{L}\mathfrak{A}I$ is a left ideal of $\mathfrak{A}$ belonging to $\mathfrak{A}I$ and containing $\mathfrak{L}$ (since $\mathfrak{L}I = \mathfrak{L}$).

If now $\mathfrak{L}\mathfrak{A}I$ contained I, then for some $X \in \mathfrak{L}$ and $A \in \mathfrak{A}$ we would have

$$I = XAI = I'XAI = I' \cdot I = I',$$

which is impossible since $I' \in \mathfrak{L}$ and $I \bar{\in} \mathfrak{L}$. Accordingly,

$$\mathfrak{L}\mathfrak{A}I \subset \mathfrak{L}.$$

Thus $\mathfrak{L}$ is either empty or is a two-sided ideal of the semigroup $\mathfrak{A}I$.

Write

$$\mathfrak{A}I \backslash \mathfrak{L} = \mathfrak{K}.$$

If $K \in \mathfrak{K}$, then $\mathfrak{A}K$ is a left ideal of $\mathfrak{A}$ lying in $\mathfrak{A}I$ and containing K. Since $K \bar{\in} \mathfrak{L}$, then $\mathfrak{A}K$ must be equal to $\mathfrak{A}I$. Conversely, if $X \in \mathfrak{A}I$ and $\mathfrak{A}X = \mathfrak{A}I$, then every left ideal of the semigroup $\mathfrak{A}I$ containing X will also contain $\mathfrak{A}IX = \mathfrak{A}X = \mathfrak{A}I$, i.e., coincide with $\mathfrak{A}I$. Therefore X cannot lie in $\mathfrak{L}$, i.e., $X \in \mathfrak{K}$.

If $K_1, K_2 \in \mathfrak{K}$, then from the given property of $\mathfrak{K}$,

$$\mathfrak{A}K_1K_2 = \mathfrak{A}K_1 \cdot K_2 = \mathfrak{A}I \cdot K_2 = \mathfrak{A}K_2 = \mathfrak{A}I,$$

i.e., $K_1K_2 \in \mathfrak{K}$. Thus $\mathfrak{K}$ is a subsemigroup of $\mathfrak{A}$. Since $I \bar{\in} \mathfrak{L}$ it follows that I belongs to $\mathfrak{K}$ and, of course, is the unit in it.

Suppose that $K \in \mathfrak{K}$. If $\mathfrak{L}$ is a two-sided ideal of $\mathfrak{A}I$, then

$$\mathfrak{A}I \cdot (\mathfrak{L} \cup \mathfrak{K}K) = (\mathfrak{L} \cup \mathfrak{K}) \cdot (\mathfrak{L} \cup \mathfrak{K}K) \subset \mathfrak{L} \cup \mathfrak{K}\mathfrak{K}K \subset \mathfrak{L} \cup \mathfrak{K}K,$$

i.e., $\mathfrak{L} \cup \mathfrak{K}K$ turns out to be a left ideal of $\mathfrak{A}I$. If $\mathfrak{L} = \varnothing$, then $\mathfrak{K}K$ is also a left ideal of $\mathfrak{K} = \mathfrak{A}I$. Also it is not contained in $\mathfrak{L}$ because it has an element $K \cdot K = \mathfrak{K}$ not belonging to $\mathfrak{L}$. By the definition of $\mathfrak{L}$ this ideal must contain I. Since $I \bar{\in} \mathfrak{L}$, for some $K' \in \mathfrak{K}$ we must have

$$K'K = I.$$

We have shown that any element K of the semigroup $\mathfrak{K}$ is a right divisor of the unit I of that semigroup. From II, 2.18, $\mathfrak{K}$ is a group. The original element cannot be contained in $\mathfrak{L}$, since either $\mathfrak{L} = \varnothing$ or $\mathfrak{L}$ is a left ideal of $\mathfrak{A}$ not containing I, while $A \cup \mathfrak{A}A$ contains I. Since $A \in \mathfrak{A}I$, A must belong to the group $\mathfrak{K}$. From III, 1.15, it therefore follows that the element A is completely regular.

3.12. Taking 3.9 into account, we have a corollary of the theorem just obtained.

Corollary. *In order that every ideal in a semigroup $\mathfrak{A}$ should contain its identity it is necessary and sufficient that $\mathfrak{A}$ should be completely regular and that every nonempty set of its idempotents should contain its unit.*

The structure of such semigroups and some of their properties were described in 3.9. In particular, it is evident that the second of the conditions of the corollary may be replaced by the requirement that all the idempotents of the semigroup should form a completely ordered set

$$I_1, I_2, \ldots, I_\xi, I_{\xi+1}, \ldots \qquad (\xi < \mu),$$

in which

$$I_\xi I_\eta = I_\eta I_\xi = I_\xi$$

for any $\eta \leqslant \xi < \mu$.

3.13. Among the semigroups we are considering there are those in particular which satisfy the stronger condition of the existence in each subsemigroup of an identity.

Theorem. *In order that each subsemigroup of the semigroup $\mathfrak{A}$ should contain its unit it is necessary and sufficient that $\mathfrak{A}$ should be a periodic completely* *regular semigroup in which every nonempty set of its idempotents contains its own unit.*

PROOF. 1) If each subsemigroup of $\mathfrak{A}$ contains its own unit, then in particular each left ideal contains its own unit and therefore by 3.11 $\mathfrak{A}$ must be a completely regular semigroup for which every nonempty set of its idempotents contains its own unit.

In addition, every element A of $\mathfrak{A}$ must generate a monogenic subsemigroup $[A]$, which must contain its unit. For this it is necessary that $[A]$ should be finite. Thus the semigroup $\mathfrak{A}$ must be periodic.

(2) Suppose that $\mathfrak{A}$ is a periodic completely regular semigroup in which each nonempty set of its idempotents contains its own unit. From 3.9, (β), $\mathfrak{A}$ is the union of a completely ordered set of groups

$$\mathfrak{A} = \bigcup_{\xi} \mathfrak{G}_{\xi}.$$

Any subsemigroup $\mathfrak{B}$ of the semigroup $\mathfrak{A}$ is equal to the union

$$\mathfrak{B} = \bigcup_{\xi} \mathfrak{G}'_{\xi}, \qquad \mathfrak{G}'_{\xi} = \mathfrak{G}_{\xi} \cap \mathfrak{B}.$$

Some of the sets $\mathfrak{G}'_{\xi}$ may turn out to be empty.

Suppose that $\mathfrak{G}'_{\nu}$ is the first nonempty set (such a set must exist, because of the complete ordering).

Let $X \in \mathfrak{G}'_{\nu}$. Since $\mathfrak{G}_{\nu}$ is a periodic group, $[X] \ni I_{\nu}$, where I_{ν} is the unit of the group $\mathfrak{G}_{\nu}$.

By property 3.9, (θ), I_{ν} is the unit of $\mathfrak{T}_{\nu}$ and is therefore also a unit of $\mathfrak{B}$ contained in $\mathfrak{T}_{\nu}$.

4. Successively Annihilating Bands

4.1. A commutative band of a semigroup determines a homomorphism onto a commutative semigroup of idempotents (1.8; 1.12). Among such semigroups one may particularly distinguish semigroups adjoint to linearly ordered sets (II, 4.11). In this connection it is natural also to distinguish the corresponding class of commutative bands. The corresponding construction was considered by A. M. Kaufman [1] under the name of successively annihilating sum. The consideration of this construction makes it possible to study the structure of the so-called holoidal semigroups. It is natural to distinguish the class of these semigroups for their interest alone. But they also play a definite role in the study of linearly ordered groups, to which we shall devote our attention in § 3 of Chapter X.

4.2. DEFINITION. *A partition of the semigroup $\mathfrak{A}$ into subsemigroups is said to be a* SUCCESSIVELY ANNIHILATING BAND *if for any two distinct components $\mathfrak{B}$ and $\mathfrak{B}'$ of this decomposition either*

$$X\mathfrak{B}' = \mathfrak{B}'X = X$$

is satisfied for each $X \in \mathfrak{B}$ *or else*

$$Y\mathfrak{B} = \mathfrak{B}Y = Y$$

is satisfied for each $Y \in \mathfrak{B}'$.

Evidently, a successively annihilating band is indeed a band in the sense of the definition of 1.6, and in addition is commutative (1.12).

4.3. It follows immediately from the definition that the set of all components of a successively annihilating band is a linearly ordered set relative to the partial ordering relation defined in the following way: $\mathfrak{B} \leqslant \mathfrak{B}'$ if $\mathfrak{B} = \mathfrak{B}'$ or

$$XX' = X'X = X'$$

for any $X \in \mathfrak{B}$, $X' \in \mathfrak{B}'$.

In addition, if for some two elements of two components, $X \in \mathfrak{B}$ and $X' \in \mathfrak{B}'$, we have $XX' = X'$ or $X'X = X'$, then evidently $\mathfrak{B} \leqq \mathfrak{B}'$.

It is immediately clear that the components of a successively annihilating band and the ordering law in their set completely determine the structure of the semigroup.

4.4. A partition of a semigroup consisting of one component, which is the whole semigroup, will be called an improper successively annihilating band.

We note some properties of successively annihilating bands.

(α) *If all the components of a successively annihilating band of a semigroup* $\mathfrak{A}$ *are commutative semigroups, then* $\mathfrak{A}$ *is also commutative.*

(β) *A successively annihilating band of semigroups with left cancellation cannot have more than two components.*

Indeed, suppose that $\mathfrak{B}_1, \mathfrak{B}_2, \mathfrak{B}_3$ are three such components of some successively annihilating band, such that

$$\mathfrak{B}_1 < \mathfrak{B}_2 < \mathfrak{B}_3$$

relative to the ordering of 4.2. Then for $X_1 \in \mathfrak{B}_1$, $X_2 \in \mathfrak{B}_2$, $X_3 \in \mathfrak{B}_3$, we have

$$X_3 = X_3X_2 = X_3X_2X_1 = X_3X_1, \qquad X_3 = X_3X_2.$$

Since $X_1 \neq X_2$, the semigroup is not a semigroup with left cancellation.

(γ) *If a semigroup with left cancellation has a proper successively annihilating band, then it has a unit which is externally adjoined.*

From (β), a proper successively annihilating band of the semigroup $\mathfrak{A}$ with left cancellation consists of two components

$$\mathfrak{B}_1 < \mathfrak{B}_2.$$

If $I, I' \in \mathfrak{B}_1$ and $X \in \mathfrak{B}_2$, then

$$XI = X, \qquad XI' = X$$

so that $I = I'$. Accordingly $\mathfrak{B}_1$ consists only of the single element I. Since $\mathfrak{B}_1$ is a subsemigroup, $I^2 = I$. Since $\mathfrak{B}_1 < \mathfrak{B}_2$, we have

$$XI = IX = X$$

for every $X \in \mathfrak{B}_2$. Accordingly, I is the unit of $\mathfrak{A}$. Since for $X, X' \in \mathfrak{B}_2$ also $XX' \in \mathfrak{B}_2$ it follows that the unit I is externally adjoined.

(δ) *If a semigroup with left cancellation has a unit which is externally adjoined, then it has a unique proper successively annihilating band.*

The first component of this band consists only of the unit, and the second of all the elements which are not units.

Indeed, suppose that $\mathfrak{B}_1 = \{E\}$ and $\mathfrak{B}_2 = \mathfrak{A} \setminus E$, where E is the unit of the semigroup $\mathfrak{A}$. If the unit E is externally adjoined, then evidently the partition of $\mathfrak{A}$ into $\mathfrak{B}_1$ and $\mathfrak{B}_2$ is a successively annihilating band. That each proper successively annihilating band of $\mathfrak{A}$ coincides with that band is proved by a repetition of the discussion of (γ).

(ε) *Suppose that $\mathfrak{n}$ is a successively annihilating band of the semigroup $\mathfrak{A}$, consisting of the components $\mathfrak{B}_\alpha, \mathfrak{B}_\beta, \ldots, \mathfrak{B}_\nu, \ldots$, and that for each of the components $\mathfrak{B}_\nu$ there is given in turn a successively annihilating band $\mathfrak{n}_\nu$ consisting of the components $\mathfrak{B}_\nu^{(\xi)}, \mathfrak{B}_\nu^{(\eta)}, \ldots$ ($\nu = \alpha, \beta, \ldots$). Then the partition of $\mathfrak{A}$ into the subsemigroups $\mathfrak{B}_\nu^{(\sigma)}$ is a successively annihilating band of $\mathfrak{A}$.*

Indeed, if for $\mathfrak{B}_\nu^{(\sigma)}$ and $\mathfrak{B}_\mu^{(\tau)}$ one has $\mathfrak{B}_\nu < \mathfrak{B}_\mu$ relative to the ordering 4.3 of the components of $\mathfrak{n}$, then

$$X\mathfrak{B}_\nu^{(\sigma)} = \mathfrak{B}_\nu^{(\sigma)}X = X$$

for every $X \in \mathfrak{B}_\mu^{(\tau)}$. If now $\nu = \mu$ but $\mathfrak{B}_\nu^{(\sigma)} < \mathfrak{B}_\mu^{(\tau)}$ relative to the ordering 4.3 of the component $\mathfrak{n}_\nu$ of the band, then

$$X\mathfrak{B}_\nu^{(\sigma)} = \mathfrak{B}_\nu^{(\sigma)}X = X$$

for every $X \in \mathfrak{B}_\mu^{(\tau)}$. It follows from this and 4.2 that our partition $\mathfrak{A}$ is a successively annihilating band.

(ζ) *If for the components of the successively annihilating band $\mathfrak{n}$ of the semigroup $\mathfrak{A}$ there is defined an equivalence relation $\mathfrak{l}$ such that*

$$\mathfrak{B}_1 \leqslant \mathfrak{B}_2 \leqslant \mathfrak{B}_3,$$
$$\mathfrak{B}_1 \sim \mathfrak{B}_3\ (\mathfrak{l})$$

always implies $\mathfrak{B}_1 \sim \mathfrak{B}_2(\mathfrak{l})$, then $\mathfrak{A}$ has the following successively annihilating band $\mathfrak{m}$. Each component of $\mathfrak{m}$ is the union of all the components of the band $\mathfrak{n}$ which are equivalent to each other modulo $\mathfrak{l}$.

In fact, the union of any set of components of $\mathfrak{n}$ evidently is a subsemigroup. Thus $\mathfrak{m}$ is a partition of $\mathfrak{A}$ into subsemigroups. Let $\mathfrak{K}$ and $\mathfrak{K}'$ be two distinct components of that partition. Let $\mathfrak{B}_1, \mathfrak{B}_2 \subset \mathfrak{K}$ and $\mathfrak{B}_1', \mathfrak{B}_1' \subset \mathfrak{K}'$ be certain components of the band $\mathfrak{n}$. If $\mathfrak{B}_1 < \mathfrak{B}_1'$, then also $\mathfrak{B}_2 < \mathfrak{B}_2'$, since if $\mathfrak{B}_2 > \mathfrak{B}_2'$ we would obtain a contradiction with the property of the equivalence $\mathfrak{l}$. Hence it follows that for $X \in \mathfrak{K}$, $X' \in \mathfrak{K}$ we always have

$$XX' = X'X = X'.$$

Hence from 4.2 it follows that $\mathfrak{m}$ is a successively annihilating band.

4.5. Let Ψ be some nonempty collection of successively annihilating bands of the semigroup $\mathfrak{A}$. Denote by $\mathfrak{k}$ their greatest lower bound and by $\mathfrak{l}$ their least upper bound (1.9). Since all the bands of Ψ are commutative it follows from 1.13 that $\mathfrak{k}$ and $\mathfrak{l}$ are also commutative bands.

Suppose that $\mathfrak{B}$ and $\mathfrak{B}'$ are two distinct components of the band $\mathfrak{k}$. From I, 5.15, for some band $\mathfrak{n} \in \Psi$ the sets $\mathfrak{B}$ and $\mathfrak{B}'$ are contained in distinct components of $\mathfrak{n}$. Therefore one necessarily has either (1) $B\mathfrak{B}' = \mathfrak{B}'B = B$ for every $B \in \mathfrak{B}$ or (2) $B'\mathfrak{B} = \mathfrak{B}B' = B'$ for every $B' \in \mathfrak{B}'$. Hence it follows that $\mathfrak{k}$ is a successively annihilating band.

Suppose that $\mathfrak{C}$ and $\mathfrak{C}'$ are two distinct components of the band $\mathfrak{l}$. From I, 5.16, relative to any band $\mathfrak{n} \in \Psi$ the sets $\mathfrak{C}$ and $\mathfrak{C}'$ lie in distinct components of $\mathfrak{n}$. Therefore it follows that either (1) $C\mathfrak{C}' = \mathfrak{C}'C = C$ for each $C \in \mathfrak{C}$ or (2) $C'\mathfrak{C} = \mathfrak{C}C' = C'$ for each $C' \in \mathfrak{C}'$. Hence it follows that $\mathfrak{l}$ is a successively annihilating band.

4.6. From what was proved in 4.5 it follows that the collection of all successively annihilating bands of a semigroup, relative to the partial ordering of I, 5.14, is a complete lattice.

4.7. Suppose that the successively annihilating band $\mathfrak{k}_0$ is the greatest lower bound of all successively annihilating bands of the semigroup $\mathfrak{A}$ (4.6). The band $\mathfrak{k}_0$ is the most refined partition of $\mathfrak{A}$ in comparison with all successively annihilating bands. By 4.3, (ε) the components of $\mathfrak{k}_0$ are semigroups which do not have proper successively annihilating bands. Furthermore, for each successively annihilating band $\mathfrak{n}$ of the semigroup $\mathfrak{A}$ we have $\mathfrak{k}_0 \leqslant \mathfrak{n}$, so that $\mathfrak{n}$ evidently may be gotten from $\mathfrak{k}_0$ by the method described in 4.3, (η).

4.8. This construction turns out to be useful, in particular, for the study of a certain class of semigroups, as follows naturally from the importance of the relation of divisibility of elements in a semigroup.

DEFINITION. *The semigroup $\mathfrak{A}$ is said to be* HOLOIDAL *if for any two different elements one is both a left and a right divisor of the other, while the second cannot be at the same time a left and right divisor of the first.*

In any semigroup one may define a relation according to which two elements A and B stand in the given relation if $A = B$ or B is divided by A both left and right. That a semigroup is holoidal means that this relation is a linear ordering relation (I, 5.11). In discussing this relation we may assume that A precedes B if $A = B$ or B is divided both left and right by A.

One must bear in mind that an individual element of a holoidal semigroup can be a proper left and right divisor of itself, i.e., it can have both left and right units. Equally well, one may have the case when there are no such units for the given element.

4.9. An infinite monogenic semigroup is obviously holoidal. As to a finite monogenic semigroup, it is holoidal if and only if its type (III, 3.7) has the form $(h, 1)$.

4.10. We note the following property of a holoidal semigroup $\mathfrak{A}$. If there is in $\mathfrak{A}$ a first element relative to a linear ordering of the holoidal semigroup, then relative to this ordering all the elements of $[X]$ precede all elements not entering into $[X]$.

Indeed, suppose that $Y \bar{\in} [X]$. We prove by induction on n that X^n precedes Y. For $n = 1$ this is valid according to the definition of X.

Suppose that X^{n-1} precedes Y, i.e., for some $U, V \in \mathfrak{A}$,

$$Y = X^{n-1}U, \qquad Y = VX^{n-1}.$$

Since $Y \bar{\in} [X]$, we have $U \neq X$, $V \neq X$ and therefore for some $U', V' \in \mathfrak{A}$,

$$U = XU', \qquad V = V'X,$$

from which we find that X^n precedes Y:

$$Y = X^{n-1}U = X^nU', \qquad Y = VX^{n-1} = V'X^n.$$

4.11. Now we consider the question of decomposition of holoidal semigroups into successively annihilating bands.

Let $\mathfrak{n}$ *be a successively annihilating band of a holoidal semigroup* $\mathfrak{A}$. *Each of its components* $\mathfrak{B}$ *is a holoidal semigroup.*

Indeed, suppose that for $B, B' \in \mathfrak{B}$ and $X, Y \in \mathfrak{A}$ we have

$$B' = XB, \qquad B' = BY.$$

X cannot belong to a component following $\mathfrak{B}$ and distinct from it (4.3), since in this case we would have $XB = X$. If now X belongs to a component preceding $\mathfrak{B}$ and differing from it, then $XB = B$, and we have $B' = B$. The situation for Y is analogous. Thus, $X, Y \in \mathfrak{B}$ and therefore B' divides B both left and right in $\mathfrak{B}$. If $B \neq B'$, then B is divided simultaneously left and right by B' in $\mathfrak{B}$, which cannot happen since it cannot happen in $\mathfrak{A}$.

4.12. There is a property inverse to 4.11. *If some semigroup* $\mathfrak{A}$ *has a successively annihilating band* $\mathfrak{n}$ *all components of which are holoidal, then* $\mathfrak{A}$ *itself is a holoidal semigroup.*

Indeed, suppose that $A, A' \in \mathfrak{A}$; $A \neq A'$; $A \in \mathfrak{B}$; $A' \in \mathfrak{B}'$, where $\mathfrak{B}$ and $\mathfrak{B}'$ are components of the band $\mathfrak{n}$. If $\mathfrak{B} \neq \mathfrak{B}'$ and $\mathfrak{B}$ precedes $\mathfrak{B}$ relative to the linear ordering of 4.3, then

$$A' = A'A, \qquad A' = AA'.$$

The relation $A = A'X$ is not possible, since the product $A'X$ must be contained in a component $\mathfrak{B}''$ of the band $\mathfrak{n}$ which follows $\mathfrak{B}$, and $A \in \mathfrak{B}$, while $\mathfrak{B} < \mathfrak{B}'$.

If $\mathfrak{B} = \mathfrak{B}'$, then in $\mathfrak{B}$, $\mathfrak{A}$ or $\mathfrak{A}'$ is divided both left and right by the other, while the second cannot be divided in $\mathfrak{B}$ by the first simultaneously left and right, since the semigroup $\mathfrak{B}$ is holoidal. For $X \in \mathfrak{A}$, belonging to a component distinct from $\mathfrak{B}$, the relation $A' = XA$ is not possible. Indeed, if the component containing X precedes $\mathfrak{B}$, then $XA = A$. If it follows $\mathfrak{B}$, then XA is contained in that component but not in $\mathfrak{B}$.

4.13. The properties of successively annihilating bands under consideration make it possible, as was shown by A. M. Kaufman [**2**], to characterize the structure and to give a complete classification of finite holoidal semigroups. This result of A. M. Kaufman is a later development of preceding papers of Klein-Barmen [**1**; **2**; **3**].

THEOREM. *Every finite holoidal semigroup decomposes into a successively annihilating band of monogenic holoidal semigroups* (4.9).

PROOF. From 4.7, a finite holoidal semigroup decomposes into a successively annihilating band of semigroups $\mathfrak{B}_1, \mathfrak{B}_2, \ldots, \mathfrak{B}_m$ which themselves do not have proper successively annihilating bands. The semigroup $\mathfrak{B}_n$ ($n = 1, 2, \ldots, m$) itself, from 4.11, is a holoidal semigroup. We shall show that the semigroup $\mathfrak{B}_n$ is monogenic. Denote by X the first element of $\mathfrak{B}_n$ relative to the linear ordering in the holoidal semigroup $\mathfrak{B}_n$. The type of the finite monogenic semigroup $[X]$ has the form $(h, 1)$ (III, 3.7), since in the contrary case there would be in $[X] \subset \mathfrak{A}$ two distinct elements which simultaneously divide each other both left and right.

Suppose that $\mathfrak{B}_n \neq [X]$. Denote by $\mathfrak{C} = \mathfrak{B}_n \backslash [X]$ and by Y the first element of $\mathfrak{C}$ relative to the ordering. Each element of $\mathfrak{C}\mathfrak{C}$ is divided by Y both left and right. Therefore it belongs to $\mathfrak{C}$, since from 4.10 all the elements of $[X]$ precede all the elements of $\mathfrak{C}$, including the element Y as well. Thus we obtain a decomposition of $\mathfrak{B}_n$ into two subsemigroups:

$$\mathfrak{B}_n = [X] \cup \mathfrak{C}, \qquad [X] \cap \mathfrak{C} = \varnothing.$$

Each element C of $\mathfrak{C}$ is divided by X^h both left and right:

$$C = UX^h, \qquad C = X^hV.$$

Since the type of $[X]$ is $(h, 1)$, then for any $X^k \in [X]$ we have

$$X^kX^h = X^hX^k = X^h.$$

Hence it follows that

$$CX^k = UX^hX^k = UX^h = C,$$

$$X^kC = X^kX^hV = X^hV = C.$$

Thus, our partition is a successively annihilating band, which contradicts the fact that $\mathfrak{B}_n$ has no proper successively annihilating bands. Thus, indeed, we must have $\mathfrak{B}_n = [X]$.

4.14. Since a monogenic holoidal semigroup evidently has no proper successively annihilating bands, the decomposition obtained in 4.13 is, for a finite holoidal semigroup $\mathfrak{A}$, the unique decomposition $\mathfrak{k}_0$ with which we were dealing in 4.7. This decomposition, from what was said in 4.3, completely determines the structure of $\mathfrak{A}$. In particular, it turns out that $\mathfrak{A}$ is necessarily commutative. Since a finite monogenic holoidal semigroup is completely characterized by the number of its elements, the structure of $\mathfrak{A}$ is completely determined by the succession of natural numbers $n_1, n_2, \ldots, n_m$, which are equal, respectively, to the number of elements in the components of the successive components of the indicated successively annihilating band. For there to be an isomorphism between two finite holoidal semigroups it is obviously necessary and sufficient that the indicated sequences of numbers should be identical for both.

4.15. It is not difficult to verify that the arguments of 4.13 and 4.14 may be extended almost without change to holoidal semigroups whose elements are ordered by the type of ordering by magnitude of the set of all natural numbers. Such semigroups also turn out to be commutative and decompose into successively annihilating bands of monogenic holoidal semigroups. The difference from the case analyzed above lies only in the fact that the number of components in the decomposition may be infinite. In the case of a finite number of components the last component is an infinite monogenic semigroup.

5. Basis Classes

5.1. In connection with the idea of studying the structure of semigroups by representing them in the form of unions of certain of their semigroups, there naturally arises the following concept, introduced by E. S. Ljapin [**11**]. In the definition presented here, as well as throughout the sequel in the present section, we shall, without special mention, understand by classes of semigroups only those classes which are closed with respect to isomorphisms, i.e., classes which contain along with a certain semigroup all the semigroups isomorphic to it.

DEFINITION. *The class Γ_0 of semigroups is said to be a* BASIS CLASS *for the class Γ of semigroups if the following conditions are satisfied.*

(α) *Each semigroup of Γ may be represented in the form of a union of subsemigroups of it belonging to the class Γ_0.*

(β) *Each semigroup, which may be represented in the form of a union of subsemigroups of it which belong to the class Γ_0, must belong to Γ.*

(γ) *If some class of semigroups Γ_1, all the semigroups of which belong to Γ_0, satisfies the conditions* (α) *and* (β) *formulated above for the class Γ_0, then Γ_1 coincides with Γ_0.*

Thus a basis class is a minimal class from which one may with the aid of the operation of taking unions of semigroups obtain all the semigroups of a given class and not go beyond it.

5.2. In condition 5.1, (γ) the reference to (β) can of course be omitted in view of the fact that for Γ_1, which is a part of Γ_0, condition (β) is automatically satisfied since it is satisfied for Γ_0.

In 5.1 condition (γ) can be replaced by the following condition:

(γ') *In each semigroup* $\mathfrak{A}_0$ *of* Γ_0 *there must exist an element* A_0 *which is not contained in any subsemigroup of the semigroup* $\mathfrak{A}_0$ *belonging to* Γ_0 *and not isomorphic to* $\mathfrak{A}_0$.

Indeed, suppose that (α), (β), (γ) of 5.1 are satisfied but (γ') is violated, i.e., in some semigroup $\mathfrak{A}_0 \in \Gamma_0$ each element X is contained in a subsemigroup $\mathfrak{B}_X$ which belongs to Γ_0 and is not isomorphic to $\mathfrak{A}_0$. Thus,

$$\mathfrak{A}_0 = \bigcup_{X \in \mathfrak{A}_0} \mathfrak{B}_X.$$

We denote by Γ_1 the class of all semigroups of Γ_0 which are not isomorphic to $\mathfrak{A}_0$. Γ_1 satisfies condition (α) since each semigroup of Γ may be represented in the form of a union of semigroups of Γ_0; and then one may replace the components entering there and isomorphic to $\mathfrak{A}_0$ by a union of subsemigroups, isomorphic to the $\mathfrak{B}_X$, which all belong to Γ_1. Condition (β), as we have already noted, is satisfied automatically and thus we find ourselves in contradiction with Definition 5.1.

Now suppose that conditions (α), (β), and (γ') are satisfied. Suppose that some class Γ_1 violates condition (γ). Suppose that $\mathfrak{A}_0$ is some semigroup of Γ_0 that is not isomorphic to any semigroup of Γ_1. Since for Γ_1 condition (α) must be satisfied, $\mathfrak{A}_0$ may be expressed in the form of a union of subsemigroups of Γ_1. The element A_0 mentioned in (γ') is contained in some component of this union, which by (γ') must be isomorphic to $\mathfrak{A}_0$. Accordingly, Γ_1 contains a semigroup isomorphic to the semigroup $\mathfrak{A}_0$, which contradicts the hypothesis.

5.3. Condition 5.2, (γ') for a class of semigroups Γ_0 is necessary and sufficient for Γ_0 to be a basis class of some class of semigroups. The necessity of this condition follows from 5.2. For the proof of sufficiency we choose a class Γ consisting of all those semigroups which may be represented in the form of unions of subsemigroups belonging to the class Γ_0. For Γ_0 and such a Γ conditions 5.1, (α) and 5.1, (β) are satisfied in an obvious way. By hypothesis 5.2, (γ') is satisfied, from which, along with 5.2, it follows that Γ_0 is a basis class for the class Γ.

5.4. It is clear that a basis class Γ_0 for a class Γ completely determines the structure of the class Γ of semigroups. However, when we have given a clear description of the structure of the semigroups of the basis class, we are usually still rather far from any clear-cut description of the structure of an arbitrary semigroup of Γ. The fact that each such semigroup may be represented as the union of subsemigroups taken from Γ_0 is far from determining its structure. Giving the structure of semigroups of the basis class may be considered as giving a description of the local structure of semigroups of Γ, since each element of a semigroup of Γ is contained in some semigroup contained in Γ_0. Semigroups of the basis class are, so to speak, elementary carriers of the properties which determine that a given subgroup belongs to the class Γ. One may say that they are the cells, the bricks, the atoms, from which the semigroups belonging to Γ are made.

5.5. In any semigroup $\mathfrak{A}$ each element A is contained in a monogenic semigroup $[A]$ generated by that element. Thus,

$$\mathfrak{A} = \bigcup_{A \in \mathfrak{A}} [A].$$

This means that the class of monogenic semigroups, relative to the class of all semigroups, satisfies condition 5.1, (α) for a basis class. Condition 5.1, (β) is satisfied in a trivial way.

It is immediately clear that condition 5.2, (γ') is also satisfied. Indeed, the element A itself, generating the monogenic semigroup $[A]$, is evidently not contained in any subsemigroup of $[A]$ different from $[A]$.

Thus, *the class of all monogenic semigroups is a basis class in the class of all semigroups.*

It is easy to prove that this is the sole basis class in the class of all semigroups. Indeed, suppose that Γ_0 is any basis class. $\mathfrak{A} = [A]$ is some monogenic semigroup. Since it must be representable in the form of a union of subsemigroups belonging to the basis class Γ_0, the element A must belong to some subsemigroup $\mathfrak{A}'$ of the semigroup $\mathfrak{A}$, belonging to Γ_0. But evidently $\mathfrak{A}' = \mathfrak{A}$, so that $\mathfrak{A}$ necessarily belongs to the class Γ_0. Since Γ_0 contains every monogenic semigroup, it must from 5.1, (γ) coincide with the class of all monogenic semigroups.

5.6. The class of all cyclic groups (II, 3.17) is easily seen to have the property 5.2, (γ'). In the finite cyclic group $[A]$ the element A is not contained in any subsemigroup different from $[A]$ itself. In the infinite cyclic group $\mathfrak{A} = \{\ldots, A^{-2}, A^{-1}, A^0, A, A^2, \ldots\}$ the element A is not contained in any finite cyclic group. Thus by (5.3) the class of all cyclic groups is a basis class for the class of semigroups which are unions of cyclic groups. Since each group is the union of its cyclic subgroups, the class just obtained may be characterized as the class of all semigroups which are unions of groups. From III, 1.15, this is the class of completely regular semigroups. We have obtained for it a basis class consisting of all cyclic groups, which may serve as a new way of defining that class.

We observe that the class of all cyclic groups is the sole basis class for the class of all completely regular semigroups. Indeed, if Γ_0 is any basis class, then it must contain every finite cyclic group $[A]$ since the latter belongs to our class, and the element A is not contained in any subsemigroup of the group $[A]$ other than $[A]$ itself. If $\mathfrak{A} = \{\ldots, A^{-2}, A^{-1}, A^0, A, A^2, \ldots\}$ is an infinite cyclic group, then A is not contained in any subgroup of $\mathfrak{A}$ other than $\mathfrak{A}$ itself. But A must be contained in a certain subsemigroup $\mathfrak{B}$ of the semigroup $\mathfrak{A}$, belonging to Γ_0. Inasmuch as $\mathfrak{B}$ is contained in the class in question, the element $A \in \mathfrak{B}$ must belong to some subgroup $\mathfrak{B}$. But this subgroup can only be $\mathfrak{A}$. Accordingly, $\mathfrak{A} = \mathfrak{B}$ belongs to Γ.

5.7. In § 5 of Chapter II we considered a certain class of semigroups Π (II, 5.2), having a certain significance for the question of the existence of a fixed point for a transformation. As E. S. Ljapin [**11**] proved, Π has a basis class. We shall construct a basis class for Π below. In preparation we shall have to consider certain auxiliary properties and constructions.

Suppose that the semigroup $\mathfrak{A}$ belongs to the class Π. Denote by $\mathfrak{A}'$ the collection of all those elements X of $\mathfrak{A}$ for which, given $XU = U$ $(U \in \mathfrak{A})$, one always has $UX^n \neq U$ for any natural number n. We also write $\mathfrak{A} \backslash \mathfrak{A}' = \mathfrak{A}''$. If X is contained in $\mathfrak{A}''$, then by the definition of $\mathfrak{A}''$ there is always a natural number n such that for some $U \in \mathfrak{A}$,

$$XU = U, \qquad UX^n = U.$$

The smallest of these numbers will be denoted by $\delta(X)$, or simply by δ. For $X \in \mathfrak{A}$ we define a set $\mathfrak{H}_X \subset \mathfrak{A}$. If $X \in \mathfrak{A}'$, then $\mathfrak{H}_X$ consists of all right zeros of the element X. If $X \in \mathfrak{A}''$, then $\mathfrak{H}_X$ is the set of all elements $U \in \mathfrak{A}$ such that

$$XU = U, \qquad UX^\delta = U.$$

5.8. LEMMA. *If* $\mathfrak{A} \in \Pi$, $X \in \mathfrak{A}$, *while* X *has finite type* (h, d) (III, 3.7), *then* $X \in \mathfrak{A}''$ *and* $\delta(X)$ *is a divisor of* d.

PROOF. Suppose that $XU = U$. From $X^{h+d} = X^h$ it follows that

$$X(UX^h) = UX^h, \qquad (UX^h)X^d = (UX^h).$$

Accordingly, $X \in \mathfrak{A}''$. Let

$$XV = V, \qquad VX^\delta = V.$$

We carry out a division with remainder, that is,

$$d = q\delta + r, \qquad 0 \leqslant r < \delta.$$

Since

$$X \cdot (VX^h) = VX^h,$$

$$(VX^h) \cdot X^r = VX^rX^h = VX^{q\delta} \cdot X^r \cdot X^h = VX^d \cdot X^h = VX^{h+d} = VX^h$$

it follows from the definition of $\delta = \delta(X)$ that r, which is less than δ, cannot differ from zero. Accordingly, d is divided by δ.

5.9. LEMMA. *If* $\mathfrak{A} \in \Pi$ *and* $X \in \mathfrak{A}$, *then* $\mathfrak{H}_X$ *is a semigroup lying in* Π.

PROOF. (1) Suppose that $X \in \mathfrak{A}'$. If $U_1, U_2 \in \mathfrak{H}_X$, then

$$X(U_1U_2) = (XU_1)U_2 = U_1U_2,$$

i.e., $U_1U_2 \in \mathfrak{H}_X$.

For $U \in \mathfrak{H}_X$ there must be in $\mathfrak{A}$ a right zero V. But then

$$XV = XUV = UV = V,$$

i.e., $V \in \mathfrak{H}_X$, and accordingly U has a right zero in $\mathfrak{H}_X$.

(2) Suppose that $X \in \mathfrak{A}''$. If $U_1, U_2 \in \mathfrak{H}_X$, i.e.,

$$U_1X^\delta = U_1, \qquad U_2X^\delta = U_2,$$

then

$$X(U_1U_2) = U_1U_2, \qquad (U_1U_2)X^\delta = U_1U_2,$$

i.e., $U_1U_2 \in \mathfrak{H}_X$.

For $U \in \mathfrak{H}_X$ there must be a right zero V in $\mathfrak{A}$. Since

$$U(VU) = VU,$$
$$X(VU) = X \cdot UV \cdot U = UV \cdot U = VU,$$
$$(VU)X^\delta = V \cdot UX^\delta = VU,$$

VU is a right zero for the element U lying in $\mathfrak{H}_X$.

5.10. Suppose that $[X]$ is a monogenic semigroup and that s is some natural number which is a divisor of d if $[X]$ is finite and has type (h, d) (III, 3.7). Observe that from $X^a = X^{a'}$, $X^b = X^{b'}$ and $a \equiv b \pmod{s}$ it follows that $a' \equiv b' \pmod{s}$.

For an infinite semigroup this is trivial. For a finite $[X]$ it follows from the definition of type that $a' \equiv a \pmod{d}$, $b \equiv b' \pmod{d}$, and therefore $a' \equiv a \pmod{s}$, $b' \equiv b \pmod{s}$.

We define in $[X]$ a relation $\mathfrak{n}_s$ by setting $X^a \sim X^b\ (\mathfrak{n}_s)$ if $a \equiv b \pmod{s}$. Evidently, $\mathfrak{n}_s$ is a two-sidedly stable equivalence. If $[X]$ is infinite, then by $\mathfrak{n}_0$ we shall understand the identity relation in $[X]$.

We observe that if $s \neq 0$ the factor-semigroup $[X]/\mathfrak{n}_s$ (VII, 2.4) is a group.

5.11. Lemma. *Suppose $H \in \mathfrak{H}_X$. If $X \in \mathfrak{A}'$, then the equality $HX^a = HX^b$ is possible only if $X^a \sim X^b\ (\mathfrak{n}_0)$, i.e., when $X^a = X^b$. If $X \in \mathfrak{A}''$, then the equality $HX^a = HX^b$ holds if and only if $X^a \sim X^b\ (\mathfrak{n}_\delta)$.*

Proof. (1) Suppose that $c = a - b > 0$ and $HX^a = HX^b$. Since evidently $HX^b \in \mathfrak{H}_X$, it follows from

$$(HX^b)X^c = HX^a = HX^b$$

that $X \in \mathfrak{A}''$. Divide c by δ with a remainder,

$$c = q\delta + r, \qquad 0 \leqslant r < \delta,$$

and suppose that $r \neq 0$. We have

$$(HX^b) = (HX^b)X^c = (HX^b)(X^\delta)^q X^r = (HX^b)X^r.$$

But since $r < \delta$ this is impossible. Accordingly, $r = 0$ and c is divisible by δ, i.e., $a \equiv b \pmod{\delta}$ and $X^a \sim X^b\ (\mathfrak{n}_\delta)$.

(2) Suppose that $X^a \sim X^b\ (\mathfrak{n}_\delta)$ and $a > b$, i.e., $a = q\delta + b$. Since $HX^\delta = H$ it follows that

$$HX^a = H(X^\delta)^q \cdot X^b = HX^b.$$

5.12. Suppose that $\mathfrak{N}$ and $\mathfrak{M}$ are two nonintersecting semigroups and that $\mathfrak{m}$ is a two-sidedly stable equivalence in $\mathfrak{M}$. In the set of all possible pairs (N, M) $(N \in \mathfrak{N}, M \in \mathfrak{M})$ we define a relation $\mathfrak{n}$, putting $(N_1, M_1) \sim (N_2, M_2)\ (\mathfrak{n})$ if $N_1 = N_2$ and $M_1 \sim M_2\ (\mathfrak{m})$. Evidently $\mathfrak{n}$ is an equivalence. The corresponding

$\mathfrak{n}$-class, containing the pair (N, M), will be denoted by $(N, M)_{\mathfrak{m}}$. We denote by $(\mathfrak{N} \times \mathfrak{M})_{\mathfrak{m}}$ the set consisting of all the elements of $\mathfrak{M}$ and of all the $\mathfrak{n}$-classes. In $(\mathfrak{N} \times \mathfrak{M})_{\mathfrak{m}}$ we define an operation, as follows. The product of two elements of $\mathfrak{M}$ will be taken to be their product obtained in $\mathfrak{M}$ by means of the operation defined in the semigroup $\mathfrak{M}$. In the three remaining cases the operation is defined in the following way:

$$(N_1, M_1)_{\mathfrak{m}} \cdot (N_2, M_2)_{\mathfrak{m}} = (N_1N_2, M_2)_{\mathfrak{m}},$$
$$(N_1, M_1)_{\mathfrak{m}} \cdot M_2 = (N_1, M_1M_2)_{\mathfrak{m}},$$
$$M_1 \cdot (N_2, M_2)_{\mathfrak{m}} = (N_2, M_2)_{\mathfrak{m}}.$$

Since $\mathfrak{m}$ is two-sidedly stable in $\mathfrak{M}$, this operation is single-valued, i.e., the result of the operation does not depend on which of the pairs (N, M) was taken as a representative of the class $(N, M)_{\mathfrak{m}}$. The associativity of the operation is verified without difficulty. The semigroup $(\mathfrak{N} \times \mathfrak{M})_{\mathfrak{m}}$ is called a *right annihilating product of the semigroups* $\mathfrak{N}$ *and* $\mathfrak{M}$.

If $\mathfrak{N}$ belongs to the class Π, then also $(\mathfrak{N} \times \mathfrak{M})_{\mathfrak{m}}$ belongs to Π. Indeed, for $M \in \mathfrak{M}$ the right zero in $(\mathfrak{N} \times \mathfrak{M})_{\mathfrak{m}}$ will be $(N', M')_{\mathfrak{m}}$, while for any $N' \in \mathfrak{N}$, $M' \in \mathfrak{M}$ and for $(N, M)_{\mathfrak{m}}$ the right zero will be $(N', M')_{\mathfrak{m}}$, where N' is a right zero of the element N in $\mathfrak{N}$.

5.13. Suppose that in a semigroup we are given a monogenic subsemigroup $\mathfrak{M} = [X]$ which is infinite or finite nonholoidal (i.e., having type (h, d), where $d > 1$) and a subsemigroup $\mathfrak{N}$ all the elements of which are right zeros for X. In addition we suppose that in $\mathfrak{N}$ for every $N_1, N_2 \in \mathfrak{N}$, $N_1 \neq N_2$ there always exists an $N_0 \in \mathfrak{N}$ such that $N_1N_0 \neq N_2N_0$. Suppose that the following condition is satisfied in $\mathfrak{A}$: $NM_1 = NM_2$ $(N \in \mathfrak{N};\ M_1, M_2 \in \mathfrak{M})$ always implies $N'M_1 = N'M_2$ for any $N' \in \mathfrak{N}$. Denote by $\mathfrak{m}$ the relation in $\mathfrak{M}$ according to which $M_1 \sim M_2\ (\mathfrak{m})$ holds if and only if $NM_1 = NM_2$. Evidently $\mathfrak{m}$ is a two-sidedly stable equivalence in $\mathfrak{M}$.

Consider the following mapping φ of the semigroup $(\mathfrak{N} \times \mathfrak{M})_{\mathfrak{m}}$ into $\mathfrak{A}$:

$$\varphi(X^k) = X^k, \qquad \varphi(N, X^k)_{\mathfrak{m}} = NX^k \qquad (k = 1, 2, \ldots).$$

This is single-valued because of the property of $\mathfrak{m}$. We shall show that φ is one-to-one. Suppose that

$$X^k = NX^l.$$

Since $XN = N$, we therefore obtain

$$X^{k+1} = XNX^l = NX^l = X^k,$$

which is not possible in a nonholoidal monogenic semigroup. Suppose that

$$N_1X^{k_1} = N_2X^{k_2}.$$

Multiplying on the right by an arbitrary element $N' \in \mathfrak{N}$, we obtain $N_1N' = N_2N'$. But the validity of this equality for any $N' \in \mathfrak{N}$, from the hypothesis made above, means that $N_1 = N_2$. In this case $X^{k_1} \sim X^{k_2}\ (\mathfrak{m})$, but then also $(N_1, X^{k_1})_{\mathfrak{m}} = (N_2, X^{k_2})_{\mathfrak{m}}$.

From the fact that the elements of $\mathfrak{N}$ are right zeros for the elements of $\mathfrak{M}$ it immediately follows that the mapping φ is a homomorphism, and therefore also an isomorphism. Thus it turns out that the set of elements of $\mathfrak{A}$ which have the forms X^k and NX^k ($N \in \mathfrak{N}$; $k = 1, 2, \ldots$) is a subsemigroup isomorphic to $(\mathfrak{N} \times \mathfrak{M})_{\mathfrak{m}}$.

5.14. Denote by Π_0 the class of semigroups consisting of all finite monogenic holoidal semigroups and of the semigroups isomorphic to the semigroups $(\mathfrak{B} \times \mathfrak{M})_{\mathfrak{m}}$, where (1) $\mathfrak{M} = [X]$ is a monogenic semigroup, infinite or finite holoidal; (2) $\mathfrak{B}$ is the unit semigroup or the semigroup $\mathfrak{A}$ (II, 5.12), or the semigroup $\mathfrak{B}$ (II, 5.13); (3) $\mathfrak{m}$ is a relation $\mathfrak{n}_s$ ($s \geqslant 0$) of type 5.10.

Since a finite monogenic semigroup has a zero, and $\mathfrak{B}$ in all three possible cases is contained in Π, it follows from 5.12 that each of the semigroups of the class Π_0 belongs to the class Π.

5.15. LEMMA. *If $\mathfrak{A} = (\mathfrak{B} \times [X])_{\mathfrak{n}_s}$ belongs to the class Π_0 and $\mathfrak{A}_0$ is a subsemigroup of it containing X and belonging to Π_0, then $\mathfrak{A}_0$ is isomorphic to $\mathfrak{A}$.*

PROOF. $\mathfrak{A}_0$ cannot be a holoidal monogenic semigroup, since every monogenic subsemigroup of such a semigroup is itself holoidal, while $\mathfrak{A}_0$ contains a nonholoidal monogenic subsemigroup $[X]$. Accordingly, $\mathfrak{A}_0$ is isomorphic to a semigroup of type

$$\overline{\mathfrak{A}} = (\overline{\mathfrak{B}} \times [\overline{X}])_{\mathfrak{n}_r}.$$

From the definition of a right annihilating product it immediately results that in $\mathfrak{A}$ the elements of $[X]$, and only those, are not right zeros for any element whatever. In $\mathfrak{A}_0$, containing $[X]$, such elements will be elements of $[X]$. In $\overline{\mathfrak{A}}$ they will be the elements of $[\overline{X}]$. Thus $[X]$ and $[\overline{X}]$ are isomorphic, while under the isomorphism of $\overline{\mathfrak{A}}$ onto $\mathfrak{A}_0$ the set $[\overline{X}]$ maps onto $[X]$.

The relation $\mathfrak{n}_s$ is completely determined by the number of different elements in the set $A \cdot [X]$ ($A \in \mathfrak{A} \backslash [X]$), while this number is the same for any choice of A. The same holds also in $\overline{\mathfrak{A}}$ for $\mathfrak{n}_r$. Since $\overline{\mathfrak{A}}$ is isomorphic to the subsemigroup $\mathfrak{A}_0$ of the semigroup $\mathfrak{A}$, we therefore conclude that the relation $\mathfrak{n}_s$ in $[X]$ is identical with the relation $\mathfrak{n}_r$ in $[\overline{X}]$, i.e., $s = r$.

According to the definition of the class Π_0, each of the semigroups $\mathfrak{B}$ and $\overline{\mathfrak{B}}$ is a unit semigroup either equal to $\mathfrak{U}$ or equal to $\mathfrak{B}$.

If $\mathfrak{B}$ is the unit semigroup, and only in that case, all the right zeros of the subsemigroup $[X]$ are idempotents. The same holds for $\overline{\mathfrak{A}}$ relative to $\overline{\mathfrak{B}}$. Inasmuch as $\overline{\mathfrak{A}}$ is isomorphic to a subsemigroup of $\mathfrak{A}$ it therefore follows that $\overline{\mathfrak{B}}$ is the unit semigroup if and only if $\mathfrak{B}$ is the unit semigroup.

Suppose that in $\mathfrak{A}$ the element P is a right zero for $[X]$ and Q is a right zero for P.

If $\mathfrak{B} = \mathfrak{U}$, then $P = (U_{\alpha_1}^{\beta_1}, X^a)_{\mathfrak{n}_s}$ and $Q = (U_{\alpha_2}^{\beta_2}, X^b)_{\mathfrak{n}_s}$, where $\alpha_1 < \alpha_2$. Therefore there exists in $\mathfrak{A}$ a pair of elements $P^* \in P \cdot [X]$ and $Q^* \in Q \cdot [X]$

(namely, $P^* = (U_{\alpha_1}^{\beta_1}, X^c)_{\mathfrak{n}_s}$ and $Q^* = (U_{\alpha_2}^{\beta_2}, X^c)_{\mathfrak{n}_s}$) such that Q^* is a two-sided zero for P^*.

If $\mathfrak{B} = \mathfrak{V}$, then no element of $Q \cdot [X]$ can be a two-sided zero for any element of $P \cdot [X]$. (This follows from the fact that in $\mathfrak{V}$ no element in general has a two-sided zero (II, 5.13).)

Since $\overline{\mathfrak{A}}$ is isomorphic to the subsemigroup $\mathfrak{A}_0$ of the semigroup $\mathfrak{A}$ containing $[X]$, and since there is in $\mathfrak{A}_0$ a right zero P for $[X]$ and a right zero Q for P, it follows from what has been said that if $\mathfrak{B} = \mathfrak{U}$ one cannot have $\overline{\mathfrak{B}} = \mathfrak{V}$, and if $\mathfrak{B} = \mathfrak{V}$ one cannot have $\overline{\mathfrak{B}} = \mathfrak{U}$.

We have shown that for $\mathfrak{A} = (\mathfrak{B} \times [X])_{\mathfrak{n}_s}$ and $\overline{\mathfrak{A}} = (\mathfrak{B} \times [\overline{X}])_{\mathfrak{n}_r}$ the semigroups $[X]$ and $[\overline{X}]$ are isomorphic and the relations $\mathfrak{n}_s$ and $\mathfrak{n}_r$ in them are identical in $\mathfrak{B} = \overline{\mathfrak{B}}$. Accordingly $\mathfrak{A}$ is isomorphic to $\overline{\mathfrak{A}}$, which is isomorphic to $\mathfrak{A}_0$.

5.16. Theorem. *The class* Π_0 (5.14) *is a basis class* (5.1) *for the class* Π.

Proof. (1) Suppose that X is any element of the semigroup $\mathfrak{A}$ belonging to the class Π. If $[X]$ is holoidal, then $[X]$ is a subsemigroup of the semigroup $\mathfrak{A}$ containing X and lying in Π_0.

Suppose that $[X]$ is not holoidal.

Since $\mathfrak{H}_X$ (5.7) belongs to Π (5.9), there exists in $\mathfrak{H}_X$, from II, 5.14, a subsemigroup $\mathfrak{B}$ which either is the unit semigroup, or is isomorphic to $\mathfrak{U}$, or is isomorphic to $\mathfrak{V}$. Denote by $\mathfrak{A}_0$ the set of all elements which belong to $[X]$ or have the form BX^k ($B \in \mathfrak{B}$; $k = 1, 2, \ldots$). Each element of $\mathfrak{B}$ will be a right zero for X. As follows from the structure of $\mathfrak{U}$ and $\mathfrak{V}$, for any $B_1, B_2 \in \mathfrak{B}$ with $B_1 \neq B_2$ there always exists a $B_0 \in \mathfrak{B}$ such that $B_1B_0 \neq B_2B_0$. If for some $B \in \mathfrak{B}$,

$$BX^a = BX^b, \qquad X^a \neq X^b,$$

then, from 5.11, $X^a \sim X^b (\mathfrak{n}_\delta)$ ($\delta = \delta(X)$). But then we would also have for any $B' \in \mathfrak{B}$ that $B'X^a = B'X^b$ since $B'X^\delta = B'$. The converse is also evident. If $X \in \mathfrak{A}''$ and $a \equiv b \pmod{\delta}$, i.e., $X^a \sim X^b (\mathfrak{n}_\delta)$, then $BX^a = BX^b$ for any $B \in \mathfrak{B}$.

In view of these properties we may apply the considerations of 5.13 to $\mathfrak{A}_0$. If $X \in \mathfrak{A}'$, then from 5.13 it follows that $\mathfrak{m}$ coincides with $\mathfrak{n}_0$ (5.10). If $X \in \mathfrak{A}''$, then $\mathfrak{m}$ coincides with $\mathfrak{n}_\delta$ (5.10). From 5.13, $\mathfrak{A}_0$ is isomorphic to the semigroup $(\mathfrak{B} \times [X])_{\mathfrak{m}}$ which belongs to the class Π_0. Thus, for Π_0 we have proved that the property 5.1, (α) is satisfied.

(2) Suppose that the semigroup $\mathfrak{A}$ is represented in the form of a union of subsemigroups belonging to the class Π_0. Each element A of $\mathfrak{A}$ is contained in some subsemigroup $\mathfrak{A}_0$ of the semigroup $\mathfrak{A}$ lying in Π_0 and therefore has a right zero since $\mathfrak{A}_0$ belongs to Π_0 (5.14).

Thus property 5.1, (β) holds for Π_0.

(3) Suppose that the semigroup $\mathfrak{A}_0$ belongs to Π_0.

If $\mathfrak{A}_0$ is a finite monogenic semigroup $\mathfrak{A} = [X]$, then the element X is not contained in any of its subsemigroups distinct from $\mathfrak{A}_0$. Suppose that $\mathfrak{A}_0$ is isomorphic to the semigroup $\overline{\mathfrak{A}}_0 = (\mathfrak{B} \times [X])_{\mathfrak{n}_s}$. From 5.15 the element X of

$\overline{\mathfrak{A}}_0$ is not contained in any subsemigroup of the semigroup $\overline{\mathfrak{A}}_0$ belonging to Π_0 and not isomorphic to $\overline{\mathfrak{A}}_0$. Accordingly, in $\mathfrak{A}_0$ itself there is an element with the analogous property.

From this it follows that the class Π_0 has property 5.2, (γ'). Taking into account what was proved in the first part of the proof, we conclude from 5.2 that Π_0 is a basis class for Π.

5.17. It should be noted that in the class Π there is not just one basis class. All the basis classes of the class Π were discovered and characterized by E. S. Ljapin [**12**]. Among them the class Π_0 considered above turned out to be the simplest in construction and the most convenient for study. The simple structure of the semigroups of the class Π_0 makes it possible to regard the local structure (5.4) of the semigroups belonging to the class Π as having been cleared up.

5.18. Of course, it is by no means true that every class of semigroups has a basis class. Indeed, a class having a basis class must evidently have the following property. It must contain every semigroup which is a union of semigroups contained in it. But many important classes of semigroups do not satisfy this condition. For example, the class of commutative semigroups does not. As a matter of fact, every semigroup is the union of certain of its commutative subsemigroups (for example, the monogenic subsemigroups), but not all semigroups are commutative.

5.19. In this connection, the following generalization of the concept of basis class is in some cases useful. Suppose that Γ_0, Γ and Σ are three classes of semigroups, where Σ contains the class Γ and the class Γ contains Γ_0. One may say that the class Γ_0 is a basis class for the class Γ relative to the class Σ if conditions 5.1, (α), 5.2, (γ') are satisfied in addition to the following condition (β').

(β') *Every semigroup of* Σ *which can be represented in the form of the union of subsemigroups belonging to the class* Γ_0 *must belong to the class* Γ.

In the sense of this generalization, the class of all monogenic semigroups, taken relative to the class of all commutative semigroups, is a basis class for the class of all commutative semigroups.

The class of all cyclic groups (II, 3.17) relative to the class of all groups is a basis class for the class of all groups.

The class of all finite cyclic groups relative to the class of all groups is a basis class for the class of all periodic groups.

5.20. The class of all inverse semigroups has no basis class in the sense of 5.1 since it evidently does not satisfy the property mentioned in 5.18. From II, 7.4 the class of all inverse semigroups is contained in the class of semigroups having the property of commutativity of idempotents.

L. M. Gluskin [**8**] showed that relative to the class of semigroups with commuting idempotents the class of all inverse semigroups has a basis class.

CHAPTER IX

RELATIONS IN SEMIGROUPS

1. Defining Systems of Relations

1.1. We have already drawn attention to the fact that the definition of elements of a semigroup in the form of a product of elements of a certain generating set of the semigroup is in general not unique. Products which are different in appearance can often be equal, i.e., represent one and the same element of the semigroup. The consideration of the corresponding relations is of evident interest. In particular, what is essential is the singling out of systems of these relations from which all other relations follow as necessary results. It is possible to define semigroups by means of systems of such relations. A special role is played by the identities, i.e., the relations which are true for all elements of the semigroup. With the notion of an identity, there is connected the notion of a free semigroup in a class. The consideration of the corresponding problems is the aim of the present chapter.

1.2. Let $\mathfrak{N}$ be an arbitrary nonempty set, which, in view of our further constructions we will call an *alphabet*. Any finite sequence of elements of $\mathfrak{N}$ written in the form of a product

$$W = X_1X_2 \dots X_n \qquad (X_1, X_2, \dots, X_n \in \mathfrak{N})$$

we will call a *word in the alphabet* $\mathfrak{N}$.

The number of members in the word (n) we will call *the length of the word W*. A word of length one X ($X \in \mathfrak{N}$) we will identify with the element X itself. We will denote the set of all words in $\mathfrak{N}$ by $\mathfrak{W}_{\mathfrak{N}}$.

The use of the term alphabet is borrowed from the theory of associative calculi, although there it is considered only for finite alphabets, while here $\mathfrak{N}$ can be infinite.

1.3. For words in an alphabet $\mathfrak{N}$ there is defined an operation of attachment, which we will call *multiplication of words*, namely,

$$U = X_1X_2 \dots X_n, \qquad V = X_{n+1}X_{n+2} \dots X_m,$$
$$W = X_1X_2 \dots X_nX_{n+1} \dots X_m,$$
$$W = UV.$$

Since the operation of multiplication of words is evidently associative, the set $\mathfrak{W}_{\mathfrak{N}}$ consisting of all words in $\mathfrak{N}$ is a semigroup with respect to this operation, which we will call the *free semigroup over* $\mathfrak{N}$. Free semigroups over various alphabets will also be simply called *free semigroups.*

1.4. Let $\mathfrak{N}$ be some subset of a semigroup $\mathfrak{A}$. To any word in $\mathfrak{N}$,

$$X_1X_2 \ldots X_n \qquad (X_1, X_2, \ldots, X_n \in \mathfrak{N}),$$

there corresponds an element $S \in \mathfrak{A}$ which is the product of the elements $X_1, X_2, \ldots, X_n$ defined by the multiplication operation in $\mathfrak{A}$. This element S is called the *value of the word* $X_1X_2 \ldots X_n$ in $\mathfrak{A}$.

Each element S from $[\mathfrak{N}]$ can be represented in $\mathfrak{A}$ in the form of a product:

$$S = X_1X_2 \ldots X_n \qquad (X_1, X_2, \ldots, X_n \in \mathfrak{N}).$$

Thus S is the value of some word in $\mathfrak{N}$. The element S is completely defined by this word.

1.5. If in the semigroup $\mathfrak{A}$ an element S belongs to $[\mathfrak{N}]$ ($\mathfrak{N} \subset \mathfrak{A}$), then it is evident that for S there can exist different words in $\mathfrak{W}_{\mathfrak{N}}$, each of which has the value S. The fact that in the semigroup $\mathfrak{A}$ the values of two words

$$X_1X_2 \ldots X_n, \qquad Y_1Y_2 \ldots Y_m \qquad (X_1, X_2, \ldots, X_n, Y_1, Y_2, \ldots, Y_m \in \mathfrak{N})$$

in $\mathfrak{W}_{\mathfrak{N}}$ are equal means that

$$X_1X_2 \ldots X_n = Y_1Y_2 \ldots Y_m$$

in $\mathfrak{A}$. Such an equality is called a *relation in* $\mathfrak{A}$ *with respect to* $\mathfrak{N}$. If it is clear what set $\mathfrak{N}$ is being discussed, the words "with respect to $\mathfrak{N}$" will usually be omitted.

It is necessary to keep in mind that by a relation with respect to $\mathfrak{N}$ we understand a connection between certain words of $\mathfrak{W}_{\mathfrak{N}}$. Each of the two parts of a relation is defined not by that one element S from $\mathfrak{A}$ which is equal in $\mathfrak{A}$ to the value of the corresponding word, but by the word itself in $\mathfrak{W}_{\mathfrak{N}}$, i.e., the sequence of factors forming the given product. Thus a relation in $\mathfrak{A}$ with respect to $\mathfrak{N}$ is a pair (W, V) of words in $\mathfrak{W}_{\mathfrak{N}}$ whose values in $\mathfrak{A}$ are equal. In view of this fact it would be more precise to write the relation not with the symbol of equality (since it is not at all the case that the two words W and V are identical) but in some other way, for example, $W \leftrightarrow V$ or the like. But usually, in view of the equality of the values of the words in $\mathfrak{A}$, the equality sign is used in writing the relation, and we will conform to this notation. In this connection, for equality of words in $\mathfrak{N}$ as elements of $\mathfrak{W}_{\mathfrak{N}}$ we will sometimes use the identity sign $\equiv$, denoting identical, or, in other words, graphical equality of words in distinction from the equality of the values of these words in $\mathfrak{A}$. It is necessary to keep in mind that in the literature such distinctions in notation are sometimes neglected, on the principle that it is clear from the context which equality is being discussed.

1.6. Let Φ be a system of relations in a semigroup $\mathfrak{A}$ with respect to a generating set $\mathfrak{K}$. We define a relation $\mathfrak{n}$ in $\mathfrak{W}_{\mathfrak{K}}$ by setting

$$W \sim V(\mathfrak{n}),$$

if the system Φ contains a relation

$$W = V$$

with respect to $\mathfrak{K}$.

1.7. Further we will make use of the notion and properties of derived relations (I, 5.20; I, 5.21; I, 5.22; I, 5.23).

If for some words $W, V \in \mathfrak{W}_{\mathfrak{K}}$ we have

$$W \sim V(\mathfrak{n}'),$$

then it is evident that the values of the words W and V in $\mathfrak{A}$ are equal, i.e., in $\mathfrak{A}$ the following relation holds with respect to $\mathfrak{K}$:

$$W = V,$$

which in this case is called an *immediate corollary of the system of relations* Φ.

About the word (or the product) W we say in this case that it can be immediately transformed into V by means of Φ. The meaning of such a statement is obvious: V can be obtained from W in the following way. W is represented in the form of a product of certain words in $\mathfrak{W}_{\mathfrak{K}}$ (1.3). In order to obtain the word V from it, one of these word-factors, which is a left or right member of some relation of Φ, is to be replaced by the other member of that relation.

1.8. If for some words $W, V \in \mathfrak{W}_{\mathfrak{K}}$ we have

$$W \sim V(\mathfrak{n}''),$$

then it is evident that the relation with respect to $\mathfrak{K}$, namely,

$$W = V,$$

is true, which in this case is called a *corollary of the system of relations* Φ.

Of the word W in this case we say that it can be transformed into V by means of Φ. This means that the word W can be transformed into V by means of a finite number of the successive transformations indicated in 1.7.

1.9. DEFINITION. *A system Φ of relations with respect to some generating set $\mathfrak{K}$ of a semigroup $\mathfrak{A}$ is a* **DEFINING SYSTEM OF RELATIONS** *if any relation with respect to $\mathfrak{K}$ is a corollary of Φ.*

Relations contained in a defining system of relations are often called defining relations.

1.10. It is obvious that in any semigroup, with respect to any of its generating sets $\mathfrak{K}$, there is a defining system of relations. Such, for example, is the system of all relations with respect to $\mathfrak{K}$.

The multiplication table of the semigroup (I, 1.6) is simply the system of all possible relations of the form $XY = Z$. It is not difficult to see that it is a defining system of relations with respect to the generating set consisting of all elements of the semigroup.

1.11. Let φ be a mapping of some set $\mathfrak{N}_1$ into $\mathfrak{N}_2$. We will say that φ induces a mapping of the words of $\mathfrak{W}_{\mathfrak{N}_1}$ into the words of $\mathfrak{W}_{\mathfrak{N}_2}$, understanding by this that the word

$$W = X_1X_2 \ldots X_n \qquad (X_1, X_2, \ldots, X_n \in \mathfrak{N}_1)$$

is mapped into the word

$$\varphi(W) = \varphi(X_1)\varphi(X_2) \ldots \varphi(X_n)$$

in $\mathfrak{N}_2$.

If $\mathfrak{N}_1$ is a subset of some semigroup $\mathfrak{A}$, then by the same token we can speak of the mapping of relations in $\mathfrak{A}$ with respect to $\mathfrak{N}_1$ induced by the map φ.

We consider the following theorem which is an immediate generalization of a well-known theorem from the theory of groups, usually called the theorem of Dyck.

THEOREM. *Let $\mathfrak{R}$ be a generating set and let Φ be a defining system of relations with respect to $\mathfrak{R}$ for a semigroup $\mathfrak{A}$. Let φ be a map of $\mathfrak{R}$ into some semigroup $\mathfrak{B}$ under which any relation of Φ is mapped into a valid relation in $\mathfrak{B}$ with respect to $\mathfrak{B}$. Then φ can be extended to a homomorphism of $\mathfrak{A}$ into $\mathfrak{B}$.*

PROOF. Any element of $\mathfrak{A}$ can be represented in the form of a product

$$A = X_1X_2 \ldots X_n \qquad (X_1, X_2, \ldots, X_n \in \mathfrak{R}).$$

We define $\psi(A)$ as the following element in $\mathfrak{B}$:

$$\varphi(X_1) \cdot \varphi(X_2) \cdot \ldots \cdot \varphi(X_n).$$

This map ψ of elements of $\mathfrak{A}$ into $\mathfrak{B}$ is single-valued. In fact, if

$$A = X_1X_2 \ldots X_n, \qquad A = Y_1 Y_2 \ldots Y_m$$
$$(X_1, X_2, \ldots, X_n, Y_1, Y_2, \ldots, Y_m \in \mathfrak{R}),$$

then

$$X_1X_2 \ldots X_n = Y_1Y_2 \ldots Y_m$$

is a relation in $\mathfrak{A}$ with respect to $\mathfrak{R}$. This relation must be a corollary of Φ. By definition, this means that there exist words $U_1, U_2, \ldots, U_k$, where

$$U_1 = X_1X_2 \ldots X_n, \qquad U_k = Y_1Y_2 \ldots Y_m$$

and the relation $U_i = U_{i+1}$ $(i = 1, 2, \ldots, n - 1)$ is an immediate corollary of Φ. By the hypothesis of the theorem about φ it follows at once that in $\mathfrak{B}$ the relations $\varphi(U_i) = \varphi(U_{i+1})$ will be true. Hence it follows that the relation

$$\varphi(X_1) \cdot \varphi(X_2) \cdot \ldots \cdot \varphi(X_n) = \varphi(Y_1) \cdot \varphi(Y_2) \cdot \ldots \cdot \varphi(Y_m)$$

is true. Thus ψ is single-valued.

If

$$A = X_1X_2 \dots X_n, \qquad A' = X_1'X_2' \dots X_m'$$
$$(X_1, X_2, \dots, X_n, X_1', X_2', \dots, X_m' \in \mathfrak{K}),$$

then evidently

$$\psi(AA') = \varphi(X_1) \cdot \varphi(X_2) \cdot \dots \cdot \varphi(X_n) \cdot \varphi(X_1') \cdot \dots \cdot \varphi(X_m') = \psi(A) \cdot \psi(A').$$

And this proves that ψ is a homomorphism.

1.12. By means of Theorem 1.11 it is easy to explain the basic meaning of the concept of a defining system of relations: A defining system of relations defines a semigroup to within isomorphism.

THEOREM. *Let $\mathfrak{K}_i$ be a generating set and let Φ_i be a defining system of relations with respect to $\mathfrak{K}_i$ for a semigroup $\mathfrak{A}_i$ $(i = 1, 2)$. If there is a one–one map φ of $\mathfrak{K}_1$ onto $\mathfrak{K}_2$ which induces a one–one map of Φ_1 onto Φ_2, then $\mathfrak{A}_1$ and $\mathfrak{A}_2$ are isomorphic.*

PROOF. By 1.11, φ can be extended to a homomorphism ψ of $\mathfrak{A}_1$ into $\mathfrak{A}_2$. Let $A, B \in \mathfrak{A}_1$, $\psi(A) = \psi(B)$. We represent A and B in the form of products of elements of $\mathfrak{K}_1$.

$$A = X_1X_2 \dots X_n, \qquad B = Y_1Y_2 \dots Y_m.$$

Since ψ is a homomorphism, from $\psi(A) = \psi(B)$ we obtain in $\mathfrak{A}_2$ the relation

$$\psi(X_1) \cdot \psi(X_2) \cdot \dots \cdot \psi(X_n) = \psi(Y_1) \cdot \psi(Y_2) \cdot \dots \cdot \psi(Y_m)$$

with respect to $\mathfrak{K}_2 = \psi(\mathfrak{K}_1)$. This relation must be a corollary of Φ_2.

Because φ induces a one–one map of Φ_1 onto Φ_2 it follows that in $\mathfrak{A}_1$ the relation

$$X_1X_2 \dots X_n = Y_1Y_2 \dots Y_m$$

with respect to $\mathfrak{K}_1$ must be a corollary of Φ_1. Thus it is shown that $\psi(A) = \psi(B)$ necessarily implies that $A = B$, i.e., the homomorphism ψ is a one–one map and hence an isomorphism.

It remains to show that $\psi(\mathfrak{A}_1) = \mathfrak{A}_2$. An arbitrary element A_2 in $\mathfrak{A}_2$ can be represented in the form of a value of some word

$$Z_1Z_2 \dots Z_k \qquad (Z_1, Z_2, \dots, Z_k \in \mathfrak{K}_2)$$

in $\mathfrak{K}_2$.

For some $X_i \in \mathfrak{K}_1$ $(i = 1, 2, \dots, k)$ we have $\varphi(X_i) = Z_i$. Since ψ is an extension of φ, for $A_1 = X_1X_2 \dots X_k \in \mathfrak{A}_1$ we obtain in $\mathfrak{A}_2$

$$\psi(A_1) = \psi(X_1X_2 \dots X_k) = \psi(X_1) \cdot \psi(X_2) \cdot \dots \cdot \psi(X_k)$$
$$= \varphi(X_1) \cdot \varphi(X_2) \cdot \dots \cdot \varphi(X_k) = Z_1Z_2 \dots Z_k = A_2.$$

1.13. We assume that canonical expressions, which appear as words in $\mathfrak{W}_{\mathfrak{K}}$ (III, 2.8), have in some way been defined for the elements of $\mathfrak{A}$ with respect to a generating set $\mathfrak{R}$. Φ is some system of relations with respect to $\mathfrak{R}$. If any word of $\mathfrak{W}_{\mathfrak{R}}$ can, by means of relations of Φ, be transformed to a canonical form, then Φ is a defining system of relations of $\mathfrak{A}$ with respect to $\mathfrak{R}$.

In fact, let $W = V$ be an arbitrary relation with respect to $\mathfrak{R}$. By assumption, the word W can by means of Φ be reduced to the canonical form U_W and V can be reduced to the canonical form U_V. Then we have the relations

$$W = V, \qquad W = U_W, \qquad V = U_V.$$

Since U_W and U_V are canonical forms of words, their values can be equal only when they coincide: $U_W \equiv U_V$. Since $W = U_W$, $V = U_V$ are corollaries of Φ, and $U_W \equiv U_V$, it follows that $W = V$ is a corollary of Φ.

1.14. As an example we consider the semigroup $\mathfrak{P}$ defined in III, 6.2; III, 6.3. The set consisting of the elements U, V, E will evidently be a generating set for $\mathfrak{P}$ (E could be omitted and then we would obtain an irreducible generating set; however, we will be able to see the relations better if we do not delete it). With respect to this generating set we have the relations

$$VU = E,$$
$$EU = U, \qquad UE = U, \qquad EV = V, \qquad VE = V, \qquad E^2 = E.$$

It is easy to see that any word in the set $\{U, V, E\}$ can be reduced by means of these relations to the form

$$U^a V^b \qquad (a, b = 0, 1, 2, \ldots)$$

(where by U^0 and V^0 is understood E). As was proven in III, 6.3, such an expression for the elements of $\mathfrak{P}$ is a canonical form with respect to the generating set $\{U, V, E\}$. Therefore, from 1.13 it follows that the indicated relations form a defining system of relations of the semigroup $\mathfrak{P}$.

2. Transformations of Defining Systems of Relations

2.1. It is evident that a semigroup, as a rule, can possess different defining systems of relations. This is connected with the fact that a semigroup usually has different generating sets. But with respect to one and the same generating set there can be different defining systems of relations. The corresponding problem in the theory of groups has been the subject of numerous investigations. In the general theory of semigroups it was considered by A. Ja. Aĭzenštat [**2**].

2.2. Let $\mathfrak{R}$ be a generating set of a semigroup $\mathfrak{A}$ and let Φ_1 be a defining system of relations with respect to $\mathfrak{R}$. Let Φ_2 be some system of relations with respect to $\mathfrak{R}$. *If any relation of* Φ_1 *is a corollary of* Φ_2, *then* Φ_2 *will also be a defining system of relations with respect to* $\mathfrak{R}$.

In fact, let W and V be words in $\mathfrak{R}$ forming the relation $W = V$. Since this relation is a corollary of Φ_1, there exist words

$$U_1 \equiv W, U_2, \ldots, U_{n-1}, \qquad U_n \equiv V$$

in $\mathfrak{R}$ such that $U_i = U_{i+1}$ is an immediate corollary of Φ_1 ($i = 1, 2, \ldots, n-1$). Therefore for some T, T', S_1, S_2 of $\mathfrak{W}_{\mathfrak{R}}$ (T and T' can be empty symbols) we have

$$U_i = TS_1T', \qquad U_{i+1} = TS_2T',$$

where $S_1 = S_2$ is a relation in Φ_1. Since $S_1 = S_2$ is a corollary of Φ_2 it follows that $U_i = U_{i+1}$ will be a corollary of Φ_2. From this it follows that $W = V$ will be a corollary of Φ_2.

2.3. Generally, the transition from one defining system of relations with respect to some generating set to any other defining system of relations with respect to the same generating set can be described in the following way.

Let $\mathfrak{R}$ be a generating set of a semigroup $\mathfrak{A}$ and let Φ be some defining system of relations with respect to $\mathfrak{R}$. If to Φ we adjoin some system Ψ of relations with respect to $\mathfrak{R}$ (each of these being a corollary of Φ), then we obtain the set $\Phi \cup \Psi$ which is evidently a defining system of relations with respect to $\mathfrak{R}$.

For the inverse transformation, let the system Φ be represented in the form of a union $\Phi = \Phi' \cup \Phi''$, where each relation of Φ'' is a corollary of Φ'. Then it is evident that Φ' is a defining system of relations for $\mathfrak{A}$ with respect to $\mathfrak{R}$.

It is easy to see that an arbitrary defining system of relations Φ with respect to $\mathfrak{R}$ can be carried into any other defining system of relations Φ_0 with respect to $\mathfrak{R}$ by means of the indicated mappings. In fact, we first take Φ into $\Phi \cup \Phi_0$ by adjoining to Φ the relations of Φ_0 (all these are corollaries of Φ). After that, we take $\Phi \cup \Phi_0$ into Φ_0, deleting the relations of $(\Phi \cup \Phi_0)\backslash\Phi_0$ (all these are corollaries of Φ_0). In case Φ and Φ_0 are finite, this can be done successively by adjoining or deleting one relation.

2.4. Now we move to the problem of modification of a defining system of relations by a modification of its generating set.

Let $\mathfrak{R}$ be a generating set of a semigroup $\mathfrak{A}$. Let $\mathfrak{R}'$ be an arbitrary subset of $\mathfrak{A}$. Let Φ be some system of relations with respect to $\mathfrak{R}$. For each $X' \in \mathfrak{R}'$ we fix some word in $\mathfrak{W}_{\mathfrak{R}}$, whose value is

$$X' = Y_1 Y_2 \ldots Y_n \qquad (Y_1, Y_2, \ldots, Y_n \in \mathfrak{R}).$$

If Φ is a defining system of relations with respect to $\mathfrak{R}$, then the system Φ' consisting of the relations belonging to Φ and the relations

$$X' = Y_1 Y_2 \ldots Y_n$$

will be a defining system of relations with respect to $\mathfrak{R} \cup \mathfrak{R}'$.

In fact, let $W = V$ be an arbitrary relation with respect to $\mathfrak{R} \cup \mathfrak{R}'$. In W and V we replace any element X' of $\mathfrak{R}'$ by the corresponding product $Y_1 Y_2 \ldots Y_n$. We obtain words W' and V' in $\mathfrak{R}$ where the relations $W = W'$ and $V = V'$ are corollaries of Φ'. The relation $W' = V'$ with respect to $\mathfrak{R}$ must be a corollary of Φ. Thus we obtain the relations in $\mathfrak{R} \cup \mathfrak{R}'$

$$W = W', \qquad W' = V', \qquad V' = V,$$

each of which is a corollary of Φ'.

Therefore the relation $W = V$ is a corollary of Φ'.

2.5. Using the notation of 2.4, it is easy to formulate and prove the converse of the assertion in 2.4. Here, however, we must agree that for each X' in $\mathfrak{K}'$ which is contained in $\mathfrak{K}$ we will take as its expression in the form of a product of elements of $\mathfrak{K}$ the identical expression $X' = X'$.

If Φ' is a defining system of relations with respect to $\mathfrak{K} \cup \mathfrak{K}'$, then Φ will be a defining system of relations with respect to $\mathfrak{K}$.

Let $W = V(W, V \in \mathfrak{W}_{\mathfrak{K}})$ be an arbitrary relation with respect to $\mathfrak{K}$. Being a relation with respect to $\mathfrak{K} \cup \mathfrak{K}'$, it is a corollary of Φ'. This shows that there exist words

$$W \equiv U_1, U_2, \ldots, U_{s-1}, \qquad U_s \equiv V$$

in $\mathfrak{K} \cup \mathfrak{K}'$ such that $U_i = U_{i+1}$ is an immediate corollary of Φ'. Replacing in all these words each element X' of $\mathfrak{K}'$ not contained in $\mathfrak{K}$ by the corresponding product $Y_1 Y_2 \ldots Y_n$, we obtain a sequence of words $T_1, T_2, \ldots, T_{s-1}, T_s$ in $\mathfrak{K}$ (the extreme words of the chain here are, of course, not changed, i.e., $T_1 \equiv W$ and $T_s \equiv V$). The words U_i and U_{i+1} can be represented in the form

$$U_i \equiv PS_1Q, \qquad U_{i+1} \equiv PS_2Q,$$

where $S_1 = S_2$ is one of the relations of Φ'. From the substitution indicated above for the words T_i and T_{i+1} we obtain the expressions

$$T_i \equiv \bar{P}\bar{S}_1\bar{Q}, \qquad T_{i+1} \equiv \bar{P}\bar{S}_2\bar{Q}.$$

If the relation $S_1 = S_2$ had the form $X' = Y_1 Y_2 \ldots Y_n$, then, replacing X' by $Y_1 Y_2 \ldots Y_n$, we obtain $T_i \equiv T_{i+1}$. But if $S_1 = S_2$ belonged to Φ, then evidently $\bar{S}_1 \equiv S_1$, $\bar{S}_2 \equiv S_2$. In this case the relation $T_i = T_{i+1}$ is a corollary of Φ. Thanks to this property of the sequence of words $T_1, T_2, \ldots, T_s$ we conclude that the relation $W = V$ is an immediate corollary of Φ.

2.6. Let $\mathfrak{K}_1$ and $\mathfrak{K}_2$ be any two generating sets of a semigroup $\mathfrak{A}$ and let Φ_1 be some defining system of relations with respect to $\mathfrak{K}_1$. We show how, starting from Φ_1, we can obtain a defining system of relations with respect to $\mathfrak{K}_2$.

We put $\tilde{\mathfrak{K}}_i = \mathfrak{K}_i \setminus (\mathfrak{K}_1 \cap \mathfrak{K}_2)$ $(i = 1, 2)$.

For each element of $\tilde{\mathfrak{K}}_1$ we fix some word in $\mathfrak{K}_2$ whose value is that element. By the same token we fix the representation of each element of $\tilde{\mathfrak{K}}_1$ in the form of a product of elements of $\mathfrak{K}_2$. We consider the equalities giving these representations as relations with respect to $\mathfrak{K}_1 \cup \mathfrak{K}_2$. The totality of these we denote by Φ_{12}.

Analogously, fixing the expressions of elements of $\tilde{\mathfrak{K}}_2$ by elements of $\mathfrak{K}_1$, we obtain a system of relations with respect to $\mathfrak{K}_1 \cup \mathfrak{K}_2$ which we denote by Φ_{21}.

If in the relations of Φ_1, we replace with the help of Φ_{12} all the elements of $\tilde{\mathfrak{K}}_1$ occurring in them by products of elements of $\mathfrak{K}_2$, then we obtain a system of relations with respect to $\mathfrak{K}_2$, which we denote by Φ_2'.

If in the relations of Φ_{21} we replace, by means of Φ_{12}, the elements of $\tilde{\mathfrak{K}}_1$ by products of elements of $\mathfrak{K}_2$, then we obtain a system of relations with respect to $\mathfrak{K}_2$, which we denote by Φ_2''.

THEOREM. *The system* $\Phi_2' \cup \Phi_2''$ *is a defining system of relations for* $\mathfrak{A}$ *with respect to the generating set* $\mathfrak{K}_2$.

PROOF. (1) Thanks to 2.4, $\Phi_1 \cup \Phi_{21}$ is a defining system of relations with respect to $\mathfrak{K}_1 \cup \mathfrak{K}_2$. Therefore by 2.3 it follows that the system

$$\Psi_1 = \Phi_1 \cup \Phi_{21} \cup \Phi_{12} \cup \Phi_2' \cup \Phi_2''$$

will be a defining system of relations with respect to $\mathfrak{K}_1 \cup \mathfrak{K}_2$.

(2) With respect to $\mathfrak{K}_1 \cup \mathfrak{K}_2$ the system of relations

$$\Psi_2 = \Phi_{21} \cup \Phi_{12} \cup \Phi_2' \cup \Phi_2''$$

is a defining system of relations.

Let $W = V$ be one of the relations of Φ_1. By means of Φ_{12} we replace the elements of $\tilde{\mathfrak{K}}_1$ in W and V by the corresponding words in $\mathfrak{W}_{\mathfrak{K}_2}$. We obtain words $\overline{W}$ and $\overline{V}$ in $\mathfrak{K}_2$. The relations $W = \overline{W}$ and $V = \overline{V}$ are corollaries of Ψ_2. The relation $\overline{W} = \overline{V}$ is one of the relations of Φ_2'. Therefore the relation $W = V$ is a corollary of Ψ_2.

Since Ψ_1 is a defining system of relations with respect to $\mathfrak{K}_1 \cup \mathfrak{K}_2$, it follows from 2.3 by the above reasoning that Ψ_2 will be a defining system of relations.

(3) The system

$$\Psi_3 = \Phi_{12} \cup \Phi_2' \cup \Phi_2''$$

is a defining system of relations with respect to $\mathfrak{K}_1 \cup \mathfrak{K}_2$.

Let $W = V$ be one of the relations of Φ_{21}. By means of Φ_{12} we replace the elements of $\tilde{\mathfrak{K}}_1$ in the words W and V by the corresponding words in $\mathfrak{W}_{\mathfrak{K}_2}$. We obtain words $\overline{W}$ and $\overline{V}$ in $\mathfrak{K}_2$, which form a relation $\overline{W} = \overline{V}$ of Φ_2''. Since the relations $W = \overline{W}$ and $V = \overline{V}$ are corollaries of Φ_{12}, the relation $W = V$ will be a corollary of Ψ_3.

Since Ψ_2 is a defining system of relations with respect to $\mathfrak{K}_1 \cup \mathfrak{K}_2$ it follows from 2.3 that Ψ_3 will be a defining system of relations.

(4) From the fact that $\Psi_3 = \Phi_{12} \cup \Phi_2' \cup \Phi_2''$ is a defining system of relations with respect to $\mathfrak{K}_1 \cup \mathfrak{K}_2$, it immediately follows from 2.5 that $\Phi_2' \cup \Phi_2''$ will be a defining system of relations with respect to $\mathfrak{K}_2$.

2.7. We note that in the case when $\mathfrak{K}_2 \subset \mathfrak{K}_1$, Φ_{21} is empty. Because of this, Φ_2'' is also empty, and hence Φ_2' will be a defining system of relations with respect to $\mathfrak{K}_2$.

2.8. We apply Theorem 2.6, together with the remark 2.7, to the semigroup $\mathfrak{P}$ considered in 1.14. Since $\{U, V\}$ is also a generating set of $\mathfrak{P}$, we obtain for it, starting from the defining system of relations with respect to $\{U, V, E\}$ deduced in 1.14, the following defining system of relations:

$$VU^2 = U, \qquad UVU = U, \qquad VUV = V, \qquad V^2U = V, \qquad VUVU = VU.$$

The last of these relations is evidently a corollary of the previous ones and can be discarded.

2.9. The preceding definitions and arguments apply to arbitrary semigroups, and in particular, to groups. However, in the theory of groups it is customary to adopt the following additional conventions. The generating set for the group is usually taken to contain the identity E and together with each element X it should contain its inverse X^{-1}. Among the relations of a defining system there must be included relations of the form

$$EX = XE = X, \qquad XX^{-1} = X^{-1}X = E.$$

Because these relations must hold in any group, they are usually not written out, nor even mentioned. So in dealing with groups, we must remember to supplement in this way the indicated generating set and the defining system of relations.

It is also necessary to mention the possibility of another approach in the theory of groups to defining systems of relations, connected with the possibility of the representation of a group as a factor group of a free group.[1] Though having well-known advantages, such an approach is in some cases less convenient. This is evident from the reasoning in 2.4 and 2.5; the particular case for the theory of groups by the indicated approach takes the form of a fairly complex theorem.

3. Semigroups Given by Defining Relations

3.1. In the preceding sections we always started from a semigroup considered as given, and examined various relations for different generating sets. Now we will approach essentially the same problem, but from a different point of view. We will start with some relations and from them construct a semigroup. The principal difficulty in such an approach is the following: as long as we have not yet finally determined the semigroup and are still considering the expressions which are to be made into elements of the semigroup, we often cannot say in advance which initially different expressions will constitute one and the same element of the semigroup under construction. This causes well-known complications for the construction of the semigroup.

3.2. Suppose we are given an initial alphabet $\mathfrak{N}$ (1.2). We take the free semigroup over it, $\mathfrak{W}_{\mathfrak{N}}$ (1.3). If in $\mathfrak{W}_{\mathfrak{N}}$ there is given some relation $\mathfrak{n}$, then for $\mathfrak{n}$ in $\mathfrak{W}_{\mathfrak{N}}$ there is defined a second derived relation $\mathfrak{n}''$ (I, 5.21). Since $\mathfrak{n}''$ is a two-sidedly stable equivalence relation (I, 5.22), we may therefore, by VII, 2.4, define the factor-semigroup $\mathfrak{W}_{\mathfrak{N}}^{\mathfrak{n}} = \mathfrak{W}_{\mathfrak{N}}/\mathfrak{n}''$, which we will call *the semigroup over the alphabet* $\mathfrak{N}$ *given by the defining relation* $\mathfrak{n}$.

An element of the semigroup $\mathfrak{W}_{\mathfrak{N}}^{\mathfrak{n}}$ is a class of words in $\mathfrak{N}$ which are equivalent to each other with respect to the second derived relation $\mathfrak{n}''$. We will usually

[1] Cf., for example, A. G. Kuroš, *The theory of groups*. Translated from the Russian and edited by K. A. Hirsch, Chelsea Publishing Co., New York, N. Y., 1955/1956.

denote it with a bar in the manner $\overline{W}$, where W is one of the words in $\mathfrak{N}$ contained in that class. Multiplication of classes in $\mathfrak{W}_{\mathfrak{N}}^{\mathfrak{n}}$ is performed by means of multiplication in $\mathfrak{W}_{\mathfrak{N}}$ (1.2) of their representatives:

$$\overline{W}_1 \cdot \overline{W}_2 = \overline{W_1 W_2}.$$

The set $\overline{\mathfrak{N}}$ consisting of all classes $\overline{X}$ ($X \in \mathfrak{N}$) containing words in $\mathfrak{N}$ of length one is a generating set of the semigroup $\mathfrak{W}_{\mathfrak{N}}^{\mathfrak{n}}$, since

$$\overline{X_1 X_2 \ldots X_n} = \overline{X}_1 \cdot \overline{X}_2 \cdot \ldots \cdot \overline{X}_n.$$

We must also keep in mind that it is entirely possible to have a case where two different elements of $\mathfrak{N}$ are equivalent with respect to $\mathfrak{n}''$, i.e., $X \not\equiv Y$ but $X \equiv Y(\mathfrak{n}'')$, i.e., $\overline{X} = \overline{Y}$ ($X, Y \in \mathfrak{N}$).

3.3. We note that in the case where $\mathfrak{n}$ is a partially identical relation, i.e., $W \sim V(\mathfrak{n})$ implies $W \equiv V$ (and also for some $W \in \mathfrak{W}_{\mathfrak{N}}$ we can have $W \not\sim W(\mathfrak{n})$), it turns out that $\mathfrak{W}_{\mathfrak{N}}^{\mathfrak{n}}$ is simply the free semigroup over $\mathfrak{N}$ (1.3). In particular, this takes place when $\mathfrak{n}$ is the identity relation (i.e., $W \sim W(\mathfrak{n})$ for all $W \in \mathfrak{W}_{\mathfrak{N}}$ and $W \not\sim V(\mathfrak{n})$ if $W \not\equiv V$) or when $\mathfrak{n}$ is the empty relation (i.e., $W \not\sim V(\mathfrak{n})$ for any $W, V \in \mathfrak{W}_{\mathfrak{N}}$).

3.4. We turn to a consideration of the semigroup $\mathfrak{W}_{\mathfrak{N}}^{\mathfrak{n}}$ for an arbitrary relation $\mathfrak{n}$ in $\mathfrak{W}_{\mathfrak{N}}$. To each element X of the alphabet $\mathfrak{N}$ we associate the element $\overline{X}$ from $\overline{\mathfrak{N}} \subset \mathfrak{W}_{\mathfrak{N}}^{\mathfrak{n}}$. Because of this, to each word $W = X_1 X_2 \ldots X_n$ in $\mathfrak{N}$ there corresponds a word $\overline{X}_1 \overline{X}_2 \ldots \overline{X}_n$ in $\overline{\mathfrak{N}}$, which we will denote by $\tilde{W}$. We note that the value of the word $\tilde{W}$ in $\mathfrak{W}_{\mathfrak{N}}^{\mathfrak{n}}$ is evidently equal to $\overline{W} \in \mathfrak{W}_{\mathfrak{N}}^{\mathfrak{n}}$. Since $W \sim V(\mathfrak{n})$ ($W, V \in \mathfrak{W}_{\mathfrak{N}}$) implies that $\overline{W} = \overline{V}$ in $\mathfrak{W}_{\mathfrak{N}}^{\mathfrak{n}}$, the pair of words W and V in $\mathfrak{W}_{\mathfrak{N}}$, which are in the relation $\mathfrak{n}$, correspond in $\mathfrak{W}_{\mathfrak{N}}^{\mathfrak{n}}$ to the relation $\tilde{W} = \tilde{V}$ with respect to $\overline{\mathfrak{N}}$. We denote the totality of all such relations in $\mathfrak{W}_{\mathfrak{N}}^{\mathfrak{n}}$ by $\Phi(\mathfrak{n})$. *$\Phi(\mathfrak{n})$ in $\mathfrak{W}_{\mathfrak{N}}^{\mathfrak{n}}$ is a defining system of relations with respect to the generating set $\overline{\mathfrak{N}}$.*

In fact, if in $\mathfrak{W}_{\mathfrak{N}}^{\mathfrak{n}}$ we have

$$\tilde{W} = \tilde{V} \qquad (W, V \in \mathfrak{W}_{\mathfrak{N}}),$$

then this means that $\overline{W} = \overline{V}$, i.e., $W \sim V(\mathfrak{n}'')$.

But $W \sim V(\mathfrak{n}'')$ implies that for the corresponding words $\tilde{W}$ and $\tilde{V}$ it is possible to obtain the relation $\tilde{W} = \tilde{V}$ as a corollary of $\Phi(\mathfrak{n})$.

3.5. Now let there be chosen in an arbitrary semigroup $\mathfrak{A}$ some generating set $\mathfrak{K}$ and let there be given some defining system Φ of relations with respect to $\mathfrak{K}$. Taking $\mathfrak{K}$ as the initial alphabet, we construct the free semigroup over it $\mathfrak{W}_{\mathfrak{K}}$. In $\mathfrak{W}_{\mathfrak{K}}$ we define the relation $\mathfrak{n}$ by putting $W \sim V(\mathfrak{n})$ ($W, V \in \mathfrak{W}_{\mathfrak{K}}$) if and only if in $\mathfrak{A}$ $W = V$ is a relation belonging to Φ. By Definitions 1.8 and 1.9 the words T and U in $\mathfrak{W}_{\mathfrak{K}}$ have the same value in $\mathfrak{A}$ if and only if $T \sim U(\mathfrak{n}'')$. But in $\mathfrak{W}_{\mathfrak{K}}^{\mathfrak{n}}$ the equality $\bar{T} = \bar{U}$ holds if and only if $T \sim U(\mathfrak{n}'')$. Thus, between the elements of $\mathfrak{A}$

and the elements of $\mathfrak{W}_{\mathfrak{N}}^{\mathfrak{n}}$ a one-to-one correspondence is established in an obvious way. This correspondence has the property of a homomorphism and is therefore an isomorphism.

Thus it follows that *any semigroup, up to within isomorphism, can be given as a semigroup over some alphabet defined by some defining relation.*

3.6. Let us return now to the question raised at the beginning of this section. The role played here by the construction of a semigroup over an alphabet defined by a defining relation consists of the following. Let there be given some set $\mathfrak{N}$ and a set Ψ of formal equalities with respect to $\mathfrak{N}$, i.e., a pair of words in $\mathfrak{N}$ connected with an equality sign. It is asked whether it is possible to construct a semigroup such that (i) each element of $\mathfrak{N}$ is an element of the semigroup, (ii) the elements of $\mathfrak{N}$ form a generating set in the semigroup and (iii) the relations of Ψ are valid in this semigroup and form a defining system of relations with respect to $\mathfrak{N}$.

3.7. For the answer to this problem we consider the free semigroup $\mathfrak{W}_{\mathfrak{N}}$ over the alphabet $\mathfrak{N}$. We shall take in it, as a defining relation, a relation $\mathfrak{n}_\Psi$ such that $W \sim V(\mathfrak{n}_\Psi)$ $(W, V \in \mathfrak{W}_{\mathfrak{N}})$ holds if and only if the pair of words W and V is connected by an equality sign contained in the given system Ψ of formal equalities. We assume that for two different elements $X, Y \in \mathfrak{N}$ we cannot have $X \sim Y(\mathfrak{n}_\Psi'')$. Then, constructing the semigroup $\mathfrak{W}_{\mathfrak{N}}^{\mathfrak{n}_\Psi}$, we identify each of its elements of the form $\bar{X}$ $(X \in \mathfrak{N})$ with the element X itself. In this semigroup the set $\mathfrak{N}$, coinciding with $\overline{\mathfrak{N}}$ in the above identification, is a generating set.

All the formal equalities of Ψ are valid relations. In this connection they form a defining system of relations with respect to $\mathfrak{N}$, since the relation $W = V$ $(W, V \in \mathfrak{W}_{\mathfrak{N}})$ in $\mathfrak{W}_{\mathfrak{N}}^{\mathfrak{n}_\Psi}$ holds if and only if $\bar{W} = \bar{V}$, i.e., if $W \sim V(\mathfrak{n}_\Psi'')$. But the latter means that $W = V$ is a corollary of Ψ.

3.8. Now let $X \sim Y(\mathfrak{n}_\Psi'')$ hold for some $X \neq Y$ $(X, Y \in \mathfrak{N})$. If $\mathfrak{N}$ is a generating set of some semigroup, in which all the formal equalities of Ψ are valid relations, then from these, as a corollary, follows the relation $X = Y$. Thus different elements of $\mathfrak{N}$ must be different elements of such a semigroup.

3.9. From the arguments of 3.7 and 3.8 follows the complete answer to the problem posed in 3.6. If the system of formal equalities Ψ is such that for the relation $\mathfrak{n}_\Psi$ (3.7) defined by it the inequality $X \neq Y$ $(X, Y \in \mathfrak{N})$ always implies $X \nsim Y(\mathfrak{n}_\Psi'')$, and only if this condition holds, then there exists a semigroup $\mathfrak{A}$ for which $\mathfrak{N}$ is a generating set and Ψ is a defining system of relations with respect to $\mathfrak{N}$.

3.10. In connection with this result we naturally make the following convention, which is always made in such cases. Let $\mathfrak{N}$ be a set and let Ψ be a system of formal equalities of certain words in $\mathfrak{W}_{\mathfrak{N}}$. The semigroup $\mathfrak{W}_{\mathfrak{N}}^{\mathfrak{n}_\Psi}$, which can also be denoted by $\Psi_{\mathfrak{N}}$, is called *the semigroup given by the defining system of relations* Ψ (or by the defining relation $\mathfrak{n}_\Psi$). Its elements are classes of words in $\mathfrak{N}$ which are equivalent to each other with respect to $\mathfrak{n}''_{\Psi}$. The class $\bar{X}$, where $X \in \mathfrak{N}$, is identified with the element X itself. If for any two distinct elements $X, Y \in \mathfrak{N}$ we have $\bar{X} = \bar{Y}$ (i.e., $X \sim Y(\mathfrak{n}''_{\Psi})$), then X and Y are regarded simply as different symbols for one and the same element of the semigroup $\mathfrak{W}_{\mathfrak{N}}^{\mathfrak{n}_{\Psi}}$. As we have already shown in (3.2), the set $\bar{\mathfrak{N}}$, and consequently $\mathfrak{N}$, if the corresponding identifications are made, is a generating set in $\mathfrak{W}_{\mathfrak{N}}^{\mathfrak{n}_{\Psi}}$, and Ψ, coinciding with $\Phi(\mathfrak{n}_{\Psi})$ (3.4) is a defining system of relations with respect to this generating set.

3.11. From the given relation $\mathfrak{n}$ in $\mathfrak{W}_{\mathfrak{N}}$, or, what is the same thing, from the form of the formal equalities Ψ, it is not immediately evident when two words from $\mathfrak{N}$ have the same value in $\mathfrak{W}_{\mathfrak{N}}^{\mathfrak{n}}$, i.e., represent one and the same element of that semigroup. The elucidation of this problem for each concrete semigroup $\mathfrak{W}_{\mathfrak{N}}^{\mathfrak{n}}$ is not only of great practical value for the investigation of its properties, but it has great theoretical importance in view of the fact that as long as it is not settled, we have no clear idea of the nature of the given semigroup, i.e., of the set of its elements. However, the solution of this problem involves profound theoretical difficulties.

The problem can be formulated more precisely for the case when the alphabet $\mathfrak{N}$ is finite and the defining relation $\mathfrak{n}$ in $\mathfrak{W}_{\mathfrak{N}}$ is given as a finite number of pairs of words in $\mathfrak{N}$, lying in the relation $\mathfrak{n}$ (or, what is the same thing, the semigroup is given by a finite number of relations). In this case we may pose the problem of finding a general algorithm by means of which for any pair of words W and V of $\mathfrak{W}_{\mathfrak{N}}^{\mathfrak{n}}$ we can decide whether or not they are in the relation $\mathfrak{n}''$ (the second derived relation with respect to $\mathfrak{n}$). But this last is equivalent to a solution of the problem of whether or not the words W and V are one and the same element of the given semigroup. The problem of the construction of such an algorithm is usually called the problem of the identity of words of the given semigroup. In some cases the solution of this problem, i.e., the finding of such an algorithm, turns out to be possible. However, as was shown for the first time by A. A. Markov [**1**], there exist semigroups over finite alphabets, defined by a finite number of relations, in which such a general algorithm as a rule does not exist. Such semigroups were constructed by A. A. Markov [**1**; **6**] and later by some other mathematicians, for example Post [**1**] and Kolmar [**1**]. In some cases the external form of the corresponding semigroups can appear quite simple. An example is the semigroup found by G. S. Ceĭtin [**1**], which is constructed over the alphabet consisting of the five elements $\{A, B, C, D, E\}$ and given by the system of the following seven relations:

$$AC = CA, \quad AD = DA, \quad BC = CB, \quad BD = DB,$$
$$ECA = AE, \quad EDB = BE,$$
$$ABAC = ABACE.$$

For this and other semigroups, the proof of the fact that the problem of identity is insolvable is very difficult and will not be taken up here. By their very character, the corresponding problems belong to a self-contained branch of mathematics, namely, the theory of algorithms (cf., for example, the book by A. A. Markov [**6**]).

3.12. It must be noted that for a semigroup over a finite alphabet, given by a finite number of relations, other algorithmic problems naturally arise. An example is the very important algorithmic problem concerned with the question of divisibility, in which it is required to construct an algorithm by means of which for any two words W and V over a given alphabet it is possible to decide whether there exists a third word U over that alphabet such that

$$WU \sim V(\mathfrak{n}'').$$

Evidently this is the problem of constructing an algorithm by means of which it would be possible to decide when one of two elements of a semigroup defined by words in a given alphabet is a left divisor of the other (and analogously for right divisibility). As was proven by A. A. Markov [**1**; **6**], for certain semigroups such an algorithm is not possible (cf. also the work of S. I. Adjan [**1**]).

3.13. It must be emphasized that the impossibility of an algorithm for the problem of identity or the problem of divisibility in these or other semigroups shows that there does not exist any general algorithm for the solution of the problem for any pair of words in the given alphabet. But this does not at all mean that there exists at least one particular pair of words for which the problem cannot be solved in principle. From the proof of the impossibility of showing for a particular pair of words W and V that they represent one and the same element, it would evidently follow that for these words we could not find a chain of words such that $W \sim V(\mathfrak{n}'')$. But this would mean that such a chain does not exist, i.e., that $W \not\sim V(\mathfrak{n}'')$. Consequently this would mean the solution of our problem.

The results mentioned above only show the absence in some cases of a single general algorithm, which would solve the problem for the entire infinite system of pairs of words in the given alphabet.

3.14. In conclusion it is necessary to note the close connection of the theory of semigroups defined over a given alphabet by means of a defining relation with the theory of associative computations (the latter is confined to the finite case in the sense indicated above). The theory of associative computations can essentially be considered as a constructive approach to the theory of such semigroups. Let us point out that a very precise exposition of the foundations of this theory and the results mentioned above about the impossibility of certain algorithms is given in the book of A. A. Markov [**6**]. In this book, as in many other works on this problem, the above results on the impossibility of certain algorithms are formulated as corresponding assertions about various associative computations.

4. Identities in Semigroups

4.1. Let T and T' be two words in a countable alphabet $\Xi = \{\xi_1, \xi_2, \ldots\}$. A relation connecting the words T and T' by the sign $\cong$ (sometimes the identity sign $\equiv$ is used, but we have already used it for other purposes),

$$T \cong T',$$

is called an *identity in the semigroup* $\mathfrak{A}$ if, for any mapping φ of Ξ into $\mathfrak{A}$, the values of the words $\varphi(T)$ and $\varphi(T')$ are equal. In other words, $T \cong T'$ is transformed, by the substitution of any of the elements of $\mathfrak{A}$ for the elements of Ξ in the words T and T', into an equality in $\mathfrak{A}$.

It is evident that the choice of elements of the alphabet Ξ does not affect the problem of identities in the semigroup. In particular, from the validity in $\mathfrak{A}$ of some identity $T \cong T'$ follows the validity of any identity $T_0 \cong T_0'$ obtained from $T \cong T'$ by a replacement of the elements of Ξ in that identity by other elements of the same alphabet. In this case we can consider that $T \cong T'$ and $T_0 \cong T_0'$ are one and the same identity.

It is possible to consider simultaneously several identities in a semigroup. We assume that we start from one and the same alphabet. Since in each of these identities there appear only a finite number of elements of the alphabet, but the number involved (for all the identities) can be unbounded, we must therefore, in order to consider all these identities, take the alphabet Ξ to be countable.

4.2. In any commutative semigroup we have the identity

$$\xi_1\xi_2 \cong \xi_2\xi_1.$$

The existence of this identity in a semigroup is the condition of its commutativity.

A semigroup of idempotents is characterized by the identity

$$\xi_1^2 \cong \xi_1.$$

The unit semigroup, i.e., the semigroup consisting of one element, is characterized by the identity

$$\xi_1 \cong \xi_2.$$

We have often drawn attention to the semigroup in which the product of any two elements is equal to the left factor. Such a semigroup is defined to within isomorphism by the cardinality of its elements. The class of such semigroups is evidently characterized by the identity

$$\xi_1\xi_2 \cong \xi_1.$$

Analogously the identity

$$\xi_1\xi_2 \cong \xi_2$$

characterizes the class of semigroups in which the product is always equal to the right factor.

4.3. It is possible to characterize each identity $T \cong T'$ by the sum of the component lengths of its words T and T'.

For the identities in which this number is equal to two, there will be the identity $\xi_1 \cong \xi_2$, which we have already considered, and the identity $\xi_1 = \xi_1$, which is trivial and true in any semigroup.

In addition to the identities considered in 4.2, in which the sum of the component lengths of their words is no more than three, only two other such identities are possible. The identity

$$\xi_1\xi_2 \cong \xi_3$$

holds only in the unit semigroup, since for any elements X and Y of the semigroup we have

$$XX = X, \qquad XX = Y.$$

Analogously the identity

$$\xi_1^2 \cong \xi_2$$

is satisfied only in the unit semigroup.

But for an increase of the sum of the lengths there are new nontrivial identities.

4.4. A semigroup $\mathfrak{A}$, in which for some natural number n we have the two identities

$$\xi_1^n \cong \xi_2^n, \qquad \xi_1^n\xi_2 \cong \xi_2$$

is a group, and the order of each of its elements is a divisor of n. In fact, for any $X \in \mathfrak{A}$ the element X^n is a left unit of $\mathfrak{A}$. Here each element $A \in \mathfrak{A}$ has the two-sided inverse A^{n-1} with respect to the unit $A^n = X^n$. Conversely, in any group, the elements of which have orders which divide n, both of the above identities evidently hold.

An important problem in the theory of groups is the so-called Burnside problem, on whether there exist infinite groups having a finite generating set, in which the order of each of the elements is a divisor of one and the same number n. For $n \leqslant 4$, it is known that such infinite groups do not exist. For $n \geqslant 72$, P. S. Novikov[2] proved the existence of such infinite groups.

As was shown in the work of Green and Rees [**1**], the Burnside problem in the theory of groups is equivalent to the following problem in the theory of semigroups: does there exist an infinite semigroup with a finite generating set in which the identity $\xi_1^{n+1} \cong \xi_1$ holds.

In one direction the connection between the two problems is evident. If for some n, any semigroup with the indicated identity and with a finite generating set is finite, then from this would follow at once the finiteness of all groups having finite generating sets, the order of whose elements is a divisor of n. This follows from the fact that the identities $\xi_1^n \cong \xi_2^n$, $\xi_1^n\xi_2 \cong \xi_2$ evidently imply the identity $\xi_1^{n+1} \cong \xi_1$.

[2] *On periodic groups*, Dokl. Akad. Nauk SSSR **127** (1959), 749–752. (Russian)

The converse is far from evident. The proof of the fact that from the existence of an infinite semigroup with finite generating set, satisfying the identity $\xi_1^{n+1} \cong \xi_1$, follows the existence of an infinite semigroup with finite generating set, satisfying the identities $\xi_1^n \cong \xi_2^n$, $\xi_1^n\xi_2 \cong \xi_1$, is not simple, and we will not give it here.

4.5. Using the concept of an identity, A. I. Mal'cev [6] was able to introduce into the theory of semigroups the notion of nilpotence, widely used in the theory of groups.[3]

We can approach the concept of nilpotence in the theory of groups from various points of view. For example, we can use the following inductive definition.

A commutative group is said to be 1-power nilpotent. A group $\mathfrak{G}$ is said to be n-power nilpotent ($n = 2, 3, 4, \ldots$) if the factor group by its center $\mathfrak{G}/\mathfrak{Z}$ is $(n - 1)$-power nilpotent.

4.6. For an alphabet $\Xi = \{\xi_1, \xi_2, \ldots\}$ we define by induction the words W_n and V_n in Ξ. We put

$$W_0 \equiv \xi_1, \qquad V_0 \equiv \xi_2.$$

Further, we put

$$W_n \equiv W_{n-1}\xi_{n+2}V_{n-1}, \qquad V_n \equiv V_{n-1}\xi_{n+2}W_{n-1} \qquad (n = 1, 2, 3, \ldots).$$

According to A. I. Mal'cev, a semigroup $\mathfrak{A}$ is said to be *n-power nilpotent* if in it we have the identity

$$W_n \cong V_n.$$

4.7. We note that evidently any n-power nilpotent semigroup will necessarily be k-power nilpotent for any $k \geqslant n$.

Commutative semigroups are all 1-power nilpotent. However, in the class of 1-power nilpotent semigroups there are some noncommutative semigroups.

In fact, the class of 1-power nilpotent semigroups evidently includes all semigroups in which the identity

$$\xi_1\xi_2\xi_3 \cong \xi_4\xi_5\xi_6$$

holds. An example of such a semigroup is the set of elements which consist of O and words of length one and two in some alphabet $\mathfrak{N}$. The operation of multiplication consists of attachment of words. If one of the factors is O or if as the result of attachment we obtain a word of length greater than two, then the product is considered to be O. In such a semigroup the indicated identity clearly holds, because the product of three factors is always O. If the alphabet $\mathfrak{N}$ contains more than one element, then the semigroup is noncommutative, since for all $X, Y \in \mathfrak{N}$ $(X \neq Y)$ the product of X by Y is the word XY, and the product of Y by X is the word YX, where by definition these words are different elements of the semigroup under consideration.

[3] Cf., for example, A. G. Kuroš, *The theory of groups*, 2nd ed., GITTL, Moscow, 1953; §62 (Russian); English transl., Chelsea, New York, 1955, 1956. See also Chapter XI, §5.5.

4.8. The value of the concept defined above lies in the fact that it is a direct generalization of the corresponding concept for the theory of groups.

THEOREM. *A group is n-power nilpotent in the sense of* 4.5 *if and only if it is an n-power nilpotent semigroup* (4.6).

PROOF. (1) We prove by induction on n that a group $\mathfrak{G}$ which is n-power nilpotent in the sense of 4.5 is an n-power nilpotent semigroup.

For $n = 1$, $\mathfrak{G}$ is commutative and is therefore a 1-power nilpotent semigroup.

Suppose $n > 1$. We denote by φ the natural homomorphism of $\mathfrak{G}$ onto the factor group of $\mathfrak{G}$ by its center: $\mathfrak{G}/\mathfrak{Z}$.

We take arbitrary elements $X_1, X_2, \ldots, X_{n+1}, X_{n+2}$ from $\mathfrak{G}$. We denote by W'_k and V'_k the words in $\mathfrak{G}$ obtained from the words W_k and V_k in Ξ (4.6) by replacing ξ_i with X_i. Using the inductive assumption, we can consider that in $\varphi(\mathfrak{G}) = \mathfrak{G}/\mathfrak{Z}$ we have the relation

$$\varphi(W'_{n-1}) = \varphi(V'_{n-1}).$$

By the property of the factor group this means that in $\mathfrak{G}$ we have

$$W'_{n-1} = V'_{n-1}Z, \qquad Z \in \mathfrak{Z}.$$

From this we obtain equalities in $\mathfrak{G}$ for the values of the words in $\mathfrak{G}$, namely,

$$\begin{aligned} W'_n = W'_{n-1}X_{n+2}V'_{n-1} &= V'_{n-1}ZX_{n+2}V_{n-1} \\ = V'_{n-1}X_{n+2}V'_{n-1}Z &= V'_{n-1}X_{n+2}W'_{n-1} = V'_n. \end{aligned}$$

The fulfilment of these equalities for arbitrary $X_1, X_2, \ldots, X_{n+2}$ in $\mathfrak{G}$ means that in $\mathfrak{G}$ we have the identity

$$W_n \cong V_n,$$

i.e., $\mathfrak{G}$ is an n-power nilpotent semigroup.

(2) Now we prove by induction on n that a group $\mathfrak{G}$ which is an n-power nilpotent semigroup will be n-power nilpotent in the sense of 4.5.

If $n = 1$, then the equality

$$X_1E_{\mathfrak{G}}X_2 = X_2E_{\mathfrak{G}}X_1,$$

obtained from the identity $W_1 \cong V_1$ by replacing ξ_1 with X_1, ξ_2 with X_2, and ξ_3 with $E_{\mathfrak{G}}$, where X_1 and X_2 are arbitrary elements of $\mathfrak{G}$, shows that $\mathfrak{G}$ is commutative.

Let $n > 1$. Using the notations of the first part, we obtain from the identity $W_n \cong V_n$, taking as X_{n+2} the identity $E_{\mathfrak{G}}$, the relation

$$W'_{n-1}V'_{n-1} = V'_{n-1}W'_{n-1}$$

in $\mathfrak{G}$. For arbitrary X_{n+2} it follows from $W_n \cong V_n$ that

$$W'_{n-1}X_{n+2}V'_{n-1} = V'_{n-1}X_{n+2}W'_{n-1}.$$

Thus we obtain

$$\begin{aligned}(V'^{-1}_{n-1}W'_{n-1})X_{n+2} &= X_{n+2}(W'_{n-1}V'^{-1}_{n-1})\\ &= X_{n+2}(V'^{-1}_{n-1}V'_{n-1}W'_{n-1}V'^{-1}_{n-1})\\ &= X_{n+2}(V'^{-1}_{n-1}W'_{n-1}V'_{n-1}V'^{-1}_{n-1}) = X_{n+2}(V'^{-1}_{n-1}W'_{n-1}).\end{aligned}$$

But this means that $(V'^{-1}_{n-1}\,W'_{n-1}) \in \mathfrak{Z}$, and therefore in $\mathfrak{G}/\mathfrak{Z}$ we obtain

$$\varphi(V'^{-1}_{n-1}W'_{n-1}) = E_{(\mathfrak{G}/\mathfrak{Z})},$$
$$\varphi(W'_{n-1}) = \varphi(V'_{n-1}).$$

This means that in $\mathfrak{G}/\mathfrak{Z}$ we have the identity $W_{n-1} \cong V_{n-1}$, i.e., $\mathfrak{G}/\mathfrak{Z}$ is an $(n-1)$-power nilpotent group in the sense of 4.5. By 4.5 it follows that $\mathfrak{G}$ is an n-power nilpotent group.

4.9. The theory of identities in semigroups is essentially a part of the theory of relations. If in the semigroup $\mathfrak{A}$ with generating set $\mathfrak{K}$ we have the identity $T \cong T'$, then in $\mathfrak{A}$ the relations with respect to $\mathfrak{K}$, which are obtained by replacing the elements of the alphabet Ξ contained in T and T' with arbitrary words in $\mathfrak{K}$, will be valid. Conversely, if in $\mathfrak{A}$ all these relations with respect to $\mathfrak{K}$ are valid, then in $\mathfrak{A}$ the identity $T \cong T'$ holds. Thus the identity $T \cong T'$ in $\mathfrak{A}$ can be considered simply as a shortened notation of the above infinite system of relations with respect to the arbitrary generating set $\mathfrak{K}$. Consequently, the different concepts and properties obtained for relations can be carried over to identities.

4.10. In view of these remarks we can independently construct a theory of identities similar to the theory of relations. An arbitrary mapping φ of the alphabet Ξ into itself maps one identity into another. If the first identity is valid in some semigroup $\mathfrak{A}$, then the second will also clearly be valid in $\mathfrak{A}$.

Let Φ be some system of identities in a semigroup $\mathfrak{A}$. We define in $\mathfrak{W}_\Xi$ the relation $\mathfrak{n}$ by setting $T \sim T'(\mathfrak{n})$, where $T, T' \in \mathfrak{W}_\Xi$, if under some mapping φ of the alphabet Ξ into itself, for some identity $T_0 = T'_0$ belonging to Φ, we obtain $\varphi(T_0) = T$, $\varphi(T'_0) = T'$. We can call an identity $T \cong T'$ a consequence of Φ if $T \sim T'(\mathfrak{n}'')$, where $\mathfrak{n}''$ is the second derived relation for $\mathfrak{n}$ (I, 5.21). It is easy to see that if all the identities of Φ are valid in some semigroup $\mathfrak{A}$, then every identity which is a consequence of Φ will also be valid in $\mathfrak{A}$.

Using the concept of consequence of a system of identities, it is possible, in the class of all identities which are valid in a given semigroup, to choose a system such that all the remaining identities are its consequences. We can consider the interrelations of various systems of this kind. We note only that for identities there is no sense in considering problems analogous to the problems of § 2, connected with a change of generating sets. In fact, by the very definition of identity the choice of the alphabet Ξ, by means of which the identities are given, makes absolutely no difference.

4.11. For the words in a countable alphabet $\Xi = \{\xi_1, \xi_2, \dots\}$ let there be given a system Φ of formal equalities of words in that alphabet, i.e., pairs of words in $\mathfrak{W}_\Xi$ joined by an equality sign. For the alphabet $\mathfrak{N}$ we can define the semigroup $\mathfrak{W}_\mathfrak{n}^\Phi$ as a semigroup over the alphabet $\mathfrak{N}$ defined by a defining system of relations, obtained from Φ by means of all possible replacements of elements of Ξ in the formal equalities of Φ by words from $\mathfrak{W}_\mathfrak{N}$ (3.10). We can call this semigroup $\mathfrak{W}_\mathfrak{N}^\Phi$ *the semigroup over* $\mathfrak{N}$ *given by the system of identities* Φ. Evidently in this semigroup all formal equalities of Φ are valid identities.

4.12. In conclusion we remark that both the concept of identity and of relation can be associated not only with semigroups, but with any multiplicative set. Here it is necessary to define words not only by a sequence of elements, but also by a system of parentheses in this sequence, indicating the proper order of performing the operation to obtain the value of the word in the given semigroup. After setting up the corresponding basis of this theory we can define a semigroup as a multiplicative set in which we have the identity

$$(\xi_1\xi_2)\xi_3 \cong \xi_1(\xi_2\xi_3).$$

5. Free Semigroups

5.1. In considering various classes of semigroups, it is natural to turn our attention to systems of elements in the semigroups of a given class between which there are no relations besides those which hold for all the elements in every semigroup of the given class. Corresponding concepts and properties were considered by Birkhoff [**1**] and E. S. Ljapin [**3**] for algebraic theories more general than the theory of semigroups.

DEFINITION. *Let a semigroup* $\mathfrak{A}$ *belong to some class* Γ *of semigroups. A nonempty subset* $\mathfrak{K}$ *of* $\mathfrak{A}$ *is said to be a* FREE SET *with respect to the class* Γ *if any mapping of the set* $\mathfrak{K}$ *into any semigroup* $\mathfrak{A}'$ *of* Γ *can be extended to a homomorphism of the semigroup* $[\mathfrak{K}]$ *into* $\mathfrak{A}'$.

Further, if $\mathfrak{A}$ *has a generating set which is a free set with respect to* Γ, *then* $\mathfrak{A}$ *is said to be* FREE *in the class* Γ.

5.2. The existence of free semigroups in a given class of semigroups proves to be essential for the study of the whole class because of the following circumstance.

If in a class Γ of semigroups there is a semigroup $\mathfrak{F}$ which is free in Γ and has a generating set $\mathfrak{K}$, free with respect to Γ, which is of cardinality $\mathfrak{m}$, then any semigroup of Γ having a generating set whose cardinality does not exceed $\mathfrak{m}$ is a homomorphic image of $\mathfrak{F}$.

In fact, let $\mathfrak{A} = [\mathfrak{K}'] \in \Gamma$, where the cardinality of $\mathfrak{K}'$ does not exceed $\mathfrak{m}$. We take an arbitrary mapping of $\mathfrak{K}$ onto $\mathfrak{K}'$. By 5.1 this extends to a homomorphism. Since the image of this homomorphism contains $\mathfrak{K}'$, it contains $[\mathfrak{K}']$. Consequently this is a homomorphism of $\mathfrak{F}$ onto $\mathfrak{A}$.

5.3. Since the cardinality of any infinite uncountable semigroup coincides with the cardinality of any one of its generating sets, it follows that for a class of semigroups Γ containing uncountable semigroups, we can state the following corollary of assertion 5.2. If in such a class there exists for each semigroup a free semigroup in Γ with no smaller cardinality, then any semigroup in Γ is a homomorphic image of some free semigroup in Γ.

If, besides the property of 5.2 the class Γ has the property that the homomorphic image of any semigroup of Γ itself belongs to Γ (i.e., Γ is closed with respect to the operation of applying homomorphisms), then the class Γ is the class of all possible homomorphic images of semigroups belonging to the class Γ_0 consisting of all free semigroups in Γ.

In this case the class Γ_0 definitely characterizes the entire class Γ. The study of the latter in principle reduces to the study of Γ_0. However, such an approach can produce a real advantage only in case we are able to form some idea of the nature of the homomorphisms of the free semigroups of Γ_0.

5.4. It is easy to see that free sets are indeed characterized by the property mentioned at the very beginning of this section. For their elements, only those relations are valid which hold for all the elements of every semigroup of the given class.

THEOREM. *A nonempty subset $\mathfrak{K}$ of a semigroup $\mathfrak{A}$ of the class Γ will be a free set with respect to Γ if and only if any relation with respect to $\mathfrak{K}$ under any mapping of $\mathfrak{K}$ into an arbitrary semigroup $\mathfrak{A}'$ of Γ is mapped onto a valid relation in $\mathfrak{A}'$.*

PROOF. (1) With respect to $\mathfrak{K} \subset \mathfrak{A}$, where $\mathfrak{A}$ belongs to the class Γ, assume that we have the relation

$$X_1X_2 \dots X_n = Y_1Y_2 \dots Y_m \qquad (X_1, X_2, \dots, X_n, Y_1, \dots, Y_m \in \mathfrak{K})$$

and let φ be an arbitrary mapping of $\mathfrak{K}$ into a semigroup $\mathfrak{A}'$ of Γ. If $\mathfrak{K}$ is a free set, then φ extends to a homomorphism ψ. With respect to ψ we have

$$\psi(X_1X_2 \dots X_n) = \psi(Y_1Y_2 \dots Y_m).$$

Using the property of the homomorphism we thus obtain in $\mathfrak{A}'$ the equality

$$\psi(X_1) \cdot \psi(X_2) \cdot \dots \cdot \psi(X_n) = \psi(Y_1) \cdot \psi(Y_2) \cdot \dots \cdot \psi(Y_m),$$

which can also be written in the form

$$\varphi(X_1) \cdot \varphi(X_2) \cdot \dots \cdot \varphi(X_n) = \varphi(Y_1) \cdot \varphi(Y_2) \cdot \dots \cdot \varphi(Y_m).$$

(2) Let every relation with respect to $\mathfrak{K} \subset \mathfrak{A}$ have the property in question and let φ be an arbitrary mapping of $\mathfrak{K}$ into $\mathfrak{A}'$, $\mathfrak{A}' \in \Gamma$. We represent an arbitrary element $Z \in [\mathfrak{K}]$ in the form of a product

$$Z = X_1X_2 \dots X_n \qquad (X_1, X_2, \dots, X_n \in \mathfrak{K}).$$

We define a mapping ψ by setting

$$\psi(Z) = \varphi(X_1) \cdot \varphi(X_2) \cdot \ldots \cdot \varphi(X_n).$$

This mapping is single-valued, i.e., it does not depend on the choice of the word in $\mathfrak{K}$ whose value is Z. In fact, if

$$Z = X_1X_2 \ldots X_n, \qquad Z = Y_1Y_2 \ldots Y_m$$
$$(X_1, X_2, \ldots, X_n, Y_1, \ldots, Y_m \in \mathfrak{K}),$$

then

$$X_1X_2 \ldots X_n = Y_1Y_2 \ldots Y_m$$

is a relation with respect to $\mathfrak{K}$. By hypothesis it is mapped onto a valid relation in $\mathfrak{A}'$, i.e., in $\mathfrak{A}'$ we have

$$\varphi(X_1) \cdot \varphi(X_2) \cdot \ldots \cdot \varphi(X_n) = \varphi(Y_1) \cdot \varphi(Y_2) \cdot \ldots \cdot \varphi(Y_m).$$

The fact that ψ has the homomorphic property is evident.

5.5. From the proof it follows that the set $\mathfrak{N}$ itself (i.e., the set of words in $\mathfrak{N}$ with length one), from the free semigroup $\mathfrak{W}_{\mathfrak{N}}$ over the alphabet $\mathfrak{N}$ (1.3), is free with respect to the class of all semigroups. In fact, with respect to this set in $\mathfrak{W}_{\mathfrak{N}}$ there are no nontrivial relations. Thus, $\mathfrak{W}_{\mathfrak{N}}$ *is a free semigroup with respect to the class of all semigroups.* Evidently any semigroup isomorphic to $\mathfrak{W}_{\mathfrak{N}}$ is free with respect to the class of all semigroups. There do not exist other free semigroups with respect to the class of all semigroups. In fact, let $\mathfrak{K} \subset \mathfrak{A}$ be a free set with respect to the class of all semigroups and let $[\mathfrak{K}] = \mathfrak{A}$. We consider the semigroup $\mathfrak{W}_{\mathfrak{N}}$ in which the alphabet $\mathfrak{N}$ has the same cardinality as $\mathfrak{K}$. Let φ be a one–one mapping of $\mathfrak{K}$ onto $\mathfrak{N}$, and let ψ be a homomorphism of $[\mathfrak{K}]$ onto $\mathfrak{W}_{\mathfrak{N}}$ extending the map φ. Let

$$\psi(Z) = \psi(Z') \qquad (Z, Z' \in \mathfrak{A}),$$
$$Z = X_1X_2 \ldots X_n, \qquad Z' = Y_1Y_2 \ldots Y_m$$
$$(X_1, X_2, \ldots, X_n, Y_1, \ldots, Y_m \in \mathfrak{K}).$$

From this follows

$$\varphi(X_1) \cdot \varphi(X_2) \cdot \ldots \cdot \varphi(X_n) = \varphi(Y_1) \cdot \varphi(Y_2) \cdot \ldots \cdot \varphi(Y_m).$$

But $\varphi(X_i)$ and $\varphi(Y_j)$ belong to $\mathfrak{N}$. By definition of $\mathfrak{W}_{\mathfrak{N}}$, the equality of values of two words in $\mathfrak{N}$ is possible only when these words are identical. Thus, the two parts of the equality so obtained must coincide identically. In view of the one–oneness of φ, this means that the words $X_1X_2 \ldots X_n$ and $Y_1Y_2 \ldots Y_m$ coincide identically. Therefore $Z = Z'$. The homomorphism ψ, being one–one, is an isomorphism.

5.6. The simple structure of free semigroups (1.3) permits us to deduce without difficulty a great number of their various properties. Those properties which have abstract character are by 5.5 properties of all semigroups which are free in the class of all semigroups. We note some of these properties.

Let $\mathfrak{F}$ be a semigroup which is free in the class of all semigroups. By 5.5 it is isomorphic to the free semigroup $\mathfrak{W}_{\mathfrak{N}}$ over some alphabet $\mathfrak{N}$. The following properties of $\mathfrak{F}$, evidently preserved by isomorphism of semigroups, can be proved for the semigroup $\mathfrak{W}_{\mathfrak{N}}$:

(α) *$\mathfrak{F}$ does not have a unit.*

(β) *$\mathfrak{F}$ is a semigroup with two-sided cancellation.*

The validity of (α) and (β) follows immediately from the rule of multiplication of elements in $\mathfrak{W}_{\mathfrak{N}}$.

(γ) *If in $\mathfrak{F}$ for elements S_1, S_2, S_3, S_4 we have*

$$S_1S_2 = S_3S_4,$$

then either $S_1 = S_3$, or S_1 is divisible from the left by S_3, or S_3 is divisible from the left by S_1.

In fact, let

$$S_i = X_1^{(i)}X_2^{(i)} \dots X_{k_i}^{(i)} \qquad (S_i \in \mathfrak{W}_{\mathfrak{N}};\ X_s^{(i)} \in \mathfrak{N};\ i = 1, 2, 3, 4).$$

By the definition of $\mathfrak{W}_{\mathfrak{N}}$ we must have $X_1^{(1)} = X_1^{(3)}$, $X_2^{(1)} = X_2^{(3)}$, etc. If $k_1 = k_3$, then $S_1 = S_3$; if $k_1 > k_3$, then S_1 is divisible from the left by S_3; if $k_1 < k_3$, then S_3 is divisible from the left by S_1.

(δ) *Each element of $\mathfrak{F}$ has only a finite number of different left divisors.*

Indeed, let

$$W = X_1X_2 \dots X_n \in \mathfrak{W}_{\mathfrak{N}} \qquad (X_1, X_2, \dots, X_n \in \mathfrak{N}).$$

We will show that besides the $n - 1$ elements $X_1X_2 \dots X_k$ $(k = 1, 2, \dots, n - 1)$, W does not have any other left divisors. Let

$$W = V_1V_2, \qquad V_i = Y_1^{(i)}Y_2^{(i)} \dots Y_{k_i}^{(i)} \qquad (Y_s^{(i)} \in \mathfrak{N};\ i = 1, 2).$$

By the definition of $\mathfrak{W}_{\mathfrak{N}}$, we must have $Y_1^{(1)} = X_1$, $Y_2^{(1)} = X_2, \dots, Y_{k_1}^{(1)} = X_{k_1}$.

(ε) *Two elements of $\mathfrak{F}$ permute if and only if they are both powers of one and the same element of $\mathfrak{F}$ (i.e., they are contained in some singly generated subsemigroup of $\mathfrak{F}$).*

In fact, let

$$WV = VW$$

in $\mathfrak{W}_{\mathfrak{N}}$.

The case where $W = V$ is trivial. Let $W \neq V$. By (γ) we can consider that for some $U \in \mathfrak{W}_{\mathfrak{N}}$, $W = VU$. But then

$$VUV = VVU$$

and, because of (β),

$$UV = VU.$$

If we apply induction on the sum of the lengths of the permutable words of $\mathfrak{W}_{\mathfrak{N}}$, we can consider that

$$U = T^k, \qquad V = T^l, \qquad T \in \mathfrak{W}_{\mathfrak{N}}.$$

But then

$$W = VU = T^{k+l}.$$

(ζ) $\mathfrak{F}$ *has a unique irreducible generating set.*

In fact, in $\mathfrak{W}_{\mathfrak{N}}$, it is evident that

$$\mathfrak{N} = \mathfrak{W}_{\mathfrak{N}} \backslash \mathfrak{W}_{\mathfrak{N}} \cdot \mathfrak{W}_{\mathfrak{N}}.$$

Consequently, $\mathfrak{N}$ is a set of indecomposable elements of $\mathfrak{W}_{\mathfrak{N}}$ and is a generating set of $\mathfrak{W}_{\mathfrak{N}}$. By III, 2.7, $\mathfrak{W}_{\mathfrak{N}}$ has no other irreducible generating sets.

(η) $\mathfrak{F}$ *has a unique generating set which is a free set with respect to the class of all semigroups.*

In $\mathfrak{W}_{\mathfrak{N}}$ the set in question is $\mathfrak{N}$. Any other generating set, according to (ζ), is reducible. This means that some of its elements can be expressed in the form of a product of some of its other elements. But, as it easily follows from 5.4, a set with such a property cannot be free with respect to the class of all semigroups.

In connection with property (η), we remark that for some classes Γ a semigroup which is free in Γ can have several generating sets which are free with respect to Γ.

5.7. From the various properties of semigroups belonging to the class of semigroups free in the class of all semigroups, we can distinguish various systems of properties characterizing that class. For example, as follows from the work of Dubreil-Jacotin [**1**], the first four properties of divisibility quoted in 5.6 form such a system.

THEOREM. *If a semigroup* $\mathfrak{A}$ *has properties* (α), (β), (γ), (δ) *of* 5.6, *then it is free in the class of all semigroups.*

PROOF. We will show that no element of $\mathfrak{A}$ can have a left or right unit in $\mathfrak{A}$. In fact, from $AX = A$ ($A, X \in \mathfrak{A}$), for arbitrary $S \in \mathfrak{A}$ we obtain $AXS = AS$ and from (β) we have $XS = S$. But then for arbitrary $S' \in \mathfrak{A}$ it follows that $S'XS = S'S$ and again by (β), that $S'X = S'$. This shows that X is a two-sided unit of $\mathfrak{A}$, which contradicts (α). The case of a left unit is analogous.

We now show that if A is a left divisor of B, then B cannot be a left divisor of A ($A, B \in \mathfrak{A}$).

In fact, from

$$A = BX, \qquad B = AY \qquad (X, Y \in \mathfrak{A})$$

we would obtain $A = AYX$, which, as we have shown above, is impossible.

We set $\mathfrak{K} = \mathfrak{A} \backslash \mathfrak{A}\mathfrak{A}$ and prove that $\mathfrak{A} = [\mathfrak{K}]$.

For arbitrary $A \in \mathfrak{A}$ we consider all its left divisors. By (δ) these are finite in number and by the property proven above there exists among them an X_1 which is not divisible from the left by any other of these left divisors:

$$A = X_1 A_1.$$

Evidently $X_1 \in \mathfrak{K}$, since for $X_1 = X_1' X_1''$ the element X_1' would be a left divisor of A which would divide X_1 from the left. Analogously A_1 has a left divisor X_1 belonging to $\mathfrak{K}$:

$$A_1 = X_2 A_2, \qquad A = X_1 X_2 A_2.$$

We continue the argument. All the elements $X_1, (X_1X_2), (X_1X_2X_3), \ldots$ must be pairwise distinct, because from

$$(X_1X_2 \ldots X_i) = (X_1X_2 \ldots X_i)(X_{i+1} \ldots X_j)$$

it would follow that $X_1X_2 \ldots X_i$ has a right unit. By (δ) our process of construction of the elements $X_i \in \mathfrak{K}$ must come to an end, i.e., for some $n \geqslant 1$ we obtain

$$A = X_1 X_2 \ldots X_n.$$

Finally, we show that there do not exist nontrivial relations with respect to the generating set $\mathfrak{K}$. Let

$$X_1X_2 \ldots X_n = Y_1Y_2 \ldots Y_m \qquad (X_1, X_2, \ldots, X_n, Y_1, \ldots, Y_m \in \mathfrak{K}).$$

If we had $X_1 \neq Y_1$, then by (γ) either X_1 would have to be divisible from the left by Y_1, or Y_1 would have to be divisible from the left by X_1, both of which are impossible since $X_1, Y_1 \in \mathfrak{K}$. But then, by ($\beta$),

$$X_2 \ldots X_n = Y_2 \ldots Y_m.$$

Repeating the argument, we obtain $X_2 = Y_2$, etc., so that finally we see that the initial relation was the identity.

The fact that there do not exist nontrivial relations with respect to $\mathfrak{K}$ shows that $\mathfrak{K}$ is a free set with respect to the class of all semigroups (5.4). Since $[\mathfrak{K}] = \mathfrak{A}$, it follows that $\mathfrak{A}$ is free in the class of all semigroups.

5.8. We will say that an identity is valid in a class of semigroups if it is valid in each of the semigroups of that class.

Let Γ be a class of semigroups which is closed with respect to isomorphisms (i.e., Γ together with each of its semigroups contains every semigroup isomorphic to it) and let Φ be a system of identities in the class such that all the other identities of that class are consequences of Φ (4.10). If the semigroup $\mathfrak{W}_{\mathfrak{N}}^{\Phi}$ (4.11) over some alphabet $\mathfrak{N}$, given by the system of identities Φ, belongs to the class Γ, then it is free in that class.

In fact, $\mathfrak{N}$ is a generating set for $\mathfrak{W}_{\mathfrak{N}}^{\Phi}$. The relations with respect to $\mathfrak{N}$ induced by the identities of Φ (4.9) form a defining system in $\mathfrak{W}_{\mathfrak{N}}^{\Phi}$ with respect to $\mathfrak{N}$ (4.9; 4.11). Since for any map of $\mathfrak{N}$ into an arbitrary semigroup $\mathfrak{A} \in \Gamma$ these relations are mapped onto valid relations in $\mathfrak{A}$, it follows from 1.11 that this map extends to a homomorphism of $[\mathfrak{N}] = \mathfrak{W}_{\mathfrak{N}}^{\Phi}$ into $\mathfrak{A}$.

We remark that if $\mathfrak{N}_1$ and $\mathfrak{N}_2$ have the same cardinality, then for two sets of identities Φ_1 and Φ_2 of the type in question the semigroups $\mathfrak{W}_{\mathfrak{N}_1}^{\Phi_1}$ and $\mathfrak{W}_{\mathfrak{N}_2}^{\Phi_2}$ are isomorphic.

5.9. The free semigroups in the considered class Γ are exhausted by the semigroups $\mathfrak{W}_{\mathfrak{N}}^{\Phi}$ and semigroups isomorphic to them.

In fact, let $\mathfrak{A} = [\mathfrak{K}]$ be a free semigroup in Γ and let $\mathfrak{K}$ be a free set with respect to Γ. By 5.4 there are no relations with respect to $\mathfrak{K}$ besides those induced by the identities which are valid in every semigroup of Γ, i.e., identities which are consequences of Φ. We consider the semigroup $\mathfrak{W}_{\mathfrak{K}}^{\Phi}$. By the definition of this semigroup, from Theorem (1.12) immediately follows the existence of an isomorphism between $\mathfrak{A}$ and $\mathfrak{W}_{\mathfrak{K}}^{\Phi}$.

5.10. The arguments in 5.8 and 5.9 give a necessary and sufficient condition for a class of semigroups which is closed with respect to isomorphisms to have free semigroups. As we have already remarked, the role of free semigroups in a class is defined by the possibility of obtaining from them by means of homomorphisms the remaining semigroups of that class. We examine this problem in detail.

Let Φ be a system of identities in an alphabet Ξ. Let $\mathfrak{M}$ be a nonempty class of cardinals such that if the cardinal number $\mathfrak{n}$ is less than the cardinal number $\mathfrak{m} \in \mathfrak{M}$, then $\mathfrak{n}$ itself must belong to $\mathfrak{M}$. We denote by $\Gamma_{\mathfrak{M}}^{\Phi}$ the class of all semigroups in which all the identities of Φ are valid and which have generating sets whose cardinal numbers belong to $\mathfrak{M}$. Of course any such class is nonempty, since it always contains the unit semigroup.

$\Gamma_{\mathfrak{M}}^{\Phi}$ always has the following three properties:

(α) $\Gamma_{\mathfrak{M}}^{\Phi}$ is closed with respect to homomorphisms, i.e., a homomorphic image of any semigroup of $\Gamma_{\mathfrak{M}}^{\Phi}$ always belongs to $\Gamma_{\mathfrak{M}}^{\Phi}$.

(β) There exist free semigroups in $\Gamma_{\mathfrak{M}}^{\Phi}$.

(γ) Each semigroup in $\Gamma_{\mathfrak{M}}^{\Phi}$ is the homomorphic image of a free semigroup in $\Gamma_{\mathfrak{M}}^{\Phi}$.

The validity of (α) follows from the fact that under a homomorphism of a semigroup onto a semigroup a generating set is carried onto a generating set and any relation is carried onto a valid relation.

(β) follows from 5.8.

Let $\mathfrak{A}$ be a semigroup in $\Gamma_{\mathfrak{M}}^{\Phi}$. Let $\mathfrak{K}$ be a generating set of $\mathfrak{A}$ whose cardinality belongs to $\mathfrak{M}$. The semigroup $\mathfrak{W}_{\mathfrak{K}}^{\Phi}$ belongs to the class $\Gamma_{\mathfrak{K}}^{\Phi}$ and is free in it (5.8). By 5.1 the identity mapping of $\mathfrak{K}$ extends to a homomorphism of $\mathfrak{W}_{\mathfrak{K}}^{\Phi}$ into $\mathfrak{A}$. In this case the homomorphic image must coincide with all of $\mathfrak{A}$.

5.11. Properties 5.10, (α), (β), (γ) show that the class $\Gamma_{\mathfrak{M}}^{\Phi}$ (5.10) consists of all semigroups obtained by means of homomorphisms from the free semigroups in that class. It turns out that no other class which is closed with respect to homomorphisms has these properties.

If some class of semigroups Σ has the properties (α), (β), (γ) formulated in 5.10 for $\Gamma_{\mathfrak{M}}^{\Phi}$, then Σ is one of the classes $\Gamma_{\mathfrak{M}}^{\Phi}$ (5.10).

We denote by $\mathfrak{M}$ the class of cardinals $\mathfrak{m}$ such that there exists a free semigroup in Σ having a generating set which is free with respect to Σ and whose cardinality is not less than $\mathfrak{m}$. By Φ we denote the system of all identities over some alphabet Ξ which are valid in all the semigroups of the class Σ. We show that $\Sigma = \Gamma_{\mathfrak{M}}^{\Phi}$.

Let $\mathfrak{A} \in \Sigma$. By (γ), $\mathfrak{A}$ is a homomorphic image of some semigroup which is free in Σ. From this it follows that $\mathfrak{A}$ has a generating set whose cardinal number belongs to $\mathfrak{M}$. Since in $\mathfrak{A}$ all the identities of Φ are valid, $\mathfrak{A}$ belongs to $\Gamma_{\mathfrak{M}}^{\Phi}$.

On the other hand, if $\mathfrak{B} \in \Gamma_{\mathfrak{M}}^{\Phi}$, then in $\mathfrak{B}$, according to (γ), there must be a generating set $\mathfrak{K}$ whose cardinal number belongs to $\mathfrak{M}$. In Σ there must exist a free semigroup $\mathfrak{A}$ having a free generating set $\mathfrak{K}'$ with respect to Σ whose cardinality is greater than or equal to the cardinality of $\mathfrak{K}$. We take any map φ of $\mathfrak{K}'$ onto $\mathfrak{K}$. By 5.4, in $\mathfrak{A}$ only those relations with respect to $\mathfrak{K}'$ which are induced by the identities of Φ are valid. Since $\mathfrak{B} \in \Gamma_{\mathfrak{M}}^{\Phi}$, all those relations are mapped onto valid relations in $\mathfrak{B}$. Because of this and 1.11, φ extends to a homomorphism. Since $\mathfrak{K}$ is contained in the image of this homomorphism, the homomorphism is a homomorphism of $\mathfrak{A}$ onto $\mathfrak{B}$. From this and (α) it follows that $\mathfrak{B} \in \Sigma$.

5.12. In an arbitrary class $\Gamma_{\mathfrak{M}}^{\Phi}$ of type 5.1, the free semigroups are, by 5.8 and 5.9, the semigroups $\mathfrak{W}_{\mathfrak{N}}^{\Phi}$ over various alphabets $\mathfrak{N}$, whose cardinalities belong to $\mathfrak{M}$, and semigroups isomorphic to them. Thus the semigroups which are free in $\Gamma_{\mathfrak{M}}^{\Phi}$, are by 5.8 completely determined to within isomorphism by the cardinality of a generating set which is free in $\Gamma_{\mathfrak{M}}^{\Phi}$.

5.13. An example of a class of type 5.10 is the class of all commutative semigroups. This is the class defined by the identity $\xi_1\xi_2 \cong \xi_2\xi_1$. A free semigroup in this class with free generating set $\mathfrak{K}$ can be defined as a set of classes of words in $\mathfrak{K}$, where each class includes the words obtained from one another by the permutation of elements.

The class of all n-power nilpotent semigroups (4.6) for any $n = 1, 2, \ldots$ is also a class of the form $\Gamma_{\mathfrak{M}}^{\Phi}$ (5.10). Namely, Φ consists of the one identity 4.6. Thus all n-power nilpotent semigroups are homomorphic images of free n-power nilpotent semigroups, given by the indicated identity and the cardinality of a free generating set.

From the indicated classes we can choose new classes of type 5.10 if in addition we impose a condition on the cardinality of the generating sets of the semigroups.

5.14. Among the classes of type 5.10 belongs also the class of idempotent semigroups, characterized by the identity $\xi_1^2 \cong \xi_1$. This class can also be considered as the class of semigroups which are bands of unit groups. We have repeatedly referred to semigroups of that class. Now we dwell on the consideration of its free semigroups. In the discussion just below, we will take Φ to be the system consisting of the single identity $\xi_1^2 \cong \xi_1$. The class of idempotent semigroups (i.e., $\Gamma_{\mathfrak{M}}^{\Phi}$, where $\mathfrak{M}$ is the class of all cardinals) will be denoted by Γ. Every semigroup which is free in Γ is isomorphic to $\mathfrak{W}_{\mathfrak{N}}^{\Phi}$, where $\mathfrak{N}$ is an arbitrary alphabet.

5.15. For each word W in $\mathfrak{N}$ we denote by $\mathfrak{H}(W)$ the system of those elements of $\mathfrak{N}$ which are contained in the word W. If in $\mathfrak{W}_{\mathfrak{N}}^{\Phi}$ the value of the word W coincides with the value of the word V, then $\mathfrak{H}(W) = \mathfrak{H}(V)$.

In fact, we can go from W to V by means of transformations which replace a part of the word of the form $X_1 X_2 \cdots X_k$ by

$$X_1 X_2 \dots X_k X_1 X_2 \dots X_k \, (X_1, X_2, \dots, X_k \in \mathfrak{N}),$$

or vice versa. It is evident that each time $\mathfrak{H}(W)$ will not be changed. Therefore it is possible to consider the set $\mathfrak{H}(W)$ as the unique characteristic of that element of $\mathfrak{W}_{\mathfrak{N}}^{\Phi}$ which is the value of the word W.

5.16. We consider some properties of the characteristic introduced in 5.15 which are connected with the properties of regular conjugacy of elements in $\mathfrak{W}_{\mathfrak{N}}^{\Phi}$, a procedure which is all the more natural since this semigroup, like all semigroups of the class Γ, is evidently completely regular.

(α) For any $S, S' \in \mathfrak{W}_{\mathfrak{N}}^{\Phi}$,

$$\mathfrak{H}(SS') = \mathfrak{H}(S) \cup \mathfrak{H}(S').$$

The validity of this immediately follows from the definition of the characteristic (5.15).

(β) If the elements $S, S' \in \mathfrak{W}_{\mathfrak{N}}^{\Phi}$ are regularly conjugate with each other, then

$$\mathfrak{H}(S) = \mathfrak{H}(S').$$

In fact, from

$$SS'S = S, \quad S'SS' = S',$$

according to (α) we obtain

$$\mathfrak{H}(S) \cup \mathfrak{H}(S') = \mathfrak{H}(S), \qquad \mathfrak{H}(S') \cup \mathfrak{H}(S) = \mathfrak{H}(S').$$

(γ) If for $S, S' \in \mathfrak{W}_{\mathfrak{N}}^{\Phi}$ we have

$$\mathfrak{H}(S) = \mathfrak{H}(S'),$$

then S and S' are regularly conjugate.

In fact, we represent S and S' in the form of values of some words in $\mathfrak{N}$:

$$S = X_1 X_2 \dots X_n,$$
$$S' = Y_1 Y_2 \dots Y_m \qquad (X_1, X_2, \dots, X_n, Y_1, \dots, Y_m \in \mathfrak{N}).$$

In view of the equality $\mathfrak{H}(S) = \mathfrak{H}(S')$ it is possible by means of permutation of the neighboring factors in the first of these words and the replacement of X by XX or XX by X ($X \in \mathfrak{N}$) to transform the first word into the second word. By VII, 6.8, (δ), under each such transformation we will obtain an element regularly conjugate with the element which is the value of the preceding word. Because of this and VII, 6.9 it follows that S and S' are regularly conjugate.

(δ) If the elements S and S' of $\mathfrak{W}_{\mathfrak{N}}^{\Phi}$ are regularly conjugate and if for some $T \in \mathfrak{W}_{\mathfrak{N}}^{\Phi}$ we have

$$\mathfrak{H}(T) \subset \mathfrak{H}(S),$$

then

$$STS' = SS'.$$

In fact, by (α),

$$\mathfrak{H}(ST) = \mathfrak{H}(S) \cup \mathfrak{H}(T) = \mathfrak{H}(S)$$

and therefore, by (γ), ST and S are regularly conjugate.

But $\mathfrak{H}(S') = \mathfrak{H}(S)$, and therefore it follows from (γ) that S' and ST are also regularly conjugate. Thus

$$S'(ST)S' = S',$$

$$SS' = SS'STS' = (SS'S)TS' = STS'.$$

5.17. Using these properties, we can prove a theorem with respect to the finiteness of semigroups which are free in Γ. This theorem is essentially a particular case of one of the theorems we have already mentioned in the work of Green and Rees [**1**]. Further, it was separately proven by McLean [**1**].

THEOREM. *If a semigroup which is free in the class of all idempotent semigroups has a finite generating free set, then it is finite.*

PROOF. The proof of the finiteness of $\mathfrak{W}_{\mathfrak{N}}^{\Phi}$ (5.14) proceeds by induction on n, the number of elements in the alphabet $\mathfrak{N}$.

If $n = 1$, then $\mathfrak{W}_{\mathfrak{N}}^{\Phi}$ is the unit semigroup.

Assume $n > 1$. From the assumption that any semigroup $\mathfrak{W}_{\mathfrak{N}'}^{\Phi}$ in which $\mathfrak{N}'$ consists of $n - 1$ elements is finite follows the existence of a natural number m such that any element of that semigroup can be represented as the value of some word in $\mathfrak{N}'$ with length at least m.

We take an arbitrary word W in $\mathfrak{N}$ with length $2m + 1$, namely,

$$W \equiv X_1X_2 \dots X_{2m+1} \qquad (X_1, X_2, \dots, X_{2m+1} \in \mathfrak{N}),$$

and starting from it we examine the structure of the word

$$U \equiv X_1X_2 \dots X_m, \qquad V \equiv X_{m+2}X_{m+3} \dots X_{2m+1}.$$

If $\mathfrak{N}' = \mathfrak{H}(U) \neq \mathfrak{N}$, then U is a word in $\mathfrak{W}_{\mathfrak{N}'}^{\Phi}$ and therefore, by assumption, there exists a word $Y_1Y_2 \dots Y_k$ ($Y_1, Y_2, \dots, Y_k \in \mathfrak{N}' \subset \mathfrak{N}$) with the same value but with smaller length ($k < m$). Thus, the value of the word W turns out to be identical with the value of the word

$$Y_1Y_2 \dots Y_kX_{m+1}X_{m+2} \dots X_{2m+1}$$

with length less than $2m + 1$.

An analogous result follows in the case $\mathfrak{H}(V) \neq \mathfrak{N}$.

If $\mathfrak{H}(V) = \mathfrak{H}(V) = \mathfrak{N}$, then by 5.16, ($\gamma$) the values of the words U and V are regularly conjugate elements. Applying 5.16, (δ) to these elements and to X_{m+1}, we obtain

$$W = UX_{m+1}V = UV.$$

Again, the value of W is equal to the value of the shorter word UV.

Thus we see that in $\mathfrak{W}_{\mathfrak{N}}^{\Phi}$ the value of any word in $\mathfrak{N}$ with length $2m + 1$ is equal to the value of some word with smaller length. From this it evidently follows that any element of $\mathfrak{W}_{\mathfrak{N}}^{\Phi}$ can be represented in the form of the value of some word in $\mathfrak{N}$ having length less than $2m + 1$. Since, in view of the finiteness of $\mathfrak{N}$, there are only finitely many such words, the number of distinct elements in $\mathfrak{W}_{\mathfrak{N}}^{\Phi}$ turns out to be finite.

5.18. Corollary. *If an idempotent semigroup has a finite generating set, then it is finite.*

In fact, for the above semigroup $\mathfrak{A} = [\mathfrak{K}]$ there exists a homomorphism of $\mathfrak{W}_{\mathfrak{K}}^{\Phi}$ onto $\mathfrak{A}$. In view of the finiteness of $\mathfrak{W}_{\mathfrak{K}}^{\Phi}$ (5.17), $\mathfrak{A}$ also turns out to be finite.

5.19. In connection with the semigroup $\mathfrak{W}_{\mathfrak{N}}^{\Phi}$ which is free in the class of all idempotent semigroups, there naturally arises the problem of a canonical form of its elements with respect to the generating set $\mathfrak{N}$. It is entirely evident that any element of $\mathfrak{W}_{\mathfrak{K}}^{\Phi}$ can be represented in the form

$$Z = X_1X_2 \dots X_n \qquad (X_1, X_2, \dots, X_n \in \mathfrak{N}),$$

where for any natural numbers p and q the equality of words

$$X_{p+1}X_{p+2} \dots X_{p+q} \equiv X_{p+q+1}X_{p+q+2} \dots X_{p+2q}$$

is impossible. However, it turns out that such a form of a given element is indeed not yet canonical. What is wrong is that although the semigroup $\mathfrak{W}_{\mathfrak{N}}^{\Phi}$ itself is finite, for a number of elements greater than two the set of words in $\mathfrak{N}$ satisfying the above condition turns out to be infinite. This circumstance was discovered by S. E. Aršon.[4] It follows also from results in the work of Morse and Hedlund [**1**].

6. Determination of Free Semigroups by the Subsemigroup Characteristic

6.1. To the properties examined in the preceding section of semigroups which are free in the class of all semigroups, we add another important property. Namely, we show that these semigroups are defined by the subsemigroup characteristic (III, 7.6). This concept and some of the ideas connected with it were examined in § 7 of the third chapter.

[4] S. E. Aršon, *Proof of the existence of n-valued infinite asymmetric sequences*, Mat. Sb. (N.S.) **2** (1937), 769–779. (Russian)

Because of the results obtained in 5.5, in what follows we can consider the free semigroup $\mathfrak{W}_{\mathfrak{N}}$ over some alphabet. The theorem about the definability of free semigroups by the semigroup characteristic is due to R. V. Petropavlovskaja [2].

6.2. For the elements X and Y of a semigroup $\mathfrak{A}$ we call the element $P \in [X, Y]$ their *special product* (R. V. Petropavlovskaja called P the maximal element for X and Y), if for any Z which is an empty symbol or an element of $[X, Y]\backslash\{X, Y, P\}$, in the semigroup $\Sigma(\mathfrak{A})$ the element $[P]$ is not a unit for $[X] \circ [Z] \circ [Y^2] \circ [Y^3]$ or for $[Y] \circ [Z] \circ [X^2] \circ [X^3]$, i.e., in other words, in $\mathfrak{A}$

$$P \bar{\in} [X, Z, Y^2, Y^3, \ldots] \cup [Y, Z, X^2, X^3, \ldots].$$

By the way, we mention (in what follows we will constantly make use of this) that for any element S it is always true that

$$[S^2] \circ [S^3] = [S^2, S^3] = \{S^2, S^3, S^4, S^5, \ldots\}.$$

From the definition it immediately follows that the special product of X and Y is also the special product of Y and X. The special product P of the elements X and Y of the semigroup $\mathfrak{A}$ will be their special product with respect to any super-semigroup of $\mathfrak{A}$ and any of its subsemigroups containing X and Y.

6.3. The terminology introduced above is justified by the following property.

LEMMA. *If P is the special product of $X, Y \in \mathfrak{A}$, then either $P = XY$ or $P = YX$.*

PROOF. (1) We suppose that $XY = X$. Then evidently any element of $[X, Y]$ has the form $Y^n X^m$ ($n, m = 0, 1, 2, \ldots$; X^0, Y^0 denote the empty symbol). Let $P = Y^p X^q$. If it were true that $p > 1$, then we would have

$$P = Y^p X^q \in [X, Y^2, Y^3, \ldots],$$

which would contradict the definition of special product.

Analogously we see that $q > 1$ is impossible.

Since $P \neq X$ and $P \neq Y$, it follows from the above that in the case considered, $P = YX$.

(2) Analogously we consider the cases $XY = Y$, $YX = X$, $YX = Y$.

(3) Since it is evident that $X^\alpha Y^\beta \in [X, XY, Y^2, Y^3]$ $(\alpha, \beta = 1, 2, \ldots)$, it follows in the case where P has the form $X^{\alpha_1} Y^{\beta_1} X^{\alpha_2} Y^{\beta_2} \ldots X^{\alpha_s} Y^{\beta_s}$ or $X^{\alpha_1} Y^{\beta_1} X^{\alpha_2} Y^{\beta_2} \ldots X^{\alpha_s}$ $(\alpha_i, \beta_i = 1, 2, \ldots)$, that

$$P \in [X, XY, Y^2, Y^3, \ldots].$$

Taking into account the definition of special product, we conclude from this that $XY \in \{X, Y, P\}$, i.e., either $P = XY$ or we have one of the two cases already discussed: $XY = X$, $XY = Y$.

Analogously we consider the case where in the expression for P in the form of a product of X and Y, the first factor is equal to Y.

6.4. COROLLARY. *For two elements there exist not more than two special products.*

6.5. It is understood that the product XY is not always equal to the special product of X and Y. One sufficient condition for this is the requirement that XY cannot be represented in a form different from XY and YX as a product of factors equal to X or Y.

In fact, for any Z which belongs to $[X, Y]\setminus\{X, Y, XY\}$ or is the empty symbol, any element of

$$[X, Z, Y^2, Y^3, \ldots] \cup [Y, Z, X^2, X^3, \ldots]$$

is evidently different from XY because of the indicated condition.

6.6. For the problem of interest to us here, concerning isomorphisms of subsemigroup characteristics, the concept of special product turns out to be essential because of the following.

THEOREM. *Let every element X_1, Y_1, P_1 of the semigroup $\mathfrak{A}_1$ have infinite type, where P_1 is the special product of X_1 and Y_1, and for some semigroup $\mathfrak{A}_2$ suppose that there exists an isomorphism φ of $\Sigma(\mathfrak{A}_1)$ onto $\Sigma(\mathfrak{A}_2)$. Then there exist unique elements X_2, Y_2, P_2 in $\mathfrak{A}_2$ such that*

$$\varphi[X_1] = [X_2], \qquad \varphi[Y_1] = [Y_2], \qquad \varphi[P_1] = [P_2],$$

where P_2 is the special product of X_2 and Y_2.

PROOF. The existence of the unique elements X_2, Y_2, P_2 connected with X_1, Y_1, P_1 in the indicated way follows from III, 7.10, (γ). We suppose that P_2 is not the special product of X_2 and Y_2 in $\mathfrak{A}_2$. Then for some Z_2, which is the empty symbol or an element of $[X_2, Y_2]\setminus\{X_2, Y_2, P\}$, we must have

$$[P_2] \subset [X_2, Z_2, Y_2^2, Y_2^3, \ldots] \cup [Y_2, Z_2, X_2^2, X_2^3, \ldots].$$

But then, applying the isomorphism φ^{-1} and taking III, 7.10, (ε), (ζ) into account, we obtain a relation which contradicts the fact that P_1 is the special product of X_1 and Y_1, namely,

$$[P_1] \subset [X_1, Z_1, Y_1^2, Y_1^3, \ldots] \cup [Y_1, Z_1, X_1^2, X_1^3, \ldots].$$

Here Z_1 is the empty symbol if Z_2 is the empty symbol. But if Z_2 belonged to $[X_2, Y_2]\setminus\{X_2, Y_2, P_2\}$, then Z_1 is an element such that $\varphi[Z_1] = [Z_2]$.

By III, 7.10, (ε), Z_1 belongs to $[X_1, Y_1]$. Since $[Z_2] \neq [X_2]$, $[Z_2] \neq [Y_2]$, $[Z_2] \neq [P_2]$, it follows that $Z_1 \neq X_1$, $Z_1 \neq Y_1$, $Z_1 \neq P_1$.

6.7. We consider the problem of special products of elements of free semigroups.

We recall (we will repeatedly need this) that in a free semigroup two elements permute if and only if they are both contained in some singly generated semigroup, i.e., they are powers of some third element (5.6, (ε)).

LEMMA. *If the elements V and W of the free semigroup $\mathfrak{W}_{\mathfrak{N}}$ do not permute, then both VW and WV are special products of V and W.*

PROOF. Since $\mathfrak{W}_{\mathfrak{N}}$ is evidently a semigroup with two-sided cancellation in which none of the elements have left or right units, from an equality of the form $VW = V^n$ we would conclude that $n > 1$ and $W = V^{n-1}$, i.e., V and W would permute. Also $VW = W^n$ is impossible. Let VW be represented in the form of a product of factors equal to V and W. By the above, V and W must occur in this product. We assume that the number of factors in this product is greater than two. V and W are words in $\mathfrak{N}$. So the word VW, with length equal to the sum of the lengths of the words V and W, turns out to be equal to a word with length greater than the sum of the lengths of V and W. But this is impossible by the definition of a free semigroup over an alphabet $\mathfrak{N}$. From what was said it follows that VW cannot be represented in the form of a product in the indicated way, except for VW or WV.

But then, according to 6.5, VW is the special product of V and W. The argument for WV is similar.

6.8. We can now proceed directly to consider isomorphisms of subsemigroup characters of free semigroups. Let $\mathfrak{W}_{\mathfrak{N}}$ be a free semigroup over $\mathfrak{N}$ and let $\mathfrak{U}$ be a semigroup for which there exists an isomorphism φ of $\Sigma(\mathfrak{W}_{\mathfrak{N}})$ onto $\Sigma(\mathfrak{U})$. For any $V \in \mathfrak{W}_{\mathfrak{N}}$ the singly generated semigroup $[V]$ is infinite. From III, 7.10, (γ), it follows that $\varphi[V]$ must be an infinite singly generated subsemigroup of $\mathfrak{U}$, that is,

$$\varphi[V] = [S] \subset \mathfrak{U}.$$

The element S, generating the infinite singly generated semigroup $[S]$, is uniquely defined by this semigroup. We set $S = \bar{V}$. From the fact that all the elements of $\mathfrak{W}_{\mathfrak{N}}$ have infinite type it follows that in $\mathfrak{U}$ the types of all the elements are infinite (III, 7.9).

For any $T \in \mathfrak{U}$ under the isomorphism φ the subsemigroup $[T]$ must be the image of some uniquely defined singly generated subsemigroup $[Q] \subset \mathfrak{W}_{\mathfrak{N}}$, i.e., $T = \bar{Q}$. Thus the isomorphism φ of $\Sigma(\mathfrak{W}_{\mathfrak{N}})$ onto $\Sigma(\mathfrak{U})$ defines a one-one transformation ψ of the set of all elements of $\mathfrak{W}_{\mathfrak{N}}$ onto the set of all elements of $\mathfrak{U}$

$$\psi(V) = \bar{V} \qquad (\varphi[V] = [\bar{V}],\ V \in \mathfrak{W}_{\mathfrak{N}}).$$

We note that, by III, 7.10, (ζ), $\varphi[V] = [\bar{V}]$ ($V \in \mathfrak{W}_{\mathfrak{N}}$) implies $\varphi(V^n) = [\bar{V}^n]$ for all natural numbers n. This means that

$$\overline{V^n} = \bar{V}^n.$$

6.9. In the free semigroup $\mathfrak{W}_{\mathfrak{N}}$ we define a relation $\mathfrak{n}$ by setting

$$V \sim W(\mathfrak{n}) \qquad (V, W \in \mathfrak{W}_{\mathfrak{N}})$$

if in $\mathfrak{U}$ we have

$$\overline{VW} = \bar{V}\bar{W}.$$

(Thus the relation $\mathfrak{n}$ depends on $\mathfrak{U}$ and φ.)

Analogously we define the relation $\mathfrak{m}$ in $\mathfrak{W}_{\mathfrak{N}}$ by setting

$$V \sim W(\mathfrak{m}) \qquad (V, W \in \mathfrak{W}_{\mathfrak{N}}),$$

if

$$\overline{VW} = \overline{W}\overline{V}.$$

Evidently the one-one transformation ψ of $\mathfrak{W}_{\mathfrak{N}}$ onto $\mathfrak{U}$ (6.8) pairing off each $V \in \mathfrak{W}_{\mathfrak{N}}$ with $\psi(V) = \overline{V} \in \mathfrak{U}$ will be an isomorphism if and only if any two elements of $\mathfrak{W}_{\mathfrak{N}}$ are related to each other with respect to $\mathfrak{n}$. ψ will be an anti-isomorphism (I, 1.17) if and only if any two elements of $\mathfrak{W}_{\mathfrak{N}}$ are related to each other with respect to $\mathfrak{m}$.

In what follows we will make use, without special mention of the fact, of the notations and results introduced here and in 6.8. Also we will constantly use the uniqueness of the expressions for elements of $\mathfrak{W}_{\mathfrak{N}}$ in the form of products of elements of $\mathfrak{N}$.

6.10. We note a few properties of the relations $\mathfrak{n}$ and $\mathfrak{m}$.

(α) If the elements V and W of $\mathfrak{W}_{\mathfrak{N}}$ are permutable, then

$$V \sim W(\mathfrak{n}), \qquad V \sim W(\mathfrak{m}).$$

In fact, V and W must both be contained in some singly generated semigroup (IX, 5.6, (ε))

$$V, W \in [U], \qquad V = U^p, \qquad W = U^q, \qquad U \in \mathfrak{W}_{\mathfrak{N}}.$$

By 6.8,

$$\overline{VW} = \overline{WV} = \overline{U^{p+q}} = \overline{U}^{p+q},$$

$$\overline{V} = \overline{U^p} = \overline{U}^p, \qquad \overline{W} = \overline{U^q} = \overline{U}^q,$$

from which follows

$$\overline{VW} = \overline{V}\overline{W} = \overline{W}\overline{V}.$$

(β) $\mathfrak{n}$ and $\mathfrak{m}$ are reflexive relations.

This immediately follows from (α), since each of the elements permutes with itself.

(γ) Any two elements of $\mathfrak{W}_{\mathfrak{N}}$ are related with respect to $\mathfrak{n}$ or with respect to $\mathfrak{m}$.

If V and W permute, then our assertion follows from (α). If V and W do not permute, then VW, by 6.7, is the special product of V and W. From this, because of 6.6, it follows that $\overline{VW}$ must be the special product of $\overline{V}$ and $\overline{W}$ in $\mathfrak{U}$, i.e., we must have $\overline{VW} = \overline{V}\overline{W}$ or $\overline{VW} = \overline{W}\overline{V}$. In the first case we have $V \sim W(\mathfrak{n})$, in the second $V \sim W(\mathfrak{m})$.

(δ) $\mathfrak{n}$ and $\mathfrak{m}$ are symmetric relations.

In the case where V and W ($V, W \in \mathfrak{W}_{\mathfrak{N}}$) permute, this immediately follows from (α). Suppose V and W do not permute. By 6.7, VW is the special product of V and W. By 6.6, $\overline{WV}$ must be the special product of $\overline{W}$ and $\overline{V}$, i.e., in $\mathfrak{U}$ we must have either $\overline{WV} = \overline{W}\overline{V}$ or $\overline{WV} = \overline{V}\overline{W}$.

If $V \sim W(\mathfrak{n})$, then $\overline{VW} = \bar{V}\overline{W}$ and therefore $\overline{WV} \neq \bar{V}\overline{W}$, since $VW \neq WV$. Consequently, $\overline{WV} = \overline{W}\bar{V}$, i.e., $W \sim V(\mathfrak{n})$.

If $V \sim W(\mathfrak{m})$, then $\overline{VW} = \overline{W}\bar{V}$ and therefore $\overline{WV} \neq \overline{W}\bar{V}$. Therefore $\overline{WV} = \bar{V}\overline{W}$, i.e., $W \sim V(\mathfrak{m})$.

(ε) If V and W do not permute, then $V \sim W(\mathfrak{n})$ and $V \sim W(\mathfrak{m})$ cannot occur simultaneously.

Indeed, if $V \sim W(\mathfrak{n})$, then by (δ), $W \sim V(\mathfrak{n})$. Therefore $\overline{WV} = \overline{W}\bar{V}$. Since $VW \neq WV$, it follows that $\overline{VW}$ cannot equal $\overline{WV}$, i.e., $V \sim W(\mathfrak{m})$ is impossible.

6.11. LEMMA. *If $X, Y \in \mathfrak{N}$ $(X \neq Y)$ and $X \sim Y(\mathfrak{n})$, then*

$$X^n \sim Y(\mathfrak{n})$$

for all positive integers n.

PROOF. If $n > 1$ and $X^{n-1} \sim XY(\mathfrak{n})$, then

$$\overline{X^nY} = \overline{X^{n-1} \cdot XY} = \overline{X^{n-1}} \cdot \overline{XY} = \bar{X}^{n-1} \cdot \bar{X} \cdot \bar{Y} = \bar{X}^n\bar{Y} = \overline{X^n}\,\bar{Y}.$$

We suppose that for some $n > 1$ we have

$$X^n \nsim Y(\mathfrak{n}), \qquad X^{n-1} \nsim XY(\mathfrak{n}).$$

Then by 6.10, (γ),

$$X^n \sim Y(\mathfrak{m}), \qquad X^{n-1} \sim XY(\mathfrak{m}),$$

and we obtain

$$\overline{XY \cdot X^{n-1}} = \overline{X^{n-1}} \cdot \overline{XY} = \bar{X}^{n-1} \cdot \bar{X} \cdot \bar{Y} = \bar{X}^n\bar{Y},$$

$$\overline{Y \cdot X^n} = \overline{X^n} \cdot \bar{Y} = \bar{X}^n\bar{Y},$$

which contradicts the fact that $XYX^{n-1} \neq YX^n$.

6.12. LEMMA. *If $X, Y \in \mathfrak{N}$ $(X \neq Y)$ and $X \sim Y(\mathfrak{n})$, then*

$$XY \sim W(\mathfrak{n})$$

for all $W \in \mathfrak{W}_{\mathfrak{N}}$.

PROOF. If XY and W permute, then our assertion follows from 6.10, (α).

Suppose XY and W do not permute. We suppose that $XY \nsim W(\mathfrak{n})$, i.e., by 6.10 (γ), $XY \sim W(\mathfrak{m})$. It will be necessary for us to consider six different cases, in each of which our assumption will reduce to a contradiction. In our arguments we will constantly use the properties of 6.10.

(1) $Y \sim W(\mathfrak{n})$, $X \sim YW(\mathfrak{n})$.

In this case we obtain

$$\overline{W \cdot XY} = \overline{XY} \cdot \overline{W} = \bar{X}\bar{Y}\overline{W},$$

$$\overline{X \cdot YW} = \overline{XY} \cdot \overline{W} = \bar{X}\bar{Y}\overline{W},$$

which contradicts the fact that $WXY \neq XYW$.

(2) $Y \sim W(\mathfrak{n})$, $X \sim YW(\mathfrak{m})$.
Then

$$\overline{W \cdot XY} = \overline{XY} \cdot \overline{W} = \overline{X}\,\overline{Y}\,\overline{W},$$

$$\overline{YW \cdot X} = \overline{X} \cdot \overline{YW} = \overline{X}\,\overline{Y}\,\overline{W},$$

and this contradicts the fact that $WXY \neq YWX$.

(3) $Y \sim W(\mathfrak{m})$, $X \sim YW(\mathfrak{n})$, $YX \sim W(\mathfrak{n})$.
Then

$$\overline{YW \cdot X} = \overline{YW} \cdot \overline{X} = \overline{W}\,\overline{Y}\,\overline{X},$$

$$\overline{W \cdot YX} = \overline{W} \cdot \overline{YX} = \overline{W}\,\overline{Y}\,\overline{X},$$

from which follows $YWX = WYX$, and therefore $YW = WY$, since $\mathfrak{W}_{\mathfrak{N}}$ has two-sided cancellation. But since $YW = WY$, we obtain $Y \sim W(\mathfrak{n})$, i.e., we arrive at one of the first two cases.

(4) $Y \sim W(\mathfrak{m})$, $X \sim YW(\mathfrak{n})$, $YX \sim W(\mathfrak{m})$.
Then

$$\overline{YX \cdot W} = \overline{W}\,\overline{YX} = \overline{W}\,\overline{Y}\,\overline{X},$$

$$\overline{YW \cdot X} = \overline{YW} \cdot \overline{X} = \overline{W}\,\overline{Y}\,\overline{X},$$

from which follows $YXW = YWX$ and therefore $XW = WX$. Since $X \in \mathfrak{N}$ does not belong to any singly generated subsemigroup of $\mathfrak{W}_{\mathfrak{N}}$ other than $[X]$, it follows that $W \in [X]$, i.e., $W = X^k$.

Using 6.11 we obtain

$$\overline{X^{k+1}Y} = X^{k+1}\overline{Y} = \overline{X}^{k+1}\overline{Y} = \overline{X}^k\overline{X}\,\overline{Y} = \overline{W}\,\overline{XY} = \overline{XYW} = \overline{XYX^k},$$

which contradicts the fact that $X^{k+1}Y \neq XYX^k$.

(5) $Y \sim W(\mathfrak{m})$, $X \sim YW(\mathfrak{m})$, $X \sim WY(\mathfrak{n})$.
Then

$$\overline{W \cdot XY} = \overline{XY}\,\overline{W} = \overline{X}\,\overline{Y}\,\overline{W},$$

$$\overline{X \cdot WY} = \overline{X}\,\overline{WY} = \overline{X}\,\overline{Y}\,\overline{W},$$

from which follows $WXY = XWY$, i.e., $WX = XW$. Repeating the arguments of the preceding case we arrive at a contradiction.

(6) $Y \sim W(\mathfrak{m})$, $X \sim YW(\mathfrak{m})$, $X \sim WY(\mathfrak{m})$.
Then

$$\overline{W \cdot XY} = \overline{XY}\,\overline{W} = \overline{X}\,\overline{Y}\,\overline{W},$$

$$\overline{WY \cdot X} = \overline{X}\,\overline{WY} = \overline{X}\,\overline{Y}\,\overline{W}.$$

This contradicts the fact that $WXY \neq WYX$.

6.13. Using the properties obtained for the above relations in free semigroups, we are able, finally, to show that a free semigroup is defined by its subsemigroup characteristic.

THEOREM. *Let $\mathfrak{U}$ be a semigroup in which the subsemigroup characteristic $\Sigma(\mathfrak{U})$ is isomorphic to the subsemigroup characteristic $\Sigma(\mathfrak{W}_{\mathfrak{N}})$ of the free semigroup $\mathfrak{W}_{\mathfrak{N}}$. Then $\mathfrak{U}$ and $\mathfrak{W}_{\mathfrak{N}}$ are isomorphic.*

PROOF. (1) If the alphabet $\mathfrak{N}$ consists of a single element, then $\mathfrak{W}_{\mathfrak{N}}$ is an infinite singly generated semigroup and the truth of the assertion of the theorem in this case immediately follows from III, 7.9. So in what follows we can assume that $\mathfrak{N}$ contains at least two distinct elements $X, Y \in \mathfrak{N}$ $(X \neq Y)$.

Let φ be the isomorphism of $\Sigma(\mathfrak{W}_{\mathfrak{N}})$ onto $\Sigma(\mathfrak{U})$. As was shown in 6.8, φ induces a one-one transformation ψ of $\mathfrak{W}_{\mathfrak{N}}$ onto $\mathfrak{U}$,

$$\psi(W) = \overline{W} \qquad (W \in \mathfrak{W}_{\mathfrak{N}}),$$

by means of which in $\mathfrak{W}_{\mathfrak{N}}$ the two relations $\mathfrak{n}$ and $\mathfrak{m}$ (6.9) are defined.

(2) First we consider the case where $X \sim Y(\mathfrak{n})$. Let V and W be any two elements of $\mathfrak{W}_{\mathfrak{N}}$. We assume that $V \sim W(\mathfrak{m})$. Applying 6.12 we obtain

$$\overline{XY \cdot VW} = \overline{XY} \cdot \overline{VW} = \overline{X}\overline{Y}\overline{W}\overline{V}.$$

If $XYW \sim V(\mathfrak{n})$, then again using 6.12 we obtain

$$\overline{XYW \cdot V} = \overline{XYW}\overline{V} = \overline{XY}\overline{W}\overline{V} = \overline{X}\overline{Y}\overline{W}\overline{V}.$$

But $\overline{XYVW} = \overline{XYWV}$ implies $XYVW = XYWV$ and since $\mathfrak{W}_{\mathfrak{N}}$ has two-sided cancellation, $VW = WV$. From this, on the basis of 6.10 (α), we conclude that $V \sim W(\mathfrak{n})$.

The case where $XYV \sim W(\mathfrak{n})$ is completely analogous and we obtain $V \sim W(\mathfrak{n})$.

If neither $XYW \sim V(\mathfrak{n})$ nor $XYV \sim W(\mathfrak{n})$ holds, then by 6.10, (γ), $XYW \sim V(\mathfrak{m})$ and $XYV \sim W(\mathfrak{m})$. In this case, using 6.12 we obtain

$$\overline{V \cdot XYW} = \overline{XYW}\overline{V} = \overline{XY}\overline{W}\overline{V} = \overline{X}\overline{Y}\overline{W}\overline{V}.$$

This would mean that $\overline{VXYW} = \overline{XYVW}$, and hence $VXYW = XYVW$, and since $\mathfrak{W}_{\mathfrak{N}}$ has two-sided cancellation, $VXY = XYV$. Since V and XY turn out to permute, they must lie in some singly generated subsemigroup of $\mathfrak{W}_{\mathfrak{N}}$. But the only singly generated subsemigroup $\mathfrak{W}_{\mathfrak{N}}$ containing XY is $[XY]$. Therefore $V = (XY)^p$. In a completely analogous way we see that in this case $W = (XY)^q$. From this it follows that V and W permute and therefore by 6.10, (α) we have $V \sim W(\mathfrak{n})$.

By 6.10, (γ), V and W must be related by $\mathfrak{n}$ or $\mathfrak{m}$. As we have shown in this case, the latter implies the former. Thus for any V, W we must have $V \sim W(\mathfrak{n})$. By 6.9, ψ is an isomorphism of $\mathfrak{W}_{\mathfrak{N}}$ onto $\mathfrak{U}$.

(3) By 6.10, (γ), there remains for us to consider only the case where $X \sim Y(\mathfrak{m})$.

We consider the one-one map ξ of $\mathfrak{W}_{\mathfrak{N}}$ onto itself,

$$\xi(X_1X_2 \ldots X_n) = X_n \ldots X_2X_1 \qquad (X_1, X_2, \ldots, X_n \in \mathfrak{N}),$$

and the one-one map η of $\Sigma(\mathfrak{W}_{\mathfrak{N}})$ onto itself, $\eta\{U_\alpha, U_\beta, \ldots\} = \{\xi(U_\alpha), \xi(U_\beta), \ldots\}$ ($\{U_\alpha, U_\beta, \ldots\} \in \Sigma(\mathfrak{W}_{\mathfrak{N}})$). Evidently ξ is an anti-automorphism of $\mathfrak{W}_{\mathfrak{N}}$ (I, 1.17) and η is an automorphism of $\Sigma(\mathfrak{W}_{\mathfrak{N}})$.

The mapping $\varphi^* = \varphi\eta$ is an isomorphism of $\Sigma(\mathfrak{W}_{\mathfrak{N}})$ onto $\Sigma(\mathfrak{U})$. For this isomorphism we define by the method of 6.8 a mapping ψ^* of $\mathfrak{W}_{\mathfrak{N}}$ onto $\mathfrak{U}$ and relations $\mathfrak{n}^*$ and $\mathfrak{m}^*$ (6.9). Setting $\psi^*(W) = \overline{\overline{W}}$, we evidently have

$$\varphi^*[W] = \varphi\eta[W] = \varphi[\xi(W)] = [\overline{\xi(W)}],$$

$$\overline{\overline{W}} = \overline{\xi(W)}.$$

Since $X \sim Y(\mathfrak{m})$, therefore

$$\overline{\overline{XY}} = \overline{\xi(XY)} = \overline{YX} = \overline{X}\,\overline{Y},$$

$$\overline{\overline{X}} = \overline{\xi(X)} = \overline{X}, \qquad \overline{\overline{Y}} = \overline{\xi(Y)} = \overline{Y}, \qquad \overline{\overline{XY}} = \overline{\overline{X}}\,\overline{\overline{Y}}.$$

Thus for φ^* we have $X \sim Y(\mathfrak{n}^*)$. By the reasoning in the preceding part of the proof it follows that ψ^* is an isomorphism of $\mathfrak{W}_{\mathfrak{N}}$ onto $\mathfrak{U}$.

CHAPTER X

EMBEDDING OF SEMIGROUPS

1. Some Cases of Embedding

1.1. One of the methods of studying a given semigroup is to embed it in some supersemigroup which has certain additional properties that make possible a more successful study of the semigroup. Such an embedding can be carried out in many different ways. Which of them to use in each separate case, and to what extent it will be used, is determined by those properties of the initial semigroup which interest us and by those properties of the enveloping supersemigroup which contribute to the study of the given properties.

The problem of the embedding of semigroups of a certain class into semigroups of another class, besides the fact that it often serves as a useful device for the study of certain properties of semigroups, is of interest in itself, since it elucidates the interrelations of semigroups of different classes.

1.2. In order to consider the problem we must bear in mind that for a proof of the possibility of embedding a semigroup $\mathfrak{A}$ into a semigroup having some property, it is often sufficient to obtain an isomorphic mapping of $\mathfrak{A}$ into some semigroup $\mathfrak{A}'$ with the given property. In fact, by the method described in III, 1.8, we can then in $\mathfrak{A}'$ replace by the elements of $\mathfrak{A}$ those elements of $\mathfrak{A}'$ onto which the elements of $\mathfrak{A}$ are mapped under the given isomorphism. As a result of this we convert $\mathfrak{A}'$ into a new semigroup $\mathfrak{A}''$, which is a supersemigroup of $\mathfrak{A}$. Of course this achieves its aim only in the case where the property we are interested in has an "abstract character" and is not destroyed as a result of the indicated substitution.

1.3. We will repeatedly use the important property of regularity (II, 6.1). As we know (II, 6.1), the semigroup $\mathfrak{S}_\Omega$ of all transformations of the set Ω is regular. In I, 3.9 it was shown that any semigroup can be isomorphically mapped into a semigroup $\mathfrak{S}_\Omega$ for some Ω. Because of Remark 1.2 it follows that any semigroup can be embedded into a regular semigroup.

We must remark that in the embedding of a semigroup it is often desirable to add to it as few new elements as possible to obtain a supersemigroup of the kind required. In this respect the embedding of $\mathfrak{A}$ into the semigroup of all transformations, as in I, 3.9, can turn out to be very uneconomical. In the case of embedding into a regular supersemigroup that we are now considering, the situation can be somewhat improved in the following way. Having embedded $\mathfrak{A}$ into a regular supersemigroup $\mathfrak{A}'$, for each $A \in \mathfrak{A}$ we fix $\bar{A} \in \mathfrak{A}'$ which is regularly adjoint to A (II, 6.6; II, 6.7) and then we consider the subsemigroup $\mathfrak{A}_1$ generated by all the elements A of $\mathfrak{A}$ and all their fixed regularly conjugate elements $\bar{A}$. We obtain a supersemigroup $\mathfrak{A}_1$ of $\mathfrak{A}$ in which all the elements of $\mathfrak{A}$ will be regular (of course $\mathfrak{A}_1$ itself may not be regular). Then for $\mathfrak{A}_1$ we construct in the very same way a supersemigroup $\mathfrak{A}_2$ in which all the elements of $\mathfrak{A}_1$ will be regular, and so on.

As was remarked in III, 1.12, the set

$$\mathfrak{B} = \bigcup_n \mathfrak{A}_n$$

is a semigroup. Each of its elements will be regular, since it belongs to some $\mathfrak{A}_n$ and is therefore regular in $\mathfrak{A}_{n+1}$.

If $\mathfrak{A}$ was countable, then evidently $\mathfrak{A}_1$ will be countable; also $\mathfrak{A}_2$ will be countable, etc. As a result $\mathfrak{B}$ will be countable. At the same time the semigroup of all transformations $\mathfrak{S}_\Omega$ for countable Ω (by the device of I, 3.9 we embed $\mathfrak{A}$ into such a semigroup) has the cardinality of the continuum and therefore from the point of view of the principle formulated above it forms a less suitable supersemigroup for embedding $\mathfrak{A}$ than the $\mathfrak{B}$ just constructed. Of course, we can apply a similar method in many other embedding problems.

1.4. In some problems the interest lies in the existence in a semigroup of a generating set of the smallest possible cardinality. If the cardinality $\mathfrak{m}$ of a semigroup $\mathfrak{A}$ is more than countable, then any generating set for $\mathfrak{A}$ has cardinality $\mathfrak{m}$ and the cardinality of a generating set of any of its supersemigroups is not less than $\mathfrak{m}$. Thus in this case the method of embedding to obtain a reduction of the cardinality of the generating set is impossible. However if $\mathfrak{A}$ is finite or countable, then, as was shown by Evans [2], it can always be embedded in a semigroup with two generators. We prove this using a construction somewhat different from that of Evans.

THEOREM. *Any finite or countable semigroup is contained in a supersemigroup with a generating set of two elements.*

PROOF. Let $A_1, A_2, A_3, \ldots$ be all the elements of the semigroup $\mathfrak{A}$ (we consider the countable case; the finite case is analogous; further, a finite semigroup can always be embedded in a countable semigroup and thereby reduced to the countable case). We define a function $\varphi(i, j)$ $(i, j = 1, 2, 3, \ldots)$ by letting

$$A_i A_j = A_{\varphi(i,j)}.$$

Clearly $\{A_1, A_2, A_3, \ldots\}$ is a generating set of $\mathfrak{A}$ and the indicated equalities form a defining system of relations with respect to this generating set.

We consider the semigroup $\mathfrak{B}$ given by the generating set $\mathfrak{N} = \{X_1, X_2, \ldots, Y, Z\}$ whose defining system of relations (IX, 3.10) consists of the relations of three types:

$$(\alpha)\ X_iX_j = X_{\varphi(i,j)}; \qquad (\beta)\ Y^2 = Y;$$

$$(\gamma)\ X_i = YZ^iY \qquad (i, j = 1, 2, 3, \ldots).$$

Since $X_i = YZ^iY$ and $Y^2 = Y$, from our relations follow the corollaries,

$$X_iY = X_i, \qquad YX_i = X_i.$$

We consider the three following special types of words:

$$(1)\ X_{\alpha_1}X_{\alpha_2}\ldots X_{\alpha_m}; \qquad (2)\ Y^p; \qquad (3)\ Z^q.$$

Each word in $\mathfrak{N}$ can be represented as a product of words (IX, 1.2) of these three types in the manner

$$W = U_1U_2\ldots U_n,$$

where no two neighboring U_i are of the same type. We denote by $\mathfrak{M}$ the totality of all words in $\mathfrak{N}$ not belonging to the second type, whose first and last factors belong to the first or second type. We note, as a result, that under any mapping by relations of our defining system, words of $\mathfrak{M}$ are carried to words of $\mathfrak{M}$.

We consider the operator σ applied to each word W of $\mathfrak{M}$ given in the indicated form. We put

$$\sigma(W) = \sigma(U_1)\cdot\sigma(U_2)\cdot\ldots\cdot\sigma(U_n),$$

where $\sigma(U_i) = U_i$ if U_i belongs to the first special type of words; $\sigma(U_i)$ is the empty symbol if U_i belongs to the second type; $\sigma(U_i) = X_q$ if $U_i = Z^q$ belongs to the third type. The values of the words W and $\sigma(W)$ in $\mathfrak{B}$ are identical. In fact, if $U_i = Z^q$, then by means of the relations in the defining system and by their above corollaries, it is possible in W, without changing its value, to replace U_{i-1} by $U_{i-1}Y$ and U_{i+1} by YU_{i+1}. After that it is possible to replace by X_q the product YZ^qY formed in the word. As a result of all this, in the word $W = U_1U_2\ldots U_i\ldots U_n$ we replace $U_i = Z^q$ by $\sigma(U_i) = X_q$. We proceed thus with all the U_i belonging to the third type. After this we can simply eliminate all the remaining Y's (i.e., replace them by empty symbols) because of the above corollaries of the defining system of relations.

As we know (IX, 3.10), from the form in which $\mathfrak{B}$ is given it is not immediately clear whether given X_i and X_j can be equal as elements of the semigroup $\mathfrak{B}$. Let us suppose that this happens for some X_i and X_j. This means that there is a sequence of words

$$X_i \equiv V_1, V_2, \ldots, V_{s-1}, \qquad V_s \equiv X_j$$

in $\mathfrak{N}$ such that $V_k = V_{k+1}$ $(k = 1, 2, \ldots, s-1)$ is an immediate corollary of the defining system of relations. Since the first word belongs to $\mathfrak{M}$, all the other words of this sequence belong to $\mathfrak{M}$. We consider the sequence of words

$$\sigma(V_1) = \sigma(X_i) = X_i, \qquad \sigma(V_2), \ldots, \sigma(V_{s-1}), \qquad \sigma(V_s) = \sigma(X_j) = X_j.$$

We consider any two of these words which are neighbors, representing them in the form indicated above as a product of words of the three special types:

$$V_k \equiv U_1'U_2' \ldots U_p', \qquad \sigma(V_k) = \sigma(U_1') \cdot \sigma(U_2') \cdot \ldots \cdot \sigma(U_p'),$$
$$V_{k+1} \equiv U_1''U_2'' \ldots U_q'', \qquad \sigma(V_{k+1}) = \sigma(U_1'') \cdot \sigma(U_2'') \cdot \ldots \cdot \sigma(U_q'').$$

V_{k+1} is obtained from V_k by a single replacement of some part of the word V_k which is one of the parts of a relation in the defining system by another part of that relation. If the relation belongs to the second or third type of relations, then clearly $\sigma(V_k)$ will entirely coincide with $\sigma(V_{k+1})$. If the relation is of the first type, then $\sigma(V_{k+1})$ is obtained from $\sigma(V_k)$ by means of a transformation by a relation of type (α).

Thus the sequence $\sigma(V_1), \sigma(V_2), \ldots, \sigma(V_s)$ represents a sequence of words in $\{X_1, X_2, \ldots\}$ in which each term is equal to the preceding term or is obtained from it by means of a transformation by some relation of the form $X_iX_j = X_{\varphi(i,j)}$. Replacing each X_k in the words of this sequence by A_k, we obtain a sequence of words in $\{A_1, A_2, \ldots\}$ in which any two neighboring words have the same value in $\mathfrak{A}$. Consequently, A_i and A_j must also have the same value. Since for $i \neq j$, A_i and A_j are different in $\mathfrak{A}$, we arrive at the result that $i = j$. Carrying this over to $\mathfrak{B}$ we conclude that X_i can equal X_j in $\mathfrak{B}$ only for $i = j$.

We consider the one-to-one mapping ψ of $\mathfrak{A}$ into $\mathfrak{B}$,

$$\psi(A_i) = X_i \qquad (i = 1, 2, \ldots).$$

If $A_iA_j = A_{\varphi(i,j)}$, then $X_iX_j = X_{\varphi(i,j)}$. Therefore ψ is an isomorphism of $\mathfrak{A}$ into $\mathfrak{B}$ with generating set $\{Y, Z\}$. Replacing in $\mathfrak{B}$ all the X_i's by the A_i's we obtain a supersemigroup of $\mathfrak{A}$ with a two-element generating set.

1.5. The problem of embedding a semigroup into a supersemigroup with given properties is often complicated by an additional condition, that under such an embedding we do not lose some other important properties of the initial semigroup. For example, if we consider the construction by which we proved Theorem 1.4, we note that the constructed supersemigroup $\mathfrak{B}$ may not have some of the intrinsic properties that $\mathfrak{A}$ has. For example, $\mathfrak{A}$ can be commutative, whereas $\mathfrak{B}$ is noncommutative (any of the relations of the defining system of $\mathfrak{B}$ maps the word Y^nZ into the word $Y^{n\pm1}Z$ and hence no sequence of maps carries YZ into ZY). $\mathfrak{A}$ can be left cancellative (right cancellative), while in $\mathfrak{B}$,

$$Y \cdot ZY = X_1, \qquad YX_1 = X_1$$

but $ZY \neq X_1$.

The problem of the existence of supersemigroups belonging to some class to which the initial semigroup also belongs can be very difficult. We mention in this connection that Evans [3] has considered the problem of embedding a semigroup into a supersemigroup having a generating set with a given number of elements under the additional condition that both the initial semigroup and the desired supersemigroup be cancellative.

1.6. Numerous investigations have been made on this problem of embedding semigroups. We mention the work of Cohn [**1; 2**] and E. G. Šutov [**1; 2; 3; 4**]. In the following section we turn our attention to the embedding of semigroups in groups. We now consider an example of a slightly different statement of the problem of embedding.

The importance of the relation of left divisibility and right divisibility was shown repeatedly above.

Let A and B be any two elements of a semigroup $\mathfrak{A}$. We ask when there exists a supersemigroup $\mathfrak{A}'$ of $\mathfrak{A}$ in which A is divisible by B from the right. It is immediately evident that the following condition is necessary for this. If such a semigroup $\mathfrak{A}'$ exists, then in $\mathfrak{A}$ for all X and Y, where X and Y are either elements of $\mathfrak{A}$ or empty symbols, $BX = BY$ must always imply $AX = AY$. In fact, by assumption, there exists $Z \in \mathfrak{A}'$ such that $A = ZB$. Therefore $BX = BY$ in $\mathfrak{A}'$ implies $ZBX = ZBY$, i.e., $AX = AY$. But this equality, being valid in $\mathfrak{A}'$, is likewise valid in $\mathfrak{A}$.

It turns out that the condition is also sufficient. In fact, let it be satisfied for A and B. We consider the semigroup of left shifts $\mathfrak{T}_{\mathfrak{A}}$ corresponding to the elements of $\mathfrak{A}$ (I, 3.9). Since in $\mathfrak{A}$, $BX = BY$ always implies $AX = AY$, therefore in $\mathfrak{T}_{\mathfrak{A}}$ for the corresponding left shifts, $T_B X = T_B Y$ always implies $T_A X = T_A Y$ (here X and Y can also be equal to the separating element I, since in the preceding equalities in $\mathfrak{A}$ the X and Y can be not only elements of $\mathfrak{A}$ but empty symbols).

$\mathfrak{T}_{\mathfrak{A}}$ is a subsemigroup of the semigroup $\mathfrak{S}_{\Lambda_{\mathfrak{A}}}$ of all transformations of the set $\Lambda_{\mathfrak{A}}$. By II, 3.1, T_A is divisible from the right by T_B in $\mathfrak{S}_{\Lambda_{\mathfrak{A}}}$. If in $\mathfrak{S}_{\Lambda_{\mathfrak{A}}}$ by the method of III, 1.8 we replace the left shifts of $\mathfrak{T}_{\mathfrak{A}}$ by the corresponding elements of $\mathfrak{A}$, we obtain a semigroup in which A is divisible from the right by B.

1.7. Similarly we see that for the existence of a supersemigroup $\mathfrak{A}''$ of $\mathfrak{A}$ in which A is divisible by B from the left, it is necessary that in $\mathfrak{A}$, $XB = YB$ always imply $XA = YA$. However, for the proof of the sufficiency of this condition it does not help us to represent $\mathfrak{A}$ by left shifts. Instead we can use the representation of $\mathfrak{A}$ by right shifts (I, 3.12). Let $XB = YB$ imply $XA = YA$ in $\mathfrak{A}$. Then for the corresponding right shifts in $\mathfrak{T}^{\circ}_{\mathfrak{A}}$, $T^{\circ}_B X = T^{\circ}_B Y$ implies $T^{\circ}_A X = T^{\circ}_A Y$.

$\mathfrak{T}^{\circ}_{\mathfrak{A}}$ is a subsemigroup of $\mathfrak{S}^{\circ}_{\Lambda_{\mathfrak{A}}}$. This semigroup is anti-isomorphic to $\mathfrak{S}_{\Lambda_{\mathfrak{A}}}$. Therefore from II, 3.1 it is easily shown that the indicated condition is satisfied for the maps T°_A and T°_B. This means that in $\mathfrak{S}^{\circ}_{\Lambda_{\mathfrak{A}}}$ the former is divisible from the left by the latter. Replacing the right shifts in $\mathfrak{S}^{\circ}_{\Lambda_{\mathfrak{A}}}$ by the corresponding elements of $\mathfrak{A}$, we obtain a supersemigroup of $\mathfrak{A}$ in which A will be divisible by B from the right. We remark that a different approach to this same problem was used by Cohn [**1; 2**].

1.8. In connection with the preceding arguments, it is interesting to note that the semigroup of all transformations $\mathfrak{S}_{\Omega}$ on some set (as well as its anti-isomorphic semigroup $\mathfrak{S}^{\circ}_{\Omega}$) has the following property. If A is not divisible by B from the right in $\mathfrak{S}_{\Omega}$ ($A, B \in \mathfrak{S}_{\Omega}$), then A will not be divisible by B from the right in any supersemigroup $\mathfrak{S}'$ of $\mathfrak{S}_{\Omega}$. The analogous statement is true for divisibility from the left.

Indeed, since A is not divisible by B from the right in $\mathfrak{S}_{\Omega}$, it follows from II, 3.1 that for some $\alpha, \beta \in \Omega$ we have $A\alpha \neq A\beta$ and $B\alpha = B\beta$. We consider the maps U_{λ} ($\lambda \in \Omega$), where $U_{\lambda}\xi = \lambda$ for all $\xi \in \Omega$. We have

$$AU_{\alpha} = U_{(A\alpha)} \neq U_{(A\beta)} = AU_{\beta},$$
$$BU_{\alpha} = U_{(B\alpha)} = U_{(B\beta)} = BU_{\beta}.$$

Therefore, by 1.6, A is not divisible by B from the right in any supersemigroup $\mathfrak{S}'$ of $\mathfrak{S}_{\Omega}$.

If A is not divisible by B from the left in $\mathfrak{S}_{\Omega}$, then by II, 3.2, there exists $\lambda \in \Omega$ such that $\mu = A\lambda \in B\Omega$. We take the identity map E and a map I which takes each $\xi \in \Omega$ ($\xi \neq \mu$) to ξ, but for which $I\mu \neq \mu$.

Since

$$IA\lambda = I\mu \neq \mu = A\lambda$$

it follows that

$$IA \neq A = EA.$$

But for any $\xi \in B\Omega$ we have $I\xi = \xi$, and therefore $IB = EB$. By 1.7, A cannot be divisible from the left by B in any supersemigroup $\mathfrak{N}'$ of $\mathfrak{S}_{\Omega}$.

2. Embedding in Groups

2.1. Among the various problems of embedding, the greatest attention has been attracted to the problem of embedding a semigroup into a group. This is entirely natural. The theory of groups is considerably better developed than the general theory of semigroups. Therefore the possibility of embedding a given semigroup into a group, allowing us to investigate not the initial semigroup but the group containing it, opens up wide possibilities for the application of results and methods of the theory of groups. On the other hand, semigroups which are embedded in groups are of obvious interest in the theory of groups. Each of them is a closed subset of some group with respect to multiplication, so that examination of it from different points of view is essential for the study of properties of that group. We also mention that the class of semigroups which are embeddable in groups can be defined as the maximal class for which the class of all groups is a universal class (III, 1.9; III, 1.10).

2.2. It must be noted, however, that it is not always expedient to pass from the study of a semigroup which is embeddable in a group to the study of this group itself. The fact is, a semigroup $\mathfrak{A}$ can frequently interest us as a subsemigroup of some supersemigroup $\mathfrak{A}'$, where it can happen that $\mathfrak{A}$ is a semigroup embeddable in a group but there does not exist any supersemigroup $\mathfrak{A}''$ of $\mathfrak{A}'$ in which there is a subgroup containing $\mathfrak{A}$. In this case it is impossible, without giving up the idea of considering $\mathfrak{A}$ as a subsemigroup of $\mathfrak{A}'$, to reduce its study to the study of any group.

We mention an example of this possibility, which was constructed by E. G. Šutov.

Let $\mathfrak{B}$ be the multiplicative semigroup of natural numbers. We consider the supersemigroup $\mathfrak{A}$ of $\mathfrak{B}$ which contains besides $\mathfrak{B}$ two elements U and V. Multiplication in $\mathfrak{A}$, besides the rule of multiplication in $\mathfrak{B}$, is defined by the conditions:

(1) $UN = U$, $VN = U$ if N is even;

(2) $UN = U$, $VN = V$ if N is not even;

(3) $UV = V$, $VU = U$, $U^2 = U$, $V^2 = V$;

(4) $NU = U$, $NV = V$ $(N \in \mathfrak{B})$.

We can show without difficulty that this operation is associative. $\mathfrak{B}$ is embeddable in a group. At the same time no supersemigroup of $\mathfrak{A}$ contains a subgroup containing $\mathfrak{B}$. In fact, if there were such a semigroup it would contain an element X such that $2 \cdot X = 1$ and we would obtain in $\mathfrak{A}$,

$$V = V \cdot 1 = V \cdot 2 \cdot X = UX = U \cdot 2 \cdot X = U \cdot 1 = U.$$

2.3. Since any group is a semigroup with two-sided cancellation it follows that any semigroup which is embeddable in a group must be a semigroup with two-sided cancellation. The solution of the converse problem, whether two-sided cancellation is sufficient for embedding a semigroup into a group, proved to be difficult. A negative answer was first obtained in 1937 by A. I. Mal'cev **[1]**, who constructed a semigroup with two-sided cancellation which was not embeddable in a group. We will consider a similar example below.

2.4. Let $\mathfrak{A}$ be a semigroup. Let $\mathfrak{K}$ be a generating set for $\mathfrak{A}$, and let $\Phi(\mathfrak{K})$ be a defining system of relations with respect to $\mathfrak{K}$. We consider two new sets $\mathfrak{K}'$ and $\mathfrak{K}''$, each of which consists of elements placed in one-to-one correspondence with the elements of $\mathfrak{K}$ (the element of $\mathfrak{K}'$ corresponding to $K \in \mathfrak{K}$ will be denoted by K' and the corresponding element of $\mathfrak{K}''$ by K''). We define the semigroup $\mathfrak{G}_{\mathfrak{A}}$ as the semigroup with generating set $\mathfrak{K}' \cup \mathfrak{K}'' \cup E$, where the element E belongs neither to $\mathfrak{K}'$ nor $\mathfrak{K}''$, and with defining system of relations (IX, 3.10) consisting of the relations $\Phi(\mathfrak{K}')$ (i.e., the relations obtained from the relations of $\Phi(\mathfrak{K})$, replacing the elements of $\mathfrak{K}$ by the corresponding elements of $\mathfrak{K}'$) and relations of the form

$$EK' = K'E = K', \qquad EK'' = K''E = K'', \qquad E^2 = E,$$
$$K'K'' = K''K' = E \qquad (K \in \mathfrak{K}).$$

$\mathfrak{G}_{\mathfrak{A}}$ is a group, since for any of its elements $X_1X_2 \dots X_n$ $(X_1, X_2, \dots, X_n \in \mathfrak{K}' \cup \mathfrak{K}'' \cup E)$ E is a two-sided unit, and $Y_nY_{n-1} \dots Y_2Y_1$ (where $Y_i = K'$ if $X_i = K'' \in \mathfrak{K}''$, $Y_i = K''$ if $X_i = K' \in \mathfrak{K}'$, and $Y_i = E$ if $K_i = E$) is the inverse element with respect to E.

We define φ mapping $\mathfrak{A}$ into $\mathfrak{G}_{\mathfrak{A}}$ by

$$\varphi(K_1K_2\dots K_n) = K_1'K_2'\dots K_n' \qquad (K_1, K_2, \dots, K_n \in \mathfrak{K}).$$

This mapping is single-valued since the image of $A \in \mathfrak{A}$ does not depend on the choice of the product of elements of $\mathfrak{K}$ by means of which A is expressed. In fact, if the products $K_1K_2\dots K_n$ and $K_{\alpha_1}K_{\alpha_2}\dots K_{\alpha_m}$ represent one and the same element $A \in \mathfrak{A}$, then the relation $K_1K_2\dots K_n = K_{\alpha_1}K_{\alpha_2}\dots K_{\alpha_m}$ is a corollary of $\Phi(\mathfrak{K})$. But then $K_1'K_2'\dots K_n' = K_{\alpha_1}'K_{\alpha_2}'\dots K_{\alpha_m}'$ is a corollary of $\Phi(\mathfrak{K}')$ and therefore $K_1'K_2'\dots K_n'$ and $K_{\alpha_1}'K_{\alpha_2}'\dots K_{\alpha_m}'$ represent one and the same element of the group $\mathfrak{G}_{\mathfrak{A}}$. Clearly φ is a homomorphism of $\mathfrak{A}$ into $\mathfrak{G}_{\mathfrak{A}}$.

The importance of the group $\mathfrak{G}_{\mathfrak{A}}$ as far as the semigroup $\mathfrak{A}$ is concerned depends on the character of the relations in $\mathfrak{G}_{\mathfrak{A}}$ between the elements of $\mathfrak{K}'$.

We suppose that in $\mathfrak{G}_{\mathfrak{A}}$, any relation between the elements of $\mathfrak{K}'$ is a corollary of the relations of $\Phi(\mathfrak{K}')$. Then φ will be one-to-one. Indeed, from

$$\varphi(K_1K_2\dots K_n) = K_1'K_2'\dots K_n',$$
$$\varphi(K_{\alpha_1}K_{\alpha_2}\dots K_{\alpha_m}) = K_{\alpha_1}'K_{\alpha_2}'\dots K_{\alpha_m}',$$
$$K_1'K_2'\dots K_n' = K_{\alpha_1}'K_{\alpha_2}'\dots K_{\alpha_m}'$$

it follows by our assumption that the latter relation can be obtained as a corollary of $\Phi(\mathfrak{K}')$. But then the relation $K_1K_2\dots K_n = K_{\alpha_1}K_{\alpha_2}\dots K_{\alpha_m}$ can be obtained as a corollary of $\Phi(\mathfrak{K})$, i.e., $K_1K_2\dots K_n$ and $K_{\alpha_1}K_{\alpha_2}\dots K_{\alpha_m}$ are one and the same element of $\mathfrak{A}$. Being a homomorphism, φ turns out in our case to be an isomorphism of $\mathfrak{A}$ into the group $\mathfrak{G}_{\mathfrak{A}}$. Thus under our assumption, $\mathfrak{A}$ is embeddable in a group.

Now we assume that $\mathfrak{A}$ can be embedded in some group $\mathfrak{F}$. We define a map ψ of the generating set $\mathfrak{K}' \cup \mathfrak{K}'' \cup E$ of the group $\mathfrak{G}_{\mathfrak{A}}$ into $\mathfrak{F}$ by

$$\psi(K') = K, \qquad \psi(K'') = K^{-1}, \qquad \psi(E) = E_{\mathfrak{F}}$$

(K^{-1} is the inverse of K in $\mathfrak{F}$). Under this map all the relations in the initial defining system of relations of the group $\mathfrak{G}_{\mathfrak{A}}$ are mapped into relations which are valid in $\mathfrak{F}$. By IX, 1.11, ψ can be extended to a homomorphism ξ of the group $\mathfrak{G}_{\mathfrak{A}}$ into $\mathfrak{F}$.

Let

$$K_1'K_2'\dots K_p' = K_{\alpha_1}'K_{\alpha_2}'\dots K_{\alpha_q}'$$

be any relation between the elements of $\mathfrak{K}'$ in the group $\mathfrak{G}_{\mathfrak{A}}$. Since ξ is a homomorphism into $\mathfrak{F}$ we must have the corresponding relation

$$K_1K_2\dots K_p = K_{\alpha_1}K_{\alpha_2}\dots K_{\alpha_q}.$$

This relation, being a relation for $\mathfrak{K}$ in $\mathfrak{A}$, must be a corollary of $\Phi(\mathfrak{K})$. But since the relations of $\Phi(\mathfrak{K}')$ hold in $\mathfrak{G}_{\mathfrak{A}}$, the relation

$$K_1'K_2'\dots K_p' = K_{\alpha_1}'K_{\alpha_2}'\dots K_{\alpha}'$$

will be a corollary of the relations of $\Phi(\mathfrak{K}')$ in $\mathfrak{G}_{\mathfrak{A}}$.

This reasoning shows that *a necessary and sufficient condition for the embedding of a semigroup $\mathfrak{A}$ into a group is that any relation in $\mathfrak{G}_{\mathfrak{A}}$ between the elements of $\mathfrak{K}'$ can be obtained as a corollary of the relations of* $\Phi(\mathfrak{K}')$.

2.5. The criterion for embedding in a group which was obtained in 2.4 is not very effective, and can be developed to some extent. For example, Pták has done research in a similar direction [**1**; **2**; **3**]. A. I. Mal'cev [**2**] penetrated deeper in this territory and has formulated a more refined criterion. However, this criterion is also far from effective and is qualitatively of the same character as the criterion of 2.4. The reasons for this situation were to a great extent explained by A. I. Mal'cev [**3**], who showed that no finite number of conditions, belonging to a class of conditions which he described and containing, in particular, the law of cancellation, can form a necessary and sufficient condition for the embedding of an arbitrary semigroup into a group.

In this connection an important role is played by various sufficient criteria which were obtained for the embedding of a semigroup into a group. These are found in a series of articles by different authors: Ore [**1**], Dubreil-Jacotin [**1**], Lambek [**1**], Doss [**1**], Tamari [**5**], Pták [**1**; **2**; **3**].

2.6. Let us consider Ore's criterion [**1**]. We examine the argument of Rees, who deduced this criterion by considering semigroups of partial transformations.

Let $\mathfrak{H}$ be a semigroup of one-to-one partial transformations of some set Ω which, for each partial transformation $X \in \mathfrak{H}$, contains the inverse $\bar{X}$ of that partial transformation (I, 4.5). We introduce in $\mathfrak{H}$ the relation $\mathfrak{n}$ by setting $X \sim Y(\mathfrak{n})$ if for $X, Y \in \mathfrak{H}$ there exists $T \in \mathfrak{H}$ such that $T \leqslant X$ and $T \leqslant Y$ (I, 4.3). This relation is clearly symmetric and reflexive. We prove that it is transitive.

Let

$$T_1 \leqslant X, \qquad T_1 \leqslant Y,$$
$$T_2 \leqslant Y, \qquad T_2 \leqslant Z \qquad (X, Y, Z, T_1, T_2 \in \mathfrak{H}).$$

We take the element

$$U = T_1\bar{T}_2T_2 \in \mathfrak{H}.$$

By the definition of multiplication of partial transformation it follows that

$$\Pi_1^{(U)} \subset \Pi_1^{(T_1)}, \qquad \Pi_1^{(U)} \subset \Pi_1^{(T_2)}.$$

For each $\alpha \in \Pi_1^{(U)}$ we have

$$U\alpha = T_1[\bar{T}_2(T_2\alpha)] = T_1\alpha = Y\alpha = T_2\alpha$$

and therefore

$$U \leqslant T_1 \leqslant X, \qquad U \leqslant T_2 \leqslant Z,$$

whence $X \sim Z(\mathfrak{n})$. The relation introduced is a two-sidedly stable equivalence. In fact, if

$$X \sim Y(\mathfrak{n}), \qquad T \leqslant X, \qquad T \leqslant Y \qquad (X, Y, T \in \mathfrak{H}),$$

then for any $Z \in \mathfrak{H}$,

$$TZ \leqslant XZ, \qquad TZ \leqslant YZ, \qquad ZT \leqslant ZX, \qquad ZT \leqslant ZY,$$

i.e., $XZ \sim YZ(\mathfrak{n})$ and $ZX \sim ZY(\mathfrak{n})$.

The semigroup $\mathfrak{H}/\mathfrak{n}$ (VII, 2.4) is a group. In fact, for any $\tilde{A}, \tilde{B} \in \mathfrak{H}/\mathfrak{n}$ (where $\tilde{S}$ denotes the class of elements of $\mathfrak{H}$ which are equivalent to S with respect to $\mathfrak{n}$) for $X = A\bar{B}$ and $Y = \bar{B}A$ we have

$$XB = A\bar{B}B \leqslant A, \qquad BY = B\bar{B}A \leqslant A,$$

whence $XB \sim A(\mathfrak{n})$, $BY \sim A(\mathfrak{n})$, and therefore

$$\tilde{X}\tilde{B} = \widetilde{XB} = \tilde{A}, \qquad \tilde{B}\tilde{Y} = \widetilde{BY} = \tilde{A}.$$

2.7. Let us use this construction to deduce one sufficient condition for embedding a semigroup into a group.

THEOREM. *Let $\mathfrak{A}$ be a semigroup with two-sided cancellation in which for any $A, B \in \mathfrak{A}$ there always exist $U, V \in \mathfrak{A}$ such that*

$$AU = BV.$$

Then $\mathfrak{A}$ is embeddable in a group.

PROOF. $\mathfrak{T}_{\mathfrak{A}}$ is the semigroup of all partial one-to-one transformations on all of $\mathfrak{A}$.

With each $A \in \mathfrak{A}$ we associate the partial transformation S_A on the set $\mathfrak{A}$ such that

$$\Pi_1^{(S_A)} = \mathfrak{A}, \qquad S_A X = AX \qquad (X \in \mathfrak{A}).$$

S_A is one-to-one since by the hypothesis of the theorem the equality $S_A X = S_A Y$ (i.e., $AX = AY$) is only possible when $X = Y$. $\Pi_1^{(S_A)}$ and $\Pi_2^{(S_A)} = A\mathfrak{A}$ are right ideals of $\mathfrak{A}$.

For any $X, Y \in \mathfrak{A}$

$$S_A(XY) = (S_A X) Y.$$

For the inverse $\bar{S}_A$ we have

$$\Pi_1^{(\bar{S}_A)} = \Pi_2^{(S_A)}, \qquad \Pi_2^{(\bar{S}_A)} = \Pi_1^{(S_A)}$$

and therefore $\Pi_1^{(\bar{S}_A)}$ and $\Pi_2^{(\bar{S}_A)}$ are also right ideals of $\mathfrak{A}$. If $X \in \Pi_1^{(\bar{S}_A)}$, i.e., $X = AZ$ for some $Z \in \mathfrak{A}$, then for arbitrary $Y \in \mathfrak{A}$

$$XY = AZY = S_A(ZY),$$

i.e.,

$$\bar{S}_A(XY) = ZY = (\bar{S}_A X) \cdot Y.$$

In $\mathfrak{T}_{\mathfrak{A}}$ we consider the subsemigroup $\mathfrak{H}$, generated by all S_A and $\bar{S}_A$ $(A \in \mathfrak{A})$. For any element of that subsemigroup

$$H = H_1 H_2 \dots H_m \qquad (H_i = S_{A_i} \text{ or } H_i = \bar{S}_{A_i};\ i = 1, 2, \dots, m)$$

we show by induction on m that $\Pi_1^{(H)}$ and $\Pi_2^{(H)}$ are right ideals of $\mathfrak{A}$. The case $m = 1$ was examined above. Let $m > 1$, let $X \in \Pi_1^{(H)}$, and let $Y \in \mathfrak{A}$. Using the inductive assumption and the property of partial transformations proven above we obtain

$$X \in \Pi_1^{(H)} \subset \Pi_1^{(H_2H_3\ldots H_m)}, \qquad XY \in \Pi_1^{(H_2H_3\ldots H_m)}$$

$$(H_2H_3\ldots H_m)(XY) = [(H_2H_3\ldots H_m)X]\cdot Y;$$

$$(H_2H_3\ldots H_m)\cdot X \in \Pi_1^{(H_1)}, \qquad [(H_2H_3\ldots H_m)X]\cdot Y \in \Pi_1^{(H_1)},$$

whence

$$(H_2H_3\ldots H_m)(XY) \in \Pi_1^{(H_1)},$$

i.e., $XY \in \Pi_1^{(H)}$. Thus $\Pi_1^{(H)}$ is either a right ideal or the empty set. In order to eliminate the second possibility, we choose elements

$$A \in \Pi_2^{(H_2H_3\ldots H_m)}, \qquad B \in \Pi_1^{(H_1)}.$$

By the hypothesis of the theorem there must exist $U, V \in \mathfrak{A}$ such that

$$AU = BV = W.$$

Since $\Pi_2^{(H_2H_3\ldots H_m)}$ and $\Pi_1^{(H_1)}$ are right ideals, we have

$$W \in \Pi_2^{(H_2H_3\ldots H_m)}, \qquad W \in \Pi_1^{(H_1)}.$$

The first of these inclusions means that

$$(H_2H_3\ldots H_m)C = W$$

for some $C \in \mathfrak{A}$.

Taking into account the second inclusion, we conclude that C belongs to $\Pi_1^{(H_1H_2\ldots H_m)} = \Pi_1^{(H)}$, which shows that $\Pi_1^{(H)}$ is nonempty. Clearly

$$\Pi_2^{(H_1H_2\ldots H_m)} = \Pi_1^{(\bar{H}_m\ldots\bar{H}_2\bar{H}_1)},$$

and hence $\Pi_2^{(H)}$ is also a right ideal of $\mathfrak{A}$. Since the partial transformation $\bar{H} = \bar{H}_m\ldots\bar{H}_2\bar{H}_1$, which is the inverse of $H = H_1H_2\ldots H_m$, also belongs to $\mathfrak{H}$, it follows that in $\mathfrak{H}$ by the method of 2.6 we can define the two-sidedly stable relation $\mathfrak{n}$. By 2.6 the factor semigroup $\mathfrak{H}/\mathfrak{n}$ is a group. We denote the natural homomorphism of $\mathfrak{H}$ onto $\mathfrak{H}/\mathfrak{n}$ by χ.

We consider the map ψ of $\mathfrak{A}$ into $\mathfrak{H}$,

$$\psi(A) = S_A \qquad (A \in \mathfrak{A}).$$

Since for any $A, B, X \in \mathfrak{A}$

$$(S_AS_B)X = S_A(S_BX) = S_A(BX) = ABX = S_{AB}X$$

it follows that ψ is a homomorphism of $\mathfrak{A}$ into $\mathfrak{H}$. We consider the product of the homomorphisms $\xi = \chi\psi$. Clearly ξ is a homomorphism of $\mathfrak{A}$ into the group $\mathfrak{H}/\mathfrak{n}$. Let us suppose that for some $A, B \in \mathfrak{A}$ we have

$$\xi(A) = \xi(B), \qquad A \neq B.$$

This means $S_A \sim S_B(\mathfrak{n})$, i.e., there exists a partial transformation $T \in \mathfrak{H}$ such that $T \leqslant S_A$ and $T \leqslant S_B$. We take some element $Z \in \Pi_1^{(T)}$. For this element we have

$$S_A Z = TZ, \qquad S_B Z = TZ.$$

But

$$S_A Z = AZ, \qquad S_B Z = BZ,$$

and the equality $AZ = BZ$ is impossible. The contradiction obtained shows that ψ is one-to-one, i.e., an isomorphism of $\mathfrak{A}$ into the group $\mathfrak{H}/\mathfrak{n}$.

2.8. Among the semigroups satisfying the condition of the theorem considered are the commutative semigroups. In fact, in a commutative semigroup for any A, B we have $AU = BV$ for $U = B$ and $V = A$. As a result we obtain the following corollary.

COROLLARY. *Any cancellative commutative semigroup is embeddable in a group.*

Of course this important result can be proven immediately.

2.9. Let us point out another particular case of the theorem, which includes the result in 2.8.

COROLLARY. *Any semigroup with two-sided cancellation satisfying the commutative condition* (IV, 6.7) *is embeddable in a group.*

In fact, for elements A and B of such a semigroup we can take $U = B$ and $V = R_{AB}$.

For such a choice we have

$$AU = BV.$$

2.10. In connection with this study of semigroups which are embeddable in groups, let us point out an interesting class of such semigroups. Let Ω be the n-dimensional Euclidean space (it is possible to take a space of more general form). Let $\mathfrak{G}$ be the group of motions of the space Ω, considered as transformations of Ω. Let Γ be some subset of Ω. We denote by $\mathfrak{A}_\Gamma$ the system of all motions X in $\mathfrak{G}$ for which $X\Gamma \subset \Gamma$. Clearly $\mathfrak{A}_\Gamma$ is a semigroup, and it is embeddable in a group ($\mathfrak{A}_\Gamma \subset \mathfrak{G}$). The algebraic properties of the semigroup $\mathfrak{A}_\Gamma$ to a certain extent characterize the geometric properties of the set Γ. The problem of deciding in what cases and to what extent it is possible to establish connections between these two types of properties has apparently not been investigated up to now. But it seems extremely probable that inclusion relations exist which will have valuable applications to the theory of semigroups.

3. Linear Ordering in Groups

3.1. In the investigation of certain classes of semigroups or of particular concrete semigroups it is sometimes necessary to take account of the fact that a partial ordering of their elements is defined in some natural way. For example, in § 7 of the second chapter a partial ordering naturally arises in inverse semigroups. In the holoidal semigroups considered in § 4 of the eighth chapter, we have a linear ordering. From the point of view of the general theory of semigroups, the interesting cases are those in which the partial ordering in the semigroup is compatible with the multiplication, i.e., it is a two-sidedly stable relation. In the first of the above cases this, as we have proven, indeed takes place. In the second case we cannot generally give an answer. However, in some particular cases, as will be proven below, two-sided stability takes place.

3.2. DEFINITION. *If in the set of elements of a semigroup $\mathfrak{A}$ there is defined a two-sidedly stable relation of partial ordering, then with respect to this partial ordering, $\mathfrak{A}$ is called a* PARTIALLY ORDERED SEMIGROUP. *If this relation is linear, then $\mathfrak{A}$ is said to be* LINEARLY ORDERED (*in both cases, we use the term ordered semigroup*).

The study of partially ordered semigroups has been carried on up to now mainly for separate classes of semigroups. Most of all, attention has been paid to partially ordered groups, especially in the case when the group is a lattice with respect to the partial ordering (such a group is called an *l*-group). Some results on partially ordered semigroups and groups are mentioned in the thirteenth and fourteenth chapters of the book of Birkhoff [3].

The problem of the possibility of introducing into a group a two-sidedly stable partial ordering, and in particular a linear ordering, deserves special attention. We shall see that this problem is connected with the study of certain subsemigroups of the group. Thus it can be included in our study of the embedding of semigroups into groups.

3.3. With a view to studying one particular case of the linear ordering of a holoidal semigroup (VII, 4.8), let us prove a preliminary lemma.

LEMMA. *Let a holoidal semigroup with two-sided cancellation have a unit. Then this unit is externally adjoined. If the element X is divisible by Y from the left, then it is divisible by Y from the right.*

PROOF. Assume for elements A and B of the given semigroup $\mathfrak{A}$ that

$$AB = E_{\mathfrak{A}}.$$

If, in the linear ordering of the holoidal semigroup, B precedes A, then for some $Z \in \mathfrak{A}$, we have $BZ = A$ (if $A = B$, then Z is the empty symbol). Thus we obtain

$$BZB = AB = E_{\mathfrak{A}}, \qquad B = E_{\mathfrak{A}} B E_{\mathfrak{A}}.$$

This means that B precedes $E_{\mathfrak{A}}$ and simultaneously $E_{\mathfrak{A}}$ precedes B. This is possible only when $B = E_{\mathfrak{A}}$. But then $A = E_{\mathfrak{A}}$. The case when A precedes B is analogous.

Let $X = YS$. If X precedes Y in the linear order relation of the holoidal semigroup, then $Y = XT$ for some $T \in \mathfrak{A}$. Hence

$$X = XTS,$$

i.e., $TS = E_{\mathfrak{A}}$. But, by what we have just proved, this is possible only when $T = S = E_{\mathfrak{A}}$, i.e., when $X = Y$. Consequently Y precedes X and therefore X is divisible by Y from the left and right.

3.4. THEOREM. *If a holoidal semigroup $\mathfrak{A}$ has a unit and has two-sided cancellation, then its linear ordering is two-sidedly stable.*

PROOF. Let B be distinct from A ($A, B \in \mathfrak{A}$) and precede it with respect to the linear order in $\mathfrak{A}$, i.e., for some $X, Y \in \mathfrak{A}$,

$$A = BX, \qquad A = YB.$$

Let us assume that for some $Z \in \mathfrak{A}$ the element ZA precedes ZB. For some $T \in \mathfrak{A}$ we must have

$$ZB = ZAT.$$

By the law of cancellation, we have $B = AT$ and therefore

$$AE_{\mathfrak{A}} = A = BX = ATX.$$

From this we obtain $TX = E_{\mathfrak{A}}$. By Lemma 3.3 this is possible only when $T = X = E_{\mathfrak{A}}$, and this contradicts the fact that $A \neq B$. Right stability is proved analogously.

3.5. We pass to partially ordered groups.

DEFINITION. *The set of all elements of a partially ordered group $\mathfrak{G}$ which follow the identity $E_{\mathfrak{G}}$ is called the* POSITIVE PART *of $\mathfrak{G}$ and is denoted by $\mathfrak{G}^+$.*

We must note that sometimes the positive part is taken to be only those X for which $X > E_{\mathfrak{G}}$.

The use of the term "positive part" is explained by the fact that in considering partially ordered groups the additive notation is very often used for the operation. In this case the identity is denoted by the symbol zero, and the fact that an element follows the identity is written by a formula which is identical in appearance with the usual condition of positiveness.

3.6. The positive part $\mathfrak{G}^+$ of a partially ordered group is a subsemigroup, since $X \geqslant E_{\mathfrak{G}}$, $Y \geqslant E_{\mathfrak{G}}$ implies

$$XY \geqslant E_{\mathfrak{G}} Y = Y \geqslant E_{\mathfrak{G}}.$$

Prescription of the positive part $\mathfrak{G}^+$ of a partially ordered group $\mathfrak{G}$ completely defines the entire partial ordering.

In order that $X \geqslant Y$ it is necessary and sufficient that $XY^{-1} \in \mathfrak{G}^+$. In fact, $X \geqslant Y$ implies

$$XY^{-1} \geqslant YY^{-1} = E_{\mathfrak{G}}.$$

In its turn, $XY^{-1} \geqslant E_{\mathfrak{G}}$ implies

$$X = XY^{-1}Y \geqslant E_{\mathfrak{G}}Y = Y.$$

Thus, prescribing a relation of partial ordering in a group is equivalent to choosing some subsemigroup in that group. Consequently, an answer to the question, which subsemigroups of the group can serve as positive parts for given two-sidedly stable relations of partial ordering, turns out to be equivalent to determining all possible two-sidedly stable relations of partial ordering in the group. Thus, the problem of transforming a given group into a partially ordered group by introducing into it a relation of partial ordering can be considered as part of the study of subsemigroups of groups.

3.7. THEOREM. *In order that a subsemigroup $\mathfrak{A}$ of a group $\mathfrak{G}$ be the positive part of the partially ordered group obtained from $\mathfrak{G}$ by introducing into it in some way a two-sidedly stable relation of partial ordering it is necessary and sufficient that $\mathfrak{A}$ be a semigroup with externally adjoined unit and that for any $X \in \mathfrak{G}$,*

$$X^{-1}\mathfrak{A}X = \mathfrak{A}.$$

PROOF. (1) If $\mathfrak{A}$ is the positive part of $\mathfrak{G}$ for some partial ordering, then $\mathfrak{A} \ni E_{\mathfrak{G}}$ and $E_{\mathfrak{G}}$ is, of course, a unit for $\mathfrak{A}$. If for some $X, Y \in \mathfrak{A}$,

$$XY = E_{\mathfrak{G}},$$

then since $X \geqslant E_{\mathfrak{G}}$ we have

$$E_{\mathfrak{G}} = XY \geqslant E_{\mathfrak{G}}Y = Y,$$

and since $Y \geqslant E_{\mathfrak{G}}$ it turns out that $Y = E_{\mathfrak{G}}$. Consequently, $X = E_{\mathfrak{G}}$.

For any $A \in \mathfrak{A}$, from $A \geqslant E_{\mathfrak{G}}$ we obtain for any $X \in \mathfrak{G}$,

$$X^{-1}AX \geqslant X^{-1}E_{\mathfrak{G}}X = E_{\mathfrak{G}},$$

whence

$$X^{-1}\mathfrak{A}X \subset \mathfrak{A}.$$

But

$$\mathfrak{A} = E_{\mathfrak{G}}\mathfrak{A}E_{\mathfrak{G}} = X^{-1}(X\mathfrak{A}X^{-1})X \subset X^{-1}\mathfrak{A}X.$$

Therefore $X^{-1}\mathfrak{A}X = \mathfrak{A}$.

(2) If a subsemigroup $\mathfrak{A}$ has the stated properties, then its unit is $E_{\mathfrak{G}}$ since it is the unique idempotent in $\mathfrak{G}$. We put $X \geqslant Y$ if $XY^{-1} \in \mathfrak{A}$. For such a relation we have

$$X \geqslant X,$$

since $XX^{-1} = E_{\mathfrak{G}} \in \mathfrak{A}$. If

$$X \geqslant Y, \qquad Y \geqslant Z,$$

then

$$XZ^{-1} = XY^{-1} \cdot YZ^{-1} \in \mathfrak{A},$$

i.e., $X \geqslant Z$.

If $X \geqslant Y$ and $Y \geqslant X$, then XY^{-1}, $YX^{-1} \in \mathfrak{A}$, and since

$$(XY^{-1}) \cdot (YX^{-1}) = E_{\mathfrak{G}},$$

and the unit $E_{\mathfrak{G}}$ is an externally adjoined zero in $\mathfrak{A}$, it follows that $XY^{-1} = YX^{-1} = E_{\mathfrak{G}}$, i.e., $X = Y$.

Finally, for $A \geqslant B$, for any $X, Y \in \mathfrak{G}$ (which can also be empty symbols) we have

$$XAY \cdot (XBY)^{-1} = XAYY^{-1}B^{-1}X^{-1} = X(AB^{-1})X^{-1} \subset X\mathfrak{A}X^{-1} = \mathfrak{A},$$

i.e., $XAY \geqslant XBY$.

3.8. In a partially ordered group $\mathfrak{G}$ it is also possible to define its negative part $\mathfrak{G}^-$ as the set of all X such that $X \leqslant E_{\mathfrak{G}}$.

In order that X belong to $\mathfrak{G}^-$ it is necessary and sufficient that $X^{-1} \in \mathfrak{G}^+$. In fact, if $X \leqslant E_{\mathfrak{G}}$, then

$$E_{\mathfrak{G}} = XX^{-1} \leqslant E_{\mathfrak{G}}X^{-1} = X^{-1}.$$

On the other hand, if $X^{-1} \geqslant E_{\mathfrak{G}}$, then

$$E_{\mathfrak{G}} = XX^{-1} \geqslant XE_{\mathfrak{G}} = X.$$

The map of a group onto itself, associating each element with its inverse, is an anti-automorphism of the group. This anti-isomorphism induces an anti-isomorphism between the positive part and the negative part of a partially ordered group.

3.9. In a partially ordered group $\mathfrak{G}$, clearly only the identity belongs to both the positive and negative parts. Therefore in a linearly ordered group all of the nonunit elements must have infinite order. In fact, for $X^n = E_{\mathfrak{G}}$ $(n > 1)$ we have $X^{-1} = X^{n-1}$, which is impossible since for $X \in \mathfrak{G}^+$ it follows from 3.6 that $X^{n-1} \in \mathfrak{G}^+$, and from 3.8 that X^{-1} belongs to $\mathfrak{G}^-$ and $\mathfrak{G}^+ \cap \mathfrak{G}^- = E_{\mathfrak{G}}$. The reasoning for $X \in \mathfrak{G}^-$ is analogous.

3.10. Let us consider linearly ordered groups in greater detail. As follows from the theorem mentioned below, which is due to A. M. Kaufman **[2]**, it is convenient here to make use of the concept of holoidal semigroups (VIII, 4.8).

THEOREM. *The positive part $\mathfrak{G}^+$ of a linearly ordered group $\mathfrak{G}$ is a holoidal semigroup, where for $X, Y \in \mathfrak{G}^+$ $(X \neq Y)$ the relation $X > Y$ holds in $\mathfrak{G}$ if and only if X is divisible by Y from the left and right in $\mathfrak{G}^+$.*

PROOF. Let $X, Y \in \mathfrak{G}^+$ $(X \neq Y)$. We must have either $X > Y$ or $Y > X$. Suppose the first holds. Then, by 3.6, $XY^{-1} \in \mathfrak{G}^+$ and by 3.7, $Y^{-1}X = Y^{-1}(XY^{-1})Y \in \mathfrak{G}^+$. Therefore X is divisible by Y from the left and right in $\mathfrak{G}^+$ i.e.,

$$X = Y \cdot (Y^{-1}X), \qquad X = (XY^{-1}) \cdot Y.$$

As regards Y, it can not be divisible by X from either side. In fact, if $Y = XS$, $S \in \mathfrak{G}^+$, it would follow that $X^{-1}Y = S \in \mathfrak{G}^+$, and by 3.7, $YX^{-1} = X(X^{-1}Y)X^{-1} \in \mathfrak{G}^+$. But this is impossible since $YX^{-1} = (XY^{-1})^{-1}$ and by 3.8, $YX^{-1} \in \mathfrak{G}^-$.

Since a linear order in $\mathfrak{G}$ induces a linear order in $\mathfrak{G}^+$, the theorem follows from the proven coincidence of partial orders in $\mathfrak{G}^+$.

3.11. From what has been said above, we can formulate in a natural way a criterion for the possibility of introducing a two-sidedly stable relation of linear ordering in a group.

THEOREM. *In order that a group $\mathfrak{G}$ can be transformed into a linearly ordered group by means of a two-sidedly stable relation of linear order it is necessary and sufficient that it be the union of two of its subsemigroups which are holoidal semigroups with intersection equal to $E_{\mathfrak{G}}$.*

PROOF. (1) If $\mathfrak{G}$ is a linearly ordered group, then

$$\mathfrak{G} = \mathfrak{G}^+ \cup \mathfrak{G}^-, \qquad \mathfrak{G}^+ \cap \mathfrak{G}^- = E_{\mathfrak{G}}.$$

$\mathfrak{G}^+$ is a holoidal semigroup by 3.10. $\mathfrak{G}^-$ is anti-isomorphic to $\mathfrak{G}^+$ by 3.8, and therefore is also a holoidal semigroup.

(2) Let

$$\mathfrak{G} = \mathfrak{A} \cup \mathfrak{A}', \qquad \mathfrak{A} \cap \mathfrak{A}' = E_{\mathfrak{G}},$$

where $\mathfrak{A}$ and $\mathfrak{A}'$ are holoidal subsemigroups of $\mathfrak{G}$.

We note that for the two elements X and X^{-1} one always belongs to $\mathfrak{A}$ and the other to $\mathfrak{A}'$.

In fact, let $X, X^{-1} \in \mathfrak{A}$ (the case when $X, X^{-1} \in \mathfrak{A}'$ is completely analogous). Since

$$X = E_{\mathfrak{G}} X E_{\mathfrak{G}}, \qquad E_{\mathfrak{G}} = XX^{-1} = X^{-1}X,$$

it follows that X precedes $E_{\mathfrak{G}}$, and $E_{\mathfrak{G}}$ precedes X in $\mathfrak{A}$. Therefore $X = E_{\mathfrak{G}}$, and in this case, $E_{\mathfrak{G}} \in \mathfrak{A}$, $E_{\mathfrak{G}}^{-1} = E_{\mathfrak{G}} \in \mathfrak{A}'$.

Let us introduce in $\mathfrak{G}$ a partial ordering relation by setting $X \geqslant Y$ if $XY^{-1} \in \mathfrak{A}$. For any $X, Y \in \mathfrak{G}$ one of the two elements XY^{-1} and $(XY^{-1})^{-1} = YX^{-1}$ belongs to $\mathfrak{A}$. Therefore either $X \geqslant Y$ or $Y \geqslant X$. Here the two relations hold simultaneously only when $XY^{-1} = E_{\mathfrak{G}}$, i.e., $X = Y$.

Let $X \geqslant Y$, $Y \geqslant Z$, i.e., XY^{-1}, $YZ^{-1} \in \mathfrak{A}$. Then we have

$$XZ^{-1} = XY^{-1} \cdot YZ^{-1} \in \mathfrak{A},$$

i.e., $X \geqslant Z$.

Thus the relation we defined is a linear ordering relation.

It remains to be shown that this relation is two-sidedly stable.

If $X \geqslant Y$, then for any $Z \in \mathfrak{G}$,

$$(XZ)(YZ)^{-1} = XZ \cdot Z^{-1}Y^{-1} = XY^{-1} \in \mathfrak{A},$$

i.e., $XZ \geqslant YZ$.

For the proof of left stability we prove as a preliminary that the condition (α) $XY^{-1} \in \mathfrak{A}$ is equivalent to the condition (β) $Y^{-1}X \in \mathfrak{A}$.

In fact, if $X, Y \in \mathfrak{A}$ then (α) means that X is divisible by Y from the right in $\mathfrak{A}$, and (β) means that X is divisible by Y from the left. By 3.3 one implies the other.

If $X, Y \in \mathfrak{A}'$ then $X^{-1}, Y^{-1} \in \mathfrak{A}$. Condition ($\alpha$) means that Y^{-1} is divisible from the left by X^{-1} in $\mathfrak{A}$, and (β) means that Y^{-1} is divisible by X^{-1} from the right. Again by 3.3 one implies the other.

If $X \in \mathfrak{A}$ and $Y \in \mathfrak{A}'$, i.e., $Y^{-1} \in \mathfrak{A}$, then (α) and (β) must be true.

If $X \in \mathfrak{A}'$ and $Y \in \mathfrak{A}$, then neither (α) nor (β) is true except in the case $X = Y = E_{\mathfrak{G}}$.

Now let $X \geqslant Y$. By definition, $XY^{-1} \in \mathfrak{A}$ and therefore, by what has been proven, $Y^{-1}X \in \mathfrak{A}$. For any $Z \in \mathfrak{G}$ we have

$$(ZY)^{-1}(ZX) = Y^{-1}Z^{-1}ZX = Y^{-1}X \in \mathfrak{A},$$

which in view of the proven equivalence gives

$$(ZX)(ZY)^{-1} \in \mathfrak{A},$$

i.e., $ZX \geqslant ZY$.

3.12. In the process of proving the second part of the theorem we have shown that in $\mathfrak{G}$ it is possible to introduce a linear order such that $\mathfrak{A}$ will be the positive part and $\mathfrak{A}'$ the negative part of the linearly ordered group so obtained. Thus, by 3.8 we have the following. If a group $\mathfrak{G}$ is the union of two of its subsemigroups which are holoidal semigroups with intersection equal to $E_{\mathfrak{G}}$, then these holoidal semigroups must be anti-isomorphic.

The representation of $\mathfrak{G}$ as such a union of two of its holoidal subsemigroups can be converted into a decomposition of $\mathfrak{G}$ since the identity $E_{\mathfrak{G}}$, by 3.3, can be picked out as a third component.

3.13. As we have seen, the positive part of a linearly ordered group is a holoidal semigroup with two-sided cancellation and a unit. It turns out that the presence of these properties in a semigroup is also sufficient for it to be the positive part of some linearly ordered group.

THEOREM. *If $\mathfrak{A}$ is a holoidal semigroup with two-sided cancellation and a unit, then there exists a linearly ordered group $\mathfrak{G}$ for which $\mathfrak{A}$ is its positive part.*

PROOF. We take a semigroup $\mathfrak{A}'$, anti-isomorphic to $\mathfrak{A}$ and having no elements in common with $\mathfrak{A}$ with the exception of a common unit $E = E_{\mathfrak{A}} = E_{\mathfrak{A}'}$. Clearly $\mathfrak{A}'$ will also be a holoidal semigroup with two-sided cancellation. The element of $\mathfrak{A}'$ corresponding to the element $X \in \mathfrak{A}$, under some anti-isomorphism of $\mathfrak{A}$ onto $\mathfrak{A}'$ which we are considering to be fixed, will be denoted by X'.

In the set $\mathfrak{G} = \mathfrak{A} \cup \mathfrak{A}'$ we define a multiplication. If two elements of $\mathfrak{G}$ both belong to $\mathfrak{A}$ or both belong to $\mathfrak{A}'$, then their product is defined according to the operation in these semigroups.

Let $A \in \mathfrak{A}$, $B' \in \mathfrak{A}'$. If A precedes B in the holoidal semigroup $\mathfrak{A}$, i.e., for some $X, Y \in \mathfrak{A}$,

$$B = AX, \qquad B = YA,$$

then we put

$$AB' = A \cdot (YA)' = A \cdot (A'Y') = Y',$$
$$B'A = (AX)' \cdot A = (X'A') \cdot A = X'.$$

In this connection we should remark that X and Y are uniquely defined, since $\mathfrak{A}$ has two-sided cancellation. If $A = B$, then $X = Y = E$. In what follows we must keep in mind Lemma 3.3, according to which for A to precede B it is sufficient that B be divisible by A from the left or right.

If B precedes A, i.e., for some $U, V \in \mathfrak{A}$,

$$A = BU, \qquad A = VB,$$

then we put

$$A \cdot B' = (VB) \cdot B' = V,$$
$$B' \cdot A = B' \cdot (BU) = U.$$

We now show that this multiplication is associative. Let $G_1, G_2, G_3 \in \mathfrak{G}$. We denote

$$S_1 = (G_1G_2)G_3, \qquad S_2 = G_1(G_2G_3).$$

If G_1, G_2, G_3 all belong either to $\mathfrak{A}$ or $\mathfrak{A}'$, then $S_1 = S_2$ by the associativity of the operations in $\mathfrak{A}$ and $\mathfrak{A}'$. Let $G_1, G_2 \in \mathfrak{A}$, $G_3 = H' \in \mathfrak{A}'$, where G_2 is divisible by H from the right in $\mathfrak{A}$, i.e., $G_2 = XH$ $(X \in \mathfrak{A})$. Then

$$S_1 = (G_1XH) \cdot H' = G_1X,$$
$$S_2 = G_1 [(XH) \cdot H'] = G_1X.$$

Let $G_1 \in \mathfrak{A}$, $G_2 \in \mathfrak{A}$, $G_3 = H' \in \mathfrak{A}'$, where H is divisible from the right by G_2 in $\mathfrak{A}$, i.e., $H = XG_2$ $(X \in \mathfrak{A})$. If G_1 is divisible by X from the right, i.e., $G_1 = UX$, then

$$S_1 = (G_1G_2) \cdot (XG_2)' = (U \cdot XG_2) \cdot (XG_2)' = U,$$
$$S_2 = G_1 \cdot (G_2 \cdot G_2'X') = UX \cdot X' = U.$$

If X is divisible from the right by G_1, i.e., $X = VG_1$, then

$$S_1 = (G_1G_2) \cdot (XG_2)' = G_1G_2 \cdot (V \cdot G_1G_2)' = (G_1G_2)[(G_1G_2)' \cdot V'] = V',$$
$$S_2 = G_1[G_2 \cdot (XG_2)'] = G_1 \cdot [G_2 \cdot (G_2X)] = G_1X' = G_1(G_1'V') = V'.$$

The case $G_1 \in \mathfrak{A}'$, $G_2 \in \mathfrak{A}$, $G_3 \in \mathfrak{A}$ is dealt with analogously.

Let $G_1 \in \mathfrak{A}$, $G_3 \in \mathfrak{A}$, and let $G_2 = H' \in \mathfrak{A}'$, where H is divisible by G_1 from the right in $\mathfrak{A}$, i.e., $G_1 = XH$ $(X \in \mathfrak{A})$. Then, whether G_3 is divisible from the left by H or H is divisible from the left by G_3, in both cases it is easy to see that

$$S_1 = (XH \cdot H')G_3 = XG_3,$$
$$S_2 = (XH) \cdot (H'G_3) = XG_3.$$

Let $G_1 \in \mathfrak{A}$, $G_3 \in \mathfrak{A}$, $G_2 = H' \in \mathfrak{A}'$, where H is divisible by G_1 from the right in $\mathfrak{A}$, i.e., $H = XG_1$ $(X \in \mathfrak{A})$, and $X = G_3 Y$ $(Y \in \mathfrak{A})$. Then

$$S_1 = [G_1 \cdot (XG_1)']G_3 = X'G_3 = Y',$$
$$S_2 = G_1[(XG_1)' \cdot G_3] = G_1 \cdot [(G_1'X') \cdot G_3] = G_1 \cdot (G_1'Y') = Y'.$$

Let $G_1 \in \mathfrak{A}$, $G_3 \in \mathfrak{A}$, $G_2 = H' \in \mathfrak{A}'$, where H is divisible from the right by G_1 in $\mathfrak{A}$, i.e., $H = XG_1$ $(X \in \mathfrak{A})$, and $G_3 = XY$ $(Y \in \mathfrak{A})$. Then, whether Y is divisible from the left by G_1 or G_1 is divisible from the left by Y, in both cases, it is easy to see that

$$S_1 = [G_1 \cdot (XG_1)'] \cdot (XY) = X' \cdot (XY) = Y,$$
$$S_2 = G_1 \cdot [(XG_1)' \cdot (XY)] = G_1[(G_1'X') \cdot (XY)] = Y.$$

We need not consider the cases when two of the factors G_1, G_2, G_3 belong to $\mathfrak{A}'$ and the third to $\mathfrak{A}$, in view of the fact that the subsemigroups $\mathfrak{A}$ and $\mathfrak{A}'$ in the multiplicative set $\mathfrak{G}$ play entirely the same role.

We have thus seen that $\mathfrak{G}$ is a semigroup. E is evidently an identity for $\mathfrak{G}$. The elements X and X' (where $X \in \mathfrak{A}$, $X' \in \mathfrak{A}'$) are inverses of each other. Therefore $\mathfrak{G}$ is a group. It is the union of two of its subsemigroups $\mathfrak{A}$ and $\mathfrak{A}'$ which are holoidal semigroups, where $\mathfrak{A} \cap \mathfrak{A}' = E$. By 3.11, $\mathfrak{G}$ can be converted into a linearly ordered group by introducing into $\mathfrak{G}$ some relation of partial ordering. As was remarked in 3.12, this can be done so that the holoidal subsemigroup $\mathfrak{A}$ is the positive part of the linearly ordered semigroup $\mathfrak{G}$.

3.14. We note that the result of Theorem 3.13 is also interesting from the point of view of the preceding section. The condition of two-sided cancellation turns out to be not only necessary but also sufficient for embedding into a group all the semigroups of a certain class, namely the holoidal semigroups with a unit.

3.15. In conclusion we remark that many of the various types of ordering are not possible in holoidal semigroups which are positive parts of linearly ordered semigroups. For example, from the theorem mentioned below it follows that the only nontrivial type of well-ordering which is possible in this respect is the type of ordering by magnitude in the set of all natural numbers.

THEOREM. *If all the elements of the positive part $\mathfrak{G}^+$ of a linearly ordered group $\mathfrak{G}$ are well-ordered with respect to a relation of linear ordering in a holoidal semigroup, then $\mathfrak{G}$ is the unit group or an infinite cyclic group.*

PROOF. The first element of $\mathfrak{G}^+$ is evidently $E_{\mathfrak{G}}$. If $\mathfrak{G} \neq E_{\mathfrak{G}}$, then by 3.8 $\mathfrak{G}^+ \neq E_{\mathfrak{G}}$. Let X be an element of $\mathfrak{G}^+$ following $E_{\mathfrak{G}}$. Put $\mathfrak{G}' = [E_{\mathfrak{G}}, X]$. We suppose that $\mathfrak{G}' \neq \mathfrak{G}^+$. We denote by Y the element of $\mathfrak{G}^+ \backslash \mathfrak{G}'$ preceding all the elements of $\mathfrak{G}^+ \backslash \mathfrak{G}'$. Since Y follows X, for some $U \in \mathfrak{G}^+$ we have

$$Y = XU \qquad (U \mathrel{\overline{\in}} \mathfrak{G}'),$$

since otherwise $Y \in \mathfrak{G}'$. Therefore for some $V \in \mathfrak{G}^+$,

$$U = VY.$$

We obtain $Y = XVY$, i.e., $XV = E_{\mathfrak{G}}$. But by 3.3 this would mean that X precedes $E_{\mathfrak{G}}$ in $\mathfrak{G}^+$. But this is impossible since $E_{\mathfrak{G}}$ is the first element. This contradiction means that

$$\mathfrak{G}^+ = [E_{\mathfrak{G}}, X] = \{E_{\mathfrak{G}}, X, X^2, \ldots, X^n, \ldots\}.$$

By 3.8,

$$\mathfrak{G}^- = \{E_{\mathfrak{G}}, X^{-1}, X^{-2}, \ldots, X^{-n}, \ldots\}$$

and therefore by 3.9, $\mathfrak{G}$ is an infinite cyclic group.

4. Potential Invertibility of Elements

4.1. In considering the properties of invertibility of elements it often happens that an element A of some semigroup $\mathfrak{A}$ is not invertible from the left in $\mathfrak{A}$ (VI, 1.1) but there exists a supersemigroup $\mathfrak{A}'$ of $\mathfrak{A}$ such that A is invertible from the left by an element of $\mathfrak{A}'$. Thus embedding $\mathfrak{A}$ into $\mathfrak{A}'$ transforms A from an element which is noninvertible from the left into an element which is invertible from the left. In view of the significant importance of the property of invertibility the problem of when this possibility occurs is worthy of attention.

DEFINITION. *An element A of a semigroup $\mathfrak{A}$ is* POTENTIALLY INVERTIBLE FROM THE LEFT *if there exists a supersemigroup $\mathfrak{A}'$ of $\mathfrak{A}$ in which A is invertible from the left.*

The situation for invertibility from the right is analogous.

4.2. The concept and term "potential" for various properties in semigroups is due to E. S. Ljapin. We can speak of various potential properties of separate elements or of subsets of semigroups, meaning by this that the given property of an element or a subset is realized in some supersemigroup of the given semigroup. E. G. Šutov [**1**; **2**; **3**; **4**] has investigated the problem of potential realization of a number of properties in semigroups. In particular, he completed and then generalized a certain problem of potential invertibility which was first taken up by E. S. Ljapin. The statement of this problem has a definite similarity to the problem of divisibility for two elements in some supersemigroup, which was considered in 1.6 and 1.7. However, the method of solving it turns out to be more complicated.

4.3. We note the following property of invertibility. Let $\mathfrak{K}$ be a generating set of a semigroup $\mathfrak{A}$ and let $A \in \mathfrak{A}$. If for each $K \in \mathfrak{K}$ there exists $Z_K \in \mathfrak{A}$ such that $Z_K A = K$, then A is invertible from the left. In fact, an arbitrary element S of $\mathfrak{A}$ can be represented in the form

$$S = K_1K_2 \ldots K_n \qquad (K_1, K_2, \ldots, K_n \in \mathfrak{K}).$$

Thus we have

$$(K_1K_2 \ldots K_{n-1}Z_{K_n}) \cdot A = K_1K_2 \ldots K_{n-1}K_n = S.$$

4.4. Let an element A of the semigroup $\mathfrak{A}$ be potentially invertible from the left; that is, in some supersemigroup $\mathfrak{A}'$ of $\mathfrak{A}$ it is invertible from the left. If for some natural number n, and for some X and Y which are elements of $\mathfrak{A}$ or empty symbols, we have $A^nX = A^nY$, then for any $S \in \mathfrak{A}$ it follows that $SX = SY$. In fact, in $\mathfrak{A}'$ there must be elements $Z_1, Z_2, \ldots, Z_{n-1}, Z_n$ such that

$$Z_1A = S, \qquad Z_2A = Z_1, \ldots, Z_nA = Z_{n-1}.$$

Therefore

$$Z_nA^nX = Z_{n-1}A^{n-1}X = \ldots = Z_2A^2X = Z_1AX = SX,$$
$$Z_nA^nY = Z_{n-1}A^{n-1}Y = \ldots = Z_2A^2Y = Z_1AY = SY,$$
$$SX = Z_nA^nX = Z_nA^nY = SY.$$

4.5. In connection with 4.4 we note that if in the semigroup $\mathfrak{A}$, for some $A \in \mathfrak{A}$, $A^2X = A^2Y$ implies $SX = SY$ for any $S \in \mathfrak{A}$, then for any natural number n, $A^nX = A^nY$ implies $SX = SY$ for any $S \in \mathfrak{A}$.

In fact, if $AX = AY$, then $A^2X = A^2Y$ and therefore $SX = SY$. If $n > 2$, then $A^nX = A^nY$ implies $A^2(A^{n-2}X) = A^2(A^{n-2}Y)$ and therefore $A(A^{n-2}X) = A(A^{n-2}Y)$, i.e., $A^{n-1}X = A^{n-1}Y$. Repeating the argument, we obtain $A^2X = A^2Y$, whence, by assumption, $SX = SY$.

4.6. In the case when an element A of a semigroup $\mathfrak{A}$ is of finite type it is at once evident that the above necessary condition for potential invertibility of A from the left is also sufficient. But more than that, this condition in the case considered simply means that A is invertible from the left in $\mathfrak{A}$.

In fact, let A satisfy condition 4.4 and

$$A^{h+d} = A^h.$$

From $A^h \cdot A^d = A^h$ it follows that for any $S \in \mathfrak{A}$ we have $SA^d = S$. Thus

$$(SA^{d-1}) \cdot A = S,$$

i.e., A is invertible from the left in $\mathfrak{A}$.

We also note that in this case $h = 1$, i.e., $[A]$ is a group. In fact, by VI, 1.8 for some $X \in \mathfrak{A}$, $XA^h = A$. Therefore from $A^{h+d} = A^h$, multiplying by X, we obtain $A^{1+d} = A$, whence it follows that $h = 1$.

4.7. Let A be an element of a semigroup $\mathfrak{A}$ such that for any natural number n and any $X, Y, S \in \mathfrak{A}$, $A^nX = A^nY$ implies $SX = SY$. We consider pairs in which the first component is an arbitrary element S of $\mathfrak{A}$ and the second is an arbitrary whole non-negative number k. Such a pair we will denote by $S^{(k)}$. The totality of all such pairs we denote by $\mathfrak{R}$. We denote by $\mathfrak{W}$ all words

$$W = S_1^{(k_1)}S_2^{(k_2)} \ldots S_s^{(k_s)}$$

in $\mathfrak{R}$ in which no two neighboring k_i and k_{i+1} are both zero. In $\mathfrak{W}$ we define multiplication by setting

$$W_1 \cdot W_2 = W_3 \qquad (W_1, W_2, W_3 \in \mathfrak{W}),$$
$$W_i = S_{1i}^{(k_{1i})}S_{2i}^{(k_{2i})} \ldots S_{s_ii}^{(k_{s_ii})} \qquad (i = 1, 2, 3),$$

where W_3 is the ordinary product of the words W_1 and W_2 if $k_{s_11} \neq 0$ or $k_{12} \neq 0$. If $k_{s_11} = k_{12} = 0$, then W_3 is obtained from such a product by replacing the neighboring terms $S_{s_11}^{(0)}$ and $S_{12}^{(0)}$ by the element $(S_{s_11} \cdot S_{12})^{(0)}$ (where $S_{s_11} \cdot S_{12}$ is the product of S_{s_11} and S_{12} in $\mathfrak{A}$).

The indicated operation in $\mathfrak{B}$ is evidently associative.

4.8. In $\mathfrak{B}$ we define the following relations:

$$W \sim W'(\mathfrak{n}_1) \qquad (W, W' \in \mathfrak{B})$$

if

$$W = S_1^{(k)} S_2^{(0)}, \qquad AS_2 = A^{k+1}R, \qquad W' = (S_1 \cdot R)^{(0)} \qquad (S_1, S_2, R \in \mathfrak{A}),$$

where $k > 0$ and R can also be the empty symbol;

$$W \sim W'(\mathfrak{n}_2),$$

if

$$W = S_1^{(k)} S_2^{(0)}, \qquad AS_2 = A^l R, \qquad W' = S_1^{(k-l+1)} R^{(0)} \qquad (k \geqslant l > 1),$$

where R again can be the empty symbol.

If $W \sim W'(\mathfrak{n}_1)$, or $W' \sim W(\mathfrak{n}_1)$, or $W \sim W'(\mathfrak{n}_2)$, or $W' \sim W(\mathfrak{n})$, or $W = W'$, we write

$$W \sim W'(\mathfrak{n}_3).$$

We note that $W \sim W'(\mathfrak{n}_i)$ implies $WS^{(0)} \sim W'S^{(0)}(\mathfrak{n}_i)$ $(i = 1, 2, 3)$ for all $S \in \mathfrak{A}$.

4.9. For words in $\mathfrak{B}$ we define an operator σ. Let

$$W = S_1^{(k_1)} S_2^{(k_2)} \ldots S_m^{(k_m)} \in \mathfrak{B}$$

and let $S_t^{(k_t)}$ be the element of this word which is farthest to the right and such that $t < m$, $k_{t+1} = 0$, and let there exist an R which is an element of $\mathfrak{A}$ or an empty symbol such that $AS_{t+1} = A^{k_t+1}$. Then $\sigma(W)$ is obtained from W by replacing $S_t^{(k_t)} S_{t+1}^{(0)}$ by $(S_t R)^{(0)}$ (if, in this connection, $k_{t-1} = 0$, then $S_{t-1}^{(0)}$ is also adjoined to these elements: $(S_{t-1} S_t R)^{(0)}$). If there is no element $S_t^{(k_t)}$ in W with the property in question, then we put $\sigma(W) = W$.

We note that the operator σ is single-valued. In fact, if $AS_{t+1} = A^{k_t+1}R = A^{k_t+1}R'$, then by the assumption on A we have $XR = XR'$ for all $X \in \mathfrak{A}$, and in particular, $S_t R = S_t R'$.

Evidently $W \sim \sigma(W)(\mathfrak{n}_1')$, where $\mathfrak{n}_1'$ is the first derived relation for $\mathfrak{n}_1$ (I, 5.20). For $X \in \mathfrak{A}$ we always have $\sigma(X^{(0)}) = X^{(0)}$.

4.10. Continuing with the notation introduced, we consider some properties of the relation $\mathfrak{n}_3$.

LEMMA. *If $W \sim W'(\mathfrak{n}_3')$, then $\sigma(W) \sim \sigma(W')(\mathfrak{n}_3')$.*

PROOF. Let

$$W = T_1 \cdot U \cdot T_2, \qquad W' = T_1 \cdot U' \cdot T_2,$$

where $U \sim U'(\mathfrak{n}_i)$ $(i = 1, 2)$ (the case $W = W'$ is trivial, and the case $U' \sim U(\mathfrak{n}_i)$ is symmetric to the ones considered). By what was stated at the end of 4.8 we can consider that in the word T_2 for the first factor $S^{(p)}$ we have $p \neq 0$.

If $\sigma(T_2) \neq T_2$, then evidently

$$\sigma(W) = T_1 \cdot U \cdot \sigma(T_2), \qquad \sigma(W') = T_1 \cdot U' \cdot \sigma(T_2)$$

and therefore $\sigma(W) \sim \sigma(W')(\mathfrak{n}_1')$.

In what follows we will assume that $\sigma(T_2) = T_2$.

If $U \sim U'(\mathfrak{n}_1)$, then, as it is easily seen, $\sigma(W) = T_1 \cdot \sigma(U) \cdot T_2$, i.e., $\sigma(W) = W'$ (since $\sigma(U) = U'$). Also, since $W' \sim \sigma(W')(\mathfrak{n}_1')$, it follows that $\sigma(W) \sim \sigma(W')(\mathfrak{n}_1')$.

Let $U \sim U'(\mathfrak{n}_2)$;

$$U = S_1^{(k)} S_2^{(0)},\ AS_2 = A^l R,\ U' = S_1^{(k-l+1)} R^{(0)} \qquad (k \geqslant l > 1).$$

If for some P which is an element of $\mathfrak{A}$ or an empty symbol, $AS_2 = A^{k+1}P$, then $\sigma(W) = T_1 \cdot (S_1 P)^{(0)} \cdot T_2$. On the other hand, $A^l R = A^{k+1} P$ implies $AR = A^{k-l+2}P$. Therefore

$$\sigma(W') = T_1 \cdot (S_1 P)^{(0)} \cdot T_2 = \sigma(W).$$

If there is no P with the indicated property, then there is no P' such that $AR = A^{k-l+2}P'$. Therefore

$$\sigma(W) = \sigma(T_1) \cdot U \cdot T_2, \qquad \sigma(W') = \sigma(T_1) \cdot U' \cdot T_2.$$

4.11. LEMMA. *If $X^{(0)} \sim Y^{(0)}(\mathfrak{n}_3'')$ $(X, Y \in \mathfrak{A})$, then $X = Y$.*

PROOF. By definition of the second derived relation there are words $V_1, V_2, \ldots, V_s \in \mathfrak{B}$ such that

$$X^{(0)} = V_1, \qquad Y^{(0)} = V_s, \qquad V_i \sim V_{i+1}(\mathfrak{n}_3') \qquad (i = 1, 2, \ldots, s-1).$$

The proof of the above assertion is by induction on s. For $s = 1$, it is trivial. Let $s > 1$. We may assume that $V_2 \neq V_1$. By 4.9 and 4.10,

$$X^{(0)} = \sigma(X^{(0)}) = \sigma(V_1), \qquad Y^{(0)} = \sigma(Y^{(0)}) = \sigma(V_s),$$
$$\sigma(V_i) \sim \sigma(V_{i+1})(\mathfrak{n}_3').$$

Since $V_2 \sim X^{(0)}(\mathfrak{n}_3')$, we evidently have $V_2 \sim V_1(\mathfrak{n}_1')$, where

$$V_2 = S_1^{(0)} S_2^{(k)} S_3^{(0)}, \qquad AS_3 = A^{k+1}R, \qquad V_1 = X^{(0)},$$
$$S_1 S_2 R = X$$

and therefore $\sigma(V_2) = (S_1 S_2 R)^{(0)} = X^{(0)}$.

By the inductive assumption for the chain $X^{(0)} = \sigma(V_2), \sigma(V_3), \ldots, \sigma(V_{s-1}), \sigma(V_s) = Y^{(0)}$ we conclude that $X = Y$.

4.12. THEOREM. *In order that an element A of a semigroup $\mathfrak{A}$ be potentially invertible from the left* (4.1) *it is necessary and sufficient that for any X and Y*

which are elements of $\mathfrak{A}$ *or empty symbols,* $A^2X = A^2Y$ *implies* $SX = SY$ *for all* $S \in \mathfrak{A}$.

PROOF. The necessity was proven in 4.4.

We now show the sufficiency (taking 4.5 into account). We construct the semigroup $\mathfrak{B}$ (4.7), considering it as a supersemigroup of $\mathfrak{A}$ (for which we identify $S^{(0)}$ with S for each $S \in \mathfrak{A}$). We consider the factor semigroup $\mathfrak{B} = \mathfrak{B}/\mathfrak{n}_3''$. In this semigroup, by 4.3, the element $\bar{A}$ (i.e., the class of words in $\mathfrak{B}$ which are equivalent to A mod $\mathfrak{n}_3''$) is left invertible, since for any $S^{(k)}$ we have

$$S^{(k+1)}A \sim S^{(k)}(\mathfrak{n}_3'')$$

and hence in $\mathfrak{B}$

$$\overline{S^{(k+1)}} \cdot \bar{A} = \overline{S^{(k)}}.$$

We consider the natural homomorphism φ of $\mathfrak{B}$ onto $\mathfrak{B} = \mathfrak{B}/\mathfrak{n}_3''$. φ when restricted to $\mathfrak{A}$ is an isomorphism, since by 4.11 for $S \neq R$ ($S, R \in \mathfrak{A}$) we have $S^{(0)} \not\sim R^{(0)}(\mathfrak{n}_3'')$, i.e., $\overline{S^{(0)}} \neq \overline{R^{(0)}}$. Here $\varphi(A) = \bar{A}$. Replacing in $\mathfrak{B}$ the elements of $\varphi(\mathfrak{A})$ by the corresponding elements of $\mathfrak{A}$, we obtain a supersemigroup of φ in which A is left invertible.

4.13. Along with the properties of potential left invertibility and potential right invertibility of elements it is natural to consider potential two-sided invertibility. Evidently an element which is potentially two-sidedly invertible is potentially invertible from the left and potentially invertible from the right. However the fulfillment of this necessary condition is, generally speaking, not sufficient for an element to be potentially two-sidedly invertible. Below we mention an example of a semigroup constructed by E. G. Šutov [**1**] patterned after a construction of A. I. Mal'cev [**1**], in which, since it is a semigroup with two-sided cancellation, each element is potentially left invertible and potentially right invertible. At the same time we shall indicate in it an element which is not potentially two-sidedly invertible. Hence, in particular, it would follow that this is a semigroup with two-sided cancellation which is not embeddable in a group, since it is clear that all the elements of any semigroup which are embeddable in a group are potentially two-sidedly invertible.

4.14. We consider the free semigroup $\mathfrak{W}_{\mathfrak{R}}$ over the alphabet $\mathfrak{R} = \{X_1, X_2, X_3, X_4, X_5, X_6\}$ (IX, 1.3). We denote

$$\mathfrak{R}_1 = \{X_1, X_2, X_3, X_4\}, \qquad \mathfrak{R}_2 = \{X_1, X_2, X_5, X_6\}.$$

The set of words X_1X_5, X_1X_6, X_2X_5, X_3X_1, X_3X_2, X_4X_1 we denote by $\mathfrak{B}$. We note that if $X_iX_j \in \mathfrak{B}$, then $X_i \in \mathfrak{R}_1$ and $X_j \in \mathfrak{R}_2$.

4.15. We denote by δ the following map of $\mathfrak{R}_1$ onto itself:

$$\delta(X_1) = X_3, \qquad \delta(X_2) = X_4, \qquad \delta(X_3) = X_1, \qquad \delta(X_4) = X_2,$$

and by τ the map of $\mathfrak{K}_2$ onto itself,

$$\tau(X_1) = X_5, \qquad \tau(X_2) = X_6, \qquad \tau(X_5) = X_1, \qquad \tau(X_6) = X_2.$$

We note that $\delta^2(X_i) = X_i$ and $\tau^2(X_i) = X_i$.

4.16. We introduce in $\mathfrak{W}_{\mathfrak{K}}$ a relation $\mathfrak{n}$ by setting

$$U \sim V(\mathfrak{n}) \qquad (U, V \in \mathfrak{W}_{\mathfrak{K}}),$$

if $U = V$ or $U = X_iX_j \in \mathfrak{B}$ and $V = \delta(X_i) \cdot \tau(X_j) \in \mathfrak{B}$. The relation $\mathfrak{n}$ is reflexive, and by the above properties of δ and τ it is symmetric.

4.17. We shall indicate a number of properties of the relation $\mathfrak{n}$. The validity of some of these follows immediately from the definition of $\mathfrak{n}$.

(α) If $X_iX_j, X_iX_k \in \mathfrak{B}$, then the equality $X_j = \delta(X_k)$ is impossible.

(β) Let

$$X_{i_1}X_{i_2} \dots X_{i_s} \sim X_{j_1}X_{j_2} \dots X_{j_s}(\mathfrak{n}'')$$

and $X_{i_1} \bar{\in} \mathfrak{K}_1$. Then $X_{j_1} = X_{i_1}$.

(γ) Let

$$X_{i_1}X_{i_2} \dots X_{i_s} \sim X_{j_1}X_{j_2} \dots X_{j_s}(\mathfrak{n}'')$$

and $X_{i_1} \in \mathfrak{K}_1$. Then either $X_{j_1} = X_{i_1}$ or $X_{j_1} = \delta(X_{i_1})$.

(δ) If

$$X_kX_{i_1}X_{i_2} \dots X_{i_s} \sim X_kX_{j_1}X_{j_2} \dots X_{j_s}(\mathfrak{n}'),$$

then $X_{i_1}X_{i_2} \dots X_{i_s} \sim X_{j_1}X_{j_2} \dots X_{j_s}(\mathfrak{n}')$.

(ε) If

$$X_kX_{i_1}X_{i_2} \dots X_{i_s} \sim X_kX_{j_1}X_{j_2} \dots X_{j_s}(\mathfrak{n}''),$$

then $X_{i_1}X_{i_2} \dots X_{i_s} \sim X_{j_1}X_{j_2} \dots X_{j_s}(\mathfrak{n}'')$.

In fact, let

$$U_1 \sim U_2(\mathfrak{n}'), \qquad U_2 \sim U_3(\mathfrak{n}'), \dots, U_{m-1} \sim U_m(\mathfrak{n}')$$
$$(U_1, U_2, \dots, U_m \in \mathfrak{W}_{\mathfrak{K}}),$$
$$U_l = X_{l0}X_{l1} \dots X_{ls} \qquad (X_{lr} \in \mathfrak{K}; l = 1, 2, \dots, m; r = 1, 2, \dots, s),$$
$$U_1 = X_kX_{i_1}X_{i_2} \dots X_{i_s}, \qquad U_m = X_kX_{j_1}X_{j_2} \dots X_{j_s}.$$

If $X_kX_{i_1} \bar{\in} \mathfrak{B}$ or $X_kX_{j_1} \bar{\in} \mathfrak{B}$, then by the definition of derived relations, the validity of the indicated relation follows immediately. Therefore, in what follows we can consider that both the indicated words belong to $\mathfrak{B}$. Thus $X_k \in \mathfrak{K}_1$.

The proof will be by induction on m. For $m = 2$ we have the case (δ).

Let $m > 2$.

If in some U_l ($l = 2, 3, \dots, m - 1$) we have $X_0 = X_k$, then by the inductive assumption applied to $U_1, U_2, \dots, U_l$ and $U_l, U_{l+1}, \dots, U_m$ we obtain

$$X_{i_1}X_{i_2} \dots X_{i_s} \sim X_{l1}X_{l2} \dots X_{ls}(\mathfrak{n}''),$$
$$X_{l1}X_{l2} \dots X_{ls} \sim X_{j_1}X_{j_2} \dots X_{j_s}(\mathfrak{n}''),$$

whence follows the indicated relation.

Here we may assume in what follows that $X_{l0} \neq X_k$ ($l = 2, 3, \ldots, m-1$), i.e., by (γ), that $X_{l0} = \delta(X_k)$. Hence it follows in particular that $X_{20} \neq X_{10}$. Therefore

$$X_{20} = \delta(X_k), \qquad X_{21} = \tau(X_{i_1}),$$

$$X_{22} = X_{i_2}, \qquad X_{23} = X_{i_3}, \ldots, X_{2s} = X_{i_s}.$$

Analogously,

$$X_{m-1,0} = \delta(X_k), X_{m-1,1} = \tau(X_{j_1}),$$

$$X_{m-1,2} = X_{j_2}, \qquad X_{m-1,3} = X_{j_3}, \ldots, X_{m-1,s} = X_{j_s}.$$

From $U_2 \sim U_{m-1}(\mathfrak{n}'')$, by the inductive assumption it follows that $X_{21}X_{22}\ldots X_{2s} \sim X_{m-1,1}X_{m-1,2}\ldots X_{m-1,s}(\mathfrak{n}'')$. By ($\beta$) and ($\gamma$) either $X_{m-1,1} = X_{21}$ or $X_{m-1,1} = \delta(X_{21})$. But $X_{20}X_{21}$ and $X_{m-1,0}X_{m-1,1}$ belong to $\mathfrak{B}$, and by (α) the second equation is impossible. Therefore $X_{m-1,1} = X_{21}$. So we can apply the inductive assumption to the words $X_{21}X_{22}\ldots X_{2s}$ and $X_{m-1,1}X_{m-1,2}\ldots X_{m-1,s}$, with the result that

$$X_{i_2}X_{i_3}\ldots X_{i_s} \sim X_{j_2}X_{j_3}\ldots X_{j_s}(\mathfrak{n}'').$$

But as was proven,

$$X_{21} = \tau(X_{i_1}), \qquad X_{m-1,1} = \tau(X_{j_1}), \qquad X_{21} = X_{m-1,1}.$$

Therefore, multiplying the relation so obtained on the left by X_{i_1}, which is equal to X_{j_1}, we obtain the required relation.

(ζ) If

$$X_{i_1}X_{i_2}\ldots X_{i_s}X_k \sim X_{j_1}X_{j_2}\ldots X_{j_s}X_k(\mathfrak{n}''),$$

then $X_{i_1}X_{i_2}\ldots X_{i_s} \sim X_{j_1}X_{j_2}\ldots X_{j_s}(\mathfrak{n}'')$.

The proof is completely analogous to (ε).

(η) $X_2X_6 \nsim X_4X_2(\mathfrak{n}'')$.

This follows from the fact that the word X_2X_6 is congruent only to itself mod $\mathfrak{n}$.

4.18. We consider the semigroup $\mathfrak{W}_{\mathfrak{K}}^{\mathfrak{n}} = \mathfrak{W}_{\mathfrak{K}}/\mathfrak{n}''$ (IX, 3.2).

The elements of $\mathfrak{W}_{\mathfrak{K}}^{\mathfrak{n}}$ being $\mathfrak{n}''$-classes, we will write them in the form $\overline{W}$ (where $\overline{W}$ is the $\mathfrak{n}''$-class containing W, $W \in \mathfrak{W}_{\mathfrak{K}}$).

From 4.17 (ε), (ζ) it follows that in $\mathfrak{W}_{\mathfrak{K}}^{\mathfrak{n}}$, $\overline{X}_k\overline{U} = \overline{X}_k\overline{V}$ implies $\overline{U} = \overline{V}$ and $\overline{U}\overline{X}_k = \overline{V}\overline{X}_k$ implies $\overline{U} = \overline{V}$. Therefore $\mathfrak{W}_{\mathfrak{K}}^{\mathfrak{n}}$ is a semigroup with two-sided cancellation.

4.19. We deduce a necessary condition for an element of a semigroup to be potentially two-sidedly invertible.

LEMMA. *If in a semigroup $\mathfrak{A}$ elements $Z_1, Z_2, Z_3, Z_4, Z_5, Z_6$ satisfy the equalities*

$$Z_1Z_5 = Z_3Z_1,$$
$$Z_2Z_5 = Z_4Z_1,$$
$$Z_1Z_6 = Z_3Z_2,$$

and if there exists a supersemigroup $\mathfrak{A}'$ *of* $\mathfrak{A}$ *in which* Z_1 *is two-sidedly invertible, then*

$$Z_2Z_6 = Z_4Z_2.$$

PROOF. In $\mathfrak{A}'$ there must exist U, V such that

$$Z_1U = Z_2, \qquad VZ_1 = Z_2.$$

Using the relations between the elements Z_i we obtain

$$Z_2Z_6 = VZ_1Z_6 = VZ_3Z_2 = VZ_3Z_1U = VZ_1Z_5U = Z_2Z_5U = Z_4Z_1U = Z_4Z_2.$$

4.20. In the semigroup with two-sided cancellation $\mathfrak{W}_{\mathfrak{R}}^{\mathfrak{n}} = \mathfrak{W}_{\mathfrak{R}}/\mathfrak{n}''$ (4.18) the elements $\bar{X}_1, \bar{X}_2, \bar{X}_3, \bar{X}_4, \bar{X}_5, \bar{X}_6$ satisfy the equalities

$$\bar{X}_1\bar{X}_5 = \bar{X}_3\bar{X}_1,$$
$$\bar{X}_2\bar{X}_5 = \bar{X}_4\bar{X}_1,$$
$$\bar{X}_1\bar{X}_6 = \bar{X}_3\bar{X}_2.$$

At the same time, by 4.17 (η) we have the inequality

$$\bar{X}_2\bar{X}_6 \neq \bar{X}_4\bar{X}_2.$$

From this, by 4.19, it follows that $\bar{X}_1$ is not potentially two-sidedly invertible.

4.21. In conclusion we note that, by IX, 3.2, 3.10, the semigroup $\mathfrak{W}_{\mathfrak{R}}^{\mathfrak{n}} = \mathfrak{W}_{\mathfrak{R}}/\mathfrak{n}''$ can be defined as a semigroup with generating set $\{X_1, X_2, X_3, X_4, X_5, X_6\}$ and with the defining system of relations

$$X_1X_5 = X_3X_1,$$
$$X_2X_5 = X_4X_1,$$
$$X_1X_6 = X_3X_2.$$

5. Free and Direct Products

5.1. Along with the problem of embedding a given semigroup into a supersemigroup with given properties there naturally arises the related problem of the embedding of entire systems of semigroups. But we must at once point out that even if we do not impose any requirements at all on the desired supersemigroup, we sometimes obtain a negative answer to the question of the possibility of such an embedding.

Consider the following example. We take the two sets of rational numbers,

$$\mathfrak{A} = \{\ldots, \tfrac{1}{16}, \tfrac{1}{8}, \tfrac{1}{4}, \tfrac{1}{2}, 1, 2, 4, 8, \ldots\},$$
$$\mathfrak{B} = \{\ldots, -\tfrac{1}{16}, -\tfrac{1}{8}, -\tfrac{1}{4}, -\tfrac{1}{2}, 1, 2, 4, 8, \ldots\}.$$

We will consider $\mathfrak{A}$ with respect to the usual operation of multiplication of rational numbers. In $\mathfrak{B}$ we define the operation, denoted by the symbol $\circ$, in the following way:

$$\alpha \circ \beta = \begin{cases} \alpha\beta & (\text{if} \quad \alpha\beta \in \mathfrak{B}), \\ -\alpha\beta & (\text{if } -\alpha\beta \in \mathfrak{B}). \end{cases}$$

Both these operations are associative.

The semigroups $\mathfrak{A}$ and $\mathfrak{B}$ have the common part $\{1, 2, 4, 8, \ldots\}$. The operations in $\mathfrak{A}$ and $\mathfrak{B}$ are compatible, i.e., the product of two elements of the common part defined according to the multiplication in $\mathfrak{A}$ coincides with the product of those elements defined according to the multiplication in $\mathfrak{B}$. Nevertheless there does not exist a common supersemigroup for both these semigroups. In fact, in such a common supersemigroup, if it existed, we would obtain

$$\tfrac{1}{2} = \tfrac{1}{2} \cdot 1 = \tfrac{1}{2} \cdot [2 \circ (-\tfrac{1}{2})] = (\tfrac{1}{2} \cdot 2) \circ (-\tfrac{1}{2}) = 1 \circ (-\tfrac{1}{2}) = -\tfrac{1}{2}.$$

Evidently the impossibility of embedding both semigroups in a common supersemigroup in the above example depends on the presence of common elements in $\mathfrak{A}$ and $\mathfrak{B}$. As we will show in 5.4, with the absence of common elements, embedding into a common supersemigroup is always possible.

5.2. Definition. *A semigroup $\mathfrak{A}$ is the* FREE PRODUCT *of its pairwise disjoint subsemigroups $\mathfrak{B}_\alpha, \mathfrak{B}_\beta, \ldots$ if $\mathfrak{B} = \mathfrak{B}_\alpha \cup \mathfrak{B}_\beta \cup \ldots$ is a generating set of $\mathfrak{A}$, where the aggregate of all the relations between the elements of $\mathfrak{B}_\alpha$, all the relations between the elements of $\mathfrak{B}_\beta$, and so forth, is a defining system of relations for $\mathfrak{A}$ with respect to $\mathfrak{B}$.*

The representation of a semigroup in the form of a free product is usually called a decomposition of the semigroup into a free product.

5.3. A semigroup which is decomposed into a free product is determined up to isomorphism by giving up to isomorphism the components of that product. This immediately follows from IX, 1.12.

5.4. Let $\mathfrak{A}$ be a free product of subsemigroups $\mathfrak{B}_\alpha, \mathfrak{B}_\beta, \ldots$. An arbitrary word W in the alphabet $\mathfrak{B} = \mathfrak{B}_\alpha \cup \mathfrak{B}_\beta \cup \ldots$ can be represented in the form

$$W = U_{\xi_1} U_{\xi_2} \ldots U_{\xi_k} \ldots U_{\xi_n},$$

where U_{ξ_i} $(i = 1, 2, \ldots, n)$ is a word in $\mathfrak{B}_{\xi_i}$ and $\mathfrak{B}_{\xi_{i-1}}$ and $\mathfrak{B}_{\xi_i}$ are always distinct. If we transform W by means of any of the relations in the defining system of relations 5.2, then W is transformed to a word of the form

$$W = U_{\xi_1} U_{\xi_2} \ldots U'_{\xi_k} \ldots U_{\xi_n},$$

where the words U_{ξ_k} and U'_{ξ_k} have the same value in $\mathfrak{B}_{\xi_k}$. As the result of a sequence of such mappings W is transformed into a word of the form

$$W = U'_{\xi_1} U'_{\xi_2} \ldots U'_{\xi_k} \ldots U'_{\xi_n},$$

where the words U_{ξ_i} and U'_{ξ_i} in $\mathfrak{B}_{\xi_i}$ have the same value in $\mathfrak{B}_{\xi_i}$ $(i = 1, 2, \ldots, n)$.

5.5. Any $A \in \mathfrak{A}$ can be given in the form of a product

$$A = B_{\xi_1} B_{\xi_2} \dots B_{\xi_n},$$

where B_{ξ_i} $(i = 1, 2, \dots, n)$ is an element of $\mathfrak{B}_{\xi_i}$, and the $\mathfrak{B}_{\xi_{i-1}}$ and $\mathfrak{B}_{\xi_i}$ are distinct. This form of an element of $\mathfrak{A}$ is canonical, since no two different products in such a form can be equal to each other. This follows immediately from 5.4.

The indicated canonical representation of elements of a free product $\mathfrak{A}$ is very convenient inasmuch as elements given in this canonical form are easily multiplied. Let

$$A = B_{\xi_1} B_{\xi_2} \dots B_{\xi_n}, \qquad A' = B'_{\eta_1} B'_{\eta_2} \dots B'_{\eta_m}.$$

If $\xi_n \neq \eta_1$, then the canonical form of the product will evidently be

$$AA' = B_{\xi_1} B_{\xi_2} \dots B_{\xi_n} B'_{\eta_1} B'_{\eta_2} \dots B'_{\eta_m}.$$

If $\xi_n = \eta_1$ and in $\mathfrak{B}_{\eta_1}$ we have $B_{\xi_n} B'_{\eta_1} = B^*_{\eta_1}$, then evidently

$$AA' = B_{\xi_1} B_{\xi_2} \dots B_{\xi_{n-1}} B^*_{\eta_1} B'_{\eta_2} \dots B'_{\eta_m}.$$

The description of $\mathfrak{A}$ as a set in the indicated canonical form compatible with the above rule for multiplication can evidently serve as a definition of free product.

5.6. It is not hard to solve the converse to the above problem of decomposing a given semigroup into a free product.

THEOREM. *For semigroups $\mathfrak{B}_\alpha, \mathfrak{B}_\beta, \dots$ which do not have pairwise common elements, there exists a common supersemigroup $\mathfrak{A}$ which is decomposed into the free product $\mathfrak{B}_\alpha, \mathfrak{B}_\beta, \dots$.*

PROOF. Let $\mathfrak{A}$ be the semigroup given by the generating set $\mathfrak{N} = \mathfrak{B}_\alpha \cup \mathfrak{B}_\beta \cup \dots$ and the defining system Ψ consisting of all relations of the semigroup $\mathfrak{B}_\alpha$, all relations of the semigroup $\mathfrak{B}_\beta$, and so forth (IX, 3.10). The relation $\mathfrak{n}_\Psi$ in the free semigroup $\mathfrak{W}_\mathfrak{N}$ corresponding to the system of relations Ψ (IX, 3.7) is such that for $X \in \mathfrak{B}_\alpha$ the condition $X \sim Y(\mathfrak{n}''_\Psi)$ $(Y \in \mathfrak{N})$ is valid only for $Y \in \mathfrak{B}_\alpha$ and $X = Y$. In fact, the transformation of X by means of any relation of Ψ gives an element which again belongs to $\mathfrak{B}_\alpha$ and is equal to X. The same result is obtained by successively applying several of these transformations. By IX, 3.7, in $\mathfrak{A}$ all the elements of $\mathfrak{N}$ are distinct. Thus $\mathfrak{A}$ turns out to be a supersemigroup for each $\mathfrak{B}_\alpha, \mathfrak{B}_\beta, \dots$ and is evidently their free product.

5.7. The construction of a free product is not simply one of the solutions of the problem of embedding given semigroups into a common supersemigroup, but it plays a special role for this problem. Let the semigroups $\mathfrak{B}_\alpha, \mathfrak{B}_\beta, \dots$ be pairwise disjoint and let $\mathfrak{A}$ be a common supersemigroup of them. We denote by $\mathfrak{A}'$ the subsemigroup of $\mathfrak{A}$ generated by $\mathfrak{B} = \mathfrak{B}_\alpha \cup \mathfrak{B}_\beta \cup \cdots$. In $\mathfrak{A}'$, $\mathfrak{B}$ is a generating set, and in $\mathfrak{A}'$ we have the validity with respect to $\mathfrak{B}$ of all the relations between the elements of $\mathfrak{B}_\alpha$, all the relations between the elements of $\mathfrak{B}_\beta$, and so forth. By IX, 1.11 there exists a homomorphism of the free product $\mathfrak{F}$ of the semigroups $\mathfrak{B}_\alpha, \mathfrak{B}_\beta, \ldots$ onto $\mathfrak{A}'$ which is an extension of the identity maps of $\mathfrak{B}_\alpha, \mathfrak{B}_\beta, \ldots$, considered as subsemigroups of $\mathfrak{F}$, onto $\mathfrak{B}_\alpha, \mathfrak{B}_\beta, \ldots$, considered as subsemigroups of $\mathfrak{A}'$.

Thus any supersemigroup of the semigroups $\mathfrak{B}_\alpha, \mathfrak{B}_\beta, \ldots$ can be obtained from their free product $\mathfrak{F}$ by means of a homomorphism of $\mathfrak{F}$, which is an extension of the identity maps of $\mathfrak{B}_\alpha, \mathfrak{B}_\beta, \ldots$ onto themselves, and a subsequent embedding of the semigroup so obtained into the arbitrary supersemigroup. That this argument is similar to the arguments about free semigroups (§ 5, Chapter IX) does not happen by chance. The fact is that, as is easily seen, free products of infinite singly generated semigroups are free semigroups in the class of all semigroups.

5.8. Let a semigroup $\mathfrak{A}$ have subsemigroups $\mathfrak{B}_\alpha, \mathfrak{B}_\beta, \ldots$, where each $\mathfrak{B}_\xi$ is given by a generating set $\mathfrak{K}_\xi$ and defining system of relations Ψ with respect to $\mathfrak{K}_\xi$ $(\xi = \alpha, \beta, \ldots)$. If $\mathfrak{K}_\alpha, \mathfrak{K}_\beta, \ldots$ are pairwise disjoint, and $\mathfrak{K} = \mathfrak{K}_\alpha \cup \mathfrak{K}_\beta \cup \ldots$ is a generating set for $\mathfrak{A}$, and $\Psi_\alpha \cup \Psi_\beta \cup \ldots$ is a defining system of relations with respect to $\mathfrak{K}$, then $\mathfrak{A}$ is the free product of $\mathfrak{B}_\alpha, \mathfrak{B}_\beta, \ldots$.

In fact, since $\mathfrak{K}$ is a generating set for $\mathfrak{A}$, the system $\mathfrak{B}_\alpha \cup \mathfrak{B}_\beta \cup \ldots$ will all the more generate $\mathfrak{A}$. For an arbitrary word W in $\mathfrak{K}_\alpha$, if we transform it by relations of $\Psi_\alpha \cup \Psi_\beta \cup \ldots$, we will always obtain a word in $\mathfrak{K}_\alpha$. From this it follows that no element of $[\mathfrak{K}_\alpha] = \mathfrak{B}_\alpha$ can equal any element of $[\mathfrak{K}_\beta] = \mathfrak{B}_\beta$ $(\alpha \neq \beta)$. Therefore $\mathfrak{B}_\alpha, \mathfrak{B}_\beta, \ldots$ are pairwise disjoint. Since $\Psi_\alpha \cup \Psi_\beta \cup \ldots$ is a defining system for $\mathfrak{A}$ with respect to $\mathfrak{K}$ it follows from IX, 2.7 that the system of all relations between the elements of $\mathfrak{B}_\alpha$, all relations between the elements of $\mathfrak{B}_\beta$, and so forth, will be a defining system of relations for $\mathfrak{A}$ with respect to $\mathfrak{B}_\alpha \cup \mathfrak{B}_\beta \cup \ldots$.

5.9. If we consider semigroups only to within isomorphism, i.e., if we do not distinguish semigroups which are isomorphic to each other, then the construction of a free product defines an operation between semigroups. The result of this operation on the semigroups $\mathfrak{A}$ and $\mathfrak{B}$ is the semigroup $\mathfrak{F}$, decomposed into the free product of its subsemigroups $\mathfrak{A}'$ and $\mathfrak{B}'$, which are isomorphic to the corresponding semigroups $\mathfrak{A}$ and $\mathfrak{B}$. By 5.6 such a semigroup always exists and by 5.3 it is completely determined up to isomorphism.

5.10. Commutativity of the operation mentioned in 5.9 follows directly from Definition 5.2, and associativity also holds, as follows from 5.8. In fact, let $\mathfrak{F}_{12}$ be the free product of the semigroups $\mathfrak{A}_1$ and $\mathfrak{A}_2$, and let $\mathfrak{F}'$ be the free product of $\mathfrak{F}_{12}$ and $\mathfrak{A}_3$. By 5.8, $\mathfrak{F}'$ is the free product of the three subsemigroups $\mathfrak{A}_1, \mathfrak{A}_2, \mathfrak{A}_3$. Similarly $\mathfrak{F}''$, which is the free product of $\mathfrak{A}_1$ and $\mathfrak{F}_{23}$, where $\mathfrak{F}_{23}$ is the free product of $\mathfrak{A}_2$ and $\mathfrak{A}_3$, turns out to be the free product of the three semigroups $\mathfrak{A}_1, \mathfrak{A}_2, \mathfrak{A}_3$.

5.11. By 5.9, 5.10, starting from given semigroups, considered to within isomorphism, we can construct a commutative semigroup $\mathfrak{T}$ whose elements are semigroups. The operation in $\mathfrak{T}$ will be the construction of the free product of two semigroups. The initial semigroups will be the elements of the generating set for $\mathfrak{T}$.

5.12. Construction of a free product is related to the construction of a direct product.

DEFINITION. *A semigroup* $\mathfrak{A}$ *is the* DIRECT PRODUCT *of its pairwise disjoint subsemigroups* $\mathfrak{B}_\alpha, \mathfrak{B}_\beta, \ldots$ *if* $\mathfrak{B} = \mathfrak{B}_\alpha \cup \mathfrak{B}_\beta \cup \ldots$ *is a generating set of* $\mathfrak{A}$, *where for any* $B_\xi \in \mathfrak{B}_\xi$, $B_\eta \in \mathfrak{B}_\eta$ $(\xi \neq \eta;\ \xi, \eta = \alpha, \beta, \ldots)$ we have

$$B_\xi B_\eta = B_\eta B_\xi$$

and the system of these relations together with all the relations between the elements of $\mathfrak{B}_\alpha$, *all the relations between the elements of* $\mathfrak{B}_\beta$, *and so forth, is a defining system of relations of* $\mathfrak{A}$ *with respect to* $\mathfrak{B}$.

5.13. If in some way we order the components of a direct decomposition of a semigroup $\mathfrak{A}$, then clearly any $A \in \mathfrak{A}$ can be represented in the form

$$A = B_{\xi_1} B_{\xi_2} \ldots B_{\xi_n}, \qquad B_{\xi_i} \in \mathfrak{B}_{\xi_i} \qquad (i = 1, 2, \ldots, n)$$

where $B_{\xi_i} B_{\xi_j} = B_{\xi_j} B_{\xi_i}$ and $\mathfrak{B}_{\xi_1} < \mathfrak{B}_{\xi_2} < \ldots < \mathfrak{B}_{\xi_n}$. By a method similar to 5.4, it is not difficult to show that such a representation of an element is unique, i.e., it is the canonical form of the elements of $\mathfrak{A}$ with respect to the generating set $\mathfrak{B}_\alpha \cup \mathfrak{B}_\beta \cup \ldots$. Multiplication of elements in such a canonical form, due to the permutability of elements of different components, is quite simple. Let

$$A = B_{\xi_1} B_{\xi_2} \ldots B_{\xi_n}, \qquad A' = B'_{\xi_1} B'_{\xi_2} \ldots B'_{\xi_n}$$

(some of the B_{ξ_i} and B'_{ξ_j} can be empty symbols, so that in the decomposition of both the elements we can formally write elements from one and the same component). Then clearly

$$AA' = (B_{\xi_1} B'_{\xi_1}) \cdot (B_{\xi_2} B'_{\xi_2}) \cdot \ldots \cdot (B_{\xi_n} B'_{\xi_n}).$$

Evidently $\mathfrak{A}$ can be given as the set of indicated formal products with the above law of multiplication.

We note at the same time that by a direct product of finitely many semigroups $\mathfrak{B}_1, \mathfrak{B}_2, \ldots, \mathfrak{B}_n$ is meant (by some authors) the set of formal products of the form $B_1 B_2 \ldots B_n$ (where B_i is always an element of $\mathfrak{B}_i$, but not the empty symbol; $i = 1, 2, \ldots, n$) with respect to the same operation indicated above. However such a semigroup in the general case will not necessarily be a supersemigroup of $\mathfrak{B}_1, \mathfrak{B}_2, \ldots, \mathfrak{B}_n$.

5.14. By reasoning similar to the reasoning in 5.6 or by starting from the method of describing a direct product indicated in 5.13, it is not hard to see that the assertion for direct products of the statement analogous to 5.6 is valid. For mutually disjoint semigroups $\mathfrak{B}_\alpha, \mathfrak{B}_\beta, \ldots$ there exists a supersemigroup $\mathfrak{A}$ which is decomposed into the direct product of $\mathfrak{B}_\alpha, \mathfrak{B}_\beta, \ldots$.

5.15. The role of a direct product in the problem of embedding of systems of semigroups is determined in the following way. Let $\mathfrak{B}_\alpha, \mathfrak{B}_\beta, \ldots$ be mutually disjoint semigroups. $\mathfrak{A}$ is a supersemigroup of each of these; moreover, elements of different semigroups $\mathfrak{B}_\xi$ permute with each other in $\mathfrak{A}$. By IX, 1.11 for the direct product $\mathfrak{N}$ of $\mathfrak{B}_\alpha, \mathfrak{B}_\beta, \ldots$ there exists a homomorphism into $\mathfrak{A}$ extending the identity maps of $\mathfrak{B}_\alpha, \mathfrak{B}_\beta, \ldots$. As in 5.7, any supersemigroup $\mathfrak{A}$ of $\mathfrak{B}_\alpha, \mathfrak{B}_\beta, \ldots$ having the indicated property of permutability can be obtained by some homomorphism of $\mathfrak{N}$ and a subsequent embedding of the semigroup so obtained into the arbitrary supersemigroup.

5.16. As in 5.9, 5.10, it is easy to see that the construction of the direct product can be considered as an operation between semigroups considered up to isomorphism. This operation is commutative and associative. With respect to it we can consider semigroups whose elements are semigroups.

5.17. Besides free products and direct products there are other analogous constructions which accomplish embeddings of systems of semigroups with similar properties.

It is also possible to consider the analogous problem and to set up different constructions for the case where the initial semigroups can have common elements. The arguments in this case are at once essentially complicated. Here the solution, as we have seen in 5.1 will in general not always be possible.

5.18. The simplest and most important case is when the semigroups $\mathfrak{B}_\alpha, \mathfrak{B}_\beta, \ldots$ all have a common unit but have no other elements in common. In this case we can define the free product of these semigroups with common unit as the semigroup in which $\mathfrak{B} = \mathfrak{B}_\alpha \cup \mathfrak{B}_\beta \cup \ldots$ is a generating set, and the system of all relations between the elements of $\mathfrak{B}_\alpha$, all the relations between the elements of $\mathfrak{B}_\beta$, and so forth, is the defining system of relations with respect to $\mathfrak{B}$.

Examination of properties analogous to the properties considered for free products of semigroups without common elements is more complicated. The interesting thing about this case is that it includes the usual construction of a free product of groups[1] (if $\mathfrak{B}_\alpha, \mathfrak{B}_\beta, \ldots$ are all groups, then their free product with common unit will evidently be a group). Also the usual direct product of groups[2] is an obvious particular case of the construction given above for the direct product of semigroups with common unit.

---

[1] Cf., for example, A. G. Kuroš, *The theory of groups*, 2nd ed., GITTL, Moscow 1953; Chapter IX (Russian); English transl., Chelsea, New York, 1955, 1956.

[2] Ibid. § 17.

The question of the existence of other constructions with analogous properties for groups with a common unit was first stated as a problem by A. G. Kuroš.[3] A positive solution is found in the work of O. N. Golovin.[4]

After that other authors considered various constructions with analogous properties. We mention in this connection the work of M. A. Fridman.[5] For collections of semigroups with common unit a construction different from a free product with common unit and different from a direct product with common unit was given by E. S. Ljapin [**4**].

Free products of semigroups with common unit and also a construction which is a free product with certain complications, were considered in the work of Griffith [**1**]. Also, R. V. Petropavlovskaja [**5**] studied and made use of a construction which she called a direct product; it is related to the direct product described above, but is not identical with it.

5.19. The general case in which the initial semigroups can have various sets of common elements has not been considered up to now in the general theory of semigroups. Only in the case of groups has some investigation been carried out on free products of groups with an arbitrary system of intersections.

6. Representations of Semigroups by Partial Transformations

6.1. As we have already shown (I, 3.9, 3.10), every semigroup is representable by transformations. An obviously important problem is to describe the structure of all representations. It is of interest to consider representations, not merely by transformations, but by partial transformations; and not just isomorphic representations, but homomorphic ones.

There are presently three different solutions of the problem. They have been given by V. V. Vagner [**6**], E. S. Ljapin and B. M. Šaĭn [**S 1, 3-6, 8**]. The latter has, indeed, carried out a number of investigations of representations. He has given special attention to representations of inverse semigroups, studying in particular the transitive representations (i.e. those such that, for any two elements α and β of the set being transformed, there exists an element of the semigroup whose image under the representation is a transformation taking α into β).

Along with representations by transformations and partial transformations, Šaĭn has also considered representations by relations (taken with respect to the operation of multiplication defined in I, 5.2).

Transitive representations of semigroups by transformations have also

[3] Ibid.

[4] O. N. Golovin, *On associative operations on a set of groups*, Dokl. Akad. Nauk SSSR **58** (1947), 1257–1260 (Russian); and subsequent papers in Mat. Sb. (Russian)

[5] M. A. Fridman, *On semi-commutative multiplications*, Dokl. Akad. Nauk SSSR **109** (1956), 710–712 (Russian); and a series of papers Učen. Zap. Glazovsk Ped. Inst. (1956). (Russian)

been investigated by Tully [S 1].

We consider below a construction, due to E. S. Ljapin [S 1], which gives all representations of a semigroup (including homomorphic representations) by partial transformations, and which is based on embedding of semigroups.

6.2. Suppose the semigroup $\mathfrak{A}$ is embedded in some supersemigroup $\mathfrak{B}$, and let $\mathfrak{L}_1 \supset \mathfrak{L}_2$ be two left ideals of $\mathfrak{B}$, where $\mathfrak{L}_1 \neq \mathfrak{L}_2$ and $\mathfrak{L}_2$ may also be the void set. Let $\Omega = \mathfrak{L}_1 \backslash \mathfrak{L}_2$. Associate with every element $A \in \mathfrak{A}$ the partial transformation σ_A of Ω such that the set $\Pi_1^{(\sigma A)}$ consists of those $X \in \Omega$ for which $AX \in \Omega$ and, for each such X,

$$\sigma_A(X) = AX.$$

The corresponding mapping of the semigroup $\mathfrak{A}$ into the semigroup $\mathfrak{P}_\Omega$ of all partial transformations of the set Ω (I, 4.2) will be denoted by $\sigma(\mathfrak{A}, \mathfrak{B}, \mathfrak{L}_1, \mathfrak{L}_2)$.

6.3. It should be kept in mind in what follows that since $\mathfrak{L}_1$ is a left ideal of $\mathfrak{B}$, it must be true for any $A \in \mathfrak{A}$ and $X \in \Omega$ that $AX \in \mathfrak{L}_1$. Thus to have $X \in \Pi_1^{(\sigma A)}$ (i.e., $AX \in \Omega$), it is necessary and sufficient that $AX \overline{\in} \mathfrak{L}_2$.

6.4. **Theorem.** *The mapping* $\sigma(\mathfrak{A}, \mathfrak{B}, \mathfrak{L}_1, \mathfrak{L}_2)$ (6.2) *of the semigroup* $\mathfrak{A}$ *into* $\mathfrak{P}_\Omega$ *is a homomorphic representation of* $\mathfrak{A}$ *by partial transformations of the set* Ω.

Proof. Suppose $A_1 A_2 = A_3 (A_1, A_2, A_3 \in \mathfrak{A})$.

If, for $X \in \Omega$, it is the case that $X \overline{\in} \Pi_1^{(\sigma A 2)}$ (i.e., if $A_2 X \in \mathfrak{L}_2$), then $X \overline{\in} \Pi_1^{(\sigma A 3)}$ (since $A_3 X = A_1 A_2 X = A_1(A_2 X) \in \mathfrak{L}_2$).

If $X \in \Pi_1^{(\sigma A 2)}$ but $A_2 X \overline{\in} \Pi_1^{(\sigma A 1)}$, then we have $X \overline{\in} \Pi_1^{(\sigma A 3)}$ (since $A_3 X = A_1(A_2 X) \in \mathfrak{L}_2$).

If $X \in \Pi_1^{(\sigma A 2)}$ and $A_2 X \in \Pi_1^{(\sigma A 1)}$, then $A_1(A_2 X) \in \Omega$ and so $A_3 X \in \Omega$, i.e., $X \in \Pi_1^{(\sigma A 3)}$; moreover,

$$\sigma_{A_3}(X) = A_3 X = A_1(A_2 X) = \sigma_{A_1}(\sigma_{A_2}(X)) = \sigma_{A_1}\sigma_{A_2}(X).$$

This proves that the mapping $\sigma(\mathfrak{A}, \mathfrak{B}, \mathfrak{L}_1, \mathfrak{L}_2)$, assigning to each $A \in \mathfrak{A}$ the partial transformation σ_A, is a homomorphism.

6.5. The significance of the construction in 6.2 lies in the fact that every homomorphic representation of a semigroup by partial transformations is essentially obtainable by means of such a construction. Indeed, given any family of representations, there exists a single supersemigroup $\mathfrak{B}$ of the semigroup $\mathfrak{A}$ such that they are all obtainable in the above fashion from the embedding of $\mathfrak{A}$ in the one semigroup $\mathfrak{B}$.

6.6. In order to state this result more precisely, we must introduce the notion of inessential distinction of homomorphic representations by partial transformations, an obvious generalization of that given earlier (§7, Chapter V) for representations by transformations.

Let φ_1 and φ_2 be two homomorphic representations of the semigroup $\mathfrak{A}$ by partial transformations of the sets Ω_1 and Ω_2. They are said to be inessentially distinct (or to be similar) if there exists a one-to-one mapping ψ of Ω_1 onto Ω_2 such that for any $A \in \mathfrak{A}$ the following conditions hold:

1) if $\xi \in \Pi_1^{(\varphi_1(A))}$, then $\psi(\xi) \in \Pi_1^{(\varphi_2(A))}$;
2) if $\xi \in \Pi_1^{(\varphi_2(A))}$, then $\psi^{-1}(\xi) \in \Pi_1^{(\varphi_1(A))}$;
3) if $\xi \in \Pi_1^{(\varphi_1(A))}$, then $\psi(\varphi_1(A)(\xi)) = \varphi_2(A)(\psi(\xi))$.

Obviously, inessential distinction means that φ_1 and φ_2 differ only in the notation for, or the intrinsic nature of, the elements of the sets Ω_1 and Ω_2.

6.7. We remark that for any family Ξ of homomorphic representations of a semigroup $\mathfrak{A}$ by partial transformations, it is obvious that one can always find a new family Ξ' of homomorphic representations which has the following properties:

1) There exists a one-to-one correspondence between the representations in Ξ and those in Ξ', such that corresponding representations are only inessentially distinct.

2) If φ_1 and φ_2 are two different members of Ξ', being representations by partial transformations of the sets Ω_1 and Ω_2 respectively, then $\Omega_1 \cap \Omega_2 = \emptyset$.

6.8. **Theorem.** *Let Ξ be an arbitrary family of homomorphic representations of a semigroup $\mathfrak{A}$ by partial transformations. Then $\mathfrak{A}$ is embeddable in a supersemigroup $\mathfrak{B}$, which has a zero O, and is such that every representation in Ξ is only inessentially distinct from some representation $\sigma(\mathfrak{A}, \mathfrak{B}, \mathfrak{L}_1, O)$* (6.2, 6.4).

Proof. For each $\varphi \in \Xi$, let Ω_φ be the set operated on by the transformations in $\varphi(\mathfrak{A})$ (thus φ is a homomorphism of $\mathfrak{A}$ into $\mathfrak{P}_{\Omega_\varphi}$). By the remark in 6.7, it is obvious that we may assume that for any two different $\varphi, \psi \in \Xi$ we have $\Omega_\varphi \cap \Omega_\psi = \emptyset$ and $\Omega_\varphi \cap \mathfrak{A} = \emptyset$.

Form the set

$$\mathfrak{B} = O \cup \mathfrak{A} \cup \Omega_\varphi \cup \Omega_\psi \cup \cdots,$$

where $\varphi, \psi, \cdots$ are all the representations in Ξ, and O is a new element not contained in either $\mathfrak{A}$ or any Ω_φ.

Define an operation of multiplication in $\mathfrak{B}$ as follows:

1) Elements of $\mathfrak{A}$ multiply according to the multiplication rule in $\mathfrak{A}$.
2) $O \cdot X = X \cdot O = O$ for every $X \in \mathfrak{B}$.
3) $\alpha_\varphi \cdot \beta_\psi = O$ for any $\alpha_\varphi \in \Omega_\varphi$, $\beta_\psi \in \Omega_\psi$.
4) $\alpha_\varphi \cdot A = \alpha_\varphi$ for any $\alpha_\varphi \in \Omega_\varphi$, $A \in \mathfrak{A}$.
5) $A \cdot \alpha_\varphi = \varphi(A)(\alpha_\varphi)$ when $\alpha_\varphi \in \Omega_\varphi$, $A \in \mathfrak{A}$ and $\alpha_\varphi \in \Pi_1^{(\varphi(A))}$ (here, the expression $\varphi(A)(\alpha_\varphi)$ designates a certain element of Ω_φ, since $\varphi(A)$ is a partial transformation of the set Ω_φ and $\alpha_\varphi \in \Pi_1^{(\varphi(A))}$).

6) $A \cdot \alpha_\varphi = O$ when $\alpha_\varphi \in \Omega_\varphi$, $A \in \mathfrak{A}$, and $\alpha_\varphi \bar{\in} \Pi_1^{(\varphi(A))}$.

We have thus defined a product for any pair of elements of Ω. Let us show that this operation is associative.

Let $Z_1, Z_2, Z_3 \in \mathfrak{B}$ and

$$U = (Z_1 Z_2) Z_3, \quad V = Z_1(Z_2 Z_3).$$

Consider all the possible ways in which the elements Z_i may belong to the several components of $\mathfrak{B}$.

1) If any two of the elements Z_1, Z_2, Z_3 belong to $\Omega_\varphi \cup \Omega_\psi \cup \cdots$, or if any one is equal to O, then it is obvious that $U = O$ and $V = O$.

2) If all three elements Z_1, Z_2, Z_3 belong to $\mathfrak{A}$, then $U = V$ because of the associativity of the operation in $\mathfrak{A}$.

3) If $Z_1 = \alpha_\varphi \in \Omega_\varphi$ and $Z_2, Z_3 \in \mathfrak{A}$, then $U = \alpha_\varphi$ and $V = \alpha_\varphi$.

4) If $Z_2 = \alpha_\varphi \in \Omega_\varphi$ and $Z_1, Z_3 \in \mathfrak{A}$, then

$$U = (Z_1 \alpha_\varphi) Z_3 = Z_1 \alpha_\varphi, \quad V = Z_1(\alpha_\varphi Z_3) = Z_1 \alpha_\varphi.$$

5) If $Z_1, Z_2 \in \mathfrak{A}$ and $Z_3 = \alpha_\varphi \in \Omega_\varphi$, with $\alpha_\varphi \in \Pi_1^{(\varphi(Z_1 Z_2))} = \Pi_1^{\varphi(Z_1) \cdot \varphi(Z_2)}$, then, by the multiplication rule for partial transformations, $\alpha_\varphi \in \Pi_1^{(\varphi(Z_2))}$ and $\varphi(Z_2)(\alpha_\varphi) \in \Pi_1^{(\varphi(Z_1))}$.

In this case, therefore,

$$U = (Z_1 Z_2) Z_3 = (Z_1 Z_2) \alpha_\varphi = \varphi(Z_1 Z_2)(\alpha_\varphi) = \varphi(Z_1)\varphi(Z_2)(\alpha_\varphi),$$

$$V = Z_1(Z_2 Z_3) = Z_1(\varphi(Z_2)(\alpha_\varphi)) = \varphi(Z_1)(\varphi(Z_2)(\alpha_\varphi)) = \varphi(Z_1)\varphi(Z_2)(\alpha_\varphi).$$

Now suppose $\alpha_\varphi \bar{\in} \Pi_1^{(\varphi(Z_1 Z_2))} = \Pi_1^{(\varphi(Z_1) \cdot \varphi(Z_2))}$. Then

$$U = (Z_1 Z_2) Z_3 = (Z_1 Z_2) \alpha_\varphi = O.$$

As for the element V, if $\alpha_\varphi \bar{\in} \Pi_1^{(\varphi(Z_2))}$, we find $Z_2 \alpha_\varphi = O$ and so $V = O$. If $\alpha_\varphi \in \Pi_1^{\varphi(Z_2)}$ and $\varphi(Z_2)(\alpha_\varphi) \bar{\in} \Pi_1^{\varphi(Z_1)}$, we find

$$V = Z_1(Z_2 Z_3) = Z_1(\varphi(Z_2)(\alpha_\varphi)) = O.$$

We have thus shown that $\mathfrak{B}$ is a semigroup. It contains $\mathfrak{A}$ as a subsemigroup, and has the element O as its zero. It follows from the multiplication rule in $\mathfrak{B}$ that $\mathfrak{L}_\varphi = \Omega_\varphi \cup O$ is a two-sided ideal of $\mathfrak{B}$.

Let us now take an arbitrary $\varphi \in \Xi$ and compare it with the representation $\sigma = \sigma(\mathfrak{A}, \mathfrak{B}, \mathfrak{L}_\varphi, O)$. Since $\Omega_\varphi = \mathfrak{L}_\varphi \setminus O$, both φ and σ are homomorphisms of $\mathfrak{A}$ into $\mathfrak{P}_{\Omega_\varphi}$.

Suppose $A \in \mathfrak{A}$ and $\alpha_\varphi \in \Omega_\varphi$. If $\alpha_\varphi \in \Pi_1^{(\varphi(A))}$ then by the multiplication rule in $\mathfrak{B}$, we have $A \cdot \alpha_\varphi = \varphi(A)(\alpha_\varphi) \in \Omega_\varphi$, so that $\alpha_\varphi \in \Pi_1^{(\sigma A)}$ and $\sigma(A)(\alpha_\varphi) = \sigma_A(\alpha_\varphi) = A \cdot \alpha_\varphi = \varphi(A)(\alpha_\varphi)$.

If $\alpha_\varphi \bar{\in} \Pi_1^{(\varphi(A))}$, then $A \cdot \alpha_\varphi = O \bar{\in} \Omega_\varphi$ and so $\alpha_\varphi \bar{\in} \Pi_1^{(\sigma A)} = \Pi_1^{(\sigma(A))}$.

We have thus shown that $\Pi_1^{(\sigma(A))} = \Pi_1^{(\varphi(A))}$ and $\sigma(A)(\alpha_\varphi) = \varphi(A)(\alpha_\varphi)$ for every $\alpha_\varphi \in \Pi_1^{\sigma(A)} = \Pi_1^{\varphi(A)}$. Consequently $\sigma(A) = \varphi(A)$ for every $A \in \mathfrak{A}$.

Therefore, finally,

$$\varphi = \sigma = \sigma(\mathfrak{A}, \mathfrak{B}, \mathfrak{L}_\varphi, O).$$

6.9. In conclusion we point out the following properties of the representation $\sigma = \sigma(\mathfrak{A}, \mathfrak{B}, \mathfrak{L}_1, \mathfrak{L}_2)$, which are immediate consequences of the definition.

(α). σ is a representation by transformations if and only if $AX \in \mathfrak{L}_1 \setminus \mathfrak{L}_2$ for every $A \in \mathfrak{A}$ and $X \in \mathfrak{L}_1 \setminus \mathfrak{L}_2$. This holds in particular if $\mathfrak{L}_2 = \emptyset$, and more generally if $\mathfrak{L}_1 \setminus \mathfrak{L}_2$ is itself a left ideal of $\mathfrak{B}$.

(β). The representation σ is an isomorphism if and only if for every pair $A_1 \neq A_2$ $(A_1, A_2 \in \mathfrak{A})$ there exists an $X \in \mathfrak{L}_1$ such that $A_1X \neq A_2X$ and at least one of the elements A_1X and A_2X is not in $\mathfrak{L}_2$.

(γ). If $\mathfrak{B}$ is a semigroup with left cancellation then σ is a homomorphic representation by one-to-one partial transformations.

(δ). If $\mathfrak{B}$ is the semigroup obtained from $\mathfrak{A}$ by external adjunction of a unit, and $\mathfrak{L}_1 = \mathfrak{B}$, $\mathfrak{L}_2 = \emptyset$, then σ is the isomorphic representation by transformations considered in I, 3.9, and called there the representation of the semigroup by left translations. We remark in passing that this designation for the representation in question is not recently in common use.

7. Translations of Semigroups

7.1. In considering various transformations of a semigroup, it is natural to give special attention to those which are in one way or another related to the operation in the semigroup. Examples of this are the endomorphisms and automorphisms considered earlier. Another important class of such transformations is that of the so-called translations. As will be seen, the notion of translation is related in an essential way to that of the embedding of semigroups. Translations were introduced by Clifford [**8**]. They were then studied by Tamura ([**9, 12**] and [**S1**]) and subsequently by a number of other mathematicians.

7.2. **Definition.** A transformation φ of a semigroup $\mathfrak{A}$ is called a *left translation* if for any elements X and Y of $\mathfrak{A}$

$$\varphi(XY) = \varphi(X) \cdot Y.$$

A transformation ψ of $\mathfrak{A}$ is called a *right translation* if for any elements X and Y of $\mathfrak{A}$

$$\psi(XY) = X \cdot \psi(Y).$$

7.3. Every element A of a semigroup $\mathfrak{A}$ determines the following transformations φ_A and ψ_A of $\mathfrak{A}$:

$$\varphi_A(X) = AX, \quad \psi_A(X) = XA \quad (X \in \mathfrak{A}).$$

φ_A is obviously a left translation of $\mathfrak{A}$. It is called the inner left translation of $\mathfrak{A}$ induced by A.

ψ_A is a right translation of $\mathfrak{A}$ and is called the inner right translation of $\mathfrak{A}$ induced by A.

Denote the set of all left translations of the semigroup $\mathfrak{A}$ by $\Phi(\mathfrak{A})$, and the set of all inner left translations by $\Phi_0(\mathfrak{A})$. Similarly, we shall use for right translations the notations $\Psi(\mathfrak{A})$ and $\Psi_0(\mathfrak{A})$, respectively.

7.4. Since translations are elements of the semigroup $\mathfrak{S}_{\mathfrak{A}}$, there is defined for them a multiplication. It is immediately clear that $\Phi(\mathfrak{A})$, $\Phi_0(\mathfrak{A})$, $\Psi(\mathfrak{A})$, $\Psi_0(\mathfrak{A})$ are semigroups with respect to this operation of multiplication of transformations.

It should be remarked that in the mathematical literature the right translations are sometimes regarded as being elements of the semigroup $\mathfrak{S}_{\mathfrak{A}}^{0}$ (I, 3.11) rather than, as here, of the semigroup $\mathfrak{S}_{\mathfrak{A}}$. This is reflected in the outward formulation of some statements.

7.5. The nature of the inner translations is essentially clearer, and the study of their properties is much simpler, than for general translations. It is natural in this connection to ask when all translations of a semigroup are inner.

Theorem. *For a semigroup $\mathfrak{A}$, $\Phi(\mathfrak{A}) = \Phi_0(\mathfrak{A})$ if and only if $\mathfrak{A}$ has a left unit.*

Proof. 1) If E is a left unit of $\mathfrak{A}$, then for every $\varphi \in \Phi(\mathfrak{A})$ and $X \in \mathfrak{A}$ we have

$$\varphi(X) = \varphi(EX) = \varphi(E) \cdot X = \varphi_{\varphi(E)}X,$$

so that $\varphi = \varphi_{\varphi(E)}$.

2) Suppose $\Phi(\mathfrak{A}) = \Phi_0(\mathfrak{A})$. The identity mapping ϵ of $\mathfrak{A}$ onto itself is certainly a left translation. $\mathfrak{A}$ must therefore contain an element Z such that $\epsilon = \varphi_Z$.

Then for every $X \in \mathfrak{A}$, we have

$$X = \epsilon(X) = \varphi_Z(X) = ZX.$$

Z is thus a left unit of $\mathfrak{A}$.

7.6. A theorem analogous to 7.5 obviously holds for right translations.

7.7. As shown by È. G. Šutov [S 1], this question is related to the notion of dense embedding (VII, 5.6).

Theorem. *If a semigroup $\mathfrak{A}$ has a unit, then every translation, both left and right, is inner, and $\mathfrak{A}$ has a two-sided ideal $\mathfrak{T}$ densely embedded in it relative to the class of semigroups containing $\mathfrak{T}$ as a two-sided ideal.*

If $\mathfrak{A}$ does not have a unit, then it has a noninner translation, and it cannot have a two-sided ideal $\mathfrak{T}$ densely embedded in it relative to the class of semigroups containing $\mathfrak{T}$ as a two-sided ideal.

Proof. 1) Suppose $\mathfrak{A}$ has a unit E. We shall show that $\mathfrak{A}$ is itself densely embedded in $\mathfrak{A}$ relative to the class of semigroups containing $\mathfrak{A}$ as a two-sided ideal.

The first condition in VII, 5.6 is trivially satisfied.

Now let $\mathfrak{A}'$ be a supersemigroup of $\mathfrak{A}$, different from $\mathfrak{A}$, in which $\mathfrak{A}$ is a two-sided ideal. Then we have $EX \in \mathfrak{A}$ for every $X \in \mathfrak{A}'$, and therefore $EX = (EX)E = EXE = E(XE) = XE$.

The mapping χ of $\mathfrak{A}'$ into $\mathfrak{A}$ such that

$$\chi(X) = EX = XE = EXE \ (X \in \mathfrak{A}')$$

obviously induces the identity isomorphism on $\mathfrak{A}$. Furthermore, χ is a homomorphism of $\mathfrak{A}'$, since we have, for any $X, Y \in \mathfrak{A}'$,

$$\chi(XY) = EXYE = EX \cdot YE = \chi(X) \cdot \chi(Y).$$

But χ is not an isomorphism of $\mathfrak{A}'$. Indeed, take any element $Z \in \mathfrak{A}' \setminus \mathfrak{A}$. Then Z and EZ are distinct, since $EZ \in \mathfrak{A}$. On the other hand,

$$\chi(Z) = EZE = EZ \cdot E = \chi(EZ).$$

The assertion concerning translations follows from 7.5 and 7.6.

2) Now suppose that $\mathfrak{A}$ does not have a unit.

Suppose $\mathfrak{A}$ has a two-sided ideal $\mathfrak{T}$ densely embedded in $\mathfrak{A}$ relative to the class of semigroups containing $\mathfrak{T}$ as a two-sided ideal, and consider the supersemigroup $\mathfrak{A}'$ of $\mathfrak{A}$ obtained from $\mathfrak{A}$ by external adjunction of a unit E.

$\mathfrak{T}$ is obviously a two-sided ideal of $\mathfrak{A}'$. Therefore, by the second condition of VII, 5.6, there must exist a homomorphism χ of $\mathfrak{A}'$ which is not an isomorphism but is an extension of the identity isomorphism of $\mathfrak{T}$. Thus for some $U, V \in \mathfrak{A}'$ we must have

$$\chi(U) = \chi(V), \quad U \neq V.$$

But by the first condition of VII, 5.6, χ induces an isomorphism on $\mathfrak{A}$. Hence at least one of the elements U or V is not in $\mathfrak{A}$. Since $\mathfrak{A}' \setminus \mathfrak{A}$ consists of the single element E, we may suppose $U = E$, $V \in \mathfrak{A}$.

Then for any $A \in \mathfrak{A}$ we have

$$\chi(A) \cdot \chi(V) = \chi(A)\chi(E) = \chi(AE) = \chi(A),$$

$$\chi(V) \cdot \chi(A) = \chi(E)\chi(A) = \chi(EA) = \chi(A).$$

It follows that $\chi(V)$ is a unit for the semigroup $\chi(\mathfrak{A})$. But this is impossible, since $\mathfrak{A}$ and $\chi(\mathfrak{A})$ are isomorphic.

As for translations, if every left and right translation of $\mathfrak{A}$ were inner, then by 7.5 and 7.6 $\mathfrak{A}$ would have a left unit E_1 and a right unit E_2. But then $E_1 = E_1E_2 = E_2$ and the element $E_1 = E_2$ would be a unit for $\mathfrak{A}$.

7.8. If not all translations of a semigroup are inner, then it is natural to try to reduce them to inner translations of some new semigroup by embedding the former as a subsemigroup of the latter. Investigations in this direction have been carried out by Tamura (see the papers referred to above).

Suppose the semigroup $\mathfrak{A}$ is embedded in a supersemigroup $\mathfrak{B}$ such that $\mathfrak{A}$ is a left ideal of $\mathfrak{B}$. Then every inner translation φ_B of $\mathfrak{B}$ ($B \in \mathfrak{B}, \varphi_B \in \Phi(\mathfrak{B})$) induces a left translation on $\mathfrak{A}$. Indeed, for every $A \in \mathfrak{A}$,

$$\varphi_B(A) = BA \in \mathfrak{A},$$

and

$$\varphi_B(AA') = B(AA') = (BA)A' = \varphi_B(A) \cdot A'.$$

Thus the restriction of φ_B to $\mathfrak{A}$ is a left translation of $\mathfrak{A}$.

7.9. Now in fact an appropriately chosen embedding does allow all the left translations of a semigroup to be obtained in this fashion from the inner left translations of the supersemigroup.

Theorem. *For every semigroup $\mathfrak{A}$ there exists a supersemigroup $\mathfrak{B}$ containing $\mathfrak{A}$ as a left ideal and such that every left translation of $\mathfrak{A}$ is induced by left multiplication in $\mathfrak{A}$ by some element of $\mathfrak{B}$* (7.8).

Proof. Let $\mathfrak{M}$ be the set of all pairs (A, φ), where $A \in \mathfrak{A}$ and $\varphi \in \Phi(\mathfrak{A})$. For uniformity of notation we shall designate the elements of $\mathfrak{A}$ also as $(A, \emptyset)$ ($A \in \mathfrak{A}$), and the elements of $\Phi(\mathfrak{A})$ as $(\emptyset, \varphi)$ ($\varphi \in \Phi(\mathfrak{A})$). The set of all such pairs, i.e. $\mathfrak{A} \cup \Phi(\mathfrak{A}) \cup \mathfrak{M}$, is to be the set $\mathfrak{B}$.

In $\mathfrak{B}$ we define an operation of multiplication by setting

$$(X_1, \varphi_1) \cdot (X_2, \varphi_2) = (X_1 \cdot \varphi_1(X_2), \varphi_2) \quad (X_2 \neq \emptyset),$$

$$(X_1, \varphi_1) \cdot (\emptyset, \varphi_2) = (X_1, \varphi_1\varphi_2)$$

(where, in the first expression on the right, if $\varphi_1 = \emptyset$ or $X_1 = \emptyset$, these are to be regarded as empty symbols; and similarly in the second expression, if $\varphi_1 = \emptyset$).

A straightforward verification shows that the operation in $\mathfrak{B}$ is associative. Thus $\mathfrak{B}$ is a semigroup.

The semigroups $\mathfrak{A}$ and $\Phi(\mathfrak{A})$ are subsets of $\mathfrak{B}$. Furthermore, the operations in both $\mathfrak{A}$ and $\Phi(\mathfrak{A})$ are consistent with the operation in $\mathfrak{B}$, i.e., coincide on both $\mathfrak{A}$ and $\Phi(\mathfrak{A})$ with the multiplication defined in $\mathfrak{B}$.

Thus $\mathfrak{B}$ is a supersemigroup for both $\mathfrak{A}$ and $\Phi(\mathfrak{A})$. Furthermore $\mathfrak{A}$ is obviously a left ideal of $\mathfrak{B}$.

Finally, in $\mathfrak{B}$, an arbitrary left translation φ of the semigroup $\mathfrak{A}$ is induced by multiplication of the elements of $\mathfrak{A}$ by the element $(\emptyset, \varphi) \in \mathfrak{B}$. Indeed, for every $A \in \mathfrak{A}$ we have, according to the multiplication rule in $\mathfrak{B}$,

$$\varphi(A) = (\varphi(A), \emptyset) = (\emptyset, \varphi) \cdot (A, \emptyset).$$

7.10. The corresponding theorem for right translations may obviously be proved by a similar construction. Thus the study of left or right translations is essentially the study of certain special embeddings of the semigroup.

The question naturally arises whether it is possible to embed a semigroup $\mathfrak{A}$ into a supersemigroup $\mathfrak{B}$ such that $\mathfrak{A}$ is a two-sided ideal of $\mathfrak{B}$, with every left translation of $\mathfrak{A}$ being induced by left multiplication of the elements of $\mathfrak{A}$ by some element of $\mathfrak{B}$, and every right translation by right multiplication by some element of $\mathfrak{B}$. Such an embedding is not always possible. This question has been studied by Tamura and Graham [S1].

7.11. The question of what semigroups, up to isomorphism, are the semigroups of all left translations of some semigroup (in other words, the problem of abstract characterization of semigroups of translations) has been solved by È. G. Šutov [S1], again by means of embedding of semigroups.

Theorem. *For every semigroup* $\mathfrak{A}$, *there exists a supersemigroup* $\mathfrak{A}'$ *such that* $\mathfrak{A}$ *is isomorphic to* $\Phi_0(\mathfrak{A}')$.

Proof. Let $\mathfrak{A}'$ be the set of all elements of the semigroup $\mathfrak{A}$, together with all pairs of the form (X, n), where either $X \in \mathfrak{A}$ or X is the symbol for the void set, and $n = 1, 2, 3, \cdots$. Any element $A \in \mathfrak{A}$ will also be written as a pair $A = (A, 0)$.

Choosing in $\mathfrak{A}$ an arbitrary fixed element T, we define a multiplication in $\mathfrak{A}'$ by setting

$$(X, k) \cdot (Y, l) = (XT^k Y, l)$$

(if $k = 0$, the factor T^0 on the right is simply to be omitted, and similarly for X or Y if it is the void set).

The operation in $\mathfrak{A}'$ is consistent with that in $\mathfrak{A}$, i.e. the two operations coincide on $\mathfrak{A}$.

Associativity of the operation in $\mathfrak{A}'$ can be verified easily. Thus $\mathfrak{A}'$ is a semigroup, and $\mathfrak{A}$ a subsemigroup of it (and in fact a left ideal).

Every element $A = (A, 0) \in \mathfrak{A}$ induces an inner left translation φ_A in $\mathfrak{A}'$. We assert that the correspondence $A \sim \varphi_A$ is one-to-one.

To see this, suppose $\varphi_X = \varphi_Y$ $(X, Y \in \mathfrak{A})$. Since

$$\varphi_X(\emptyset, 1) = (X, 0) \cdot (\emptyset, 1) = (X, 1),$$

$$\varphi_Y(\emptyset, 1) = (Y, 0) \cdot (\emptyset, 1) = (Y, 1),$$

the equality $\varphi_X = \varphi_Y$ implies $(X, 1) = (Y, 1)$, i.e., $X = Y$.

Furthermore, if $AB = C$ $(A, B, C \in \mathfrak{A})$ then for every $X \in \mathfrak{A}'$ we have

$$\varphi_A \varphi_B(X) = A(BX) = (AB)X = CX = \varphi_C(X),$$

so that $\varphi_A \varphi_B = \varphi_C$.

We have thus shown that the mapping of $\mathfrak{A}$ into $\Phi_0(\mathfrak{A}')$ which assigns to each $A \in \mathfrak{A}$ the inner left translation φ_A of the semigroup $\mathfrak{A}'$ is an isomorphism. It remains to show that every inner left translation φ_Z of $\mathfrak{A}'$ is of the form φ_A for some $A \in \mathfrak{A}$.

Let $Z = (X, n)$, where we may assume $n \neq 0$, since $Z = (X, 0)$ is already an element of $\mathfrak{A}$, for $n = 0$. Then for any $(S, k) \in \mathfrak{A}'$ we have

$$\begin{aligned} \varphi_Z(S, k) &= (X, n)(S, k) = (XT^n S, k) \\ &= (XT^n, 0)(S, k) = \varphi_{(XT^n)}(S, k). \end{aligned}$$

Thus, $\varphi_Z = \varphi_{(XT^n)}$, where $XT^n \in \mathfrak{A}$.

7.12. One important embedding of semigroups of a quite extensive class can be realized by means of translations. This embedding has been used by Clifford [8] to study the so-called ideal extensions of semigroups, and by L. M. Gluskin [S 1] to study dense embeddings.

Let φ be a left translation of a semigroup $\mathfrak{A}$, and ψ a right translation. φ and ψ are said to be linked if for any $X, Y \in \mathfrak{A}$ we have

$$X \cdot \varphi(Y) = \psi(X) \cdot Y.$$

In particular, the inner left and right translations determined by the same element are linked:

$$X \cdot \varphi_A(Y) = X(AY), \quad \psi_A(X) \cdot Y = (XA)Y.$$

7.13. Given the semigroup $\mathfrak{A}$, let $\mathfrak{B}(\mathfrak{A})$ be the set of all pairs (φ, ψ), where $\varphi \in \Phi(\mathfrak{A})$ and $\psi \in \Psi(\mathfrak{A})$. Define an operation in $\mathfrak{B}$ by setting

$$(\varphi_1, \psi_1) \cdot (\varphi_2, \psi_2) = (\varphi_1\varphi_2, \psi_2\psi_1).$$

This operation is obviously associative. Thus $\mathfrak{B}(\mathfrak{A})$ is a semigroup.

7.14. The subset $\mathfrak{B}_1(\mathfrak{A})$ of $\mathfrak{B}(\mathfrak{A})$, consisting of those (φ, ψ) such that φ and ψ are linked (7.12), is a subsemigroup of $\mathfrak{B}(\mathfrak{A})$.

Indeed, suppose φ_1 and ψ_1 are linked and φ_2 and ψ_2 are linked. Then for any $X, Y \in \mathfrak{A}$,

$$\begin{aligned} X \cdot \varphi_1\varphi_2(Y) &= X \cdot \varphi_1[\varphi_2(Y)] = \psi_1(X) \cdot \varphi_2(Y) \\ &= \psi_2[\psi_1(X)] \cdot Y = \psi_2\psi_1(X) \cdot Y. \end{aligned}$$

In other words, $\varphi_1\varphi_2$ and $\psi_2\psi_1$ are linked.

7.15. Let $\mathfrak{B}_0(\mathfrak{A})$ be the subset of the semigroup $\mathfrak{B}(\mathfrak{A})$ consisting of those pairs of inner translations (φ_A, ψ_A) determined by the same element $A \in \mathfrak{A}$. It is easily seen that $\mathfrak{B}_0(\mathfrak{A})$ is a subsemigroup of $\mathfrak{B}(\mathfrak{A})$, and that $\mathfrak{B}_0(\mathfrak{A}) \subset \mathfrak{B}_1(\mathfrak{A})$.

7.16. Following Clifford [8], a semigroup $\mathfrak{A}$ is called weakly reductive (one also uses the term: semigroup without equi-operational elements) if for any $A_1 \neq A_2$ $(A_1, A_2 \in \mathfrak{A})$ there exists an $X \in \mathfrak{A}$ such that either $XA_1 \neq XA_2$ or $A_1X \neq A_2X$. This means that $\mathfrak{A}$ does not contain two elements which induce the same left and right translations.

7.17. If $\mathfrak{A}$ is a weakly reductive semigroup, its mapping χ into $\mathfrak{B}(\mathfrak{A})$:

$$\chi(A) = (\varphi_A, \psi_A),$$

is a one-to-one mapping of $\mathfrak{A}$ onto $\mathfrak{B}_0(\mathfrak{A})$.

It is easily seen that χ is an isomorphism. Thus, identifying the elements $A \in \mathfrak{A}$ with the corresponding elements $\chi(A) \in \mathfrak{B}_0(\mathfrak{A})$, we obtain an embedding of $\mathfrak{A}$ into $\mathfrak{B}_1(\mathfrak{A})$. In various situations this embedding is quite useful and important.

For example, L. M. Gluskin [S1] proved that if $\mathfrak{A}$ is a weakly reductive semigroup, the subsemigroup $\mathfrak{B}_0(\mathfrak{A})$ is densely embedded in $\mathfrak{B}_1(\mathfrak{A})$ relative to the class of all semigroups containing $\mathfrak{B}_0(\mathfrak{A})$ as a two-sided ideal (VII, 5.6). Thus every weakly reductive semigroup is densely embedded in some supersemigroup relative to the class of semigroups containing it as a two-sided ideal.

This result is of importance because, as shown in Gluskin [S1], and in a number of other papers by him and other authors, dense embeddings play an important part in the theory of semigroups (and even beyond its limits, as is evident from the constructions of L. M. Gluskin [S3]). In particular, by considering dense embeddings it has been possible to obtain abstract characterizations for a wide class of semigroups of transformations, similar to that described in VII, 5.9, 5.10.

CHAPTER XI

SUBDIRECT PRODUCTS

1. The Notion of Subdirect Product

1.1. The notion and general theory of subdirect products are a concern of general algebra (see, e.g., Birkhoff [**3**]). The properties of subdirect products are used in various algebraic disciplines.

The specific study of subdirect products of semigroups was begun by Thierrin [**S 1**]. They were then investigated independently by B. M. Saĭn [**S 2, 10**]. Subsequently subdirect products have been used as a means of studying one or another property of semigroups. The question of approximation, so-called, also enters, as we shall see, into the domain of the theory of subdirect decomposition of semigroups.

1.2. Suppose given a family of semigroups $\mathfrak{B}_\alpha, \mathfrak{B}_\beta, \cdots, \mathfrak{B}_\xi, \cdots (\xi \in \Gamma)$. Consider the set $\mathfrak{M}$ of which an element is any family of elements taken one from each of the semigroups $\mathfrak{B}_\xi$:

$$M = (B_\xi)_{\xi \in \Gamma} = (B_\alpha, B_\beta, \cdots, B_\xi, \cdots) \ (B_\xi \in \mathfrak{B}_\xi),$$
$$\mathfrak{M} = \{M\}.$$

The notation for the components of the element M will sometimes indicate this $M : B_\xi = B_\xi^{(M)}$.

Define an operation in $\mathfrak{M}$ as follows:

If

$$M = (B_\xi^{(M)})_{\xi \in \Gamma} = (B_\alpha^{(M)}, B_\beta^{(M)}, \cdots, B_\xi^{(M)}, \cdots)$$

and

$$M' = (B_\xi^{(M')})_{\xi \in \Gamma} = (B_\alpha^{(M')}, B_\beta^{(M')}, \cdots, B_\xi^{(M')}, \cdots),$$

then

$$MM' = (B_\xi^{(M)} B_\xi^{(M')})_{\xi \in \Gamma} = (B_\alpha^{(M)} \cdot B_\alpha^{(M')}, B_\beta^{(M)} \cdot B_\beta^{(M')}, \cdots, B_\xi^{(M)} \cdot B_\xi^{(M')}, \cdots).$$

It is obvious that the operation in $\mathfrak{M}$ is associative and $\mathfrak{M}$ is a semigroup. We shall call $\mathfrak{M}$ the *Cartesian product* of the semigroups $\mathfrak{B}_\xi$ $(\xi \in \Gamma)$ and write $\mathfrak{M} = \times_{\xi \in \Gamma} \mathfrak{B}_\xi$. $\mathfrak{M}$ is sometimes called the direct product of the semigroups $\mathfrak{B}_\xi$ or their full direct product. It must be kept in mind that, despite the similarity and a certain connection, this construction is different from the direct product defined in §5 of the tenth chapter.

If a semigroup $\mathfrak{A}$ is isomorphic to the above semigroup $\mathfrak{M}$, one says that $\mathfrak{A}$ decomposes into the Cartesian product of the semigroups $\mathfrak{B}_\xi$ $(\xi \in \Gamma)$. In this case, as usual in considering semigroups up to isomorphism ("from the abstract point of view"), $\mathfrak{A}$ is often simply identified with $\mathfrak{M}$.

1.3. Let φ be a mapping of a semigroup $\mathfrak{A}$ into $\mathfrak{M} = \times_{\xi \in \Gamma} \mathfrak{B}_\xi$. It induces natural mappings φ_ξ $(\xi \in \Gamma)$ of $\mathfrak{A}$ into $\mathfrak{B}_\xi$:

$$\varphi(A) = (B_\alpha^{(A)}, B_\beta^{(A)}, \cdots, B_\xi^{(A)}, \cdots),$$

$$\varphi_\xi(A) = B_\xi^{(A)}.$$

φ is obviously a homomorphism if and only if every φ_ξ is.

Each homomorphism φ_ξ determines a two-sidedly stable equivalence $\mathfrak{n}_\xi$ in $\mathfrak{A}$ (VII, 2.1). The set of all equivalences $\mathfrak{n}_\xi$ $(\xi \in \Gamma)$ has a greatest lower bound $\mathfrak{m}$, namely, their intersection (I, 5.15). $\mathfrak{m}$ is itself two-sidedly stable (VII, 2.9).

We assert that $\mathfrak{m}$ *is the equivalence corresponding to the homomorphism* φ. Indeed, let $\mathfrak{l}$ be the equivalence corresponding to φ.

Then if $A \sim A'(\mathfrak{l})$, i.e. $\varphi(A) = \varphi(A')$, we have $\varphi_\xi(A) = \varphi_\xi(A')$ for all $\xi \in \Gamma$, i.e. $A \sim A'(\mathfrak{n}_\xi)$. This means that $\mathfrak{l} \leqq \mathfrak{n}_\xi$ $(\xi \in \Gamma)$, i.e. $\mathfrak{l} \leqq \mathfrak{m}$.

On the other hand, if $A \sim A'(\mathfrak{m})$ then $A \sim A'(\mathfrak{n}_\xi)$, i.e. $\varphi_\xi(A) = \varphi_\xi(A')$ $(\xi \in \Gamma)$. But then obviously $\varphi(A) = \varphi(A')$, i.e. $A \sim A'(\mathfrak{l})$. We have therefore shown that $\mathfrak{m} \leqq \mathfrak{l}$. Thus $\mathfrak{m} = \mathfrak{l}$.

The homomorphism φ determines a natural isomorphism of the factor-semigroup $\mathfrak{A}/\mathfrak{m}$ into $\mathfrak{M}$.

1.4. Suppose the homomorphism φ of $\mathfrak{A}$ into $\mathfrak{M} = \times_{\xi \in \Gamma} \mathfrak{B}_\xi$ is an isomorphism, and $\varphi_\xi(\mathfrak{A}) = \mathfrak{B}_\xi$ for every $\xi \in \Gamma$. Then one says that $\mathfrak{A}$ *decomposes into a subdirect product* of the semigroups $\mathfrak{B}_\xi$ $(\xi \in \Gamma)$. One sometimes also says that $\mathfrak{A}$ *is* a subdirect product of the $\mathfrak{B}_\xi$ $(\xi \in \Gamma)$.

Observe that, by 1.3, the condition that φ be an isomorphism amounts to the condition that the equivalence $\mathfrak{m}$ be the identity equivalence, i.e., that $X \sim Y(\mathfrak{m})$ only when $X = Y$.

1.5. Thus every decomposition of $\mathfrak{A}$ into a subdirect product determines a set of two-sidedly stable equivalences $\mathfrak{n}_\xi$ $(\xi \in \Gamma)$, whose greatest lower bound is the identity equivalence $\mathfrak{e}$. Moreover the decomposition of $\mathfrak{A}$ in question is into a subdirect product of the semigroups $\mathfrak{A}/\mathfrak{n}_\xi$ $(\xi \in \Gamma)$, since $\mathfrak{B}_\xi$ is isomorphic to $\mathfrak{A}/\mathfrak{n}_\xi$

On the other hand, suppose given in $\mathfrak{A}$ an arbitrary set of two-sidedly stable equivalences $\mathfrak{n}_\xi$ $(\xi \in \Gamma)$ whose greatest lower bound is the identity equivalence $\mathfrak{e}$. Consider the Cartesian product

$$\mathfrak{M} = \times_{\xi \in \Gamma} \mathfrak{A}/\mathfrak{n}_\xi.$$

Let φ_ξ be the natural homomorphism of $\mathfrak{A}$ onto the factor-semigroup $\mathfrak{A}/\mathfrak{n}_\xi$ (VII, 2.5). Define the mapping φ of $\mathfrak{A}$ into $\mathfrak{M}$ such that for every $A \in \mathfrak{A}$

$$\varphi(A) = (\varphi_\alpha(A), \varphi_\beta(A), \cdots, \varphi_\xi(A), \cdots).$$

Then φ is obviously an isomorphism and the φ_ξ are the homomorphisms defined for it in 1.3. Hence φ determines a decomposition of $\mathfrak{A}$ into a subdirect product of the semigroups $\mathfrak{A}/\mathfrak{n}_\xi$ $(\xi \in \Gamma)$.

It is also obvious that for this decomposition the given equivalences $\mathfrak{n}_\xi$ are the same as those defined above for every subdirect decomposition.

1.6. Observe that if in the decomposition of $\mathfrak{A}$ into a subdirect product of the $\mathfrak{B}_\xi$ $(\xi \in \Gamma)$, one has $\mathfrak{n}_\xi \leqq \mathfrak{n}_\eta$ for some $\xi \neq \eta$ $(\xi, \eta \in \Gamma)$, then the component $\mathfrak{B}_\eta$ can be omitted, since the greatest lower bound of the remaining equivalences is still the identity equivalence. The entire family of components can be similarly dealt with, under a suitable corresponding condition. In particular, all unit components can be omitted.

1.7. The notion of subdirect decomposition is closely related to that of a complete system of representations of the semigroup $\mathfrak{A}$ in a class of semigroups (VII, 1.6). We note in this connection that one says that a semigroup $\mathfrak{A}$ can be *approximated* within a class of semigroups Ξ if it has a complete system of representations by semigroups in Ξ.

The possibility of approximation within the class of all finite semigroups is obviously of fundamental importance. One says in this case that the semigroup is finitely approximable. A. I. Mal′cev [S1] has shown that a commutative semigroup which has a finite generating set is finitely approximable.

It is easily seen that any semigroup approximable within the class of groups must be a semigroup with two-sided cancellation.

Indeed, let $\mathfrak{A}$ be a semigroup such that there exists a complete system of representations of $\mathfrak{A}$ by groups. Suppose $SX = SY$ for some elements $X, Y, S \in \mathfrak{A}$, with $X \neq Y$ (a similar argument applies if $XS = YS$). The complete system of representations includes a homomorphism φ_0 of $\mathfrak{A}$ into a group $\mathfrak{G}$ such that $\varphi_0(X) \neq \varphi_0(Y)$. Then in $\mathfrak{G}$ we find that $\varphi_0(S) \cdot \varphi_0(X) = \varphi_0(S) \cdot \varphi_0(Y)$, which is impossible for $\varphi_0(X) \neq \varphi_0(Y)$.

From the fact that every cancellative commutative semigroup is embeddable in a group (X, 2.8), it follows that for commutative semigroups the cancellability condition is not only necessary but sufficient for approximability within the class of groups.

1.8. The possibility of approximation within one or another class is directly related to decomposition of the semigroup into a subdirect product.

Theorem. *A semigroup* $\mathfrak{A}$ *is approximable within a class of semigroups*

Ξ *if and only if* $\mathfrak{A}$ *decomposes into a subdirect product of semigroups which are imbedded in semigroups of the class* Ξ.

Proof. 1) Suppose $\mathfrak{A}$ is approximable within the class Ξ, and let $\varphi_\alpha, \varphi_\beta, \cdots, \varphi_\xi, \cdots$ be a complete system of representations of $\mathfrak{A}$ in Ξ, with $\mathfrak{n}_\alpha, \mathfrak{n}_\beta, \cdots, \mathfrak{n}_\xi, \cdots$ the corresponding two-sidedly stable equivalences, and $\mathfrak{m}$ their greatest lower bound.

For every pair of distinct elements $X, Y \in \mathfrak{A}$, there exists a φ_ξ such that $\varphi_\xi(X) \neq \varphi_\xi(Y)$, i.e. $X \nsim Y\ (\mathfrak{n}_\xi)$. Since $\mathfrak{m} \leqq \mathfrak{n}_\xi$, we have $X \nsim Y\ (\mathfrak{m})$. This says that the equivalence $\mathfrak{m}$ is the identity equivalence. It follows by 1.5 that $\mathfrak{A}$ decomposes into a subdirect product of the semigroups $\mathfrak{A}/\mathfrak{n}_\alpha$, $\mathfrak{A}/\mathfrak{n}_\beta, \cdots, \mathfrak{A}/\mathfrak{n}_\xi, \cdots$. But the semigroup $\varphi_\xi(\mathfrak{A})$, which is embedded in some semigroup of Ξ, is isomorphic with $\mathfrak{A}/\mathfrak{n}_\xi$. Thus $\mathfrak{A}$ decomposes into a subdirect product of the semigroups $\varphi_\alpha(\mathfrak{A}), \varphi_\beta(\mathfrak{A}), \cdots, \varphi_\xi(\mathfrak{A}), \cdots$ embedded in semigroups of the class Ξ.

2) Suppose $\mathfrak{A}$ decomposes into a subdirect product of semigroups $\mathfrak{B}_\xi$ $(\xi \in \Gamma)$, where each $\mathfrak{B}_\xi$ is embedded in a semigroup $\mathfrak{B}'_\xi$ of the class Ξ. The homomorphisms φ_ξ $(\xi \in \Gamma)$ of $\mathfrak{A}$ onto the $\mathfrak{B}_\xi$, corresponding to this decomposition, are then homomorphisms of $\mathfrak{A}$ into semigroups $\mathfrak{B}'_\xi$ of Ξ. Furthermore, these homomorphisms constitute a complete system of representations. Indeed, if $\varphi_\xi(X) = \varphi_\xi(Y)$ for some $X, Y \in \mathfrak{A}$ $(X \neq Y)$ and all $\xi \in \Gamma$, this would imply that $X \sim Y\ (\mathfrak{m})$, where $\mathfrak{m}$ is the greatest lower bound of the equivalences corresponding to the homomorphisms φ_ξ $(\xi \in \Gamma)$.

2. Subdirectly Indecomposable Semigroups

2.1. As usual, in considering the different forms of the decompositions of various algebraic systems, among the first questions to arise are those concerning indecomposable systems and decomposition into indecomposable components. In the theory of semigroups, the first results concerning subdirectly indecomposable semigroups are obtained by direct application of the general theory of subdirect products of arbitrary algebras. The subsequent specific consideration of subdirectly indecomposable semigroups has been carried out in the papers referred to in §1.

2.2. Suppose that for a subdirect decomposition of a semigroup $\mathfrak{A}$, given by a mapping φ into

$$\mathfrak{M} = \mathop{\times}_{\xi \in \Gamma} \mathfrak{B}_\xi,$$

one of the homomorphisms φ_{ξ_0} is an isomorphism (i.e. that $\mathfrak{n}_{\xi_0} = \mathfrak{e}$). Such a subdirect decomposition is called *improper*. From the point of view described in 1.6, an improper decomposition can be reduced, by making eliminations as there indicated, to a decomposition consisting of a single component isomorphic to the original semigroup. If none of the homomorphisms φ_ξ is an isomorphism (i.e., if every $\mathfrak{n}_\xi$ is different from $\mathfrak{e}$), the decomposition is called *proper*.

If the semigroup $\mathfrak{A}$ has no proper subdirect decompositions, it is called *subdirectly indecomposable*.

2.3. Consider, for a semigroup $\mathfrak{A}$, the set of all two-sidedly stable equivalences other than $\mathfrak{e}$. Let $\mathfrak{e}'$ be the greatest lower bound of this set (VII, 2.9). It may happen that $\mathfrak{e}' = \mathfrak{e}$. In some semigroups, however, $\mathfrak{e}'$ will itself belong to the set in question, i.e. $\mathfrak{e}' \neq \mathfrak{e}$. This latter is equivalent with the condition that $\mathfrak{A}$ have a smallest nonidentity two-sidedly stable equivalence.

Theorem. *A semigroup $\mathfrak{A}$, other than the unit semigroup, is subdirectly indecomposable if and only if $\mathfrak{e}' \neq \mathfrak{e}$.*

Proof. 1) Let φ be an isomorphism of $\mathfrak{A}$ into a Cartesian product $\mathfrak{M} = \times_{\xi \in \Gamma} \mathfrak{A}/\mathfrak{n}_\xi$ which yields a proper subdirect decomposition of $\mathfrak{A}$. All the equivalences $\mathfrak{n}_\xi$ are then different from $\mathfrak{e}$. But by 1.5, their greatest lower bound is $\mathfrak{e}$. Since, on the one hand, we must therefore have, for the greatest lower bound $\mathfrak{e}'$ of *all* nonidentity two-sidedly stable equivalences, that $\mathfrak{e}' \leqq \mathfrak{e}$, and, on the other hand, $\mathfrak{e}$ precedes all equivalences whatever, it follows that $\mathfrak{e}' = \mathfrak{e}$.

2) Suppose $\mathfrak{e}' = \mathfrak{e}$. Let $\mathfrak{n}_\alpha, \mathfrak{n}_\beta, \cdots, \mathfrak{n}_\xi, \cdots$ $(\xi \in \Gamma)$ be all the nonidentity two-sidedly stable equivalences of $\mathfrak{A}$. Since their greatest lower bound is $\mathfrak{e}$, it follows by 1.5 that $\mathfrak{A}$ decomposes into a subdirect product of the semigroups $\mathfrak{A}/\mathfrak{n}_\xi$ $(\xi \in \Gamma)$, the decomposition being proper since all the equivalences $\mathfrak{n}_\xi$ are different from the identity.

2.4. **Corollary.** *A subdirectly indecomposable semigroup, other than the unit semigroup, has a smallest nonzero two-sided ideal.*

Proof. Consider the set of all nonzero two-sided ideals $\mathfrak{T}_\alpha, \mathfrak{T}_\beta, \cdots, \mathfrak{T}_\xi, \cdots$ of a subdirectly indecomposable nonunit semigroup $\mathfrak{A}$, and let $\mathfrak{n}_\alpha, \mathfrak{n}_\beta, \cdots, \mathfrak{n}_\xi, \cdots$ be the corresponding ideal equivalences (VII, 4.15). By 2.3, their greatest lower bound $\mathfrak{m} \neq \mathfrak{e}$. Consequently, $\mathfrak{A}$ contains two elements $X \neq Y$ such that $X \sim Y$ $(\mathfrak{m})$. For these elements we have $X \sim Y$ $(\mathfrak{n}_\xi)$ for all $\mathfrak{n}_\xi$. It follows from the definition of ideal equivalence that X and Y are contained in all the ideals $\mathfrak{T}_\xi$, and hence in their intersection. Thus, the intersection of all the $\mathfrak{T}_\xi$ is nonvoid and contains at least two elements. It is then obviously the smallest nonzero two-sided ideal.

2.5. **Corollary.** *Let $\mathfrak{n}$ be a nonidentity two-sidedly stable equivalence of a subdirectly indecomposable semigroup $\mathfrak{A}$, and $\mathfrak{T}$ a nonzero two-sided ideal of $\mathfrak{A}$. Then $\mathfrak{n}$ induces a nonidentity equivalence on $\mathfrak{T}$.*

Proof. Consider the ideal equivalence $\mathfrak{m}$ corresponding to the ideal $\mathfrak{T}$. Let $\mathfrak{l}$ be the greatest lower bound of $\mathfrak{n}$ and $\mathfrak{m}$. By 2.3, $\mathfrak{l}$ is not the identity equivalence. Consequently, there exist $X, Y \in \mathfrak{A}$ such that

$$X \sim Y\ (\mathfrak{l}),\ X \neq Y.$$

Since $\mathfrak{m} \geqq \mathfrak{l}$, $X \sim Y\ (\mathfrak{m})$. It follows, by the definition of ideal equivalence, that $X, Y \in \mathfrak{T}$. In addition, since $\mathfrak{n} \geqq \mathfrak{l}$, we have $X \sim Y\ (\mathfrak{n})$.

2.6. The role of subdirectly indecomposable semigroups is indicated by the following theorem.

THEOREM. *Every semigroup decomposes into a subdirect product of subdirectly indecomposable semigroups.*

PROOF. For each pair of distinct elements X and Y of the semigroup $\mathfrak{A}$, let $\Gamma(X, Y)$ be the set of all two-sidedly stable equivalences $\mathfrak{n}$ of $\mathfrak{A}$ such that $X \nsim Y\ (\mathfrak{n})$ (such equivalences exist, e.g. the identity equivalence). Thinking of $\Gamma(X, Y)$ as a partially ordered set (VII, 2.9), we shall apply to it the theorem of II, 4.17. Indeed, suppose Γ' is a nonvoid linearly ordered subset of $\Gamma(X, Y)$. Define the relation $\mathfrak{m}$ in $\mathfrak{A}$ by the condition that $X \sim Y\ (\mathfrak{m})$ if there exists an $\mathfrak{n} \in \Gamma'$ such that $X \sim Y\ (\mathfrak{n})$. $\mathfrak{m}$ is obviously reflexive and symmetric, and is transitive because, if $X \sim Y\ (\mathfrak{n})$ and $Y \sim Z\ (\bar{\mathfrak{n}})$, where $\mathfrak{n}, \bar{\mathfrak{n}} \in \Gamma'$, we have $X \sim Y\ (\mathfrak{n}^*)$, where $\mathfrak{n}^*$ is that one of $\mathfrak{n}$ and $\bar{\mathfrak{n}}$ which follows $\mathfrak{n}$ and $\bar{\mathfrak{n}}$. Since all $\mathfrak{n} \in \Gamma'$ are two-sidedly stable, so also is $\mathfrak{m}$. Furthermore, $\mathfrak{m}$ belongs to $\Gamma(X, Y)$, since if $X \nsim Y(\mathfrak{n})$ for all $\mathfrak{n} \in \Gamma(X, Y)$, then also $X \nsim Y(\mathfrak{m})$.

Thus by the theorem of II, 4.17 there exists a maximal equivalence in $\Gamma(X, Y)$. Select one such and call it $\mathfrak{l}_{X,Y}$.

The greatest lower bound of all the $\mathfrak{l}_{X,Y}$ over all pairs (X, Y) is the identity equivalence e, since for any pair $X \neq Y$ we have $X \nsim Y\ (\mathfrak{l}_{X,Y})$, so that X and Y are necessarily nonequivalent with respect to the greatest lower bound. It follows by 1.5 that $\mathfrak{A}$ decomposes into a subdirect product of the semigroups $\mathfrak{A}/\mathfrak{l}_{X,Y}$.

We assert now that each semigroup $\mathfrak{A}/\mathfrak{l}_{X,Y}$ is subdirectly indecomposable.

To see this, consider the set of all two-sidedly stable equivalences $\bar{\mathfrak{n}}$ in $\mathfrak{A}/\mathfrak{l}_{X,Y}$ other than the identity equivalence. Each such equivalence $\bar{\mathfrak{n}}$ determines an equivalence $\mathfrak{n}$ in $\mathfrak{A}$ by the condition that $X \sim Y\ (\mathfrak{n})$ if $\bar{X} \sim \bar{Y}\ (\bar{\mathfrak{n}})$, where $\bar{Z}\ (Z \in \mathfrak{A})$ means the element of $\mathfrak{A}/\mathfrak{l}_{X,Y}$ which is the equivalence class, with respect to $\mathfrak{l}_{X,Y}$, of elements of $\mathfrak{A}$ which contains Z. It is easily seen that $\mathfrak{n}$ is a two-sidedly stable equivalence of $\mathfrak{A}$. Furthermore, $\mathfrak{n} > \mathfrak{l}_{X,Y}$. But, recalling that $\mathfrak{l}_{X,Y}$ is maximal among all two-sidedly stable equivalences $\mathfrak{k}$ such that $X \nsim Y\ (\mathfrak{k})$, we conclude that $X \sim Y\ (\mathfrak{n})$. Since this holds for all $\mathfrak{n}$, it must also be the case for their greatest lower bound $\mathfrak{n}_0$ that $X \sim Y\ (\mathfrak{n}_0)$. Hence $\mathfrak{n}_0 > \mathfrak{l}_{X,Y}$. The equivalence $\mathfrak{n}_0$ determines in an obvious fashion an equivalence $\bar{\mathfrak{n}}_0$ in $\mathfrak{A}/\mathfrak{l}_{X,Y}$, which is then, of course, the greatest lower bound of the original set of two-sidedly stable equivalences $\bar{\mathfrak{n}}$ in $\mathfrak{A}/\mathfrak{l}_{X,Y}$. But $\bar{\mathfrak{n}}_0$ cannot be the identity equivalence; this follows from the fact that $\overline{X} \sim \overline{Y}$ ($\overline{\mathfrak{n}}_0$), whereas $X \nsim Y$ ($\mathfrak{l}_{X,Y}$), so that $\overline{X}$ and $\overline{Y}$ are distinct elements of the semigroup $\mathfrak{A}/\mathfrak{l}_{X,Y}$. We conclude, by 2.3, that $\mathfrak{A}/\mathfrak{l}_{X,Y}$ is subdirectly indecomposable.

2.7. The following construction can be regarded as a generalization of a construction in the theory of rings.

Let $\mathfrak{N}$ be a nonvoid subset of a semigroup $\mathfrak{A}$, consisting of certain elements which commute with all the elements of $\mathfrak{A}$ (central elements, in group-theoretical terminology). Let $\mathfrak{T}_{\mathfrak{N}}$ be the set of elements T such that $NT = T$ for all $N \in \mathfrak{N}$. If $\mathfrak{T}_{\mathfrak{N}}$ is nonvoid, then it is obviously a two-sided ideal of $\mathfrak{A}$. Let $\mathfrak{m}_{\mathfrak{N}}$ be the ideal equivalence corresponding to $\mathfrak{T}_{\mathfrak{N}}$.

Define a relation $\mathfrak{l}_{\mathfrak{N}}$ in $\mathfrak{A}$ by setting $X_1 \sim X_2$ ($\mathfrak{l}_{\mathfrak{N}}$) if $X_1N = X_2N$ for all $N \in \mathfrak{N}$. $\mathfrak{l}_{\mathfrak{N}}$ is obviously a two-sidedly stable equivalence.

If $T_1, T_2 \in \mathfrak{T}_{\mathfrak{N}}$ and $T_1 \sim T_2$ ($\mathfrak{l}_{\mathfrak{N}}$), then for any $N \in \mathfrak{N}$ we have

$$NT_1 = T_1, \quad NT_2 = T_2, \quad T_1N = T_2N,$$

so that $T_1 = T_2$. Thus the greatest lower bound of $\mathfrak{m}_{\mathfrak{N}}$ and $\mathfrak{l}_{\mathfrak{N}}$ is the identity equivalence. Hence if $\mathfrak{T}_{\mathfrak{N}}$ is nonvoid it follows by 1.5 that $\mathfrak{A}$ decomposes into a subdirect product of the semigroups $\mathfrak{A}/\mathfrak{m}_{\mathfrak{N}}$ and $\mathfrak{A}/\mathfrak{l}_{\mathfrak{N}}$.

2.8. **Theorem.** *Suppose the semigroup $\mathfrak{A}$ is subdirectly indecomposable, and let I be an idempotent which commutes with all elements of $\mathfrak{A}$. Then I is either the unit of $\mathfrak{A}$ or the zero of $\mathfrak{A}$.*

Proof. Consider the construction of 2.7, taking for $\mathfrak{N}$ the set consisting of the single element I. Then $\mathfrak{T}_{\mathfrak{N}}$ is nonvoid, since $I \in \mathfrak{T}_{\mathfrak{N}}$. By 2.2, one of the equivalences $\mathfrak{m}_{\mathfrak{N}}$ or $\mathfrak{l}_{\mathfrak{N}}$ must be the identity equivalence.

The ideal equivalence $\mathfrak{m}_{\mathfrak{N}}$ is the identity only if $\mathfrak{T}_{\mathfrak{N}}$ consists of a single element, which is then the zero of $\mathfrak{A}$. But $I \in \mathfrak{T}_{\mathfrak{N}}$. In this case, consequently, I is the zero of $\mathfrak{A}$.

On the other hand, since for any $X \in \mathfrak{A}$, $XI = (XI)I$, i.e., $X \sim XI$ ($\mathfrak{l}_{\mathfrak{N}}$), we have, if $\mathfrak{l}_{\mathfrak{N}}$ is the identity equivalence, that for any $X \in \mathfrak{A}$, $X = XI$ (and similarly $X = IX$); i.e., I is then the unit of $\mathfrak{A}$.

2.9. **Corollary.** *If a subdirectly indecomposable semigroup $\mathfrak{A}$ does not have a zero and is not a group, then the intersection of all its ideals is void.*

Proof. Let $\mathfrak{T} \neq \emptyset$ be the intersection. By V, 1.7, $\mathfrak{T}$ is a two-sided ideal which is a group. By V, 1.8, the identity $E_{\mathfrak{T}}$ of this group is an element which commutes with every element of $\mathfrak{A}$. Since $E_{\mathfrak{T}}$ is not the zero of the semigroup, it follows by 2.8 that $E_{\mathfrak{T}}$ is the unit of $\mathfrak{A}$. But by V, 1.8, $E_{\mathfrak{T}}$ is divisible on the left and right by every element of $\mathfrak{A}$. This implies that $\mathfrak{A}$ is a group.

2.10. Let us mention that in [**S2, 10**] B. M. Šaĭn gave a description of all subdirectly indecomposable commutative semigroups.

3. Restrictive Semigroups

3.1. The subdirect product construction is not only of interest in its own right, but is also a useful tool for studying various questions in the theory of semigroups. A good example of the latter use is afforded by the study of the so-called restrictive semigroups that has been made by B. M. Šaĭn [**S** 7, **9**].

Restrictive semigroups had been investigated by V. V. Vagner [**S 1**] in considering a certain operation on partial transformations, which will be described below. The name is also due to him. They were earlier studied by Yamada and Kimura [**1**].

3.2. **Definition.** A semigroup $\mathfrak{A}$ is called *restrictive* if the following identities hold in it:

$$\xi_1\xi_2\xi_3 = \xi_2\xi_1\xi_3, \quad \xi_1^2 = \xi_1.$$

Such semigroups are sometimes called left-restrictive, the corresponding notion of right-restrictive semigroups being defined in the obvious fashion.

3.3. We note the following properties of restrictive semigroups.

(α) Every subsemigroup of a restrictive semigroup is restrictive.

(β) The homomorphic image of a restrictive semigroup is a restrictive semigroup.

(γ) If a restrictive semigroup $\mathfrak{A}$ decomposes into a subdirect product of semigroups $\mathfrak{B}_\xi$ ($\xi \in \Gamma$), then every $\mathfrak{B}_\xi$ is a restrictive semigroup.

(δ) If a semigroup $\mathfrak{A}$ decomposes into a subdirect product of restrictive semigroups $\mathfrak{B}_\xi$ ($\xi \in \Gamma$), it is itself a restrictive semigroup.

(α) and (β) follow immediately from the Definition 3.2. Indeed, every identity in a semigroup is satisfied in every subsemigroup and in every homomorphic image. (γ) follows from (β).

Let us prove (δ). Let X, Y, Z be any elements of $\mathfrak{A}$. We have

$$\varphi(X) = (B_\xi^{(X)})_{\xi\in\Gamma}, \ \varphi(Y) = (B_\xi^{(Y)})_{\xi\in\Gamma}, \ \varphi(Z) = (B_\xi^{(Z)})_{\xi\in\Gamma}.$$

Since every $\mathfrak{B}_\xi$ ($\xi \in \Gamma$) is restrictive, we obtain for the elements of $\mathfrak{A}$

$$X^2 = (B_\xi^{(X)} \cdot B_\xi^{(X)})_{\xi\in\Gamma} = (B_\xi^{(X)})_{\xi\in\Gamma} = X,$$

$$XYZ = (B_\xi^{(X)} \cdot B_\xi^{(Y)} \cdot B_\xi^{(Z)})_{\xi\in\Gamma} = (B_\xi^{(Y)} \cdot B_\xi^{(X)} \cdot B_\xi^{(Z)})_{\xi\in\Gamma} = YXZ,$$

so that the required identities are satisfied in $\mathfrak{A}$.

3.4. The simplest restrictive semigroups are the unit semigroup $\mathfrak{R}_0$ and the following three semigroups:

$$\mathfrak{R}_1 = \{O, E\},$$

$$EE = E, \ OE = EO = OO = O;$$

$$\mathfrak{R}_2 = \{U, V\},$$
$$UU = VU = U, \ UV = VV = V;$$

$\mathfrak{R}_3 = \{U, V, O\}$, the semigroup obtained from $\mathfrak{R}_2$ by external adjunction of a zero O. That each of these semigroups is restrictive is easily verified.

3.5. The reason for singling out the above particular semigroups is explained by the following role they play from the point of view of the theory of subdirect products.

Theorem. *The restrictive semigroups $\mathfrak{R}_0$, $\mathfrak{R}_1$, $\mathfrak{R}_2$, $\mathfrak{R}_3$ (3.4) are subdirectly indecomposable, and every subdirectly indecomposable restrictive semigroup is isomorphic to one of them.*

Proof. That $\mathfrak{R}_0$, $\mathfrak{R}_1$, $\mathfrak{R}_2$, $\mathfrak{R}_3$ are subdirectly indecomposable can be verified straightforwardly without difficulty.

Now let $\mathfrak{A}$ be a subdirectly indecomposable restrictive semigroup other than the unit semigroup. By 2.4, $\mathfrak{A}$ has a smallest nonzero two-sided ideal $\mathfrak{T}$.

For any $X \in \mathfrak{A}$, the set $X\mathfrak{A}$ is not only a right ideal, but a left as well, since, $\mathfrak{A}$ being restrictive, we obtain for all A, $Y \in \mathfrak{A}$ that

$$Y(XA) = YXA = XYA \subset X\mathfrak{A}.$$

If X is not the zero of $\mathfrak{A}$ then since $X\mathfrak{A} \ni X^2 = X$ it follows that $X\mathfrak{A}$ is a nonzero two-sided ideal of $\mathfrak{A}$. Hence $\mathfrak{T} \subset X\mathfrak{A}$, and so for any $T \in \mathfrak{T}$ there exists an $X' \in \mathfrak{A}$ such that $T = XX'$. It follows that

$$XT = XXX' = XX' = T.$$

Next we define, for any element $A \in \mathfrak{A}$, a relation $\mathfrak{n}_A$ in $\mathfrak{A}$ by setting

$$X \sim Y \ (\mathfrak{n}_A)$$

if $AX = AY$. This relation is obviously an equivalence. In a restrictive semigroup it is two-sidedly stable, since for any $Z \in \mathfrak{A}$ we have, if $X \sim Y \ (\mathfrak{n}_A)$,

$$A(ZX) = ZAX = ZAY = A(ZY),$$
$$A(XZ) = (AX)Z = (AY)Z = A(YZ).$$

Now suppose $\mathfrak{T} \neq \mathfrak{A}$. Pick an element $S \in \mathfrak{A}$ which is not in $\mathfrak{T}$, and an element $T \in \mathfrak{T}$ which is not the zero of $\mathfrak{A}$.

Since S is not in the ideal $\mathfrak{T}$, S is not the zero of $\mathfrak{A}$. Hence, using what was proved above, we obtain

$$T = ST = STT = TST.$$

It follows that TS is likewise not the zero of $\mathfrak{A}$. The equivalence $\mathfrak{n}_{TS}$ is not the identity equivalence, since although $S \neq TS$ (as follows from the fact that $S \bar{\in} \mathfrak{T}$, while $TS \in \mathfrak{T}$) we have $(TS)S = TS$ and $(TS)(TS) = TS$, i.e. $S \sim TS \ (\mathfrak{n}_{TS})$. Consequently, by 2.5, $\mathfrak{n}_{TS}$ must induce a nonidentity equivalence in $\mathfrak{T}$. But from what was proved above, it follows, since TS is not zero, that for any $T_1, T_2 \in \mathfrak{T}$,

$$(TS)\,T_1 = T_1, \quad (TS)\,T_2 = T_2.$$

Thus, the relation $T_1 \sim T_2\ (\mathfrak{n}_{TS})$ can occur only if $T_1 = T_2$. The assumption that $\mathfrak{T} \neq \mathfrak{A}$ has therefore led to a contradiction.

Since $\mathfrak{T} = \mathfrak{A}$ we have, by what was proved above, that for any nonzero elements $A, B \in \mathfrak{A}$,

$$AB = B.$$

Suppose $\mathfrak{A}$ has three distinct nonzero elements A_1, A_2, A_3. Consider the following two equivalences $\mathfrak{m}_1$ and $\mathfrak{m}_2$ in $\mathfrak{A}$: $X \sim Y\ (\mathfrak{m}_1)$ if and only if $X = Y$ or one of the elements X and Y is A_1 and the other A_3, and $X \sim Y\ (\mathfrak{m}_2)$ if and only if $X = Y$ or one of the elements X and Y is A_2 and the other A_3.

Given the fact that $A_1 \sim A_3\ (\mathfrak{m}_1)$, we observe, for any $Z \in \mathfrak{A}$, that $ZA_1 = A_1$ and $ZA_3 = A_3$ if Z is different from zero, while $ZA_1 = Z$ and $ZA_3 = Z$ if Z is the zero of $\mathfrak{A}$. Furthermore, $A_1Z = Z$, $A_3Z = Z$. It follows that $\mathfrak{m}_1$ is two-sidedly stable. Similarly for $\mathfrak{m}_2$.

As we know (2.3), the set of nonidentity two-sidedly stable equivalences of $\mathfrak{A}$ contains a smallest member, $\mathfrak{k}$. But it follows from the inequalities $\mathfrak{k} \leqq \mathfrak{m}_1$ and $\mathfrak{k} \leqq \mathfrak{m}_2$ that if $X \sim Y\ (\mathfrak{k})$, then $X \sim Y\ (\mathfrak{m}_1)$ and $X \sim Y\ (\mathfrak{m}_2)$, which is possible only if $X = Y$.

The contradiction thus arrived at shows that $\mathfrak{A}$ can contain no more than two distinct nonzero elements. Now recall that in $\mathfrak{A}$, for any nonzero elements A and B, we have $AB = B$. It follows that if $\mathfrak{A}$ has no zero, then $\mathfrak{A} = \{X, Y\}$, with $XY = Y$ and $YX = X$, i.e., $\mathfrak{A}$ is isomorphic to $\mathfrak{R}_2$.

If $\mathfrak{A}$ consists of a zero and a single nonzero element, then $\mathfrak{A}$ is obviously isomorphic to $\mathfrak{R}_1$.

If $\mathfrak{A}$ consists of a zero and two nonzero elements, it is obviously isomorphic to $\mathfrak{R}_3$.

3.6. **Corollary.** *Every nonunit restrictive semigroup $\mathfrak{A}$ decomposes into a subdirect product of semigroups isomorphic to $\mathfrak{R}_1, \mathfrak{R}_2, \mathfrak{R}_3$ (3.4).*

Indeed, by 2.6 $\mathfrak{A}$ decomposes into a subdirect product of nonunit subdirectly indecomposable semigroups $\mathfrak{B}_\xi (\xi \in \Gamma)$. By 3.3 ($\gamma$), every $\mathfrak{B}_\xi$ is restrictive. Thus, by 3.5, every $\mathfrak{B}_\xi$ is isomorphic to one of the semigroups $\mathfrak{R}_1, \mathfrak{R}_2, \mathfrak{R}_3$.

3.7. Observe that by 3.3 (δ) the converse of 3.6 is also true. A semigroup decomposable into a subdirect product of semigroups isomorphic to $\mathfrak{R}_1$, $\mathfrak{R}_2$, $\mathfrak{R}_3$ is restrictive.

3.8. These results allow us to explain the relation between the notion of restrictive semigroup and a certain operation for partial transformations. This relation was first obtained by V. V. Vagner by a different method from the one presented here, the latter being due to B. M. Šaĭn.

We define an operation on the set of all partial transformations of a set Ω, by setting

$$X \nabla Y = Z$$

if $\Pi_1^{(Z)} = \Pi_1^{(X)} \cap \Pi_1^{(Y)}$ and $Z\xi = Y\xi$ for every $\xi \in \Pi_1^{(Z)}$. The operation is associative, since clearly both $X \nabla (Y \nabla T)$ and $(X \nabla Y) \nabla T$ are the partial transformation W such that $\Pi_1^{(W)} = \Pi_1^{(X)} \cap \Pi_1^{(Y)} \cap \Pi_1^{(T)}$ and $W\xi = T\xi$ for every $\xi \in \Pi_1^{(W)}$. The set of all partial transformations of Ω constitutes a semigroup with respect to this operation, which we shall denote by $\mathfrak{P}_\Omega^\nabla$.

We assert that $\mathfrak{P}_\Omega^\nabla$ is restrictive. Firstly, if $X \nabla X = X'$ $(X \in \mathfrak{P}_\Omega^\nabla)$, then $\Pi_1^{(X')} = \Pi_1^{(X)} \cap \Pi_1^{(X)} = \Pi_1^{(X)}$ and $X'\xi = X\xi$ $(\xi \in \Pi_1^{(X')})$, i.e. $X \nabla X = X$.

Secondly, if $X, Y, Z \in \mathfrak{P}_\Omega^\nabla$, $X \nabla Y \nabla Z = U$, $Y \nabla X \nabla Z = V$, then

$$\Pi_1^{(U)} = \Pi_1^{(X)} \cap \Pi_1^{(Y)} \cap \Pi_1^{(Z)}, \quad \Pi_1^{(V)} = \Pi_1^{(Y)} \cap \Pi_1^{(X)} \cap \Pi_1^{(Z)},$$
$$U\xi = Z\xi, \quad V\xi = Z\xi \quad (\xi \in \Pi_1^{(U)} = \Pi_1^{(V)}).$$

3.9. It follows from 3.3 (α) that every subsemigroup of $\mathfrak{P}_\Omega^\nabla$ is restrictive, and consequently so is every semigroup isomorphic to one such. The theorem proved below asserts that the class of all restrictive semigroups is exhausted by such semigroups. Thus the restrictiveness condition yields an abstract characterization of semigroups of partial transformations taken with respect to the above operation.

3.10. **Theorem.** *Every restrictive semigroup is isomorphic to a semigroup of one-to-one partial transformations taken with respect to the operation* ∇ (3.8).

Proof. Let $\mathfrak{A}$ be an arbitrary nonunit restrictive semigroup (for the unit semigroup the assertion of the theorem is trivial). By 3.6, $\mathfrak{A}$ decomposes into a subdirect product of semigroups $\mathfrak{B}_\xi$ $(\xi \in \Gamma)$, each isomorphic to one of the semigroups $\mathfrak{R}_1, \mathfrak{R}_2, \mathfrak{R}_3$. Consider a family of disjoint sets Ω_ξ $(\xi \in \Gamma)$, where each Ω_ξ consists of three elements $\alpha_\xi, \beta_\xi, \gamma_\xi$.

If $\mathfrak{B}_\xi = \{O_\xi, E_\xi\}$ is isomorphic to $\mathfrak{R}_1$, define a mapping φ_ξ of $\mathfrak{B}_\xi$ into $\mathfrak{P}_{\Omega_\xi}^\nabla$ by setting

$$\varphi_\xi(O_\xi) = \begin{pmatrix} \alpha_\xi \\ \alpha_\xi \end{pmatrix}, \quad \varphi_\xi(E_\xi) = \begin{pmatrix} \alpha_\xi & \beta_\xi & \gamma_\xi \\ \alpha_\xi & \beta_\xi & \gamma_\xi \end{pmatrix}$$

(the elements of $\mathfrak{P}_{\Omega_\xi}^\nabla$ are here written as permutations, the upper row being the set Π_1).

If $\mathfrak{B}_\xi = \{U_\xi, V_\xi\}$ is isomorphic to $\mathfrak{R}_2$, define φ_ξ by setting

$$\varphi_\xi(U_\xi) = \begin{pmatrix} \alpha_\xi & \beta_\xi & \gamma_\xi \\ \alpha_\xi & \beta_\xi & \gamma_\xi \end{pmatrix}, \quad \varphi_\xi(V_\xi) = \begin{pmatrix} \alpha_\xi & \beta_\xi & \gamma_\xi \\ \alpha_\xi & \gamma_\xi & \beta_\xi \end{pmatrix}.$$

If $\mathfrak{B}_\xi = \{U_\xi, V_\xi, O_\xi\}$ is isomorphic to $\mathfrak{R}_3$, define φ_ξ by setting

$$\varphi_\xi(U_\xi) = \begin{pmatrix} \alpha_\xi & \beta_\xi & \gamma_\xi \\ \alpha_\xi & \beta_\xi & \gamma_\xi \end{pmatrix}, \quad \varphi_\xi(V_\xi) = \begin{pmatrix} \alpha_\xi & \beta_\xi & \gamma_\xi \\ \alpha_\xi & \gamma_\xi & \beta_\xi \end{pmatrix}, \quad \varphi_\xi(O_\xi) = \begin{pmatrix} \alpha_\xi \\ \alpha_\xi \end{pmatrix}.$$

It is easily seen that in each case φ_ξ is an isomorphism with a subsemigroup of $\mathfrak{P}^{\nabla}_{\Omega_\xi}$ consisting of one-to-one transformations.

Using these isomorphisms, we now assign to each $A \in \mathfrak{A}$ a partial transformation $\psi(A)$ of the set $\Omega = \bigcup_\xi \Omega_\xi$. In the decomposition of $\mathfrak{A}$, let the element A correspond to $(B_\xi^{(A)})_{\xi \in \Gamma}$. Set

$$\Pi_1^{\psi(A)} \cap \Omega_\xi = \Pi_1^{\varphi_\xi(B_\xi(A))}$$

and, for $\lambda_\xi \in \Pi_1{}^{\psi(A)} \cap \Omega_\xi$,

$$\psi(A)\,\lambda_\xi = \varphi_\xi(B_\xi{}^{(A)})\,\lambda_\xi.$$

Then ψ is a homomorphism of $\mathfrak{A}$ into $\mathfrak{P}^{\nabla}_{\Omega}$.

Indeed, for $A_1, A_1 \in \mathfrak{A}$, it is readily seen that for every $\xi \in \Gamma$,

$$\Pi_1{}^{\psi(A_1A_2)} \cap \Omega_\xi = \Pi_1{}^{\psi(A_1)} \cap \Pi_1{}^{\psi(A_2)} \cap \Omega_\xi,$$

so that $\Pi_1{}^{\psi(A_1A_2)} = \Pi_1{}^{\psi(A_1)} \cap \Pi_1{}^{\psi(A_2)}$. Furthermore, if $\lambda_\xi \in \Pi_1{}^{\psi(A_1A_2)} \cap \Omega_\xi$, then

$$\begin{aligned} \psi(A_1A_2)\lambda_\xi &= \varphi_\xi(B_\xi{}^{(A_1A_2)})\lambda_\xi = \varphi_\xi(B_\xi{}^{(A_1)} \cdot B_\xi{}^{(A_2)})\lambda_\xi \\ &= [\varphi_\xi(B_\xi{}^{(A_1)}) \nabla \varphi_\xi(B_\xi{}^{(A_2)})]\lambda_\xi = \varphi_\xi(B_\xi{}^{(A_2)})\lambda_\xi = \psi(A_2)\lambda_\xi. \end{aligned}$$

Thus $\psi(A_1A_2) = (\psi A_1) \nabla (\psi A_2)$.

Secondly, ψ is an isomorphism. For if $A_1 \neq A_2$ then $B_\xi{}^{(A_1)} \neq B_\xi{}^{(A_2)}$ for some $\xi \in \Gamma$, and therefore, since φ_ξ is an isomorphism, $\varphi_\xi(B_\xi{}^{(A_1)}) \neq \varphi_\xi(B_\xi{}^{(A_2)})$. This implies that $\psi(A_1) \neq \psi(A_2)$.

Finally, since each $\varphi_\xi(A)$ was a one-to-one partial transformation of the set Ω_ξ, the partial transformation $\psi(A)$ is obviously also one-to-one.

4. Weakly Cancellative Semigroups

4.1. A subdirect product of two-sidedly cancellative semigroups is easily seen to be itself a two-sidedly cancellative semigroup.

Consider the class of semigroups consisting of those semigroups with two-sided cancellation, together with those obtained from the latter by external adjunction of a zero. As shown by I. E. Burmistrovič [S1], the semigroups which decompose into a subdirect product of members of this class constitute a class of semigroups which is of interest from the point of view of the theory of bands of semigroups. In the light of their properties, which were studied by him, I. E. Burmistrovič called such semigroups weakly cancellative.

4.2. **Definition.** A semigroup $\mathfrak{A}$ is called *weakly cancellative* if it satisfies the following two conditions:

(α) for $X, Y \in \mathfrak{A}$, if

$$X^2 = XY, \quad Y^2 = YX,$$

then $X = Y$;

(β) for $X, Y \in \mathfrak{A}$, if

$$X^2 = YX, \quad Y^2 = XY,$$

then $X = Y$.

For commutative semigroups, each of these conditions is equivalent with that of Schwarz mentioned in VIII, 1.3.

4.3. We consider here several properties related to conditions (α) and (β) of 4.2. Some of them have been examined previously by Thierrin [**24**].

(α) If a semigroup $\mathfrak{A}$ satisfies condition 4.2 (α), then for any $X, Y, Z \in \mathfrak{A}$ the equality

$$XY = XZ$$

implies $YX = ZX$.

Indeed, if $XY = XZ$, then

$$Y(XY)X = Y(XZ)X, \quad Z(XZ)X = Z(XY)X,$$

i.e.,

$$(YX)^2 = (YX)(ZX), \quad (ZX)^2 = (ZX)(YX).$$

It follows by 4.2 (α) that $YX = ZX$.

(β) If a semigroup $\mathfrak{A}$ satisfies condition 4.2 (β), then for any $X, Y, Z \in \mathfrak{A}$ the equality

$$YX = ZX$$

implies $XY = XZ$.

The demonstration is symmetric to that of 4.3 (α).

(γ) If a semigroup $\mathfrak{A}$ satisfies condition 4.2 (α), then for any $X, Y, Z \in \mathfrak{A}$ the equality

$$X^2Y = X^2Z$$

implies $YX = ZX$.

Indeed, suppose $X^2Y = X^2Z$, i.e. $X(XY) = X(XZ)$. Then by 4.3 (α), $(XY)X = (XZ)X$, whence

$$YX \cdot YX = YX \cdot ZX, \quad ZX \cdot YX = ZX \cdot ZX.$$

It follows by 4.2 (α) that

$$YX = ZX.$$

(δ) If a semigroup $\mathfrak{A}$ satisfies both conditions 4.2 (α) and 4.2 (β), then, for any $X, Y, Z \in \mathfrak{A}$ the equality

$$X^2 Y = X^2 Z$$

implies $XY = XZ$.

Indeed, if $X^2Y = X^2Z$, then by 4.3 (γ), $YX = ZX$; but then by 4.3 (β), $XY = XZ$.

(ϵ) If a semigroup $\mathfrak{A}$ satisfies both conditions 4.2 (α) and 4.2 (β), then for any $X, Y, A, B \in \mathfrak{A}$ the equality

$$XYA = XYB$$

implies $YXA = YXB$.

Indeed, suppose $XYA = XYB$. Then

$$X \cdot YAY = X \cdot YBY,$$

and by 4.3 (α),

$$YAYX = YBYX.$$

Multiplying on the right by A, we have

$$Y \cdot AYXA = Y \cdot BYXA,$$

whence, by 4.3 (α),

$$AYXAY = BYXAY,$$

and so

$$AYXAYX = BYXAYX,$$

i.e.

$$(AYX)^2 = (BYX)(AYX).$$

From the equality $YAYX = YBYX$ obtained above we have, multiplying on the right by B,

$$Y \cdot AYXB = Y \cdot BYXB.$$

It follows by 4.3 (α) that

$$AYXBY = BYXBY,$$

and so

$$AYXBYX = BYXBYX,$$

i.e.

$$(AYX)(BYX) = (BYX)^2.$$

From this and the previous equality $(AYX)^2 = (BYX)(AYX)$, we have, by 4.2 (β),

$$AYX = BYX,$$

whence, by 4.3 (β), it finally follows that

$$YXA = YXB.$$

4.4. The following theorem, due to I. E. Burmistrovič, was obtained earlier for the commutative case by Hewitt and Zuckerman [**3**].

Theorem. *A semigroup $\mathfrak{A}$ is a commutative band of semigroups, each with two-sided cancellation, if and only if $\mathfrak{A}$ is weakly cancellative* (4.2).

Proof. 1) Suppose $\mathfrak{A}$ is a band of semigroups as described. Let X and Y be elements such that $X^2 = XY$ and $Y^2 = YX$. Let $\mathfrak{B}$ be the component of the band that contains XY. Since the band is commutative we have $YX \in \mathfrak{B}$ as well. Thus $X^2, Y^2 \in \mathfrak{B}$. It follows by the remark in VIII, 1.6 that $X, Y \in \mathfrak{B}$. But $\mathfrak{B}$ is a semigroup with cancellation, and therefore the equality $XX = XY$ implies $X = Y$. A similar argument applies if $X^2 = YX$ and $Y^2 = XY$.

2) Suppose $\mathfrak{A}$ is weakly cancellative. Define a relation $\mathfrak{n}$ on $\mathfrak{A}$ by setting $X \sim Y$ $(\mathfrak{n})$ if $XA = XB$ $(A, B \in \mathfrak{A})$ always implies $YA = YB$ and conversely. Then $\mathfrak{n}$ is an equivalence. We show, first of all, that it is two-sidedly stable.

Suppose $X \sim Y$ $(\mathfrak{n})$ and, for some $Z \in \mathfrak{A}$, $(XZ)A = (XZ)B$; then by definition of $\mathfrak{n}$, $YZA = YZB$. The converse holds similarly. This proves stability on the right.

Now suppose $ZXA = ZXB$. Then by 4.3 (α), $XAZ = XBZ$, and since $X \sim Y$ $(\mathfrak{n})$, this implies $YAZ = YBZ$; whence, by 4.3 (β), $ZYA = ZYB$. Similarly, $ZYA = ZYB$ implies $ZXA = ZXB$. This proves stability on the left.

By 4.3 (δ), $X \sim X^2$ $(\mathfrak{n})$ for every $X \in \mathfrak{A}$. It follows by the remark in VIII, 1.6 that $\mathfrak{n}$ is a band. By 4.3 (ϵ) the band is commutative.

Let $\mathfrak{B}$ be a component of the band and suppose that $XA = XB$, where $X, A, B \in \mathfrak{B}$. Since $X \sim A$ $(\mathfrak{n})$ and $X \sim B$ $(\mathfrak{n})$ it follows that

$$A^2 = AB, \quad BA = B^2.$$

But then $A = B$ by 4.2 (α).

If $AX = BX$ it follows by 4.3 (β) that $XA = XB$, and by what we have just proved, $A = B$.

Thus every component of the band $\mathfrak{n}$ is a semigroup with two-sided cancellation.

4.5. Weakly cancellative semigroups have a natural characterization from the point of view of the theory of subdirect products.

THEOREM. *A semigroup is weakly cancellative if and only if it decomposes into a subdirect product of semigroups which either are semigroups with two-sided cancellation or are obtained from such by external adjunction of a zero.*

PROOF. 1) Suppose $\mathfrak{A}$ decomposes into a subdirect product of semigroups $\mathfrak{B}_\xi$ $(\xi \in \Gamma)$ of the indicated type. We show that $\mathfrak{A}$ satisfies condition 4.2 (α) (condition 4.2 (β) is verified similarly).

Suppose $X, Y \in \mathfrak{A}$ and

$$X^2 = XY, \quad Y^2 = YX.$$

Then for $\varphi(X) = (B_\xi^{(X)})_{\xi \in \Gamma}$ and $\varphi(Y) = (B_\xi^{(Y)})_{\xi \in \Gamma}$ the same equalities hold, so that, for each $\xi \in \Gamma$,

$$(B_\xi^{(X)})^2 = B_\xi^{(X)} \cdot B_\xi^{(Y)}, \quad (B_\xi^{(Y)})^2 = B_\xi^{(Y)} \cdot B_\xi^{(X)}.$$

If either $B_\xi^{(X)}$ or $B_\xi^{(Y)}$ is the zero of the semigroup $\mathfrak{B}_\xi$, then both $(B_\xi^{(X)})^2$ and $(B_\xi^{(Y)})^2$ are zero, and since the zero is externally adjoined this implies that both $B_\xi^{(X)}$ and $B_\xi^{(Y)}$ are zero, and $B_\xi^{(X)} = B_\xi^{(Y)}$.

If neither $B_\xi^{(X)}$ nor $B_\xi^{(Y)}$ is the zero of $\mathfrak{B}_\xi$, then since the set of all nonzero elements of $\mathfrak{B}_\xi$ is a semigroup with two-sided cancellation, the equality $B_\xi^{(X)} \cdot B_\xi^{(X)} = B_\xi^{(X)} \cdot B_\xi^{(Y)}$ implies $B_\xi^{(X)} = B_\xi^{(Y)}$.

Since therefore $B_\xi^{(X)} = B_\xi^{(Y)}$ for all $\xi \in \Gamma$ it follows that $\varphi(X) = \varphi(Y)$ and so $X = Y$.

2) Suppose $\mathfrak{A}$ is weakly cancellative. By 4.4, $\mathfrak{A}$ is a commutative band of semigroups, each with two-sided cancellation. Denote by $\mathfrak{H}_T$ the component of the band that contains any given element $T \in \mathfrak{A}$. In accordance with the defining property of a commutative band, we have $\mathfrak{H}_{T_1 \cdot T_2} = \mathfrak{H}_{T_2 \cdot T_1}$ and $\mathfrak{H}_{T_1} \cdot \mathfrak{H}_{T_2} \subset \mathfrak{H}_{T_1 \cdot T_2}$ for any $T_1, T_2 \in \mathfrak{A}$.

For each $T \in \mathfrak{A}$, let $\mathfrak{N}_T$ be the set of all $X \in \mathfrak{A}$ such that $XZ \bar{\in} \mathfrak{H}_T$ for every $Z \in \mathfrak{H}_T$, or, equivalently, $ZX \bar{\in} \mathfrak{H}_T$. Let $\mathfrak{M}_T$ be the set of all $X \in \mathfrak{A}$ such that $XZ \in \mathfrak{H}_T$ for every $Z \in \mathfrak{H}_T$, and so also $ZX \in \mathfrak{H}_T$. Since $\mathfrak{H}_T$ is a component of the band, it suffices for this that $XZ_0 \in \mathfrak{H}_T$ for some one $Z_0 \in \mathfrak{H}_T$. Hence $\mathfrak{A} = \mathfrak{N}_T \cup \mathfrak{M}_T$, $\mathfrak{N}_T \cap \mathfrak{M}_T = \emptyset$.

We assert that $\mathfrak{N}_T$, provided it is nonvoid, is a two-sided ideal. To see this, suppose that for some $X \in \mathfrak{N}_T$ and $S \in \mathfrak{A}$ we have $XS \bar{\in} \mathfrak{N}_T$, i.e. for some $Z \in \mathfrak{H}_T$ we have $Z' = (XS)Z \in \mathfrak{H}_T$. Since $X^2 \in \mathfrak{H}_X$, the fact that $XSZ \in \mathfrak{H}_T$ implies that $X^2SZ \in \mathfrak{H}_T$, i.e. that $XZ' \in \mathfrak{H}_T$, which is impossible since $X \in \mathfrak{N}_T$ and $Z' \in \mathfrak{H}_T$. In an entirely similar fashion, we may show it is impossible that $SX \bar{\in} \mathfrak{N}_T$.

Next we assert that $\mathfrak{M}_T$ is a subsemigroup of $\mathfrak{A}$. Indeed, if $X, Y \in \mathfrak{M}_T$, then for any $Z \in \mathfrak{H}_T$ we have $YZ \in \mathfrak{H}_T$, and so also $(XY)Z = X(YZ) \in \mathfrak{H}_T$, i.e. $XY \in \mathfrak{M}_T$.

We now define a relation $\mathfrak{l}_T$ in $\mathfrak{A}$ by the condition that, for $X, Y \in \mathfrak{M}_T$,

$X \sim Y\ (\mathfrak{l}_T)$ if $XZ = YZ$ for $Z \in \mathfrak{H}_T$; and in addition $X \sim Y\ (\mathfrak{l}_T)$ whenever X and Y both belong to $\mathfrak{N}_T$. $\mathfrak{l}_T$ is obviously an equivalence, and $\mathfrak{N}_T$ is one of the equivalence classes.

The equivalence $\mathfrak{l}_T$ is two-sidedly stable.

To see this, suppose, on the one hand, that $X \sim Y\ (\mathfrak{l}_T)$ with $X, Y \in \mathfrak{N}_T$. Then the fact that $\mathfrak{N}_T$ is a two-sided ideal of $\mathfrak{A}$ implies that $SX \sim SY\ (\mathfrak{l}_T)$ and $XS \sim YS\ (\mathfrak{l}_T)$ for any $S \in \mathfrak{A}$.

If on the other hand $X \sim Y\ (\mathfrak{l}_T)$ with $X, Y \in \mathfrak{M}_T$, then the equality $XZ = YZ\ (Z \in \mathfrak{H}_T)$ implies $(SX)Z = (SY)Z$, whence $SX \sim SY\ (\mathfrak{l}_T)$. In turn, we conclude from the equality $SXZ = SYZ$, by 4.3 (ϵ), that $XSZ = YSZ$, and thus $XS \sim YS\ (\mathfrak{l}_T)$.

If $\mathfrak{N}_T$ is nonvoid, then $\mathfrak{N}_T$, being a two-sided ideal and one of the equivalence classes for $\mathfrak{l}_T$, is the zero of the factor-semigroup $\mathfrak{A}/\mathfrak{l}_T$. Since $\mathfrak{A} = \mathfrak{N}_T \cup \mathfrak{M}_T$, this zero is externally adjoined, and the semigroup of all nonzero elements of $\mathfrak{A}/\mathfrak{l}_T$ is isomorphic to the semigroup $\mathfrak{M}_T/\mathfrak{l}_T$.

Next we show that $\mathfrak{M}_T/\mathfrak{l}_T$ is a semigroup with two-sided cancellation. Observe that $\mathfrak{H}_T$ is a two-sided ideal of $\mathfrak{M}_T$. Suppose $\underline{U}\underline{A} = \underline{V}\underline{A}$, where $\underline{U}, \underline{V}, \underline{A}$ are equivalence classes with respect to $\mathfrak{l}_T$ containing respectively the elements $U, V, A \in \mathfrak{M}_T$, i.e., elements of the factor-semigroup $\mathfrak{M}_T/\mathfrak{l}_T$. Then $UA = VA$, so that $Z_1UAZ_2 = Z_1VAZ_2$ for any $Z_1, Z_2 \in \mathfrak{H}_T$. But $(Z_1U), (Z_1V), (AZ_2) \in \mathfrak{H}_T$, and since $\mathfrak{H}_T$ is a semigroup with two-sided cancellation, we have $Z_1U = Z_1V$, whence, by 4.3 (α), $UZ_1 = VZ_1$. This implies that $U \sim V\ (\mathfrak{l}_T)$, i.e., $\underline{U} = \underline{V}$.

If $\underline{A}\underline{U} = \underline{A}\underline{V}$, then $AU = AV$ and so $Z_1AUZ_2 = Z_1AVZ_2$ for any $Z_1, Z_2 \in \mathfrak{H}_T$. It follows that $UZ_2 = VZ_2$, i.e. $U \sim V\ (\mathfrak{l}_T)$ and $\underline{U} = \underline{V}$.

Now let $\mathfrak{m}$ be the greatest lower bound of all the equivalences $\mathfrak{l}_T\ (T \in \mathfrak{A})$.

Consider an arbitrary pair of distinct elements $X, Y \in \mathfrak{A}$.

Suppose first that X and Y are in the same component $\mathfrak{H}_X = \mathfrak{H}_Y$ of the band. Then $X, Y \in \mathfrak{M}_X$. Since $\mathfrak{H}_X$ is a semigroup with two-sided cancellation, we cannot have $XZ = YZ$ for $Z \in \mathfrak{H}_X$. But $XZ \neq YZ$ implies that $X \nsim Y\ (\mathfrak{l}_X)$, so that $X \nsim Y\ (\mathfrak{m})$.

Next suppose that $\mathfrak{H}_X \neq \mathfrak{H}_Y$, with $\mathfrak{H}_X\mathfrak{H}_Y \subset \mathfrak{H}_W$ but $\mathfrak{H}_W \neq \mathfrak{H}_Y$. Then $Y \in \mathfrak{M}_Y$. But $XY \in \mathfrak{H}_X\mathfrak{H}_Y \subset \mathfrak{H}_W$, and since $\mathfrak{H}_W \cap \mathfrak{H}_Y = \emptyset$, we have $X \in \mathfrak{N}_Y$. Hence $X \nsim Y\ (\mathfrak{l}_Y)$, so that again $X \nsim Y\ (\mathfrak{m})$.

Finally, suppose $\mathfrak{H}_X \neq \mathfrak{H}_Y$, with $\mathfrak{H}_X\mathfrak{H}_Y \subset \mathfrak{H}_Y$. Then $\mathfrak{H}_Y\mathfrak{H}_X \subset \mathfrak{H}_Y$, where $\mathfrak{H}_Y \neq \mathfrak{H}_X$, so that, by the preceding paragraph, $X \nsim Y\ (\mathfrak{l}_X)$ and therefore $X \nsim Y\ (\mathfrak{m})$.

We have thus shown that $\mathfrak{m}$ is the identity equivalence. It follows, by 1.5, that $\mathfrak{A}$ decomposes into a subdirect product of the semigroups $\mathfrak{A}/\mathfrak{l}_T$ $(T \in \mathfrak{A})$. If $\mathfrak{N}_T$ is void then $\mathfrak{A}/\mathfrak{l}_T$ coincides with $\mathfrak{M}_T/\mathfrak{l}_T$, which is a semigroup with two-sided cancellation. If $\mathfrak{N}_T \neq \emptyset$ then $\mathfrak{A}/\mathfrak{l}_T$ is a semigroup with externally adjoined zero, its subsemigroup of nonzero elements being isomorphic to $\mathfrak{M}_T/\mathfrak{l}_T$, which is a semigroup with two-sided cancellation.

5. Approximation Properties

5.1. The idea of approximation of a semigroup within a certain class of semigroups (XI, 1.7, 1.8; VII, 1.6) is closely related to the construction of subdirect product, and admits a natural development that gives rise to a number of kindred problems.

In addition to approximation effected by arbitrary homomorphisms, it is of interest to consider approximation by means of homomorphisms of a specified type.

Suppose given a certain class of homomorphisms Φ. It is natural to say that the semigroup $\mathfrak{A}$ *can be approximated within the class of semigroups* Ξ *by homomorphisms in* Φ if for any $A \neq B$ ($A, B \in \mathfrak{A}$) there exists a homomorphism $\varphi \in \Phi$ of $\mathfrak{A}$ into a semigroup in Ξ such that $\varphi(A) \neq \varphi(B)$.

5.2. Among the various homomorphisms of semigroups, especially simple and well understood are the ideal homomorphisms (VII, 4.15, 4.19). For convenience of formulation, we shall in the present context include among ideal homomorphisms the identity isomorphism (regarding it, so to speak, as the ideal homomorphism corresponding to the "void ideal").

The question of finite approximability (i.e., approximability within the class of all finite semigroups) by ideal homomorphisms has been examined by M. M. Lesohin [**S1**].

5.3. We shall say that an element B of the semigroup $\mathfrak{A}$ is a divisor of the element $A \in \mathfrak{A}$ if $A = XBY$, where X and Y are either elements of $\mathfrak{A}$ or empty symbols.

5.4. **Theorem.** *In order that a semigroup* $\mathfrak{A}$ *be finitely approximable by ideal homomorphisms, it is necessary and sufficient that every nonzero element have only finitely many divisors.*

Proof. 1) Suppose $\mathfrak{A}$ is finitely separable by ideal homomorphisms. Assume that $\mathfrak{A}$ contains an element Z with infinitely many divisors. For an arbitrary $A \in \mathfrak{A}$ different from Z, there must exist an ideal homomorphism φ such that $\varphi(A) \neq \varphi(Z)$. Furthermore, $\varphi(\mathfrak{A})$ is finite. Let $\mathfrak{T}$ be the two-sided ideal corresponding to φ. Since $\varphi(A) \neq \varphi(Z)$, at least one of A or Z is not contained in $\mathfrak{T}$; call it Y. Since $Y \overline{\in} \mathfrak{T}$, and $\mathfrak{T}$ is a two-sided ideal, no element of $\mathfrak{T}$ can be a divisor of Y. But $\varphi(\mathfrak{A})$ is finite, so that the set of elements of $\mathfrak{A}$ that are outside $\mathfrak{T}$ is finite. Consequently, Y has only finitely many divisors. Hence $Y \neq Z$, i.e., $Y = A$. We have thus shown that the number of divisors of an arbitrary element $A \in \mathfrak{A}$ different from Z is finite.

For any $X \in \mathfrak{A}$, every divisor of the element Z is also a divisor of the element XZ. The set of divisors of XZ is therefore infinite. In view of the uniqueness, just proved, of an element with this property, we obtain

$XZ = Z$. Similarly, $ZX = Z$. Consequently, the only element of $\mathfrak{A}$ with infinitely many divisors is the zero of the semigroup.

2) Suppose the set of divisors of any nonzero element of $\mathfrak{A}$ is finite. If $A \neq B$ $(A, B \in \mathfrak{A})$, we can assume that the set of divisors of the element A is finite. Let $\mathfrak{T}$ be the set of all elements that are not divisors of A. For any $X \in \mathfrak{A}$ and $T \in \mathfrak{T}$, the products XT and TX are not divisors of A, since otherwise T would be a divisor of A. Consequently $\mathfrak{T}$ is a two-sided ideal in $\mathfrak{A}$. Let φ be the corresponding ideal homomorphism. The semigroup $\varphi(\mathfrak{A})$ is finite, since there are only finitely many elements of $\mathfrak{A}$ that are outside $\mathfrak{T}$, i.e., that are divisors of A. If $B \in \mathfrak{T}$, then $\varphi(A) \neq \varphi(B)$, since $A \overline{\in} \mathfrak{T}$. If $B \overline{\in} \mathfrak{T}$, then the definition of ideal homomorphism implies, since $A, B \overline{\in} \mathfrak{T}$ and $A \neq B$, that $\varphi(A) \neq \varphi(B)$.

5.5. In the recent theory of semigroups, the term "nilpotent semigroup" has come to be applied in a sense different from that of Mal′cev, as used in IX, 4.5 –4.8. Accordingly, we shall now apply it only in the following sense.

DEFINITION. *A semigroup $\mathfrak{A}$ is called nilpotent if it has a zero and there exists an integer n such that the product of any n elements of $\mathfrak{A}$ is zero.*

The smallest integer n with this property relative to $\mathfrak{A}$ is called the *nilpotency degree* or the *nilpotency index* of the semigroup $\mathfrak{A}$. A nilpotent semigroup for which the nilpotency index does not exceed n is called *n-nilpotent.*

It is immediate that every subsemigroup of an n-nilpotent semigroup, as well as every homomorphic image, is itself an n-nilpotent semigroup.

The simplest case is a nilpotent semigroup with nilpotency index equal to two. Such a semigroup amounts to a set with a distinguished element O, in which the product of any two elements is equal to O.

5.6. For approximation within the class of nilpotent semigroups, it turns out that it suffices to use only ideal homomorphisms. A study of this approximation has been carried out by M. M. Lesohin [S3]. Also related to the question is a previous investigation by G. Thierrin [S1].

Denote by $\mathfrak{A}^{(n)}$ $(n = 1, 2, 3, \cdots)$ the set of all elements of the semigroup $\mathfrak{A}$ that are factorable into a product of n factors; and let $\mathfrak{A}^{(\infty)} = \bigcap_{n=1}^{\infty} \mathfrak{A}^{(n)}$.

Each $\mathfrak{A}^{(n)}$, as well as $\mathfrak{A}^{(\infty)}$, is obviously a two-sided ideal in $\mathfrak{A}$ (it may be that $\mathfrak{A}^{(\infty)} = \emptyset$).

If $\mathfrak{A}^{(\infty)}$ consists of a single element, the element is the zero of the semigroup.

5.7. THEOREM. *If a semigroup $\mathfrak{A}$ is approximable within the class of nilpotent semigroups, it is approximable within the same class by means of ideal homomorphisms. For such approximation, it is necessary and sufficient that $\mathfrak{A}^{(\infty)}$ consist of at most one element.*

PROOF. 1) Suppose $\mathfrak{A}$ is approximable within the class of nilpotent semigroups.

For $A \neq B$ $(A, B \in \mathfrak{A})$, there exists a homomorphism φ of $\mathfrak{A}$ onto a nilpotent semigroup $\varphi(\mathfrak{A})$, with $\varphi(A) \neq \varphi(B)$. We can suppose that $\varphi(A) \neq O_{\varphi(\mathfrak{A})}$. If n is the nilpotency degree of $\varphi(\mathfrak{A})$, then $A \overline{\in} \mathfrak{A}^{(n)}$. Indeed, we should otherwise have $A = X_1 X_2 \cdots X_n$ and $\varphi(A) = \varphi(X_1) \cdot \varphi(X_2) \cdot \cdots \cdot \varphi(X_n) = O_{\varphi(\mathfrak{A})}$.

Let ψ be the ideal homomorphism of $\mathfrak{A}$ corresponding to the two-sided ideal $\mathfrak{A}^{(n)}$. If $B \in \mathfrak{A}^{(n)}$, then $\psi(B) = O_{\psi(\mathfrak{A})}$, whereas $\psi(A) \neq O_{\psi(\mathfrak{A})}$ because $A \overline{\in} \mathfrak{A}^{(n)}$. If $B \overline{\in} \mathfrak{A}^{(n)}$, then since neither element A or B lies in $\mathfrak{A}^{(n)}$, and ψ is an ideal homomorphism, we have $\psi(A) \neq \psi(B)$.

Now assume that $Z, Z' \in \mathfrak{A}^{(\infty)}$, with $Z \neq Z'$. Then there exists a homomorphism ξ of $\mathfrak{A}$ onto a nilpotent semigroup $\xi(\mathfrak{A})$ such that $\xi(Z) \neq \xi(Z')$. Let m be the nilpotency degree of the semigroup $\xi(\mathfrak{A})$. Since $Z, Z' \in \mathfrak{A}^{(\infty)}$, we have $Z = X_1 X_2 \cdots X_m$ and $Z' = X_1' X_2' \cdots X_m'$. But then $\xi(Z) = \xi(X_1) \cdot \xi(X_2) \cdot \cdots \cdot \xi(X_m) = O_{\xi(\mathfrak{A})}$. Similarly, $\xi(Z') = O_{\xi(\mathfrak{A})}$, and we have a contradiction.

2) Suppose $\mathfrak{A}^{(\infty)}$ consists of at most one element. If $A \neq B$ $(A, B \in \mathfrak{A})$, we can assume that $A \overline{\in} \mathfrak{A}^{(\infty)}$. This means that $A \overline{\in} \mathfrak{A}^{(n)}$ for some n. Under the ideal homomorphism φ corresponding to the two-sided ideal $\mathfrak{A}^{(n)}$, we have $\varphi(A) \neq O_{\varphi(\mathfrak{A})}$. If $B \in \mathfrak{A}^{(n)}$, then $\varphi(B) = O_{\varphi(\mathfrak{A})}$. If $B \overline{\in} \mathfrak{A}^{(n)}$, then since neither A nor B belongs to $\mathfrak{A}^{(n)}$, and φ is an ideal homomorphism, we have $\varphi(A) \neq \varphi(B)$.

5.8. The approximation considered so far can be called approximation with respect to the equality relation between elements of the semigroup. The same idea extends in a natural way to other relations and properties of a semigroup.

We say that a semigroup $\mathfrak{A}$ is approximable within a class of semigroups Ξ with respect to inclusion of elements in subsets if for any $A \in \mathfrak{A}$ and $\mathfrak{M} \subset \mathfrak{A}$ $(\mathfrak{M} \neq \emptyset)$ such that $A \overline{\in} \mathfrak{M}$, there exists a homomorphism φ of $\mathfrak{A}$ into a semigroup in the class Ξ such that $\varphi(A) \overline{\in} \varphi(\mathfrak{M})$. It is immediate that if $\mathfrak{A}$ is approximable within Ξ with respect to inclusion of elements in subsets, then $\mathfrak{A}$ is approximable within Ξ with respect to equality of elements.

A condition for finite approximability (i.e., approximability within the class of finite semigroups) with respect to inclusion of elements in subsets has been obtained by È. A. Golubov [**S1, S2**]. His procedure makes essential use of a construction considered earlier by P. Dubreil [**1**].

5.9. Denote by $\mathfrak{A}'$ the semigroup obtained from a semigroup $\mathfrak{A}$ by external adjunction of a unit E.

For any pair of elements A, B of $\mathfrak{A}$, we define the set $[A:B]$ to consist of the pairs (U, V), where $U, V \in \mathfrak{A}'$, such that $A = UBV$.

Theorem. *In order that a semigroup $\mathfrak{A}$ be finitely approximable with respect to inclusion of elements in subsets, it is necessary and sufficient that for each element $A \in \mathfrak{A}$ the family of all sets of the form $[A : X]$, where $X \in \mathfrak{A}$, be finite.*

Proof. 1) Suppose that for $A \in \mathfrak{A}$ the family of sets $[A : X]$ $(X \in \mathfrak{A})$ is finite, and that $A \overline{\in} \mathfrak{M}$ $(\mathfrak{M} \subset \mathfrak{A},\ \mathfrak{M} \neq \emptyset)$.

On $\mathfrak{A}$ define a relation $\mathfrak{n}$ by putting $X \sim Y(\mathfrak{n})$ if $[A : X] = [A : Y]$.

Suppose $X \sim Y(\mathfrak{n})$. For any $Z \in \mathfrak{A}$, the set $[A : XZ]$ consists of the pairs (U', V') such that $A = U'XZV'$. This means that $(U', ZV') \in [A : X] = [A : Y]$, and therefore $A = U'YZV'$, i.e., $(U', V') \in [A : YZ]$. Thus, $[A : XZ] \subset [A : YZ]$. Similarly, $[A : YZ] \subset [A : XZ]$. Hence $[A : XZ] = [A : YZ]$, i.e., $XZ \sim YZ(\mathfrak{n})$. The relation $\mathfrak{n}$ is thus right stable. Similarly, it is left stable.

Consider the natural homomorphism φ of $\mathfrak{A}$ onto the factor-semigroup $\mathfrak{A}/\mathfrak{n}$. The latter is finite, as follows from the finiteness of the family of all sets $[A : X]$ $(X \in \mathfrak{A})$.

Since $[A : A] \ni (E, E)$, but $[A : X] \overline{\ni} (E, E)$ for any $X \neq A$, the element A alone constitutes one of the $\mathfrak{n}$-classes.

Hence $\varphi(A) \neq \varphi(X)$ for any $X \neq A$ $(X \in \mathfrak{A})$. But then $A \overline{\in} \mathfrak{M}$ implies $\varphi(A) \overline{\in} \varphi(\mathfrak{M})$.

If the property indicated at the beginning of the proof is enjoyed by every element $A \in \mathfrak{A}$, then the required approximability holds.

2) Suppose $\mathfrak{A}$ is finitely approximable with respect to inclusion of elements in subsets, but for some $A \in \mathfrak{A}$ there is a countable set of elements $X_1, X_2, \cdots$ in $\mathfrak{A}$ such that $[A : X_i] \neq [A : X_j]$ for $i \neq j$ $(i, j = 1, 2, 3, \cdots)$. Take for $\mathfrak{M}$ the set $\mathfrak{M} = \mathfrak{A} \setminus \{A\}$. Under some homomorphism φ of $\mathfrak{A}$ onto a finite semigroup $\varphi(\mathfrak{A})$, we must have $\varphi(A) \overline{\in} \varphi(\mathfrak{M})$.

Since $\varphi(\mathfrak{A})$ is finite, there is a pair $p \neq q$ such that $\varphi(X_p) = \varphi(X_q)$.

Since $[A : X_p] \neq [A : X_q]$, we can suppose that $[A : X_p]$ is not contained in $[A : X_q]$. This means that there exists a pair (U, V) such that

$$UX_pV = A, \qquad UX_qV \neq A.$$

Since $UX_qV \neq A$, we have $UX_qV \in \mathfrak{M} = \mathfrak{A} \setminus \{A\}$, and therefore $\varphi(UX_qV) \in \varphi(\mathfrak{M})$.

But the equality $\varphi(X_p) = \varphi(X_q)$ implies

$$\varphi(A) = \varphi(U) \cdot \varphi(X_p) \cdot \varphi(V) = \varphi(U) \cdot \varphi(X_q) \cdot \varphi(V) = \varphi(UX_qV),$$

and we have a contradiction, since $\varphi(A) \overline{\in} \varphi(\mathfrak{M})$.

5.10. Among the approximation properties of semigroups that have been studied should be mentioned also the problem of separability of elements from subsemigroups. This is similar to 5.8, but the sets $\mathfrak{M}$ are taken to be the subsemigroups of the semigroup $\mathfrak{A}$ that is being approximated. Results relating to this question have been obtained by D. M. Smirnov [**S1**], M. M. Lesohin [**S1, S2**], E. A. Golubov [**S1, S2**], and S. G. Mamikonjan [**S1**].

Also studied has been approximation with respect to the property of divisibility. A semigroup $\mathfrak{A}$ is called approximable within a class of semigroups Ξ with respect to the relation of divisibility if for any $A, B \in \mathfrak{A}$ such that $A \overline{\in} B\mathfrak{A}$ and $A \overline{\in} \mathfrak{A}B$, there exists a homomorphism φ of $\mathfrak{A}$ into some semigroup in Ξ such that $\varphi(A) \overline{\in} \varphi(B) \cdot \varphi(\mathfrak{A})$ and $\varphi(A) \overline{\in} \varphi(\mathfrak{A}) \cdot \varphi(B)$.

For this type of approximation, results have been obtained by M. M. Lesohin [**S2, S3**] and É. A. Golubov (Golubov and Lesohin [**S1**]).

CHAPTER XII

IDENTITIES

1. The Operators Θ and Π

1.1. The notion of identity has been introduced in §4 of Chapter IX. There, as elsewhere, we have encountered various particular identities, and we have had the opportunity more than once to observe the usefulness of this notion. We shall consider now some further results relating to the line of investigation dealing with identities in semigroups.

1.2. By $\Xi = \{\xi_1, \xi_2, \cdots\}$ we shall always mean a certain fixed countable alphabet. Its elements $\xi_1, \xi_2, \cdots$ will be called letters. By an identity we mean a pair of words in the alphabet Ξ connected by the symbol $\cong$. Not infrequently we shall denote an identity by a single letter:

$$\sigma = (\xi_{i_1}\xi_{i_2} \cdots \xi_{i_p} \cong \xi_{j_1}\xi_{j_2} \cdots \xi_{j_q}).$$

For each letter $\xi_k \in \Xi$ there is defined in an obvious way the number of its occurrences in the first and in the second word of $\sigma : p_k, q_k$.

Obviously

$$\sum_k p_k = p, \qquad \sum_k q_k = q.$$

Important characteristics of the identity σ are the following.

The pair of lengths of the two words constituting the identity: (p, q).

The number of letters ξ_k actually present in σ (i.e., those ξ_k for which $p_k + q_k > 0$).

The nonnegative integer $d(\sigma)$ which is the greatest common divisor of the set of all numbers $|p_k - q_k|$.

If $p_k = q_k$ for each $k = 1, 2, \cdots$, the identity is called *balanced*. For a balanced identity, $d(\sigma) = 0$.

1.3. In what follows, as earlier (e.g., IX, 1.11), when we are given a mapping φ of one set $\mathfrak{N}_1$ into another, $\mathfrak{N}_2$, we shall automatically extend the mapping (using the same notation for the extended mapping) to the set of words $\mathfrak{W}_{\mathfrak{N}_1}$:

$$W = X_1 X_2 \cdots X_m \quad (X_1, X_2, \cdots, X_m \in \mathfrak{N}_1),$$

$$\varphi(W) = \varphi(X_1)\varphi(X_2) \cdots \varphi(X_m) \quad (\varphi(X_1), \varphi(X_2), \cdots, \varphi(X_m) \in \mathfrak{N}_2).$$

Keeping in mind the ordinary point of view regarding identities, we shall usually, if φ is a one-to-one mapping of the alphabet Ξ into itself, consider the identities $U \cong V$ and $\varphi(U) \cong \varphi(V)$ to be the same identity.

1.4. The definition of what is meant by saying that a semigroup satisfies a certain identity was given in IX, 4.1.

As we have more than once had occasion to observe, the fact that one or another identity is satisfied by a given semigroup has an essential connection with the structure and properties of the semigroup.

1.5. The following two operators Θ and Π will play a fundamental role in our exposition.

Let Γ be a given class of semigroups, and Φ a given family of identities.

$\Theta(\Gamma)$ *is the family of all identities satisfied by all the semigroups in* Γ.

$\Pi(\Phi)$ *is the class of all semigroups that satisfy all the identities in* Φ.

1.6. $\Theta(\Gamma)$ is always nonvoid, since it contains all the trivial identities $\xi_{i_1}\xi_{i_2}\cdots\xi_{i_p} \cong \xi_{i_1}\xi_{i_2}\cdots\xi_{i_p}$.

$\Pi(\Phi)$ is always nonvoid, since it contains every one-element semigroup $\{E\}$, as follows from the equality

$$\underbrace{E\cdot E\cdot\cdots\cdot E}_{p} = \underbrace{E\cdot E\cdot\cdots\cdot E}_{q}.$$

For the void class of semigroups, $\Theta(\emptyset)$ is obviously the family of all identities. For a one-element semigroup $\{E\}$, the family $\Theta(E)$ likewise consists of all identities.

If the class Γ contains a semigroup $\mathfrak{A}$ with more than one element, then $\Theta(\Gamma)$ does not contain all identities. For example, $\Theta(\Gamma)$ obviously does not contain the identity $\xi_1 \cong \xi_2$. In addition, if for two elements $X, Y \in \mathfrak{A}$, $X \neq Y$, we have $XY = X$, then $\mathfrak{A}$ fails to satisfy the identity $\xi_1\xi_2 \cong \xi_2$; and if $XY \neq X$, then $\mathfrak{A}$ fails to satisfy the identity $\xi_1\xi_2 \cong \xi_1$.

The trivial identity $U \cong U$ $(U \in \mathfrak{W}_\Xi)$ is satisfied by every semigroup. Hence $\Pi(U \cong U)$, as well as $\Pi(\emptyset)$, is the class of all semigroups.

If a family of identities Φ contains a nontrivial identity $U \cong V$ (i.e., one in which U and V are different words of $\mathfrak{W}_\Xi$), then $\Pi(\Phi)$ is a proper subclass of the class of all semigroups. This follows from the existence of semigroups that satisfy no nontrivial identity whatever.

An important example of such a semigroup is the free semigroup $\mathfrak{W}_\mathfrak{N}$ of all words over an alphabet $\mathfrak{N}$ containing at least two elements (IX, 1.3).

Indeed, suppose the word U has in the kth position the letter $\xi_r \in \Xi$, while for the word V either the kth position is occupied by a different letter $\xi_s \neq \xi_r$, or else the length of V is actually less than k. If φ is a mapping of the alphabet Ξ into the alphabet $\mathfrak{N}$ such that $\varphi(\xi_r) = X_1$ and $\varphi(\xi_s) = X_2$ $(X_1, X_2 \in \mathfrak{N};\ X_1 \neq X_2)$, then the words $\varphi(U)$ and $\varphi(V)$ are different elements of the semigroup $\mathfrak{W}_{\mathfrak{N}}$. Consequently, $(U \cong V) \bar{\in} \Theta(\mathfrak{W}_{\mathfrak{N}})$.

As for the case that $\mathfrak{N}$ consists of just one element: $\mathfrak{N} = \{X\}$, it is obvious that the semigroup $\mathfrak{W}_X$, which in this case is infinite monogenic: $\mathfrak{W}_X = [X]$, does satisfy nontrivial identities. In fact, $\Theta(\mathfrak{W}_X)$ is precisely the family of all balanced identities (1.2). On the one hand, any balanced identity is clearly satisfied by any commutative semigroup, and so by $\mathfrak{W}_X$. On the other hand, suppose the identity $U \cong V$ is not balanced. Then some letter ξ_k occurs r times in U and s times in V. We can obviously assume $r < s$. Consider a mapping φ of Ξ into $\mathfrak{W}_X$ such that $\varphi(\xi_i) = X$ $(i \neq k)$ and $\varphi(\xi_k) = X^m$, where m is sufficiently large. By a suitable choice of m we can evidently ensure that, for $\varphi(U) = X^c$ and $\varphi(V) = X^d$, we obtain $c < d$; i.e., $\varphi(U) \neq \varphi(V)$. Consequently, $(U \cong V) \bar{\in} \Theta(\mathfrak{W}_X)$.

1.7. We shall often find ourselves considering the *n-nilpotency identity*:

$$\xi_1 \xi_2 \cdots \xi_n \cong \xi_{n+1} \xi_{n+2} \cdots \xi_{2n}.$$

THEOREM. *A semigroup* $\mathfrak{A}$ *belongs to* $\Pi(\xi_1 \xi_2 \cdots \xi_n \cong \xi_{n+1} \xi_{n+2} \cdots \xi_{2n})$ *if and only if it is n-nilpotent* (XI, 5.5).

PROOF. 1) If $\mathfrak{A}$ is n-nilpotent, then for any $X_i \in \mathfrak{A}$ $(i = 1, 2, \cdots, 2n)$ we have

$$X_1 X_2 \cdots X_n = O, \qquad X_{n+1} X_{n+2} \cdots X_{2n} = O,$$
$$X_1 X_2 \cdots X_n = X_{n+1} X_{n+2} \cdots X_{2n},$$

so that the n-nilpotency identity is satisfied.

2) Suppose $\mathfrak{A} \in \Pi(\xi_1 \xi_2 \cdots \xi_n \cong \xi_{n+1} \xi_{n+2} \cdots \xi_{2n})$.

For arbitrary $X_1, X_2, \cdots, X_n, Y \in \mathfrak{A}$, consider a mapping φ such that $\varphi(\xi_k) = X_k$ $(k = 1, 2, \cdots, n-1)$, $\varphi(\xi_n) = X_n Y$, and $\varphi(\xi_l) = X_{l-n}$ $(l = n+1, n+2, \cdots, 2n)$. Then corresponding to φ we must have in $\mathfrak{A}$ the relation

$$X_1 X_2 \cdots X_{n-1} X_n Y = X_1 X_2 \cdots X_n.$$

Since Y was an arbitrary element of $\mathfrak{A}$, this means that $X_1 X_2 \cdots X_n$ is a left zero of $\mathfrak{A}$. Similarly, $X_1 X_2 \cdots X_n$ is a right zero. Hence $O = X_1 X_2 \cdots X_n$ is a two-sided zero, and it is equal to the product of any n elements of the semigroup.

1.8. The operators Θ and Π defined in 1.5 satisfy the following properties, which will be so frequently used that we shall generally not even refer to them.

Let $\{\Phi_\lambda\}$ be an arbitrary collection of families of identities, and $\{\Gamma_\mu\}$ an arbitrary collection of classes of semigroups.

(α) If $\Phi_1 \subset \Phi_2$, then $\Pi(\Phi_1) \supset \Pi(\Phi_2)$.

(β) $\Phi_1 \subset \Theta\Pi(\Phi_1)$.

(γ) $\Theta(\bigcup_\mu \Gamma_\mu) = \bigcap_\mu \Theta(\Gamma_\mu)$.
(δ) $\Theta\Pi\Theta(\Gamma_1) = \Theta(\Gamma_1)$.
(α') If $\Gamma_1 \subset \Gamma_2$, then $\Theta(\Gamma_1) \supset \Theta(\Gamma_2)$.
(β') $\Gamma_1 \subset \Pi\Theta(\Gamma_1)$.
(γ') $\Pi(\bigcup_\lambda \Phi_\lambda) = \bigcap_\lambda \Pi(\Phi_\lambda)$.
(δ') $\Pi\Theta\Pi(\Phi_1) = \Pi(\Phi_1)$.

Property (α) follows immediately from definitions.

If $\mathfrak{A} \in \Pi(\Phi_1)$, then $\Theta(\mathfrak{A}) \supset \Phi_1$. Since this holds for every $\mathfrak{A} \in \Pi(\Phi_1)$, we have the inclusion (β).

If $\sigma \in \Theta(\bigcup_\mu \Gamma_\mu)$, then σ is satisfied by every semigroup in any Γ_μ; i.e., $\sigma \in \Theta(\Gamma_\mu)$ for every Γ_μ. But then $\sigma \in \bigcap_\mu \Theta(\Gamma_\mu)$. Hence $\Theta(\bigcup_\mu \Gamma_\mu) \subset \bigcap_\mu \Theta(\Gamma_\mu)$.

On the other hand, if $\sigma \in \bigcap_\mu \Theta(\Gamma_\mu)$, then σ is satisfied by every semigroup belonging to any Γ_μ. But then $\sigma \in \Theta(\bigcup_\mu \Gamma_\mu)$. Consequently, $\bigcap_\mu \Theta(\Gamma_\mu) \subset \Theta(\bigcup_\mu \Gamma_\mu)$. This gives the equality (γ).

The proofs of (α'), (β') and (γ') are similar.

By (β), we have $\Theta(\Gamma_1) \subset \Theta\Pi(\Theta(\Gamma_1))$. But by ($\beta'$), $\Pi\Theta(\Gamma_1) \supset \Gamma_1$, and applying Θ we obtain by (α) that $\Theta\Pi\Theta(\Gamma_1) \subset \Theta(\Gamma_1)$. This gives the equality (δ). The argument for (δ') is similar.

1.9. Let $\mathfrak{A}$ be an arbitrary 2-nilpotent semigroup (1.7; XI, 5.5) with more than one element. We assert that $\Theta(\mathfrak{A})$ consists of the trivial identities together with all identities $U \cong V$ for which the lengths of U and V are greater than one.

Indeed, on the one hand, every such identity is satisfied in $\mathfrak{A}$, since for any mapping φ of Ξ into $\mathfrak{A}$ both $\varphi(U)$ and $\varphi(V)$ are equal to the zero O of $\mathfrak{A}$.

On the other hand, if $\mathfrak{A}$ satisfied a nontrivial identity

$$\xi_1 \cong \xi_{j_1}\xi_{j_2} \cdots \xi_{j_q} \quad (q \geqq 2),$$

then for any mapping φ we should have

$$\varphi(\xi_1) = \varphi(\xi_{j_1}) \cdot \varphi(\xi_{j_2}) \cdot \cdots \cdot \varphi(\xi_{j_q}) = O.$$

This would mean that the arbitrary element $\varphi(\xi_1)$ of $\mathfrak{A}$ is equal to O; i.e., $\mathfrak{A}$ would consist of a single element.

1.10. Let $\mathfrak{A}$ be an arbitrary commutative semigroup of idempotents with more than one element. Then $\Theta(\mathfrak{A})$ consists of all identities $U \cong V$ for which every letter $\xi_k \in \Xi$ that occurs in one of the words U or V also occurs in the other.

Indeed, let

$$\xi_{i_1}\xi_{i_2} \cdots \xi_{i_p} \cong \xi_{j_1}\xi_{j_2} \cdots \xi_{j_q}$$

be such an identity, and φ a mapping from Ξ to $\mathfrak{A}$.

The product $\varphi(U) = \varphi(\xi_{i_1}) \cdot \varphi(\xi_{i_2}) \cdot \cdots \cdot \varphi(\xi_{i_p})$ is equal, since $\mathfrak{A}$ is

commutative and its elements idempotent, to the product of its different factors, each taken just once. Their order is irrelevant. But $\varphi(V)$ is equal to the same product. Consequently, the identity $U \cong V$ is satisfied in $\mathfrak{A}$.

Now suppose that for some identity $U \cong V$ in $\Theta(\mathfrak{A})$ the letter ξ_1 occurs in U but not in V. For $A, B \in \mathfrak{A}$, take the mapping for which $\varphi(\xi_1) = A$ and $\varphi(\xi_k) = B$ $(k = 2, 3, \cdots)$. Then in view of the commutativity and idempotency, we obtain one of the equalities $A = B$ or $AB = B$. In the latter case, we can similarly obtain $AB = A$. Thus, in any case, we should have $A = B$ for any elements A, B of $\mathfrak{A}$.

1.11. Let $\mathfrak{A}$ be an arbitrary semigroup in $\Pi(\xi_1\xi_2 = \xi_1)$ with more than one element.

If an identity $U \cong V$ has both words U and V beginning with the same letter ξ_k, it is satisfied in $\mathfrak{A}$. Indeed, for any mapping φ of Ξ into $\mathfrak{A}$, the validity in $\mathfrak{A}$ of the identity $\xi_1\xi_2 \cong \xi_1$ implies $\varphi(U) = \varphi(\xi_k)$. Similarly, $\varphi(V) = \varphi(\xi_k)$.

On the other hand, if in the identity $U \cong V$ the word U begins with the letter ξ_k and the word V with the letter ξ_l, where $\xi_l \neq \xi_k$, then under any mapping φ for which $\varphi(\xi_k) \neq \varphi(\xi_l)$ we have $\varphi(U) = \varphi(\xi_k) \neq \varphi(\xi_l) = \varphi(V)$. Consequently, the identity is not satisfied in $\mathfrak{A}$.

1.12. Let $\mathfrak{A}$ be an arbitrary commutative *group* with more than one element, in which the order of every nonunit element is a prime number p.

If for the identity $U \cong V$ the number $d(U \cong V)$ (1.2) is divisible by p, then under any mapping φ of Ξ into $\mathfrak{A}$, each element will occur on the two sides of the relation $\varphi(U) = \varphi(V)$ to powers that differ only by a multiple of p. Since $\mathfrak{A}$ is commutative, and $A^p = E$ for any $A \in \mathfrak{A}$, where E is the unit of $\mathfrak{A}$, we conclude that the relation $\varphi(U) = \varphi(V)$ is satisfied in $\mathfrak{A}$. This means that $\mathfrak{A}$ satisfies the identity $U \cong V$.

Now suppose $d(U \cong V)$ is not divisible by p; i.e., that some letter ξ_k occurs r_k times in U and q_k times in V, with $r_k - q_k$ not divisible by p.

Consider the mapping φ of Ξ into $\mathfrak{A}$ such that $\varphi(\xi_k) = A$ and $\varphi(\xi_i) = E$ for all other letters $\xi_i \in \Xi$. Under this mapping, $\varphi(U) = A^{r_k}$ and $\varphi(V) = A^{q_k}$.

Since $A^{r_k} \neq A^{q_k}$, it follows that in this opposite case the identity $U \cong V$ is not satisfied in $\mathfrak{A}$.

1.13. Among the identities receiving particularly thorough study should be mentioned the so-called permutation identities, of the form

$$\xi_1\xi_2 \cdots \xi_n \cong \xi_{j_1}\xi_{j_2} \cdots \xi_{j_n},$$

where $\{j_1, j_2, \cdots, j_n\} = \{1, 2, \cdots, n\}$.

Some special cases of these identities have been studied individually. In addition, they have also been studied in the aggregate.

The work of M. Yamada [S1] has brought out a number of properties common to all those regular semigroups that satisfy one or another nontrivial permutation identity.

1.14. The question of existence of identities satisfied by a given semigroup has been studied for the case of semigroups specified by a generating set and a defining system of relations. For a certain large class of semigroups, with a system of relations of a special form, S. I. Adjan has shown [S1] that when a semigroup in the class satisfies at least one nontrivial identity, it must be a group, provided it is not the infinite monogenic semigroup or the semigroup $\mathfrak{P}$ considered in III, 6.2, 6.3. These two occupy a special position. The first satisfies the commutativity identity. For the second, Adjan discovered the identity

$$\xi_1\xi_2^2\xi_1\xi_2\xi_1^2\xi_2^2\xi_1 \cong \xi_1\xi_2^2\xi_1^2\xi_2\xi_1\xi_2^2\xi_1.$$

In addition, E. I. Grindlinger [S1] has described certain conditions on the defining system of relations which imply that the semigroup satisfies no nontrivial identity whatever, since it then contains a subsemigroup isomorphic to $\mathfrak{W}_{\mathfrak{N}}$, where $\mathfrak{N}$ consists of two elements. As shown in 1.6, such a semigroup has nontrivial identities.

2. The Relation of Consequentiality between Identities

2.1. In the study of identities in the theory of semigroups, the relation of consequentiality between them is of basic importance. This notion has different nonequivalent definitions. The definition below in 2.2 seems to be the most natural. We shall from now on understand consequentiality only in the sense of 2.2. It is different from the definition in IX, 4.10. The latter will not be used in what follows.

2.2. DEFINITION. *Let σ be an identity, Φ a family of identities, and Γ a class of semigroups. We shall say that σ is a consequence of Φ in Γ if every semigroup in Γ that satisfies all the identities in Φ also satisfies the identity σ.*

When Γ is the class of all semigroups, one usually speaks simply of a consequence of Φ, without mentioning the class.

2.3. By definition 2.2, the family of all consequences of Φ in the class Γ is

$$\Theta[\Pi(\Phi) \cap \Gamma].$$

For the case that Γ is the class of all semigroups, the corresponding family of consequences of Φ is $\Theta\Pi(\Phi)$.

2.4. We list some elementary properties of the relation of consequentiality. Their validity is obvious.

(α) *If in a given class* Γ *all the identities in a family* Φ_1 *are consequences of a family* Φ_2, *and all the identities in* Φ_2 *are consequences, in* Γ, *of a family* Φ_3, *then in* Γ *all the identities in* Φ_1 *are consequences of* Φ_3.

(β) *If* $\Gamma_1 \subset \Gamma_2$, *and the identity* σ *is a consequence of the family* Φ *in the class* Γ_2, *it is a consequence of* Φ *in the class* Γ_1.

(γ) *If the identity* σ *holds for all semigroups in the class* Γ, *then its consequences in* Γ *are all the identities that hold for all semigroups in* Γ.

(δ) *If the identity* σ *holds for no semigroup whatever in the class* Γ, *then in* Γ *all identities are consequences of* σ.

(ϵ) *If all the identities* $(U_1 \cong U_2), (U_2 \cong U_3), \cdots, (U_{m-1} \cong U_m)$ *are consequences of the family* Φ *in the class* Γ, *then the identity* $(U_1 \cong U_m)$ *is a consequence of* Φ *in* Γ.

2.5. Let $U \cong V$ be a member of a certain family of identities Φ, and ψ a mapping of the alphabet Ξ into $\mathfrak{W}_\Xi$. Then the identity $\psi(U) \cong \psi(V)$ is a consequence of Φ.

Indeed, let φ be a mapping of Ξ into a semigroup $\mathfrak{A} \in \Pi(\Phi)$. Then $\varphi\psi$ is likewise a mapping of Ξ into $\mathfrak{A}$. Since $\mathfrak{A} \in \Pi(\Phi)$, it must be that $\mathfrak{A}$ satisfies $\varphi\psi(U) \cong \varphi\psi(V)$. Consequently, $\psi(U) \cong \psi(V)$ is satisfied in $\mathfrak{A}$.

It follows that for any C and C' that are either words in Ξ or empty words, $\mathfrak{A}$ will also satisfy the identity $C\psi(U)C' \cong C\psi(V)C'$.

Thus *the identity* $C\psi(U)C' \cong C\psi(V)C'$ *is a consequence of* Φ. Such a consequence will be called an *immediate consequence* of Φ.

2.6. **Definition.** *Two families of identities* Φ_1 *and* Φ_2 *are called equivalent in the class of semigroups* Γ *if every identity in* Φ_i *is a consequence in* Γ *of* Φ_j $(i,j = 1,2)$.

By 2.4 (α), the condition for equivalence amounts to

$$\Theta[(\Pi(\Phi_1) \cap \Gamma] = \Theta[\Pi(\Phi_2) \cap \Gamma].$$

In the case of equivalence in the class of all semigroups, we shall usually speak simply of equivalence, without mentioning the class. In this case, by 1.8 (δ'), the equivalence condition $\Theta\Pi(\Phi_1) = \Theta\Pi(\Phi_2)$ amounts to the condition $\Pi(\Phi_1) = \Pi(\Phi_2)$.*

2.7. Any balanced identity (1.2) is obviously satisfied by every commutative semigroup. This means that all balanced identities are consequences of the commutativity identity, i.e., belong to $\Theta\Pi(\xi_1\xi_2 \cong \xi_2\xi_1)$.

---

* *Translator's note*: In fact, since

$$\Pi\Theta[\Pi(\Phi) \cap \Gamma] \cap \Gamma = \Pi(\Phi) \cap \Gamma,$$

the condition for arbitrary Γ also has an equivalent formulation:

$$\Pi(\Phi_1) \cap \Gamma = \Pi(\Phi_2) \cap \Gamma.$$

On the other hand, if an identity is not balanced, then, as shown in 1.6, the infinite monogenic semigroup, which is commutative, fails to satisfy it.

It follows that the family of all balanced identities is the family of all consequences of the identity $\xi_1\xi_2 \cong \xi_2\xi_1$, i.e., is equal to $\Theta\Pi(\xi_1\xi_2 \cong \xi_2\xi_1)$.

Since the identity $\xi_1\xi_2 \cong \xi_2\xi_1$ is itself balanced, the family of all balanced identities is equivalent to the commutativity identity $\xi_1\xi_2 \cong \xi_2\xi_1$.

2.8. How to construct all identities that are consequences, in the class of all semigroups, of a given family of identities, is to a considerable extent explained by the following theorem, first published by A. P. Birjukov [**S1, S3**]. Other algebraists have later independently arrived at the same result.

Theorem. *In order that an identity $\sigma = (U \cong V)$ be a consequence, in the class of all semigroups, of the family of identities Φ, it is necessary and sufficient that there exist a sequence of words $U \equiv T_1, T_2, \cdots, T_n \equiv V$ in the alphabet Ξ such that $T_i \cong T_{i+1}$ $(i = 1, 2, \cdots, n-1)$ is an immediate consequence of Φ* (2.5).

Proof. 1) If the sequence exists, then, by 2.4 (ϵ), $T_1 \cong T_n$ is a consequence of Φ.

2) Suppose σ is a consequence of Φ.

Consider the free semigroup $\mathfrak{W}_{\mathfrak{N}}$ of words over a countable alphabet $\mathfrak{N} = \{X_1, X_2, \cdots\}$ (IX, 1.3). In $\mathfrak{W}_{\mathfrak{N}}$, let $\mathfrak{n}$ be the relation for which $S \sim S'(\mathfrak{n})$ if there exists a sequence of words in $\mathfrak{W}_{\mathfrak{N}}$: $S \equiv P_1, P_2, \cdots, P_n \equiv S'$, such that under the mapping δ, where $\delta(X_i) = \xi_i$ $(i = 1, 2, \cdots)$, the identities $\delta(P_j) \cong \delta(P_{j+1})$ $(j = 1, 2, \cdots, n-1)$ are immediate consequences of Φ.

It is easily seen that $\mathfrak{n}$ is an equivalence. Furthermore, it is two-sidedly stable. Indeed, if $S \sim S'(\mathfrak{n})$ as above, then for RS and RS' the requisite sequence is obviously $RP_1, RP_2, \cdots, RP_n$. Similarly for SR and $S'R$.

Consider the factor-semigroup $\mathfrak{W}' = \mathfrak{W}_{\mathfrak{N}}/\mathfrak{n}$ (VII, 2.4). Every identity $\xi_{s_1}\xi_{s_2}\cdots\xi_{s_p} \cong \xi_{r_1}\xi_{r_2}\cdots\xi_{r_q}$ in Φ is satisfied by $\mathfrak{W}'$. Indeed, let φ be an arbitrary mapping of the alphabet Ξ into $\mathfrak{W}'$. It can obviously be represented in the form $\varphi = \varphi'\varphi''\delta^{-1}$, where φ'' is a mapping of $\mathfrak{N}$ into $\mathfrak{W}_{\mathfrak{N}}$, and φ' is the natural homomorphism of $\mathfrak{W}_{\mathfrak{N}}$ onto $\mathfrak{W}' = \mathfrak{W}_{\mathfrak{N}}/\mathfrak{n}$ (VII, 2.5). Since $\delta\varphi''\delta^{-1}$ is a mapping of Ξ into $\mathfrak{W}_{\Xi}$, the identity $\delta\varphi''\delta^{-1}(\xi_{s_1}\xi_{s_2}\cdots\xi_{s_p}) \cong \delta\varphi''\delta^{-1}(\xi_{r_1}\xi_{r_2}\cdots\xi_{r_q})$ is an immediate consequence of Φ. Hence

$$\varphi''\delta^{-1}(\xi_{s_1}\xi_{s_2}\cdots\xi_{s_p}) \sim \varphi''\delta^{-1}(\xi_{r_1}\xi_{r_2}\cdots\xi_{r_q})(\mathfrak{n}).$$

Applying the mapping φ', we obtain $\varphi(\xi_{s_1}\xi_{s_2}\cdots\xi_{s_p}) = \varphi(\xi_{r_1}\xi_{r_2}\cdots\xi_{r_q})$, and so

$$\varphi(\xi_{s_1}) \cdot \varphi(\xi_{s_2}) \cdot \cdots \cdot \varphi(\xi_{s_p}) = \varphi(\xi_{r_1}) \cdot \varphi(\xi_{r_2}) \cdot \cdots \cdot \varphi(\xi_{r_q}).$$

Since all the identities in Φ are satisfied by the semigroup $\mathfrak{W}' = \mathfrak{W}_{\mathfrak{N}}\,\mathfrak{n}$, so must be the identity $\sigma = (U \cong V)$. Using the construction of the factor-semigroup, it follows that in $\mathfrak{W}_{\mathfrak{N}}$ we must have $\delta^{-1}(U) \sim \delta^{-1}(V)(\mathfrak{n})$.

By definition of $\mathfrak{n}$, this means there exists a sequence of words in $\mathfrak{W}_{\mathfrak{N}}$: $\delta^{-1}(U) \equiv P_1, P_2, \cdots, P_n \equiv \delta^{-1}(V)$ such that each identity $\delta(P_j) \cong \delta(P_{j+1})$ $(j = 1, 2, \cdots, n-1)$ is an immediate consequence of Φ. But then the words $\delta(P_1), \delta(P_2), \cdots, \delta(P_n)$ in the alphabet Ξ are the requisite words for the identity $\sigma = (U \cong V)$.

2.9. By using 2.8, we can show that the important n-nilpotency identity (1.7)

$$\xi_1\xi_2 \cdots \xi_n \cong \xi_{n+1}\xi_{n+2} \cdots \xi_{2n}$$

is equivalent to the following family Φ, consisting of two identities but containing fewer letters:

$$\xi_1\xi_2 \cdots \xi_n \cong \xi_1\xi_2 \cdots \xi_n\xi_{n+1}, \qquad \xi_1\xi_2 \cdots \xi_n \cong \xi_{n+1}\xi_1\xi_2 \cdots \xi_n.$$

Using a mapping ψ_1 such that $\psi_1(\xi_i) = \xi_i$ $(i = 1, 2, \cdots, n)$, $\psi_1(\xi_j) = \xi_{j-n}$ $(j = n+1, n+2, \cdots, 2n-1)$ and $\psi_1(\xi_{2n}) = \xi_n\xi_{n+1}$, we find that the first identity in Φ is a consequence of the original n-nilpotency identity. Similarly, using a mapping ψ_2 such that $\psi_2(\xi_i) = \xi_i$ $(i = 1, 2, \cdots, n, n+1)$, $\psi_2(\xi_j) = \xi_{j-n-1}$ $(j = n+2, n+3, \cdots, 2n-1)$ and $\psi_2(\xi_{2n}) = \xi_{n-1}\xi_n$, we find that the second identity is a consequence.

On the other hand, using a mapping ψ_3 such that $\psi_3(\xi_i) = \xi_i$ $(i = 1, 2, \cdots, n)$ and $\psi_3(\xi_{n+1}) = \xi_{n+1}\xi_{n+2} \cdots \xi_{2n}$, we obtain as a consequence of the first identity in Φ the identity $\xi_1\xi_2 \cdots \xi_n \cong \xi_1\xi_2 \cdots \xi_n\xi_{n+1}\xi_{n+2} \cdots \xi_{2n}$. Similarly, from the second identity, using a mapping ψ_4 such that $\psi_4(\xi_i) = \xi_{i+n}$ $(i = 1, 2, \cdots, n)$ and $\psi_4(\xi_{n+1}) = \xi_1\xi_2 \cdots \xi_n$, we obtain the consequence $\xi_{n+1}\xi_{n+2} \cdots \xi_{2n} \cong \xi_1\xi_2 \cdots \xi_n\xi_{n+1}\xi_{n+2} \cdots \xi_{2n}$. From these two consequences of Φ, we conclude at once that the n-nilpotency identity is a consequence of Φ.

2.10. The relation of consequentiality in a miscellaneous class of semigroups is in a sense determined by the relation of consequentiality in the class of all semigroups. Indeed, for any class of semigroups Γ and any family of identities Φ, there always exists a family Ψ such that

$$\Theta[\Pi(\Phi) \cap \Gamma] = \Theta\Pi(\Psi).$$

For example, we can obviously take $\Psi = \Theta[\Pi(\Phi) \cap \Gamma]$.

In general, this method of reduction by itself leaves something to be desired. An important case is when Ψ can be taken as a family whose composition is determined from Γ and Φ in a simpler fashion.

2.11. **Theorem.** *Let Φ and Ψ be two families of identities. Then an* *identity σ is a consequence of Φ in the class $\Pi(\Psi)$ if and only if σ is a consequence of $\Phi \cup \Psi$ in the class of all semigroups.*

PROOF. That σ is a consequence of Φ in the class $\Pi(\Psi)$ means that σ belongs to $\Theta[\Pi(\Phi) \cap \Pi(\Psi)]$.

That σ is a consequence of $\Phi \cup \Psi$ in the class of all semigroups means that σ belongs to $\Theta\Pi(\Phi \cup \Psi)$.

But by 1.8 (γ'), $\Theta[\Pi(\Phi) \cap \Pi(\Psi)] = \Theta\Pi(\Phi \cup \Psi)$.

2.12. Special mention should be made of the case that the family of consequences of the family of identities Φ in the class of semigroups Γ coincides with the family of consequences of Φ in the class of all semigroups:

$$\Theta[\Pi(\Phi) \cap \Gamma] = \Theta\Pi(\Phi).$$

Note that this is equivalent, obviously, to the equality

$$\Pi\Theta[\Pi(\Phi) \cap \Gamma] = \Pi(\Phi).$$

Below (in 4.14, 4.15, 4.16) we shall examine the question of its validity for the class of all finite semigroups.

3. Irreducible Families of Identities

3.1. DEFINITION. *A family of identities Φ is called irreducible in a given class of semigroups Γ if no identity in Φ is a consequence, in Γ, of the remaining identities in Φ.*

A finite subset Ψ of a family of identities Φ is sometimes called a *basis* of Φ in the class Γ if, with respect to Γ, Ψ is irreducible and equivalent to Φ.

As always, if the class in question is the class of all semigroups, it is usually left unmentioned.

3.2. A finite family of identities always has a basis in any class Γ. To obtain it, one simply discards in succession identities that are consequences of the remaining ones.

Among infinite families of identities, there are some that have bases in one class or another, and some that do not. The existence of families of identities without bases in the class of all semigroups was discovered by A. P. Birjukov [S4] and A. K. Austin [S1].

As was seen subsequently, there exist infinite families of identities that are even irreducible. Such families have been constructed independently by several algebraists (V. L. Murskiĭ [S1], T. Evans [S1], E. S. Ljapin [S5]). The above-mentioned family of identities of Birjukov [S4] is also such a family. We shall describe below the construction given by Ljapin.

3.3. If a family of identities Φ_1 is equivalent, in the class $\Pi(\Phi_2)$, to a finite family Φ_3, then Φ_1 has a basis in $\Pi(\Phi_2)$.

Indeed, by 2.11, every identity in Φ_3 is then a consequence of $\Phi_1 \cup \Phi_2$. By 2.8, every identity in Φ_3 is a consequence of a finite subset of the family $\Phi_1 \cup \Phi_2$. Since Φ_3 is finite, it follows that there exists a finite family $\Phi_4 \subset \Phi_1$ such that every identity in Φ_3 is a consequence of $\Phi_2 \cap \Phi_4$. Every identity in Φ_1 is by 2.11 a consequence of $\Phi_2 \cap \Phi_3$, and every identity in $\Phi_2 \cup \Phi_3$ is a consequence of $\Phi_2 \cup \Phi_4$. Hence every identity in Φ_1 is a consequence of $\Phi_2 \cap \Phi_4$. In other words, every identity in Φ_1 is a consequence of Φ_4 in $\Pi(\Phi_2)$.

But the finite family Φ_4 has a basis Φ_5 in $\Pi(\Phi_2)$, and every identity in Φ_1 is a consequence of Φ_5 in $\Pi(\Phi_2)$. Hence Φ_5 is a basis for Φ_1 in $\Pi(\Phi_2)$.

3.4. Consider the free semigroup $\mathfrak{W}_{\mathfrak{N}}$ over a two-letter alphabet $\mathfrak{N} = \{X_1, X_2\}$ (IX, 1.3).

Every word in $\mathfrak{W}_{\mathfrak{N}}$ can be written in the form

$$U \equiv X_{i_1} X_{i_2} \cdots X_{i_n} = X_{j_1}^{k_1} X_{j_2}^{k_2} \cdots X_{j_m}^{k_m}$$
$$(i_1, i_2, \cdots, i_n, j_1, j_2, \cdots, j_m = 1, 2;\ j_{s-1} \neq j_s).$$

We shall write $l_1(U) = n$, $l_2(U) = m$ and $l_3(U) = t$, where t is the number of exponents $k_1, k_2, \cdots, k_m$ exceeding 1.

Let k be an arbitrary fixed natural number greater than 1. We shall denote by $\mathfrak{T}_k$ the family of all words $U \in \mathfrak{W}_{\mathfrak{N}}$ that satisfy at least one of the conditions

$$l_1(U) > k + 3, \qquad l_2(U) > 4, \qquad l_3(U) > 1.$$

From the nature of the multiplication of elements in $\mathfrak{W}_{\mathfrak{N}}$ it follows at once that for any $U \in \mathfrak{T}_k$ and $W \in \mathfrak{W}_{\mathfrak{N}}$ one has $UW, WU \in \mathfrak{T}_k$; i.e., $\mathfrak{T}_k$ is a two-sided ideal in the semigroup $\mathfrak{W}_{\mathfrak{N}}$.

Note that the factor-semigroup $\mathfrak{W}_{\mathfrak{N}}/\mathfrak{T}_k$ is finite, since any nonzero element U of $\mathfrak{W}_{\mathfrak{N}}/\mathfrak{T}_k$ is a word in $\mathfrak{W}_{\mathfrak{N}}$ that must satisfy the condition $l_1(U) \leqq k + 3$.

In $\mathfrak{W}_{\mathfrak{N}}$, define a relation $\mathfrak{n}_k$ for which $U \sim V(\mathfrak{n}_k)$ $(U, V \in \mathfrak{W}_{\mathfrak{N}})$ if $U = V$ or if $U, V \in \mathfrak{T}_k$, and also

$$X_i X_j X_i^s X_j \sim X_j X_i^s X_j X_i (\mathfrak{n}_k),$$
$$X_j X_i^s X_j X_i \sim X_i X_j X_i^s X_j (\mathfrak{n}_k)$$
$$(i, j = 1, 2;\ s = 2, 3, \cdots, k-1, k+1, k+2, \cdots).$$

The relation $\mathfrak{n}_k$ is reflexive and symmetric. It is easy to see that it is also transitive. Indeed, if $U \sim V(\mathfrak{n}_k)$ and $U \in \mathfrak{T}_k$, then obviously $V \in \mathfrak{T}_k$. Hence to prove transitivity it suffices to consider the case

$$U_1 \sim U_2(\mathfrak{n}_k), \qquad U_2 \sim U_3(\mathfrak{n}_k),$$

when $U_1 \neq U_2$, $U_2 \neq U_3$ and none of U_1, U_2, U_3 belongs to $\mathfrak{T}_k$. But then

U_2 is such that there exists just one $U' \in \mathfrak{W}_{\mathfrak{N}}$ for which $U_2 \sim U'(\mathfrak{n}_k)$, $U' \neq U_2$. Hence $U_1 = U'$, $U_3 = U'$, and $U_1 \sim U_3(\mathfrak{n}_k)$.

We assert that the equivalence relation $\mathfrak{n}_k$ is two-sidedly stable. Indeed, suppose $U \sim V(\mathfrak{n}_k)$. If $U = V$ or $U, V \in \mathfrak{T}_k$, then for any $W \in \mathfrak{W}_{\mathfrak{N}}$ we have obviously $WU \sim WV(\mathfrak{n}_k)$ and $UW \sim VW(\mathfrak{n}_k)$. As for $X_iX_jX_i^sX_j$ and $X_jX_i^sX_jX_i$ $(i, j = 1, 2)$, multiplication of either of these by any element of $\mathfrak{W}_{\mathfrak{N}}$ gives a product that belongs to $\mathfrak{T}_k$ (since for this product either l_2 is greater than 4 or l_3 is greater than 1).

We shall want to consider the factor-semigroups $\mathfrak{B}_k = \mathfrak{W}_{\mathfrak{N}}/\mathfrak{n}_k$ $(k = 2, 3, \cdots)$.

The element of $\mathfrak{B}_k = \mathfrak{W}_{\mathfrak{N}}/\mathfrak{n}_k$ which is the $\mathfrak{n}_k$-class containing the element $U \in \mathfrak{W}_{\mathfrak{N}}$ will be denoted by $\overline{U}$. We have always $\overline{U}_1 \cdot \overline{U}_2 = \overline{U_1U_2}$.

3.5. Denote by σ_s $(s = 2, 3, \cdots)$ the identity

$$\xi_1\xi_2\xi_1^s\xi_2 \cong \xi_2\xi_1^s\xi_2\xi_1.$$

We show now that the semigroup $\mathfrak{B}_k$ of 3.4 satisfies every identity σ_s $(s = 2, 3, \cdots, k-1, k+1, \cdots)$.

Let φ be a mapping from Ξ to $\mathfrak{B}_k$.

If $\varphi(\xi_1) = \overline{X}_i$ and $\varphi(\xi_2) = \overline{X}_j$, then the relation

$$\overline{X}_i\overline{X}_j\overline{X}_i^s\overline{X}_j = \overline{X}_j\overline{X}_i^s\overline{X}_j\overline{X}_i,$$

obtained from σ_s via φ is satisfied in $\mathfrak{B}_k$, since its left-hand side is equal to $\overline{X_iX_jX_i^sX_j}$ and its right to $\overline{X_jX_i^sX_jX_i}$, and $X_iX_jX_i^sX_j \sim X_jX_i^sX_jX_i(\mathfrak{n}_k)$.

If for either $i = 1$ or $i = 2$ we have $\varphi(\xi_i) = \overline{X_{t_1}X_{t_2}\cdots X_{t_r}}$, where $r > 1$, then it is easy to see that $\varphi(\xi_1\xi_2\xi_1^s\xi_2) = \overline{U}_s$, where $U_s \in \mathfrak{T}_k$ (since either $l_2(U_s) > 4$ or $l_3(U_s) > 1$). Similarly, $\varphi(\xi_2\xi_1^s\xi_2\xi_1) = \overline{U}_s'$, where $U_s' \in \mathfrak{T}_k$. Hence $U_s \sim U_s'(\mathfrak{n}_k)$ and $\overline{U}_s = \overline{U}_s'$. Thus again the relation obtained from σ_s via φ is satisfied in $\mathfrak{B}_k$.

3.6. As for the identity σ_k, it is not satisfied in $\mathfrak{B}_k$, since under a mapping φ for which $\varphi(\xi_1) = \overline{X}_1$ and $\varphi(\xi_2) = \overline{X}_2$, the left- and right-hand sides map into the elements

$$\overline{X}_1\overline{X}_2\overline{X}_1^k\overline{X}_2, \qquad \overline{X}_2\overline{X}_1^k\overline{X}_2\overline{X}_1.$$

But these elements of $\mathfrak{B}_k$ are unequal, since in $\mathfrak{W}_{\mathfrak{N}}$ the elements $X_1X_2X_1^kX_2$ and $X_2X_1^kX_2X_1$ are nonequivalent with respect to $\mathfrak{n}_k$.

3.7. The fact that for each σ_k we have found a semigroup $\mathfrak{B}_k$ that satisfies every σ_s $(s \neq k)$, but not σ_k, means that σ_k is not a consequence of the family of identities $\{\sigma_2, \sigma_3, \cdots, \sigma_{k-1}, \sigma_{k+1}, \cdots\}$. This means that the infinite family $\{\sigma_2, \sigma_3, \cdots\}$ is irreducible. In fact, it is irreducible not only in the class of all semigroups, but in the class of finite semigroups, since the semigroups $\mathfrak{B}_k$ are all finite. This is evident from the fact, already noted in 3.4, that the semigroups $\mathfrak{W}_{\mathfrak{N}}/\mathfrak{T}_k$ are finite.

3.8. Suppose an identity τ_1 is included in an infinite family of identities $\{\tau_1, \tau_2, \tau_3, \cdots\}$ which is irreducible in the class of all semigroups. Then by 2.11, the family $\{\tau_2, \tau_3, \cdots\}$ is irreducible in the class $\Pi(\tau_1)$.

This implies that the preceding construction provides an infinite set of different classes of semigroups, in each of which there is an infinite irreducible family of identities.

3.9. At the same time, in certain important classes of semigroups matters are otherwise. For example, P. Perkins has shown [S1] that in the class of all commutative semigroups $\Pi(\xi_1\xi_2 \cong \xi_2\xi_1)$, every family of identities is equivalent to a finite family. By 3.3, this means that in $\Pi(\xi_1\xi_2 \cong \xi_2\xi_1)$ every family of identities has a basis.

It follows by 3.8 that the commutativity identity $\xi_1\xi_2 \cong \xi_2\xi_1$ cannot be included in any infinite family of identities which is irreducible in the class of all semigroups, and so in any infinite family irreducible in any other class of semigroups.

In order to obtain Perkins' result, we must first carry out some preliminaries. Repeated use will be made of Theorem 2.11.

3.10. We shall denote here by Σ_n the family of all sequences of n non-negative integers.

If two sequences $E = (m_1, m_2, \cdots, m_n)$ and $E' = (m_1', m_2', \cdots, m_n')$ in Σ_n satisfy the equalities $m_i \leqq m_i'$ $(i = 1, 2, \cdots, n)$, we shall write $E \leqq E'$.

Lemma. *For any infinite sequence*

$$E_1, E_2, \cdots, E_k, \cdots \quad (E_k \in \Sigma_n;\ k = 1, 2, \cdots)$$

there exists an infinite subsequence

$$E_{i_1} \leqq E_{i_2} \leqq \cdots \leqq E_{i_l} \leqq \cdots \quad (i_1 < i_2 < \cdots).$$

Proof. The proof is by induction on n.

For $n = 1$, the family Σ_1 is simply the family of all nonnegative integers, for which the assertion is obviously valid.

Suppose $n > 1$. For $E_k \in \Sigma_n$, let $E_k' \in \Sigma_{n-1}$ have as components the first $n - 1$ components of E_k.

By the induction hypothesis, there exists a subsequence

$$E_{j_1}' \leqq E_{j_2}' \leqq \cdots \leqq E_{j_k}' \leqq \cdots \quad (j_1 < j_2 < \cdots).$$

If the nth components of all the E_{j_k} $(k = 1, 2, \cdots)$ are unbounded, this sequence has a subsequence $E_{i_1}' \leqq E_{i_2}' \leqq \cdots$ for which in the corresponding $E_{i_1}, E_{i_2}, \cdots$ the nth components form an increasing sequence. Hence $E_{i_1} \leqq E_{i_2} \leqq \cdots$.

On the other hand, if the nth components of all the E_{j_k} are bounded, there exists a subsequence $E'_{i_1} \leqq E'_{i_2} \leqq \cdots$ for which in the corresponding $E_{i_1}, E_{i_2}, \cdots$ all the nth components are equal. Hence again $E_{i_1} \leqq E_{i_2} \leqq \cdots$.

3.11. **Lemma.** *If a family of identities Φ includes an unbalanced identity, then the family $\Theta\Pi(\Phi)$ includes an identity of the form $\xi_1^m \cong \xi_1^l$ $(m > l)$.*

Proof. Suppose $\sigma \in \Theta\Pi(\Phi)$ is unbalanced. Some letter ξ_k occurs on the left-hand side p_k times and on the right q_k times, where, say, $p_k > q_k$. A mapping φ such that $\varphi(\xi_k) = \xi_1^s$ and $\varphi(\xi_i) = \xi_1$ $(i = 1, 2, \cdots, k-1, k+1, \cdots)$ determines a consequence of Φ of the form $\xi_1^m \cong \xi_1^l$. Since $p_k > q_k$, a sufficiently large choice of s will obviously ensure that $m > l$.

3.12. If in one of the words in the identity $U \cong V$ we interchange two adjacent letters, we obtain a new identity, which is equivalent to the original in the class of all commutative semigroups.

Let σ be an arbitrary identity, and $\sigma_0 = (\xi_1^m \cong \xi_1^l)$ $(m > l)$. If in one of the words in σ we replace ξ_k^m by ξ_k^l, we obtain a new identity σ', with $\{\sigma, \sigma_0\}$ obviously equivalent to $\{\sigma', \sigma_0\}$.

Putting these facts together, we conclude that for any σ the family $\{\sigma, \sigma_0\}$ is equivalent, in the class of all commutative semigroups, to a family consisting of σ_0 and an identity of the form

$$\xi_1^{p_1}\xi_2^{p_2}\cdots\xi_r^{p_r} \cong \xi_1^{q_1}\xi_2^{q_2}\cdots\xi_r^{q_r} \quad (p_i, q_i < m).$$

Here we allow $p_i = 0$ or $q_i = 0$ (but not simultaneously $p_i = q_i = 0$), regarding ξ_i^0 as an empty symbol.

3.13. Consider the pairs of nonnegative integers (p, q) with $0 \leqq p, q < m$, $p + q > 0$. Assign to the family of all $m^2 - 1$ such pairs a linear order.

To each identity σ of the type indicated in 3.12 associate the sequence

$$E^{(\sigma)} = (e_{10}^{(\sigma)}, e_{01}^{(\sigma)}, \cdots, e_{pq}^{(\sigma)}, \cdots) \in \Sigma_{m^2-1}.$$

Here $e_{pq}^{(\sigma)}$ is the number of letters $\xi_k \in \Xi$ for which $p_k = p$ and $q_k = q$.

From 3.12 we conclude that when $E^{(\sigma_1)} = E^{(\sigma_2)}$, the identities σ_1 and σ_2 are equivalent in the class of all commutative semigroups.

3.14. **Lemma.** *If $E^{(\sigma_1)} \leqq E^{(\sigma_2)}$, then in the class of commutative semigroups the identity σ_2 is a consequence of the identity σ_1.*

Proof. We prove this by induction on $s = s(\sigma_1, \sigma_2)$, where s is the number of components for which $e_{pq}^{(\sigma_1)} < e_{pq}^{(\sigma_2)}$.

If $s = 0$, then σ_1 and σ_2 are equivalent.

Suppose $s = 1$. We pass from σ_1 and σ_2 to identities σ_1' and σ_2' equivalent to them in the class of commutative semigroups and of the form

$$\sigma_1' = (\xi_1^p \xi_2^p \cdots \xi_k^p U \cong \xi_1^q \xi_2^q \cdots \xi_k^q V),$$
$$\sigma_2' = (\xi_1^p \xi_2^p \cdots \xi_k^p \xi_{k+1}^p \cdots \xi_{k+r}^p U \cong \xi_1^q \xi_2^q \cdots \xi_k^q \xi_{k+1}^q \cdots \xi_{k+r}^q V),$$

where in the words U and V the letters $\xi_1, \xi_2, \cdots, \xi_{k+r}$ do not occur. Applying to σ_1' the mapping φ for which $\varphi(\xi_k) = \xi_k \xi_{k+1} \cdots \xi_{k+r}$ and $\varphi(\xi_i) = \xi_i$ $(i = 1, 2, \cdots, k-1, k+1, \cdots)$, we see that in the class of commutative semigroups the identity σ_2' is a consequence of the identity σ_1'. Hence σ_2 is a consequence of σ_1.

If $s > 1$, there exists a σ_3 such that $E^{(\sigma_1)} < E^{(\sigma_3)} < E^{(\sigma_2)}$, where $s(\sigma_1, \sigma_3)$, $s(\sigma_3, \sigma_2) < s(\sigma_1, \sigma_2)$. By the induction hypothesis, σ_3 is a consequence of σ_1, and σ_2 of σ_3. Hence σ_2 is a consequence of σ_1.

3.15. **Theorem.** *In the class of all commutative semigroups, any family of identities is equivalent to a finite family.*

Proof. For the rest of the argument, in speaking of consequence and equivalence we shall mean consequence and equivalence in the class of all commutative semigroups.

If all the identities in Φ are balanced, then by 2.7 and 2.11 Φ is equivalent to the identity $\xi_1 \xi_2 \cong \xi_2 \xi_1$.

Now suppose the family $\Phi = \{\sigma_1, \sigma_2, \cdots\}$ is infinite and contains an unbalanced identity. By 3.11, among the consequences of Φ is an identity $\sigma_0 = (\xi_1^m \cong \xi_1^l)$ $(m > l)$.

Consider the family $\Psi_1 = \{\sigma_0, \sigma_1, \sigma_2, \cdots\}$. The family Φ is equivalent to Ψ_1, and Ψ_1 is equivalent by 3.12 to $\Psi_2 = \{\sigma_0, \sigma_1', \sigma_2', \cdots\}$, where $\sigma_1', \sigma_2', \cdots$ are all identities of the form indicated in 3.12.

If, from some t on, every σ_i' $(i > t)$ is a consequence of $\{\sigma_0, \sigma_1', \cdots, \sigma_{i-1}'\}$, then Ψ_2 is equivalent to the finite family $\{\sigma_0, \sigma_1', \cdots, \sigma_t'\}$. But then Φ is likewise equivalent to this finite family.

Suppose, on the other hand, that this is not the case. Then there exists an infinite sequence $\sigma_0, \sigma_{i_1}', \sigma_{i_2}', \cdots$ in which none of the σ_{i_k}' is a consequence of the preceding identities. If we apply 3.10 to the sequence consisting of the $E^{(\sigma_{i_s}')}$, we find that for some σ_{i_c}' and σ_{i_d}' $(1 < c < d)$ we must have $E^{(\sigma_{i_c}')} \leqq E^{(\sigma_{i_d}')}$. But by 3.14 this contradicts the fact that σ_{i_d}' is not a consequence of σ_{i_c}'.

4. Varieties of Semigroups

4.1. Among the important classes of semigroups, particular attention is due those defined by families of identities. We have already had frequent occasion to deal with such classes.

Definition. *A class of semigroups Γ is called a variety if there exists a family of identities Φ such that $\Gamma = \Pi(\Phi)$.*

The theory of varieties has been extensively developed in the theory of groups (see, e.g., H. Neumann [S1]). It must be recognized, however, that the theory of group varieties is not directly contained in the theory of semigroup varieties. This stems from the fact that when we consider identities in the theory of groups, the groups are regarded as algebraic systems with two operations—multiplication and taking the inverse of an element.

Varieties in the theory of semigroups have been studied in a number of papers. A broad survey appears in a paper of T. Evans [S3], which includes an extensive bibliography.

4.2. One of the important varieties that we have already encountered is the variety $\Pi(\xi_1\xi_2 \cong \xi_2\xi_1)$ of all commutative semigroups.

The identity $\xi_1^2 \cong \xi_1$ determines the variety $\Pi(\xi_1^2 \cong \xi_1)$ of all idempotent semigroups.

The class

$$\Pi(\xi_1\xi_2\cdots\xi_n \cong \xi_{n+1}\xi_{n+2}\cdots\xi_{2n})$$
$$= \Pi(\xi_1\xi_2\cdots\xi_n \cong \xi_1\xi_2\cdots\xi_n\xi_{n+1}, \xi_1\xi_2\cdots\xi_n \cong \xi_{n+1}\xi_1\xi_2\cdots\xi_n)$$

is the variety of all n-nilpotent semigroups (XI, 5.5; XII, 1.7, 2.9). The class $\Pi(\xi_1\xi_2\xi_3 \cong \xi_2\xi_1\xi_3,\ \xi_1^2 \cong \xi_1)$ is the variety of all restrictive semigroups (XI, 3.2).

4.3. **Theorem.** *In order that a class of semigroups* Γ *be a variety, it is necessary and sufficient that* $\Pi\Theta(\Gamma) = \Gamma$.

Proof. If the equality holds, then Γ is a variety, since the family Φ of Definition 4.1 can be taken as $\Phi = \Theta(\Gamma)$.

On the other hand, suppose $\Gamma = \Pi(\Phi)$. Then, applying the operators Θ and Π, we obtain $\Pi\Theta(\Gamma) = \Pi\Theta\Pi(\Phi) = \Pi(\Phi) = \Gamma$.

4.4. For a variety Γ, there obviously exist in general a number of different families of identities Φ defining the variety: $\Gamma = \Pi(\Phi)$. In studying a given variety Γ, it can be important to describe all the families Φ defining Γ. By 2.6, this is equivalent to the problem of describing all the families of identities that are equivalent to Φ.

4.5. It follows from 4.3 that one of the families of identities defining the variety Γ is the family $\Theta(\Gamma)$. Among all families Φ such that $\Gamma = \Pi(\Phi)$, this is the largest. Indeed, by 1.8 (β),

$$\Theta(\Gamma) = \Theta\Pi(\Phi) \supset \Phi.$$

4.6. In connection with 4.4, note that the condition $\Theta(\Gamma) \supset \Phi$, for an arbitrary variety Γ and family of identities Φ, means that the variety $\Pi(\Phi)$ contains the variety Γ.

Indeed, if $\Theta(\Gamma) \supset \Phi$, then, applying the operator Π, we obtain $\Pi\Theta(\Gamma) \subset \Pi(\Phi)$. But $\Gamma \subset \Pi\Theta(\Gamma)$.

On the other hand, if $\Pi(\Phi) \supset \Gamma$, then, applying Θ, we obtain $\Theta\Pi(\Phi) \subset \Theta(\Gamma)$. But $\Phi \subset \Theta\Pi(\Phi)$.

4.7. Let $\mathfrak{N}$ be an alphabet, and $\mathfrak{W}_{\mathfrak{N}}$ the free semigroup of words over it (IX, 1.3). In this semigroup, $\mathfrak{N}$ is a free generating set with respect to the class of all semigroups.

For a family of identities Φ, we have constructed (IX, 4.11) a semigroup $\mathfrak{W}^{\Phi}_{\mathfrak{N}}$ with $\mathfrak{N}$ as generating set, the defining system of relations consisting of all relations $\varphi(U) = \varphi(V)$, where $(U \cong V) \in \Phi$ and φ is any mapping of Ξ into $\mathfrak{W}_{\mathfrak{N}}$.

Clearly $\mathfrak{W}^{\Phi}_{\mathfrak{N}} \in \Pi(\Phi)$, and $\mathfrak{N}$ is a free set with respect to the variety $\Pi(\Phi)$. Indeed, under any mapping φ of the set $\mathfrak{N}$ into any semigroup $\mathfrak{A} \in \Pi(\Phi)$, every relation in the above-described defining system maps into a relation that holds in $\mathfrak{A}$, since $\mathfrak{A} \in \Pi(\Phi)$. By IX, 1.11, it follows that φ extends to a homomorphism of $\mathfrak{W}^{\Phi}_{\mathfrak{N}}$ into $\mathfrak{A}$.

Thus, the semigroup $\mathfrak{W}^{\Phi}_{\mathfrak{N}}$ is free in the variety $\Pi(\Phi)$.

As follows from the argument of §5, Chapter IX, the semigroups of this type exhaust all the free semigroups in the class $\Pi(\Phi)$. Every semigroup in $\Pi(\Phi)$ can be obtained from one of these by a homomorphism.

4.8. Let Γ be a class of semigroups. One says of the variety $\Pi\Theta(\Gamma)$ that it is generated by the class Γ. The usual notation is

$$\operatorname{var}\Gamma = \Pi\Theta(\Gamma).$$

$\Pi\Theta(\Gamma)$ is the intersection of all varieties containing Γ. Thus, it is the smallest variety in the class of all such varieties. Indeed, suppose $\Gamma \subset \Pi(\Phi)$. Then

$$\Pi\Theta(\Gamma) \subset \Pi\Theta\Pi(\Phi) = \Pi(\Phi).$$

4.9. For a variety Γ, the question arises of describing all the classes Γ_0 that generate it: $\Gamma = \Pi\Theta(\Gamma_0)$. Especially important is the case that Γ_0 is as small as possible. In particular, a case of interest is that $\Gamma = \Pi\Theta(\mathfrak{A})$ for some semigroup $\mathfrak{A}$ that by itself generates the whole variety Γ. This is especially convenient if the structure of $\mathfrak{A}$ is as simple as possible.

4.10. Let $\mathfrak{N}$ be a countable alphabet. For a variety $\Gamma = \Pi(\Phi)$, consider the semigroup $\mathfrak{W}^{\Phi}_{\mathfrak{N}}$, which by 4.7 is free in the variety. It satisfies all the identities in Φ. On the other hand, if $\mathfrak{W}^{\Phi}_{\mathfrak{N}}$ satisfies a certain identity σ, the corresponding relation is satisfied for elements of $\mathfrak{N}$. But any relation between elements of $\mathfrak{N}$, which is a free set with respect to $\Pi(\Phi)$ (IX, 5.1), must also be satisfied between any elements of any semigroup in Γ (IX, 5.4), i.e., corresponds to an identity in $\Theta(\Gamma)$. Thus, $\Theta(\mathfrak{W}^{\Phi}_{\mathfrak{N}}) = \Theta(\Gamma) = \Theta\Pi(\Phi)$.

Hence

$$\Pi\Theta(\mathfrak{W}^{\Phi}_{\mathfrak{N}}) = \Pi\Theta\Pi(\Phi) = \Pi(\Phi) = \Gamma.$$

This means that $\mathfrak{W}^{\Phi}_{\mathfrak{N}}$ generates the variety Γ.

4.11. Suppose a variety $\Gamma = \Pi(\Phi)$ is generated by a semigroup $\mathfrak{A}$: $\Gamma = \Pi\Theta(\mathfrak{A})$.

Consider the case that the family of identities Φ is such that every identity obtained by dropping out any one letter $\xi_k \in \Xi$ from an identity in Φ either also belongs to Φ or else is trivial.

Let $\mathfrak{A}'$ be the semigroup obtained from $\mathfrak{A}$ by external adjunction to $\mathfrak{A}$ of a unit E.

Since $\mathfrak{A}' \supset \mathfrak{A}$, we have $\Theta(\mathfrak{A}') \subset \Theta(\mathfrak{A})$.

On the other hand, suppose $\sigma \in \Theta(\mathfrak{A})$. Under a mapping φ of the set Ξ into $\mathfrak{A}'$, certain letters ξ_i may map into $\mathfrak{A}$ and others onto the element E. Drop out from σ those ξ_i that map into E. This gives a new identity σ', which by assumption also belongs to Φ (or is trivial). But the relation in $\mathfrak{A}'$ obtained from σ via the mapping φ is obviously essentially the same as the relation obtained from σ' (differing only in factors equal to E). This latter relation is valid, since $\sigma' \in \Phi$. Consequently, the identity σ also becomes a valid equality under the mapping φ.

It follows that $\Theta(\mathfrak{A}) \subset \Theta(\mathfrak{A}')$. Thus, $\Theta(\mathfrak{A}') = \Theta(\mathfrak{A})$, and so $\Pi\Theta(\mathfrak{A}') = \Pi\Theta(\mathfrak{A}) = \Gamma$.

4.12. The argument of 4.11 shows that for a family Φ of the type in question, the variety $\Gamma = \Pi(\Phi)$ is generated not only by the free semigroup $\mathfrak{W}^{\Phi}_{\mathfrak{N}}$ (4.10), but also by the semigroup $\mathfrak{W}'$ obtained from it by external adjunction of a unit E. We assert that except for the trivial case that Γ consists only of one-element semigroups, $\mathfrak{W}'$ is never free in $\Pi(\Phi)$.

Indeed, the unit E of $\mathfrak{W}'$ must be a member of any generating set. If $\mathfrak{W}'$ were free, E would be a member of the free generating set $\mathfrak{M}$ of $\mathfrak{W}'$. Then for any $\mathfrak{A} \subset \Pi(\Phi)$ and $A, B \in \mathfrak{A}$, a mapping φ such that $\varphi(E) = A$ and $\varphi(M) = B$ for some $M \in \mathfrak{M}$, $M \neq E$, would extend to a homomorphism of $\mathfrak{W}'$ into $\mathfrak{A}$. Since in $\mathfrak{W}'$ we have $EM = ME = M$, in $\mathfrak{A}$ we should have $AB = BA = B$. The arbitrary element A of the semigroup $\mathfrak{A}$ is thus a unit in $\mathfrak{A}$. Hence $\mathfrak{A}$ is a one-element semigroup.

4.13. As an example of a family of identities with the property of 4.11, we can mention the family consisting of a single identity with just one letter: $\xi_1^n \cong \xi_1^m$ (in particular, $\xi_1^2 \cong \xi_1$). Other examples: the family consisting of the commutativity identity $\xi_1\xi_2 \cong \xi_2\xi_1$; and the family consisting of all balanced identities in which the number of letters does not exceed a fixed n.

4.14. Consideration of the free semigroup $\mathfrak{W}^{\Phi}_{\mathfrak{N}}$ in the variety $\Pi(\Phi)$, for a countable $\mathfrak{N} = \{X_1, X_2, \cdots\}$, allows us to clarify for the important class Γ_f of all finite semigroups the question posed in 2.12.

As indicated there, the condition in question is equivalent to the equality

$$\Pi\Theta(\Pi(\Phi) \cap \Gamma_f) = \Pi(\Phi).$$

The equality says that the variety $\Pi(\Phi)$ is generated, in the obvious sense, by its finite semigroups.

As T. Evans [**S2**] observed, this is connected with finite approximability (XI, 1.7, 1.8, and §5).

4.15. **Theorem.** *The equality*

$$\Theta(\Pi(\Phi) \cap \Gamma_f) = \Theta\Pi(\Phi)$$

holds if and only if the semigroup $\mathfrak{W}_{\mathfrak{N}}^{\Phi}$ (4.14) *is finitely approximable.*

Proof. 1) Suppose the family Φ satisfies the equality.

Consider any two distinct elements of the semigroup $\mathfrak{W}_{\mathfrak{N}}^{\Phi}$:

$$X_{i_1}X_{i_2}\cdots X_{i_p} \neq X_{j_1}X_{j_2}\cdots X_{j_q} \quad (X_{i_k}, X_{j_l} \in \mathfrak{N}).$$

Since $\mathfrak{W}_{\mathfrak{N}}^{\Phi} \in \Pi(\Phi)$, the identity

$$\xi_{i_1}\xi_{i_2}\cdots\xi_{i_p} \cong \xi_{j_1}\xi_{j_2}\cdots\xi_{j_q}$$

does not belong to $\Theta\Pi(\Phi)$. In view of the hypothesized equality, the class $\Pi(\Phi)$ must contain a finite semigroup $\mathfrak{A}$ in which the identity also fails to be satisfied, i.e., in which for appropriate elements we have

$$A_{i_1}A_{i_2}\cdots A_{i_p} \neq A_{j_1}A_{j_2}\cdots A_{j_q}.$$

Since $\mathfrak{W}_{\mathfrak{N}}^{\Phi}$ is free in $\Pi(\Phi)$, and $\mathfrak{N}$ is a free generating set, a mapping φ for which

$$\varphi(X_{i_r}) = A_{i_r}, \quad \varphi(X_{j_s}) = A_{j_s} \quad (r = 1, 2, \cdots, p;\ s = 1, 2, \cdots, q)$$

extends to a homomorphism of $\mathfrak{W}_{\mathfrak{N}}^{\Phi}$ into $\mathfrak{A}$. For this homomorphism,

$$\varphi(X_{i_1}X_{i_2}\cdots X_{i_p}) \neq \varphi(X_{j_1}X_{j_2}\cdots X_{j_q}).$$

This means that the semigroup $\mathfrak{W}_{\mathfrak{N}}^{\Phi}$ is finitely approximable.

2) Suppose $\mathfrak{W}_{\mathfrak{N}}^{\Phi}$ is finitely approximable.

Since $\Pi(\Phi) \cap \Gamma_f \subset \Pi(\Phi)$, we have $\Theta(\Pi(\Phi) \cap \Gamma_f) \supset \Theta\Pi(\Phi)$.

Suppose equality fails to hold, i.e., that there exists an identity

$$\xi_{i_1}\xi_{i_2}\cdots\xi_{i_p} \cong \xi_{j_1}\xi_{j_2}\cdots\xi_{j_q}$$

belonging to $\Theta(\Pi(\Phi) \cap \Gamma_f)$ but not to $\Theta\Pi(\Phi)$. This means that in $\Pi(\Phi)$ there exists a semigroup $\mathfrak{A}$ in which for appropriate elements one has the inequality

$$A_{i_1}A_{i_2}\cdots A_{i_p}\neq A_{j_1}A_{j_2}\cdots A_{j_q}.$$

But then in $\mathfrak{W}^*_{\mathfrak{N}}$ we must also have

$$X_{i_1}X_{i_2}\cdots X_{i_p}\neq X_{j_1}X_{j_2}\cdots X_{j_q}.$$

Indeed, if we had equality, we could take a mapping φ of $\mathfrak{N}$ into $\mathfrak{A}$ such that $\varphi(X_{i_r})=A_{i_r}$ and $\varphi(X_{j_s})=A_{j_s}$ $(r=1,2,\cdots,p;\ s=1,2,\cdots,q)$. Since $\mathfrak{N}$ is a free set with respect to $\Pi(\Phi)$, the mapping extends to a homomorphism. Hence equality of the products of elements $X_k\in\mathfrak{N}$ would imply equality for the corresponding elements $A_k\in\mathfrak{A}$, whereas we have inequality.

In view of the finite approximability of the semigroup $\mathfrak{W}^*_{\mathfrak{N}}$, there must exist for the unequal elements $X_{i_1}X_{i_2}\cdots X_{i_p}$ and $X_{j_1}X_{j_2}\cdots X_{j_q}$ a finite semigroup $\mathfrak{B}$ and a homomorphism ψ of $\mathfrak{W}^*_{\mathfrak{N}}$ onto $\mathfrak{B}$ such that

$$\psi(X_{i_1}X_{i_2}\cdots X_{i_p})\neq\psi(X_{j_1}X_{j_2}\cdots X_{j_q}).$$

Since all the identities in Φ are satisfied in $\mathfrak{W}^*_{\mathfrak{N}}$, they must obviously be satisfied in $\mathfrak{B}$, the latter being obtained from $\mathfrak{W}^*_{\mathfrak{N}}$ by a homomorphism: $\mathfrak{B}=\psi(\mathfrak{W}^*_{\mathfrak{N}})$. Consequently, $\mathfrak{B}\in\Pi(\Phi)$. But $\mathfrak{B}$ is finite. Hence $\mathfrak{B}\in\Pi(\Phi)\cap\Gamma_f$. The fact that in $\mathfrak{B}$ we have

$$\psi(X_{i_1})\psi(X_{i_2})\cdots\psi(X_{i_p})\neq\psi(X_{j_1})\psi(X_{j_2})\cdots\psi(X_{j_q})$$

now contradicts the assumption that the identity $\xi_{i_1}\xi_{i_2}\cdots\xi_{i_p}\cong\xi_{j_1}\xi_{j_2}\cdots\xi_{j_q}$ belongs to $\Theta(\Pi(\Phi)\cap\Gamma_f)$.

4.16. **Theorem.** *Let Φ be a family of balanced identities. Then*

$$\Theta(\Pi(\Phi)\cap\Gamma_f)=\Theta\Pi(\Phi).$$

Proof. By 4.15, it suffices to prove that the semigroup $\mathfrak{W}^*_{\mathfrak{N}}$ (4.14) is finitely approximable.

Consider any two unequal elements of $\mathfrak{W}^*_{\mathfrak{N}}$:

$$S=X_{i_1}X_{i_2}\cdots X_{i_p},\quad S'=X_{j_1}X_{j_2}\cdots X_{j_q},\quad S\neq S'.$$

They are represented as products of elements of $\mathfrak{N}$.

Take $n>p,q$. Let $\mathfrak{T}$ be the family of all elements $T\in\mathfrak{W}^*_{\mathfrak{N}}$ that are representable as a product of elements of $\mathfrak{N}$: $T=X_{k_1}X_{k_2}\cdots X_{k_t}$, in which either $t\geqq n$, or else some X_{k_c} is different from all the X_{i_r} and X_{j_l}.

It is obvious that $\mathfrak{T}$ is a two-sided ideal in $\mathfrak{W}^*_{\mathfrak{N}}$. Let φ be the ideal homomorphism corresponding to $\mathfrak{T}$ (VII, 4.19).

The semigroup $\mathfrak{W}^*_{\mathfrak{N}}$ is by definition the semigroup given by the generating set $\mathfrak{N}$ and a certain defining system of relations with respect to $\mathfrak{N}$. These relations are obtained in a prescribed fashion from identities that are all balanced. This immediately implies the following. Suppose we are given for some $T \in \mathfrak{T}$ an expression $T = X_{k_1} X_{k_2} \cdots X_{k_t}$ of the form described above. When we transform this product by using any of the defining system of relations, the new expression $T = X_{l_1} X_{l_2} \cdots X_{l_t}$ is again of the same nature.

In view of what it means for a semigroup to be given by a defining system of relations, we conclude that any expression whatever of an element of $\mathfrak{T}$ as a product of factors in $\mathfrak{N}$ must be of the above nature. Since the original expressions for S and S' do not meet this condition, we conclude that $S, S' \,\overline{\in}\, \mathfrak{T}$. Since $S \neq S'$, and $S, S' \,\overline{\in}\, \mathfrak{T}$, it follows, by the properties of an ideal homomorphism, that $\varphi(S) \neq \varphi(S')$.

It is evident from the definition of $\mathfrak{T}$ that the corresponding ideal factor-semigroup $\varphi(\mathfrak{W}^{\Phi}_{\mathfrak{N}}) = \mathfrak{W}^{\Phi}_{\mathfrak{N}}/\mathfrak{T}$ is finite. This shows that $\mathfrak{W}^{\Phi}_{\mathfrak{N}}$ is finitely approximable.

4.17. Consider a variety $\Gamma = \Pi(\Phi)$.

For a semigroup $\mathfrak{A} \in \Gamma = \Pi(\Phi)$, it is obvious that every subsemigroup $\mathfrak{B}$, as well as every semigroup $\varphi(\mathfrak{A})$ obtained from $\mathfrak{A}$ by a homomorphism, satisfies all the identities in Φ; i.e., $\mathfrak{B}$, $\varphi(\mathfrak{A}) \in \Gamma = \Pi(\Phi)$.

Furthermore, if a semigroup $\mathfrak{M}$ is a Cartesian product (XI, 1.2) of semigroups in Γ, it is easily seen that also $\mathfrak{M} \in \Gamma = \Pi(\Phi)$.

Thus, as one says briefly, every variety is closed with respect to the operations of taking subsemigroups, homomorphisms, and construction of Cartesian products.

4.18. It is of interest to observe that the converse also holds. Every class of semigroups closed with respect to these operations is a variety.

This is a special case of a more general assertion concerning general algebraic systems, first obtained by Birkhoff [**1**]. The proof is given in, e.g., the books of P. M. Cohn [**S1**], G. Grätzer [**S1**], and A. I. Mal'cev [**S2**].

It should be remarked that in these, as well as in a number of other works dealing with general algebraic systems and universal algebras, many of the general concepts and results are of value in the special case of their application to semigroups.

5. The Lattice of Varieties of Semigroups

5.1. We shall denote by Σ the set of all varieties of semigroups. We regard this set as partially ordered by inclusion.

Since every variety is given by a family of identities over a fixed alphabet Ξ, the elements of Σ could be regarded not as the classes of semigroups (varieties) themselves, but as equivalence classes of families of identities. This would avoid the dangers associated with dealing with such notions as the class of all semigroups satisfying certain identities. But in this book we are never involved in the subtle distinctions between set and class. As far as our exposition goes, there is no necessity for this, and we nowhere run the danger of entanglement with the corresponding set-theoretical difficulties.

If $\Gamma_1 \subset \Gamma_2$ $(\Gamma_1, \Gamma_2 \in \Sigma)$, the variety Γ_1 is called a *subvariety* of the variety Γ_2.

For varieties $\Gamma_1 = \Pi(\Phi_1)$ and $\Gamma_2 = \Pi(\Phi_2)$, the relation $\Gamma_1 \subset \Gamma_2$ holds if and only if all the identities in Φ_2 are consequences of Φ_1: $\Phi_2 \subset \Theta\Pi(\Phi_1)$.

This means that the relation of precedence in the ordered set Σ determines the relation of consequentiality for families of identities.

5.2. Every variety is given by a family of identities. The set of identities over a fixed countable alphabet is countable. Hence the cardinality of the class of all families of identities is that of the continuum. It follows that the cardinality of the set Σ is no greater than that of the continuum. We cannot immediately conclude that it is equal to that of the continuum, since different families of identities can determine the same variety. But this is nevertheless the case, as shown by T. Evans [**S1**].

To see this, consider any infinite (and so countable) irreducible family of identities Φ. For example, it can be the family in 3.5–3.7. The cardinality of the set of all subsets of Φ is equal to that of the continuum. Furthermore, any two different subsets $\Psi_1, \Psi_2 \subset \Phi$ $(\Psi_1 \neq \Psi_2)$ determine different varieties: $\Pi(\Psi_1) \neq \Pi(\Psi_2)$.

Indeed, suppose $\sigma \in \Psi_1$, $\sigma \overline{\in} \Psi_2$. Then $\sigma \in \Theta\Pi(\Psi_1)$, but $\sigma \overline{\in} \Theta\Pi(\Psi_2)$. The latter follows from the fact that σ is not a consequence of the family of all identities in Φ other than σ, so that σ is all the more not a consequence of Ψ_2.

It follows that the cardinality of the set of varieties determined by all possible families of identities in Φ is equal to that of the continuum. Consequently, the cardinality of the set of all varieties is that of the continuum.

5.3. **Theorem.** *The partially ordered set* Σ (5.1) *is a complete lattice. The greatest lower bound of any family of varieties in* Σ *is their intersection.*

Proof. Let Σ' be any nonvoid family of varieties, $\Sigma' = \{\Gamma_\alpha\}_\alpha$.

By 1.8 (γ'), the intersection $\Gamma_0 = \bigcap_\alpha \Gamma_\alpha$ is itself a variety. Obviously Γ_0 is the greatest lower bound for Σ'.

The greatest lower bound of the set of all varieties that contain every variety in Σ' (there is at least one, namely, the variety of all semigroups) is obviously the least upper bound for Σ'.

5.4. The structure of the lattice Σ is to a large extent still obscure. Certain lattice properties of Σ have nevertheless been obtained (see, e.g., T. Evans [**S3**]).

As we shall see shortly, the following varieties are elements of Σ that play special roles in Σ.

$$\Gamma_l = \Pi(\xi_1\xi_2 \cong \xi_1),$$
$$\Gamma_r = \Pi(\xi_1\xi_2 \cong \xi_2),$$
$$\Gamma_s = \Pi(\xi_1^2 \cong \xi_1, \xi_1\xi_2 \cong \xi_2\xi_1),$$
$$\Gamma_o = \Pi(\xi_1\xi_2 \cong \xi_3\xi_4),$$
$$\Gamma_p = \Pi(\xi_1^p\xi_2 \cong \xi_2, \xi_1\xi_2 \cong \xi_2\xi_1) \ (p \text{ a prime number}).$$

The structure of the semigroups in Γ_l and Γ_r is extremely plain and simple. Γ_s consists of the semigroups conjugate to semilattices (II, 4.3).

The structure of the semigroups in Γ_o has been described in XI, 5.5 and XII, 1.7. Every semigroup in Γ_o has a zero, and the product of any two elements is zero.

Since, as is easily seen, the identities $\xi_1^p\xi_2 \cong \xi_2$ and $\xi_1\xi_2 \cong \xi_2\xi_1$ have as consequence the identity $\xi_1^p \cong \xi_2^p$, every semigroup in Γ_p is, by IX, 4.4, a commutative group in which the order of any element other than the unit is equal to p. Conversely, every such group belongs to Γ_p. The structure of these groups is very simple and well known.

5.5. **Lemma.** *For any semigroup $\mathfrak{A}$ with more than one element, there exists a semigroup $\mathfrak{A}' \in \Pi\Theta(\mathfrak{A})$ which either consists of two elements and belongs to one of the varieties Γ_l, Γ_r, Γ_s or Γ_o, or else consists of p elements and belongs to the variety Γ_p.*

Proof. 1) Suppose $\mathfrak{A}$ contains an element A of infinite type. The infinite monogenic semigroup $[A]$, being a subsemigroup of $\mathfrak{A}$, is by 4.17 a member of $\Pi\Theta(\mathfrak{A})$. The mapping ψ of $[A]$ onto the cyclic group $\mathfrak{A}' = \{E, G\} \in \Gamma_2$ of order two, given by $\psi(A^{2k-1}) = G$, $\psi(A^{2k}) = E$ $(k = 1, 2, 3, \cdots)$, is a homomorphism. By 4.17 again, $\mathfrak{A}' \in \Pi\Theta(\mathfrak{A})$.

2) Suppose $\mathfrak{A}$ contains an element A of type (h, d), where $d > 1$. Then $\mathfrak{A}$ contains a commutative subgroup of order d (III, 3.11). As is well known, this group contains in turn, for any prime p dividing d, a cyclic subgroup $\mathfrak{A}'$ of order p. Then $\mathfrak{A}' \in \Pi\Theta(\mathfrak{A})$ (4.17) and $\mathfrak{A}' \in \Gamma_p$.

3) Suppose $\mathfrak{A}$ contains an element A of type $(h, 1)$, $h \neq 1$. Then $\mathfrak{A}' = \{A^{h-1}, A^h\}$ is a subsemigroup of $\mathfrak{A}$. We have $\mathfrak{A}' \in \Pi\Theta(\mathfrak{A})$ (4.17) and, as is easily seen, $\mathfrak{A}' \in \Gamma_o$.

4) Suppose $\mathfrak{A}$ is commutative and all its elements are idempotent. Take $A, B \in \mathfrak{A}$, $A \neq B$; we can assume $AB \neq A$. Then $\mathfrak{A}' = \{A, AB\}$ is a subsemigroup of $\mathfrak{A}$ that belongs to Γ_s, and $\mathfrak{A}' \in \Pi\Theta(\mathfrak{A})$.

5) Suppose $\mathfrak{A}$ is noncommutative and all its elements are idempotent. If for $A, B \in \mathfrak{A}$ we have $AB \neq BA$ and $ABA = BA$, then $\mathfrak{A}' = \{AB, BA\}$ is a subsemigroup of $\mathfrak{A}$ that belongs to Γ_r, and $\mathfrak{A}' \in \Pi\Theta(\mathfrak{A})$.

If $ABA \neq BA$, then $\mathfrak{A}' = \{ABA, BA\}$ is a subsemigroup of $\mathfrak{A}$ that belongs to Γ_l, and $\mathfrak{A}' \in \Pi\Theta(\mathfrak{A})$.

5.6. **Lemma.** *Let* $\mathfrak{A}$ *be a semigroup with more than one element, contained in one of the varieties* Γ_i $(i = l, r, s, o, p)$. *Then* $\Theta(\mathfrak{A}) = \Theta(\Gamma_i)$.

Proof. By 1.9–1.12 (and the fact that the case $i = r$ is similar to $i = l$), in each of the cases $i = l, r, s, o, p$ the set $\Theta(\mathfrak{A})$ is the same for all semigroups $\mathfrak{A} \in \Gamma_i$ with more than one element. For the one-element semigroup $\{E\}$, the set $\Theta(E)$ is the set of all identities. It follows at once that $\Theta(\Gamma_i) = \Theta(\mathfrak{A})$ for any $\mathfrak{A} \in \Gamma_i$ with more than one element.

5.7. **Corollary.** $\Theta(\Gamma_l)$ *is the family of all identities of the form* $\xi_1 U_1 \cong \xi_1 U_2$ (*where* U_1 *and* U_2 *are either words in* Ξ *or empty symbols*).

$\Theta(\Gamma_r)$ *is the family of all identities of the form* $U_1\xi_1 \cong U_2\xi_1$.

$\Theta(\Gamma_s)$ *is the family of all identities* $V_1 \cong V_2$ *for which every letter* $\xi_k \in \Xi$ *that occurs in one of the words* V_1 *or* V_2 *also occurs in the other.*

$\Theta(\Gamma_o)$ *is the family of all identities* $V_1 \cong V_2$ *for which the lengths of* V_1 *and* V_2 *are both greater than one, together with the trivial identities.*

$\Theta(\Gamma_p)$ *is the family of all identities* σ *for which* $d(\sigma)$ *is divisible by* p.

By 1.9–1.12, this follows immediately from 5.6.

5.8. One of the important characteristics of a family of identities Φ, due to A. P. Birjukov [S1], is the nonnegative integer $d(\Phi)$ which is the greatest common divisor of the numbers $d(\sigma)$ for all $\sigma \in \Phi$ (1.2).

It is evident at once that for an identity τ that is an immediate consequence of Φ, $d(\tau)$ is divisible by $d(\Phi)$. It follows by 2.8 that the same is true for any consequence τ of Φ.

5.9. From 5.7, using 4.5, we have for an arbitrary variety $\Pi(\Phi)$ the following.

Corollary. $\Pi(\Phi)$ *contains* Γ_l *if and only if all the identities in* Φ *are of the form*

$$\xi_1 U_1 \cong \xi_1 U_2.$$

$\Pi(\Phi)$ *contains* Γ_r *if and only if all the identities in* Φ *are of the form*

$$U_1\xi_1 \cong U_2\xi_1.$$

$\Pi(\Phi)$ *contains* Γ_s *if and only if, for every identity* $V_1 \cong V_2$ *in* Φ, *any letter* ξ_k *that occurs in one of the words* V_1 *or* V_2 *occurs in the other.*

$\Pi(\Phi)$ *contains* Γ_o *if and only if, for every nontrivial identity* $V_1 \cong V_2$ *in* Φ, *the lengths of both words* V_1 *and* V_2 *are greater than* 1.

$\Pi(\Phi)$ *contains* Γ_p *if and only if* $d(\Phi)$ *is divisible by* p.

5.10. In an arbitrary ordered set $\mathfrak{M}$, an element X is said to *cover* an element Y if $X > Y$ and there is no element Z of $\mathfrak{M}$ such that $X > Z > Y$.

If S is a lower bound for all of $\mathfrak{M}$ (sometimes called the least element), the elements covering S are called the *atoms*. They evidently have a special role in $\mathfrak{M}$.

5.11. In the lattice Σ, the lower bound (least element) is the variety consisting of all one-element semigroups:

$$\Gamma_E = \Pi(\xi_1 \cong \xi_2) = \Pi(\Phi_0),$$

where Φ_0 is the family of all identities (1.6). We shall continue to denote by Γ_E this smallest variety.

5.12. The atoms of the lattice Σ have been determined by J. Kalicky and D. Scott [**S1**].

Theorem. *The atoms of the lattice Σ are the varieties Γ_l, Γ_r, Γ_s, Γ_o, and Γ_p (5.4). Every variety, except for the least, contains an atom of Σ.*

Proof. 1) Suppose a variety Γ, other than Γ_E, is contained in one of the varieties Γ_i $(i = l, r, s, o, p)$. Let $\mathfrak{A}$ be a semigroup in Γ with more than one element.

Since $\mathfrak{A} \in \Gamma \subset \Gamma_i$, we have $\Theta(\mathfrak{A}) \supset \Theta(\Gamma) \supset \Theta(\Gamma_i)$. But by 5.6, $\Theta(\mathfrak{A}) = \Theta(\Gamma_i)$. Hence $\Theta(\Gamma) = \Theta(\Gamma_i)$. Since Γ and Γ_i are varieties, applying the operator Π gives the equality $\Gamma = \Gamma_i$.

This shows that Γ_i is an atom in Σ.

2) Now let Γ be an arbitrary variety other than Γ_E.

Let $\mathfrak{A}$ be any semigroup in Γ with more than one element. By 5.5, there is a semigroup $\mathfrak{A}' \in \Pi\Theta(\mathfrak{A})$, with more than one element, that belongs to one of the varieties Γ_i $(i = l, r, s, o, p)$.

Since $\mathfrak{A} \in \Gamma$, we have $\Pi\Theta(\mathfrak{A}) \subset \Pi\Theta(\Gamma) = \Gamma$. Hence $\mathfrak{A}' \in \Gamma$.

The intersection $\Gamma \cap \Gamma_i$ contains $\mathfrak{A}'$, and by 5.3 is a variety. Since $\Gamma \cap \Gamma_i \subset \Gamma_i$, what has already been proved in the first part implies that $\Gamma \cap \Gamma_i = \Gamma_i$. It follows that Γ contains the atom Γ_i. It follows also that if $\Gamma \neq \Gamma_i$, then Γ is not an atom in Σ.

5.13. **Corollary.** *In order that a variety $\Pi(\Phi)$ be the least variety: $\Pi(\Phi) = \Gamma_E$ (5.11), it is necessary and sufficient that $d(\Phi) = 1$ and that Φ include identities of the following form:*

1) $\xi_1 U_1 \cong \xi_2 U_2$;

2) $V_1\xi_1 \cong V_2\xi_2$ *(U_1, U_2, V_1, V_2 words in Ξ or empty symbols)*;

3) $\xi_1 \cong W_0$ *(W_0 a word in Ξ different from ξ_1)*;

4) $W_1 \cong W_2$, *where W_1 and W_2 are words in Ξ such that there is a letter $\xi_k \in \Xi$ that occurs in one but not the other.*

Proof. By 5.12, the variety $\Pi(\Phi)$ is equal to Γ_E if and only if it contains none of the atoms Γ_i $(i = l, r, s, o, p)$ of Σ. By 5.9, this amounts to the above conditions on Φ.

5.14. In the theory of semigroups, individual classes of semigroups are of special interest and deserving of special study. In studying a given class of semigroups Γ, it can be important to single out these subclasses that are determined by one or another family of identities.

Any family of identities Φ determines in Γ the subclass

$$\Pi(\Phi) \cap \Gamma.$$

In the theory of semigroups belonging to the class Γ, such a class can be called a variety in the class Γ.

If the class Γ is itself a variety, $\Gamma = \Pi(\Psi)$, then the subclass is a variety in the class of all semigroups:

$$\Pi(\Phi) \cap \Gamma = \Pi(\Phi) \cap \Pi(\Psi) = \Pi(\Phi \cup \Psi).$$

In the general case, this is of course not always true.

5.15. Special attention is due the case that $\Pi(\Phi) \cap \Gamma = \Pi(\Phi)$, i.e., $\Pi(\Phi) \subset \Gamma$.

For a number of important classes of semigroups Γ, A. P. Birjukov has found [S2] conditions that must be satisfied by Φ in order that $\Pi(\Phi) \subset \Gamma$. We give some of these results below, deriving them by means of Theorem 5.12 concerning the atoms of the lattice Σ.

5.16. **Theorem.** *In order that a variety $\Pi(\Phi)$ consist of semigroups of idempotents:*

$$\Pi(\Phi) \subset \Pi(\xi_1^2 \cong \xi_1),$$

it is necessary and sufficient that $d(\Phi) = 1$ and that Φ include an identity of the form $\xi_1 \cong W$ $(W \in \mathfrak{W}_{\Xi},\ W \neq \xi_1)$.

Proof. 1) Suppose $\Pi(\Phi) \subset \Pi(\xi_1^2 \cong \xi_1)$. Since Γ_o and Γ_p are obviously not contained in $\Pi(\xi_1^2 \cong \xi_1)$, it follows from 5.9 that Φ has the indicated properties.

2) If Φ has the indicated properties, then by 5.9 it contains neither Γ_o nor Γ_p. Since Γ_o and Γ_p are atoms, it follows that $\Pi(\Phi) \cap \Gamma_o = \Pi(\Phi) \cap \Gamma_p = \Gamma_E$.

For any $\mathfrak{A} \in \Pi(\Phi)$ and $A \in \mathfrak{A}$, we have $[A] \in \Pi(\Phi)$. If $[A]$ had more than one element, the argument used in the proof of Lemma 5.5 would show that $\Pi(\Phi)$ contains a semigroup, with more than one element, that belongs to Γ_o or Γ_p. Since this is not the case, $[A]$ must consist of just one element; i.e., any element A of any semigroup $\mathfrak{A} \in \Pi(\Phi)$ must be idempotent.

5.17. **Theorem.** *In order that a variety* $\Pi(\Phi)$ *consist of inverse semigroups, it is necessary and sufficient that* Φ *include identities of the following form:*

1) $\xi_1 U_1 \cong \xi_2 U_2$;
2) $V_1 \xi_1 \cong V_2 \xi_2$ (U_1, U_2, V_1, V_2 *are words in* Ξ *or empty symbols*);
3) $\xi_1 \cong W_0$ (W_0 *a word in* Ξ *different from* ξ_1).

Proof. 1) Suppose all the semigroups in $\Pi(\Phi)$ are inverse. Since obviously no semigroup in Γ_l, Γ_r or Γ_o with more than one element is inverse, $\Pi(\Phi)$ cannot contain Γ_l, Γ_r or Γ_o. It follows by 5.9 that Φ must include identities of the three indicated types.

2) Suppose $\Pi(\Phi) \neq \Gamma_E$, and Φ has identities of the indicated types. By 5.9, this means that $\Pi(\Phi)$ contains none of the atoms Γ_l, Γ_r or Γ_o, and so has in common with them no semigroup with more than one element.

For the identity in Φ of the third type, the length of the word W_0 must be greater than 1, since $\Pi(\Phi) \neq \Gamma_E$. A consequence of this identity is one of the form $\xi_1 \cong \xi_1^n$ $(n > 1)$, obtained from it by the mapping $\varphi(\xi_i) = \xi_1$ $(i = 1, 2, \cdots)$. Thus, the identity $\xi_1 \cong \xi_1^n$ is satisfied in $\Pi(\Phi)$.

Consider first the case $n = 2$.

If $\Pi(\Phi)$ contained a noncommutative semigroup $\mathfrak{A}$, the argument in the last part of the proof of Lemma 5.5 would exhibit a two-element semigroup $\mathfrak{A}' \in \Pi(\Phi)$ that belongs to either Γ_r or Γ_l; and this, as we have seen, is impossible.

Thus, every semigroup in $\Pi(\Phi)$ is a commutative semigroup of idempotents. But any such semigroup is inverse.

Now consider the case $n > 2$.

The identity $\xi_1^n \cong \xi_1$ $(n > 2)$ implies that every element A in any semigroup $\mathfrak{A} \in \Pi(\Phi)$ is regular and has as regular conjugate the element A^{n-2}:

$$AA^{n-2}A = A^n = A, \quad A^{n-2}AA^{n-2} = A^nA^{n-3} = AA^{n-3} = A^{n-2}$$

(here and in what follows, A^{n-3} means the empty symbol if $n = 3$).

On the other hand, suppose $B \in \mathfrak{A}$ is a regular conjugate of A.

If $AB \neq A^{n-1}$, it is easy to see that the two-element family $\{AB, A^{n-1}\}$ is a semigroup and belongs to Γ_r. As already observed, this is impossible, since $\{AB, A^{n-1}\} \in \Pi(\Phi)$.

Hence $AB = A^{n-1}$. Similarly, $AB = B^{n-1}$, $BA = B^{n-1}$ and $BA = A^{n-1}$. It follows that $AB = BA$.

Using this equality, we obtain

$$B = BAB = BA^nB = A^{n-1}BAB = A^{n-1}B = A^{n-3}ABA = A^{n-2}.$$

We have thus shown that any element A in $\mathfrak{A}$ has only one regular conjugate. Hence $\mathfrak{A}$ is inverse.

5.18. In studying a lattice, of equal interest with atoms (5.10) are the so-called anti-atoms: those elements that are covered by the upper bound of the lattice. But the lattice Σ has no such elements. This is shown by the following.

Theorem. *For any variety Γ other than the variety Γ_a of all semigroups, there exists a variety Γ' such that*

$$\Gamma \subset \Gamma' \subset \Gamma_a, \qquad \Gamma' \neq \Gamma, \qquad \Gamma' \neq \Gamma_a.$$

Proof. Consider a nontrivial identity $(U \cong V) \in \Theta(\Gamma)$, (it exists, since $\Gamma \neq \Gamma_a$). Suppose it involves the letters $\xi_1, \xi_2, \cdots, \xi_n$. Let m be a number greater than the length of either word U or V. Let φ be a mapping for which

$$\varphi(\xi_k) = \xi_k \xi_{n+1} \cdots \xi_{n+m} \quad (k = 1, 2, \cdots, n).$$

Since the identity $U \cong V$ is nontrivial, i.e., U is different from V, the same must obviously be the case for $\varphi(U) \cong \varphi(V)$. The latter is a consequence of $U \cong V$, and so a consequence of Φ. Hence for the variety $\Gamma' = \Pi(\varphi(U) \cong \varphi(V))$ we have $\Gamma \subset \Gamma' \subset \Gamma_a$. Furthermore, $\Gamma' \neq \Gamma_a$, since the identity $\varphi(U) \cong \varphi(V)$ is nontrivial and therefore not in $\Theta(\Gamma_a)$.

Since both the words $\varphi(U)$ and $\varphi(V)$ have length greater than m, it follows immediately from 2.8 that in every identity $U' \cong V'$ which is a nontrivial consequence of $\varphi(U) \cong \varphi(V)$ the lengths of the words U' and V' must also exceed m. Consequently, the identity $U \cong V$ cannot be a consequence of $\varphi(U) \cong \varphi(V)$. Hence $\Theta(\Gamma)$ and $(\varphi(U) \cong \varphi(V))$ are not equivalent. It follows that $\Gamma \neq \Gamma'$.

5.19. For any variety Γ, the family of all subvarieties, which we here denote by $\Sigma(\Gamma)$, is a complete lattice. This lattice $\Sigma(\Gamma)$ has been studied for certain varieties Γ. Especially far advanced at present is the study of the lattice $\Sigma(\Gamma_3)$ for the variety Γ_3 of semigroups of idempotents. Several investigations of this variety, and in particular of its subvarieties, have been carried out by N. Kimura [**S1, S2**] and T. Tamura [**S2**].

A full description of the lattice $\Sigma(\Gamma_3)$ was first published by A. P. Birjukov [**S5, S6**]. It was also obtained independently, apparently, by C. F. Fennemore [**S1, S2**] and J. A. Gerhard [**S1**]. The lattice $\Sigma(\Gamma_3)$ can be very well described by a diagram whose law of formation is quite explicit. In particular, $\Sigma(\Gamma_3)$ is countable. Its atoms are Γ_l, Γ_r and Γ_s. The atom Γ_l is covered by the varieties

$$\Pi(\xi_1\xi_2\xi_3 \cong \xi_1\xi_3\xi_2, \xi_1^2 = \xi_1) \quad \text{and} \quad \Pi(\xi_1\xi_2\xi_1 \cong \xi_1, \xi_1^2 \cong \xi_1);$$

and symmetrically for Γ_r. The atom Γ_s is covered by the varieties

$$\Pi(\xi_1\xi_2\xi_3 \cong \xi_1\xi_3\xi_2, \xi_1^2 \cong \xi_1) \quad \text{and} \quad \Pi(\xi_1\xi_2\xi_3 \cong \xi_2\xi_1\xi_3, \xi_1^2 \cong \xi_1).$$

Every subvariety of Γ_3 can be specified by a family consisting of the identity $\xi_1^2 \cong \xi_1$ and one other identity. These identities are described in the papers cited.

The description of the lattice gives a complete picture of what identities are consequences in the class Γ_3 of what families of identities.

6. Identical Inclusions

6.1. The theory of semigroup identities can be regarded as part of a more general line of investigation in the theory of semigroups.

Suppose given certain words in the countable alphabet $\Xi = \{\xi_1, \xi_2, \cdots\}$:

$$U, V_1, V_2, \cdots \in \mathfrak{W}_\Xi.$$

We can say that the semigroup $\mathfrak{A}$ satisfies identically the inclusion

$$U \tilde{\in} \{V_1, V_2, \cdots\}$$

if for any mapping φ of Ξ into $\mathfrak{A}$ the values of the corresponding words satisfy the inclusion

$$\varphi(U) \in \{\varphi(V_1), \varphi(V_2), \cdots\}.$$

When the family $\{V_1, V_2, \cdots\}$ consists of a single word V, we obtain as a special case the identity (identical equality) $U \cong V$, with the corresponding condition that it be satisfied in the given semigroup (IX, 4.1).

Except for results concerning identities, this line of investigation has at present only a small number of results.

6.2. We shall say that a semigroup $\mathfrak{A}$ is a *semigroup with unique generating set* if no proper subset is a generating set.

6.3. A semigroup $\mathfrak{A}$ is a semigroup with unique generating set if and only if every nonvoid subset of $\mathfrak{A}$ is a subsemigroup.

Indeed, if every nonvoid subset of $\mathfrak{A}$ is a subsemigroup, then for any $\mathfrak{N} \subset \mathfrak{A}$ ($\mathfrak{N} \neq \emptyset$, $\mathfrak{N} \neq \mathfrak{A}$) we have $[\mathfrak{N}] = \mathfrak{N} \neq \mathfrak{A}$, i.e., $\mathfrak{N}$ is not a generating set.

On the other hand, if a subset $\mathfrak{N} \subset \mathfrak{A}$ ($\mathfrak{N} \neq \emptyset$, $\mathfrak{N} \neq \mathfrak{A}$) is not a subsemigroup, then for some $X, Y \in \mathfrak{N}$ we have $Z = XY \bar{\in} \mathfrak{N}$. Then for the set $\mathfrak{M} = \mathfrak{A} \setminus \{Z\}$ we have $[\mathfrak{M}] \supset \mathfrak{M} = \mathfrak{A} \setminus \{Z\}$ and $[\mathfrak{M}] \supset Z$, since $\mathfrak{M} \supset \mathfrak{N}$. Hence $[\mathfrak{M}] = \mathfrak{A}$, i.e., $\mathfrak{M}$ is a generating set for $\mathfrak{A}$. But $\mathfrak{M} \neq \mathfrak{A}$.

6.4. The significance of this class of semigroups is obvious, both from the point of view of the notion of generating set and from the point of view of studying the system of subsemigroups of a given semigroup.

6.5. These semigroups can be characterized in the following way by means of the notion defined in 6.1.

6.6. A semigroup $\mathfrak{A}$ is a semigroup with unique generating set if and only if it satisfies identically the inclusion

$$\{\xi_1\xi_2\} \mathrel{\widetilde{\in}} \{\xi_1, \xi_2\}.$$

Otherwise put, this means that the product of any two elements of the semigroup is always equal to one of the factors.

The equivalence of this condition with that in 6.3 is immediate.

6.6. The structure of semigroups with unique generating set has been described in more than one way. Below, we describe their structure by means of the notion of successively annihilating band. In this form the result appears in a paper of E. S. Ljapin [**S9**].

6.7. **Lemma.** *The semigroups with unique generating set that are indecomposable into a proper successively annihilating band* (VIII, 4.2–4.4) *are those semigroups that satisfy the identity* $\xi_1\xi_2 \cong \xi_1$ *and those that satisfy the identity* $\xi_1\xi_2 \cong \xi_2$.

Proof. Let A be an element of a semigroup $\mathfrak{A}$ with unique generating set that is indecomposable into a proper successively annihilating band.

Denote by $\mathfrak{B}_0$ the family of all $X \in \mathfrak{A}$ such that $XA = AX = X$, $X \neq A$; by $\mathfrak{B}_1$ the family of all $X \in \mathfrak{A}$ such that $XA = AX = A$, $X \neq A$; by $\mathfrak{B}_l$ the family of all $X \in \mathfrak{A}$ such that $XA = X$ and $AX = A$, $X \neq A$; and by $\mathfrak{B}_r$ the family of all $X \in \mathfrak{A}$ such that $XA = A$ and $AX = X$, $X \neq A$.

For $B_0 \in \mathfrak{B}_0$ and $X \in \mathfrak{B}_l \cup \mathfrak{B}_1$, we cannot have $B_0X = X$, since multiplying this equality on the left by A would give $B_0X = A$, contradicting $X \neq A$. Likewise, $XB_0 = X$ is impossible, since again multiplying on the left by A would give $B_0 = A$. Hence $B_0X = XB_0 = B_0$.

Similarly, we obtain $B_0X = XB_0 = B_0$ for any $X \in \mathfrak{B}_r$.

It follows that if $\mathfrak{B}_0$ were nonvoid, we should have a decomposition of $\mathfrak{A}$ into a proper successively annihilating band:

$$\mathfrak{A} = (\mathfrak{B}_1 \cup \mathfrak{B}_l \cup \mathfrak{B}_r \cup \{A\}) \cup \mathfrak{B}_0.$$

In the same way, for $B_1 \in \mathfrak{B}_1$ and $X \in \mathfrak{B}_l \cup \mathfrak{B}_0$, neither of the equalities $XB_1 = B_1$ or $B_1X = B_1$ is possible, since multiplying each of them on the right by A would give an invalid equality. Hence $XB_1 = B_1X = X$.

Similarly, again, we obtain $B_1X = XB_1 = X$ for any $X \in \mathfrak{B}_r$.

It follows that if $\mathfrak{B}_1$ were nonvoid, we should again have a decomposition of $\mathfrak{A}$ into a proper successively annihilating band:

$$\mathfrak{A} = \mathfrak{B}_1 \cup (\mathfrak{B}_l \cup \mathfrak{B}_r \cup \mathfrak{B}_0 \cup \{A\}).$$

Now suppose $X \in \mathfrak{B}_l$ and $Y \in \mathfrak{B}_r$.

If $XY = X$, multiplying this equality on the left by A gives $Y = A$.

If $XY = Y$, multiplying on the right by A gives $X = A$. Hence either $\mathfrak{B}_l$ or $\mathfrak{B}_r$ must be empty.

Thus, finally, for the semigroup $\mathfrak{A}$ we must have either $\mathfrak{A} = \mathfrak{B}_l \cup \{A\}$ or $\mathfrak{A} = \mathfrak{B}_r \cup \{A\}$.

For any $X, Y \in \mathfrak{B}_l$, we have $XAY = (XA)Y = XY$, and also $XAY = X(AY) = XA = X$. Hence $XY = X$. Since furthermore $XA = X$ and $AX = A$, it follows that $\mathfrak{B}_l \cup \{A\} \in \Pi(\xi_1\xi_2 \cong \xi_1)$.

Similarly, for $\mathfrak{B}_r$ we find that $\mathfrak{B}_r \cup \{A\} \in \Pi(\xi_1\xi_2 \cong \xi_2)$.

In the opposite direction, it is obvious that there is no decomposition into a proper successively annihilating band for any semigroup in either $\Pi(\xi_1\xi_2 \cong \xi_1)$ or $\Pi(\xi_1\xi_2 \cong \xi_2)$.

6.8. Now let $\mathfrak{A}$ be an arbitrary semigroup with unique generating set. By VIII, 4.7, it splits in a unique fashion into a successively annihilating band of subsemigroups which themselves have no further such decomposition. Each is also, obviously, itself a semigroup with unique generating set. Hence by 6.7, each belongs to either $\Pi(\xi_1\xi_2 \cong \xi_1)$ or $\Pi(\xi_1\xi_2 \cong \xi_2)$.

The structure of $\mathfrak{A}$ is completely described by specifying the order type of the set of components in this decomposition into a successively annihilating band, together with the structure of each component. The structure of a component is determined by its cardinality as a set and by one of the symbols l or r, indicating whether it belongs to $\Pi(\xi_1\xi_2 \cong \xi_1)$ or $\Pi(\xi_1\xi_2 \cong \xi_2)$.

Conversely, every semigroup of this form is obviously a semigroup with unique generating set.

6.9. Equally well with identical relations (identities, identical inclusions), which involve only "variables," it is meaningful to consider relations of a more general form that include not only variables but also certain fixed elements of the semigroup. Some investigations concerning identities with distinguished elements have been published by A. P. Birjukov [S7].

6.10. M. Sasaki [S1] has considered semigroups that can be characterized by the property that they have exactly two generating sets. For any such semigroup $\mathfrak{A}$, the generating sets are $\mathfrak{A}$ itself and $\mathfrak{A} \setminus \{Z_0\}$, where Z_0 is a certain element of $\mathfrak{A}$. Indeed, if the complement of a generating set contained at least two elements Z_1 and Z_2, we should have three different generating sets for $\mathfrak{A} : \mathfrak{A}, \mathfrak{A} \setminus \{Z_1\}$, and $\mathfrak{A} \setminus \{Z_1, Z_2\}$.

Semigroups with at most two generating sets can obviously be characterized in the following fashion. For a fixed element Z_0, such a semigroup must satisfy identically the inclusion

$$\{\xi_1\xi_2\} \mathrel{\widetilde{\in}} \{\xi_1, \xi_2, Z_0\}.$$

6.11. E. S. Ljapin has considered [S9] elements U of a semigroup called unit ideal elements, for which UX is equal to either X or U itself for any $X \in \mathfrak{A}$, and similarly XU is equal to either X or U. From the point of view above, the definition of unit ideal element can be formulated as the requirement that a pair of identical inclusions be satisfied for the fixed element U:

$$U\xi_1 \mathrel{\tilde{\in}} \{U, \xi_1\}, \quad \xi_1 U \mathrel{\tilde{\in}} \{U, \xi_1\}.$$

6.12. An even further extension can be made of this notion. The semigroup $\mathfrak{A}$ can have distinguished in it a fixed subset $\mathfrak{K}$. From this point of view, the notion of left ideal can be formulated as a condition of identical inclusion: $\xi_1 \mathfrak{K} \mathrel{\tilde{\subset}} \mathfrak{K}$. Similarly for right and two-sided ideals.

6.13. Identical inclusions can also be used to define certain generalizations of the notion of ideal.

Thus, for example, for a distinguished subset $\mathfrak{K} \subset \mathfrak{A}$, the requirement that $\mathfrak{A}$ satisfy the identical inclusion

$$\xi_1 \mathfrak{K} \mathrel{\tilde{\subset}} \{\mathfrak{K} \cup \xi_1\}, \qquad \mathfrak{K}\xi_1 \mathrel{\tilde{\subset}} \{\mathfrak{K} \cup \xi_1\}$$

defines a unit ideal subsemigroup $\mathfrak{K}$ of the semigroup $\mathfrak{A}$. This notion has been given by E. S. Ljapin [**S5, S7, S8**] in connection with the study of independence of subsemigroups of a given semigroup.

A family of subsemigroups $\{\mathfrak{A}_\alpha\}_\alpha$ is called independent if none of them is contained in the union of the rest, and if no nontrivial equality is satisfied of the form

$$X_1 X_2 \cdots X_p = Y_1 Y_2 \cdots Y_q,$$

where all the elements X_i and Y_j belong to $\bigcup_\alpha \mathfrak{A}_\alpha$ and no two adjacent ones belong to the same $\mathfrak{A}_\alpha$.

Two independent subsemigroups $\mathfrak{A}_1$ and $\mathfrak{A}_2$ can intersect. Their intersection, when it is nonvoid, is a unit ideal subsemigroup in both of them.

BIBLIOGRAPHY

ADJAN, S. I.

1. *On the problem of divisibility in semi-groups,* Dokl. Akad. Nauk SSSR **103** (1955), 747-750. (Russian) MR **19**, 117.
2. *The role of the cancellation law in presenting cancellation semi-groups by means of defining relations,* Dokl. Akad. Nauk SSSR **113** (1957), 1191-1194. (Russian) MR **20** # 1716.

AĬZENŠTAT, A. JA.

1. *On the semigroup of all* 1-1 *mappings of the set of natural numbers into itself,* Učen. Zap. Vyborgsk. Ped. Inst. **2** (1957), 15-24. (Russian)
2. *Defining relations of finite symmetric semigroups,* Mat. Sb. **45** (87) (1958), 261-280. (Russian) MR **21** # 88.
3. *Defining relations in symmetric semigroups,* Leningrad. Gos. Ped. Inst. Učen. Zap. **166** (1958), 121-142. (Russian) MR **21** # 86.
4. *On a class of periodic semigroups,* Leningrad. Gos. Ped. Inst. Učen. Zap. **183** (1958), 241-249. (Russian)

ALIMOV, N. T.

1. *On ordered semigroups,* Izv. Akad. Nauk SSSR Ser. Mat. **14** (1950), 569-576. (Russian) MR **12**, 480.

AMITSUR, A. S.

1. *Semi-group rings,* Riveon Lematematika **5** (1951), 5-9. (Hebrew) MR **13**, 7.

ANDREOLI, G.

1. *Sulla teoria della sostituzioni generalizzate e dei loro gruppi generalizzati,* Rend. Accad. Sci. Fis. Mat. Napoli (4) **10** (1940), 115-127. MR **8**, 439.

ARNOL'D, I. V.

1. *Ideale in kommutativen Halbgruppen,* Mat. Sb. **36** (1929), 401-408.

ASANO, K. AND MURATA, K.

1. *Arithmetical ideal theory in semigroups,* J. Inst. Polytech. Osaka City Univ. Ser. A. Math. **4** (1953), 9-33. MR **15**, 502.

AUBERT, K.

1. *On the ideal theory of commutative semigroups,* Math. Scand. **1** (1953), 39-54. MR **15**, 7.
2. *Some characterizations of valuation rings,* Duke Math. J. **21** (1954), 517-525. MR **16**, 8.
3. *Une théorie générale des idéaux,* Séminaire Dubreil et Pisot 1955/56, Exposé 8, Secrétariat mathématique, Paris, 1956. MR **19**, 522.
4. *Un théorème de représentation dans la théorie des idéaux,* C. R. Acad. Sci. Paris **242** (1956), 320-322. MR **17**, 583.
5. *Généralisation de la théorie des r-idéaux de Prüfer-Lorenzen,* C. R. Acad. Sci. Paris **238** (1954), 2214-2216. MR **15**, 848.

BAER, R.

1. *Free sums of groups and their generalizations. An analysis of the associative law,* Amer. J. Math. **71** (1949), 706-742. MR **11**, 78.

BAER, R. AND LEVI, F.

1. *Vollständige irreduzibele Systeme von Gruppenaxiomen,* S.-B. Heidelberg. Akad. Wiss. Abh. **2** (1932), 1-12.

BALLIEU, R.

1. *Sur les groupes de parties d'un demi-groupe*, Ann. Soc. Sci. Bruxelles Sér. I **64** (1950), 139-147. MR **12**, 586.
2. *Une relation d'équivalence dans les groupoîdes et son application à une classe de demi-groupes*, IIIe Congrès National des Sciences (Bruxelles, 1950), vol. 2, Fédération Belge des Sociétés Scientifiques, Bruxelles, pp. 46-50. MR **17**, 126.

BELL, E. T.

1. *Unique decomposition*, Amer. Math. Monthly **37** (1930), 400-418.

BIRKHOFF, G.

1. *On the structure of abstract algebras*, Proc. Cambridge Philos. Soc. **31** (1935), 433-454.
2. *An ergodic theorem for general semi-groups*, Proc. Nat. Acad. Sci. U.S.A. **25** (1939), 625-627. MR **1**, 148.
3. *Lattice theory*, Amer. Math. Soc. Colloq. Publ., vol. 25, rev. ed., Amer. Math. Soc., Providence, R. I., 1948; Russian transl., IL, Moscow, 1952. MR **10**, 673.

BOCCIONI, D.

1. *P-gruppoide dei quozienti di un gruppoide con operatori*, Rend. Sem. Mat. Univ. Padova **25** (1956), 176-195. MR **18**, 283.
2. *$\mathfrak{A}$-modulo supplementare di un S-semigruppo commutativo*, Rend. Sem. Mat. Univ. Padova **27** (1957), 48-59. MR **19**, 940.
3. *Q-pseudogruppi complementarizzabili*, Rend. Sem. Mat. Univ. Padova **26** (1956), 85-123. MR **18**, 638.

BRANDT, H.

1. *Über eine Verallgemeinerung des Gruppenbegriffes*, Math. Ann. **96** (1927), 360-366.
2. *Über die Axiome des Gruppoids*, Vierteljschr. Naturforsch. Ges. Zürich **85** (1940), 95-104. MR **2**, 218.

BRUCK, R.

1. *A survey of binary systems*, Ergebnisse der Mathematik und ihrer Grenzgebiete, N.F., Heft 20. Reihe: Gruppentheorie, Springer-Verlag, Berlin, 1958. MR **20** # 76.

CEĬTIN, G. S.

1. *Associative calculus with insoluble equivalence problem*, Dokl. Akad. Nauk SSSR **107** (1956), 370-371. (Russian) MR **18**, 103.

CHAMBERLIN, E. AND WOLFE, J.

1. *Multiplicative homomorphisms of matrices*, Proc. Amer. Math. Soc. **4** (1953), 37-42. MR **14**, 611.

CHEHATA, C.

1. *On an ordered semigroup*, J. London Math. Soc. **28** (1953), 353-356. MR **14**, 944.

CLIFFORD, A. H.

1. *A system arising from a weakened set of group postulates*, Ann. of Math. (2) **34** (1933), 865-871.
2. *Arithmetic and ideal theory of commutative semigroups*, Ann. of Math. (2) **39** (1938), 594-610.
3. *Partially ordered abelian groups*, Ann. of Math. (2) **41** (1940), 465-473. MR **2**, 4.
4. *Semigroups admitting relative inverses*, Ann. of Math. (2) **42** (1941), 1037-1049. MR **3**, 199.
5. *Matrix representations of completely simple semigroups*, Amer. J. Math. **64** (1942), 327-342. MR **4**, 4.
6. *Semigroups containing minimal ideals*, Amer. J. Math. **70** (1948), 521-526. MR **10**, 12.
7. *Semigroups without nilpotent ideals*, Amer. J. Math. **71** (1949), 834-844. MR **11**, 327.
8. *Extensions of semigroups*, Trans. Amer. Math. Soc. **68** (1950), 165-173. MR **11**, 499.
9. *A class of d-simple semigroups*, Amer. J. Math. **75** (1953), 547-556. MR **15**, 98.

10. *Naturally totally ordered commutative semigroups,* Amer. J. Math. **76** (1954), 631-646. MR **15**, 930.
11. *Bands of semigroups,* Proc. Amer. Math. Soc. **5** (1954), 499-504. MR **15**, 930.
12. *Ordered commutative semigroups of the second kind,* Proc. Amer. Math. Soc. **9** (1958), 682-687. MR **20** # 3223.
13. *Totally ordered commutative semigroups,* Bull. Amer. Math. Soc. **64** (1958), 305-316. MR **20** # 7070.
14. *Connected ordered topological semigroups with idempotent endpoints.* I, Trans. Amer. Math. Soc. **88** (1958), 80-98. MR **20** # 1727.

Clifford, A. H. and Miller, D. D.
1. *Semigroups having zeroid elements,* Amer. J. Math. **70** (1948), 117-125. MR **9**, 330.

Climescu, Al. C.
1. *Sur les quasicycles,* Bul. Politehn. Gh. Asachi. Iaşi **1** (1946), 5-14. MR **8**, 134.

Cohn, P. M.
1. *Embeddings in semigroups in semigroups with one-sided division,* J. London Math. Soc. **31** (1956), 169-181. MR **18**, 14.
2. *Embeddings in sesquilateral division semigroups,* J. London Math. Soc. **31** (1956), 181-191. MR **18**, 14.

Conrad, P. F.
1. *Generalized semigroup rings,* J. Indian Math. Soc. **21** (1957), 73-95. MR **20** # 887.

Cotlar, M. and Zarantonello, E.
1. *Semiordered groups and Riesz-Birkhoff L-ideals,* Fac. Ci. Mat. Univ. Nac. Litoral. Publ. Inst. Mat. 8 (1948), 105-192. (Spanish) MR **10**, 99.

Croisot, R.
1. *Holomorphies d'un semi-groupe,* C. R. Acad. Sci. Paris **226** (1948), 1134-1136. MR **10**, 353.
2. *Autre généralisation de l'holomorphie dans un semi-groupe,* C. R. Acad. Sci. Paris **227** (1948), 1195-1197. MR **10**, 430.
3. *Propriétés des complexes forts et symétriques des demi-groupes,* Bull. Soc. Math. France **80** (1952), 217-223. MR **14**, 842.
4. *Demi-groupes et axiomatique des groupes,* C. R. Acad. Sci. Paris **237** (1953), 778-780. MR **15**, 195.
5. *Demi-groupes inversifs et demi-groupes réunions de demi-groupes simples,* Ann. Sci. École Norm. Sup. (3) **70** (1953), 361-379. MR **15**, 680.
6. *Automorphismes intérieurs d'un semi-groupe,* Bull. Soc. Math. France **82** (1954), 161-194. MR **16**, 215.
7. *Demi-groupes simples inversifs à gauche,* C. R. Acad. Sci. Paris **239** (1954), 845-847. MR **16**, 215.
8. *Sur la classification des demi-groupes,* Proc. Internat. Congress Math. (Amsterdam, 1954), vol. 2, pp. 13-14.
9. *Applications résiduées,* Ann. Sci. École Norm. Sup. (3) **73** (1956), 453-474. MR **18**, 790.
10. *Équivalences principales bilatères définies dans un demi-groupe,* J. Math. Pures Appl. (9) **36** (1957), 373-417. MR **19**, 1037.
11. *Une interprétation des relations d'équivalence dans un ensemble,* C. R. Acad. Sci. Paris **226** (1948), 616-617. MR **9**, 406.
12. *Condition suffisante pour l'égalité des longueurs de deux chaînes de mêmes extrémités dans une structure. Application aux relations d'équivalence et aux sous-groupes,* C. R. Acad. Sci. Paris **226** (1948), 767-768. MR **9**, 406.

Day, M. M.
1. *Means for the bounded functions and ergodicity of the bounded representations of semi-groups,* Trans. Amer. Math. Soc. **69** (1950), 276-291. MR **13**, 357.

DICKSON, L. E.
1. *Definitions of a group and a field by independent postulates,* Trans. Amer. Math. Soc. **6** (1905), 198-204.
2. *On semigroups and the general isomorphism between infinite groups,* Trans. Amer. Math. Soc. **6** (1905), 205-208.

DOSS, R.
1. *Sur l'immersion d'un semi-groupe dans un groupe,* Bull. Sci. Math. (2) **72** (1948), 139-150. MR **10**, 591.

DRAGUNOVA, T. E.
1. *Structure of right ideals of a semigroup with unit element and with left cancellation,* Uspehi Mat. Nauk **12** (1957), no. 4 (76), 285-288. (Russian) MR **19**, 730.

DUBREIL, P.
1. *Contribution à la théorie des demi-groupes,* Mém. Acad. Sci. Inst. France (2) **63** (1941), no. 3, 52 pp. MR **8**, 15.
2. *Sur les problèmes d'immersion et la théorie des modules,* C. R. Acad. Sci. Paris **216** (1943), 625-627. MR **5**, 144.
3. *Algèbre.* Tome I. *Équivalences, opérations, groupes, anneaux, corps,* Gauthier-Villars, Paris, 1946. MR **8**, 192.
4. *Contribution à la théorie des demi-groupes.* II, Univ. Roma Ist. Naz. Alta Mat. Rend. Mat. e Appl. (5) **10** (1951), 183-200. MR **14**, 12.
5. *Contribution à la théorie des demi-groupes.* III, Bull. Soc. Math. France **81** (1953), 289-306. MR **15**, 680.
6. *Introduction à la théorie des demi-groupes ordonnés,* Convegno italo-francese di algebra astratta (Padova, 1956), Edizioni Cremonese, Rome, 1957, pp. 1-33. MR **20** # 3228.

DUBREIL, P. ET DUBREIL-JACOTIN, M.-L.
1. *Équivalences et opérations,* Ann. Univ. Lyon. Sect. A (3) **3** (1940), 7-23. MR **8**, 254.

DUBREIL-JACOTIN, M.-L.
1. *Sur l'immersion d'un semi-groupe dans un groupe,* C. R. Acad. Sci. Paris **225** (1947), 787-788. MR **9**, 174.
2. *Quelques propriétés arithmétiques dans un demi-groupe demi-réticulé entier,* C. R. Acad. Sci. Paris **232** (1951), 1174-1176. MR **12**, 670.

DUBREIL-JACOTIN, M.-L. ET CROISOT, R.
1. *Sur les congruences dans les ensembles où sont définies plusieurs opérations,* C. R. Acad. Sci. Paris **233** (1951), 1162-1164. MR **13**, 620.

DYNKIN, E. B.
1. *Markov processes and semi-groupes of operators,* Teor. Verojatnost. i Primenen. **1** (1956), 25-37. (Russian) MR **19**, 469.

EMELIČEV, V. A.
1. *Commutative semigroups with one defining relation,* Šuiskogo Gos. Univ. Učen. Zap. **6** (1958), 227-242. (Russian)

EVANS, T.
1. *A note on the associative law,* J. London Math. Soc. **25** (1950), 196-201. MR **12**, 75.
2. *Embedding theorems for multiplicative systems and projective geometries,* Proc. Amer. Math. Soc. **3** (1952), 614-620. MR **14**, 347.
3. *An embedding theorem for semigroups with cancellation,* Amer. J. Math. **76** (1954), 399-413. MR **15**, 681.

FAUCETT, W. M.
1. *Compact semigroups irreducibly connected between two idempotents,* Proc. Amer. Math. Soc. **3** (1955), 741-747. MR **17**, 173.

FAUCETT, W. M., KOCH, R. J. AND NUMAKURA, K.

1. *Complements of maximal ideals in compact semigroups*, Duke Math. J. **22** (1955), 655-661. MR **17**, 282.

FELLER, W. K.

1. *The parabolic differential equations and the associated semi-groups of transformations*, Ann. of Math. (2) **55** (1952), 468-519. MR **13**, 948.

FORSYTHE, G. E.

1. SWAC *computes* 126 *distinct semigroups of order* 4, Proc. Amer. Math. Soc. **6** (1955), 443-447. MR **16**, 1085.

FUCHS, L.

1. *On semigroups admitting relative inverses and having minimal ideals*, Publ. Math. Debrecen **1** (1950), 227-231. MR **12**, 473.

FULLERTON, R. E.

1. *On a semi-group of subsets of a linear space*, Proc. Amer. Math. Soc. **1** (1950), 440-442. MR **12**, 188.

GARDAŠNIK, M. F.

1. *Ideals in associative systems*, Nauk. Zap. Poltavs′k. Ped. Inst. **8** (1955), 57-60. (Russian)

GELBAUM, B. R., KALISCH, G. K. AND OLMSTED, J. M. H.

1. *On the embedding of topological semigroups and integral domains*, Proc. Amer. Math. Soc. **2** (1951), 807-821. MR **13**, 206.

GLUSKIN, L. M.

1. *An associative system of square matrices*, Dokl. Akad. Nauk SSSR **97** (1954), 17-20. (Russian) MR **16**, 4.
2. *Homomorphisms of unilaterally simple semigroups on groups*, Dokl. Akad. Nauk SSSR **102** (1955), 673-676. (Russian) MR **17**, 237.
3. *Simple semigroups with zero*, Dokl. Akad. Nauk SSSR **103** (1955), 5-8. (Russian) MR **17**, 237.
4. *Automorphisms of multiplicative semigroups of matrix algebras*, Uspehi Mat. Nauk **11** (1956), no. 1 (67), 199-206. (Russian) MR **17**, 936.
5. *Extension of homomorphisms of semigroups*, Har′kov. Gos. Univ. Učen. Zap. **18** (1956), 33-39. (Russian)
6. *Completely simple semigroups*, Har′kov. Gos. Univ. Učen. Zap. **18** (1956), 41-55. (Russian)
7. *Semigroups of matrices with non-negative elements*, Zap. Mat. Otd. Fiz.-Mat. Fak. i Har′kov. Mat. Obšč. **25** (1957), 167-173. (Russian)
8. *Elementary generalized groups*, Mat. Sb. **41 (83)** (1957), 23-36. (Russian) MR **19**, 836.
9. *Semigroups of nonsingular matrices with non-negative elements*, Har′kov. Gos. Univ. Učen. Zap. **21** (1957), 81-98. (Russian)
10. *Normal series of completely simple semigroups*, Har′kov. Gos. Univ. Učen. Zap. **21** (1957), 99-106. (Russian)
11. *Matricial semigroups*, Izv. Akad. Nauk SSSR Ser. Mat. **22** (1958), 439-488. (Russian) MR **20** # 2386.
12. *The semigroup of homeomorphic mappings of an interval*, Mat. Sb. **49 (91)** (1959), 13-28; English transl., Amer. Math. Soc. Transl. (2) **30** (1963), 273-290. MR **22** # 2663; **27** # 5238.
13. *Ideals of semigroups of transformations*, Mat. Sb. **47 (89)** (1959), 111-130. (Russian) MR **21** # 4197.
14. *Semigroups of topological mappings*, Dokl. Akad. Nauk SSSR **125** (1959), 699-702. (Russian) MR **21** # 4412.

15. *Semigroups and rings of linear transformations*, Dokl. Akad. Nauk SSSR **127** (1959), 1151-1154. (Russian) MR **26** # 1382.
16. *Transitive semigroups of transformations*, Dokl. Akad. Nauk SSSR **129** (1959), 16-18. (Russian) MR **21** # 7260.

GOTÔ, M. AND KIMURA, N.
1. *Semigroup of endomorphisms of a locally compact group*, Trans. Amer. Math. Soc. **87** (1958), 359-371. MR **20** # 3229.

GREEN, J. A.
1. *On the structure of semigroups*, Ann. of Math. (2) **54** (1951), 163-172. MR **13**, 100.

GREEN, J. A. AND REES, D.
1. *On semigroups in which $x^r = x$*, Proc. Cambridge Philos. Soc. **38** (1952), 35-40. MR **13**, 720.

GRIFFITHS, H. B.
1. *Infinite products of semi-groups and local connectivity*, Proc. London Math. Soc. (3) **6** (1956), 455-480. MR **18**, 192.

HALEZOV, E. A.
1. *Automorphisms of matrix subgroups*, Dokl. Akad. Nauk SSSR **96** (1954), 245-248. (Russian) MR **16**, 333.
2. *Isomorphisms of matrix semigroups*, Ivanov, Gos. Ped. Inst. Učen. Zap. Fiz.-Mat. Nauki **5** (1954), 42-56. (Russian) MR **17**, 825.

HALL, M. JR.
1. *The word problem for semigroups with two generators*, J. Symbolic Logic **14** (1949), 115-118. MR **11**, 1.

HASHIMOTO, H.
1. *On a generalization of groups*, Proc. Japan Acad. **30** (1954), 548-549. MR **16**, 564.
2. *On the kernel of semigroups*, J. Math. Soc. Japan **7** (1955), 59-66. MR **16**, 670.
3. *On the structure of semigroups containing minimal left ideals and minimal right ideals*, Proc. Japan Acad. **31** (1955), 264-266. MR **17**, 459.

HEWITT, E.
1 & 2. *Compact monothetic semigroups*, Duke Math. J. **23** (1956), 447-457. MR **17**, 1225.

HEWITT, E. AND ZUCKERMAN, H. S.
1. *Finite dimensional convolution algebras*, Acta Math. **93** (1955), 67-119. MR **17**, 1048.
2. *Arithmetic and limit theorems for a class of random variables*, Duke Math. J. **22** (1955), 595-615. MR **17**, 754.
3. *The l_1-algebra of a commutative semigroup*, Trans. Amer. Math. Soc. **83** (1956), 70-97. MR **18**, 465.
4. *The irreducible representations of a semigroup related to the symmetric group*, Illinois J. Math. **1** (1957), 188-213. MR **19**, 249.

HILLE, E.
1. *Lie theory of semi-groups of linear transformations*, Bull. Amer. Math. Soc. **56** (1950), 89-114. MR **12**, 10.
2. *Functional analysis and semigroups*, Amer. Math. Soc. Colloq. Publ., vol. 31, Amer. Math. Soc., Providence, R. I., 1948. MR **9**, 594.

HILLE, E. AND PHILLIPS, R. S.
1. *Functional analysis and semigroups*, rev. ed., Amer. Math. Soc. Colloq. Publ., vol. 31, Amer. Math. Soc., Providence, R. I., 1957. MR **19**, 664.

HILLE, E. AND ZORN, M.
1. *Open additive semi-groups of complex numbers*, Ann. of Math. (2) **44** (1943), 554-561. MR **5**, 40.

HION, JA. V.
1. *Ordered semigroups*, Izv. Akad. Nauk SSSR Ser. Mat. **21** (1957), 209-222. (Russian) MR **19**, 530.

HÖLDER, O.
1. *Die Axiome der Quantität und die Lehre vom Mass,* Ber. Verh. Sächs. Wiss. Leipzig Math.-Phys. Kl. **51** (1901), 1-64.

HUNTINGTON, E. V.
1. *Note on the definitions of abstract groups and fields by sets of independent postulates,* Trans. Amer. Math. Soc. **6** (1905), 181-197.

ISÉKI, K.
1. *Sur les demi-groupes,* C. R. Acad. Sci. Paris **236** (1953), 1524-1525. MR **14**, 842.
2. *On compact abelian semi-groups,* Michigan Math. J. **3** (1954), 59-60. MR **15**, 933.
3. *Sur un théorème de M. G. Thierrin concernant demi-groupe limitatif,* Proc. Japan Acad. **31** (1955), 54-55. MR **17**, 173.
4. *Contribution to the theory of semi-groups.* I, Proc. Japan Acad. **32** (1956), 174-175. MR **17**, 1184.
5. *Contribution to the theory of semi-groups.* II, Proc. Japan Acad. **32** (1956), 225-227. MR **18**, 282.
6. *Contributions to the theory of semi-groups.* III, Proc. Japan Acad. **32** (1956), 323-324. MR **18**, 282.
7. *Contributions to the theory of semi-groups.* IV, Proc. Japan Acad. **32** (1956), 430-435. MR **18**, 282.
8. *Contributions to the theory of semi-groups.* V, Proc. Japan Acad. **32** (1956), 560-561. MR **18**, 872.
9. *Contributions to the theory of semi-groups.* VI, Proc. Japan Acad. **33** (1957), 29-30. MR **19**, 1158.
10. *A characterisation of regular semi-group,* Proc. Japan Acad. **32** (1956), 676-677. MR **18**, 717.
11. *On compact semi-groups,* Proc. Japan Acad. **32** (1956), 221-224. MR **18**, 282.
12. *Generalisation of a theorem by Š. Schwarz on semigroup,* Portugal. Math. **15** (1956), 71-72.

IVAN, J.
1. *On the direct product of semigroups,* Mat.-Fyz. Časopis. Sloven. Akad. Vied **3** (1953), 57-66. (Slovak) MR **16**, 9.
2. *On the decomposition of simple semigroups into a direct product,* Mat.-Fyz. Časopis. Sloven. Akad. Vied **4** (1954), 181-202. (Slovak) MR **17**, 346.
3. *On the matrix representations of simple semigroups,* Mat.-Fyz. Časopis. Sloven. Akad. Vied **8** (1958), 27-39. (Russian) MR **20** #7073.

JAFFARD, P.
1. *Théorie axiomatique des groupes définies par des systèmes de générateurs,* Bull. Sci. Math. (2) **75** (1951), 114-128. MR **13**, 430.

KACMAN, A. D.
1. *On some properties of a semigroup invariant in a group,* Uspehi Mat. Nauk **11** (1956), 179-183. (Russian) MR **18**, 378.
2. *On the question of generating and nongenerating elements of a semigroup invariant in a group,* Učen. Zap. Ural′sk. Univ. **19** (1956), 43-50. (Russian)

KATTSOFF, L. O.
1. *The independence of the associative law,* Amer. Math. Monthly **65** (1958), 620-623. MR **21** #91.

KAUFMAN, A. M.
1. *Successively annihilating sums of associative systems,* Leningrad. Gos. Ped. Inst. Učen. Zap. **86** (1949), 145-165. (Russian)
2. *The structure of certain haloids,* Leningrad. Gos. Ped. Inst. Učen. Zap. **86** (1949), 167-182. (Russian)

3. *Associative systems with an ideally solvable series of length two,* Leningrad. Gos. Ped. Inst. Učen. Zap. **89** (1953), 67-93. (Russian) MR **17**, 942.

KAWASA, Y. AND KONDÔ, K.

1. *Idealtheorie in nicht kommutativen Halbgruppen,* Jap. J. Math. **16** (1939), 37-45. MR **1**, 164.

KEMPERMAN, J. H. B.

1. *On complexes in a semigroup,* Nederl. Akad. Wetensch. Proc. Ser. A **59** (1956), 247-254. MR **19**, 13.

KIMURA, N.

1. *Maximal subgroups of a semigroup,* Kōdai Math. Sem. Rep. **1954**, 85-88. MR **16**, 443.
2. *On some examples of semigroups,* Kōdai Math. Sem. Rep. **1954**, 89-92. MR **16**, 443.
3. *Note on idempotent semigroups.* I, Proc. Japan Acad. **33** (1957), 642-645. MR **20** # 4602.
4. *Note on idempotent semigroups.* III, Proc. Japan Acad. **34** (1958), 113-114. MR **20** # 4604.
5. *The structure of idempotent semi-groups.* I, Pacific J. Math. **8** (1958), 257-275. MR **21** # 1354.
6. *Note on idempotent semigroups.* IV, Proc. Japan Acad. **34** (1958), 121-123. MR **20** # 4604.

KIMURA, N. AND TAMURA, T.

1. *Compact mob with a unique left unit,* Math. J. Okayama Univ. **5** (1956), 115-119. MR **17**, 1107.
2. *Counter examples to Wallace's problem,* Proc. Japan Acad. **31** (1955), 499-500. MR **17**, 643.

KLEIN-BARMEN, F.

1. *Über gewisse Halbverbände und kommutative Semigruppen.* I, Math. Z. **48** (1942), 275-288. MR **5**, 31.
2. *Über gewisse Halbverbände und kommutative Semigruppen.* II, Math. Z. **48** (1943), 715-734. MR **5**, 31.
3. *Ein Beitrag zur Theorie der linearen Holoide,* Math. Z. **51** (1948), 355-366. MR **10**, 353.
4. *Zur Theorie der Operative und Assoziative,* Math. Ann. **126** (1953), 23-30. MR **15**, 95.
5. *Zur Axiomatik der Semigruppen,* Bayer Akad. Wiss. Math.-Nat. Kl. S.-B. **1956**, 287-294. MR **20** # 3227.
6. *Verallgemeinerung des Verbandsbegriffs durch Abschwächung des Axioms der Idempotenz,* Math. Z. **70** (1958), 38-51. MR **21** # 1356.
7. *Über eine weitere Verallgemeinerung des Verbandsbegriffes,* Math. Z. **46** (1940), 472-480. MR **1**, 327.

KOCH, R. J.

1. *Remarks on primitive idempotents in compact semigroups with zero,* Proc. Amer. Math. Soc. **5** (1954), 828-833. MR **16**, 447.
2. *On monothetic semigroups,* Proc. Amer. Math. Soc. **8** (1957), 397-401. MR **19**, 290.

KOCH, R. J. AND WALLACE, A. D.

1. *Maximal ideals in compact semigroups,* Duke Math. J. **21** (1954), 681-685. MR **16**, 112.
2. *Stability in semigroups,* Duke Math. J. **24** (1957), 193-195. MR **18**, 907.
3. *Admissibility of semigroup structures on continua,* Trans. Amer. Math. Soc. 88 (1958), 277-287. MR **20** # 1729.

KOLIBIAROVÁ, B.

1. *On the semigroups, every subsemigroup of which has a left unit element,* Mat.-Fyz. Časopis. Sloven. Akad. Vied **7** (1957), 177-182. (Slovak) MR **20** # 7072.

KOLMAR, L.

1. *Another proof of the Markov-Post theorem*, Acta Math. Acad. Sci. Hungar. **3** (1952), 1-25.

KONTOROVIČ, P. G.

1. *On the theory of semigroups in a group*, Kazan. Gos. Univ. Učen. Zap. **114** (1954), no. 8, 35-43. (Russian) MR **17**, 942.
2. *On the theory of semigroups in a group*, Dokl. Akad. Nauk SSSR **93** (1953), 229-231. (Russian) MR **15**, 681.
3. *Some questions in the theory of semigroups in groups*, Uspehi Mat. Nauk **11** (1956), no. 1 (67), 255-258. (Russian)
4. *On the theory of semigroups in a group*. II, Učen. Zap. Ural′sk. Univ. **19** (1956), 3-20. (Russian)

KONTOROVIČ, P. G. AND KACMAN, A. D.

1. *Some types of elements of a semigroup invariant in a group*, Uspehi Mat. Nauk **11** (1956), no. 3 (69), 145-150. (Russian) MR **18**, 193.

KRUMING, P. D.

1. *On divisibility in topological semigroups*, Leningrad. Gos. Ped. Inst. Učen. Zap. **183** (1958), 271-275. (Russian)

LAMBEK, J.

1. *The immersibility of a semigroup into a group*, Canad. J. Math. **3** (1951), 34-43. MR **12**, 481.
2. *Initial segments of positive semigroups*, Trans. Roy. Soc. Canad. Sect. III (3) **50** (1956), 41-46. MR **20** # 6473.

LEFEBVRE, P.

1. *Demi-groupes admettant des complexes nets à droite minimaux*, C. R. Acad. Sci. Paris **247** (1958), 393-396. MR **20** # 1719.

LESIEUR, L.

1. *Théorèmes de décomposition dans certains demi-groupes réticulés satisfaisant à la condition de chaîne descendante affaiblie*, C. R. Acad. Sci. Paris **234** (1952), 2250-2252. MR **13**, 906.
2. *Sur les idéaux irréductibles d'un demi-groupe*, Rend. Sem. Mat. Univ. Padova **24** (1955), 29-36. MR **17**, 347.

3 & 4. *Sur les demi-groupes réticulés satisfaisants à une condition de chaîne*, Bull. Soc. Math. France **83** (1955), 161-193. MR **17**, 584.

5. *Sur les idéaux irréductibles d'un demi-groupe*, Rend. Sem. Mat. Univ. Padova **24** (1955), 29-36. MR **17**, 347.

LESOHIN, M. M.

1. *Some properties of generalized characters of semigroups*, Leningrad. Gos. Ped. Inst. Učen. Zap. **183** (1958), 277-286. (Russian)

LEVI, F. W.

1. *A problem on rigid motion*, Math. Student **8** (1940), 1-10. MR **2**, 128.
2. *On semigroups*, Bull. Calcutta Math. Soc. **36** (1944), 141-146. MR **6**, 202.
3. *On semigroups*. II, Bull. Calcutta Math. Soc. **38** (1946), 123-124. MR **8**, 368.

LEVITZKI, J.

1. *On powers with transfinite exponents*, Riveon Lematematika **1** (1946), 8-13. (Hebrew) MR **8**, 193.
2. *On powers with transfinite exponents*. II, Riveon Lematematika **2** (1947), 1-7. (Hebrew) MR **9**, 173.

LIBER, A. E.

1. *On symmetric generalized groups*, Mat. Sb. **33** (**75**) (1953), 531-544. (Russian) MR **15**, 502.

2. *On the theory of generalized groups,* Dokl. Akad. Nauk SSSR **97** (1954), 25-38. (Russian) MR **16**, 9.

LJAPIN, E. S.

1. *Free systems with an infinite univalent operation,* C. R. (Dokl.) Acad. Sci. URSS **51** (1946), 493-496. MR **8**, 501.
2. *The kernels of homomorphisms of associative systems,* Mat. Sb. **20 (62)** (1947), 497-515. (Russian) MR **9**, 134.
3. *Algebraic systems with several operations,* Leningrad. Gos. Ped. Inst. Učen. Zap. **64** (1948), 53-72. (Russian)
4. *Complete operations in classes of associative systems and groups,* Leningrad. Gos. Ped. Inst. Učen. Zap. **86** (1949), 93-106. (Russian)
5. *Normal complexes of associative systems,* Izv. Akad. Nauk SSSR Ser. Mat. **14** (1950), 179-192. (Russian) MR **11**, 575.
6. *Simple commutative associative systems,* Izv. Akad. Nauk SSSR Ser. Mat. **14** (1950), 275-282. (Russian) MR **12**, 5.
7. *Semisimple commutative associative systems,* Izv. Akad. Nauk SSSR Ser. Mat. **14** (1950), 367-380. (Russian) MR **12**, 154.
8. *Canonical form of elements of an associative system given by defining relations,* Leningrad. Gos. Ped. Inst. Učen. Zap. **89** (1953), 45-54. (Russian) MR **17**, 942.
9. *Increasing elements of associative systems,* Leningrad. Gos. Ped. Inst. Učen. Zap. **89** (1953), 55-65. (Russian) MR **17**, 942.
10. *Associative systems of all partial transformations,* Dokl. Akad. Nauk SSSR **88** (1953), 13-15; errata, **92** (1953), 692. (Russian) MR **15**, 395.
11. *Semigroups in all of whose representations the operators have fixed points.* I, Mat. Sb. **34 (76)** (1954), 289-306. (Russian) MR **15**, 850.
12. *Semigroups in all of whose representations the operators have fixed points.* II, Mat. Sb. **36 (78)** (1955), 111-124. (Russian) MR **16**, 792.
13. *Abstract characterization of certain semigroups of transformations,* Leningrad. Gos. Ped. Inst. Učen. Zap. **103** (1955), 5-29. (Russian)
14. *On the existence and uniqueness of solutions of a general type of equation in connection with reversibility in semigroups of transformations,* Dokl. Akad. Nauk SSSR **116** (1957), 552-555. (Russian) MR **20** # 3923.
15. *Inversion of elements in semi-groups,* Leningrad. Gos. Ped. Inst. Učen. Zap. **166** (1958), 65-74. (Russian) MR **21** # 4984.

LO LI-PO AND WANG SHIH-CHIANG

1. *Finite associative systems and finite groups.* I, Advancement in Math. **3** (1957), 268-270. (Chinese) MR **20** # 3226.

LOEWY, A.

1. *Über abstrakt definierte Transmutationssysteme oder Mischgruppe,* J. Reine Angew. Math. **157** (1927), 239-254.

LORENZEN, P.

1. *Abstrakte Begründung der multiplikativen Idealtheorie,* Math. Z. **45** (1939), 533-553. MR **1**, 101.

LOS, J.

1. *On the existence of linear order in a group,* Bull. Acad. Polon. Sci. Cl. III **2** (1954), 21-23. MR **16**, 564.

McLEAN, D.

1. *Idempotent semigroups,* Amer. Math. Monthly **61** (1954), 110-113. MR **15**, 681.

MACKENZIE, R. E.

1. *Commutative semigroups,* Duke Math. J. **21** (1954), 471-477. MR **16**, 8.

MAL'CEV, A. I.

1. *On the immersion of an algebraic ring into a field,* Math. Ann. **113** (1937), 686-691.
2. *Über die Einbettung von assoziativen Systemen in Gruppen,* Mat. Sb. **6** **(48)** (1939), 331-336. (Russian) MR **2**, 7.
3. *Über die Einbettung von assoziativen Systemen in Gruppen.* II, Mat. Sb. **8** **(50)** (1940), 251-264. (Russian) MR **2**, 128.
4. *Symmetric groupoids,* Mat. Sb. **31** **(73)** (1952), 136-151. (Russian) MR **14**, 349.
5. *Multiplicative congruences of matrices,* Dokl. Akad. Nauk SSSR **90** (1953), 333-335. (Russian) MR **14**, 1057.
6. *Nilpotent semigroups,* Ivanov. Gos. Ped. Inst. Učen. Zap. Fiz.-Mat. Nauki **4** (1953), 107-111. (Russian) MR **17**, 825.
7. *On the general theory of algebraic systems,* Mat. Sb. **35** **(77)** (1954), 3-20; English transl., Amer. Math. Soc. Transl. (2) **27** (1963), 125-142. MR **16**, 440; **27** # 1401.

MANN, H. B.

1. *On certain systems which are almost groups,* Bull. Amer. Math. Soc. **50** (1944), 879-881. MR **6**, 147.

MANTEUFFEL, K.

1. *Bemerkung zur Zerlegung und Erzeugung endlicher Gruppoide,* Wiss. Z. Hochsch. Schwermaschinenbau Magdeburg **1** (1956/57), 1-2. MR **23** # A955.

MARKOV, A. A.

1. *On the impossibility of certain algorithms in the theory of associative systems,* Dokl. Akad. Nauk SSSR **55** (1947), 583-586. MR **8**, 558.
2. *The impossibility of certain algorithms in the theory of associative systems.* II, Dokl. Akad. Nauk SSSR **58** (1947), 353-356. (Russian) MR **9**, 321.
3. *The impossibility of certain algorithms in the theory of associative systems,* Dokl. Akad. Nauk SSSR **77** (1951), 19-20. (Russian) MR **12**, 661.
4. *The impossibility of algorithms for the recognition of certain properties of associative systems,* Dokl. Akad. Nauk SSSR **77** (1951), 953-956. (Russian) MR **13**, 4.
5. *The theory of algorithms,* Trudy Mat. Inst. Steklov. **38** (1951), 176-189; English transl., Amer. Math. Soc. Transl. (2) **15** (1960), 1-14. MR **13**, 811; **22** # 5572.
6. *The theory of algorithms,* Trudy Mat. Inst. Steklov. **42** (1954); English transl., Israel Program for Scientific Translations, Jerusalem, 1961. MR **17**, 1039; **24** # A2527.

MAURY, G.

1. *Une caractérisation des demi-groupes noethériens intégralement clos,* C. R. Acad. Sci. Paris **247** (1958), 254-255. MR **20** # 6474.

MILLER, D. D. AND CLIFFORD, A. H.

1. *Regular D-classes in semigroups,* Trans. Amer. Math. Soc. **82** (1956), 270-280. MR **17**, 1184.

MIYANAGA, J.

1. *A generalisation of Wallace theorem on semi-groups,* Proc. Japan Acad. **32** (1956), 254.

MOLINARO, I.

1. *Demi-groupes résidutifs,* Séminare P. Dubreil et Ch. Pisot, 9e année 1955/56, Secrétariat mathématique, Paris, 1956, pp. 1-16. MR **19**, 522.

MOORE, E. H.

1. *A definition of abstract groups,* Trans. Amer. Math. Soc. **3** (1902), 485-492.

MORSE, M. AND HEDLUND, G. A.

1. *Unending chess, symbolic dynamics and a problem in semigroups,* Duke Math. J. **11** (1944), 1-7. MR **5**, 202.

MOSTERT, P. S. AND SHIELDS, A. L.

1. *On continuous multiplications on the two-sphere,* Proc. Amer. Math. Soc. **7** (1956), 942-947. MR **18**, 317.

2. *On a class of semi-groups on E_n*, Proc. Amer. Math. Soc. **7** (1956), 729-734. MR **17**, 1224.
3. *On the structure of semi-groups on a compact manifold with boundary*, Ann. of Math. (2) **65** (1957), 117-143. MR **18**, 809.

MUNN, W. D.

1. *On semigroup algebras*, Proc. Cambridge Philos. Soc. **51** (1955), 1-15. MR **16**, 561.
2. *Matrix representations of semigroups*, Proc. Cambridge Philos. Soc. **53** (1957), 5-12. MR **18**, 489.
3. *The characters of the symmetric inverse semigroup*, Proc. Cambridge Philos. Soc. **53** (1957), 13-18. MR **18**, 465.
4. *Semigroups satisfying minimal conditions*, Proc. Glasgow Math. Assoc. **3** (1957), 145-152. MR **22** #9543.

MUNN, W. D. AND PENROSE, R.

1. *A note on inverse semigroups*, Proc. Cambridge Philos. Soc. **51** (1955), 396-399. MR **17**, 10.

MURATA, K.

1. *On the quotient semi-group of a noncommutative semi-group*, Osaka Math. J. **2** (1950), 1-5. MR **12**, 155.

NAKADA, O.

1. *Partially ordered Abelian semigroups*. I. *On the extension of the strong partial order defined on Abelian semigroups*, J. Fac. Sci. Hokkaido Univ. Ser. I **11** (1951), 181-189. MR **13**, 817.
2. *Partially ordered Abelian semigroups*. II. *On the strongness of the linear order defined on Abelian semigroups*, J. Fac. Sci. Hokkaido Univ. Ser. I **12** (1952), 73-86. MR **14**, 945.

NUMAKURA, K.

1. *On bicompact semigroups with zero*, Bull. Yamagata Univ. **1951**, no. 4, 405-412. MR **16**, 447.
2. *On bicompact semigroups*, Math. J. Okayama Univ. **1** (1952), 99-108. MR **14**, 18.
3. *A note on the structure of commutative semigroups*, Proc. Japan Acad. **30** (1954), 262-265. MR **16**, 214.
4. *Prime ideals and idempotents in compact semigroups*, Duke Math. J. **24** (1957), 671-680. MR **19**, 967.
5. *Naturally totally ordered compact semigroups*, Duke Math. J. **25** (1958), 639-645. MR **20** #7077.

OGANESJAN, V. A.

1. *Invariant and normal subsystems of a symmetric system of partial substitutions*, Akad. Nauk Armjan. SSR Dokl. **21** (1955), 49-56. (Russian) MR **17**, 943.
2. *A theorem on the structure of a system of substitutions*, Sb. Naučn. Trudy Armjan. Gos. Zaočn. Ped. Inst. **3** (1956), no. 2, 9-17. (Russian)
3. *A system of substitutions and the generalized Cayley theorem*, Sb. Naučn. Trudy Armjan. Gos. Zaočn. Ped. Inst. **3** (1956), no. 2, 19-31. (Russian)
4. *On the semisimplicity of a system algebra*, Akad. Nauk Armjan. SSR Dokl. **21** (1955), 145-147. (Russian) MR **18**, 465.

ORE, O.

1. *Linear equations in non-commutative fields*, Ann. of Math. (2) **32** (1931), 463-477.

PAŘÍZEK, B.

1. *Note on structure of multiplicative semigroup of residue classes*, Mat.-Fyz. Časopis. Sloven. Akad. Vied **7** (1957), 183-185. (Slovak) MR **21** #85.

PARKER, E. T.

1. *On multiplicative semigroups of residue classes*, Proc. Amer. Math. Soc. **5** (1954), 612-612. MR **16**, 9.

PEREL, W. M.
1. *Principal representations in commutative semigroups,* Proc. Amer. Math. Soc. **9** (1957), 957-960. MR **20** # 3922.

PETROPAVLOVSKAJA, R. V.
1. *On the decomposition into a direct sum of the structure of subsystems of an associative system,* Dokl. Akad. Nauk SSSR **81** (1951), 999-1002. (Russian) MR **13**, 525.
2. *Structural isomorphisms of free associative systems,* Mat. Sb. **28** **(70)** (1951), 589-602. (Russian) MR **13**, 100.
3. *On the determination of a group by the structure of its subsystems,* Mat. Sb. **29** **(71)** (1951), 63-78. (Russian) MR **13**, 203.
4. *Associative systems that are lattice-isomorphic to a given group.* I, Vestnik Leningrad. Univ. **11** (1956), no. 13, 5-26. (Russian) MR **18**, 555.
5. *Associative systems that are lattice-isomorphic to a given group.* II. Vestnik Leningrad. Univ. **11** (1956), no. 19, 80-99. (Russian) MR **20** # 3799a.
6. *Semigroups lattice-isomorphic to groups.* III, Vestnik Leningrad. Univ. **12** (1957), no. 19, 5-19. (Russian) MR **20** # 3799b.

PIERCE, R. S.
1. *Homomorphisms of semi-groups,* Ann. of Math. (2) **59** (1954), 287-291. MR **15**, 930.

PODDERJUGIN, V. D.
1. *Conditions for orderability of a group,* Izv. Akad. Nauk SSSR Ser. Mat. **21** (1957), 199-208. (Russian) MR **20** # 911.

PONIZOVSKIĬ, I. S.
1. *On matrix representations of associative systems,* Mat. Sb. **38** **(80)** (1956), 241-260. (Russian) MR **18**, 378.
2. *On irreducible representations of finite associative systems,* Učen. Zap. Kemorovskogo Ped. Inst. **1** (1956), 245-250. (Russian)
3. *On irreducible matrix representations of finite semigroups,* Uspehi Mat. Nauk **13** (1958), no. 6 (84), 139-144. (Russian) MR **21** # 2015.

POOLE, A. R.
1. *Finite ova,* Amer. J. Math. **59** (1937), 23-32.

POST, E. L.
1. *Recursive unsolvability of a problem of Thue,* J. Symbolic Logic **12** (1947), 1-11. MR **8**, 558.

POST, K. A.
1. *Rank functions on semigroups,* Nederl. Akad. Wetensch. Proc. Ser. A **61** = Indag. Math. **20** (1958), 332-334. MR **20** # 7074.

PRESTON, G. B.
1. *Inverse semi-groups,* J. London Math. Soc. **29** (1954), 396-403. MR **16**, 215.
2. *Inverse semi-groups with minimal right ideals,* J. London Math. Soc. **29** (1954), 404-411. MR **16**, 215.
3. *Representations of inverse semi-groups,* J. London Math. Soc. **29** (1954), 411-419. MR **16**, 216.
4. *The structure of normal inverse semigroups,* Proc. Glasgow Math. Assoc. **3** (1956), 1-9. MR **18**, 717.
5. *Matrix representations of semigroups,* Quart. J. Math. Oxford Ser. (2) **9** (1958), 169-176. MR **20** # 5814.
6. *A note on representations of inverse semigroups,* Proc. Amer. Math. Soc. **8** (1957), 1144-1147. MR **19**, 941.

PTAK, V.
1. *Immersibility of semigroups,* Acta Fac. Nat. Univ. Carol. (Prague) No. 192 (1949), 16 pp. MR **12**, 155.

2. *Immersibility of semigroups,* Czechoslovak Math. J. **2** (**77**) (1952), 247-271. (Russian) MR **15**, 598.

3. *Immersibility of semigroups,* Časopis Pěst. Mat. **78** (1953), 259-261. (Czech) MR **18**, 110.

RAUTER, H.

1. *Abstrakte Kompositionssysteme oder Übergruppen,* J. Reine Angew. Math. **159** (1928), 229-237.

RÉDEI, L.

1. *Die Verallgemeinerung der Schreierschen Erweiterungstheorie,* Acta Sci. Math. (Szeged) **14** (1952), 252-273. MR **14**, 614.

RÉDEI, L. AND STEINFELD, O.

1. *Über Ringe mit gemeinsamer multiplikativen Halbgruppe,* Comment. Math. Helv. **26** (1952), 146-151. MR **14**, 10.

REES, D.

1. *On semi-groups,* Proc. Cambridge Philos. Soc. **36** (1940), 387-400. MR **2**, 127.
2. *Note on semi-groups,* Proc. Cambridge Philos. Soc. **37** (1941), 434-435. MR **3**, 199.
3. *On the ideal structure of a semi-group satisfying a cancellation law,* Quart. J. Math. Oxford Ser. **19** (1948), 101-108. MR **9**, 567.
4. *On the group of a set of partial transformations,* J. London Math. Soc. **22** (1947), 281-284. MR **9**, 568.

RICH, R. P.

1. *Completely simple ideals of a semigroup,* Amer. J. Math. **71** (1949), 883-885. MR **11**, 327.

RYBAKOFF, L.

1. *Sur une classe de semi-groupes commutatifs,* Mat. Sb. **5** (**47**) (1939), 521-536. (Russian) MR **1**, 164.

SAITÔ, T. AND HORI, S.

1. *On semigroups with minimal left ideals and without minimal right ideals,* J. Math. Soc. Japan **10** (1958), 64-70. MR **20** # 1717.

SCHENKMAN, E.

1. *A certain class of semigroups,* Publ. Math. Debrecen **4** (1956), 262-275.
2. *A certain class of semigroups,* Amer. Math. Monthly **63** (1956), 242-243. MR **17**, 1055.

SCHIEFERDECKER, E.

1. *Die fastperiodischen Funktionen einer Oreschen Halbgruppe,* Arch. Math. **6** (1955), 428-438. MR **17**, 346.
2. *Zur Einbettung metrischer Halbgruppen in ihre Quotientenhalbgruppen,* Math. Z. **62** (1955), 443-468. MR **17**, 384.
3. *Einbettungssätze für topologische Halbgruppen,* Math. Ann. **131** (1956), 372-384. MR **18**, 658.
4. *Fastperiodische Fortsetzung von Funktionen auf Halbgruppen,* Math. Nachr. **14** (1955), 253-261. MR **19**, 118.

SCHMIDT, F.

1. *Bemerkungen zum Brandtschengruppoid, Beitrage zur Algebra,* S.-B. Heidelberg. Akad. Wiss. Mat. Nat. Kl. **1927**, 91-103.

SCHREIER, J.

1. *Über Abbildungen einer abstrakten Menge auf ihre Teilmenge,* Fund. Math. **28** (1936), 261-264.

SCHÜTZENBERGER, M.-P.

1. *Une théorie algébrique du codage,* C. R. Acad. Sci. Paris **242** (1956), 862-864. MR **17**, 702.

2. *Sur une représentation des demi-groupes,* C. R. Acad. Sci. Paris **242** (1956), 2907-2908. MR **18**, 13.
3. *Sur deux représentations des demi-groupes finis,* C. R. Acad. Sci. Paris **243** (1956), 1385-1387. MR **18**, 282.
4. *D-représentation des demi-groupes,* C. R. Acad. Sci. Paris **244** (1957), 1994-1996. MR **19**, 249.
5. *Applications des $\bar{D}$-représentations à l'étude des homomorphismes des demi-groupes,* C. R. Acad. Sci. Paris **244** (1957), 2219-2221. MR **19**, 249.
6. *Sur une propriété combinatoire des demi-groupes,* C. R. Acad. Sci. Paris **245** (1957), 16-18. MR **19**, 528.
7. *Sur la représentation monomiale des demi-groupes,* C. R. Acad. Sci. Paris **246** (1958), 865-867. MR **20** # 2384.
8. *Sur les homomorphismes d'un demi-groupe sur un groupe,* C. R. Acad. Sci. Paris **246** (1958), 2442-2444. MR **20** # 1720.
9. *Sur certains treillis gauches,* C. R. Acad. Sci. Paris **224** (1947), 776-778. MR **8**, 432.

SCHWARZ, Š.

1. *Zur Theorie der Halbgruppen,* Sbornik Prác Prirodovedeckej Fakulty Slovenskej Univerzity v Bratislave No. 6 (1943), 64 pp. (Slovak) MR **10**, 12.
2. *On various generalizations of the notion of a group,* Časopis Pěst. Mat. Fys. **74** (1949), 95-112. (Czech) MR **12**, 389.
3. *On the structure of simple semigroups without zero,* Czechoslovak Math. J. **1 (76)** (1951), 41-53. MR **14**, 12.
4. *On semigroups having a kernel,* Czechoslovak Math. J. **1 (76)** (1951), 229-264. MR **14**, 444.
5. *Contribution to the theory of torsion semigroups,* Czechoslovak Math. J. **3 (78)** (1953), 7-21. (Russian) MR **15**, 850.
6. *On maximal ideals in the theory of semigroups.* I, Czechoslovak Math. J. **3 (78)** (1953), 139-153. (Russian) MR **15**, 850.
7. *On maximal ideals in the theory of semigroups.* II, Czechoslovak Math. J. **3 (78)** (1953), 365-383. (Russian) MR **15**, 850.
8. *Maximal ideals and the structure of semigroups,* Mat.-Fyz. Časopis. Sloven. Akad. Vied **3** (1953), 17-30. (Slovak) MR **16**, 335.
9. *The theory of characters of commutative semigroups,* Czechoslovak Math. J. **4 (79)** (1954), 219-247. (Russian) MR **16**, 1085.
10. *Characters of commutative semigroups as class functions,* Czechoslovak Math. J. **4 (79)** (1954), 291-295. (Russian) MR **16**, 1086.
11. *On a Galois connexion in the theory of characters of commutative semigroups,* Czechoslovak Math. J. **4 (79)** (1954), 296-313. (Russian) MR **16**, 1086.
12. *On Hausdorff bicompact semigroups,* Czechoslovak Math. J. **5 (80)** (1955), 1-23. (Russian) MR **17**, 643.
13. *Characters of bicompact semigroups,* Czechoslovak Math. J. **5 (80)** (1955), 24-28. (Russian) MR **17**, 643.
14. *Topological semigroups with one-sided units,* Czechoslovak Math. J. **5 (80)** (1955), 153-163. (Russian) MR **17**, 643.
15. *Remark on the theory of bicompact semigroups,* Mat.-Fyz. Časopis. Sloven. Akad. Vied **5** (1955), 86-89. (Slovak) MR **17**, 1107.
16. *On expanding elements in the theory of semigroups,* Dokl. Akad. Nauk SSSR **102** (1955), 697-698. (Russian) MR **17**, 173.
17. *Semigroups satisfying some weakened forms of the cancellation law,* Mat.-Fyz. Časopis. Sloven. Akad. Vied **6** (1956), 149-158. (Slovak) MR **19**, 940.

18. *The theory of characters of commutative Hausdorff bicompact semigroups*, Czechoslovak Math. J. **6** **(81)** (1956), 330-364. MR **19**, 1063.

19. *On the existence of invariant measures on certain types of bicompact semigroups*, Czechoslovak Math. J. **7** **(82)** (1957), 165-182. (Russian) MR **19**, 663.

20. *On the structure of the semigroup of measures on a finite semigroup*, Czechoslovak Math. J. **7** **(82)** (1957), 358-373. MR **20** # 1721.

21. *An elementary semigroup theorem and a congruence relation of Rédei*, Acta Sci. Math. (Szeged) **19** (1958), 1-4. MR **20** # 3222.

SCOTT, W. R.

1. *Half-homomorphisms of groups*, Proc. Amer. Math. Soc. **8** (1957), 1141-1144. MR **20** # 2388.

SEMIGROUPS

1. *Séminaire A. Châtelet et P. Dubreil de la Faculté des Sciences de Paris*, 7e année complémentaire 1953/54, Secrétariat mathématique, Paris, 1956. MR **18**, 282.

SHIELDS, A. L.

1. *The n-cube as a product semigroup*, Michigan Math. J. **4** (1957), 165-166. MR **19**, 1064.

SIVERCEVA, N.

1. *On the simplicity of the associative system of singular square matrices*, Mat. Sb. **24** **(66)** (1949), 101-106. (Russian) MR **10**, 508.

SKOLEM, T.

1. *Some remarks on semi-groups*, Norske Vid. Selsk. Forh. (Trondheim) **24** (1951), 42-47. MR **13**, 906.

2. *Theorems of divisibility in some semi-groups*, Norske Vid. Selsk. Forh. (Trondheim) **24** (1951), 48-53. MR **13**, 906.

3. *Theory of divisibility in some commutative semi-groups*, Norske Mat. Tidsskr. **33** (1951), 82-88. MR **13**, 430.

4. *A theorem on some semigroups*, Norske Vid. Selsk. Forh. (Trondheim) **25** (1952), 72-77. MR **14**, 842.

SONNEBORN, L. M.

1. *On the arithmetic structure of a class of commutative semigroups*, Amer. J. Math. **77** (1955), 783-790. MR **17**, 459.

STEINFELD, O.

1. *Über die Quasiideale von Halbgruppen*, Publ. Math. Debrecen **4** (1956), 262-275. MR **18**, 790.

2. *Über die Quasiideale von Halbgruppen mit eigentlichem Suschkewitsch-Kern*, Acta Sci. Math. (Szeged) **18** (1957), 235-242. MR **20** # 1718.

STOLL, R. R.

1. *Representations of finite simple semigroups*, Duke Math. J. **11** (1944), 251-265. MR **5**, 229.

2. *Homomorphisms of a semigroup onto a group*, Amer. J. Math. **73** (1951), 475-481. MR **12**, 799.

STOLT, B.

1. *Über Axiomensysteme die eine abstrakte Gruppe bestimmen*, Thesis, University of Uppsala, 1953, Almquist & Wiksells, Uppsala, 1953, 99 pp. MR **15**, 99.

2. *Abschwächung einer klassischen Gruppendefinition*, Math. Scand. **3** (1955), 303-305. MR **17**, 1052.

3. *Über eine besondere Halbgruppe*, Ark. Mat. **3** (1956), 275-286. MR **17**, 942.

4. *Zur Axiomatik des Brandtschen Gruppoids*, Math. Z. **70** (1958), 156-164. MR **21** # 90.

SUŠKEVIČ, A. K.

1. *Über die Darstellung der eindeutig nicht umkehrbaren Gruppen mittels der verallgemeinerten Substitutionen*, Mat. Sb. **33** (1926), 371-374.

2. *Sur quelques cas des groupes finis sans la loi de l'inversion univoque,* Soobšč. Har'kov. Mat. Obšč. (4) **1** (1927), 17-24.
3. *Über die endlichen Gruppen ohne das Gesetz der eindeutigen Umkehrbarkeit,* Math. Ann. **99** (1928), 30-50.
4. *Untersuchungen über verallgemeinerte Substitutionen,* Atti Congr. Internat. Mat. (Bologna, 1928), vol. I, pp. 147-157.
5. *Über die Matrizendarstellung der verallgemeinerten Gruppen,* Soobšč. Har'kov. Mat. Obšč. i Ukrain Naučno-Issled. Inst. Mat. Meh. (4) **6** (1933), 27-38.
6. *Über Semi-gruppen,* Zap. Inst. Mat. Meh. (4) **8** (1934), 25-28.
7. *Über einen merkwürdigen Typus der verallgemeinerien unendlichen Gruppen,* Soobšč. Har'kov. Mat. Obšč. i Naučno-Issled. Inst. Mat. Meh. Har'kov. Gos. Univ. (4) **9** (1934), 39-46.
8. *On certain properties of a type of generalized groups,* Učen. Zap. Har'kov. Univ. **2** (1935), 23-25. (Russian)
9. *On an extension of semigroups to groups,* Soobšč. Har'kov. Mat. Obšč. i Naučno-Issled. Inst. Mat. Meh. Har'kov. Gos. Univ. (4) **12** (1936), 81-88. (Russian)
10. *Über eine Verallgemeinerung der Semi-gruppen,* Soobšč. Har'kov. Mat. Obšč. i Naučno-Issled. Inst. Mat. Meh. Har'kov. Gos. Univ. (4) **12** (1936), 89-98.
11. *Sur quelques propriétés des semi-groupes généralisés,* Soobšč. Naučno-Issled. Inst. Mat. Meh. Har'kov. Gos. Univ. i Har'kov. Mat. Obšč. (4) **13** (1936), 29-33.
12. *Theory of generalized groups,* GNITI, Kharkov, 1937. (Russian)
13. *On certain types of singular matrices,* Učen. Zap. Har'kov. Univ. **10** (1937), 5-16. (Russian)
14. *Investigations on infinite substitutions,* Memorial Volume Dedicated to D. A. Grave, Moscow, 1940, pp. 245-253. (Russian) MR **2**, 217.
15. *Groupes généralisés des matrices singulières,* Zap. Naučno-Issled. Inst. Mat. Meh. i Har'kov. Mat. Obšč. (4) **16** (1940), 3-11. (Russian) MR **3**, 99.
16. *Groupes généralisés de quelques types des matrices infinies,* Zap. Naučno-Issled. Inst. Mat. Meh. i Har'kov. Mat. Obšč. (4) **16** (1940), 115-120. (Ukrainian). MR **3**, 99.
17. *Über einen Typus der verallgemeinerten Semigruppen,* Zap. Naučno-Issled. Inst. Mat. Meh. i Har'kov. Mat. Obšč. (4) **17** (1940), 19-28. (Russian) MR **3**, 37.
18. *Untersuchungen über unendliche Substitutionen,* Zap. Naučno-Issled. Inst. Mat. Meh. i Har'kov. Mat. Obšč. (4) **18** (1940), 27-37. (Russian) MR **3**, 99.
19. *Direct products of certain types of generalized groups,* Naučn. Zap. Har'kov. Inst. Sov. Torgovli **3** (1941), 11-14. (Russian)
20. *Über den Zusammenhang der Rauterschen Übergruppe mit den gewöhnlichen Gruppen,* Math. Z. **38** (1934), 643-649.

ŠUTOV, È. G.

1. *Potential divisibility of elements in semigroups,* Leningrad. Gos. Ped. Inst. Učen. Zap. **166** (1958), 75-103. (Russian) MR **21** #2018.
2. *Potential conjugacy of elements in semigroups,* Leningrad. Gos. Ped. Inst. Učen. Zap. **166** (1958), 105-119. (Russian) MR **21** #2698.
3. *Potential stationarity of elements of semigroups,* Učen. Zap. Udmurtsk. Ped. Inst. **12** (1958), 16-23. (Russian)
4. *Potential one-sided invertibility of elements of semigroups,* Učen. Zap. Udmurtsk. Ped. Inst. **12** (1958), 24-36. (Russian)

SZÁSZ, G.

1. *Die Unabhängigkeit der Assoziativitätsbedingungen,* Acta Sci. Math. (Szeged) **15** (1953), 20-28. MR **15**, 95.
2. *Über die Unabhängigkeit der Assoziativitätsbedingungen kommutativer multiplikativer Strukturen,* Acta Sci. Math. (Szeged) **15** (1954), 130-142. MR **15**, 773.

3. *Die Translationen der Halbverbände,* Acta Sci. Math. (Szeged) **17** (1956), 165-169. MR **18**, 555.

SZÁSZ, G. AND SZENDREI, J.

1. *Über die Translationen der Halbverbände,* Acta Sci. Math. (Szeged) **18** (1957), 44-47. MR **19**, 389.

SZÉP, J.

1. *Zur Theorie der Halbgruppen,* Publ. Math. Debrecen **4** (1956), 344-346. MR **18**, 110.

TAMARI, D.

1. *Caractérisation des semi-groupes à un paramètre,* C. R. Acad. Sci. Paris **228** (1949), 1092-1094. MR **10**, 508.
2. *Groupoïdes reliés et demi-groupes ordonnés,* C. R. Acad. Sci. Paris **228** (1949), 1184-1186. MR **10**, 508.
3. *Groupoïdes ordonnés. L'ordre lexicographique pondéré,* C. R. Acad. Sci. Paris **228** (1949), 1909-1911. MR **11**, 9.
4. *Ordres pondérés. Caractérisation de l'ordre naturel comme l'ordre du semi-groupe multiplicatif des nombres naturels,* C. R. Acad. Sci. Paris **229** (1949), 98-100. MR **11**, 80.
5. *Les images homomorphes des groupoïdes de Brandt et l'immersion des semi-groupes,* C. R. Acad. Sci. Paris **229** (1949), 1291-1293. MR **11**, 327.
6. *Sur l'immersion d'un semi-groupe topologique dans un groupe topologique,* Algèbre et Théorie des Nombres, Colloq. Internat. du Centre National de la Recherche Scientifique, no. 24, Centre National de la Recherche Scientifique, Paris, 1950, pp. 217-221. MR **12**, 670.
7. *On a certain classification of rings and semigroups,* Bull. Amer. Math. Soc. **54** (1948), 153-158. MR **9**, 490.

TAMURA, T.

1. *Characterization of groupoids and semilattices by ideals in a semigroup,* J. Sci. Gakugei Fac. Tokushima Univ. **1** (1950), 37-44. MR **13**, 430.
2. *Some remarks on semi-groups and all types of semi-groups of order* 2, 3, J. Gakugei Tokishima Univ. **3** (1953), 1-11. MR **15**, 7.
3. *On compact one-idempotent semigroups,* Kōdai Math. Sem. Rep. **1954**, 17-21. MR **15**, 933.
4. *Supplement to the paper "On compact one-idempotent semigroups",* Kōdai Math. Sem. Rep. **1954**, 96. MR **16**, 335.
5. *On finite one-idempotent semigroups.* I, J. Gakugei Tokushima Univ. **4** (1954), 11-20. MR **15**, 850.
6. *On a monoid whose submonoids form a chain,* J. Gakugei Tokushima Univ. Math. **5** (1954), 8-16. MR **16**, 1085.
7. *Notes on finite semigroups and determination of semigroups of order* 4, J. Gakugei Tokushima Univ. Math. **5** (1954), 17-27. MR **16**, 1085.
8. *Note on unipotent inversible semigroups,* Kōdai Math. Sem. Rep. **1954**, 93-95. MR **16**, 443.
9. *One-sided bases and translations of a semigroup,* Math. Japon. **3** (1955), 137-141. MR **17**, 1184.
10. *Indecomposable completely simple semigroups except groups,* Osaka Math. J. **8** (1956), 35-42. MR **18**, 282.
11. *The theory of construction of finite semigroups.* I, Osaka Math. J. **8** (1956), 243-261. MR **18**, 717.
12. *On translations of a semigroup,* Kōdai Math. Sem. Rep. **7** (1955), 67-70. MR **18**, 318.

13. *The theory of construction of finite semigroups*. II, Osaka Math. J. **9** (1957), 1-42; errata, 242. MR **19**, 940.
14. *Supplement to my paper "The theory of construction of finite semigroups. II"*, Osaka Math. J. **9** (1957), 235-237. MR **19**, 940.
15. *On the system of semigroup operations defined in a set*, J. Gakugei Coll. Tokushima Univ. **2** (1952), 1-12. MR **14**, 616.
16. *Commutative nonpotent archimedean semigroup with cancellation law*. I, J. Gakugei Tokushima Univ. **8** (1957), 5-11. MR **20** # 3224.

TAMURA, T. AND KIMURA, N.
1. *On decompositions of a commutative semigroup*, Kōdai Math. Sem. Rep. **1954**, 109-112. MR **16**, 670.
2. *Existence of greatest decomposition of a semigroup*, Kōdai Math. Sem. Rep. **7** (1955), 83-84. MR **18**, 192.

TARSKI, A.
1. *Remarks on direct products of commutative semigroups*, Math. Scand. **5** (1957), 218-223. MR **21** # 7168.

TESSIER, M.
1. *Sur les équivalences régulières dans les demi-groupes*, C. R. Acad. Sci. Paris **232** (1951), 1987-1989. MR **12**, 799.
2. *Sur la théorie des idéaux dans les demi-groupes*, C. R. Acad. Sci. Paris **234** (1952), 386-388. MR **13**, 620.
3. *Sur l'algèbre d'un demi-groupe fini simple*. II. *Cas général*, C. R. Acad. Sci. Paris **234** (1952), 2511-2513. MR **14**, 10.
4. *Sur quelques propriétés des idéaux dans des demi-groupes*, C. R. Acad. Sci. Paris **235** (1952), 767-769. MR **14**, 445.
5. *Sur les demi-groupes admettant l'existence du quotient d'un côté*, C. R. Acad. Sci. Paris **236** (1953), 1120-1122. MR **14**, 721.
6. *Sur les demi-groupes ne contenant pas d'élément idempotent*, C. R. Acad. Sci. Paris **237** (1953), 1375-1377. MR **15**, 598.
7. *Sur les demi-groupes*, C. R. Acad. Sci. Paris **236** (1953), 1524-1525.
8. *Sur l'algèbre d'un demi-groupe fini simple*, C. R. Acad. Sci. Paris **234** (1952), 2413-2414. MR **14**, 10.

TETSUYA, K., HASHIMOTO, T., AKAZAWA, T., SHIBATA, R., INUI, T. AND TAMURA, T.
1. *All semigroups of order at most* 5, J. Gakugei Tokushima Univ. Mat. Sci. Math. **6** (1955), 19-39. MR **17**, 1184.

THIERRIN, G.
1. *Sur une condition nécessaire et suffisante pour qu'un semigroupe soit un groupe*, C. R. Acad. Sci. Paris **232** (1951), 376-378. MR **12**, 389.
2. *Sur les éléments inversifs et les éléments unitaires d'un demi-groupe inversif*, C. R. Acad. Sci. Paris **234** (1952), 33-34. MR **13**, 621.
3. *Sur une classe de demi-groupes inversifs*, C. R. Acad. Sci. Paris **234** (1952), 177-179. MR **13**, 621.
4. *Sur une classe de transformations dans les demi-groupes inversifs*, C. R. Acad. Sci. Paris **234** (1952), 1015-1017. MR **13**, 621.
5. *Sur les demi-groupes inversés*, C. R. Acad. Sci. Paris **234** (1952), 1336-1338. MR **14**, 12.
6. *Sur les homogroupes*, C. R. Acad. Sci. Paris **234** (1952), 1519-1521. MR **13**, 902.
7. *Sur quelques classes de demi-groupes*, C. R. Acad. Sci. Paris **236** (1953), 33-35. MR **14**, 616.
8. *Sur quelques équivalences dans les demi-groupes*, C. R. Acad. Sci. Paris **236** (1953), 565-567. MR **14**, 616.

9. *Quelques propriétés des équivalences réversibles généralisées dans un demi-groupe D,* C. R. Acad. Sci. Paris **236** (1953), 1399-1401. MR **14**, 842.
10. *Sur une équivalence en relation avec l'équivalence réversible généralisée,* C. R. Acad. Sci. Paris **236** (1953), 1723-1725. MR **14**, 944.
11. *Quelques propriétés des sous-groupoïdes consistants d'un demi-groupe abélien D,* C. R. Acad. Sci. Paris **236** (1953), 1837-1839. MR **14**, 944.
12. *Sur la caractérisation des équivalences régulières dans les demi-groupes,* Acad. Roy. Belg. Bull. Cl. Sci. (5) **39** (1953), 942-947. MR **15**, 680.
13. *Sur quelques classes de demi-groupes possédant certaines propriétés des semi-groupes,* C. R. Acad. Sci. Paris **238** (1954), 1765-1767. MR **15**, 849.
14. *Sur la caractérisation des groupes pars leurs équivalences régulières,* C. R. Acad. Sci. Paris **238** (1954), 1954-1956. MR **15**, 850.
15. *Sur la caractérisation des groupes par leurs équivalences simplifiables,* C. R. Acad. Sci. Paris **238** (1954), 2046-2048. MR **15**, 850.
16. *Sur quelques propriétés de certaines classes de demi-groupes,* C. R. Acad. Sci. Paris **239** (1954), 1335-1337. MR **16**, 443.
17. *Sur la caractérisation des groupes par certaines propriétés de leurs relations d'ordre,* C. R. Acad. Sci. Paris **239** (1954), 1453-1455. MR **16**, 443.
18. *Demi-groupes inversés et rectangulaires,* Acad. Roy. Belg. Bull. Cl. Sci. (5) **41** (1955), 83-92. MR **17**, 10.
19. *Contribution à la théorie des équivalences dans les demi-groupes,* Bull. Soc. Math. France **83** (1955), 103-159. MR **17**, 584.
20. *Sur une propriété caractéristique des demi-groupes inversés et rectangulaires,* C. R. Acad. Sci. Paris **241** (1955), 1192-1194. MR **17**, 459.
21. *Sur la théorie des demi-groupes,* Comment. Math. Helv. **30** (1956), 211-223. MR **17**, 711.
22. *Sur les automorphismes intérieurs d'un demi-groupe réductif,* Comment. Math. Helv. **31** (1956), 145-151. MR **18**, 872.
23. *Demi-groupes réductifs,* Séminaire P. Dubriel et Ch. Pisot, 9e année 1955/56, Secrétariat mathématique, Paris, 1956, pp. 1-9. MR **19**, 522.
24. *Contribution à la théorie des anneaux et des demi-groupes,* Comment. Math. Helv. **32** (1957), 93-112. MR **19**, 1158.
25. *Sur la structure des demi-groupes,* Publ. Sci. Univ. Alger. Sér. A **3** (1956), 161-171. MR **20** # 7071.

THURSTON, H. A.
1. *Equivalences and mappings,* Proc. London Math. Soc. (3) **2** (1952), 175-182. MR **14**, 241.
2. *Some properties of partly-associative operations,* Proc. Amer. Math. Soc. **5** (1954), 487-497. MR **16**, 443.

TURING, A. M.
1. *The word problem in semi-groups with cancellation,* Ann. of Math. (2) **52** (1950), 491-505. MR **12**, 239.

VAGNER, V. V.
1. *On the theory of partial transformations,* Dokl. Akad. Nauk SSSR **84** (1952), 653-656. (Russian) MR **14**, 10.
2. *Generalized groups,* Dokl. Akad. Nauk SSSR **84** (1952), 1119-1122. (Russian) MR **14**, 12.
3. *The theory of generalized heaps and generalized groups,* Mat. Sb. **32** **(74)** (1953), 545-632. (Russian) MR **15**, 501.
4. *Generalized heaps reducible to generalized groups,* Scientific Yearbook of Saratov University for 1954, Saratov, 1955, pp. 668-669. (Russian)

5. *Generalized heaps reducible to generalized groups,* Ukrain. Mat. Ž. **8** (1956), 235-253. (Russian) MR **20** # 3924.
6. *Representation of ordered semigroups,* Mat. Sb. **38 (80)** (1956), 203-240; English transl., Amer. Math. Soc. Transl. (2) **36** (1964), 295-336. MR **17**, 942.
7. *Semigroups of partial mappings with a symmetric transitivity relation,* Izv. Vysš. Učebn. Zaved. Matematika **1957**, no. 1, 81-88. (Russian) MR **28** # 4049.

VAKSELJ, A.
1. *Eine neue Form der Gruppenpostulate und eine Erweiterung des Gruppenbegriffes,* Publ. Math. Univ. Belgrade **3** (1934), 195-211.
2. *Faktorhalbgruppen,* Glasnik Mat.-Fiz. Astronom. Društvo Mat. Fiz. Hrvatske Ser. II **12** (1957), 9-16. MR **20** # 4605.

VANDIVER, H. S.
1. *On the imbedding of one semi-group in another, with application to semi-rings,* Amer. J. Math. **62** (1940), 72-78. MR **1**, 105.
2. *The elements of a theory of abstract discrete semi-groups,* Vierteljschr. Naturforsch. Ges. Zürich **85** (1940), 71-86. MR **2**, 310.

VANDIVER, H. S. AND WEAVER, M. W.
1. *A development of associative algebra and algebraic theory of numbers.* III, Math. Mag. **29** (1956), 135-151. MR **17**, 825.

VINOGRADOV, A. A.
1. *On the theory of ordered semigroups,* Ivanov. Gos. Ped. Inst. Učen. Zap. Fiz.-Mat. Nauki **4** (1953), 19-21. (Russian) MR **17**, 710.

VOROB'EV, N. N.
1. *Normal subsystems of finite symmetric associative systems,* Dokl. Akad. Nauk SSSR **58** (1947), 1877-1879. (Russian) MR **9**, 330.
2. *Defect ideals of associative systems,* Leningrad. Gos. Univ. Učen. Zap. Ser. Mat. Nauk **16** (1949), 47-53. (Russian)
3. *On ideals of associative systems,* Dokl. Akad. Nauk SSSR **83** (1952), 641-644. (Russian) MR **14**, 10.
4. *Associative systems of which every subsystem has a unity,* Dokl. Akad. Nauk SSSR **88** (1953), 393-396. (Russian) MR **14**, 718.
5. *On symmetric associative systems,* Leningrad. Gos. Ped. Inst. Učen. Zap. **89** (1953), 161-166. (Russian) MR **17**, 943.
6. *On the theory of ideals of associative systems,* Leningrad. Gos. Ped. Inst. Učen. Zap. **103** (1955), 31-74. (Russian)
7. *On canonical representations of elements of symmetric associative systems,* Leningrad. Gos. Ped. Inst. Učen. Zap. **103** (1955), 75-82. (Russian)
8. *On associative systems every left ideal of which has a unit element,* Leningrad. Gos. Ped. Inst. Učen. Zap. **103** (1955), 83-90. (Russian)

WALLACE, A. D.
1. *A note on mobs,* An. Acad. Brasil. Ci. **24** (1952), 329-334. MR **14**, 724.
2. *A note on mobs.* II, An. Acad. Brasil. Ci. **25** (1953), 335-336. MR **15**, 854.
3. *Indecomposable semigroups,* Math. J. Okayama Univ. **3** (1953), 1-3. MR **15**, 933.
4. *Cohomology, dimension and mobs,* Summa Brasil. Math. **3** (1953), 43-55. MR **15**, 336.
5. *Inverses in Euclidean mobs,* Math. J. Okayama Univ. **3** (1953), 23-28. MR **15**, 933.
6. *Topological invariance of ideals in mobs,* Proc. Amer. Math. Soc. **5** (1954), 866-868. MR **16**, 568.
7. *The structure of topological semigroups,* Bull. Amer. Math. Soc. **61** (1955), 95-112. MR **16**, 796.
8. *The position of C-sets in semigroups,* Proc. Amer. Math. Soc. **6** (1955), 639-642. MR **17**, 172.

9. *The Rees-Suschkewitsch structure theorem for compact simple semigroups*, Proc. Nat. Acad. Sci. U.S.A. **42** (1956), 430-432. MR **18**, 14.
10. *The Gebietstreue in semigroups*, Nederl. Akad. Wetensch. Proc. Ser. A **59** = Indag. Math. **18** (1956), 271-274. MR **18**, 14.
11. *Ideals in compact connected semigroups*, Nederl. Akad. Wetensch. Proc. Ser. A **59** = Indag. Math. **18** (1956), 535-539. MR **18**, 490; 1118.
12. *Retractions in semigroups*, Pacific J. Math. **7** (1957), 1513-1517. MR **20** #2400.

WEAVER, M. W.
1. *Cosets in a semi-group*, Math. Mag. **25** (1952), 125-136. MR **14**, 12.
2. *On the imbedding of a finite commutative semigroup of idempotents in a uniquely factorable semigroup*, Proc. Nat. Acad. Sci. U.S.A. **42** (1956), 772-775. MR **18**, 283.

WIEGANDT, R.
1. *On complete semi-groups*, Acta Sci. Math. (Szeged) **19** (1958), 93-97. MR **20** #2387.

WIEGMANN, N. A.
1. *Some generalizations of Burnside's theorem*, Canad. J. Math. **9** (1957), 336-346. MR **19**, 634.

WOIDISLAWSKI, M.
1. *Ein konkreter Fall einiger Typen der verallgemeinerten Gruppen*, Zap. Inst. Mat. Meh. (4) **17** (1940), 127-144. (Russian) MR **3**, 37.

WRIGHT, F. B.
1. *Semigroups and submodular functions*, Michigan Math. J. **3** (1955/56), 169-172. MR **18**, 493.
2. *Semigroups in compact groups*, Proc. Amer. Math. Soc. **7** (1956), 309-311. MR **17**, 1107.

YAMADA, M.
1. *A note on middle unitary semigroups*, Kōdai Math. Sem. Rep. **7** (1955), 49-52. MR **17**, 585.
2. *On the greatest semilattice decomposition of a semigroup*, Kōdai Math. Sem. Rep. **7** (1955), 59-62. MR **17**, 584.
3. *Compositions of semigroups*, Kōdai Math. Sem. Rep. **8** (1956), 107-111. MR **19**, 118.
4. *Regularly totally ordered semi-groups*. I, Bull. Shimane Univ. Nat. Sci. **1957**, no. 7, 14-23.

YAMADA, M. AND KIMURA, N.
1. *Note on idempotent semigroups*. II, Proc. Japan Acad. **34** (1958), 110-112. MR **20** #4603.

ZARECKII, K. A.
1. *An abstract characterization of the semigroup of all binary relations*, Leningrad. Gos. Ped. Inst. Učen. Zap. **183** (1958), 251-263. (Russian)
2. *An abstract characterization of the semigroup of all reflexive binary relations*, Leningrad. Gos. Ped. Inst. Učen. Zap. **183** (1958), 265-269. (Russian)
3. *On ideals of semigroups*, Uspehi Mat. Nauk **14** (1959), no. 6 (90), 173-174. (Russian) MR **22** #736.

SUPPLEMENTARY BIBLIOGRAPHY (1968)

BURMISTROVIČ, I. E.

S1. *Commutative bands of cancellative semigroups,* Sibirsk. Mat. Ž. **6** (1965), 284-299. (Russian) MR **30** # 3929.

GLUSKIN, L. M.

S1. *Ideals of semigroups,* Mat. Sb. **55** (**97**) (1961), 421-448. (Russian) MR **25** # 3106.

S2. *Semi-groups of isotone transformations,* Uspehi Mat. Nauk **16** (1961), no. 5 (101), 157-162. (Russian) MR **24** # A1336.

S3. *On dense embeddings,* Mat. Sb. **61** (**103**) (1963), 175-206. (Russian) MR **27** # 2570.

HEDRLÍN, Z. AND PULTR, A.

S1. *Relations (graphs) with given finitely generated semigroups,* Monatsh. Math. **68** (1964), 213-217. MR **29** # 5942.

S2. *Relations (graphs) with given infinite semigroups,* Monatsh. Math. **68** (1964), 421-425. MR **30** # 1076.

KALMANOVIČ, A. M.

S1. *Semigroups of partial endomorphisms of a graph,* Dopovīdī Akad. Nauk Ukraïn. RSR **1965**, 147-150. (Ukrainian) MR **31** # 264.

LJAPIN, E. S.

S1. *On representations of semigroups by partial transformations,* Mat. Sb. **52** (**94**) (1960), 589-596; English transl., Amer. Math. Soc. Transl. (2) **27** (1963), 289-296. MR **22** # 12155; **27** # 1522.

S2. *Some endomorphisms of an ordered set,* Izv. Vysš. Učebn. Zaved. Matematika **1962**, no. 2 (27), 87-94. (Russian) MR **27** # 4877.

S3. *Some representations of semigroups by mappings,* In Memoriam: N. G. Čebotarev, Izdat. Kazan. Univ., Kazan, 1964, pp. 60-66. (Russian) MR **33** # 5766.

S4. *An abstract characterization of the class of semigroups of endomorphisms of systems of general type,* Mat. Sb. **70** (**112**) (1966), 172-179; English transl., Amer. Math. Soc. Transl. (2) **94** (1970), 211-219. MR **33** # 2746.

MAL'CEV, A. I.

S1. *Homomorphisms onto finite groups,* Ivanov. Gos. Ped. Inst. Učen. Zap. **18** (1958), 49-60. (Russian)

POPOVA, L. M.

S1. *On a semigroup of partial endomorphisms of a set with a relation,* Leningrad. Gos. Ped. Inst. Učen. Zap. **238** (1962), 49-77. (Rusian) MR **31** # 1313.

S2. *Semigroups of partial endomorphisms of a set with a relation,* Sibirsk. Mat. Ž. **4** (1963), 309-317. (Russian) MR **27** # 3726.

ŠAIN, B. M.

S1. *Representation of semigroups by binary relations,* Dokl. Akad. Nauk SSSR **142** (1962), 808-811 = Soviet Math. Dokl. **3** (1962), 188-191. MR **26** # 3810.

S2. *On subdirectly irreducible semigroups,* Dokl. Akad. Nauk SSSR **144** (1962), 999-1002 = Soviet Math. Dokl. **3** (1962), 868-871. MR **27** # 4885.

S3. *Representations of generalized groups,* Izv. Vysš. Učebn. Zaved. Matematika **1962**, no. 3 (28), 164-176. (Russian) MR **25** # 3105.

S4. *Representation of ordered semigroups,* Uspehi Mat. Nauk **17** (1962), no. 6 (108), 200-201. (Russian)

S5. *Representation of semigroups by means of binary relations*, Mat. Sb. **60** (**102**) (1963), 293-303. (Russian) MR **27** # 3721.

S6. *Transitive representations of semigroups*, Uspehi Mat. Nauk **18** (1963), no. 3 (111), 215-222. (Russian) MR **28** # 150.

S7. *On the theory of restrictive semigroups*, Izv. Vysš. Učebn. Zaved. Matematika **1963**, no. 2 (33), 152-154. (Russian) MR **27** # 1527.

S8. *The representation of ordered semigroups*, Mat. Sb. **65** (**107**) (1964), 188-197. (Russian) MR **30** # 2093.

S9. *Restrictive bisemigroups*, Izv. Vysš. Učebn. Zaved. Matematika **1965**, no. 1 (44), 166-179; English transl., Amer. Math. Soc. Transl. (2) **100** (1972), 293-307. MR **30** # 4847.

S10. *Homomorphisms and subdirect decompositions of semigroups*, Pacific J. Math. **17** (1966), 529-547. MR **33** # 5768.

Šutov, È. G.

S1. *Translations of semigroups*, Uspehi Mat. Nauk **19** (1964), no. 4 (118), 215-218. (Russian) MR **29** # 5946.

Tamura, T.

S1. *Notes on translations of a semigroup*, Kōdai Math. Sem. Rep. **10** (1958), 9-26. MR **20** # 6475.

Tamura, T. and Graham, N.

S1. *Certain embedding problems of semigroups*, Proc. Japan. Acad. **40** (1964), 8-13. MR **29** # 1280.

Thierrin, G.

S1. *Sur la structure des demi-groupes*, Publ. Sci. Univ. Alger. Sér. A **3** (1965), 161-171. MR **20** # 7071.

Tulley, E. J., Jr.

S1. *Representation of a semigroup by transformations acting transitively on a set*, Amer. J. Math. **83** (1961), 533-541. MR **25** # 135.

Vagner, V. V.

S1. *Restrictive semigroups*, Izv. Vysš. Učebn. Zaved. Matematika **1962**, no. 6 (31), 19-27. (Russian) MR **26** # 5086.

Zybina, L. D.

S1. *On relations determining the order of certain endomorphisms of an ordered set*, Leningrad. Gos. Ped. Inst. Učen. Zap. **274** (1965), 106-121. (Russian) MR **33** # 5526.

SUPPLEMENTARY BIBLIOGRAPHY (1972)

ADJAN, S. I.

S1. *Defining relations and algorithmic problems for groups and semigroups,* Trudy Mat. Inst. Steklov. **85** (1966) = Proc. Steklov Inst. Math. **85** (1966). MR **34** # 4340; **36** # 1520.

AUSTIN, A. K.

S1. *A closed set of laws which is not generated by a finite set of laws,* Quart. J. Math. Oxford Ser. (2) **17** (1966), 11-13. MR **34** # 4341.

BIRJUKOV, A. P.

S1. *Semigroups given by identities,* Učen. Zap. Novosibirsk. Gos. Ped. Inst. **18** (1963), 139-169. (Russian)

S2. *On some identities in semigroups,* Učen. Zap. Novosibirsk. Gos. Ped. Inst. **18** (1963), 170-183. (Russian)

S3. *Semigroups defined by identities,* Dokl. Akad. Nauk SSSR **149** (1963), 230-232 = Soviet Math. Dokl. **4** (1963), 333-336. MR **26** # 6281.

S4. *On infinite aggregates of identities in semi-groups,* Algebra i Logika Sem. **4** (1965), no. 2, 31-32. (Russian) MR **35** # 2980.

S5. *The lattice of varieties of idempotent semigroups,* Eighth All-Union Colloq. on General Algebra, Abstract of Reports, Riga, 1967, pp. 16-18. (Russian)

S6. *Varieties of idempotent semigroups,* Algebra i Logika **9** (1970), no. 3, 255-273 = Algebra and Logic **9** (1970), 153-164. MR **45** # 6949.

S7. *Semigroups determined by identities with a distinguished element,* Mat. Sb. **67** (**109**) (1965), 461-473; English transl., Amer. Math. Soc. Transl. (2) **82** (1969), 209-223. MR **33** # 2740.

COHN, P. M.

S1. *Universal algebra,* Harper & Row, New York, 1965; Russian transl., "Mir", Moscow, 1968. MR **31** # 224.

EVANS, T.

S1. *The number of semigroup varieties,* Quart. J. Math. Oxford Ser. (2) **19** (1968), 335-336. MR **38** # 257.

S2. *Some connections between residual finiteness, finite embeddability and the word problem,* J. London Math. Soc. (2) **1** (1969), 399-403. MR **40** # 2589.

S3. *The lattice of semigroup varieties,* Semigroup Forum **2** (1971), 1-43.

FENNEMORE, C.

S1. *All varieties of bands,* Semigroup Forum **1** (1970), 172-179. MR **42** # 6140.

S2. *All varieties of bands,* Math. Nachr. **48** (1971), 237-252, 253-262.

GERHARD, J. A.

S1. *The lattice of equational classes of idempotent semigroups,* J. Algebra **15** (1970), 195-224. MR **41** # 8552.

GOLUBOV, È. A.

S1. *On finite separability in semigroups,* Dokl. Akad. Nauk SSSR **189** (1969), 20-22 = Soviet Math. Dokl. **10** (1969), 1321-1323. MR **41** # 368.

S2. *Finite separability in semigroups,* Sibirsk. Mat. Ž. **11** (1970), 1247-1263 = Siberian Math. J. **11** (1970), 920-931. MR **43** # 376.

GOLUBOV, E. A. AND LESOHIN, M. M.

S1. *Finite approximability relative to the divisibility of commutative naturally totally ordered semigroups,* Ural. Gos. Univ. Mat. Zap. **6** (1968), tetrad′ 3, 12-17. (Russian) MR **42** # 4460.

GRÄTZER, G.

S1. *Universal algebra,* Van Nostrand, Princeton, N. J., 1968. MR **40** # 1320.

GRINDLINGER, E. I.

S1. *On the question of identities in semigroups,* Izv. Vysš. Učebn. Zaved. Matematika **1967**, no. 5 (60), 6-12. (Russian) MR **35** # 6769.

KALICKY, J. AND SCOTT, D.

S1. *Equational completeness of abstract algebras,* Nederl. Akad. Wetensch. Proc. Ser. A **58** = Indag. Math. **17** (1955), 650-659. MR **17**, 571.

KIMURA, N.

S1. *Note on idempotent semigroups.* IV, Proc. Japan Acad. **34** (1958), 121-123. MR **20** # 4604.

S2. *The structure of idempotent semi-groups.* I, Pacific J. Math. **8** (1958), 257-275. MR **21** # 1354.

LESOHIN, M. M.

S1. *Finitary approximability of semigroups and finite separability of subsemigroups,* Leningrad. Gos. Ped. Inst. Učen. Zap. **387** (1968), 134-158. (Russian) MR **39** # 5731.

S2. *On approximations of semigroups with respect to predicates.* I, II, Leningrad. Gos. Ped. Inst. Učen. Zap. **404** (1971), 191-219; **464** (1971), 56-108. (Russian)

S3. *On some algorithmic questions in semigroups,* Leningrad. Gos. Ped. Inst. Učen. Zap. **541** (1972), 118-128.

LJAPIN, E. S.

S5. *The independence of subsemigroups of a semigroup,* Dokl. Akad. Nauk SSSR **185** (1969), 1229-1231 = Soviet Math. Dokl. **10** (1969), 492-494. MR **40** # 259.

S6. *A certain infinite irreducible collection of semigroup identities,* Mat. Zametki **7** (1970), 545-549 = Math. Notes **7** (1970), 330-332. MR **42** # 396.

S7. *Intersections of independent subsemigroups of a semigroup,* Izv. Vysš. Učebn. Zaved. Matematika **1970**, no. 4 (95), 67-73. (Russian) MR **43** # 7526.

S8. *Unique ideal subsemigroups of semigroups,* Ural. Gos. Univ. Mat. Zap. **7** (1970), no. 3, 119-126. (Russian)

S9. *Unique ideal elements of semigroups,* Theory of Semigroups and its Appl., vol. II, Izdat. Saratov Univ., Saratov, 1971, pp. 41-50. (Russian)

MAL′CEV, A. I.

S2. *Algebraic systems,* "Nauka", Moscow, 1970; English transl., Die Grundlehren der math. Wissenschaften, Band 192, Springer-Verlag, New York, 1973.

MAMIKONJAN, S. G.

S1. *The approximation of semigroups that satisfy a commutator condition,* Izv. Vysš. Učebn. Zaved. Matematika **1970**, no. 1 (92), 70-72. (Russian) MR **41** # 7006.

MURSKIĬ, V. L.

S1. *Some examples of varieties of semigroups,* Mat. Zametki **3** (1968), 663-670 = Math. Notes **3** (1968), 423-427. MR **38** # 256.

NEUMANN, H.

S1. *Varieties of groups,* Ergebnisse der Mathematik und ihrer Grenzgebiete, Band 37, Springer-Verlag, New York, 1967; Russian transl., "Mir", Moscow, 1969. MR **35** # 6734; **40** # 1460.

PERKINS, P.

S1. *Bases for equational theories of semigroups,* J. Algebra **11** (1969), 293-314. MR **38** # 2232.

SASAKI, M.

S1. *Semigroups whose arbitrary subsets containing a definite element are subsemigroups,* Proc. Japan Acad. **39** (1963), 628-633. MR **30** # 3169.

SMIRNOV, D. M.

S1. *On the theory of finitely approximable groups,* Ukrain. Mat. Ž. **15** (1963), 453-457. (Russian) MR **28** # 2151.

TAMURA, T.

S2. *Attainability of systems of identities of semigroups,* J. Algebra **3** (1966), 261-276. MR **32** # 7668.

YAMADA, M.

S1. *Regular semigroups whose idempotents satisfy permutation identities,* Pacific J. Math. **21** (1967), 371-392. MR **37** # 2887.

INDEX

The numbers in parentheses are page numbers.

INDEX

The numbers in parentheses are page numbers.